THE BIOLOGY OF HAGFISHES

THE BIOLOGY OF HAGFISHES

JØRGEN MØRUP JØRGENSEN, JENS PETER LOMHOLT, ROY E. WEBER and HANS MALTE

Department of Zoophysiology
Institute of Biological Sciences
University of Aarhus
Denmark

Springer-Science+Business Media, B.V.

Published by Chapman and Hall, an imprint of Thomson Science, 2–6 Boundary Row, London SE1 8HN, UK

Thomson Science, 2–6 Boundary Row, London SE1 8HN, UK

Thomson Science, 115 Fifth Avenue, New York, NY 10003, USA

Thomson Science, Suite 750, 400 Market Street, Philadelphia, PA 19106, USA

Thomson Science, Pappelallee 3, 694469 Weinheim, Germany

First edition 1998

Originally published by Chapman & Hall Ltd in 1998
Softcover reprint of the hardcover 1st edition 1998

Typeset in 10/12pt Palatino by Cambrian Typesetters, Frimley, Surrey

ISBN 978-94-010-6465-1 ISBN 978-94-011-5834-3 (eBook)
DOI 10.1007/978-94-011-5834-3

A catalogue record for this book is available from the British Library

Library of Congress Catalog Card Number: 97-69524

Printed on acid-free text paper, manufactured in accordance with ANSI/NISO Z39.48–1992 (Permanence of Paper).

CONTENTS

LIST OF CONTRIBUTORS

Richard J. Aldridge
Department of Geology
University of Leicester
University Road
Leicester LE1 7RH
United Kingdom
E-mail: ra12@leicester.ac.uk

Karl H. Andres
Abt. für Neuroanatomie
Institut für Anatomie
Ruhr-Universität Bochum
Universitätsstrasse 150
D-44780 Bochum
Deutschland (Germany)
E-mail: Karl.h.andres@rz.ruhr-uni-bochum.de

David Bardack
Department of Biological
Sciences (M/C 066)
University of Illinois at Chicago
845 West Taylor Street
Chicago, Illinois 60607-7060
USA
E-mail: DBardack@uic.edu

Helmut Bartels
Anatomische Anstalt
Universität München
Pettenkoferstrasse 11
D-80336 München
Deutschland (Germany)
Fax: +49 89 51604857
E-mail: bartels@anat.med.uni-muenchen.de

Nicholas J. Bernier
Department of Biology
University of Ottawa
30 Marie Curie
PO Box 450 STN A
Ottawa ON K1N 6N5
Canada
E-mail: nbernier@oreo.uottawa.ca

Christopher B. Braun
Parmly Hearing Institute
Loyola University Chicago
6525 N. Sheridan Rd
Chicago IL 60626
USA
E-mail: cbraun@luc.ed4

Simone Büchl
Anatomische Anstalt, Lehrstuhl II
Pettenkoferstrasse 11
Ludwig-Maximilians-Universität München
D-80336
Deutschland (Germany)

Akira Chiba
Department of Histology
Nippon Dental University
1–8 Hamaura-Cho
Niigata, 951
Japan

M. Edwin DeMont
Department of Biology
St Francis Xavier University
Antigonish, Nova Scotia B2G 2W5
Canada
E-mail: edemont@juliet.stfx.ca

Niels Dohn
Department of Zoophysiology
Institute of Biological Sciences
University of Aarhus
Universitetsparken, building 131
DK-8000 Aarhus C
Danmark (Denmark)

Philip C.J. Donoghue
School of Earth Sciences
University of Birmingham
Edgbaston
Birmingham B15 2TT
United Kingdom
E-mail: donoghue@ers.birmingham.ac.uk

Monika von Düring
Abt. für Neuroanatomie
Institut für Anatomie
Ruhr-Universität Bochum
Universitätsstrasse 150 MA6/162
D-44780 Bochum
Deutschland (Germany)
Fax: +49 234 709 4457
E-mail: Monika.U.Duering@rz.ruhr-uni-bochum-de

Kjell B. Døving
Avdeling for generell fysiologi
Biologisk Institut
Universitetet i Oslo
PO Box 1051
N-0316 Oslo
Norge (Norway)
Fax: +47 2285 4664
E-mail: kjelld@whitney.ufl.edu

Rainer Erlinger
Anatomische Anstalt, Lehrstuhl II
Ludwig-Maximilians-Universität München
Pettenkoferstrasse 11
D-80336 München
Deutschland (Germany)

Angela Fago
Department of Zoophysiology
Institute of Biological Sciences
University of Aarhus
Universitetsparken, building 131
DK-8000 Aarhus C
Danmark (Denmark)
E-mail: angela@bio.aau.dk

Sture Falkmer
Institute of Morphology and Pathology
Regionsykehuset i Trondheim
N-7006 Trondheim
Norge (Norway)
E-mail: Sture.Falkmer@medisin.ntnu.no

Lüder M. Fels
Abteilung für Nephrologie
Arbeitsbereich Experimentelle Nephrologie
Medizinische Hochschule Hannover
Carl-Neuberg-Strassel
D-30625 Hannover
Deutschland (Germany)
Fax: +49 511 5323 780

Bo Fernholm
Sektionen för Vertebratzoologi
Naturhistoriska Riksmuseet
Box 50007
S-104 05 Stockholm
Sverige (Sweden)
E-mail: VE-Bo@NRM.SE

Per R. Flood
Zoologisk Institut
Allégaten 41
N-5007 Bergen
Norge (Norway)
E-mail: Per.Flood@zoo.uib.no

Malcolm E. Forster
Department of Zoology
University of Canterbury
Private Bag 4800
Christchurch
New Zealand
E-mail: FORSTER@zool.canterbury.ac.nz

Frida Franko-Dossar
Department of Zoophysiology
Institute of Biological Sciences
University of Aarhus
Universitetsparken, building 131
DK-8000 Aarhus C
Danmark (Denmark)

Ragnar Fänge
Storängsgatan 24
S-413 19 Göteborg
Sverige (Sweden)

Gerolf Gros
Abteilung Vegetative Physiologie
Zentrum Physiologie
Medizinische Hochschule Hannover
Postfach 61 01 80
D-30623 Hannover
Deutschland (Germany)
Fax: +49 511 532 2938
E-mail: Gros.Gerolf@MH-Hannover.de

Susanne Holmgren, Dr
Zoofysiologiska Avdelningen
Zoologiska Institutionen
Göteborgs Universitet
Medicinaregatan 18
S-413 90 Göteborg
Sverige (Sweden)
E-mail: S.Holmgren@zool.gu.se

Yoshiharu Honma, Dr
3460-55 Inarimachi
Niigata, 951
Japan

Jørgen Mørup Jørgensen
Department of Zoophysiology
Institute of Biological Sciences
University of Aarhus
Universitetsparken, building 131
DK-8000 Aarhus C
Danmark (Denmark)
E-mail: jmj@bio.aau.dk

Sabine Kastner
Abteilung für Nephrologie
Arbeitsbereich Experimentelle Nephrologie
Medizinische Hochschule Hannover
Carl-Neuberg-Strasse 1
D-30625 Hannover
Deutschland (Germany)
Fax: +49 511 5323 780

Fred W. Keeley
Division of Cardiovascular Research
Research Institute
Hospital for Sick Children
Toronto, Ontario M5G 1X8
Canada
E-mail: fwk@resunix.sickkids.on.ca

Elizabeth A. Koch
Department of Biological Chemistry
The Chicago Medical School
Finch University of Health Sciences
3333 Green Bay Road
North Chicago, Illinois 60064-3095
USA
E-mail: koche@mis.finchcms.edu

Sei-ichi Kohno
Department of Biology
Faculty of Science
Toho University
Miyama 2-2-1, Funabashi,
Chiba 274
Japan
E-mail: skohno@toho-u.ac.jp

Souichirou Kubota
Department of Biology
Faculty of Science
Toho University
Miyama 2-2-1, Funabashi,
Chiba 274
Japan

N. Adam Locket
Department of Anatomy and Histology
University of Adelaide
South Australia 5005
Australia
E-mail:
AdamLocket@medicine.ccmail.adelaide.edu.au

Jens Peter Lomholt
Department of Zoophysiology
Institute of Biological Sciences
University of Aarhus
Universitetsparken, building 131
DK-8000 Aarhus C
Danmark (Denmark)
Fax: +45 8619 4186

Hans Malte
Department of Zoophysiology
Institute of Biological Sciences
University of Aarhus
Universitetsparken, building 131
DK-8000 Aarhus C
Danmark (Denmark)
E-mail: hans@bio.aau.dk

Frederic H. Martini
School of Ocean and Earth Sciences and Technology
University of Hawaii
5071 Hana Hwy
Haiku, Hawai 96708
USA
E-mail: martini@maui.net

Alistair R. McVean
School of Biological Sciences
University of London
Royal Holloway
Egham, Surrey TW20 0EX
United Kingdom
E-mail: a.mcvean@rhbnc.ac.uk

Yasuharu Nakai
Safety Research Department
Pharmaceuticals Development Research Laboratories
Teijin Limited
Asahigaoka 4-3-2
Hino, Tokyo, 191
Japan

Stefan Nilsson
Zoofysiologiska Avdelningen
Zoologiska Institutionen
Göteborgs Universitet
Medicinaregatan 18
S-413 90 Göteborg
Sverige (Sweden)
E-mail: S.Nilsson@zoofys.gu.se

R. Glenn Northcutt
Department of Neurosciences, 0201
University of California, San Diego
9500 Gilman Drive
La Jolla, California 92093-0201
USA
Fax: +1 619 534 5622

Robert A. Patzner
Zoologisches Institut
Universität Salzburg
Hellbrunnerstrasse 34
A-5020 Salzburg
Österreich (Austria)
E-mail: robert.patzner@sbg.ac.at

Steve F. Perry
Department of Biology
University of Ottawa
30 Marie Curie
PO Box 450 Stn A
Ottawa
Ontario 6N5
Canada

Thomas Peters
Abteilung Vegetative Physiologie
Zentrum Physiologie
Medizinische Hochschule Hannover
Postfach 61 01 80
D-30623 Hannover
Deutschland (Germany)
Fax: +49 511 532 2938
E-mail: Peters.Thomas@MH-Hannover.de

Ian C. Potter
School of Environmental and Life Sciences
Murdoch University
Murdoch
Western Australia, 6150
Australia

Robert L. Raison
Department of Cell and Molecular Biology
University of Technology, Sydney
PO Box 123 Broadway
New South Wales 2007
Australia
E-mail: Robert.Raison@uts.edu.au

Nicholas J. dos Remedios
Department of Cell and Molecular Biology
University of Technology, Sydney
PO Box 123 Broadway
New South Wales 2007
Australia

Jay A. Riegel
Department of Zoology
University of Cambridge
Downing Street
Cambridge CB2 3EJ
United Kingdom
Fax: +44 223 336 676

Mark Ronan
Physiology Section
School of Medicine
Indiana University
Meyers Hall 263
Bloomington, Indiana 47405-4201
USA
Fax: +1 812 855 4436

Robert H. Spitzer
Department of Biological Chemistry
The Chicago Medical School
Finch University of Health Sciences
3333 Green Bay Road
North Chicago, Illinois 60064-3095
USA
E-mail: koche@mis.finchcms.edu

Hilmar Stolte
Abteilung für Nephrologie
Arbeitsbereich Experimentelle Nephrologie
Medizinische Hochschule Hannover
Carl-Neuberg-Strassel
D-30625 Hannover
Deutschland (Germany)
E-mail: 101711.1144@Compuserve.com

Michael C. Thorndyke
School of Biological Sciences
University of London
Royal Holloway
Egham, Surrey TW20 0EX
United Kingdom
E-mail: m.thorndyke@rhbnc.ac.uk

Udo Tusch
Klinikum der Johann Wolfgang Goethe-Universität
Dr. Senckenbergische Anatomie
Theodor-Stern-Kai 7
D-60590 Frankfurt/Main
Deutschland (Germany)

Roy E. Weber
Department of Zoophysiology
Institute of Biological Sciences
University of Aarhus
Universitetsparken, building 131
DK-8000 Aarhus C
Danmark (Denmark)
E-mail: rw@bio.aau.dk

Ulrich Welsch
Lehrstuhl Anatomie II
Anatomische Anstalt
Ludwig-Maximilians-Universität München
Pettenkoferstrasse 11
D-80336 München
Deutschland (Germany)
Fax: +49 89 5160 4897
E-mail: Welsch@anat.med.uni-muenchen.de

Helmut Wicht
Klinikum der Johann Wolfgang Goethe-Universität
Dr. Senckenbergische Anatomie
Theodor-Stern-Kai 7
D-60590 Frankfurt am Main
Deutschland (Germany)
E-mail: wicht@em.uni-frankfurt.de

Glenda M. Wright
Department of Anatomy and Physiology
Atlantic Veterinary College
University of Prince Edward Island
550 University Avenue
Charlottetown, Prince Edward Island
Canada C1A 4P3
E-mail: gwright@upei.ca

EDITORS' PREFACE

The hagfishes comprise a uniform group of some 60 species inhabiting the cool or deep parts of the oceans of both hemispheres. They are considered the most primitive representatives of the group of craniate chordates, which – apart from the hagfishes that show no traces of vertebrae – includes all vertebrate animals. Consequently the hagfishes have played and still play a central role in discussions concerning the evolution of the vertebrates. Although most of the focus on hagfishes may be the result of their being primitive, it should not be forgotten that, at the same time, they are specialized animals with a unique way of life that is interesting in its own right.

It is now more than 30 years since a comprehensive treatise on hagfishes was published. *The Biology of Myxine*, edited by Alf Brodal and Ragnar Fänge (Universitetsforlaget, Oslo, 1963), provided a wealth of information on the biology of hagfishes, and over the years remained a major source of information and inspiration to students of hagfishes.

The three decades since the publication of that volume have witnessed a surge of new information and insight in traditional as well as in newly developed areas of biological investigation. It is the aim of the present book to make this new information on hagfish biology available in a single volume for readers interested in a group of animals that is fascinating not only because of the available knowledge, but also because of the fact that major aspects of their life – such as reproduction and development – largely remain closely guarded secrets.

An important step in the production of this book was a symposium organized for the contributors in July 1996. At the symposium speakers addressed an audience from all walks of biology but with a shared interest in hagfishes. It is our hope that this has resulted in a book that will be of use to readers with a general interest in hagfishes, while at the same time reflecting recent progress in hagfish research.

The symposium was held at the Kristineberg Marine Biological Station near Göteborg, Sweden – a classic locality for hagfish research. We express our gratitude for the hospitality shown by the Director, Professor Jarl-Ove Strömberg and the staff at Kristineberg. The symposium was supported by the Danish Natural Science Research Council and the Carlsberg Foundation.

Jørgen Mørup Jørgensen, Jens Peter Lomholt,
Roy E. Weber, Hans Malte

INTRODUCTION: EARLY HAGFISH RESEARCH

R. Fänge

SUMMARY

The Atlantic hagfish was briefly described by Kalm (1753). Linnaeus (1754, 1758) gave it the name *Myxine glutinosa* and classified it as a worm, but anatomical studies by Abildgaard (1792) and others proved that it was related to fishes. Research during the nineteenth century by J. Müller, A. Retzius, F. Nansen, G. Retzius and others centred on anatomy and histology, but during the twentieth century physiological, biochemical and ecological aspects, etc., have become important. Several of the early investigators were Scandinavians. Innumerable questions remain to be solved, for instance concerning the phylogenetic relationships and the reproduction. Some features unique to hagfishes are listed.

EARLIEST OBSERVATIONS

Our knowledge of the Atlantic hagfish goes back to Pehr Kalm, a disciple of Linnaeus. In 1747 Kalm started a journey to North America. He left Göteborg in November on board a ship bound for England, but he had to wait many weeks at Grimstad in Norway for the ship to be repaired after a storm. As a naturalist of the Enlightenment he used the compulsory stay to study plants, animals, agriculture, fishery and other things in the surrounding areas. One day in January a local fisherman, 'Pehr i Haven', brought him a fish which he had never seen before. It was called 'Pihrål' by the local people and seemed to be a kind of *Petromyzon*, or 'neijnögon' (Swedish for lamprey, literally meaning 'nine-eyes'). It secreted enormous amounts of slime and was disliked for devouring and destroying fishes caught by hook or net. Linnaeus knew the 'Pihrål' from Kalm's travel report (1753) and pictured and described it in a magnificent catalogue of the private Natural History Museum at Ulriksdal near Stockholm which was owned by the king, Adolf Fredrik. Linnaeus (1754) introduced the name *Myxina glutinosa* (Gr. *myxa*, slime; Lat. *gluten*, glue). He thought that the hagfish was recognized by contemporary ichthyologists as a blind lamprey (*Lampetra caeca, Enneophthalmos caecus*), but this was probably wrong (Bloch, 1793). However, disregarding any similarity to lampreys, Linnaeus classified the hagfish among the worms (*Vermes intestina*) in the tenth edition of the *Systema Naturae* (1758) and changed the genus name to *Myxine*.

The Norwegian bishop and naturalist Gunnerus (1763) dissected *Myxine* and observed teeth, jaws, oesophagus, intestine, a two-lobed liver, gall bladder, heart and large yolk-containing eggs, etc. However, the longitudinal lingual muscle he mistook for a cartilaginous trachea connected with 12 so-called lungs. In spite of the anatomical results, Gunnerus followed Linnaeus in calling the animal 'Sleep-Mark', i.e. slime worm. In 1790 A.J. Retzius, professor of natural history at Lund, Sweden, in a paper entitled 'Anmärkningar vid slägtet *Myxine*' (Comments on the genus *Myxine*), wondered what had induced the sharp-sighted Linnaeus to classify *Myxine glutinosa* among the order of worms (Vermes). Both the literature and inspection of preserved specimens had convinced Retzius that *Myxine* was related to lampreys or snakes. Two years later Abildgaard (1792), zoologist and father of veterinary medicine in Denmark, on the basis of an anatomical investigation concluded that *Myxine* was not a worm but a fish. This was confirmed by Bloch (1793), who named the hagfish *Gastrobranchus coecus* (Gr. *gaster*, belly; *branchus*, gill; Lat. *caecus*, blind; the genus name alludes to ventral gill openings).

RESEARCH DURING THE NINETEENTH CENTURY

In 1822 and 1824 A.A. Retzius, the son of A.J. Retzius, carried out detailed anatomical studies on the hagfish. The animal was difficult to get hold of, but Retzius had obtained a few specimens preserved in alcohol from the amateur ichthyologist, baron Nils Gyllenstierna at Krapperup, Scania. He investigated the circulatory system by injecting mercury and observed the subcutaneous blood sinus and other vascular structures, the pronephros, the ureters, the muscles of the tongue, different kinds of cartilage, etc. For several years the two brief communications by A.A. Retzius were the main source of knowledge on the anatomy of the hagfish, but during the period 1836–45 the German anatomist, Johannes Müller, published a series of well-illustrated articles on the comparative anatomy of myxinoids. To Müller, as to Retzius, the hagfish was a good vertebrate, more specifically a fish. Müller classified it among the Cyclostomata, which were divided into Hyperoartia (lampreys) and Hyperotreta (hagfishes) (Gr. *hypero*, palate; *artios*, whole; *tretos*, perforated; Hyperoartia have a whole palate whereas, in the Hyperotreta, the palate is penetrated by the nasopharyngeal duct). Müller described with expertise the macroscopic anatomy of the main organ systems including a few microscopic details such as blood cells. However, when first investigating the kidney of *Myxine*, Müller (1836) failed to recognize the true structure of the renal corpuscles. The English physician Bowman (1842) showed in a comparative anatomical study that the capsules of the Malpighian corpuscles of the vertebrate kidney are blind endings of urinary canaliculi and concluded that the glomeruli function by separating water from blood. Müller (1845) in continued work, almost embarrassed, could not but confirm Bowman's anatomical results. He marvelled at the extreme simplicity of the hagfish renal system but abstained from speculating about the function.

In 1841 Müller and A.A. Retzius, who were friends, worked together at Kristineberg in Sweden on the circulation of blood in tiny living lancelets (*Amphioxus*). On that occasion Müller also studied *Myxine glutinosa* and discovered that a part of the portal vein, previously recognized as a special anatomic structure by Retzius (1822), performed heart-like rhythmic contractions (Müller, 1845). One of several researchers, who worked on the hagfish during the late part of the nineteenth century, was Gustaf Retzius. He was the son of A.A. Retzius and grandson of A.J. Retzius, and among other activities he was for some years professor of histology at the Karolinska Institute in Stockholm. For many years he intensively investigated the morphology of *Myxine glutinosa*, following up a more than 100-year-old family tradition. Most of his results appeared in the *Biologische Untersuchungen*, a magnificent journal which was owned, edited, distributed and mainly written by himself. A younger contemporary of his, the Norwegian zoologist, arctic explorer and diplomat Fridtjof Nansen, also studied the hagfish. To a large extent the achievements of these two scientists concerned the nervous system. Nansen (1887) observed that the nerve fibres of the hagfish spinal cord lacked myelin sheaths, and by the newly invented Golgi technique he discovered that the posterior spinal root axons branched dichotomously, a structural arrangement later detected in all vertebrates. Retzius made careful anatomical observations of nervous and sensory structures. Before dissecting he macerated the hagfishes in 20% nitric acid, which enabled preparation of delicate nerve branches (Figure 1, upper left), and for microscopic work he used mainly the Ehrlich methylene blue vital staining method (Figure 1, lower left) but also the Golgi method (Figure 1, lower right). He studied many different tissues in addition to the nervous system and discovered and described the function of the caudal heart (Retzius, 1890a). Nansen and Retzius were talented artists. Nansen's illustrations were 'drawn under the camera lucida, from the microscope directly upon the stone' (Nansen, 1887), (Figure 1, upper right). Retzius made drawings and watercolours himself and also sought assistance by professional artists and engravers. The two researchers corresponded with each other and were friends. Nansen (1888) presented the much debated hypothesis that *Myxine* is a protandric hermaphrodite. He was visionary and speculative, but he soon gave up his zoological studies for arctic research, and most of his theories on nerve structure are now obsolete or forgotten. Retzius was an accurate recorder of facts who avoided theorizing, and many of his scientific illustrations are still appreciated. An extensive review of investigations on hagfishes up to approximately 1900 is found in Lönnberg (1924).

THE BEGINNING OF THE TWENTIETH CENTURY

Since the latter half of the nineteenth century biology has been strongly influenced by the theory of evolution. *Myxine* was often regarded as a sort of intermediate link between the lancelet (*Amphioxus* or *Branchiostoma*) and true vertebrates. By investigating the hagfish anatomists hoped to get clues to understanding morphologic details in higher vertebrates. In Scandinavia the first-year medical students dissected the hagfish as a model of a primitive vertebrate (Müller, 1922), and research on *Myxine* was performed at both zoological and medical institutes. Schreiner, professor of human anatomy in Oslo, Norway, published results from extensive studies on *Myxine* carried out over almost 60 years (e.g. Schreiner, 1898, 1955 and 1957), and colleagues at the same university institute followed his choice of research object. When Cole (1925) finished a series of monographs on the morphology of *Myxine* by publishing a survey of the vascular system, the anatomy and histology of *Myxine* had become relatively well known, but progress in other sectors long lagged behind. However, since around 1950–60 problems and techniques from physiology, biochemistry, genetics, molecular biology, immunology, pathology, ecology and palaeontology have become increasingly important. New and more complete and nuanced informations on the myxinoid vertebrates are continuously accumulating.

THE UNIQUENESS OF HAGFISHES

Although in their organization hagfishes follow a vertebrate pattern, they are in many respects unique. This is even more evident if other than morphological properties are considered. Some of these points of uniqueness of the hagfishes are listed below.

Liver and biliary system

The liver is remarkably large and has a well-developed gall bladder, but unlike the liver of other vertebrates it is built as a tubular gland. The metabolism of the hagfish liver has not

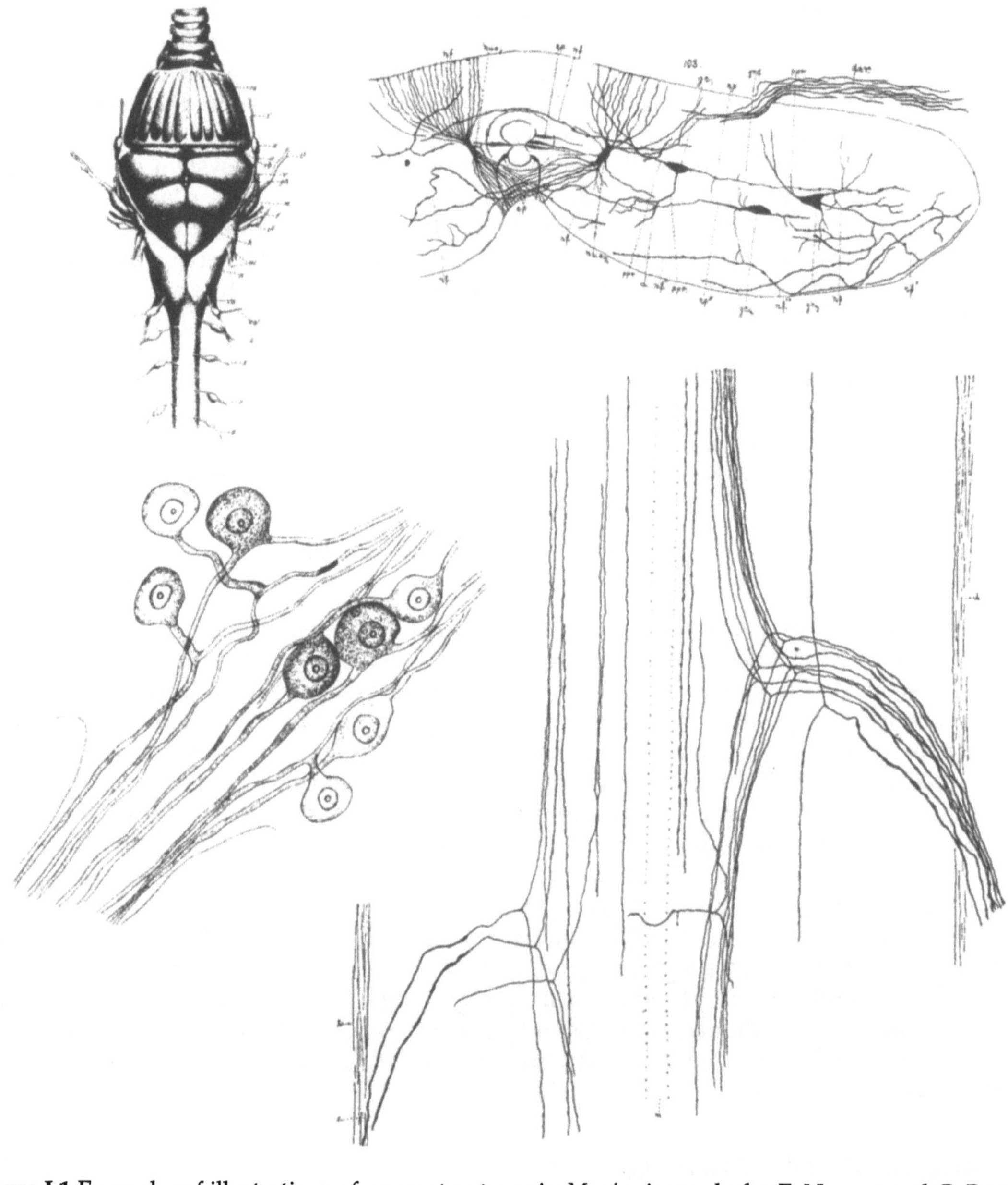

Figure I.1 Examples of illustrations of nerve structures in *Myxine* in works by F. Nansen and G. Retzius. Upper left: dorsal view of the brain (probably prepared by the nitric acid method; see the text), drawn by Ingrid Andersson on behalf of Retzius. Upper right: part of a transverse section through the spinal cord showing neurones (nf), glia cells (gc) and a dorsal nerve root (dnr), stained by the Golgi method, drawn by Nansen 'under the camera lucida'. Lower left: bipolar and unipolar neurons in a spinal nerve ganglion. Methylene blue staining, drawn by G. Retzius. Lower right: Y-shaped branching of dorsal root nerve fibres at their entrance into the spinal cord. Golgi staining, drawn by G. Retzius. Similar pictures are found in Nansen (1887). (After Nansen, 1887, Retzius, 1890b, and Retzius, 1893; the sizes are slightly changed from the originals.)

been much investigated but probably to a considerable extent it resembles that of other vertebrates. For instance it produces a cytochrome P-450 (Andersson and Nilsson, 1989). The gall bladder is controlled by cholinergic nerves, not by the hormone cholecystokinin, and the bile contains a bile alcohol (myxinol disulphate) instead of bile acids (Haslewood, 1966).

Pancreas
A specific variety of insulin is produced by cells around the bile duct (Cutfield *et al.*, 1979) and exocrine pancreatic substances, including colipase (Sternby *et al.*, 1983), are produced by intestinal cells.

Blood and immune system
The haemoglobin of the blood is monomeric with unusual composition and function. The defence system produces no immunoglobulins but a complement-like factor may be formed in response to antigenes.

Circulatory system
Hagfishes have a low pressure circulation, and the arterial walls contain an unusual type of elastic fibres (Welsch and Potter, 1994). Pulsatory activities and vascular tone are influenced by hormones rather than nerves. The heart, and other tissues, have a remarkable capacity of anaerobic metabolism.

Nervous and sensory systems
Eyes, lateral line system, inner ear and pineal complex of the brain are poorly developed, but the skin, especially in the tentacles of the head, has a rich sensory innervation. The physiologic implications of the lack of myelin in the nervous system (Peters, 1964) are not known.

Excretory system
Several authors assume that hagfishes have always been living in salt water. However, peculiarities of the kidney of hagfishes may indicate that their ancestors were once adapted to fresh or brackish water. The renal corpuscles are relatively enormous, resembling in size those of freshwater fishes (Nash, 1931) or amphibians. Reabsorption of sodium from the urine, a capacity which is important for freshwater life, occurs at a low rate in the hagfish kidney (McInerney, 1974). Moreover, hagfishes possess a prominent pronephros and in certain vertebrates (larval Urodela) the pronephros is an osmoregulatory organ eliminating water from the body (Fox, 1963). In myxinoids the pronephros has the appearance of a ciliated apparatus pumping fluid from the body cavity (peritoneal fluid) into the venous blood, and the pronephric duct has regressed. The myxinoid pronephros possibly represents a vestigial water excretory structure. H.W. Smith and A.S. Romer both defended the theory that vertebrates in general originated in fresh water (Smith, 1959a; Romer, 1955). Griffith (1994) assumes that vertebrates once were anadromous, migrating from salt water into fresh water to spawn. The ancestral hagfishes might have returned to a purely marine life very early during evolution, whereas the lampreys, and to a large extent the gnathostomes, remained anadromous.

Reproduction
The reproduction biology remains a mystery. The relatively small size of sperm-producing organs may indicate internal fertilization, but no copulation organs have been observed and

Fernholm (1975) supports the hypothesis that hagfishes of both sexes deposit the gonadal products in burrows in the mud. The number of myxinoid embryos which have until now been discovered and examined is extremely limited (Wicht and Northcutt, 1995; Wicht and Tusch, this volume). It is remarkable that so little is known on the embryology and reproduction in spite of the extensive commercial fishery of hagfishes now going on (Gorbman *et al.*, 1990; Honma, this volume).

Some features of the hagfishes may be originally primitive, inherited from extinct vertebrates or protovertebrates, but others, such as the poor development of certain sensory organs, could be secondary adaptations to a burrowing habit. Only few features, if any, indicate affinity to worms (annelids) or other invertebrates contrary to the imaginations of some of the early investigators. Thus the tooth apparatus has sometimes been compared with a molluscan radula. Liljeqvist *et al.* (1982) noticed a similarity in the primary structure of haemoglobin from *Myxine* and from an insect, *Chironomus*. Future sequence analyses of homeotic genes, etc., may reveal similarities in gene structure not yet anticipated.

REFERENCES

Abildgaard, P.C. (1792) Kurze anatomische Beschreibung des Säugers (*Myxine glutinosa* Linn.). *Schriften der Gesellschaft naturforschender Freunde zu Berlin*, **10**, 193–200 (1 pl.).

Andersson, T. and Nilsson, E. (1989) Characterization of cytochrome P-450-dependent activities in hagfish, dogfish, perch and spectacle caiman. *Comparative Biochemistry and Physiology*, **94B**, 99–105.

Bloch, M.E. (1793) *Naturgeschichte der ausländischen Fische* (7. Theil, 66–74, Pl. 413), J. Morino & Comp., Berlin, pp. 66–74.

Bowman, Sir W. (1842) On the structure and use of the Malpighian bodies of the kidney, with observations on the circulation through that gland. *Philosophical Transactions of the Royal Society London*, **132**, 57 (cit. from Smith, 1959b).

Cole, F.J. (1925) A monograph on the general morphology of the myxinoid fishes, based on a study of *Myxine*. VI. The morphology of the vascular system. *Transactions of the Royal Society of Edinburgh*, **54**(II), 309–42, 5 Pl.

Cutfield, J.F., Cutfield, S.M., Dodson, E.J., Dodson, G.G., Emdin, S.F. and Reynolds, C.D. (1979) Structure and biological activity of hagfish insulin. *Journal of Molecular Biology*, **132**, 85–100.

Fernholm, B. (1975) Ovulation and eggs of the hagfish *Eptatretus burgeri*. *Acta Zoologica* (*Stockholm*), **56**, 199–204.

Fox, H. (1963) The amphibian pronephros. *Quarterly Review of Biology*, **38**, 1–25.

Gorbman, A., Kobayashi, H., Honma, Y. and Matsuyama, M. (1990) The Hagfishery of Japan. *Fisheries*, **15**, 12–18.

Griffith, R.W. (1994) The life of the first vertebrates. *BioScience*, **44**, 408–17.

Gunnerus, J.E. (1763) Om Sleep-Marken. *Det Trondhiemske Selskabs Skrifter*, **2**, 250–7 (1 Pl.).

Haslewood, G.A.D. (1966) Comparative studies on bile salts. Myxinol disulfate, the principal bile salt of hagfish, *Myxine glutinosa* L. *Biochemical Journal*, **100**, 133–7.

Kalm, P. (1753) *En Resa till Norra AMERICA* (New edn 1966, ed. M. Kerkkonen, *Resejournal över resan till Norra Amerika*. Svenska Litteratursällskapet, Helsingfors).

Liljeqvist, G., Paléus, S. and Braunitzer, G. (1982) Hemoglobins, XLVIII. The primary structure of a monomeric hemoglobin from the hagfish, *Myxine glutinosa* L. Evolutionary aspects. *Journal of Molecular Evolution*, **18**, 102–8.

Linnaeus, C. (1754) *Museum S:ae R:ae M:tis Adolphi Friderici regis* (90–92, Pl. VIII, 4). Typographia Regia, Holmiae.

Linnaeus, C. (1758) *Systema Naturae. Regnum Animale*. Stockholm. (A photographic facsimile of the first volume of the tenth edition. British Museum, Natural History, 1956.)

Lönnberg, E. (1924) Cyclostomi, in *Dr. H.G. Bronn's Klassen and Ordnungen des Tier-Reichs*, 6. Band, I. Abteilung. Pisces (Fische), I. Buch. Leipzig: Akademische Verlagsgesellschaft, pp. 250–336.

McInerney, J.E. (1974) Renal absorption in the hagfish, *Eptatretus stouti*. *Comparative Biochemistry and Physiology*, **49A**, 273.

Müller, E. (1922) *Lärobok i ryggradsdjurens jämförande anatomi* (2nd edn), A. Bonnier, Stockholm.

Müller, J. (1836) Vergleichende Anatomie der Myxinoiden. I. Osteologie und Myologie. *Abhandlungen der königlichen Akademic der Wissenschaften zu Berlin*, **1834**, 65–340.

Müller, J. (1839) Ueber den eigenthümlichen Bau des Gehörorganes bei den Cyclostomen, mit Bemerkungen über die ungleiche Ausbildung der Sinnesorgane bei den Myxinoiden. *Abhandlungen der königlichen Akademie der Wissenschaften zu Berlin*, **1837**, 15–48.

Müller, J. (1839) Vergleichende Neurologie der Myxinoiden. *Abhandlungen der königlichen Akademie der Wissenschaften zu Berlin*, **1838**, 171–251.

Müller, J. (1841) Vergleichende Anatomie der Myxinoiden. III. Ueber das Gefässsystem. *Abhandlungen der königlichen Akademie der Wissenschaften zu Berlin*, **1839**, 175–304.

Müller, J. (1845) Untersuchungen über die Eingeweide der Fische. Schluss der vergleichenden Anatomie der Myxinoiden. I. Ueber die Eingeweide der Myxinoiden. *Abhandlungen der königlichen Akademie der Wissenschaften zu Berlin*, **1843**, 109–70.

Nansen, F. (1887) The structure and combination of the histological elements of the central nervous system. *Bergens Museums Aarsberetning (for 1886)*, 29–214.

Nansen, F. (1888) A protandric hermaphrodite (*Myxine glutinosa*, L) amongst the vertebrates. *Bergens Museums Aarsberetning (for 1887)*, 3–34.

Nash, J. (1931) The number and size of glomeruli in the kidney of fishes, with observations on the morphology of the renal tubules of fishes. *American Journal of Anatomy*, **47**, 425–45.

Peters, A. (1964) An electron microscope study of the peripheral nerves of the hagfish (*Myxine glutinosa*, L.). *Quarterly Journal of Experimental Physiology*, **49**, 35–42.

Retzius, A.J. (1790) Anmärkningar vid Slägtet MYXINE. *Kongliga Vetenskapsacademiens Nya Handlingar, Stockholm*, **11**, 110–117 (1 pl.).

Retzius, A. (1822) Bidrag till Åder-och Nerfsystemets Anatomie hos *Myxine glutinosa*. *Kongliga Vetenskapsacademiens Handlingar, Stockholm*, 233–47 (1 pl.).

Retzius, A. (1824) Ytterligare bidrag till anatomien af Myxine glutinosa. *Kongliga Vetenskapsakademiens Handlingar. Stockholm*, 408–31 (1 pl.).

Retzius, G. (1890a) Ein sogenanntes Caudalherz bei *Myxine glutinosa*. *Biologische Untersuchungen N.F.*, **1**, 94–6 (Pl. 18).

Retzius, G. (1890b) Ganglienzellen der Cerebrospinalganglien und über subcutane Ganglienzellen der Myxine glutinosa. *Biologische Untersuchungen N.F.*, **1**, 97–8. (Pl. 18).

Retzius, G. (1893) Das Gehirn und das Auge von Myxine. *Biologische Untersuchungen, N.F.*, **5**, 55–63 (Pl. 24–6).

Romer, A.S. (1955) Fish origins – fresh or salt water? *Papers in marine Biology and Oceanography. Deep Sea Research, Suppl.*, **3**, 261–80.

Schreiner, K. (1898) Zur Histologie des Darmkanals bei *Myxine glutinosa*. *Bergens Museum Aarbog*, **1**, 1–16.

Schreiner, K.E. (1955) Studies on the gonad of *Myxine glutinosa L. Universitetet i Bergen, Årbok 1955. Naturvitenskapelig rekke*, **8**, 1–40.

Schreiner, K.E. (1957) The pancreas-like organ of *Myxine glutinosa*. *Avhandlinger utgitt av Det Norske Vitenskapsakademi i Oslo*, **1**, 1–19.

Smith, H.W. (1959a) *From Fish to Philosopher* (2nd edn), CIBA Pharmaceutical Products, Summit, N.J.

Smith, H.W. (1959b) Highlights in the history of renal physiology. *The Bulletin-Georgetown Medical Center*, **13**, 1–48.

Sternby, B., Larsson, A. and Bergström, B. (1983) Evolutionary studies on pancreatic colipase. *Biochimica et Biophysica Acta*, **750**, 340–5.

Welsch, U. and Potter, I. (1994) Variability in the presence of elastic fibre-like structures in the ventral aorta of agnathans (hagfishes and lampreys). *Acta Zoologica*, **75**, 323–7.

Wicht, H. and Northcutt, R.G. (1995) Ontogeny of the head of the Pacific hagfish (*Eptatretus stouti*, Myxinoidea): development of the lateral line system. *Philosophical Transactions of the Royal Society London B*, **349**, 119–34.

PART ONE

Evolution, Taxonomy and Ecology

1
RELATIONSHIPS OF LIVING AND FOSSIL HAGFISHES

David Bardack

SUMMARY

The fossil record of agnathans now includes representatives of the myxinoids (hagfish) as well as lampreys. The hagfish is from late Pennsylvanian age deposits and part of the taxonomically diverse Mazon Creek biota in northeastern Illinois. The fossil comes from shallow marine sediments. The specimen shows tentacles, parts of the head skeleton and some soft tissues. The fossil differs from living hagfish in the position of gills, feeding apparatus and relatively well developed eyes but is otherwise quite similar to extant hagfish. Fossils of other jawless fishes, both skeletal and soft-bodied forms, are discussed.

1.1 INTRODUCTION

It is more than 30 years since contributors to Brodal and Fänge (1963) explored the biology of *Myxine*, and since then knowledge of agnathan fishes and early chordates has grown significantly (Forey and Janvier, 1993, 1994). In this period, the first fossils of lampreys (Bardack and Zangerl, 1968) and hagfishes (Bardack, 1991) were discovered and intact specimens recognizable as the long sought conodont animal (Aldridge *et al.*, 1993; see Aldridge and Donoghue, this volume) were found. Also, new and better preserved specimens of previously known agnathan fossils were collected. But more importantly, during these years entirely new material referable to new orders of the jawless fishes with hard external skeletons (ostracoderms) were discovered and studied (Janvier, 1996b). Some of these were reported from continents where they had been unknown. Finally during the last 30 years overall knowledge of the biology of living agnathans coupled with new techniques and characteristics for analysing relationships (Maisey, 1986) among groups has revitalized interest in jawless fishes (Hardisty, 1982; Forey, 1995; Janvier, 1993, 1996a, b).

The existence of living jawless fishes, lampreys and hagfishes, provides material which on the one hand is helpful to elucidate the origin of jawed fishes and on the other hand anatomy and other biological characteristics of living jawless fishes provide useful starting points for understanding relationships of fossil jawless forms. We will examine what is known about the single fossil hagfish, its anatomy and relationships, then briefly review its position among fossil agnathans.

1.2 THE SINGLE FOSSIL HAGFISH

Fossils of jawless fishes are recorded from the Paleozoic, primarily Ordovician to Devonian. With the inclusion of conodonts (Aldridge *et al.*, 1993) and some isolated scales (Smith *et al.*, 1996) among the jawless fishes, the fossil record of such 'fishes' may be extended back to the Cambrian and up to the end of the Triassic. The fossil evidence for agnathans similar to extant ones is confined to the Mississippian and Pennsylvanian. Except for the lampreys and hagfishes (and conodonts),

The Biology of Hagfishes. Edited by Jørgen Mørup Jørgensen, Jens Peter Lomholt, Roy E. Weber and Hans Malte. Published in 1998 by Chapman & Hall, London. ISBN 0 412 78530 7.

most Paleozoic agnathans are characterized by bony head shields and unusually thick body scales. In the nineteenth century the chordate affinities of many ostracoderms were unknown. But with new approaches to studying the material which were developed early in this century, especially in Sweden, and particularly through the reconstruction of the fossils using serially sectioned individuals and living agnathans as models (Stensiö, 1964), the anatomy of fossil agnathans was determined and proposals of relationships among these early fishes emerged.

While fossils comparable with living agnathans are not common, they are sufficient to provide conclusive evidence for their identification, the extension of the geologic range of living groups and information on their basic anatomy. There are two genera of fossil lampreys, *Mayomyzon* from the Westphalian D, Pennsylvanian of northeastern Illinois (Bardack and Zangerl, 1968) about 300 million years ago and *Hardistiella* from the Namurian A, Mississippian of eastern Montana (Janvier and Lund, 1983; Lund and Janvier, 1986) about 320 million years ago. *Hardistiella* is based on two specimens preserved in lateral view showing the external outline and a few other features. *Mayomyzon* was established on the basis of half-a-dozen specimens but there are now many additional individuals although most show few internal features. Most are preserved in lateral view and a few in addition to the holotype show internal cartilaginous and soft structures. One specimen is preserved in dorsoventral view (Bardack and Zangerl, 1971). *Myxinikela siroka*, the single genus and species of a fossil hagfish and also its only known specimen (Bardack, 1991) is from the same geologic horizon and the same general locality as *Mayomyzon*. Anatomical features of *Myxinikela* are almost as well preserved as those of *Mayomyzon*.

The Pennsylvanian localities yielding *Mayomyzon* and *Myxinikela* represent low-lying areas spanning a large delta similar to that of the Mississippi River today (Baird *et al.*, 1985). The region lay close to the then (Pennsylvanian) Equator. The geography and environment of the area were characterized by a complex series of channels with fresh to brackish water succeeded by and intertongued with marine limestones. Thus there are at some sites, fossils representing terrestrial and freshwater aquatic animals and plants and at others purely marine types. But in some of the latter sites, there are fossils of freshwater/terrestrial origin among the predominantly marine forms. Weather patterns in the delta were no doubt similar to those in many tropical and subtropical systems today. Animals and plants of freshwater or terrestrial origin were swept into the saline waters by floods of upstream origin or carried to these waters on receding storm surges. Fossils from all of the sites are termed the Mazon Creek Biota. This diversified fauna and flora includes upwards of 600 named genera and more species. The fossils are found on the bedding planes of ironstone concretions which formed diagenetically around the organisms. The fossil hagfish and lamprey appear as darker markings representing cartilaginous and soft tissues on these bedding planes.

The deposits yielding the Mazon Creek biota have produced fossils since the 1840s. Initially these were from the Mazon Creek itself and later from shaft mines reaching to coal beds underlying the fossiliferous rocks. Finally, during the middle of this century, recovery of fossils peaked with collections from vast strip mines. The latter were developed in areas predominantly of marine deposition. Unfortunately the coal in these deposits has been exhausted. A nuclear energy plant has been built where some of the strip mines were and on the rest of the land, a cooling lake constructed. Thus it is unlikely that more material, particularly of the less common fossils, will be obtained. It should be noted that all concretions do not contain fossils. The core of many concretions is a mineral but most finds of organic origin are fragments of plants. Animals are much less common and among these, vertebrates still less so. Estimates of the number of concretions which have been broken apart and examined in the last 150 years are in the 10–15 million range. Thus the discovery of one hagfish and a handful of lampreys must be considered in this perspective.

The Mazon Creek biota is one of the more diverse fossil assemblages known. It includes the major carboniferous plant groups and about 10 animal phyla. Lists of the fossil groups appear in Maples and Schultze (1988), Bardack (1989) and Shabica and Hay (1997).

Study of hagfish and lampreys from the Mazon Creek biota primarily involves optical microscopy with rotation of the light source to obtain a series of contrasting shadows which make more visible the slight ridges or grooves on the concretion surface. Emphasizing these markings on photographs produces a pattern which can be related to known structures in living jawless fishes. For comparative purposes, illustrations of cranial and other structures of living myxinoids particularly from Parker (1884) and Marinelli and Strenger (1956) have been essential in reconstructing the anatomy of *Myxinikela*.

1.3 DESCRIPTION

The single specimen (Figures 1.1 and 1.2) of *Myxinikela* is 7.2 cm in total length with a

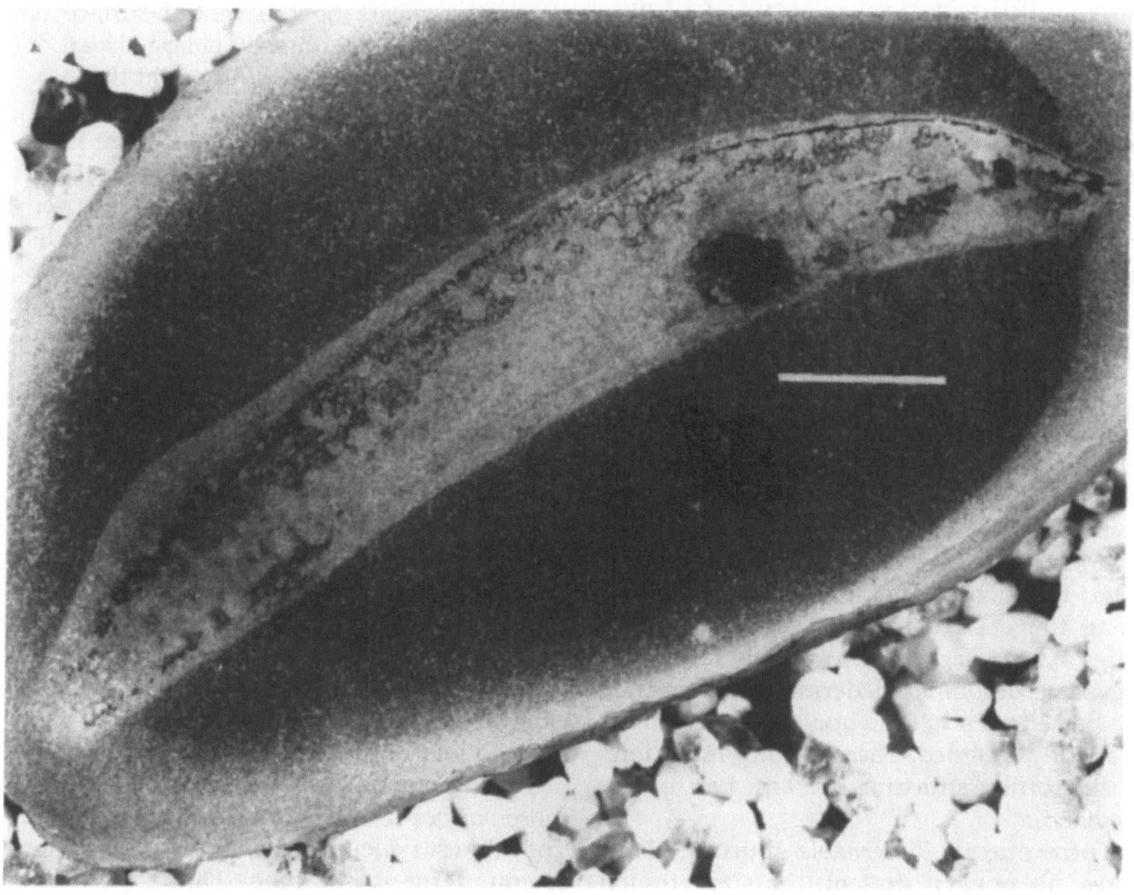

Figure 1.1 *Myxinikela siroka* Bardack. Whole specimen, scale bar = 1 cm.

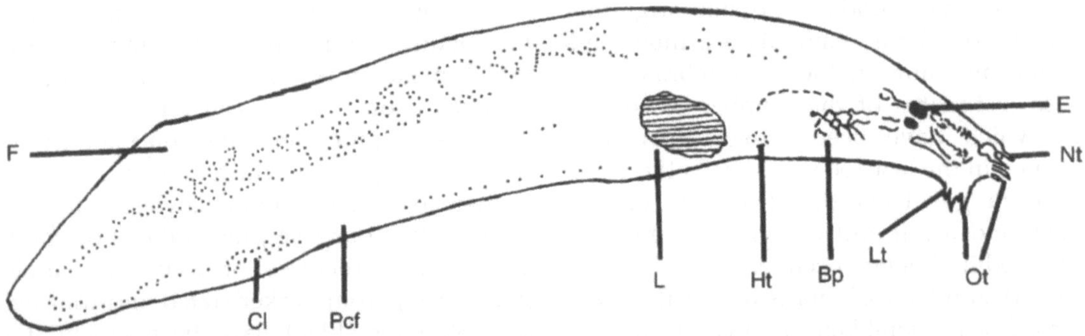

Figure 1.2 *Myxinikela siroka* Drawing of whole specimen. Abbreviations: Bp, branchial pouches; Cl, opening of cloaca; E, eye; F, dorsal part of median fin; Ht, heart; L, liver; Lt, labial tentacles; Nt, nasal tentacles; Ot, oral tentacles; Pcf, precloacal part of median fin.

maximum body depth of 1.2 cm. The anterior part of the specimen is bent slightly downward so that the head is seen dorsolaterally. While the three-dimensional body shape is not directly determinable from the specimen, the slightly down-twisted head suggests a tubular animal. Although the body is of uniform height for most of its length, the posterior fifth tapers rapidly to the caudal end. A distinct external feature is a continuous median fin extending along about one-half of the dorsal surface. It appears to arise somewhat anterior to mid-body, extend around the caudal end and then anteriorly on the ventral surface about as far as a prominent dark stain representing part of the digestive organs. Where the body tapers to the caudal end, the dorsal part of the median fin becomes one-and-a-half times higher than anteriorly. Demarcation of the median fin from the body is emphasized, especially along the posterior fifth of the body by a series of irregular blotches possibly representing some decomposed dermal pigments. At the posterior end of the body, the median fin is slightly rounded. There are no indications of supporting structures for any part of the median fin.

There are no discernable mucous gland pores, although if present, the size of such pores and the generally lateral preservation of the body might make it difficult to see them. A slight swelling also marked by some darker pigments appears on the ventral surface of the body below the dorsally enlarged part of the median fin and should represent the location of the cloacal opening.

At the head end of the animal (Figure 1.3) a series of tentacles is seen. These features alone confirm the identification of the specimen as a myxiniform. The four pairs lie peripheral to the nasal tube or mouth. Starting ventrolaterally from the mouth a pair of short, stubby structures represent the labial tentacles. Anterior to these the more elongate oral tentacles seem to extend somewhat more posteriorly from the mouth than in living hagfishes. As preserved, these are the largest of the tentacles. Dorsally two pairs of nasal tentacles extend anteriorly from the opening of the nasal tube. The lower tentacle projects from the same cartilage that supports the oral tentacles while the upper pair of nasal tentacles, much shorter than the lower pair, is supported by processes extending from the elongate subnasal cartilage. The mouth opens at an oblique angle to the body axis.

Several cranial skeletal structures besides those supporting the tentacles are preserved. The darkest skeletal markings are parts which surround the nasal tube (Figure 1.3). The subnasal cartilage forms a long dark, thin in

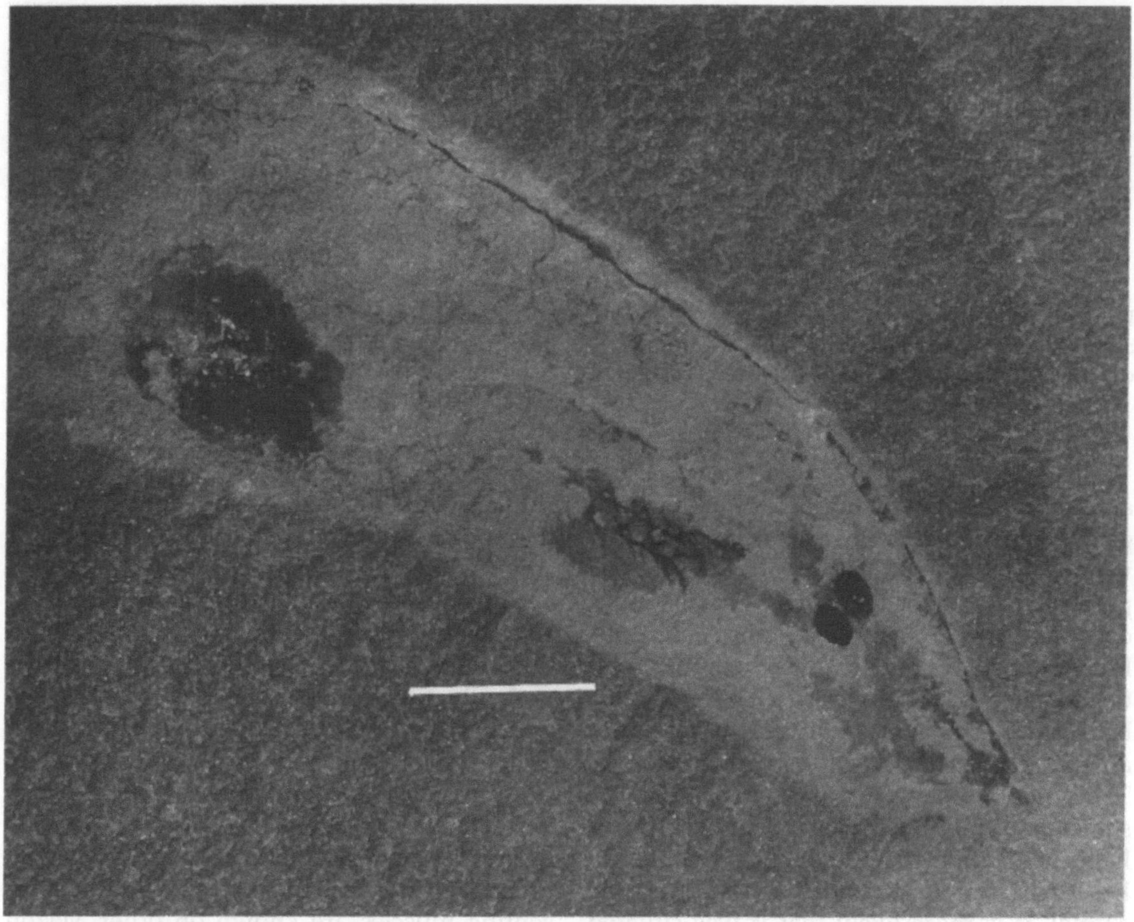

Figure 1.3 *Myxinikela siroka* View of head. Scale bar = 0.5 cm.

dorsal view, median band. Remnants of the series of circular nasal tube cartilages are suggested by extensions to either side of the subnasal cartilage. At its anterior end the nasal tube cartilage is broad and dense with several perforations. Such thickening is also seen anteriorly in *Eptatretus* (Figure 1.4) where the nasal tube cartilages are thicker and wider than in *Myxine*. About halfway between the anterior end of the head and the eyes, a broad stain represents the nasal capsule and marks the posterior end of the nasal tube canal. A series of elongate openings in this cartilage are similar to those in living hagfishes.

Other skeletal structures are less well developed or preserved and their outlines indistinct. They appear as discrete, continuous stains and no doubt were parts of the cartilaginous chondrocranium (Figure 1.5). A small, elongate stain, possibly a remnant of the palatine cartilage, lies below and anterior to the nasal capsule. Two relatively massive cartilages appear behind the mouth. These should be the anterior lingual and dentigerous cartilages. Their borders and shapes are indistinct. In living hagfishes, there are teeth on the dentigerous cartilage. Several discrete, particularly dark stains on this structure suggest the presence of these teeth including

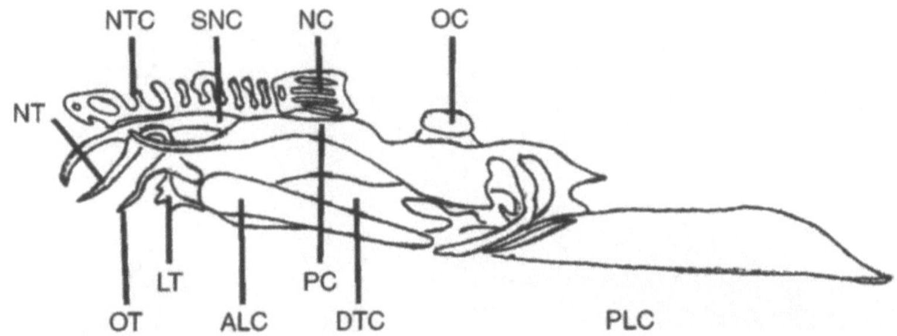

Figure 1.4 *Eptatretus* sp. Head skeleton. Abbreviations: ALC, anterior lingual cartilage; DTC, dentigerous cartilage; LT, labial tentacle; NC, nasal cartilage; NT, nasal tentacles; NTC, nasal tube cartilage; OC, otic capsule; OT, oral tentacles; PC, palatine cartilage; PLC, posterior lingual cartilage; SNC subnasal cartilage.

at least one seemingly in the form of an elongate curve. There is no reason to think this would be an ethmoid tooth which, as preserved, would have to have been turned over and backwards. The more posterior lingual cartilages are not preserved or not yet (?ontogenetically) developed. They may also be formed somewhat differently than the anterior cartilages (Wright *et al.*, this volume). The anterior lingual cartilage appears to lie somewhat more anterior than in living hagfishes. While one possibility is that it may have been drawn forward after death of the animal, it is more likely that it was situated more anteriorly in this extinct form where the lingual cartilage complex is generally less developed either ontogenetically or phylogenetically in comparison with living forms. Behind the eyes are two elongate parts of the chondrocranium probably representing pieces of the *pila anterior* and *pila posterior*. The otic capsules, at least one of which is seen as a darker ring, lie on top of this part of the chondrocranium.

The eyes are a dominant feature of the fossil (Figures 1.3 and 1.5). They lie close together, closer to the otic capsules than nasal capsule and are separated from each other by a space less than the width of the eye. The eyes, longer than wide, are about 1.2 mm long and 0.9 mm wide. An external view of *Myxine* does not show eyes but they are dimly noted below unpigmented skin on *Eptatretus* (Fernholm and Holmberg, 1974). There is a small white spot close to the center of each eye and seen better on the left than right eye. The material is a secondary clay deposit in the outer surface of the eye. This material probably fills a small opening into the eye which may or may not have held a small lens. Eyes with such clay-filled openings are preserved similarly on other Mazon Creek vertebrates, e.g. bony fishes known to have lenses. There is no evidence of extrinsic eye muscles. Impressions and swellings representing muscles are known on other Mazon Creek fossils (e.g. *Gilpichthys*) but eye muscles in such small specimens as the hagfish would probably be too small to be preserved as recognizable structures.

Behind the otic capsules and anterior to the digestive organs a pair of broadly elongate and apparently thickened areas can be recognized. In length these are about equal to the distance between the mouth and the otic capsule. These swollen areas contain the branchial apparatus, its enveloping musculature, related blood vessels and posteriorly the heart. A pattern of seemingly irregularly distributed black, intersecting lines is seen at the top and side of these areas and represents parts of the complex branchial circulatory system. These lines separate a series of clear rectangular and oval areas. Interpretation of

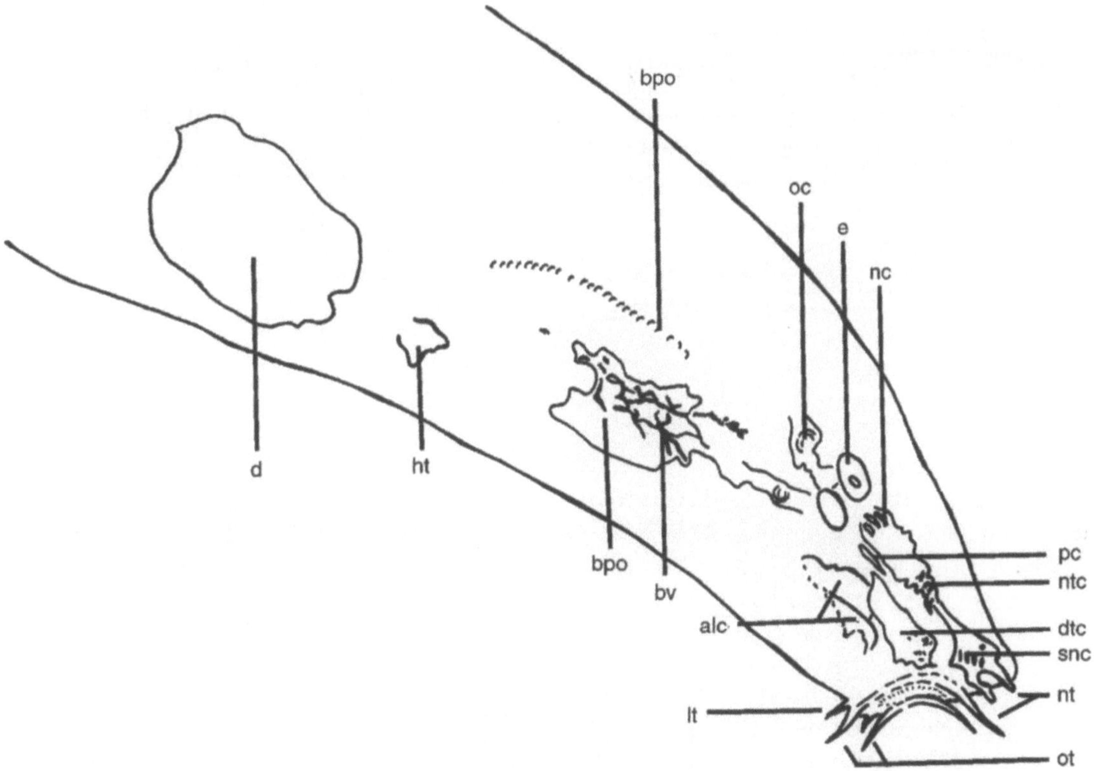

Figure 1.5 *Myxinikela siroka* drawing of head skeleton and associated structures. Abbreviations: alc, anterior lingual cartilage; bpo, branchial pouches and outline of branchial area; bv, branchial blood vessels, d, digestive organ (primarily liver); dtc, dentigerous cartilage; e, eye; ht, heart; lt, labial tentacles; nc, nasal capsule; nt, nasal tentacles; ntc, nasal tube cartilage; oc, otic capsule; ot, oral tentacles; pc, palatine cartilage; and snc, subnasal cartilage.

structures here is difficult but one can make out some parts of the respiratory and related circulatory features based on comparison with *Myxine* (Marinelli and Strenger, 1956; Hardisty, 1979). A median dorsal aorta seems to lie above the pharynx with branches (efferent aortae) extending to and surrounding the afferent branchial ducts before entering the branchial pouches (Figure 1.6). The latter appear as a series of slightly bulbous blocks. Smaller branches emerge from the efferent aortae and extend longitudinally along the pharynx. These are probably parts of the lateral aorta. Although, as preserved, these black intersecting lines are similar in coloration to those of some of the head skeletal structures, their shapes, arrangement and number preclude their being part of the branchial skeleton. There is no indication of a branchial skeleton *per se*. The few parts of the branchial skeleton in living myxinoids are quite small. Also, in living myxinoids, the branchial apparatus lies further posteriad to the head skeleton, a separation that is emphasized during the ontogenetic process (Neumayer, 1938; Holmgren, 1943).

Just behind the branchial pouch area a somewhat diffusely marked region represents the position of the heart. Posterior to the heart, a large dark mass with irregular boundaries must be a part of the digestive system, presumably the liver. There is no

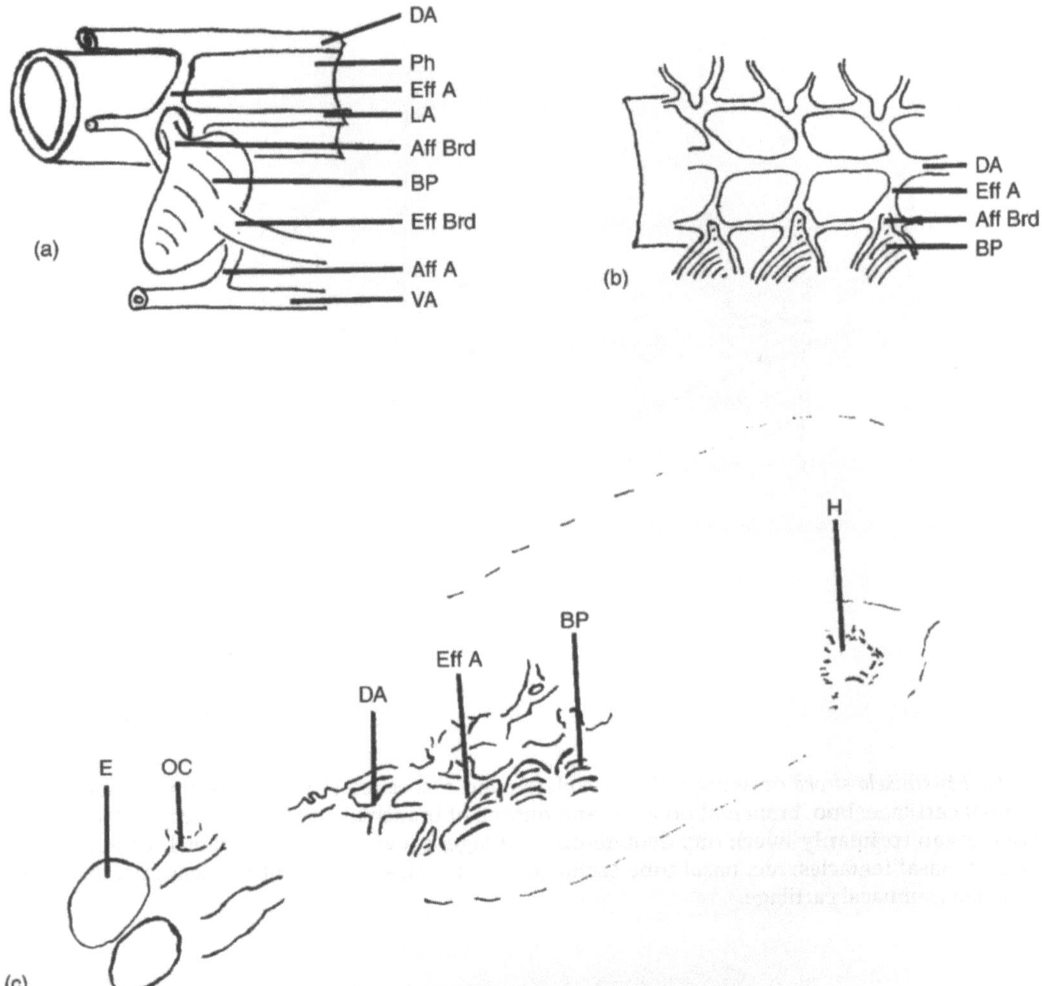

Figure 1.6 Branchial pouches and related circulatory structures. (a) *Myxine glutinosa* adapted from Marinelli and Strenger (1956). (b) *Myxinikela siroka*. Restoration of dorsal view of branchial circulation. (c) *Myxinikela siroka*. Sketch of branchial circulation, branchial pouches and adjacent structures. Abbreviations: Aff A, afferent artery; Aff Brd, Afferent branchial duct; BP, branchial pouch; DA, dorsal aorta; E, eye; Eff A, efferent artery; Eff Brd, efferent branchial duct; H, heart; LA, lateral aorta; OC, otic capsule; Ph, pharynx; VA, ventral aorta.

evidence of other body organs or any axial skeleton.

1.4 DISCUSSION

What is the significance of the fossil hagfish? First, with the discovery of a fossil myxinoid, the two extant groups of jawless fishes, hagfishes and lampreys, are now represented in the fossil record. Second, the fossils demonstrate that animals of each type, each of which is closely similar to its living jawless relatives, have been in existence for well over 300 million years. Third, the record also shows that living jawless fishes had a long, conservative evolutionary history which, to judge from their distinct characteristics when they appeared (Pennsylvanian for myxinoids and Mississippian for lampreys), must have had an even older, still unrecorded geologic history. Fourth, no features of *Myxinikela* suggest a common ancestry with *Mayomyzon* or the ostracoderms.

Some differences between *Myxinikela* and living adult hagfishes include the position of the branchial structures which lie closer to the head in *Myxinikela*, the position of the oral tentacles lying possibly somewhat more posteriad in *Myxinikela*, the possibly minimal development of skeletal elements in the fossil compared to *Myxine*, the relatively large eyes of *Myxinikela*, and the somewhat more anterior position of the feeding apparatus. Together these features suggest that the specimen of *Myxinikela* may be a juvenile individual. Indeed its length, about 7 cm, is certainly within the lower end of the size range of living juvenile myxinoids. In showing possibly juvenile features when compared to adult living hagfishes, *Myxinikela* duplicates the structural contrasts that the fossil lamprey *Mayomyzon* shows to living adult lampreys (Bardack and Zangerl, 1971). Also, a preponderant number of Mazon Creek biota vertebrates (including actinopterygians and sarcopterygians) are juveniles rather than adults (Schultze and Bardack, 1987). It may be that the near shore, deltaic region from which many of the Mazon Creek biota specimens come represents a nursery type of environment and that the habitat appropriate for adults has not been sampled or preserved. The water depth where the hagfish and lampreys were found had to be relatively shallow compared to the usual depth at which hagfishes at least are found today. Perhaps the hagfish and lampreys matured elsewhere into adults not too different from living forms. The extreme rarity of hagfishes may reflect the fact that these lived primarily elsewhere. But it is also possible that *Myxinikela* is an adult. Those features which characterize living adult hagfishes may have evolved phylogenetically later in the *Myxinikela* lineage or in a different hagfish lineage leading directly to modern forms.

Studies of relationships between hagfishes and lampreys, and between these and the ostracoderms have produced a large literature with several, often quite different, cladograms (Blieck, 1992; Forey, 1984, 1995; Forey and Janvier, 1993, 1994; Gagnier, 1995; Halstead, 1982; Janvier, 1996a, 1996b). Also, the discovery of intact conodont animals with some of their soft anatomy preserved is stimulating examination of the phylogenetic connections of the conodonta with jawless fishes (Aldridge *et al.*, 1993; Aldridge and Donoghue, this volume). The fossil hagfish being so similar to living forms does not alter the position of myxinoids on the cladograms of agnathan relationships which have been developed in recent years. Consequently only a brief perspective on the relationship of hagfish to other groups of agnathans is presented here.

Seven different groups of ostracoderm fishes comprise a fossil record extending over 125 million years. Among these are the osteostracans (Janvier, 1993), heterostracans (Elliott, 1987; Blieck *et al.*, 1991) and anaspids (Arsenault and Janvier, 1991) all of which have been known for many years. Another group, the thelodonts (Turner, 1991), also long known but of uncertain affinities, may possibly be larval forms of one or more of the groups named above. They have generated interest in recent years with new discoveries in northern Canada (Wilson and Caldwell, 1993). In only the last 20 years, discovery of three new groups, arandaspids of the Early Ordovician, South America and Australia (Gagnier, 1993), galeaspids of Silurian and Devonian, China and Vietnam (Wang, 1991) and pituriaspids of Early Devonian, Australia (Young, 1991), have raised still more interest in the anatomy, evolution and biogeography of jawless fishes.

For much of this century living jawless fishes were considered a closely related group, the cyclostomes. Recently, arguments based on ribosomal RNA (Stock and Whitt, 1992), other molecular evidence (Goodman *et al.*, 1987), morphologic position of the gill apparatus (Schaeffer and Thomson, 1980) and lingual and feeding apparatus (Mallatt, 1984; Yalden, 1985) supported this position. In the last two decades however, the majority of studies recognize that many characteristics of different organ systems and their biochemistry (Hardisty, 1982) distinguish lampreys from hagfishes and move the latter on a cladogram to a position of sister group to all other craniates. These cladograms place lampreys among some of the ostracoderms and give them a position closer to the gnathostomes (Forey and Janvier, 1993, 1994). More recently several additional cladograms utilizing different out groups, including or not including conodonts, including or not including non-mineralized fossil forms have been published and the reader is referred to these for more details (Forey, 1995; Janvier, 1996a, b). The primary differences among these trees with their often large numbers of unresolved branches lie among the ostracoderms and in some cases the lampreys but not the hagfish.

Finally, two other genera of jawless fishes should be noted here. Among the unusual soft bodied jawless fishes from the same fossil assemblage yielding the lamprey and the hagfish are two genera, *Pipiscius* and *Gilpichthys*. These two were assigned (Bardack and Richardson, 1977) to the Agnatha. Janvier (1993, 1996a) suggested that *Pipiscius* might be related to lampreys and *Gilpichthys* to myxinoids (Janvier, 1981, 1993, 1996a). The primary distinguishing feature of *Pipiscius* and the one which led Janvier to suggest an affinity with lampreys is its complex oral sucking disc which he thought 'recalls' *Ichthyomyzon*. With regard to *Gilpichthys*, Janvier said that its body shape (elongate) and its unmineralized teeth were suggestive of myxinoids. The case for both being chordates and agnathans is presented in Bardack and Richardson (1977). Both are characterized by unique, distinctive feeding structures. *Pipiscius* has a segmented muscular 'double-decker' mouth apparatus showing no evidence of teeth. In *Gilpichthys* there is an elongate oropharyngeal apparatus made up of a linear series of muscular block segments of decreasing size. Anteriorly these blocks seem to have supported fibrous teeth. It is unlikely that either *Pipiscius* or *Gilpichthys* merits placement among the known living or fossil agnathans. But they both demonstrate that agnathans were more diversified in the past than the few fossil forms with unmineralized skeletons or the numerous ostracoderm types would suggest. The Mazon Creek biota where so many soft bodied animals are preserved provides an unusual glimpse of the diversity of life of the past. In this regard, there are at least two additional fossils from the Mazon Creek biota which suggest that they may represent fossil agnathans but neither shows sufficient characteristics to warrant description.

REFERENCES

Aldridge, R.J., Briggs, D.E.G., Smith, M.P., Clarkson, E.N.K. and Clark, N.D.L. (1993) The anatomy of conodonts. *Philosophical Transactions of the Royal Society, London Ser B*, **340**, 405–21.

Arsenault, M. and Janvier, P. (1991) The anaspid-like craniates of the Escuminac formation (Upper Devonian) from Miguasha (Quebec, Canada) with remarks on anaspid-petromyzontid relationships, in *Early Vertebrates and Related Problems of Evolutionary Biology* (eds M-m Chang, Y-h Liu and G-r Zhang), Science Press, Beijing, pp. 19–40.

Baird, G., Shabica, C.W., Anderson, J.C. and Richardson, E.S. Jr (1985) Biota of a Pennsylvanian muddy coast. Habitats within the mazonian delta complex. *Journal of Paleontology*, **59**(2), 253–81.

Bardack, D. (1989) Some comments on the Mazon Creek Biota. *Acta Musei Reginaehradecensis*, ser. *Scientiae Naturales*, **22**, 53–9.

Bardack, D. (1991) First fossil hagfish (myxinoidea): a record from the Pennsylvanian of Illinois. *Science*, **254**, 701–3.

Bardack, P. and Richardson, E.S. Jr (1977) New agnathous fishes from the Pennsylvanian of Illinois. *Fieldiana Geology*, **33**(26), 489–510.

Bardack, P. and Zangerl, R. (1968) First fossil lamprey: a record from the Pennsylvanian of Illinois. *Science*, **162**, 1265–7.

Bardack, P. and Zangerl, R. (1971) Lampreys in the fossil record, in *Biology of Lampreys* (eds M.W. Hardisty and I.C. Potter), New York, Academic Press, Vol. 1, pp. 67–84.

Blieck, A. (1992) At the origin of the chordates, *Geobios* **25**, 101–113.

Blieck, A., Elliott, D. and Gagnier, P-Y. (1991) Some questions concerning the phylgenetic relationships of heterostracans, Ordovician and Devonian jawless vertebrates, in *Early Vertebrates and Related Problems of Evolutionary Biology* (eds M-m Chang, Y-h Liu and G-r Zhang), Science Press, Beijing, pp. 1–17.

Brodal, A. and Fänge, R. (eds) (1963) *The Biology of Myxine*, Universitatsforlaget, Oslo.

Elliott, D. (1987) A reassessment of Astraspis desiderata, the oldest North American vertebrate. *Science*, **237**, 190–2.

Fernholm, B.K. and Holmberg, K. (1974) The eyes in three genera of hagfish (*Eptatretus*, *Paramyxine*, and *Myxine*) a case of degenerative evolution. *Vision Research*, **15**, 253–79.

Forey, P.L. (1984) Yet more reflections on agnatha–gnathostome relationships. *Journal of Vertebrate Paleontology*, **4**(3), 330–43.

Forey, P.L. (1995) Agnathans recent and fossil, and the origin of jawed vertebrates. *Reviews in Fish Biology and Fisheries*, **5**, 267–303.

Forey, P.L. and Janvier, P. (1993) Agnathans and the origin of jawed vertebrates. *Nature*, **361**, 129–34.

Forey, P.L. and Janvier, P. (1994) Evolution of the early vertebrates. *American Scientist*, **82**, 554–65.

Gagnier, P-Y. (1993) *Sacabambaspis janvieri*, vertébrés ordovicien de Bolivie. *Annales Paléontologie*, **79**, 119–66.

Gagnier, P-Y. (1995) Ordovician vertebrates and agnathan phylogeny. *Bulletin Museum Histoire Naturelle, Paris*, ser. **17**, sec. C, 1–37.

Goodman, M., Miyamoto, M.M. and Czelusniak, J. (1987) Patterns and process in vertebrate phylogeny revealed by coevolution of molecules and morphology, in *Molecules and Morphology in Evolution: Conflict or Compromise?* (ed. C. Patterson), pp. 141–76.

Halstead, L.B. (1982) Evolutionary trends and the phylogeny of the agnatha, in *Problems of Phylogenetic Reconstruction* (eds K.A. Joysey and A. Friday), Systematics Association Special Papers, Academic Press, London, pp. 159–96.

Hardisty, M.W. (1979) *Biology of the Cyclostomes* Chapman and Hall, London, pp. 293–333.

Hardisty, M.W. (1982) Lampreys and hagfishes: analysis of cyclostome relationships, in *The Biology of Lampreys*, Vol. 4b (eds M.W. Hardisty and I.C. Potter), Academic Press, New York, pp. 165–260.

Holmgren, N. (1943) On two embryos of *Myxine glutinosa*. *Acta Zoologica*, **27**, 1–90.

Janvier, P. (1981) The phylogeny of craniata with particular reference to the significance of fossil 'agnathans'. *Journal of Vertebrate Paleontology*, **2**(4), 121–59.

Janvier, P. (1993) Patterns of diversity in the skull of jawless fishes, in *The Skull*, Vol. 2 (eds J. Hanken and B.K. Hall), University of Chicago Press, Chicago, pp. 131–88.

Janvier, P. (1996a) The dawn of the vertebrates: characters versus common ascent in the rise of current vertebrate phylogenies. *Paleontology*, **39**, 259–287.

Janvier, P. (1996b) Early vertebrates. *Oxford Monographs in Geology and Geophysics*, **33**, 393 pp.

Janvier, P. and Lund, R. (1983) *Hardestiella montanensis* n. gen et sp. (petromyzontida) from the Lower Carboniferous of Montana with remarks on the affinities of the lampreys. *Journal of Vertebrate Paleontology*, **2**(4), 407–13.

Lund, R. and Janvier, P. (1986) A second lamprey from the Lower Carboniferous (Namurian) of Bear Gulch, Montana (U.S.A.). *Geobios*, **19**(5), 647–52.

Maisey, J. (1986) Heads and tails: a chordate phylogeny. *Cladistics*, **2**(3), 201–56.

Mallatt, J. (1984) Early vertebrate evolution: pharyngeal structure and the origin of gnathostomes. *Journal of the Zoological Society London*, **204**, 169–183.

Maples, C.G. and Schultze, H-P. (1988) Preliminary comparison of Pennsylvanian assemblage of Hamilton, Kansas with marine and non-marine contemporaneous assemblages, in *Regional Geology and Paleoecology of Upper Paleozoic Hamilton Quarry area in Southeastern Kansas* (eds G. Mapes and R.H. Mapes), Kansas Geological Survey Guidebook, series 6, pp. 253–73.

Marinelli, W. and Strenger, A. (1956) *Myxine glutinosa* (L.). *Vergleichende Anatomie und*

Morphologie der Wirbeltiere, Franz Deuticke, Vienna, Bd II, pp. 81–172.

Neumayer, L. (1938) Die Entwicklung des Kopfskelettes von *Bdellostoma stoutii. Archivio italiano de anatomia e embriologia*, suppl. **40**, 1–122.

Parker, W.K. (1884) On the skeleton of the marsipobranch fishes. Part 1. The myxinoids (*Myxine* and *Bdellostoma*). *Philosophical Transactions of the Royal Society, London*, **174**, 373–409.

Sansom, I.J., Smith, M.P. and Smith, M.M. (1994) Dentine in conodonts. *Nature*, **368**, 591.

Schaeffer, B. and Thomson, K.S. (1980) Reflections on the agnathan–gnathostome relationship, in *Aspects of Vertebrate History* (ed. L.I. Jacobs), Museum of Northern Arizona, Flagstaff, pp. 19–33.

Schultze, H-P. and Bardack, D. (1987) Diversity and size changes in palaeonisciform fishes (Actinopterygii, Pisces) from the Pennsylvanian Mazon Creek fauna, Illinois (U.S.A.). *Journal of Vertebrate Paleontology*, **7**(1), 1–23.

Shabica, C. and Hay, A.A. (eds) (1997) *Richardson's Guide to the Fossil Fauna of Mazon Creek. Northeastern Illinois* (in press).

Smith, M.P., Sansom, I.J. and Repetski, J.E. (1996) Histology of the first fish. *Nature*, **380**, 702–4.

Stensiö, E.A. (1964) Les cyclostome fossiles ou ostracodermes, in *Traité de Paléontologie* (ed. J. Piveteau), Masson, Paris, pp. 96–382.

Stock, D.W. and Whitt, G.S. (1922) Evidence from 18S ribosomal RNA sequences that lampreys and hagfishes form a natural group. *Science*, **257**, 787–9.

Turner, S. (1991) Monophyly and interrelationships of the thelodonti, in *Early Vertebrates and Related Problems of Evolutionary Biology* (eds M-m Chang, Y-h Liu and G-r Zhang), Science Press, Beijing, pp. 87–119.

Wang, N-z (1991) Two new Silurian galeaspids (jawless craniates) from Zhejiang Province, China, with a discussion of galeaspid–gnathostome relationships, in *Early Vertebrates and Related Problems of Evolutionary Biology* (eds M-m Chang, Y-h Liu and G-r Zhang), Science Press, Beijing, pp. 41–65.

Wilson, M.V.H. and Caldwell, M.W. (1993) New Silurian and Devonian forked tailed 'thelodonts' are jawless vertebrates with stomachs and deep bodies. *Nature*, **361**, 442–4.

Yalden, D.W. (1985) Feeding mechanisms as evidence for cyclostome monophyly. *Zoological Journal of the Linnean Society, London* **84**, 291–300.

Young, G. (1991) First armoured agnathan vertebrates from the Devonian of Australia, in Early Vertebrates and Related Problems of Evolutionary Biology (eds M-m Chang, Y-h Liu and G-r Zhang), Science Press, Beijing, pp. 67–85.

2

CONODONTS: A SISTER GROUP TO HAGFISHES?

Richard J. Aldridge and Philip C.J. Donoghue

SUMMARY

Conodonts are an extinct group of naked agnathan fishes which range in age from Cambrian to Triassic. The conodont animal is almost exclusively represented in the fossil record by the phosphatic elements of the feeding apparatus, which was the only mineralized component of the skeleton. Only 12 specimens have been found which preserve the soft tissue anatomy of the animal.

The animal possessed a notochord, myomeres, caudal fin, paired sensory organs (optic and possibly otic) and extrinsic eye musculature; these characters indicate that the animal was a vertebrate. Just posterior of the eyes and ventral of the notochord lay a feeding apparatus of varying complexity that acted bilaterally as in hagfishes, differing from the dorsoventral arrangement and action of gnathostome jaws.

The hard tissues from which the feeding apparatus is composed are comparable with those of vertebrates, particularly other fossil agnathans and corroborate the phylogenetic position established on the basis of the soft tissue anatomy. Although conodont soft tissues suggest a relationship to hagfishes, the elements cannot be homologized with hagfish lingual 'teeth' because of fundamental differences in the modes of growth of these structures.

2.1 INTRODUCTION

Conodonts are an extinct group of chordates, represented in the fossil record almost exclusively by the phosphatic elements of their feeding apparatuses. They possessed no other biomineralized skeleton, and remained enigmatic until the discovery of the first of a number of fossils with preserved soft tissues in 1982 (Briggs *et al.*, 1983). Conodont soft tissues are now known from three separate localities: the Ordovician Soom Shale of South Africa (Aldridge and Theron, 1993; Gabbott *et al.*, 1995), the Silurian Brandon Bridge dolomite of Wisconsin, USA (Mikulic *et al.*, 1985a, b; Smith *et al.*, 1987), and the Carboniferous Granton Shrimp Bed of Edinburgh, Scotland (Briggs *et al.*, 1983; Aldridge *et al.*, 1986, 1993). The single Silurian specimen from Wisconsin is very poorly preserved and provides little information about conodont anatomy, but the Soom and Granton specimens preserve several features of the trunk and head. It must be emphasized, however, that the preservation of particular tissues and organs has been highly selective, and the processes of replacement that led to the preservation of non-biomineralized tissues are currently poorly understood. Replacement of muscles by calcium phosphate, as displayed by the Granton specimens, has been replicated in the laboratory by Briggs *et al.* (1993), but the preservation of

The Biology of Hagfishes. Edited by Jørgen Mørup Jørgensen, Jens Peter Lomholt, Roy E. Weber and Hans Malte. Published in 1998 by Chapman & Hall, London. ISBN 0 412 78530 7.

muscle fibres by clay minerals, evident in the Soom Shale, is problematic, although it may involve an intermediate phase of phosphate replacement (Gabbott *et al.*, 1995). Whatever the preservational history of these specimens, it is clear that each exhibits only part of the soft anatomy of the original organism, biased by the particular characteristics of the chemical and microbiological environment in which it died and decayed. Using information gleaned from several specimens, however, it has proved possible to reconstruct many of the characters of the living conodont animal, although details of features of low preservation potential remain obscure.

2.2 CONODONT SOFT-TISSUE ANATOMY

Ten specimens from the Granton Shrimp Bed exhibit features of the trunk of the animal (Figure 2.1); two of these also preserve the tail, and two show structures in the head (Aldridge *et al.*, 1993). A single giant specimen (Figure 2.2) from the Soom Shale displays part of the trunk and head region (Gabbott *et al.*, 1995), while at least 40 have been found in which lobate structures, interpreted as eye cartilages by Aldridge and Theron (1993), are associated with complete feeding apparatuses. All of these fossils were subject to some decay before the processes of replacement which preserved the tissues commenced, but experimental examination of the pattern of decay in extant primitive chordates, principally *Branchiostoma* (Briggs and Kear, 1994), provides a basis for interpretation of the structures that remain. These features can be compared with those of living and fossil cephalochordates and agnathans to develop hypotheses regarding the phylogenetic position of the Conodonta in relation to the Myxinoidea.

2.2.1 General features

The conodont animal specimens from the Ordovician and Carboniferous are all from taxa that possessed complicated feeding apparatuses comprising pectiniform and ramiform elements. These taxa represent at least two conodont orders: seven of the specimens from the Granton Shrimp Bed can be assigned to *Clydagnathus*, an ozarkodinid (Aldridge *et al.*, 1993), and the Soom Shale specimen is of the prioniodontid genus *Promissum* (Gabbott *et al.*, 1995). General features of the anatomy are remarkably constant; all are elongate with a short head and a laterally compressed trunk made up of somites (Figure 2.1(1)). These are apparently V-shaped in all specimens, although preservation may be incomplete; they are thus simpler than the W-shaped myomeres of adult hagfishes and lampreys, but comparable with the chevron muscle blocks of *Branchiostoma* and some fossil agnathans (e.g. *Sacabambaspis*, see Gagnier *et al.*, 1986; *Mayomyzon*, see Bardack and Zangerl, 1968, 1971; *Gilpichthys and Pipiscius*, see Bardack and Richardson, 1977). The *Clydagnathus* specimens are all small, with the largest a little over 55 mm in total length (Aldridge *et al.*, 1993), whereas the preserved portion of the *Promissum* specimen is 109 mm and the entire length may have approximated 400 mm (Gabbott *et al.*, 1995).

2.2.2 The trunk

Paired axial lines occur along the trunk of most of the Granton specimens (Figure 2.1(3)) and represent the margins of the notochord (Aldridge *et al.*, 1993); the notochord of *Branchiostoma* is one of the most decay-resistant features of this animal and collapses to a pair of lateral ridges comparable with those shown by the fossil conodonts (Briggs and Kear, 1994). Preferential preservation of the notochord is also apparent in a number of fossil agnathans from other deposits, including *Gilpichthys* (Bardack and Richardson, 1977) and *Mayomyzon* (Bardack and Zangerl, 1968, 1971) from the Carboniferous Mazon Creek fauna. The notochord is not preserved

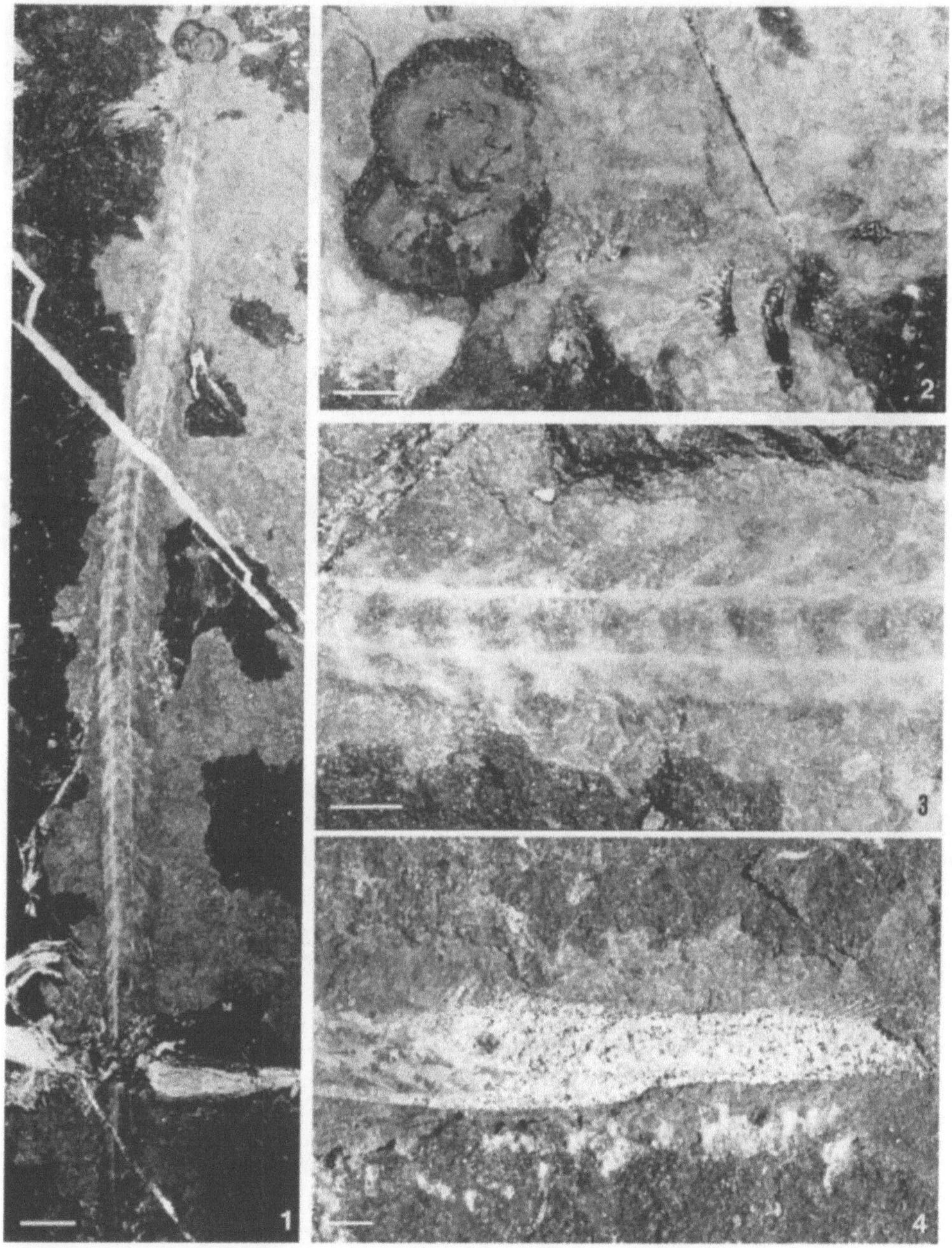

Figure 2.1 *Clydagnathus windsorensis* (Globensky); (1–3) RMS GY 1992.41.1 (refigured from Aldridge *et al.*, 1993, with permission); (4) IGSE 13822 (refigured from Briggs *et al.*, 1983, with permission). (1) Complete specimen, anterior at top and ventral to left; scale bar 2000μm. (2) Anterior portion showing eye cartilages, feeding apparatus (only partially uncovered) and anterior part of trunk with notochord; scale bar 500μm. (3) Detail of trunk at mid-length showing the notochordal sheath and shrunken myotomes; scale bar 500μm. (4) Posterior portion of trunk and tail showing closely set ray supports and tail asymmetry; scale bar 500μm.

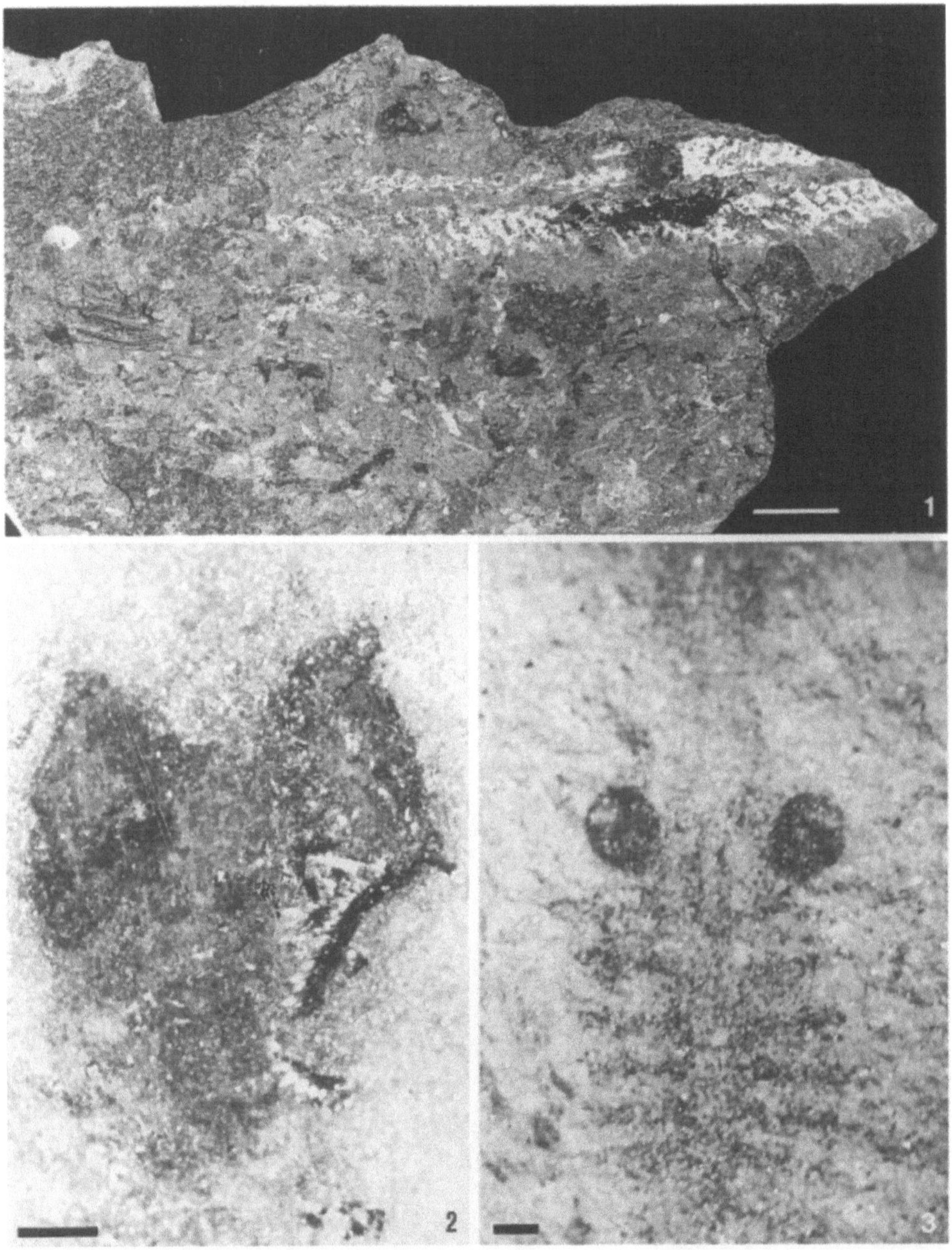

Figure 2.2 (1) *Promissum pulchrum* Kovàcs-Endrödy (GSSA C721a; refigured from Gabbott *et al.*, 1995, with permission), complete specimen (counterpart) showing the trunk, extrinsic eye musculature and feeding apparatus; anterior to left; scale bar 10 mm. (2) Head of C. *windsorensis* (IGSE 13822; refigured from Briggs *et al.* 1983, with permission) showing eye capsules, otic capsules, possible traces of gill pouches, and feeding apparatus, anterior at top (soft tissues preserved in dorsoventral orientation); scale bar 500 µm. (3) Head of *Myomazon pieckoensis* Bardack (FMNH PF 8167) a fossil lamprey from the Carboniferous Mazon Creek lagerstätte, showing nasal and eye capsules, gill pouches and trace of notochord, preserved orientation as (2); scale bar 500 µm.

in *Promissum*, but its position is indicated by a 2 mm gap in preservation within the myomeres (Gabbott *et al.*, 1995). A dorsal nerve cord may be represented on two of the Granton specimens by a medial darker trace apparent along the anterior portion of one wall of the notochord, although this interpretation remains equivocal (Aldridge *et al.*, 1993). Details of the structure of the trunk muscles are best preserved in the *Promissum* specimen, in which each myomere displays sets of fibril bundles, together with possible sarcolemmic membranes and collagenous connective tissues (Gabbott *et al.*, 1995). The fibres do not show the extreme flattening characteristic of *Branchiostoma*, and appear more circular in cross-section than those of agnathans and fishes; their size (5 μm in diameter) is consistent with their being slow muscle fibres (Gabbott *et al.*, 1995). Larger, fast muscle fibres have not been recognized, although these may be present outside the plane along which the fossil has split.

2.2.3 The tail

Closely spaced fin rays are apparent at the posterior end of two of the Granton fossils (Briggs *et al.*, 1983, figs 4, 5; Aldridge *et al.*, 1986, fig. 4; Figure 2.1(4)), but their configuration is not clear on either. From one of the specimens it is evident that fins occur on both the dorsal and ventral margins; more examples are required before we can ascertain for certain if the disposition is symmetrical, or if the apparent extension of the fin further on one of the margins is genuine. There is no evidence of articulating musculature at the base of the fin rays, suggesting that they resemble the unsupported fin folds of myxinoids (Aldridge *et al.*, 1993).

2.2.4 The head

Apart from the phosphatic feeding apparatus, the most commonly preserved features of the head are the two lobate structures interpreted by Aldridge and Theron (1993) to represent sclerotic cartilages which surrounded the eyes (Figures 2.1(2) and 2.2(2)). In specimens from Granton and from South Africa, these are evident as apparently carbonized impressions, commonly thickened marginally and with some phosphatization; they can be reconstructed as deep, inwardly tapering hollow rings (Aldridge *et al.*, 1993). They are positioned above and immediately anterior to the feeding apparatus and are closely comparable morphologically with structures that have been interpreted as eye capsules in fossil agnathans, for example *Jamoytius* (Ritchie, 1968), and as altered retinal pigments in the hagfish *Myxinikela* (Bardack, 1991), the lampreys *Mayomyzon* (Figure 2.2) and *Hardistiella* (Bardack, this volume), and larval gnathostomes such as *Esconichthys* (Bardack, 1974), *Bandringa* and *Rhabdoderma* (Richardson and Johnson, 1971). Optic capsules in living craniates are embryologically derived from ectodermal placodes (Gans and Northcutt, 1983).

The most complete *Promissum* specimen displays solid white oval patches anterior to and above the feeding apparatus, in a similar position to the sclerotic rings on other specimens (Figure 2.2(1)). These patches have a fibrous texture and were interpreted as representing extrinsic eye musculature by Gabbott *et al.* (1995); the development of such muscles is entirely patterned by connective tissue derived from neural crest (Noden, 1991; Couly *et al.*, 1992).

Other than indistinct and indecipherable patches, only the first specimen discovered from Granton has additional soft-tissue features in the head region (Briggs *et al.*, 1983, figs 2C and 3A; Aldridge *et al.*, 1993; Figure 2.2). A pair of small subcircular dark patches behind the sclerotic rings may represent the otic capsules, similar to those reported in the Carboniferous lamprey *Mayomyzon* (Bardack and Zangerl, 1971) and hagfish *Myxinikela* (Bardack, 1991). The presence of otic capsules is further supported by the occurrence of a phosphatic structure, strongly resembling the statoliths of modern lampreys, in the vicinity of the feeding apparatus in the head of another of the Granton conodont animals (Figure 2.4(2)). Transverse traces posterior to the eyes of the first specimen may be branchial structures, comparable with features so interpreted in *Mayomyzon* (Bardack and Zangerl, 1968, 1971). There is no preserved evidence of pharyngeal slits.

The relative arrangement of the structures in the head of the conodont animal is closely comparable with that shown by fossil and recent lampreys and hagfishes (Figures 2.2(2) and 2.2(3)). The otic capsules are positioned just posterior of the optic capsules, and the putative gill pouches are located very close to the head structures, as in *Mayomyzon*. The first Granton conodont animal also preserves an indistinct organic trace surrounding the head structures that resembles the unmineralized head cartilage of the fossil lampreys from the Mazon Creek fauna.

The feeding apparatus is only partly exposed in most of the specimens from Granton, but its architecture has been reconstructed using additional evidence from undisturbed assemblages of elements found occasionally on Carboniferous shale surfaces. The apparatus was bilaterally symmetrical, comprising a set of 11 ramiform elements that formed an anterior basket, behind which lay two pairs of pectiniform elements with their long axes directed dorsoventrally (Aldridge *et al.*, 1987; Purnell and Donoghue, in press; Figure 2.3). The anterior half of the apparatus has been interpreted as an oral raptorial array (Aldridge and Briggs, 1986; Purnell and von

Figure 2.3 Model of the conodont (ozarkodinid) feeding apparatus in oblique anteroventral view. From Purnell and Donoghue (*in press*).

Bitter, 1992), but this has recently been challenged (Mallatt, 1996). Mallatt contended that the position of this portion of the apparatus posterior of the eyes suggests that it lay in the pharynx. However, if the interpretation of gill pouches in the first Granton specimen is correct (Figure 2.2(2)), the position of the anterior array, anterior of the first gill pouch, and the pectiniform elements posterior, implies that the two portions of the apparatus were located in the oral cavity and pharynx respectively, thus falling into the 'old' and 'new' mouths of Mallatt (1996).

The *Promissum* apparatus was similar to that of ozarkodinids, but more complicated, with 11 ramiform elements positioned below an array of four pairs of pectiniform elements (Aldridge *et al.*, 1995). Both types of apparatus are more complex than those found in any other agnathan, and they do not compare with the jaws of fishes. However, more primitive conodonts, and their putative ancestors the paraconodonts (Szaniawski and Bengtson, 1993), had simpler apparatuses made up of conical elements which may be more readily comparable with the lingual and palatal teeth of hagfishes. The multicuspid lingual laminae of some lampreys (Potter and Hilliard, 1987) also bear a broad resemblance to some ramiform conodont elements.

2.2.5 Phylogenetic interpretations

Of the preserved soft tissues, the notochord and the chevron-shaped myomeres clearly show that the conodonts belong within the euchordates (Cephalochordata + Craniata) (although for a contrary view see Dzik, 1995). Their precise affinities are controversial, with some authorities still maintaining that they are closest to the protochordates (Urochordata + Cephalochordata) (Kemp and Nicoll, 1995), although the radials in the caudal fin, the presence of eyes and the termination of the notochord behind them, the bilaterally operative feeding apparatus, and the phosphatic skeletal biomineralization are all craniate characters (Aldridge *et al.*, 1993; Janvier, 1995).

The possession of paired external sensory organs and a distinct head anterior of the notochord are also indications of vertebrate grade. The 'new head' hypothesis for the origin of the vertebrates (Gans and Northcutt, 1983) recognizes that most of the functional and morphological differences between vertebrates and other chordates are located in the head, and contends that the vertebrate head is a new structure. Most of the new structures in the vertebrate head are embryologically derived from neural crest and ectodermal placodes.

More recently, a single colinear cluster of *Hox* genes has been identified in *Amphioxus*, the traditional proxy for a vertebrate ancestor, matching four paralogous clusters in gnathostomes (Garcia-Fernàndez and Holland, 1994; Holland and Garcia-Fernàndez, 1996). The expression of these clusters in mice never occurs more anteriorly than the rhombomeres of the hindbrain, and expression of *Hox* genes in *Amphioxus* too has distinct anterior limits, indicating a significant portion of the animal equivalent to the craniate head. Furthermore, the single cluster in *Amphioxus* also points to a gene duplication at the acraniate–craniate transition, emphasizing the fundamental importance of this event in chordate evolution. Determination of *Hox* gene clusters in hagfish and lampreys is at a preliminary stage, but multiple clusters, up to four in number, appear to be present in each group (Holland and Garcia-Fernàndez, 1996).

Much of the opposition to the interpretation of conodonts as vertebrates stems from the lack of consensus over what constitutes a vertebrate or a craniate; many workers consider these to be synonymous (Kardong, 1995; Nielsen, 1995; Young, 1995). The 'new head' hypothesis for the origin of the vertebrates places myxinoids as the first crown-group vertebrates (Gans, 1993). Janvier (1981, 1993), however, considered the lack of arcualia in hagfishes to exclude them from the vertebrates, placing them in the craniates; the lampreys were regarded to be crown-group vertebrates. On this basis, much of the controversy surrounding the interpretation of conodont affinities becomes semantic. In the present context, it is pertinent to assess the evidence for and against a close relationship between the conodonts and the hagfishes.

Aldridge *et al.* (1986) forwarded two possible phylogenetic positions for the conodonts on the basis of the soft tissues characters: as a sister group to the Myxinoidea, or immediately crownwards of them. Other placements have been suggested (see Aldridge and Purnell, 1996), including immediately anti-crownwards of the Myxinoidea, as stem-group craniates (Peterson, 1994). Conodonts differ from myxinoids in having eyes with apparent extrinsic musculature and in bearing phosphatic, not keratinous, oral elements. Large eyes with extrinsic eye muscles are a vertebrate characteristic, but their absence in myxinoids may be degenerate rather than primitive (Northcutt, 1985). The lack of a phosphatic skeleton in hagfish may also be secondary, or it might be argued that the development of phosphatic structures in conodonts was a separate, convergent feature, unrelated to the origin of skeletons in other craniates. The mode of growth of conodont elements and the nature of their phosphatic tissues are of crucial importance in resolving this particular question.

2.3 CONODONT HARD TISSUES

A typical euconodont ('true conodont') element is constructed of two structurally distinct components, a basal body and an overlying crown, which grew by the addition of calcium phosphate on their outer surfaces (Furnish, 1938; Hass, 1941). Post-Devonian elements do not have a basal body, suggesting that its function was fulfilled by unmineralized tissue in more derived forms. The crown is composed of a crystalline, hyaline tissue punctuated by numerous incremental growth lines (Figure 2.4(5)); in most conodonts the crown also includes areas of opaque tissue, traditionally known as 'white matter' because it appears albid in incident light. The cores and tips of the cusps and denticles of conodont elements are commonly composed of this white matter (Figure 2.4(4)), which is relatively fine-grained and massive, but contains numerous cavities and fine tubules (see Lindström and Ziegler, 1981). The basal body is also finely crystalline, but much more variable in structure; it commonly displays growth increments and may show spherical or tubular features (Figure 2.4(6)).

2.3.1 Lamellar crown tissue

A homology between conodont crown tissue and the enamel of vertebrates has been suggested several times (e.g. Schmidt and Müller, 1964; Dzik, 1986; Sansom *et al.*, 1992). Although only a few taxa have as yet been examined in detail, there is considerable variability in the orientation of crystallites in the hyaline lamellae with respect to the incremental growth lines. In most, the crystallites are more or less perpendicular to the growth increments (*contra* Schultze, 1996), as in true enamel, whereas one area of the crown tissue figured by Sansom *et al.* (1992) from *Parapanderodus* (fig. 3F) showed crystallites arranged at a shallower angle and this was considered outside the range of known enamel types by Forey and Janvier (1993). However, crystallite arrangement in enamel is known to vary, particularly in primitive forms of prismatic enamel (Smith, 1989, 1992), and Sansom (1996) has described a prismatic form of lamellar crown tissue that compares directly to primitive prismatic enamel from the teeth of a sarcopterygian fish. As is the case with enamel, the lamellar crown of conodont elements exhibits variation in crystallite arrangement within a single specimen.

The interpretation of the conodont tissue as enamel has been contested by Kemp and Nicoll (1995, 1996) on the grounds that etched

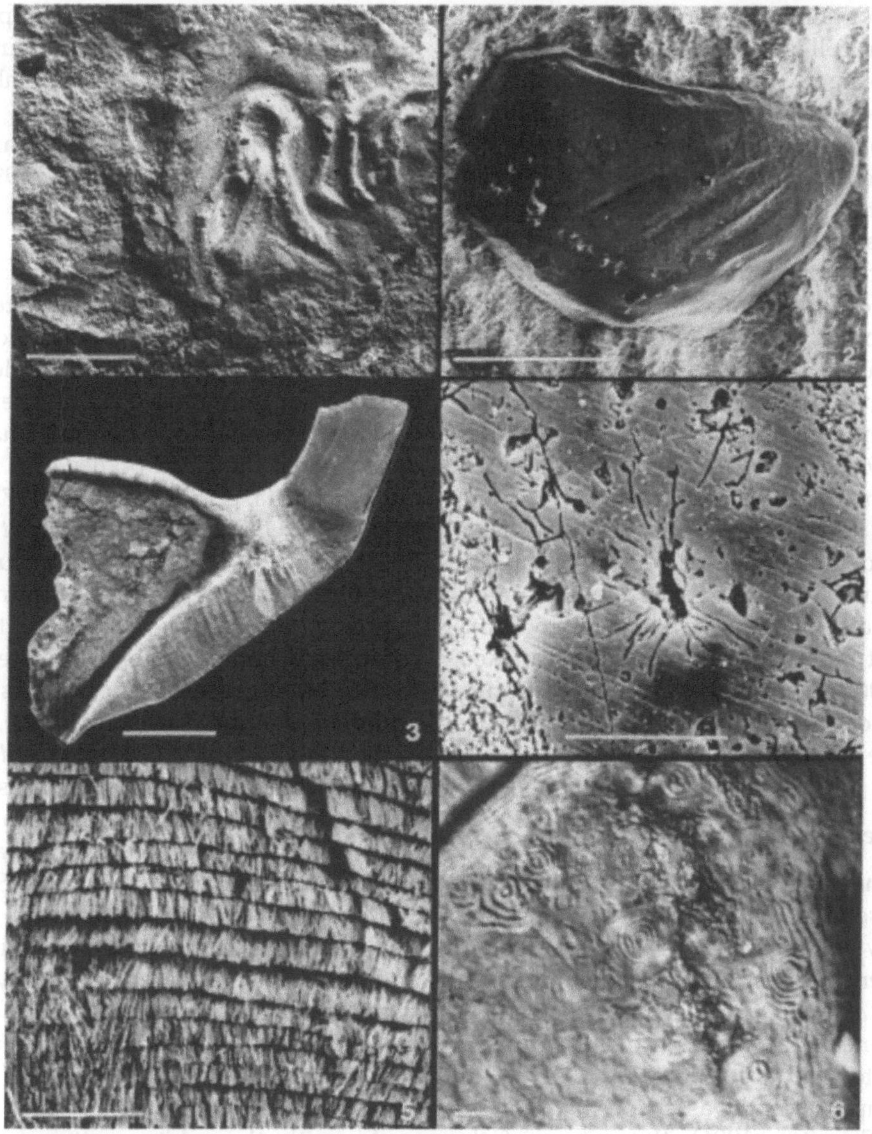

Figure 2.4 (1) Feeding apparatus from one of the Scottish conodont animals (RMS GY 1992.41.3), still partially covered by matrix. The small black asymmetric structure to the upper left of the frame is the putative statolith; scale bar 1000 µm. (2) Close-up of possible statolith, concentric grooves may represent the limit of annual growth increments; scale bar 100 µm. (3) Thin section of *Cordylodus*, a Lower Ordovician conodont, Maardu Beds, Estonia (BU 2614), micrograph taken using differential interference contrast, showing crown (to right) and basal body (to left). The hyaline crown tissue incorporates 'white matter' upper right; scale bar 100 µm. (4) SEM micrograph of an etched thin section through an element of *Ozarkodina* Upper Silurian, Gotland (BU 2615), showing fine grained ground mass and enclosed cell and cell-process spaces characteristic of white matter; scale bar 10 µm. (5) SEM micrograph of an etched thin section through the enamel crown tissue of an element of *Scaliognathus* Carboniferous, North America (BU 2613), showing incremental growth lines and crystallites organized into protoprisms; scale bar 10 µm. (6) Micrograph of detail of section 3 taken using differential interference contrast, showing lamellar and spheroidal structures in the basal body; scale bar 10 µm.

surfaces are stained by picrosirius red, a stain specific for collagen. True enamel does not contain collagen. However, the validity of such histochemical tests on fossil material remains to be established, as they have not been applied to unequivocal fossil vertebrate material. It is possible that the etching of the element surface increases porosity and permits retention of the stain which 'fixes' by electrostatic attraction; further work is required to test the results of this technique. The presence of fibrous tissues, claimed to be collagen, in conodont elements has also been reported by Fåhraeus and Fåhraeus-van Ree (1987, 1993), who demineralized Silurian conodont hard tissues then fixed, dehydrated, sectioned and stained the residue. They commented that the most remarkable result of their study was that tissue more than 400 million years old could remain histochemically intact (Fåhraeus and Fåhraeus-van Ree, 1987, p. 106). There is, however, no certainty as to which of the conodont hard tissues housed the soft tissue they recovered.

2.3.2 White matter

Many of the vacuoles within the white matter closely resemble the lacunae of odontocytes or osteocytes (Figure 2.4(4)), and together with evidence of associated canaliculi this led Sansom *et al.* (1992) to interpret this tissue as dermal bone. The vesicles are ubiquitous in white matter and are repeatedly observed in thin sections (*contra* the assertion that they are artefacts, Schultze, 1996); the nature of the tissue is different from cellular dermal bone in other vertebrates, and it is likely that white matter represents a tissue unique to conodonts.

Histochemical staining of the white matter with picrosirius red failed to indicate the presence of collagen (Kemp and Nicoll, 1995, 1996), which is present in the dentine and bone of extant vertebrates. However, it is unusual for any fossil bone or dentine to preserve collagen, which normally disintegrates shortly after death, leaving at best degradation products in the form of amino acids (Fåhraeus and Fåhraeus-van Ree, 1987). Detectable amino acids have been reported in conodont elements by Pietzner *et al.* (1968) and Savage *et al.* (1990), but see Collins *et al.* (1995).

2.3.3 The basal body

Schmidt and Müller (1964) suggested that the basal body of conodonts was homologous with the dentine of vertebrate sclerites, and branched or unbranched tubules representing different forms of dentine have been described in the basal bodies of a number of Ordovician taxa (Barnes *et al.*, 1973; Barskov *et al.*, 1982; Dzik, 1986; Sansom *et al.*, 1994). Basal bodies of other species, including most post-Ordovician elements examined, show regular lamination without tubuli or comprise a homogenous alaminate mass, the former having been interpreted as a form of atubular dentine (Sansom, 1996). In some early conodonts, for example *Cordylodus*, the basal material comprises a mass of fused spherical bodies and this has been compared with the globular calcified cartilage of the Ordovician vertebrate *Eriptychius* (Smith *et al.*, 1987; Sansom *et al.*, 1992), although it is just as likely to be an atubular dentine (Figure 2.4(3) and (6)). Such apparent diversity of tissue types in conodonts is unexpected, but parallels experimentation with different tissue combinations by other coeval agnathans (Halstead, 1987).

As with the white matter, basal bodies examined by Kemp and Nicoll (1995, 1996) failed to stain positively for collagen, although they tested positive for mucopolysaccharides.

2.3.4 Histogenesis of conodont elements

Published ontogenetic studies of conodont elements have concentrated on the development of the lamellar crown and basal body, which are known to have grown synchronously (Müller and Nogami, 1971). The pattern of divergent appositional growth between the basal body and the crown is comparable with that of the dentine and enamel of extant vertebrate teeth, and Schmidt and Müller (1964), Dzik (1976, 1986) and Smith *et al.* (1996) have argued for a homology between conodont elements and vertebrate odontodes. Odontodes are the basic building blocks of the dermal skeleton in vertebrates and are formed by interaction of the epithelium, which forms the enamel, and ectomesenchymal cells, derived from the neural crest, which ultimately form the dentine, dermal bone and cartilage. Odontodes are almost exclusively composed of a complex of enamel, dentine and underlying bone of attachment. The bone of attachment is absent in conodonts, but this is also the case in the dermal denticles of thelodonts and in the oral teeth and skin denticles of chondrichthyans (Smith *et al.*, 1996).

2.3.5 Comparison with hagfish toothlets

Both conodonts and myxinoids possess a feeding apparatus comprising a bilaterally symmetrical array of cuspate elements, and a homology between conodont elements and hagfish lingual toothlets has been proposed by Krejsa *et al.* (1990a, b). Evidence comes from a similarity in overall morphology between simple conodont elements and myxinoid teeth, and an overlap in size range between conodont elements and juvenile hagfish toothlets. The hypothesis requires that the phosphatic lamellar crown of conodont elements should be a mineralized homologue of keratin, with the pores in white matter interpreted as moribund remnants of pokal cells. The basal body is considered to be a developing replacement tooth (Krejsa *et al.*, 1990a, b).

Evidence such as analogous morphology and similarity in size is regarded as weak and circumstantial, and this interpretation of conodont elements has been severely criticized (Szaniawski and Bengtson, 1993; Smith *et al.*, 1996). Histogenetic and ontogenetic studies of conodont elements show that the crown and basal body of conodonts grew synchronously, with appositional growth increments passing confluently between the two structures (Müller and Nogami, 1971); the basal body is clearly not a replacement tooth. Indeed, except in the simplest of conical conodont elements, the upper surface of the basal body bears no morphological resemblance to the upper surface of the crown it would putatively replace.

2.4 DISCUSSION

Possible phylogenetic positions of the conodonts relative to the extant euchordates are illustrated in Figure 2.5.

Kemp and Nicoll (1995, 1996) contended that their histochemical tests prove that the hard tissues of conodont elements are not homologous with those of vertebrates, and concluded that conodonts were therefore more closely related to cephalochordates than to craniates. This is not a necessary conclusion from their arguments, even if they were correct. If conodont hard tissues were developed independently from those of vertebrates, then this could have happened at any stage in early chordate history, for example as an offshoot from the myxinoids or from the petromyzontids. The evidence from conodont soft tissues suggests that either of these positions would be more parsimonious than a sister group relationship with the cephalochordates.

How strong, then, is the evidence for the Conodonta to be considered as a sister group to the Myxinoidea? There are two hypotheses to be examined here: either conodont characters are plesiomorphic for this group and have been secondarily lost in the hagfish, or conodont hard and soft tissue features are derived and synapomorphous for the Conodonta. The latter proposal does not seem parsimonious; not only would the enamel- and dentine-like skeletal tissues of conodonts

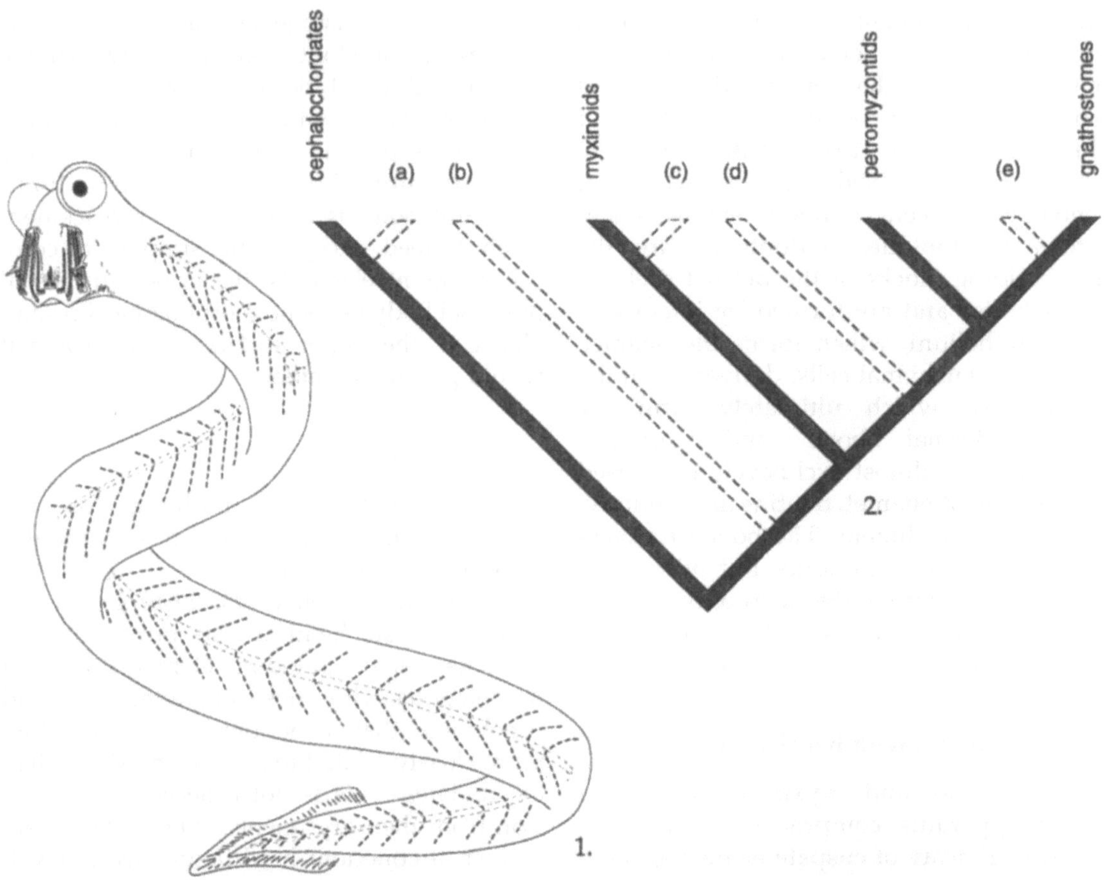

Figure 2.5 (1) Reconstruction of a conodont animal, based on current evidence; the feeding apparatus is shown in the 'everted' position according to Purnell and Donoghue (in press). (2) Cladogram of possible conodont relationships; solid branches represent phylogenetic positions of extant groups, dashed lines represent some of the proposed positions of conodonts: (a) Kemp and Nicoll (1995, 1996), (b) Peterson (1994), (c) Krejsa (1990a, b), Aldridge *et al.* (1986), (d) Aldridge *et al.* (1986, 1993) and (e) Gabbott *et al.* (1995), Janvier (1995).

represent a completely independent and fortuitously analogous development from that in other craniates, but the development of eyes with extrinsic muscles would be similarly homoplastic.

It is perhaps more likely that myxinoids separated from the conodonts by secondary loss of conodont characters. The eyes of hagfishes are connected to the brain and sensitive to light (Wicht and Northcutt, 1995) and are probably responsible for the entrainment of the circadian rhythm (Ooka-Souda *et al.*, 1993). They are nevertheless of very limited function, and despite the formation of a lens placode during development, a lens, iris, extrinsic musculature and associated nerves fail to develop (Wicht and Northcutt, 1995). The most likely interpretation is that hagfishes represent a condition degenerate from that of, for example, conodonts, with characters lost in response to their specialized mode of life (Fernholm and Holmberg, 1975; Northcutt, 1985). However, it is quite feasible that the failure of the lens placode to form a lens is a primitive condition (Wicht and Northcutt, 1995), since the placode is ultimately responsible for the formation of a number of other structures, including the cornea, which are present in hagfishes.

The conversion of phosphatic hard parts to keratin may be more problematic. The recognition of enamel-like antigens (putatively enamelin) in the pokal cell cone beneath the tip of the keratin toothlet of hagfishes (Slavkin *et al.*, 1983) may be of relevance here, and Kresja *et al.* (1990a, b) used this to support a proposal that conodonts were ancestral hagfishes that switched from secreting mineralized keratin to keratin. The occurrence of the fossil hagfish *Myxinikela* in the Mazon Creek (Bardack, 1991) shows that the two groups were distinct by the Carboniferous. The case is weakened, however, by the lack of any demonstrable homology between conodont elements and hagfish toothlets. Smith *et al.* (1996) also cited the lack of developmental support for a switch from an apatitic system to one secreting keratin.

Relatively few myxinoid embryos have been recovered, and there is little evidence to indicate the degree of neural crest involvement in the formation of hagfish skeletal head structures. Conel (1942), however, suggested that neural crest played no role in hagfish cranial skeletal development. If hagfish neural crest is indeed restricted to neuronal derivatives (Langille, 1987), the evidence for neural-crest derived tissues in conodonts would indicate that the conodonts are the more derived.

The placement of the conodonts immediately anticrownwards of myxinoids (Peterson, 1994) suffers from similar drawbacks, involving loss in the myxinoids of the phosphatic tissues, the muscularized eyes and migratory neural crest, and their re-evolution in the post-myxinoid craniates. A position crownwards of the myxinoids poses fewest problems with current evidence of conodont soft and hard tissues (Aldridge *et al.*, 1993). Their precise placement will be influenced by resolution of the controversial relationships between extant and fossil agnathans. If hagfish and lampreys form a natural group (Yalden, 1985; Stock and Whitt, 1992) or if the hagfish and lampreys are successive paraphyletic groups (Forey and Janvier, 1994; Forey 1995), then conodonts may well occupy a position crownwards of both myxinoids and petromyzontids (Gabbott *et al.*, 1995). However, Langille (1987) has described neural crest involvement in the formation of the head skeleton of lampreys, and the ability of petromyzontids to mineralize their skeleton has been demonstrated by *in vivo* (Bardack and Zangerl, 1971) and *in vitro* studies (Langille and Hall, 1993). This evidence, and the possession of arcualia by lampreys, suggests that of the two groups, the conodonts are the more primitive.

Janvier (1996) recently completed the first full cladistic analysis of the agnatha to incorporate conodonts. Lack of soft tissue characters and equivocation over the interpretation of some characters largely resulted in tree imbalance. However, Janvier's text-fig. 5c, the best resolved of the relevant trees, places the conodonts as a sister group to lampreys, with which they form a sister group to all other agnathans with a mineralized exoskeleton; hagfishes are a sister group to all other craniates. This intriguing solution awaits testing by additional cladistic analyses. Further, the speculation by Janvier (1995, 1996) that conodonts might be closer to the gnathostomes than all the ostracoderms apart from the osteostracans currently seems difficult to sustain, as it would involve secondary loss in the conodonts of the exoskeleton and the paired fins.

Whatever the final position of conodonts within craniate phylogeny, they have clearly influenced recent debates on vertebrate origin and generated a new impetus into long-standing controversies regarding the origin and early evolution of the vertebrate skeleton.

2.5 CONODONTS AS LIVING ANIMALS

Aldridge *et al.* (1993) reconstructed the conodont animal as an elongate and laterally compressed eel-shaped agnathan (Figure 2.5) capable of an anguilliform mode of swimming. Lack of muscle fibres of a size comparable with 'fast' white muscle in the Soom specimen may indicate that conodonts were adapted to sustained swimming and incapable of rapid bursts (Gabbott *et al.*, 1995). However, examination of the musculature in this specimen is at a preliminary stage, and other muscle tissue may be present.

Conodonts had a complex feeding array which performed a number of tooth functions (Aldridge and Briggs, 1986; Purnell and von Bitter, 1992; Purnell, 1995; Purnell and Donoghue, in press). Early forms possessed only conical elements which were capable of grasping and perhaps slicing food (Aldridge and Briggs, 1986; Purnell, 1995); later forms developed more highly differentiated feeding apparatuses which separated grasping from slicing and crushing elements. The great variation in conodont apparatuses suggests that the group adopted a number of different ecological strategies, although with their locomotive capability and differentiated nervous system (including eyes with associated musculature) many would have made effective hunters (Purnell *et al.*, 1995).

ACKNOWLEDGEMENTS

The organizers and fellow participants of the symposium on the biology of hagfishes are thanked for discussion. Dr Mark Purnell also provided lively discussion and made valuable comments on the manuscript. Specimens figured here are reposited at: the (BU) Lapworth Museum, School of Earth Sciences, University of Birmingham, UK; (FMNH) Field Museum of Natural History, Chicago; (GSSA) Geological Survey of South Africa, Pretoria; and (RSM) Royal Scottish Museums, Edinburgh.

REFERENCES

Aldridge, R.J. and Briggs, D.E.G. (1986) Conodonts, in *Problematic Fossil Taxa. Oxford Monographs on Geology and Geophysics No. 5* (eds A. Hoffman and M.H. Nitecki), Oxford University Press, New York, pp. 227–39.

Aldridge, R.J., Briggs, D.E.G., Clarkson, E.N.K. and Smith, M.P. (1986) The affinities of conodonts – new evidence from the Carboniferous of Edinburgh, Scotland. *Lethaia*, **19**, 279–91.

Aldridge, R.J., Briggs, D.E.G., Smith, M.P., Clarkson, E.N.K. and Clark, N.D.L. (1993) The anatomy of conodonts. *Philosophical Transactions of the Royal Society of London, Series B*, **340**, 405–21.

Aldridge, R.J. and Purnell, M.A. (1996) The conodont controversies. *Trends in Ecology and Evolution*, **11**, 463–8.

Aldridge, R.J., Purnell, M.A., Gabbott, S.E. and Theron, J.N. (1995) The apparatus architecture and function of *Promissum pulchrum* Kovács-Endrödy (Conodonta, Upper Ordovician), and the prioniodontid plan. *Philosophical Transactions of the Royal Society of London, Series B*, **347**, 275–91.

Aldridge, R.J., Smith, M.P., Norby, R.D. and Briggs, D.E.G. (1987) The architecture and function of Carboniferous polygnathacean conodont apparatuses, in *Palaeobiology of Conodonts* (ed. R.J. Aldridge), Ellis Horwood, Chichester, pp. 63–76.

Aldridge, R.J. and Theron, J.N. (1993) Conodonts with preserved soft tissue from a new Upper Ordovician *Konservat-Lagerstätte*. *Journal of Micropalaeontology*, **12**, 113–17.

Bardack, D. (1974) A larval fish from the Pennsylvanian of Illinois. *Journal of Paleontology*, **48**, 988–93.

Bardack, D. (1991) First fossil hagfish (Myxinoidea): a record from the Pennsylvanian of Illinois. *Science*, **254**, 701–703.

Bardack, D. and Richardson Jr, E.S. (1977) New agnathous fiches from the Pennsylvanian of Illinois. *Fieldiana Geology*, **33**, 489–510.

Bardack, D. and Zangerl, R. (1968) First fossil lamprey: a record from the Pennsylvanian of Illinois. *Science*, **162**, 1265–7.

Bardack, D. and Zangerl, R. (1971) Lampreys in the fossil record, in *The Biology of Lampreys* (eds M.W. Hardisty and I.C. Potter), Academic Press, pp. 67–84.

Barnes, C.R., Sass, D.B. and Monroe, E.A. (1973) Ultrastructure of some Ordovician conodonts, in *Conodont Paleozoology* (ed. F.H.T. Rhodes), Geological Society of America, Boulder, pp. 1–30.

Barskov, I.S., Moskalenko, T.A. and Starostina, L.P. (1982) New evidence for the vertebrate nature of the conodontophorids. *Paleontological Journal*, **1982**, 82–90.

Briggs, D.E.G., Clarkson, E.N.K. and Aldridge, R.J. (1983) The conodont animal. *Lethaia*, **16**, 1–14.

Briggs, D.E.G. and Kear, A. (1994) Decay of *Branchiostoma*: implications for soft-tissue preservation in conodonts and other primitive chordates. *Lethaia*, **26**, 275–87.

Briggs, D.E.G., Kear, A.J., Martill, D.M. and Wilby, P.R. (1993) Phosphatization of soft-tissue in experiments and fossils. *Journal of the Geological Society, London*, **150**, 1035–8.

Conel, J.L. (1942) The origin of the neural crest. *Journal of Comparative Neurology*, **76**, 191–215.

Couly, G., Coltey, P.M. and Le Douarin, N.M. (1992) The development of the cephalic mesoderm in quail-chick chimeras. *Development*, **114**, 1–15.

Collins, M.D., Stern, B., Abbott, G.D., Walton, D., Riley, M.S., Von Wallmenich, T., Savage, N.M., Armstrong, H.A. and Westbroek, P. (1995) 'Intracrystalline' organic matter in biominerals, in *Organic Geochemistry* (eds J.O. Grimalt and C. Dorronsoro), AIGOA, pp. 702–6.

Dzik, J. (1976) Remarks on the evolution of Ordovician conodonts. *Acta Palaeontologica Polonica*, **21**, 395–455.

Dzik, J. (1986) Chordate affinities of the conodonts, in *Problematic Fossil Taxa*. Oxford Monographs on Geology and Geophysics No. 5 (eds A. Hoffman and M.H. Nitecki), Oxford University Press, New York, pp. 240–54.

Dzik, J. (1995) *Yunnanozoon* and the ancestry of the vertebrates. *Acta Palaeontologica Polonica*, **40**, 341–60.

Fåhraeus, L.E. and Fåhraeus-van Ree, G.E. (1987) Soft tissue matrix of decalcified pectiniform elements of *Hindeodella confluens* (Conodonta, Silurian), in *Palaeobiology of Conodonts* (ed. R.J. Aldridge), Ellis Horwood Ltd, Chichester, pp. 105–10.

Fåhraeus, L.E. and Fåhraeus-van Ree, G.E. (1993) Histomorphology of sectioned and stained 415 Ma old soft-tissue matrix from internal fluorapatite skeletal elements of an extinct taxon, Conodontophorida (Conodonta), in *Structure, Formation and Evolution of Fossil Hard Tissues* (eds I. Kobayashi, H. Mutvei and A. Sahni), Tokai University Press, Tokyo, pp. 107–12.

Fernholm, B. and Holmberg, K. (1975) The eyes in three genera of hagfish (*Eptatretus, Paramyxine* and *Myxine*) – a case of degenerative evolution. *Vision Research*, **15**, 253–9.

Forey, P.L. (1995) Agnathans recent and fossil, and the origin of jawed vertebrates. *Reviews in Fish Biology and Fisheries*, **5**, 267–303.

Forey, P.L. and Janvier, P. (1993) Agnathans and the origin of jawed vertebrates. *Nature*, **361**, 129–34.

Forey, P.L. and Janvier, P. (1994) Evolution of the early vertebrates. *American Scientist*, **82**, 554–65.

Furnish, W.M. (1938) Conodonts of the Prairie du Chien (Lower Ordovician) beds of the Upper Mississippian Valley. *Journal of Paleontology*, **12**, 318–40.

Gabbott, S.E., Aldridge, R.J. and Theron, J.N. (1995) A giant conodont with preserved muscle tissue from the Upper Ordovician of South Africa. *Nature*, **374**, 800–3.

Gagnier, P.Y., Blieck, A. and Rodrigo, G. (1986) First Ordovician vertebrate from South America. *Geobios*, **19**, 629–34.

Gans, C. (1993) Evolutionary origin of the vertebrate skull, in *The Skull* (eds J. Hanken and B.K. Hall), University of Chicago Press, Chicago, pp. 1–35.

Gans, C. and Northcutt, R.G. (1983) Neural crest and the origin of the vertebrates: a new head. *Science*, **220**, 268–74.

Garcia-Fernàndez, J. and Holland, P.W.H. (1994) Archetypal organization of the amphioxus *Hox* gene cluster. *Nature*, **370**, 563–6.

Globensky, Y. (1967) Middle and Upper Mississippian conodonts from the Windsor Group of the Atlantic Provinces of Canada. *Journal of Paleontology*, **41**, 432–48.

Halstead, L.B. (1987) Evolutionary aspects of neural crest-derived skeletogenic cells in the earliest vertebrates, in *Developmental and Evolutionary Aspects of the Neural Crest* (eds P.F.A. Maderson), John Wiley & Sons, pp. 339–58.

Hass, W.H. (1941) Morphology of conodonts. *Journal of Paleontology*, **15**, 71–81.

Holland, P.W.H. and Garcia-Fernàndez, J. (1996) *Hox* genes and chordate evolution. *Developmental Biology*, **173**, 382–95.

Janvier, P. (1981) The phylogeny of the Craniata, with particular reference to the significance of fossil 'agnathans'. *Journal of Vertebrate Paleontology*, **1**, 121–59.

Janvier, P. (1993) Patterns of diversity in the skull of jawless fishes, in *The Skull* (eds J. Hanken and B.K. Hall), University of Chicago Press, Chicago, pp. 131–88.

Janvier, P. (1995) Conodonts join the club. *Nature*, **374**, 761–2.

Janvier, P. (1996) The dawn of the vertebrates: characters versus common ascent in the rise of current vertebrate phylogenies. *Palaeontology*, **39**, 259–87.

Kardong, K.V. (1995) *Vertebrates, Comparative Anatomy Function Evolution*, Wm. C. Brown Publishers, Dubuque, Iowa.

Kemp, A. and Nicoll, R.S. (1995) Protochordate affinities of conodonts. *Courier Forschungsinstitut Senckenberg*, **182**, 235–45.

Kemp, A. and Nicoll, R.S. (1996) Histology and histochemistry of conodont elements. *Modern Geology*, **20**, 287–302.

Kovàcs-Endrödy, E. (1987) The earliest known vascular plant, or a possible ancestor of vascular plants in the flora of the Lower Silurian Cedarberg Formation, Table Mountain Group, South Africa. *Annals of the Geological Survey of South Africa*, **20**, 93–118.

Krejsa, R.J., Bringas, P. and Slavkin, H.C. (1990a) The cyclostome model: an interpretation of conodont element structure and function based on cyclostome tooth morphology, function, and life history. *Courier Forschungsinstitut Senckenberg*, **118**, 473–92.

Krejsa, R.J., Bringas, P. and Slavkin, H.C. (1990b) A neontological interpretation of conodont elements based on agnathan cyclostome tooth structure, function, and development. *Lethaia*, **23**, 359–78.

Langille, R.M. (1987) The neural crest and the development of the skeleton in lower vertebrates. Unpublished PhD thesis, Dalhousie University, Halifax, Nova Scotia. 157pp.

Langille, R.M. and Hall, B.K. (1993) Calcification of cartilage from the lamprey *Petromyzon marinus* (L.) *in vitro*. *Acta Zoologica*, **74**, 31–41.

Lindström, M. and Ziegler, W. (1981) Surface micro-ornamentation and observations on internal composition, in *Treatise on Invertebrate Paleontology, Part W, Miscellanea, Supplement 2, Conodonta* (ed. R.A. Robison), Geological Society of America and University of Kansas, Lawrence, pp. W41–W103.

Mallatt, J. (1996) Ventilation and the origin of jawed vertebrates: a new mouth. *Zoological Journal of the Linnean Society*, **117**, 329–404.

Mikulic, D.G., Briggs, D.E.G. and Kluessendorf, J. (1985a) A Silurian soft-bodied biota. *Science*, **228**, 715–17.

Mikulic, D.G., Briggs, D.E.G. and Klussendorf, J. (1985b) A new exceptionally preserved biota from the Lower Silurian of Wisconsin, USA. *Philosophical Transactions of the Royal Society of London, Series B*, **311**, 78–85.

Müller, K.J. and Nogami, Y. (1971) Über die Feinbau der Conodonten. *Memoirs of the Faculty of Science, Kyoto University, Series of Geology and Mineralogy*, **38**, 1–87.

Nielsen, C. (1995) *Animal Evolution: Interrelationships of the Living Phyla*, Oxford University Press, New York.

Noden, D.M. (1991) Vertebrate craniofacial development: the relation between ontogenetic process and morphological outcome. *Brain, Behaviour, and Evolution*, **38**, 190–225.

Northcutt, R.G. (1985) The brain and sense organs of the earliest vertebrates: reconstruction of a morphotype, in *Evolutionary Biology of Primitive Fishes* (eds R.E. Foreman, A. Gorbman, J.M. Dodd and R. Olsson), Plenum Press, New York, pp. 81–112.

Ooka-Souda, S., Kadota, T. and Kabasawa, H. (1993) The preoptic nucleus: the probable location of the circadian pacemaker of the hagfish, *Eptatretus burgeri*. *Neuroscience Letters*, **164**, 33–6.

Peterson, K.J. (1994) The origin and early evolution of the Craniata, in *Major Features of Vertebrate Evolution* (eds D.R. Prothero and R.M. Schoch), The Paleontological Society, pp. 14–37.

Pietzner, H., Vahl, J., Werner, H. and Ziegler, W. (1968) Zur chemischen zusammensetzung und mikromorphologie der conodonten. *Palaeontographica Abt. A*, **128**, 115–52.

Potter, I.C. and Hilliard, R.W. (1987) A proposal for the functional and phylogenetic significance of differences in the dentition of lampreys (Agnatha: Petromyzontiformes). *Journal of Zoology*, **212**, 713–37.

Purnell, M.A. (1995) Microwear on conodont elements and macrophagy in the first vertebrates. *Nature*, **374**, 798–800.

Purnell, M.A., Aldridge, R.J., Donoghue, P.C.J. and Gabbott, S.E. (1995) Conodonts and the first vertebrates. *Endeavour*, **19**, 20–7.

Purnell, M.A. and Donoghue, P.C.J. (in press) Skeletal architecture and functional morphology of ozarkodinid conodonts. Philosophical Transactions of the Royal Society of London, series B.

Purnell, M.A. and von Bitter, P.H. (1992) Blade-shaped conodont elements functioned as cutting teeth. *Nature*, **359**, 629–31.

Richardson, E.S. and Johnson, R.G. (1971) The Mazon Creek faunas, in *Proceedings of the North American Paleontological Convention* (ed E.L. Yochelson), Allen Press Inc., Vol. 2, pp. 1222–35.

Ritchie, A. (1968) New evidence on *Jamoytius kerwoodi* White, an important ostracoderm from the Silurian of Lanarkshire, Scotland. *Palaeontology*, **11**, 21–39.

Sansom, I.J. (1996) *Pseudooneotodus*, an important Palaeozoic vertebrate lineage. *Zoological Journal of the Linnean Society*, **118**, 47–57.

Sansom, I.J., Smith, M.P., Armstrong, H.A. and Smith, M.M. (1992) Presence of the earliest vertebrate hard tissues in conodonts. *Science*, **256**, 1308–11.

Sansom, I.J., Smith, M.P. and Smith, M.M. (1994) Dentine in conodonts. *Nature*, **368**, 591.

Savage, N.M., Lindorfer, M.A. and McMillen, D.A. (1990) Amino acids from Ordovician conodonts. *Courier Forschungsinstitut Senckenberg*, **118**, 267–75.

Schmidt, H. and Müller, K.J. (1964) Weitere Funde von Conodonten-Gruppen aus dem oberen karbon des Sauerlandes. *Paläontologische Zeitschrift*, **38**, 105–35.

Schultze, H.-P. (1996) Conodont histology: an indicator of vertebrate relationship? *Modern Geology*, **20**, 275–86.

Slavkin, H.C., Graham, E., Zeichner-David, M. and Hildemann, W. (1983) Enamel-like antigens in hagfish: possible evolutionary significance. *Evolution*, **37**, 404–12.

Smith, M.M. (1989) Distribution and variation in enamel structure in the oral teeth of Sarcopterygians: its significance for the evolution of a protoprismatic enamel. *Historical Biology*, **3**, 97–126.

Smith, M.M. (1992) Microstructure and evolution of enamel amongst osteichthyan and early tetrapods, in *Structure, Function and Evolution of Teeth* (ed. P. Smith), Jerusalem, Israel, Proceedings of the 8th International Symposium on Dental Morphology (1989), pp. 1–19.

Smith, M.M., Sansom, I.J. and Smith, M.P. (1996) 'Teeth' before armour: the earliest vertebrate mineralized tissues. *Modern Geology*, **20**, 303–20.

Smith, M.P., Briggs, D.E.G. and Aldridge, R.J. (1987) A conodont animal from the lower Silurian of Wisconsin, U.S.A., and the apparatus architecture of panderodontid conodonts, in *Palaeobiology of Conodonts* (ed. R.J. Aldridge), Ellis Horwood, Chichester, pp. 91–104.

Stock, D.W. and Whitt, G.S. (1992) Evidence from 18S ribosomal RNA sequences that lampreys and hagfishes form a natural group. *Science*, **257**, 787–9.

Szaniawski, H. and Bengtson, S. (1993) Origin of euconodont elements. *Journal of Paleontology*, **67**, 640–54.

Wicht, H. and Northcutt, R.G. (1995) Ontogeny of the head of the Pacific hagfish (*Eptatretus stouti*, Myxinoidea): development of the lateral line system. *Philosophical Transactions of the Royal Society of London, Series B*, **349**, 119–34.

Yalden, D.W. (1985) Feeding mechanisms as evidence of cyclostome monophyly. *Zoological Journal of the Linnean Society*, **84**, 291–300.

Young, J.Z. (1995) *The Life of Vertebrates*, Clarendon Press, Oxford.

3
HAGFISH SYSTEMATICS

Bo Fernholm

SUMMARY

Hagfish occur in all oceans except the Polar Seas. They are absent from warm oceans or occur only in deep and cool parts.

Extant hagfish in the family Myxinidae are monophyletic. Systematically useful characters within Myxinidae are number and pattern of gill pouches, gill apertures, slime pores and dental cusps. Close to sixty valid species of hagfish in five genera are listed. The genus *Paramyxine* is problematical and its distinction from *Eptatretus* cannot be upheld. In anticipation of revisionary work *Paramyxine* is tentatively synonymized with *Eptatretus*. It makes up the subfamily Eptatretinae with 35 species, having more than one pair of gill openings. *Myxine* with 19 species, *Notomyxine*, 1 species, *Neomyxine*, 1 species and *Nemamyxine*, 2 species have only one pair of gill openings and constitute the subfamily Myxininae.

3.1 INTRODUCTION AND HISTORY

The common Atlantic species, *Myxine glutinosa* was described by Linnaeus (1758) as *Myxina glutinosa* and placed by him among the worms in the tenth and later editions of his *Systema Naturae*. He gave the succinct description: *Intrat et devorat Pisces; aquam in gluten mutat* (enters into and eats fishes; changes water into slime). About ten species of hagfish in six genera from the seas around all continents except the Antarctic were named during the following century. The remainder of the approximately 60 valid species were named during the twentieth century, the majority of them after 1963 when Adam and Strahan (1963) summarized the known species (Tables 3.1–3.4).

Hagfishes are bottom-dwelling, exclusively marine, cold water animals, feeding on dead or moribund fish and invertebrates. They are cartilaginous with two sets of laterally everting and biting-scraping keratinous cusps (or teeth) attached to a dental plate, in turn attached to the anterior end of the dental (lingual) muscle. Hagfishes have vestigial eyes embedded in the flesh of the head and covered by integument with, in several species, conspicuous transparent (whitish in preserved specimens) eye spots. Internal gill pouches are supplied with

Table 3.1 Geographical distribution of genera of hagfish

	Europe	Africa	Japan	Taiwan	N. America	S. America	Australia, N.Z.
Eptatretus		X	X	X	X	X	X
(Paramyxine)			X	X			
Myxine	X	X	X		X	X	
Nemamyxine						X	X
Neomyxine							X
Notomyxine						X	

The Biology of Hagfishes. Edited by Jørgen Mørup Jørgensen, Jens Peter Lomholt, Roy E. Weber and Hans Malte. Published in 1998 by Chapman & Hall, London. ISBN 0 412 78530 7.

Table 3.2 Recognized species of *Eptatretus*, including Japanese and Taiwanese *Paramyxine*. Abbreviations: gp – number of pairs of gill pouches; afc – number of fused cusps of anterior tooth row; pfc – number of fused cusps of posterior tooth row; tot c – total number of dental cusps on both sides; tail sp – number of slime pores on the left side of tail region (including slime pores above the cloaca and posteriorly); tot sp – total number of slime pores on the left side; tl mm – total length in mm; orig gen – genus of original description

Species	Author	gp	afc	pfc	tot c	tail sp	tot sp	tl mm	orig gen	Locality	Depth, m
E. atami	Dean (1904)	6	3	3	47	8–14	68–79	610	*Paramyxine*	Japan, Sagami and Suruga Bay	300–536
E. bischoffli	Schneider (1880)	10	3	3	48	12–16	74–83	550	*Bdellostoma*	Off Chile	8–50
E. burgeri	Girard (1855)	6	3	2	37	11–14	82–90	590	*Bdellostoma*	Japan, Korea, China, Taiwan	10–270
E. caribbeaus	Fernholm (1982)	7	3	3	56	11–13	79–85	385		Caribbean Sea	365–500
E. carlhubbsi	McMillan and Wisner (1984)	7	2	3	64–71	12–16	93–110	1160		Hawaii, Wake Island, Guam	481–1574
E. cheni	Shen and Tao (1975)	5	3	3	52	7–8	78	377	*Paramyxine*	SW Taiwan	180
E. chinensis	Kuo and Mok (1994)	6	3	3	52	11–14	72–83	375		South China Sea	600
E. cirrhatus	Forster in Bloch and Schneider (1801)	7	3	3	43–51	11–13	79–90	830		New Zealand, S. and E. Australia	40–700
E. deani	Evermann and Goldsborough (1907)	11	3	2	42	9–15	76	523	*Polistotrema*	West coast of N. Am.	107–2743
E. decatrema	Regan (1912)	10	3	3	48			480	*Heptatretus*	Jr synonym to E. Bischoffi, Chile	
E. eos	Fernholm (1991)	5	3	2	34	26–27	128–130	665		Tasman Sea	991–1013
E. fernholmi	Kuo *et al.* (1994)	6	3	2	42–50	6–11	64–71	295	*Paramyxine*	Taiwan	200–400
E. fritzi	Wisner and McMillan (1990)	11	3	2	42	8–15	79	592		Guadelope Island, Mexico	18–2743
E. hexatrema	Müller (1834)	6	3	2	46	11–14	98	720		Off Cape of Good Hope	10–45
E. indrambaryai	Wongratana (1984)	8	3	2	47	10–13	77–82	437		Andaman Sea	267–400
E. laurahubbsae	McMillan and Wisner (1984)	7	2	2	61–68	14–16	97–105	375		Juan Fernandez Islands	2400
E. longipinnis	Strahan (1975)	6	3	2	30	8–9	106	422		SE Indian Ocean off S. Australia	40
E. mcconnaugheyi	Wisner and McMillan (1990)	13	3	2	42	8–13	67–84	470		Off S. Calif., lower Gulf of Calif.	43–415
E. mendozai	Hensley (1985)	6	3	3	56–61	12–15	77–82	450		Caribbean Sea	720–1100
E. minor	Fernholm and Hubbs (1981)	6	3	3	50	11–14	74–82	395		Gulf of Mexico	300–400
E. multidens	Fernholm and Hubbs (1981)	6	3	3	55	15	87–91	655		Caribbean Sea	510–770
E. nanii	Wisner and McMillan (1988)	13	3	3	52	11–15	72–82	664		Off Valparaiso, Chile	274
E. nelsoni	Kuo *et al.* (1994)	4	3	2	40	8	62	190	*Paramyxine*	SW Taiwan	50–200
E. octatrema	Barnard (1923)	8	3	2	40			300	*Heptatretus*	Agulhas Bank, South Africa	46–73
E. okinoseanus	Dean (1904)	8	3	2	45	13	95	800	*Homea*	Japan, Taiwan	300–1020
E. polytrema	Girard (1855)	14	3	3	48	12–17	72–79	460	*Bdellostoma*	Off Valparaiso, Chile	10–350
E. profundus	Barnard (1923)	5	3	2	42			620	*Heptatretus*	Off Cape Point, South Africa	732
E. sheni	Kuo *et al.* (1994)	6	3	3		8–12	64–74	380	*Paramyxine*	SW Taiwan	200–800
E. sinus	Wisner and McMillan (1990)	10	3	2	40	7–14	74	481		Gulf of California, Mexico	198–1330
E. springeri	Bigelow and Schroeder (1952)	6	3	2	50	9–13	84–92	590	*Paramyxine*	Gulf of Mexico	400–730
E. stoutii	Lockington (1878)	12	3	2	40	8–14	79	468	*Bdellostoma*	West coast of N. Am	16–633
E. strahani	McMillan and Wisner (1984)	7	3	3	47–52	10–12	76–80	520		S. China Sea, Philippines	189
E. taiwanae	Shen and Tao (1975)	6	3	2	38	6–9	60–68	334	*Paramyxine*	NE Taiwan	20–50
E. wisneri	Kuo *et al.* (1994)	6	3	2		6–11	63–72	335	*Paramyxine*	SE Taiwan	200
E. yangi	Teng (1958)	5	3	2	38	8–11	66–78	296	*Paramyxine*	NE and SW Taiwan	20–50

Table 3.3 Recognized species of *Myxine*. Abbreviations as in Table 3.2

Species	Author	gp	afc	pfc	tot c	tail sp	tot sp	tl mm	Locality	Depth, m
M. affinis	Günther (1870)	6	2	2	42	12	112	659	Straits of Magellan	shallow?
M. australis	Jenyns (1842)	6	2	2	34	11	102	394	Straits of Magellan	shallow
M. capensis	Regan (1913a)	7	2	2	40	11	99	310	Cape of Good Hope, S. Africa	173–457
M. circifrons	Garman (1899)	5	3	2	50	10	90	650	E. Pacific	700–1860
M. debueni	Wisner and McMillan (1995)	6	3	2	37	7	104	570	Straits of Magellan	300
M. dorsum	Wisner and McMillan (1995)	6	2	2	36	11	109	490	SW Atlantic Ocean	112–650
M. fernholmi	Wisner and McMillan (1995)	6	3	2	36	8	116	846	S. Centr. Chile, Falkland Isl.	135–1480
M. garmani	Jordan and Snyder (1901)	6	3	2	45	12	94	540	Japan	500–800
M. glutinosa	Linnaeus (1758)	6	2	2	34	12	97	400	Europe; Mediterran. to Murmansk	40–1200
M. hubbsi	Wisner and McMillan (1995)	6	2	2	36	11	100	522	W. coast of N. and S. America	1100–2440
M. hubbsoides	Wisner and McMillan (1995)	6	2	2	36	12	114	826	Central Chile	735–880
M. ios	Fernholm (1981)	7	2	2	48	8	110	570	W. Africa, Ireland	614–1625
M. knappi	Wisner and McMillan (1995)	6	2	2	38	12	109	565	Falkland Islands	630–650
M. limosa	Girard (1859)	6	2	2	36	12	103	720	W.N. Atlantic	55–1006
M. mccoskeri	Wisner and McMillan (1995)	5	3	2	41	11	84	286	S. Caribbean Sea	439–567
M. mcmillanae	Hensley (1991)	6	2	2	45	11	107	473	Caribbean Sea	700–1500
M. paucidens	Regan (1913a)	6	2	2	26	10	89–93	305	Japan	630
M. pequenoi	Wisner and McMillan (1995)	7	2	2	27	9	83	183	Valdivia, Chile	185
M. robinsorum	Wisner and McMillan (1995)	5	3	2	57	9	98	540	S. Caribbean Sea	783–1768

Table 3.4 Recognized species of *Nemamyxine*, *Neomyxine* and *Notomyxine*. Abbreviations as in Table 3.2

Species	Author	gp	afc	pfc	tot c	tail sp	tot sp	tl mm	orig gen	Locality	Depth, m
Nemamyxine	Richardson (1958)										
N. elongata	Richardson (1958)		2	2	36	17	200	614		Bay of Plenty, N.Z.	
N. kreffti	McMillan and Wisner (1982)	8	3	2	36	16	140	400		Off Argentina, S. Brazil	140–800
Neomyxine	Richardson (1958)										
N. biniplicata	Richardson and Jowett (1951)	7	2	2	34	21	164	412	*Myxine*	Cook Strait, Kaikoura, N.Z.	73
Notomyxine	Nani and Gneri (1951)										
N. tridentiger	Garman (1899)	6	3	2	42	9	91	575	*Myxine*	S. coasts of S. America	11–106

water entering the single nostril above the mouth and through the nasopharyngeal duct. Discharge is via one or more pairs of adjacent gill openings (branchial apertures) and the pharyngocutaneous duct.

A row of mucus-secreting slime glands with pores occur on each side, usually one per segment. Many hagfishes live in burrows formed in mud from shallow waters to below 2700 m (Wisner and McMillan, 1990). At least some are actively searching for food at night (Fernholm, 1974). Some shallow water species occur among rocks (Wisner and McMillan, 1988).

Dean (1904) made an important contribution when he described two new Japanese species, erected the genus *Paramyxine* and discussed useful characters for hagfish taxonomical work. He stressed that it is necessary to base species determinations upon the average characters of as great a number of individuals as practicable. He further noted that the fusion of the two or three cusps at the median end of each row of cusps appears to be moderately constant for the species (see Figure 3.2). Total number of cusps, relative position of the posterior end of the tongue muscle as well as the relative position of the bifurcation of the ventral aorta were also stated to be useful features. The position of the external gill apertures relative to the internal gill pouches was used by Dean to describe the new genus *Paramyxine*. This feature has continued to be used by hagfish systematists and problems associated with this usage are discussed below.

Nani and Gneri (1951), discussing South American species, based much of their work on Dean (1904). They erected the new genus *Notomyxine* based on the separate opening of the *ductus pharyngocutaneous* posterior to the left gill aperture. The common hagfish pattern is that the duct opens into the posteriormost left gill aperture.

Two further genera of hagfishes occurring exclusively in the southern hemisphere were described relatively late. They are *Neomyxine* (Richardson, 1953) and *Nemamyxine* (Richardson, 1958) characterized by slender shape and high numbers of slime glands.

Strahan made important contributions to East Asian and Australian hagfish systematics. He discussed the genus *Paramyxine* and was first to suggest it to be considered a junior synonym of *Eptatretus* (Strahan, 1975).

Recently Taiwanese ichthyologists (Kuo *et al.*, 1994; Kuo and Mok, 1994) have described several small-sized Taiwanese hagfishes. They were all described as belonging to *Paramyxine* by having the feature of crowded gill apertures.

Methodological improvements and a consistent use of a standardized set of characteristics introduced by Carl Hubbs and a group of ichthyologists inspired by him (Fernholm and Hubbs, 1981; McMillan and Wisner, 1984; Wisner and McMillan, 1995) have been important for the increased stability and comparability in the recent surge of new descriptions.

Today we have a much improved picture of hagfish abundance and distribution compared to that available to Adam and Strahan (1963) although much still remains regarding basic taxonomic work and modern phylogenetic studies of the group.

3.2 MATERIAL AND METHODS

Most factual information in this review is derived from published sources, but especially for East Asian hagfishes, some unpublished material has been used.

3.3 SYSTEMATICALLY USEFUL CHARACTERS OF MYXINIDAE

There is no doubt that recent hagfishes in the family Myxinidae (Chordata) are monophyletic. A great number of synapomorphic features testify to that fact: notochord with no traces of vertebrae, single nostril leading respiratory water through the nasopharyngeal duct to pharynx and gill pouches, unique thread cells in numerous slime glands, one semicircular canal, adenohypophysis with little cellular differentiation and not divided into distinct regions, barbels around laterally biting mouth structures, somatic chromosome reduction, near isosmotic blood, pharyngocutaneous duct on the left side opening to the exterior.

The relationship of Myxinidae to other taxa is problematical and further discussed in Chapter 2. The old idea of a close relationship to lampreys is essentially abandoned.

Regarding the phylogenetic relationship among recent hagfishes, regrettably little work has been done. Nelson (1994) divides Myxinidae into two subfamilies: Eptatretinae characterized by one branchial duct for each of the numerous gill pouches and Myxininae characterized by the efferent branchial ducts opening by a common external aperture on each side (Figure 3.1).

The most useful features for systematic work in hagfishes are gill apertures and their position relative to the gill pouches, dental cusps and pattern of fused cusps, slime pores, relative position of ventral aorta bifurcation and gill pouches, fin fold and body proportions. It is also useful to note depth distribution in addition to geographical distribution when trying to distinguish among hagfish species. For a detailed description of characteristics see Fernholm and Hubbs (1981) and Wisner and McMillan (1995).

The likely primitive stage of gill apertures is one in which the many gill pouches each has its own short direct opening to the exterior. A tendency for the anteriormost openings to move backwards into a gradually more crowded posterior location can be seen in hagfishes from different continents. The end result of such a development is one where all gill pouches open into a common posterior pair of apertures as in Myxininae.

Hagfishes have keratinous cusps on two pairs of laterally biting dental plates. The number of fused cusps are 3 or 2 in each of

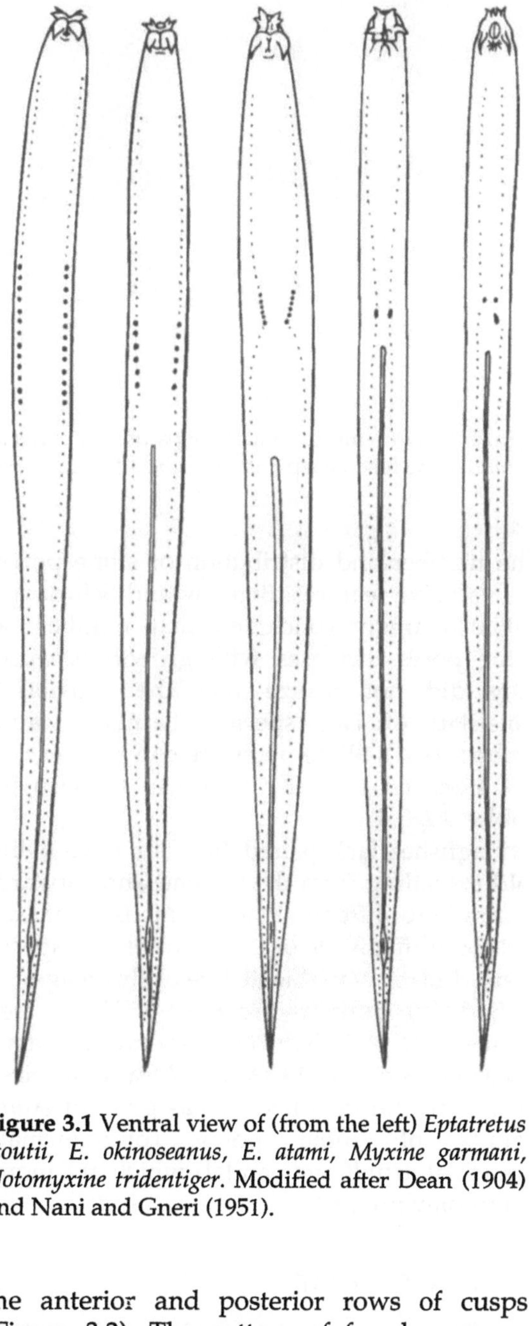

Figure 3.1 Ventral view of (from the left) *Eptatretus stoutii, E. okinoseanus, E. atami, Myxine garmani, Notomyxine tridentiger*. Modified after Dean (1904) and Nani and Gneri (1951).

the anterior and posterior rows of cusps (Figure 3.2). The pattern of fused cusps as well as the total number of cusps is of systematic value.

Slime glands and slime production is a conspicuous characteristic of hagfishes frequently commented on by field observers.

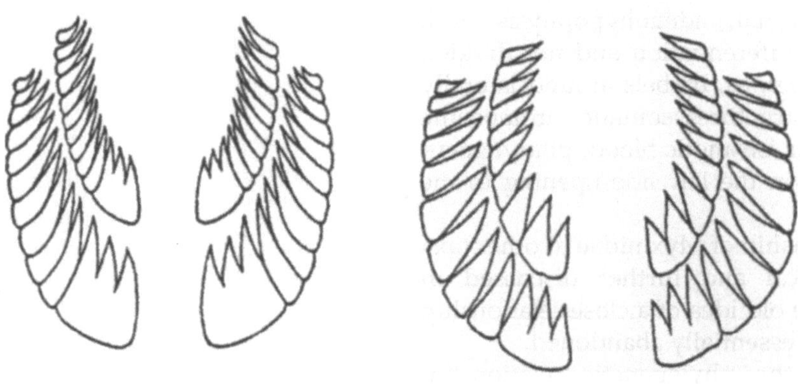

Figure 3.2 Drawing of dental cusps of *Eptatretus atami* (A) and *E. okinoseanus* (B). From Dean (1904). *E. atami* has the fused cusp pattern of 3/3 and *E. okinoseanus* 3/2.

The number and distribution of slime pores are valuable features. Bigelow and Schroeder (1952) wrongly concluded that number of slime pores increases with growth because they did not realize that their material consisted of two species (Fernholm and Hubbs, 1981). Slime pores are counted on the left side and typical ranges can be seen in Tables 3.2–3.4.

Hagfishes lack paired fins but have a fin fold extending from the tail end and forward ventrally to different extent. This is a characteristic difficult to use on preserved specimens due to a variable degree of shrinkage.

Body proportions are often of little use. Wisner and McMillan (1995) found no difference between 12 of 14 studied *Myxine* species. However, different body areas give different counts of slime pores (prebranchial, branchial, trunk and caudal) which are more commonly utilized.

3.4 EPTATRETINAE

For discussions of the genus name *Eptatretus*, see Fernholm and Hubbs (1981).

Dean (1904) diagnosed the genus *Paramyxine* as having the anterior efferent branchial duct several times longer than the posterior. Strahan (1975) pointed out that *Eptatretus burgeri* has similarly elongated anterior branchial ducts and suggested that *Paramyxine* be regarded as a junior synonym of *Eptatretus*. Another feature that has been utilized to characterize *Paramyxine* is the absence of slime pores in the gill aperture area. However, one or two slime pores are found in *E. atami* in the gill aperture area and some of the Taiwanese hagfish described as *Paramyxine* also have a few slime pores in between the crowded gill apertures e.g. *E. sheni* (Kuo *et al.*, 1994). I agree with Strahan (1975) that it is not possible to retain *Paramyxine* with the criteria given by Dean and I choose to treat all hagfish species with several pairs of gill openings as belonging to *Eptatretus*.

Paramyxine atami Dean, 1904, the type species for the genus *Paramyxine*, is rare and has, unfortunately, generally been confused with other as yet undescribed Japanese species (work in progress). Dean (1904) noted that the famous collector Kumakichi Aoki, who collected the specimen from 490 m in Sagami Bay, had never seen that species before during his many years of fishing and collecting. I have been able to locate and study only 19 specimens of the species, all from the Sagami and Suruga Bays (unpublished). *P. atami* is a deep water species (300–536 m) with a fused cusp pattern of 3/3 as described by Dean (Figure 3.2). It has been confused with one or two similar, more widespread, but yet undescribed species. They have the pattern of fused cusps 3/2 and are more easily collected in shallower water (100–200 m) around Japan. They are commonly and erroneously called *P. atami* (Honma and others, in this volume).

Another difficulty for the early discussion of *Paramyxine* was that Bigelow and Schroeder (1952) had described an Atlantic (Gulf of Mexico) species *P. springeri* which was a composite of two species, *Eptatretus springeri* and *E. minor* (Fernholm and Hubbs, 1981). Fernholm and Hubbs, after having referred *P. springeri* to *Eptatretus*, believed it was possible to retain *Paramyxine* as an exclusively East Asian genus.

Recently several species with crowded gill apertures have been described from Taiwan and adjacent waters as *Paramyxine*. Mitochondrial DNA studies on Taiwanese hagfish (Huang *et al.*, 1994) indicate that Taiwanese *Paramyxine* species are monophyletic while *Eptatretus* is suggested to be paraphyletic, *E. okinoseanus* being the sister group of *Paramyxine* and *E. burgeri*.

Protein electrophoretic studies (Jansson *et al.*, 1995) also indicate that *Eptatretus* is paraphyletic, the eastern Pacific *Eptatretus stoutii* and *E. deani* being the sister group of the Japanese *E. burgeri* and *Paramyxine* species. The dataset of Jansson *et al.* (1995) was coded for presence of equally migrating isozymes (Table 3.5). The exhaustive search option of Hennig 86 (Farris, 1988), a computer program for parsimony analysis of cladistic relationships resulted in one tree (Figure 3.3) with minimum steps 43 and c.i. 0.72.

It may well be that the character of having gill apertures in a crowded position has arisen independently in Japanese and Taiwanese hagfishes, perhaps by separate speciation from the widespread species *E. burgeri*. An observation that may support this speculation is the fact that, in contrast to the Japanese species, all Taiwanese *Paramyxine* are dwarfed forms (personal observation). The tendency for fusion of crowded gill apertures is illustrated by the finding of specimens having fewer apertures than gill pouches, e.g. specimens of *E. taiwanae* having five gill apertures on the right side but six gill pouches (Shen and Tao, 1975).

It is obvious that major revisions are needed of *Eptatretus* and *Paramyxine* to determine their generic delimitations. In anticipation of this, I

Table 3.5 Matrix resulting from coding of presence of electrophoretically similar isozymes used for producing the cladogram in Figure 3.3. Data from Jansson *et al.* (1995)

Myxine glutinosa	10000100100101100010100100100000
Myxine circifrons	00000110100100100001010000000100
Myxine sp.	10000100000101000010100100100000
Eptatretus stoutii	01010010010000001000000000101110
Eptatretus deani	01010010010000001000101001100010
Eptatretus burgeri	01101001001010000100001000001001
Paramyxine sp. A	01101000000010001100001010011100
Paramyxine sp. B	01101001001000000100000100001001

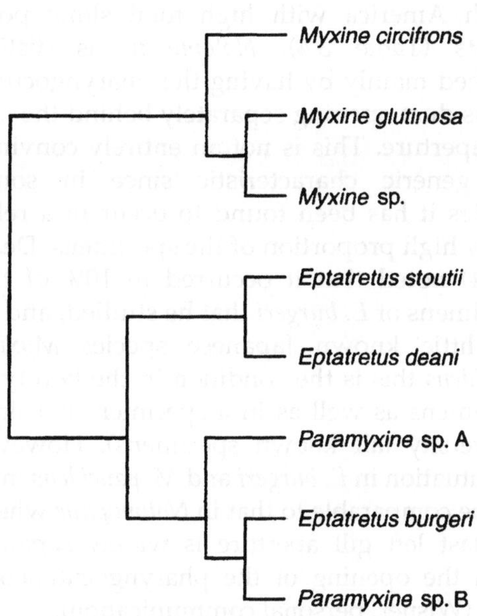

Figure 3.3 Cladogram of hagfish species based on isozyme data as presented in Table 3.5. *Myxine* sp. may represent *M. hubbsi*. *Paramyxine* spp. A and B represent common undescribed species of Japanese paramyxinids.

find it most helpful at present not to distinguish *Paramyxine* but include it in *Eptatretus* (Table 3.2).

3.5 MYXININAE

This subfamily is defined as having one pair of common gill apertures. It comprises the genera *Myxine*, *Nemamyxine*, *Neomyxine* and *Notomyxine* (Tables 3.3 and 3.4). *Nemamyxine* and *Neomyxine* are quite distinct southern hemisphere forms from New Zealand and South America with high total slime pore counts (Table 3.4). *Notomyxine* is distinguished mainly by having the pharyngocutaneous duct opening separately behind the left gill aperture. This is not an entirely convincing generic characteristic since in some species it has been found to occur in a relatively high proportion of the specimens. Dean (1904) noted that it occurred in 10% of the specimens of *E. burgeri* that he studied, and in the little known Japanese species *Myxine paucidens* this is the condition in the two type specimens as well as in a specimen in Tokyo University (all known specimens). However the situation in *E. burgeri* and *M. paucidens* may not be comparable to that in *Notomyxine* where the last left gill aperture is widely separate from the opening of the pharyngocutaneous duct (Wisner, personal communication).

3.6 BIOGEOGRAPHY

The recent increase in knowledge of hagfish species has demonstrated that they occur in all oceans except the Polar Seas (Table 3.1). They cannot survive in warm water and seem to be absent from the warm waters of the Red Sea and Gulf of Thailand and of rare occurrence in the Mediterranean (Dieuzeidae, 1956). In equatorial waters hagfishes occur only in deep, cool water. For at least one species (*E. burgeri*) living in shallow water with seasonal warming, a seasonal migration to deeper water occurs during the warm season (Fernholm, 1974).

Hagfishes occur at all depths down to at least 2700 m (Wisner and McMillan, 1990). The group does not occur in the Antarctic but is plentiful around the southern parts of Africa, New Zealand and South America and around the Falkland Islands. Hagfishes have not been found in the North Polar Sea. *M. glutinosa* occurs off the Norwegian north coast and off Murmansk but not further east. This may be related to the reduced salinity over large areas of the shallow coastal sea receiving water from the Russian rivers.

As in lampreys, a few genera are restricted to the southern hemisphere (*Notomyxine*, *Nemamyxine* and *Neomyxine*) with *Nemamyxine kreffti* off Argentina and Brazil (Mincarone, in press) and *Nemamyxine elongata* off New Zealand indicating earlier continental connections.

3.7 SPECIES LIST

In Tables 3.2–3.4 all recognized species (58) are listed alphabetically with a few selected characteristics of cusp and slime pore counts in addition to maximum known length, original description, distribution and depth of occurrence. In addition, some notes about selected species are given below.

3.7.1 New World Myxine and Notomyxine

M. affinis Günther, 1870: Straits of Magellan; frequently taken with *M. australis* but distinguished by higher unicusp and trunk slime pore counts.

M. australis Jenyns, 1842; syn. *M. acutifrons* Garman, 1899: Straits of Magellan where it was obtained by Charles Darwin by hook amongst the kelp in Tierra del Fuego, but also further north off Argentina and Chile; redescribed by Wisner and McMillan (1995) who state that although the multicusp pattern and number of gill pouches is the same as in *M. affinis* it can be distinguished by fewer unicusps, fewer trunk and total slime pores and shorter total length of mature specimens.

There is no type in the British Museum of Natural History but Norman (1937) states that the type, 285 mm, is in the Zoological Museum in Cambridge. The fusion of three cusps is given by Günther (1870) as a characteristic of *M. australis* but occurs in only one specimen from Sandy Point, Straits of Magellan. That 460 mm specimen was later selected as the type for *M. tridentiger* by Regan (1913a) and used by Nani and Gneri (1951) to establish the genus *Notomyxine* (see below).

M. circifrons: Caught close to the equator at 7°N at a depth of 1336 m in water of 3.6°C; occurring in Eastern Pacific from San Francisco to north-central Chile. Redescribed by Wisner and McMillan (1995).

M. debueni: Known only from type material in the Straits of Magellan.

M. dorsum: Known only from two specimens from southwestern Atlantic.

M. fernholmi: Known only from the type material off Chile and the Falkland Islands.

M. hubbsi: Deep water species; of 150 specimens investigated 76% were females, 23% hermaphroditic and 6% males (Wisner and McMillan, 1995).

M. hubbsoides: Known only from type material off central Chile.

M. knappi: Known only from type material near the Falkland Islands.

M. limosa Girard, 1859; syn *M. atlanticus* Regan, 1913: Grand Manan Isle, Bay of Fundy; often synonymized with *M. glutinosa* but recognized by Wisner and McMillan (1995) mainly on the basis of coloration (reddish-brown instead of greyish-pink, and is further characterized by a narrow pale streak in the dorsal midline) and size (larger); western Atlantic from the Davis Strait (66°N) to Florida in south; 75–1006 m; other recent studies, however, failed to distinguish between the two, except that the Gulf of Maine population seemed to be aberrant (Martini pers. comm.). *M. glutinosa* occurs off Greenland and Iceland and clinal variation cannot be ruled out. *M. limosa* is included here mainly to stimulate systematic work with this difficult complex.

M. mccoskeri: South Caribbean Sea; similarities with an undescribed species from off Suriname (Shimizu, 1983).

M. mcmillanae: Known only from the west and southwest coasts of Puerto Rico and St Croix, Virgin Islands, USA; white headed and similar to *M. ios* from Ireland except for gill pouch number (6 vs 7).

M. pequenoi: Known only from type material south of Valdivia, Chile; with seven gill pouches like *M. capensis* and *M. ios* but with significantly different slime pore numbers; a dwarfed species with low number of unicusps like *M. paucidens* but differs in gill pouch numbers (7 vs 6) and wide geographical separation.

M. robinsorum: Name corrected according to ICZN (1985), Article 32c(ii). Known only from the type material from the southern Caribbean Sea; *M. robinsorum*, *M. circifrons*, and *M. mccoskeri* may be closely related having five gill pouches and pattern of fused cusps 3/2 (Wisner and McMillan, 1995).

Notomyxine tridentiger (Garman, 1899): Generic character given by Nani and Gneri (1951) is one pair of branchial apertures separated by a distance of 7.5–13.5 mm from the posteriorly located opening of the pharyngocutaneous duct; holotype from Sandy Point, Straits of Magellan.

3.7.2 Old World Myxine

M. capensis: Type 310 mm, Cape of Good Hope, common around the tip of South Africa.

M. garmani: Described from three specimens from Sagami Bay, Japan; fished commercially around Japan.

M. glutinosa: Only species in the northeast Atlantic. Specific delimitation to *M. limosa* and *M. australis* is problematical.

M. ios: One of several deep sea *Myxine* with whitish head; off West Africa between 0 and 25°N and south of Ireland.

3.7.3 Nemamyxine and Neomyxine

Nemamyxine Richardson (1958): Slender myxinids having one pair of external branchial apertures, median ventral fin fold extending between the branchial apertures terminating shortly behind the head; mucous glands numerous, segmentally arranged.

N. elongata: New Zealand, caught upstream in the Kaituna River in a whitebait net in an estuary situation; known only from the type specimen; extremely elongated, black-pigmented species.

N. kreffti: Only known from type specimen off Argentina and recently from another specimen off southern Brazil (Mincarone, in press), less elongated than *N. elongata*.

Neomyxine Richardson (1953): Based on *M. biniplicata* Richardson and Jowett (1951); slender myxinids with a pair of short lateral fin folds extending anteriorly from closely behind the branchial apertures; mucous glands close together and more numerous than body segments. One species, New Zealand.

N. biniplicata: Only species in the genus, New Zealand.

3.7.4 Eptatretus

E. atami: Described from a single specimen, 550 mm female off Cape Manazuru in Sagami Bay at 494 m; notoriously confused with other unnamed species around Japan.

E. carlhubbsi: Named in honour of Carl Hubbs, who from 1965 to his death 1979 collected material and information for a worldwide revision of hagfishes; largest known hagfish reaching more than 1000 mm in total length; caught in self-rising trap in deep water in the Pacific.

E. caribbeaus: Only known seven-gilled *Eptatretus* in the Atlantic.

E. cheni: In original description pattern of fused cusps given as 3/2, later corrected to 3/3 (Kuo *et al.*, 1994).

E. eos: Only known from one type specimen; conspicuously pink coloration and peculiar elongated nostril distinguishes this species perhaps at generic level.

E. fernholmi: Taiwanese species with similarities to unnamed Japanese paramyxinid species (Kuo *et al.*, 1994). However, the unnamed Japanese species that have been confounded with *E. atami* are larger and with a higher number of slime pores.

E. springeri: Original description misleading because two species (*E. springeri* and *E. minor*) were mixed (Fernholm and Hubbs, 1981).

3.8 CONCLUDING REMARK

Although I am aware of about ten species under description I do not believe that we will see another threefold increase in the number of known species in the next two decades as we have experienced since *The Biology of Myxine* (see Adam and Strahan, 1963) was published. Clearly, however, basic taxonomic work is not finished and phylogenetic studies are much needed.

REFERENCES

Adam, H. and Strahan, R. (1963) Systematics and geographical distribution of myxinoids, in *The Biology of Myxine* (eds Alf Brodal and Ragnar Fänge), Universitetsforlaget, Oslo, pp. 1–8.

Barnard, K.H. (1923) Diagnoses of new species of marine fishes from South African waters. *Annals of the South African Museum*, **13**, 439–445.

Bigelow, H.B. and Schroeder, W.C. (1952) A new species of the cyclostome genus *Paramyxine* from the Gulf of Mexico. *Breviora*, Museum of Comparative Zoology, Cambridge, Mass., **8**, 1–10.

Bloch, M.E. and Schneider, J.G. (1801) *Systema ichthyologiae iconibus ex illustratium. Post obitum auctoris opus inchoatum absolvit, correxit, I interpolovit Jo.* Gotlob Schneider Saxo, Berlin, pp. 1–584.

Dean, B. (1904) Notes on Japanese myxinoids. *Journal of the College of Science*, Imperial University of Tokyo, Japan, **19**(2), 1–23.

Dieuzeidae, R. (1956) Les myxines en Méditerranée. *Bulletin des Travaux*. Publiés par la station d'Aquaculture et de Peche de Castiglione, **8**.

Evermann, B.W. and Goldsborough, E.L. (1907) The fishes of Alaska. *U.S. Bulletin Bureau of Fisheries*, **26** (for 1906), 219–360.

Farris, J.S. (1988) Hennig86, program and documentation. Distributed by A.G. Kluge, Museum of Zoology, University of Michigan, Ann Arbor, MI 48104 USA.

Fernholm, B. (1974) Diurnal variations in the behaviour of the hagfish *Eptatretus burgeri*. *Marine Biology*, **27**, 351–6.

Fernholm, B. (1981) A new species of hagfish of the genus *Myxine*, with notes on other eastern Atlantic myxinids. *Journal of Fish Biology*, **19**, 73–82.

Fernholm, B. (1982) *Eptatretus carribbeaus*: A new species of hagfish (Myxinidae) from the *Caribbean. Bulletin of Marine Science*, **32**(2), 434–8.

Fernholm, B. (1991) *Eptatretus eos*: a new species of hagfish (Myxinidae) from the Tasman Sea. *Japanese Journal of Ichthyology*, **18**(2), 115–18.

Fernholm, B. and Hubbs, C.L. (1981) Western Atlantic hagfishes of the genus *Eptatretus* (Myxinidae) with description of two new species. *Fishery Bulletin*, **79**(1), 69–83.

Garman, S. (1899) *The Fishes*. Reports on an exploration of the west coast of Mexico, Central and South America, and off the Galapagos Islands, in charge of Alexander Agazziz, by the U.S. Fish Commission Steamer *Albatross* during 1891, Lieut. Commander Z.L. Tanner, Commanding. *Memoirs, Museum Comparative Zoology, Harvard College*, **24**, 1–431.

Girard, C.F. (1854) Abstract of a report to Lieut. Jas. M. Gilliss, U.S.N. upon the fishes collected during the U.S.N. Astronomical Expedition to Chili. *Proceedings. Philadelphia Academy Natural Sciences*, **7**, 197–9.

Girard, C.F. (1855) Fishes. Pp. 230–253 in Lieut. J.M. Gilliss, Superintendent. The U.S. Naval Astronomical Expedition to the Southern Hemisphere during the years 1849–52, Contributions to the fauna of Chile, Vol. 2. 33rd Congress, 1st Session, Executive Document No. 121, pp. 207–262.

Girard, C. (1859) Ichthyological notices. *Proceedings of the Academy of Natural Sciences of Philadelphia*, **11**, 223–5.

Günther, A. (1870) *Catalogue of the Fish* British Museum, 510–513.

Hensley, D.A. (1985) *Eptatretus mendozai*, a new species of hagfish (Myxinidae) from off the southwest coast of Puerto Rico. *Copeia* (4), 865–869.

Hensley, D. (1991) *Myxine mcmillanae*, a new species of hagfish (Myxinidae) from Puerto Rico and the U.S. Virgin Islands. *Copeia* (4), 1040–1043.

Huang, K-F., Mok, H-K. and Huang, P-C. (1994) Hagfishes of Taiwan (II): taxonomy as inferred from mitochondrial DNA diversity. *Zoological Studies*, **33**(3), 186–91.

ICZN (1985) *International Code of Zoological Nomenclature* (3rd edn), International Trust for Zoological Nomenclature, British Museum (Natural History), London.

Jansson, H., Wyöni, P-I., Fernholm, B. *et al*. (1995) Genetic relationships among species of hagfish revealed by protein electrophoresis. *Journal of Fish Biology*, **47**, 599–608.

Jenyns, L. (1842) Cyclostomi, in C. Darwin, *The Zoology of the voyage of HMS Beagle* under the command of captain Fitzroy, R.N. during the years 1832 to 1836. Part IV, *Fish*. Smith, Elder & Co., London.

Jordan, D.S. and Snyder, J.O. (1901) A review of the lancelets, hagfishes and lampreys of Japan with a description of two new species. *Proceedings of U.S. National Museum*, **23**, 726–38.

Kuo, C-H., Huang, K-F. and Mok, H-K. (1994) Hagfishes of Taiwan (I): a taxonomic revision with description of four new *Paramyxine* species. *Zoological Studies*, **33**(2), 126–39.

Kuo, C-H. and Mok, H-K. (1994) *Eptatretus chinensis*: a new species of hagfish (Myxinidae; Myxiniformes) from the South China Sea. *Zoological Studies*, **33**(4), 246–50.

Linnaeus, C. (1758) *Systema Naturae* (10th edn) *Holmiae: Laurentii Salvii.*

Lockington, W.N. (1878) Walks round San Francisco. *American Naturalist*, **12**, 786–93.

McMillan, C.B. and Wisner, R.L. (1982) Results of the research cruises of FRV Walther Herwig to South America LX. *Nemamyxine kreffti*, a new species of hagfish from off Argentina. *Arch. Fisch Wiss*. **32**(1/3), 33–8.

McMillan, C.B. and Wisner, R.L. (1984) Three new species of seven-gilled hagfishes (Myxinidae, *Eptratretus*) from the Pacific Ocean. *Proceedings of the California Academy of Sciences*, **43**(16), 249–67.

Mincarone, M.M. (in press) Inclusão de classe Myxini (Agnatha) na ictiofauna do Brasil, com base na segunda ocorrência de *Nemamyxine kreffti* McMillan & Wisner, 1982 (Myxinformes: Myxinidae). Brazilian Ichthyology Congress, abstract.

Müller, J. (1834) Vergleichende Anatomie der Myxinoiden, der Cyclostomen mit durchbohrtem Gaumen. Erster Teil. Osteologie und Myologie. *Abhandlungen Akademie Wissenschaften, Berlin*, pp. 65–340.

Nani, A. and Gneri, F.S. (1951) Estudio de los mixinoideos sudamericanos. *Revista del Inst. Nac. de Invest. de las Ciencias Naturales*, pp. 184–223.

Nelson, J.S. (1994) *Fishes of the World*, John Wiley & Sons, pp. 26–9.

Norman, J.R. (1937) Coast fishes, Part II. The Patagonia region, *Discovery Reports*, **16**, 1–150.

Regan, C.T. (1912) A synopsis of the myxinoids of the genus *Heptatretus* or *Bdellostoma*. *Annual Magazine of Natural History*, **9**, 534–6.

Regan, C.T. (1913a) Revision of the myxinoids of the genus Myxine. *Annual Magazine of Natural History*, **11**(8), 395–8.

Regan, C.T. (1913b) Note on *Myxine capensis*. *Annual Magazine of Natural History*, **12**(8), 217.

Richardson, L.R. (1953) *Neomyxine* n.g. (Cyclostomata) based on *Myxine biniplicata*. *Royal Society of New Zealand*, **81**(3), 379–83.

Richardson, L.R. (1958) A new genus and species of Myxinidae/Cyclostomata). *Royal Society of New Zealand*, **85**(2), 283–7.

Richardson, L.R. and Jowett, J.P. (1951) A new species of *Myxine* (Cyclostomata) from Cook Strait. *Victoria University College, Wellington*, **12**, 1–5.

Schneider, A. (1880) Über die Arten von *Bdellostoma*. *Archiv für Naturgeschichte*, **46**(1), 115–16.

Shen, S-C. and Tao, H-J. (1975) Systematic studies on the hagfish (Eptatretidae) in the adjacent waters around Taiwan with description of two new species. *Chinese Bioscience*, **11**(8), 66–78.

Shimizu, T. (1983) Fishes trawled off Suriname and French Guiana. *Japan Marine Fishery Resource Center*, p. 519.

Strahan, R. (1975) *Eptatretus longipinnis*, n. sp. A new hagfish (family Eptatretidae) from South Australia, with a key to the 5–7 gilled Eptatretidae. Australian Zoology, 18(3), 138–48.

Teng, H-T. (1958) A new species of Cyclostomata from Taiwan. *Chinese Fisheries Monthly*, **66**, 3–6.

Wisner, R.L. and McMillan, C.B. (1988) A new species of hagfish, genus *Eptatretus* (Cyclostomata, Myxinidae) from the Pacific Ocean near Valparaiso, Chile, with new data on *E. bischoffii* and *E. polytrema*. *Transactions of the San Diego Society of Natural History*, **21**(14), 227–44.

Wisner, R.L. and McMillan, C.B. (1990) Three new species of hagfishes, genus *Eptatretus* (Cyclostomata, Myxinidae), from the Pacific Coast of North American with new data on *E. deani* and *E. stoutii*. *Fishery Bulletin*, **88**, 787–804.

Wisner, R.L. and McMillan, C.B. (1995) Review of new world hagfishes of the genus *Myxine* (Agnatha, Myxinidae) with descriptions of nine new species. *Fishery Bulletin*, **93**, 530–50.

Wongratana, T. (1984) *Eptatretus indrambaryai*, a new species of hagfish (Myxinidae) from the Andaman Sea. *Natural History Bulletin of the Siam Society*, **31**(2), 139–50.

4
ASIAN HAGFISHES AND THEIR FISHERIES BIOLOGY

Yoshiharu Honma

SUMMARY

The classification and distribution of 13 species of Asian hagfishes, consisting of two species of *Myxine*, three of *Eptatretus* and eight of *Paramyxine*, are reviewed. The occurrence of *Eptatretus chinensis* Kuo and Mok (1994) in the South China Sea is noticed. The recent decline of hagfishery on the Pacific coast of Japanese Islands, probably due to overfishing, is discussed.

4.1 INTRODUCTION

According to Nelson (1994), the hagfishes (Order Myxiniformes) comprise a single family, including 6 genera and 43 species, living only in sea water. Subsequently, Kuo and Mok (1994) recorded one more species belonging to the genus *Eptatretus*, and Kuo *et al.* (1994) described four additional species of *Paramyxine* from Taiwanese and adjacent waters, raising the total number of species recognized at present to 48. Hagfishes are distributed world wide, occurring in subarctic waters as well as in the temperate regions of both hemispheres. Fernholm recognizes 59 species (Chapter 3, this volume).

Figure 4.1 Ronald Strahan (University of Hong Kong, now Australian Museum, Sydney), counting and measuring brown hagfish, *Paramyxine atami*, at Teradomari Town Aquarium in October 1959.

Because of their anatomically interesting organization, the author and co-workers undertook histological and immunohistochemical studies of the neuroendocrine organs of a Japanese hagfish, *Paramyxine atami*, and the arctic lamprey, *Lampetra* (= *Lethenteron*) *japonica*, both being collected in Niigata district, Sea of Japan (Honma, 1960, 1969; Chiba and Honma, 1986a, b, 1992; Chiba *et al.*, 1993). In addition to this, a taxonomic study was made on *P. atami*, with special

The Biology of Hagfishes. Edited by Jørgen Mørup Jørgensen, Jens Peter Lomholt, Roy E. Weber and Hans Malte. Published in 1998 by Chapman & Hall, London. ISBN 0 412 78530 7.

reference to *Paramyxine* species (Strahan and Honma, 1960, 1961) (Figure 4.1).

At that time, Strahan suggested that the author make an investigation of the hagfishery, operating in Sado Strait and off the shore of Niigata, the largest city on the west coast of Honshu. The previously successful fishery in the Niigata district, probably being a reflection of the economic conditions at the time, had obviously declined. Although hagfishery in Niigata operated primarily for food, the possibility of hagfish being processed for leather was also considered by Kanebo Co. Ltd (Tokyo) in 1939, using specimens collected off the southern coast of Korea.

During his boyhood, the author witnessed the hagfishery and leather manufacture in Niigata. However, since a good account of the fishery in Japan and Korea has already been given by Gorbman *et al.* (1990), the present account will be restricted to some supplementary remarks. The following includes short descriptions of and keys to the Asian hagfishes and some recent fisheries information.

4.2 TAXONOMY AND DISTRIBUTION OF ASIAN HAGFISHES

The Asian hagfishes were first reviewed by Jordan and Snyder (1901), who recognized two species, *Eptatretus burgeri* (Girard) and *Myxine garmani* Jordan and Snyder. Later, Dean (1904) reported two additional new species, *Homea* (= *Eptatretus*) *okinoseana*, and *Paramyxine atami*; the latter was erected as a new genus owing to the six gills having separate efferent branchial ducts of distinctly unequal length, the most anterior being several times the length of the most posterior. However, taxonomic characters and the description of *Paramyxine* were based only on one specimen, collected from deep water off cape Manazuru near Atami, Izu Peninsula, on the Pacific coast of Honshu. Regan (1913) reviewed nine species of *Myxine*, and described a Japanese species, *M. paucidens*, as new to science. Subsequently Matsubara (1937) investigated intraspecific variation in *P. atami* including disposition and shape of branchial apertures, position of the base of the tongue muscle and dental formula, comparing 14 specimens caught in deep water off Kumano-nada, Pacific coast of central Honshu (Figures 4.2 and 4.3) to Dean's type specimen.

During the Second World War, the fishery of *Paramyxine* rapidly expanded in the Sea of Japan, off the Niigata district. Okada *et al.* (1948) surveyed the hagfishery as well as the morphology and ecology of *P. atami*, using almost 1000 specimens from Niigata, although their primary aim was related to the importance of the commercial operation in relation to the exhaustion of more commonly used natural resources during wartime. Therefore, their account did not include observations of taxonomic significance.

A study distinguishing *Paramyxine* species from the related genera, *Myxine* and *Eptatretus*, was made by Strahan and Honma (1960), who also recognized the decline of the hagfishery based on *P. atami*. The study examined variation in several taxonomic characters of *P. atami*, resulting in a revision of the genus *Paramyxine*, for which three species were recognized, *viz. P. atami*, *P. springeri* Bigelow and Schroeder (1952) from the Gulf of Mexico and *P. yangi* Teng (1958) from southern Taiwan. Dean's type specimen of *P. atami* was considered to be an unusual (atypical) specimen, after comparison with material from both the Pacific and the Sea of Japan populations.

The Taiwanese hagfish, *P. yangi*, with five pairs of gills and approximately half the body length of the two other known species, *P. atami* and *P. springeri*, was reported from the subtropical Pacific Ocean by Teng (1958). *P. atami* and *Eptatretus burgeri* were later collected from northern Taiwan (Yang, 1972). Shen and Tao (1975) documented two known species and two new species, *P. taiwanae* from northeastern and southwestern Taiwan and

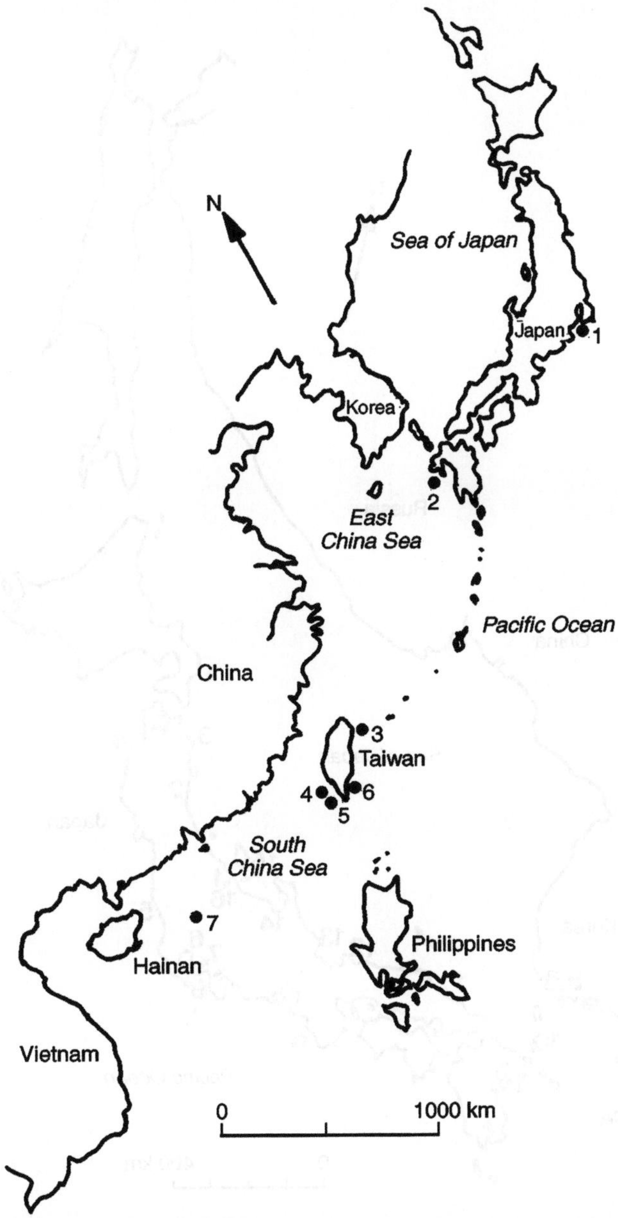

Figure 4.2 Map of western Pacific, including Japan, Taiwan and Hainan, showing type localities of the Asian hagfishes described as new to science. (1) *Myxine garmani* Jordan and Snyder, 1901; *Paramyxine atami* Dean, 1904; *Eptatretus okinoseanus* Dean, 1904. (2) *Eptatretus burgeri* (Girard, 1850). (3) *Paramyxine taiwanae* Shen and Tao (1975). (4, 5) *Paramyxine yangi* Teng (1958); *P. cheni* Shen and Tao (1975): *P. sheni* Kuo *et al.* (1994); *P. nelsoni* Kuo et *al.* (1994); *P. fernholmi* Kuo *et al.* (1994). (6) *P. wisneri* Kuo *et al.* (1994). (7) *Eptatretus chinensis* Kuo and Mok (1994).

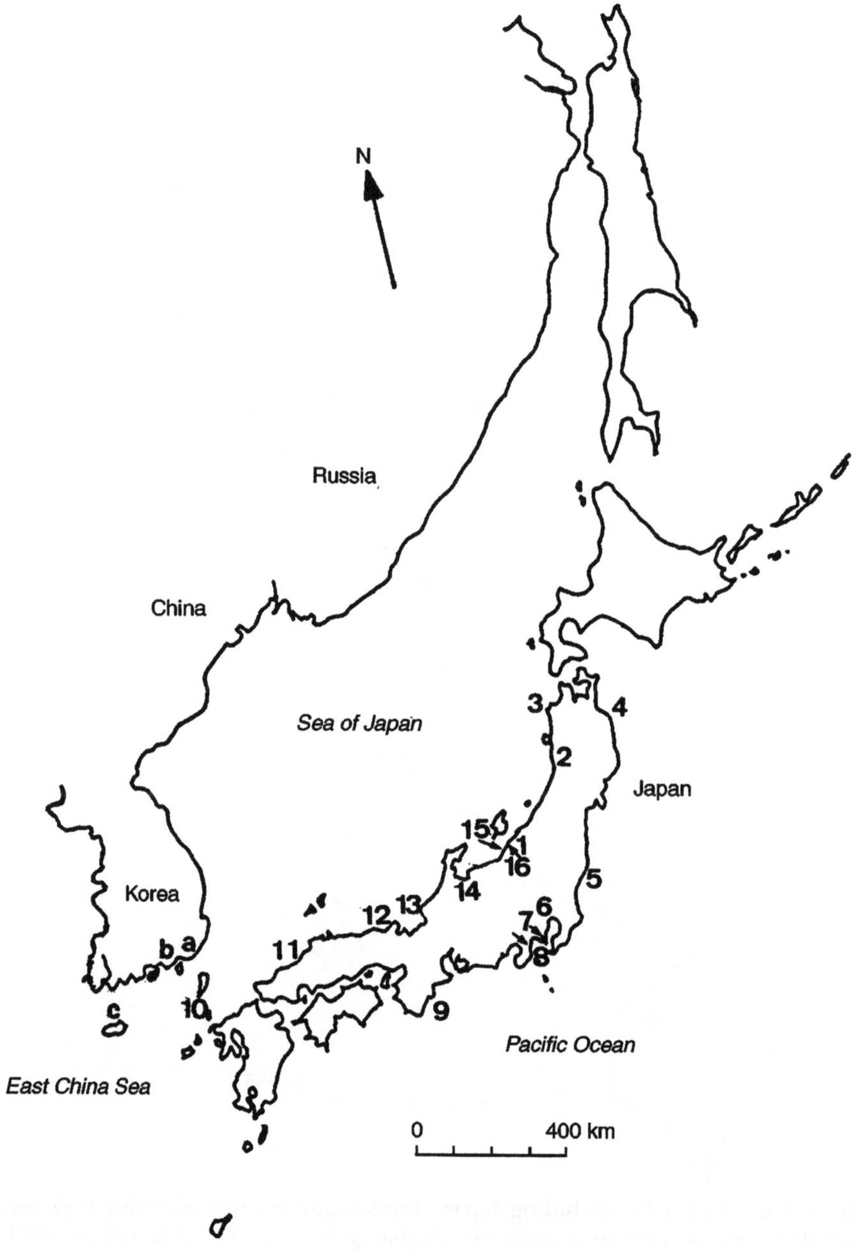

Figure 4.3 Map of Japan and Korea showing locations of hagfishing industry and collecting sites. (1) Niigata; (2) Akita, (3) Fukaura, (4) Hachino-he, (5) Onahama, (6) Tokyo, (7) Yokosuka, (8) Manazuru, (9) Kumanonada, (10) Tsushima, (11) Shimane and San'in, (12) Kasumi, (13) Fukui, (14) Toyama, (15) Izumozaki, (16) Teradomari; (a) Pusan, (b) Cheiju, (c) Kyongnan.

P. cheni from southwestern Taiwan. The size of these was intermediate between *P. atami* and *P. yangi*. Recently, Kuo *et al.* (1994) described four new *Paramyxine* from Taiwan and added the Japanese species *Eptatretus okinoseanus*, to the agnathan fauna of Taiwan. *Paramyxine nelsoni, P. sheni* and *P. wisneri* from the southwestern region of Taiwan, and *P. fernholmi* from the east coast, are all smaller than the species occurring in Japanese waters. Kuo and Mok (1994) reported a new species, *Eptatretus chinensis*, from deep water off the coast of east Hainan in the tropical South China Sea (Figure 4.2).

Huang *et al.*'s 1994 study of the mitochondrial DNA of six species (two *Eptatretus* and four *Paramyxine*) confirmed the earlier study of Huang (1989), concluding that *Paramyxine* was monophyletic and *Eptatretus* probably paraphyletic. Among the *Paramyxine* spp. they considered *P. yangi* and *P. nelsoni* to constitute a sister group, with *P. taiwanae* being the more primitive outgroup. The sympatrically occurring *P. sheni* and *P. wisneri* were shown to have homologous mitochondrial genomes, although the former showed a high frequency of heteroplasy. They concluded that *Paramyxine* only comprised hagfishes restricted to the western Pacific, thus supporting Fernholm and Hubbs (1981), who included only Asian species in *Paramyxine* and referred *P. springeri*, found in the northeastern Gulf of Mexico, to the genus *Eptatretus*. This contrasts with the earlier opinion of Strahan (1975) who believed that *Paramyxine* should be treated as a junior synonym of *Eptatretus*, owing to an apparent *Paramyxine* branchial duct characteristic being shared by the Atlantic *E. springeri* and Asian *E. burgeri*. Fernholm (Chapter 3, this volume) is of the opinion that all *Paramyxine* species should be treated as belonging to the genus *Eptatretus*.

In the light of the above and other studies, including Jordan *et al.* (1913), Matsubara (1955), Tchang (1940), Chu *et al.* (1963), Lindberg and Legeza (1959), Okamura (1986), Okamura *et al.* (1982), Okamura and Kitajima (1984) and Nakabo (1993), the classification of Asian hagfishes is as follows:

CLASS MYXINI

ORDER MYXINIFORMES

Family Myxinidae

Subfamily Myxininae

Efferent branchial ducts open by a common external aperture on each side (i.e. only one pair of branchial openings). The pharyngocutaneous duct, which leads from the pharynx behind the gills, is present only on the left side and probably allows flushing of the pharynx, thus clearing particles too large for the afferent branchial ducts.

Genus *Myxine*

No. of gill apertures 1, anal fin ending posterior to branchial aperture; 5 to 7 pairs of gill pouches.

1a Lateral teeth (= cuspids) 10–13 /10–12; fused teeth 3/2, mucous glands (26–27) + (57–61) + (12–13) = 95–101. Body length up to 50 cm. 200–1000 m depth, off Pacific coast of Iwate Prefecture, Okinawa . . . .*Myxine garmani*

1b Lateral teeth 6/7, fused teeth 2, mucous glands 26 + (53–57) + 10 = 89–93. Supposed length of head 3.4 in total length. Body length 24.0–30.5 cm. 600 m depth, Japan. Apart from the very simple original description, no information is available. . . . . *M. paucidens*

Subfamily Eptatretinae

Efferent branchial ducts open separately to the exterior with 4–16 external gill openings.

Genus *Eptatretus*

Gill apertures 6–8, arranged in a regular line on each side.

Gill apertures 6 (rarely 7). 1

Gill apertures 8 2

1a Lateral teeth 8–11/9–11, fused teeth 3/2, mucous glands (18–23) + (46–51) + (10–13) = 74–89. A white band on median dorsal ridge. Max. body length 60 cm. Shallow waters, central and southern Japan, South Korea, northeast Taiwan .... *Eptatretus burgeri*

1b Lateral teeth 10/10, fused teeth 3/3, mucous glands (15–19) + (4–5) + (42–45) + (11–14) = 72–83. No white band on median dorsal ridge, prominent eye spots present. Max. body length 40 cm, 600 m depth, South China Sea .... *E. chinensis*

2a Lateral teeth 11–13/10–12, fused teeth 3/2, mucous glands 13 + (7–8) + (54–56) + 13 = 87–92. No white band on median dorsal ridge. Max. body length 80 cm. 200–600 m depth, Pacific coast of central and southern Japan to northeast Taiwan. .... *E. okinoseanus*

Genus *Paramyxine* (= *Eptatretus* by Fernholm, Chapter 3, this volume)

Gill apertures 4–6 arranged basically in a crowded, irregular line, some in a regular line on each side.

Gill apertures 4 1
Gill apertures 5 2
Gill apertures 6 3

1a Lateral teeth 11/9, fused teeth 3/2, mucous glands 19 + 0 + 35 + 8 = 62. No light ocular patch, no white band on median dorsal ridge. Less than 20 cm. 50–200 m depth, southwest Taiwan .... *Paramyxine nelsoni*

2a Lateral teeth 9–11/9–11, fused teeth 3/2, mucous glands (16–23) + 0 + (42–47) + (8–11) = 66–78, gill apertures crowded. Max. body length 30 cm. 20–50 m depth, northeast and southwest Taiwan .... *P. yangi*

2b Lateral teeth 12–14/13, fused teeth 3/3, mucous glands 26 + 0 + (45–47) + (7–8) = 78–81, gill apertures in a regular line, white rings absent around gill aperture. Body length less than 40 cm. 180 m depth, southwest Taiwan .... *P. cheni*

3a Lateral teeth 8–13/9–13, fused teeth 3/2–3, mucous glands (16–21) + 0 + (41–49) + (9–13) = 66–83. Less than 60 cm. 50–400 m depth, Pacific and Sea of Japan coasts of Honshu and Kyushu, South Korea, northeast Taiwan .... *P. atami*

3b Lateral teeth 13–15/12–14, fused teeth 3/3, mucous glands (13–18) + (0–2) + (39–46) + (8–12) = 64–74, gill apertures arranged in a regular line, each gill aperture surrounded by a white ring. Max. body length 45 cm. 300–450 m depth, east and southwest Taiwan .... *P. sheni*

3c Lateral teeth 9–11/8–11 = 38–42, fused teeth 3/2, mucous glands (16–19) + 0 + (36–42) + (6–9) = 60–68, each gill aperture not surrounded by a single white area on both sides of ventral surface, irregularly crowded together. Max. body length 32 cm. Northeast and southwest Taiwan ... . *P. taiwanae*

3d Lateral teeth 11–13/10–12 = 42–50, fused teeth 3/2, mucous glands (16–23) + 0 + (38–44) + (6–11) = 64–72, 1 or 2 rows of gill apertures, each aperture closely spaced, surrounded by a white ring. External opening of pharyngocutaneous duct not confluent with most posterior left aperture, branchial length 2.66% of total length, entire ventral surface white. Max. body length 30 cm. 300 m depth, east and southeast Taiwan .... *P. fernholmi*

3e Lateral teeth 10–13/10–11, fused teeth 3/2, mucous glands (15–20) + (0–1) + (36–44) + (6–11) = 63–72, gill apertures in a straight line, external opening of pharyngocutaneous duct confluent with the most posterior left gill aperture, branchial length 4.58% in total length, only ventral fin fold white. Max. body length 33.5 cm. 200 m depth, southwest Taiwan .... *P. wisneri*

4.3 FISHERIES BIOLOGY

Hagfisheries in North America and the western Pacific have already been discussed by Okada *et al.* (1948), Strahan and Honma (1960), and, in a major analysis, Gorbman *et al.* (1990). Consequently, the present chapter seeks to avoid repetition and concentrates on aspects not previously covered.

Before the Second World War, local people in fishing villages along the Sea of Japan and the northeastern Pacific coasts of Honshu utilized hagfish for food following traditional customs, in addition to use as longline baits for red sea bream, *Pagrus major*, and crimson sea bream, *Evynnis japonica*.

It is well known that hagfish are carnivorous, carrion feeders and/or scavengers (Malm, 1877; Worthington, 1905; Gustafson, 1935; Okada *et al.* 1948; Honma, 1960, 1961, 1983; Strahan and Honma, 1960; Strahan, 1963; Jensen, 1966; Kobayashi *et al.*, 1972; Shelton, 1978). According to these reports, the diets of hagfish are dead and moribund fishes, invertebrate benthic animals, including annelid worms and crustaceans, and probably detritus. The narratives of fishermen's personal experiences in the Niigata district support the above (Honma, 1960, 1961, 1968). The habitat and behaviour of hagfishes have been discussed by Gustafson (1935), Jensen (1966), Foss (1968), McInerney and Evans (1970) and Patzner (1977). Hagfishes inhabit dark, muddy bottoms, consisting of silt rather than sand, in sea water of low temperature and high salinity; they have never been recorded from rivers or bodies of fresh water. Although these are burrowing animals with remarkably sluggish habits, their serpentine swimming movements can, in fact, be relatively fast, judged from aquarium observations.

In the Niigata district, during the Second World War, Mr T. Tomioka influenced the hagfishery so as to further exploit the natural resources available. Thus, *Paramyxine atami* was utilized in diverse ways, such as food, a substitute for leathercrafts and even a cleaner

Figure 4.4 Hagfish prepared for broiling (seen only at Teradomari fish market).

Table 4.1 Official hagfish catches from Korean waters in 1993

Offshore long line	6 t	Pusan	77 t
Coastal long line	12 t	Kyongnam	53 t
Coastal trap	130 t	Cheju	18 t
Totals	148 t		148 t

Table 4.2 Official monthly hagfish catches in Korean waters in 1993

January	1 t	July	17 t
February	3	August	23
March	0	September	19
April	6	October	16
May	22	November	12
June	19	December	3
Total			148 t

was made from the slime. In those days, Niigata was the largest hagfishing centre and the centre for collection of catches of *P. atami*. The leather goods were processed and sold through Mitsui Bussan (= Products) Co. Ltd. Other active hagfisheries were maintained at Kasumi, Hyogo Prefecture (San'in district), western Honshu, and Fukaura, Aomori Prefecture, northern Honshu, in addition to trial fishing undertaken at, for example, Akita, Toyama, Fukui and Shimane Prefectures, along the west coast of Honshu. However, in all these localities, the hagfish catch progressively declined, with operations

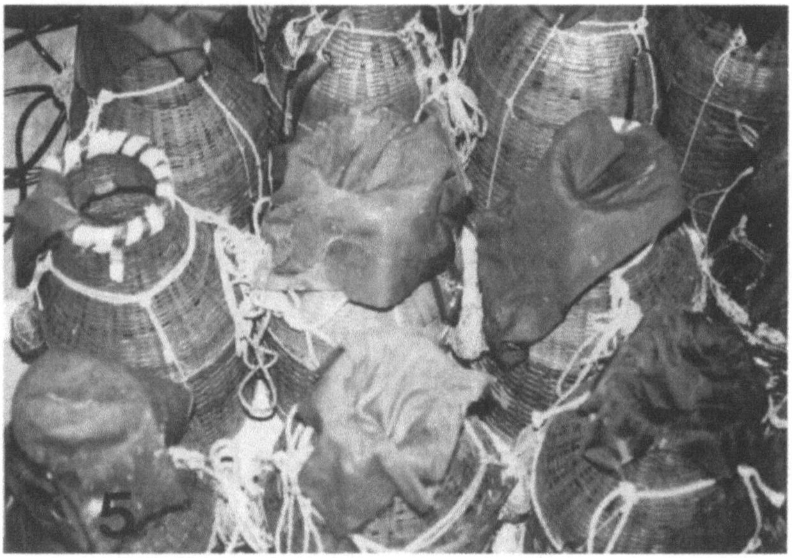

Figure 4.5 Woven bamboo traps showing open ends with canvas and round stone weights.

Figure 4.6 Hagfishing boat, *Zenpoh-maru*, 4.9 tons, at anchor in Izumozaki harbour, Niigata Prefecture, Sea of Japan.

Table 4.3 Importation of *Myxine garmani* and *Eptatretus burgeri* at Teradomari Market, supplementary to local produce, *Paramyxine atami*, landed at Izumozaki

Year	Weight imported (kg)	Purchase locality
Kakujoe Fish Dealer		
1992	37 315	Hachino-he, Aomori Prefecture
1993	23 415	Hachino-he
	4 300	Yokosuka, Kanagawa Prefecture
1994	11 749	Yokosuka
1995	11 080	Yokosuka
	1 980	unknown (?), Shimane Prefecture
Hamayaki Center (Fish Dealer)		
1994	2 775	Onahama, Fukushima Prefecture

Table 4.4 *Paramyxine atami* landed at Izumozaki* 1988–95†

Year	Number of boats	Catch (kg)	Income (Yen)
1988	1	1225	765 500
1989	1	4619	3 887 358
1990	1	3927	4 758 600
1991	1	4320	4 266 755
1992	1	1885	2 623 822
1993	1	3304	4 563 930
1994	1	2566	3 962 925
1995	1	2777	4 270 895

* Izumozaki Cooperative Fisheries Association, Niigata Prefecture.
† Landings in earlier years given by Gorbman *et al.* (1990) (Table 1).
‡ *Zenpoh-maru*, 4.9 tons (Figure 4.6).

ceasing towards the end of the Second World War due to overfishing (Figure 4.3).

In contrast to the declining hagfishery in the Sea of Japan, Onahama (Pacific coast of Fukushima Prefecture), and Tsushima Islands (Nagasaki Prefecture) and Pusan (Korea), maintained active hagfisheries although the principal target species differed from those in the northeastern province of the Sea of Japan: off the Pacific coast *Myxine garmani*, and in and near Tsushima Strait, including off San'in district, *Eptatretus burgeri* were caught. Contrary to the Niigata practice of using hagfish chiefly as food, hagfish in other centres were used exclusively as a light-leather substitute.

Because of the paucity of the local source of roast hagfish, *Paramyxine atami*, sold in Teradomari near Niigata, a large amount of *Myxine garmani* were imported from Onahama to Teradomari until 1994 in order to supplement the *Paramyxine* catch. Very recently, however, the hagfish catch off Onahama has declined rapidly, perhaps due to overfishing both by Japanese and Korean boats off the coast of Fukushima Prefecture. As a result the inshore hagfish, *Eptatretus burgeri*, has been recently imported from Yokosuka (Kanagawa Prefecture) and *Myxine garmani* from Hachino-he (Aomori Prefecture) (Table 4.3). At Teradomari Market, hagfish broiled over charcoal was sold as a speciality of the locality under the name Anago-yaki ('ana' meaning 'hole' and 'ko or go' meaning 'child' due to the burrowing nature of the hagfish; 'yaki' indicated the method of cooking) (Figure 4.4).

Baited traps currently used in Japan and Korea were discussed by Gorbman *et al.* (1990) and Japanese hagfishing methods and woven bamboo trap (Figure 4.5) illustrated by Kaneda (1977). Several Korean books noted that plastic traps used for hagfishes were also used for catching conger eels, *Conger myriaster* (National Fisheries Research and Development Agency, 1989; Li, 1988, etc.).

As shown in Tables 4.1 and 4.2 (Ministry of

Agriculture, Forestry and Fisheries, Korea, 1994), the 1993 hagfish catch in Korea was 148 t, most fish being caught by traps operating off the coast. Landings were concentrated in Pusan (77 t) and Kyongnam (53 t), with the remainder in Cheju (18 t). Gorbman *et al.* (1990) reported the hagfish catch by one boat at Onahama, during 1979–82, and noted that four boats had been operating in that area in April 1988 at the time of his visit. Because of the many Korean boats (Figure 4.6) operating in waters close to the Japanese mainland, the official total catch in 1993 (148 t) probably did not include data (no catch records available) from Japanese waters.

Despite the local importance of the hagfish industry, life history aspects of hagfish, including growth rate, age determination, reproductive biology, such as incidence of mating and spawning, number of eggs rejected per spawning and larval size at hatching, and the lifespan of any hagfish species are unknown. It seems likely that the significant decreases in the hagfish landed at Niigata in the past and at Izumozaki at present are the result of overfishing (Table 4.4). Therefore, the risk exists that other hagfish resources, especially the underexploited populations along the Pacific coast of North America, may similarly decline under increased fishing pressure.

ACKNOWLEDGEMENTS

The author wishes to express his sincere thanks to Dr C.-H. Kuo, Tunghai University, Taiwan, and Mr A. Aoyagi, Teradomari Town Aquarium, Niigata Prefecture, Japan, for preparing the photographs and drawings. Thanks are also due to Mr Y.C. Sohn, Pusan Fisheries University, Korea, for information on Korean hagfisheries.

REFERENCES

Bigelow, H.B. and Schroeder, W.C. (1952) A new species of the cyclostome genus *Paramyxine* from the Gulf of Mexico. *Brev. Mus. Comp. Zool., Harvard*, (8), 1–10.

Chiba, A. and Honma, Y. (1986a) Fine structure of the granulocytes occurring in the hypothalamic-hypophyseal ventricle and neuro-hypophysis of the hagfish, *Paramyxine atami. Japan J. Ichthyol.*, **33**, 262–8.

Chiba, A. and Honma, Y. (1986b) Comparative anatomy of the brain-ventricular system of the lamprey and hagfish, in *Indo-Pacific Fish Biology (Proc. 2nd Intern. Conf. Indo-Pacific Fishes)*, pp. 76–85.

Chiba, A., Honma, Y. and Oka, S. (1993) Immunocytochemical localization of neuropeptide Y-like substance in the brain of the hagfish, *Paramyxine atami. Cell Tissue Res.*, **271**, 289–95.

Chu, Y.-T., Chan, C. and Chen, C. (eds) (1963) *Fish of the East China Sea*, Sci. Press, Peking, 642 pp.

Dean, B. (1904) Notes on Japanese myxinoids. A new genus *Paramyxine*, and a new species *Homea okinoseana*. Reference also to their eggs. J. Coll. Sci., Imp. Univ. Tokyo, **19**, 1–23, 2 pls.

Fernholm, B. (1974) Diurnal variation in the behaviour of the hagfish *Eptatretus burgeri*, Mar. Biol., **27**, 351–6.

Fernholm, B. and Hubbs, C.L. (1981) Western Atlantic hagfishes of the genus *Eptatretus* (Myxinidae) with description of two new species. *Fish. Bull*, **79**, 69–83.

Foss, G. (1968) Behaviour of *Myxine glutinosa* L. in natural habitat. Investigation of the mud biotope by a suction technique. *Sarsia*, **31**, 1–31.

Gorbman, A., Kobayashi, H., Honma, Y. and Matsuyama, M. (1990) The hagfishery of Japan. *Fisheries (Amer. Fish. Soc.)*, **15**, 12–18.

Gustafson, G. (1935) On the biology of *Myxine glutinosa* L. *Ark. Zool.*, **28**A, 1–8.

Honma, Y. (1960) A story of the fisheries of hagfish and sea lamprey in the waters of Sado and Awashima Islands, the Sea of Japan. *Collect. Breed.*, **22**, 34–6 (in Japanese).

Honma, Y. (1961) Notes on *Paramyxine atami* Dean, and a description of a specimen with defective caudal fin. *Collect. Breed.*, **23**(6), 182–3 (in Japanese).

Honma, Y. (1969) Some evolutionary aspects of the morphology and role of the adenohypophysis in fishes. *Gunma Symp. Endocrin.*, **6**, 19–37.

Honma, Y. (1983) A story of the fisheries of hagfish and sea lamprey in the waters of Sado and Awashima islands, the Sea of Japan. *Canad. Trans. Fish. Aquat. Sci.*, No. 5033, 1–6.

Huang, K.-F. (1989) Studies on mitochondrial DNA and systematics of hagfish from Taiwan waters. Master's thesis, National Sun Yat-sen University.

Huang, K.-F., Mok, H.-K. and Huang, P.-C. (1994) Hagfishes of Taiwan (II): taxonomy as inferred from mitochondrial DNA diversity. *Zool. Stud.*, **33**, 186–91.

Jensen, D. (1966) The hagfish. *Sci. Amer.*, **214**(2), 82–90.

Jordan, D.S. and Snyder, J.O. (1901) A review of the lancelets, hagfishes, and lampreys of Japan, with a description of two new species. *Proc. U.S. Natn. Mus.*, **23**, 725–34, 1 pl.

Jordan, D.S., Tanaka, S. and Snyder, J.O. (1913) A catalogue of the fishes of Japan. *J. Coll. Sci., Imp. Univ. Tokyo*, **33**, 1–497.

Kaneda, Y. (1977) *An Illustrated Dictionary of Japanese Fishing Gear and Methods*, Seizando, Tokyo, 674pp.

Kobayashi, H., Ichikawa, T., Suzuki, H. and Sekimoto, M. (1972) Seasonal migration of the hagfish, *Eptatretus burgeri*. Japan. J. Ichthyol., **19**, 191–4 (in Japanese with English summary).

Kuo, C.-H., Huang, K.-F. and Mok, H.-K. (1994) Hagfishes of Taiwan (I): a taxonomic revision with description of four new *Paramyxine* species. *Zool. Stud.*, **33**, 126–39.

Kuo, C.-H. and Mok, H.-K. (1994) *Eptatretus chinensis*: a new species of hagfish (Myxinidae; Myxiniformes) from the South China Sea. *Zool. Stud.*, **33**, 246–50.

Li, S. (1988) *Dictionary of Aquatic Animals and Plants Name*, Modern Oceanography, Seoul, 324pp.

Lindberg, G.U. and Legeza, M.J. (1959) *Fishes of the Sea of Japan and its Adjacent Waters of Okhotsk Sea and Yellow Sea*, Vol. 1, Acad. Nauk., Moscow 207pp. (in Russian).

Malm, A.W. (1877) *Goteborgs och Bohuslans Fauna*, Ryggradsdjuren.

Matsubara, K. (1937) Studies on the deep sea fishes of Japan. III. On some remarkable variation found in *Paramyxine atami* Dean with special reference to its taxonomy. *J. Imp. Fish. Inst.*, **32**, 13–15.

Matsubara, K. (1955) *Fish Morphology and Hierarchy*, 3 vols, Ishizaki-shoten, Tokyo, 1605pp, 135 pls.

McInerney, J.E. and Evans, D.O. (1970) Habitat characteristics of the Pacific hagfish, *Polistotrema stouti*. *J. Fish. Res. Bd. Canad.*, **27**, 966–8.

Ministry of Agriculture, Forestry and Fisheries (ed.) (1994) *Statistical Year Book of Agriculture, Forestry and Fisheries*, Republic of Korea, 498pp.

Nakabo, T. (ed.) (1993) *Fishes of Japan with Pictorial Keys to the Species*, Tokai Univ. Press, Tokyo, 1474pp.

National Fisheries Research and Development Agency (1989) *Modern Fishing Gear of Korea*, Kyong-Nam, Republic of Korea, 624pp.

Nelson, J.S. (1994) *Fishes of the World* (3rd edn), John Wiley & Sons, Inc., New York, 600pp.

Okada, Y., Kuronuma, K. and Tanaka, M. (1948) Studies on *Paramyxine atami* Dean found in the Japan Sea, near Niigata and Sado Island. I & II. *Mis. Rep. Res. Inst. Nat. Resour.*, (11), 7–10; (12), 17–20 (in Japanese with English summary).

Okamura, O. (ed.) (1986) *Fishes of the East China Sea and the Yellow Sea*, Seikai Reg. Fish. Res. Lab., Nagasaki (in Japanese), 501pp.

Okamura, O., Amaoka, K. and Mitani, F. (eds) (1982) *Fishes of the Kyushu-Palau Ridge and Tosa Bay*, Japan Fish. Resour. Conserv. Assoc., Tokyo (in Japanese), 435pp.

Okamura, O. and Kitajima, T. (1984) *Fishes of Okinawa Trough and the Adjacent Waters*, Japan Fish. Resour. Conserv. Assoc., Tokyo :in Japanese), 414pp.

Patzner, R.A. (1977) Befunde über Aktivitätsphasen beim Schleimaal *Eptatretus burgeri* (Cyclostomata). *Sitzungsber. Osterr. Akad. Wiss., Math.-Naturwiss. Kl. Abt. I.*, **186**, Bd. 6, 421–4.

Regan, C.T. (1913) A revision of the myxinoids of the genus Myxine *Ann. Mag. Nat. Hist., Ser. 8*, 11, 395–8.

Shelton, R.G.J. (1978) On the feeding of the hagfish *Myxine glutinosa* in the North Sea. *J. Mar. Biol. Assoc. UK*, **58**, 81–6.

Shen, S.-C., and Tao, H.-J. (1975) Systematic studies on the hagfish (Eptatretidae) in the adjacent waters around Taiwan with description of two new species. *Chinese Biosci.*, **2**, 65–79.

Strahan, R. (1963) The behaviour of myxinoids. *Acta Zool.*, **44**, 73–102.

Strahan, R. (1975) *Eptatretus longipinnis*, n. sp., a new hagfish (Family Eptatretidae) from South Australia, with a key to the 5–7 gilled Eptatretidae. *Aust. Zool.*, **18**, 137–48.

Strahan, R. and Honma, Y. (1960) Notes on *Paramyxine atami* Dean (Fam. Myxinidae) and its fishery in Sado Strait, Sea of Japan. *Hong Kong Univ. Fish. J.*, (3), 27–35.

Strahan, R. and Honma, Y. (1961) Variation in *Paramyxine*, with a redescription of *P. atami* Dean and *P. springeri* Bigelow and Schroeder. *Bull. Mus. Comp. Zool., Harvard Univ.*, **125**, 323–42.

Tchang, T.L. (1940) On a hagfish from Foochow. *Biol. Bull. Fukien Chr. Univ.*, **2**, 99–100.

Teng, H.T. (1958) A new species of cyclostomes, *Paramyxine yangi*, found in Taiwan. *China Fish.*, (66), 3–6 (in Chinese).

Teng, H.T. (1959) A new species of cyclostomes. *Rep. Taiwan Fish. Inst.*, (5), 83–7 (in Chinese).

Yang, H.C. (1972) The discovery of cyclostomes – Notes on collection of Taiwan fishes. *Fish Mag.*, **8**(8), 42–5 (in Japanese).

5
THE ECOLOGY OF HAGFISHES

Frederic H. Martini

SUMMARY

In areas where hagfish are found, they are ecologically important for the following reasons:

1. Hagfish may be one of the most abundant groups of demersal fishes in many areas, in terms of numbers and/or biomass.
2. Where present at high densities, hagfish burrowing and feeding activities have a significant impact on substrate turnover.
3. Hagfish are significant as predators on benthic invertebrates and, in some cases, mesopelagic invertebrates and vertebrates.
4. Hagfish represent one of the most important mechanisms for the rapid cleanup and processing of carrion-falls. In areas subject to intensive commercial fisheries, hagfish probably play a key role in the removal and recycling of discarded by-catch.
5. Hagfish adults, juveniles, and eggs can represent a significant prey item for marine mammals and large predatory invertebrates.

5.1 INTRODUCTION

The lifestyles of hagfishes and the types of habitat they select pose serious logistical challenges for investigators. However, over the last three decades the widespread availability of technologies such as SCUBA diving, remote-operated vehicles (ROVs), camera sleds, and submersibles, has created new opportunities for hagfish research in the field. Although we still lack a complete picture of the biology and ecological physiology of any hagfish species, significant details are now available concerning the life history and ecology of *Eptatretus stoutii*, *E. cirrhatus*, *E.* (= *Paramyxine*) *atami*, *E. hexatrema* and *Myxine glutinosa*. The data indicate that these hagfish species have important roles in their benthic ecosystems. The fact remains, however, that the majority of hagfish species are known only from a few specimens captured in trawls or traps at great depth. Even the most basic aspects of their life histories, such as growth rate, age at sexual maturity and longevity, remain a mystery.

This chapter summarizes available information concerning the ecological preferences, impact and importance of hagfishes. The focus will be on ecological aspects of broad significance within the group. (*For information on development and reproduction of hagfishes, see Wicht and Tusch, this volume, and Patzner, this volume.*)

5.2 ENVIRONMENTAL REQUIREMENTS OF HAGFISHES

Hagfishes have been reported from the Atlantic, Pacific, Indian, Arctic, and Antarctic Oceans, and from the Bering, Mediterranean and Caribbean Seas. The latitudinal range of hagfish distribution extends from inside the Arctic Circle (Wisner and McMillan, 1995) to the South Shetland Islands off the Scotia Sea near Antarctica (Andriyashev, 1965; Fernholm, 1990). (*For a review of hagfish*

The Biology of Hagfishes. Edited by Jørgen Mørup Jørgensen, Jens Peter Lomholt, Roy E. Weber and Hans Malte. Published in 1998 by Chapman & Hall, London. ISBN 0 412 78530 7.

systematics and distribution, see Fernholm, this volume.)

Limiting factors that determine the distribution of hagfish species appear to be (1) salinity, (2) temperature, (3) depth and (4) substrate preference. All known hagfish species require a high salinity, and most thrive at low temperatures. Although individual species may have characteristic depth distributions, the ranges are so broad that depth per se is probably not the primary limiting factor. Substrate preference varies from species to species, and even among individuals within a species.

5.2.1 Salinity

Salinity is a limiting factor for all known species of hagfish. Palmgren (1927) was able to maintain *M. glutinosa* in samples of bottom water, but not in lower-salinity surface waters at the same temperature. Salinities of 20–25 ppt are rapidly lethal to *M. glutinosa* (Gustafson, 1935) and at salinities of 29–31 ppt they may survive for weeks but do not feed (Bigelow and Schroeder, 1953). When suddenly exposed to salinities below 31 ppt, individuals will struggle violently, slime copiously, and then become moribund (Adam and Strahan, 1963; personal observations). M. glutinosa have been held in aquaria for months at salinities of 32–34 ppt (Gustafson, 1935), with periodic episodes of feeding.

Strahan (1962) investigated the salinity tolerance of *E. atami* and reported that after 30 hours in 27 ppt salinity seawater all of the experimental animals were dead or moribund. Although incursions into an estuarine environment were reported for *Nemamyxine elongata* (Richardson, 1958), the hagfish may have been moving below a permanent pycnocline, and hagfish were not previously or subsequently collected at that site. The upper limit of salinity tolerance in hagfish is not known, but *Eptatretus* sp. have been reported at deep brine seeps in the Gulf of Mexico at depths of 650 m (MacDonald *et al.*, 1990).

5.2.2 Temperature

Sudden changes in temperature can have debilitating or lethal effects, but the range of temperature tolerance varies within and among species. *M. glutinosa* in the eastern Atlantic have been maintained at temperatures of 10°C by Gustafson (1935), and Palmgren (1927) reported that his animals could tolerate temperatures of 15°C. There is general agreement that *M. glutinosa* can be held in captivity for extended periods at lower temperatures (0–4°C).

E. hexatrema have been held for over two years at temperatures of 14–17°C (Kench, 1989). Worthington (1905) held *Eptatretus stoutii* for months at 22°C, and reported that they tolerated brief exposure to temperatures as high as 30°C. From November to May, *E. burgeri* feed actively within Koajiro Bay, Japan. From June to October, when temperatures in the bay rise above 20°C, the hagfish move out of the bay and into deeper, cooler waters. The summer temperatures within bay waters reach 24–28°C, probably at or near the lethal limits for this species, and hagfish do not return until temperatures fall to 21°C or below (Fernholm, 1974).

5.2.3 Depth

Depth *per se* does not appear to be a limiting factor for hagfishes as a group, although individual species have characteristic depth ranges. These ranges can be quite broad. Given acceptable temperatures and salinities, hagfish enter shallow coastal waters and may even feed at the surface or at the shoreline. For example, *Eptatretus cirrhatus* was reported feeding on immersed portions of whale carcasses at shore-based whaling stations off Cook Straits (Strahan, 1962), and both *E. bischoffii* and *E. hexatrema* may swim in surface waters, and have been collected in tidepools (Kench, 1989; Wisner and McMillan, 1988; Gosliner, personal communication). *M. glutinosa*, which may be found at

30 m depth in the northern Gulf of Maine (Bigelow and Schroeder, 1948), has been collected at depths of 1100 m off the edges of the North American continental shelf (Bigelow and Schroeder, 1953). In tropical oceans, hagfishes are often collected at depths of 400 m or more (Bigelow and Schroeder, 1952; Fernholm, 1982; Fernholm and Hubbs, 1981; Wisner and McMillan, 1995), and hagfishes (*Eptatretus* sp.) have been photographed at depths of over 5000 m in the tropical Pacific (Sumich, 1992).

As noted above, the depth distribution of *E. burgeri* changes in the course of the year, ranging from around 10 m (October–May) to 50–100 m (June–September) (Fernholm, 1974). Although this is the only known seasonal migration among hagfishes, there are indications that several populations show variation in depth distribution as a function of age, size or sex. For example, large male *E. stoutii* are most common at shallow depths (100 m), whereas large females predominate at the lower limit of their depth range (500 m). In shallow waters the sex ratio between males and females approaches 1:1, but in deep water that ratio is 1:2. Small hagfishes are most common at intermediate depths (250 m). The distribution data, reported by Johnson (1994) suggested to him that the habitat was being partitioned by age and sex, perhaps to reduce intraspecific competition. He also suggested that a depth migration of adult males and females would be necessary prior to reproduction. This study also reported that *E. deani* were unevenly distributed along the shelf slope, with smaller individuals common at 750 m, and large individuals predominant at 1000 m.

5.2.4 Substrate preference

In general, it appears that *Eptatretus* species occupy a broader range of substrate types than *Myxine* species. Cailliet *et al.* (1991) have done extensive work on habitat utilization by

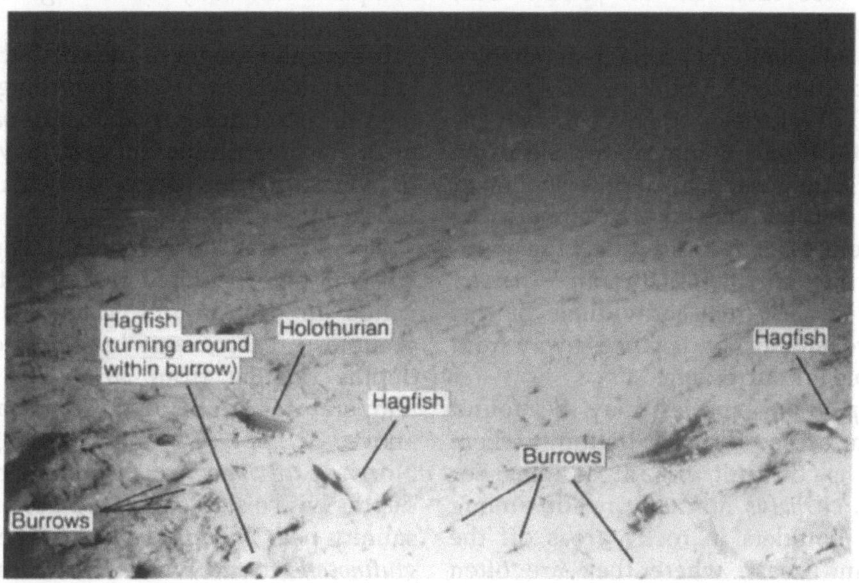

Figure 5.1 Representative habitat of *Eptatretus deani*. Several hagfish are visible in burrows; there are many other apparently empty burrows in the field of view. (600 m depth, off Point Sur, California; photograph courtesy of Dr Waldo W. Wakefield.)

E. stoutii off the California coast. All *E. stoutii* burrows and the majority of sightings were from areas of sand or mud bottom. The area of mud or sand may not be extensive, however, as burrowing may occur in isolated patches of mud. The nature of the substrate preferred for burrowing in this species has been described by McInerney and Evans (1970) off Vancouver Island, British Columbia. The bottom sediments consisted of 74.5% silt (diameter <0.625 mm), 24% clay (diameter <0.0039 mm), and 1.5% fine-to-medium coarse sand (>0.125 mm).

Despite their frequent association with soft-bottom substrates, significant numbers of *E. stoutii* can be encountered among other substrate types. Cailliet *et al.* (1991) found 30% of observed *E. stoutii* in areas of mixed substrates, such as sand/gravel or boulder/sand, and almost 15% of sightings were where the substrate consisted of massive granite formations.

As part of a larger study on demersal fishes, Wakefield (1990) investigated the habitat preference and distribution of *E. deani* at depths of 400–1200 m off Point Sur, California. Although *E. deani* were common throughout this depth range, hagfish burrows were found only at depths of 600–800 m (Figure 5.1). Wakefield suggested that the upper limit of burrowing may reflect the presence of coarse sand at depths shallower than 600 m, and the sharp decrease in burrow incidence below 800 m could reflect the abundance of the burrowing urchin *Brisaster latifrons*. It is clear that individuals of this species travel significant distances away from areas suitable for burrowing.

Other *Eptatretus* species may be found *primarily* in areas of hard bottom, where burrowing is difficult or impossible. For example, *E. cirrhatus* normally reside among ledges and boulders in rocky areas off the New Zealand coast, where they are often caught in commercial crayfish traps. *E. bischoffii* and *E. hexatrema* are often found in shallow rocky areas, and even in intertidal pools and crevices (T. Gosliner, personal communication; Kench, 1989; Wisner and McMillan, 1988).

Given suitable salinity and temperature, the common factor in the distribution among or within *Eptatretus* species may be the need to avoid exposure and detection by potential predators. Burrowing presumably offers the most protection, followed by curling among rocks or coiling. Although Worthington (1905) noted that *E. stoutii* in aquaria appeared to prefer nestling among rocks to lying or burrowing in soft substrates, the sandy substrate provided may have been too shallow for effective burrowing, as suggested by Fernholm (1974). If other forms of cover are not available, algae fronds may suffice; during daylight surveys, Fernholm (1974) encountered 37 *E. burgeri*, 31 of them (84%) within burrows. Of the six animals resting on the substrate, three were hidden by drifting algae, and two were dead or dying. Thus only one animal of 37 was both healthy and fully exposed. That individual was coiled, a position commonly seen among *Eptatretus* species in aquaria: the body forms a tight spiral, with the ventral surface in contact with the substrate and the tip of the tail at the inside of the coil (Strahan, 1963; Worthington, 1905) (Figure 5.2). Such a posture probably maximizes tactile stimulation and may make the individual harder to see, particularly against a variegated substrate.

The substrate preferences of *Myxine glutinosa* can be considered representative for the genus. In the Gulf of Maine, *M. glutinosa*, occupies areas of soft bottom sediments at depths ranging from 30 to 200 m, where temperatures are below 4°C and salinities are above 32 ppt. In the Hardanger Fjord, Norway, this species is found at comparable depths with a stable temperature of 7°C and a salinity near 35 ppt (Tambs-Lyche, 1969). *M. glutinosa* is found where the bottom is classified on sounding charts as clay, silt, sand or gravel; however, there is usually a layer of soft, flocculent, muddy sediment of variable

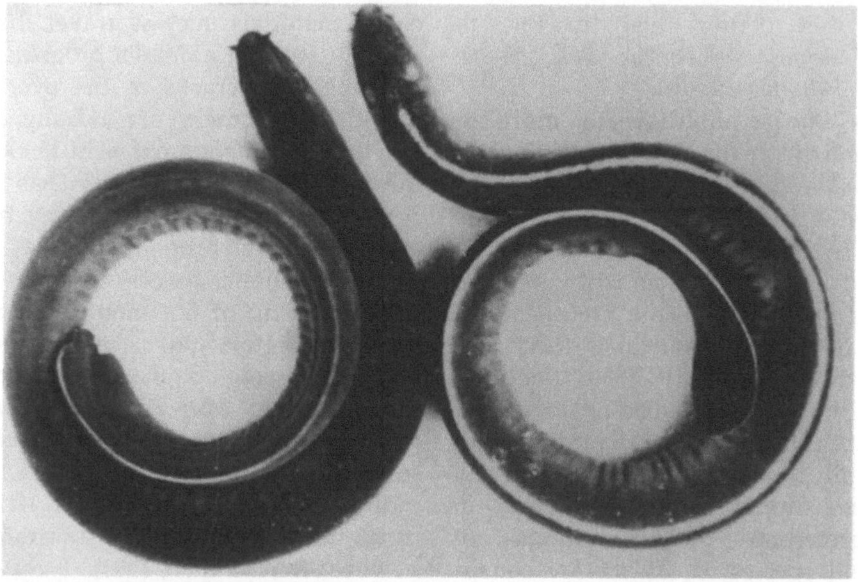

Figure 5.2 Coiling in *Eptatretus*. Individual *Eptatretus* usually rest in the coiled position when on the surface of the substrate. The tightness of the coil can vary. (*E. burgeri* and *Eptatretus* sp.; photograph courtesy of Dr Bo Fernholm.)

thickness covering these substrates. It is this superficial layer that *M. glutinosa* inhabits, occupying shallow, transient burrows that parallel the surface. The area of muddy substrate need not be large, as *M. glutinosa* has been reported in small patches of mud between rocks (Cole, 1913). In the Gulf of Maine, *M. glutinosa* is found in benthic communities characterized by Cerianthid anemones, tube worms, colonial tunicates, an abundance of errant polychaetes and nemerteans, and the edible shrimp, *Pandalus borealis* (Martini and Heiser, 1989). The investigations by Foss (1962, 1968), Palmgren (1927), Gustafson (1935), Howard (1982) and Shelton (1978a, b) indicate that *M. glutinosa* in the eastern Atlantic occupies a comparable substrate type. Many other myxinids, including *M. capensis*, *M. australis*, and *M. eos*, have been collected in similar habitats. Unlike the situation among the *Eptatretus* species discussed above, *Myxine* have not been reported in areas of hard substrate, nor seen over soft bottom substrates except where burrowing occurs.

5.2.5 Burrowing

Hagfishes have many morphological characters that are consistent with adaption to life within burrows. A partial list includes (1) the smooth, elongate body form, (2) the smooth skin with a coating of mucus, (3) the degenerate eyes, (4) an extensive reduction (*Eptatretus*) or complete elimination (*Myxine*) of the lateral line system, (5) a velar pumping mechanism for gill ventilation, (6) a tendency towards posterior migration of the efferent branchial ducts, (7) the presence of a pharyngocutaneous duct, a direct connection from the pharynx to the exterior through which silts or debris can be ejected, (8) a very low basal metabolic rate, which may be important for an opportunistic predator/scavenger in environments where nutrient sources are limited and feeding is episodic, and (9) a variety of physiological adaptations that permit them to tolerate hypoxia (*for details, see Forster, this volume; Malte and Lomholt, this volume*).

Species of *Myxine* show greater morphological specialization for burrowing than do those of *Eptatretus*. Examples of further specialization include: (1) the degenerate eyes are covered by muscle tissue, rather than by unpigmented skin characteristic of the 'eyespots' of *Eptatretus*; (2) the presence of a single pair of efferent branchial ducts that open to the exterior roughly 25% of the body length from the snout, whereas *Eptatretus* sp. have separate efferent ducts associated with individual gill pouches; and (3) a relatively thin, delicate, slippery skin. Differences in substrate preference between species of *Eptatretus* and species of *Myxine* are consistent with the degree of morphological specialization. Burrowing of *M. glutinosa* in the eastern Atlantic has been observed in aquaria (Gustafson, 1935; Strahan, 1963) and in the field using underwater television (Strahan, 1963). Observations in the western Atlantic (Gulf of Maine) have been made through ROV and manned submersible dives at depths of 120–150 m (Martini and Heiser, 1989; Martini *et al*., 1997). The descriptions are all consistent as to the stages and process of burrowing in this species (Figure 5.3a–d).

When preparing to burrow, the animal assumes an angle of 45–90° to the bottom and swims vigorously, driving the head into the substrate (Figure 5.3a). The swimming movements continue as the head moves into the substrate following a sinusoidal path that roughly parallels the surface. When one-third to one-half of the animal has entered the substrate, swimming movements cease (Figure 5.3b). Sinusoidal progression of the head and anterior trunk continue, punctuated by longitudinal contractions that pull the immobile posterior portion of the animal into the burrow in a series of pulses (Figure 5.3c). Complete disappearance of the animal into the substrate may take 5 or more minutes. When burrowing into sediments containing coarse materials, such as gravel, the burrow is shallow and the animal's progress is marked by superficial cracks in the overlying sediment. These cracks are usually not visible when the animals burrow in thick, soft sediments. Once the animal has burrowed completely, there is seldom any trace of the entry site. Over time (as much as an hour) a burrowed animal usually becomes positioned so that the tip of the snout projects from the substrate (Figure 5.3d). This may make it easier to detect passing odors and to maintain a respiratory current over the gills. (*For additional information on respiratory function and burrowing, see Malte and Lomholt, this volume.*) Burrowed individuals may disappear for 10 minutes or more, travelling within the substrate to reach a new location (Strahan, 1963).

While in the substrate the hagfish fills the width of the burrow, and observed burrows never contained more than one individual. There is no evidence that mucus produced by the slime glands plays a role in the maintenance of the burrow. After the emergence of its resident, the burrow collapses, often leaving a sinusoidal depression in the sea floor (Figure 5.3e). If stimulated to seek cover, the animal creates a new burrow rather than seeking out the remnants of a former one. However, in coarse substrates any discontinuity in the surface, whether a depression downstream of a small rock or the depression created by a recently vacated burrow, may make it easier to drive the head into the bottom and initiate the burrowing process. Foss (1962) reported observing *M. glutinosa* entering volcano-shaped mounds on the floor of Ulvikpollen, Hardanger Fjord, Norway, and suggested that these were the entrances to hagfish burrows. Despite later expeditions (Foss, 1962, 1968) she was ultimately unable to determine whether these mounds were created by hagfishes or by invertebrates that may be hagfish prey. However, in the process she did observe hagfishes burrowing as detailed above. Although comparable mounds were observed during ROV and submersible work in the Gulf of Maine, none was associated with hagfishes or their burrows.

Burrowed animals may emerge on detection of food or following mechanical stimulation (Gustafson, 1935), or in response to a reduction in the oxygen tension of the water (Strahan, 1963). In the Gulf of Maine, on those rare instances where *M. glutinosa* was encountered

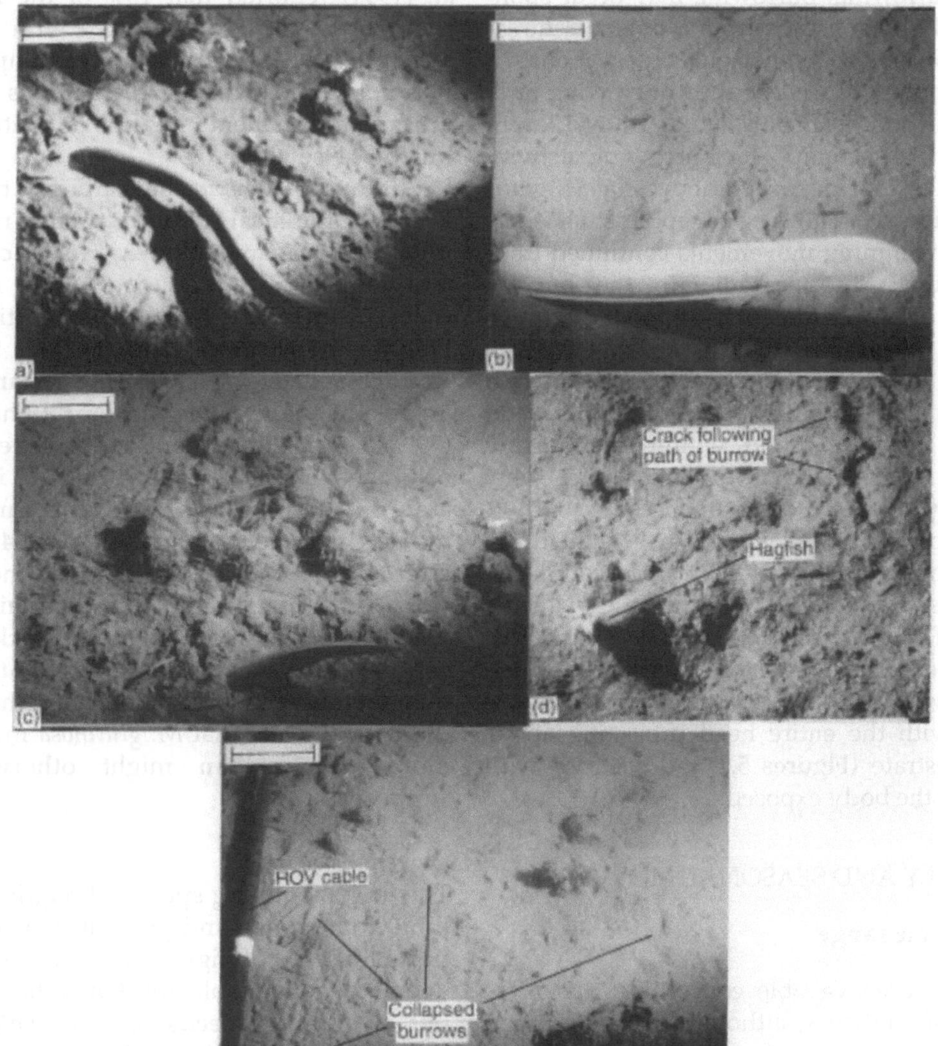

Figure 5.3 Burrowing in *Myxine glutinosa*. (a) Burrowing begins as the animal swims vigorously, forcing the head against the substrate. (b) When one-third to one-half of the animal has moved into the substrate, the exposed portion of the body relaxes. It then moves into the substrate in a series of pulses. (c) Burrowing usually takes 5 minutes or more to complete. Note the crack in the surface of the bottom that follows the path of the burrow. These cracks are easily seen when the superficial layer contains coarse materials and the burrows are shallow. (d) The tip of the snout or the entire head emerges some time after burrowing has been completed. (e) The burrows collapse when the hagfish emerge completely. This leaves distinctive sinusoidal traces in the bottom sediments.

on or above the surface of the substrate, the animals were either (1) resting on the substrate, with the ventral surface in contact with the substrate and the body in one or more S-curves, (2) swimming within 3 m of the bottom, or (3) burrowing. Burrowing may occur spontaneously or to escape from a potential threat (Foss, 1962, 1968, and personal observation).

Fernholm (1974) described burrowing in *E. burgeri* in soft sediments at a depth of 10 m in Koajiro Bay, Japan. The burrowing activity resembled that described above for *M. glutinosa*, with the following exceptions: (1) sinusoidal swimming movements continued until the animal disappeared into the burrow (Figure 5.4a); (2) the entry and exit openings remain patent, and the passageway is large enough to let the animal reverse its position without emerging (Figure 5.4b); (3) several hagfish may be found within one burrow (Figure 5.4c); (4) burrows persist over time, and may be used more than once.

Burrowing in *E. burgeri* as detailed by Fernholm is consistent with observations of burrows made by *E. stoutii* off the Pacific coast of North America (Cailliet *et al.*, 1991). Burrowed animals may be invisible, or positioned with the head near an opening (Figure 5.4b), with the entire head projecting above the substrate (Figures 5.1 and 5.4d) or with most of the body exposed.

5.3 DAILY AND SEASONAL MOVEMENTS

5.3.1 Home range

No data are available concerning the home range of hagfishes, although the assumption has been that they form relatively isolated populations with small home ranges. Tagging studies are needed to verify this assumption, especially in light of recent data concerning the distribution of *E. deani* and *E. stouti* off the California coast. Wakefield (1990) found *E. deani* to be common at 400–1200 m, but burrows were present only at depths of 600–800 m. Many of the burrows were empty, and it remains to be seen whether the hagfishes at other depths were associated with the empty burrows, or whether they were permanent residents at depths where burrowing did not occur. Similarly, Cailliett *et al.* (1991) reported that 15% of the *E. stouti* they observed were associated with massive substrates where burrowing was impossible. It is not known whether the animals in these areas are permanent residents or visitors with burrows elsewhere.

Although *Myxine* occupies transient burrows, the local community may retain its identity over time. Walvig (1967) captured and tagged 53 *M. glutinosa* and released them 2 km away from the collection site. Despite the distance, and the fact that the capture and release sites were separated by an area of deeper water (200 m), 18 hagfishes (33.9%) were subsequently recaptured at the original site. The recapture was not immediate – the first animals were recaptured nine months after release, and the last 4.5 years after release – but fishing effort in the region was low. Thus it remains uncertain as to whether the animals migrated back to the capture site, or whether over that period they merely wandered into it. In either event the data suggest that *M. glutinosa* may roam more widely than might otherwise be supposed.

Current velocity

The low swimming speeds of hagfishes may limit home range and prevent their penetration into areas of high current velocity. Field data are unavailable for *Eptatretus* species, but swimming speeds of *E. cirrhatus* have been reported in laboratory studies of cardiovascular function. Wells *et al.* (1986) reported that *E. cirrhatus* tolerated sustained swimming at 20 cm s^{-1}, but exhaustion in this species soon develops at swimming speeds of 30 cm s^{-1} (Davison *et al.*, 1990). Burst-speed limits have not been reported for *Eptatretus*. Adam (1960) reported a swimming speed in *M. glutinosa*

Figure 5.4 Burrowing in *Eptatretus burgeri*. (a) Burrowing in progress. Although much of the animal is already within the substrate, vigorous swimming movements continue. (b) A hagfish near the opening of a burrow, whose diameter is considerably larger than that of the hagfish. (c) Two hagfishes sharing a single burrow. (d) A hagfish with its head protruding from its burrow; this posture is very similar to that of *E. deani* (Figure 5.1). (Photographs courtesy of Dr Bo Fernholm.)

of 25 cm s^{-1}. However, the burst-speed of a free-swimming *M. glutinosa* (chased by divers) is approximately 1 m s^{-1} (Foss, 1968). *M. glutinosa* in the Gulf of Maine is found in areas where currents are normally below 15 cm s^{-1} (personal observation).

5.3.2 Diurnal rhythm

Hagfish in aquaria and in shallow waters have been reported to show nocturnal patterns of activity. Nocturnal cycles of activity in aquaria have been reported for *E. stoutii* by Worthington (1905) and for *E. burgeri* and *E. atami* by Ooka-Souda *et al.* (1991). Similar patterns were seen in the field for *E. burgeri* at depths of 10 m (Fernholm, 1974). During night dives he saw 79 hagfishes, and 55 (70%) of them were free-swimming or exposed. In contrast, during daylight dives he observed 37 hagfishes, and only 6 (16%) were partially or completely exposed. He suggested that the nocturnal cycle of activity might be the result of changes in illumination, since comparable activity cycles were not reported for *E. burgeri* in deeper waters (>200 m). This observation has been supported by the work of others (Ooka-Souda *et al.*, 1991; Kabasawa and Ooka-Souda, 1991), who concluded that in *Eptatretus* the eyes, although degenerate, are still responsible for the entrainment of circadian activity patterns.

Gustafson (1935), Palmgren (1927) and Strahan (1963) reported that aquarium-held *M. glutinosa* were most active at night. Unlike the situation with *Eptatretus*, cutaneous photoreceptors, rather than the eyes, are responsible for the reactions of *Myxine* to illumination (Newth and Ross, 1955). However, the depths normally inhabited by *M. glutinosa* may not permit sufficient illumination to trigger daily activity cycles. No day/night activity differences were seen in the field by Foss (1962, 1968) during SCUBA surveys in the eastern Atlantic or during trap surveys, ROV work, or manned submersible dives within the Gulf of Maine (pers. obs.).

5.3.3 Seasonal movements

Migrations in response to habitat shifts or reproductive function have been reported only for *E. burgeri* (Fernholm, 1974). However, there have been reports of seasonal changes in population distribution for *E. deani* and *E. stoutii*. Wakefield (1990) reported changes in the distribution of *E. deani* during the year. Although these hagfish were found year-round at depths of 400–1200 m, hagfish densities at 600 m were lowest in October, February and May, and highest in July, November and March. These cycles closely paralleled density changes in thornyheads (*Sebastolobus* sp.), the most common demersal fish in this area. The causes of these distribution shifts and the linkage between the two species are unknown.

5.4 POPULATION DENSITY

Hagfish are easily overlooked in the environment because they are usually burrowed or otherwise covered, and generally immobile. For this reason standard visual surveys or trawl surveys may drastically underestimate their abundance. The release of bait in an area showing few if any hagfish on superficial survey can result in the emergence of hundreds of animals in a matter of minutes. Concerning *M. glutinosa*, Cole (1913) stated that 'On their own grounds, they may be as plentiful as earthworms', and Nansen (1887) stated that in the neighbourhood of Bergen 'I think they are more common than any other fish'. These assessments are probably correct.

Wakefield (1990) found that, over the depth range of 400–1200 m, *E. deani* were certainly the second most abundant species in terms of individuals and biomass. Because hagfishes were counted only when visible, and 'empty' burrows were ignored, the actual abundance may be even greater. Hagfish distribution was patchy, with densities ranging from 0 to 592 000 km^{-2} (1:1.7 m^2). The average density year-round at a depth of 600–800 m was estimated to be 325 000 km^{-2} (1:3.1 m^2).

The density at 600 m was 176 300 km^{-2} (1:5.7 m^2) with a biomass of 11 800 kg km^{-2}. By comparison, trawling in the same area yielded a hagfish biomass estimate of 700 kg km^{-2}. The underestimate probably reflects the fact that the majority of animals (over 80%) were in burrows, and inaccessible by trawling. Over the depth range of 600–800 m, *E. deani* was the most abundant species, comprising 61% of the fishes in this region.

The peak densities reported in this study are in close agreement with those of Martini *et al.* (1996), who estimated densities of *M. glutinosa* based on arrival times at bait stations in known current velocities. The calculated peak density was 500 000 km^{-2} (1:2 m^2). The density average for several sites was 59 700 km^{-2} (1:16.75 m^2), and the estimated biomass was 8119 kg km^{-2}. The energy demands of hagfish populations at these densities are considerable, and suggests that, at least where population densities are high, hagfishes are predators first and opportunistic scavengers second.

5.4.1 Changes in distribution and abundance

The distribution and abundance of hagfishes can change over time in response to:

1. *Changes in the local ecosystem.* Bigelow and Schroeder (1953) noted that *M. glutinosa* was extremely common in the Gulf of Maine between Boon Island and the Isles of Shoals; however, no hagfishes have been reported from hagfish surveys or commercial fisherman close around the Isles of Shoals in at least the last 25 years. It is not known whether this shift reflects alterations in benthic community structure related to the precipitous decline in regional groundfish stocks, or occurred for other factors as yet unknown.
2. *Substrate alteration.* Cole (1913) found *M. glutinosa* in soft dredge spoils dumped 10 km inshore from their normal range.
3. *Fishing pressure.* The hagfish fishery has gradually expanded, from the area of Korea and Japan to the Pacific coast of North America and, as of 1990, the New England coast (Gulf of Maine). The expansion is at least in part driven by the susceptibility of hagfish populations to intensive fishing pressure (Gorbman *et al.*, 1990; Honma, this volume). A decline in abundance and a decrease in catch per unit effort with intense fishing activity has been reported for eptatretids off the California coast (Johnson 1994; Lindley, 1988), and for *M. glutinosa* within the Gulf of Maine (Hall-Arber, 1996). This is not surprising given the limited reproductive potential of hagfishes (*see Patzner, this volume*).
4. *Appearance of supplemental food resources.* In some areas, the number of hagfishes may have increased as a result of human activities, such as gill net and longline operations and the release of by-catch in trawling operations. Shelton (1978a) suggested that trawling by-catch could be an important supplement to the diet of hagfishes in the eastern Atlantic, and this is equally likely in the western Atlantic, especially within the Gulf of Maine. Bigelow and Schroeder (1948) considered a catch of 11 hagfishes in an hour or less to be evidence that they were 'plentiful' in the area sampled. They qualified that statement with 'But we question whether they ever occur in American waters in such numbers as seen in the fjords of western Sweden and southern Norway, where catches of 100 are usual in eel pots set overnight on suitable bottom ...' Until the onset of commercial hagfishing this did not seem to be the case, given the density calculations noted above. In many areas of the Gulf of Maine, catches of 100 or more animals were common in traps set for 30 minutes to 1 hour.

5.5 FEEDING

Although often considered as scavengers, it is extremely unlikely that the observed densities could be supported by scavenging alone. It is much more likely that hagfishes are predators who prey on a variety of invertebrates, but who are opportunistic scavengers on invertebrate and vertebrate remains. Table 5.1 provides an overview of the literature concerning the diets of various hagfish species.

Johnson (1994) found that the diet of *E. stoutii* and *E. deani* was dominated by the remains of mesopelagic inhabitants, including sergestids, cephalopods, euphausiids and small fishes. Fragments of benthic polychaetes

Table 5.1 Diet of hagfishes

Species	Food item	Reference
M. glutinosa	Polychaetes (*Eumenia, Lumbrinereis, Nereis*)	Gustafson (1935)
	Bird, mammal remains	Shelton (1978a)
	Shrimp (*Pandalus*), fish, epibenthic and burrowing invertebrates (polychaetes, nemerteans)	Shelton (1978b)
	Limpet, blackjack (*Pollachius virens*)	Cole (1913)
	Hermit crabs (*Calocaris, Eupagurus*), shrimp (*Spirontocharis*), herring and sprat (*Clupea harengus* and *C. sprattus*), mackerel (*Scomber scombrus*), priapuloids (*Priapulus horridus*)	Strahan (1963)
	Cod (*Gadus callarias*), ling (*Molva molva*)	Goode and Bean (1895)
	Haddock (*Melanogrammus aeglefinus*)	Bigelow and Schroeder (1948)
	Mackerel sharks, sturgeon	Bigelow and Shroeder (1953)
	Mammalian viscera, bird remains	Putnam (1874)
	Hagfish eggs (male *M. glutinosa* only)	Holmgren (1946)
	Dogfish (*Squalus acanthias*), cod (*Gadus callarias*), herring (*Clupea harrengus*), pollack (*Pollachuas virens*), bird and mammal remains	Personal observation
E. stoutii	Mesopelagic invertebrates, including sergestids, cephalopods, euphausiids; small fishes; polychaetes were 14.8% of diet	Johnson (1994)
	Hagfish eggs (male *E. stoutii* only)	Worthington (1905)
E. deani	Mesopelagic sergestids, cephalopods, euphausiids, small fishes; polychaetes were 22.8% of diet	Johnson (1994)
	Fish, brittlestars, tube worms, crustaceans	Wakefield (1990)
E. burgeri	Fish in floating live-cars	Fernholm (1974)
	Electric ray (*Astrape dipterygia*)	Dean (1904)
E. longipinnis	Burrowed into abdomen of live lobster (*Jasus novaehollandiae*)	Strahan (1975)
E. cirrhatus	Whale flesh	Strahan (1962)
E. atami	Saury (*Scomberesox* sp.), sardine (*Sardina*)	Strahan and Honma (1960)

also comprised a significant proportion of the diet in *E. stoutii* (14.8%) and *E. deani* (22.8%). The occasional presence of hair and feathers within the gut contents support the idea that these hagfishes will scavenge on carrion-falls. In deepwater environments, carrion-falls may represent a more important dietary resource (Smith, 1985).

Gut content analyses of *M. glutinosa* have shown that their primary diet consists of invertebrate organisms. Gustafson (1935) reported setae of the polychaetes *Eumenia crassa, Lumbrinereis fragilis,* and *Nereis* sp. Strahan (1963) examined faecal wastes and identified polychaete setae mixed with the remains of herring (*Clupea harengus*), hermit crabs, shrimp, and priapulids. In the most detailed study of the dietary habits of *M. glutinosa,* Shelton (1978a) examined the gut contents of 129 trawled specimens in the North Sea. He reported that the gut contents were dominated by remains of the edible shrimp, *Pandalus borealis*. It is not known whether the shrimp are seized when alive, but *Myxine* in the wild certainly appear active enough to accomplish this. The orientation of the remains in the gut indicated that the abdominal segments had been consumed first, and the author suggested that these segments may have been easier to ingest. Alternatively, the shrimp may have been attacked from behind or below, by a burrowed hagfish. There is circumstantial evidence that feeding on polychaetes and other invertebrates may occur while the animals are within the substrate (Strahan, 1963).

Shrimp were the dominant gut content in *M. glutinosa* collected in April, May, July, October and November (Shelton, 1978a). Several different year classes of this protandric hermaphroditic shrimp may be preyed upon, since juveniles remain inshore year-round, whereas adult males remain offshore and reproductively mature females move inshore during the spawning season from autumn through the winter months (Shumway *et al.*, 1985).

In addition to *Pandalus* remains, Shelton (1978a) reported finding the remains of small fishes and a combination of epibenthic and burrowing invertebrates. He suggested that the discard of trawling by-catch could enhance the feeding opportunities for hagfishes in the area, and account for the presence of fish remains in hagfish digestive tracts.

Larger invertebrates may also fall victim to hagfish predation; the first known specimen of *Eptatretus longipinnis* was found burrowed into the abdominal segments of a live Pacific spiny lobster 'and was still embedded when the lobster was brought to the surface' (Strahan, 1975).

5.5.1 Feeding behaviour

When feeding on tough material, such as vertebrate tissues, a hagfish swims vigorously, forcing the head against the target while the toothplates are everted. When a superficial irregularity has been grasped by retraction and closure of the toothplates, knotting occurs, and as the knot reaches the head a portion of tissue is torn away. This process is awkward and slow, and when feeding on dead or injured fishes, hagfishes often seek access to the relatively tender viscera by entering the mouth or anus. When feeding on soft material, such as a soft-bodied worm, knotting does not occur. First, the tooth plates are everted. As they are retracted, they grasp the object and pull it into the mouth. The toothplates are then everted once again. As this movement occurs, the object is held in place by the median fang on the inferior surface of the palate. The everted toothplates grasp the object distally, and the cycle is repeated. In this way an elongate item, such as a polychaete worm, can be ingested in a series of rapid grasp-retract-release-extend cycles. This behaviour, which can easily be elicited in the laboratory, has also been observed in the field by Strahan (1963).

5.5.2 Feeding aggregations

The local hagfish density increases markedly when a food resource becomes available. In the open ocean, these conditions may develop following the settling of a large carcass, such as that of a whale. In coastal waters, human activities such as the trapping of groundfish in gillnets, the discard of trawling by-catch, or the disruption of benthic communities by mobile fishing gear provide sudden, large-scale feeding opportunities for hagfishes. When drawn to a food resource, hagfishes converge on their target en masse, with no indication of organized schooling (Figure 5.5a). The result is a writhing swarm of hagfishes whose movements disturb the surrounding sediments and whose slime soon covers the bait and any associated equipment (Figure 5.5b). Strahan (1963) described a

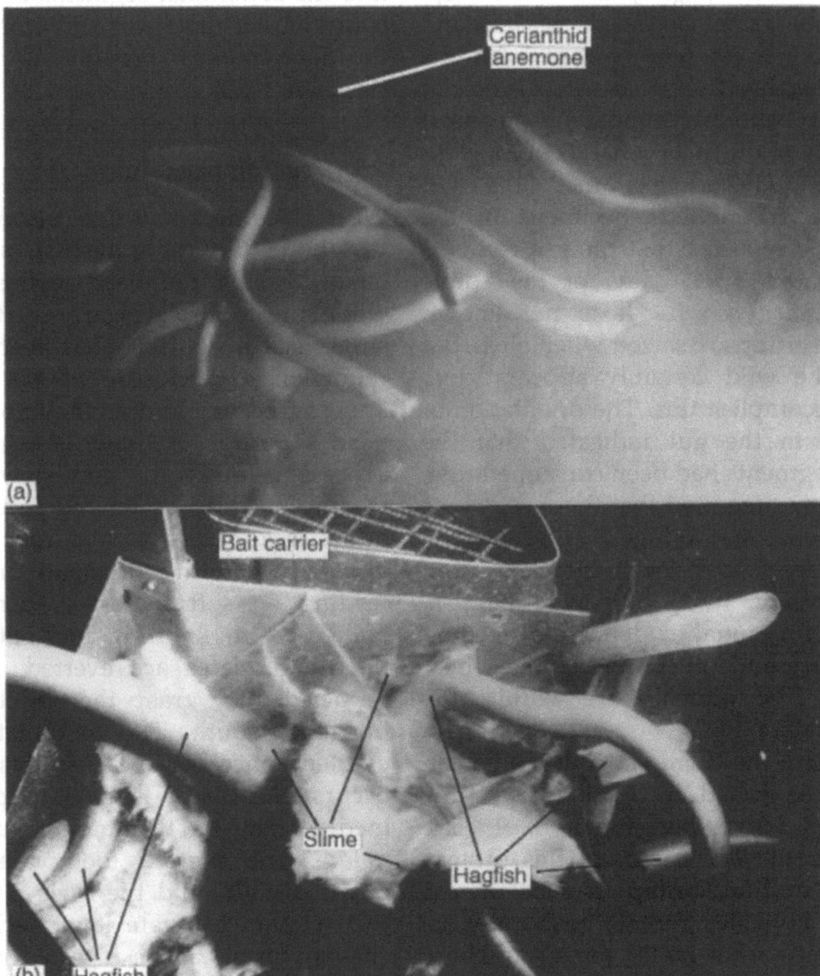

Figure 5.5 A feeding aggregation of *Myxine glutinosa*. (a) A swarm of hagfishes approaching a baited trap at a depth of 125 m in the Gulf of Maine. (b) Hagfishes feeding at a bait station. Streamers and layers of slime cover the bait. This may help deter other benthic scavengers.

rather lethargic locomotory style for aroused *M. glutinosa* in the eastern Atlantic. The animals swam against the current towards the odour source, and if passing it they would become limp, and drift until they were once again within the odour plume. Nothing of the sort has been observed in recent submersible or ROV work in the Gulf of Maine (personal observation). Instead, the aroused animals swam rapidly following a linear or zig-zag track, and they arrived from all points of the compass without regard to the prevailing current.

Smith (1985) placed bait parcels at a depth of 1310 m, below the depth where *E. deani* is abundant (Johnson, 1994; Wakefield, 1990). Nonetheless, *E. deani* at the bait reached peak densities of over 100 m^{-2} after 10–24 h, whereas the density in the absence of bait was estimated to be 1:37 000 m^2. It is worth noting that, despite their low density in the environment, they still managed to acquire an estimated 90% of the energy content of the bait parcel over a period of three weeks. Slime production appeared to discourage the activities of other scavengers over this period. The speed and efficiency of carrion-removal at depths of 600–800 m, where peak hagfish densities occur, must be far greater.

5.5.3 Feeding periodicity

In the study by Smith (1985), individual hagfishes fed to satiation and then burrowed within 30 m of the bait. Comparable feeding behaviour has been observed among *M. glutinosa* lured to a bait station (personal observation). Arriving individuals feed actively for up to an hour before moving away and burrowing, usually within a few metres of the bait. This behaviour is common among opportunistic scavengers in deepwater environments (Smith, 1985).

The normal interval between bouts of feeding is unknown. A colony of *E. hexatrema* was maintained for over five years with feeding every two weeks (Kench, 1989). Worthington (1905) felt that the natural cycle for aquarium held *E. stoutii* was 2–3 weeks, and that between feedings the animals showed little interest in food. Gustafson (1935) fed captive *M. glutinosa* on a weekly basis, but noted that sometimes animals would not feed for two weeks or more. In cases where feeding was required for experimental purposes, animals have been force-fed without difficulty (Adam and Strahan, 1963). It is not clear, however, what feeding regimens in aquaria tell us about requirements in the field.

In fact, there may not be a 'normal' periodicity for feeding. Cunningham (1886/7) was first to suggest that the lethargic habits and adipose tissue reserves of hagfishes could account for their ability to tolerate long periods of starvation. This may be important in colonizing deepwater areas where resource distribution is patchy or episodic. Hagfishes appear to spend most of their time immobile, either within burrows or hiding while emerged, arousing only to feed or to avoid unpleasant stimuli (Gustafson, 1935; Newth and Ross, 1955; Strahan, 1963; Worthington, 1905). This immobility, coupled with their unusually low basal metabolic rates (Forster, 1990; Lesser *et al.*, 1996; Munz and Morris, 1965), may allow them to survive for prolonged periods without feeding.

The deepwater hagfish *E. deani* has the lowest basal metabolic rate of any hagfish known. Based on energetic calculations, a single *E. deani* could obtain its entire annual basal metabolic demand in roughly 1.5 hours of feeding at a bait station (Smith, 1985). Natural carrion-falls of substantial size (>1 kg) may occur at this depth in sufficient quantity to support a sizeable dispersed population of *E. deani*; survey work in that study indicated an incidence of 1 carcass per 4000 m^2 at this depth.

The ability of hagfishes to tolerate prolonged fasting may also help them survive the rigours of captivity. Amid the destruction and confusion of the 1989 earthquake, aquarium-held *E. stoutii* in the rubble of the Moss

Landing Marine Lab survived for 14 weeks without feeding (Johnson, 1994).

5.5.4 Energetic requirements

The substantial numbers of hagfishes in many softbottom communities would suggest that these animals may be playing a crucial role in substrate turnover and nutrient recycling. The density estimates noted above permit a rough estimate of the feeding requirements for *E. deani* and *M. glutinosa* at a temperature of 4–6°C. Smith and Hessler (1974) measured the resting oxygen consumption for burrowed *E. deani*. Resting oxygen consumption was 0.10 μmol O_2 g^{-1} h^{-1}. This is an extremely low oxygen demand, even for a hagfish; this may be important for an opportunistic predator/scavenger in environments where nutrient sources are limited and feeding is episodic. At the average density and biomass reported for 600 m by Wakefield (1990), the resting energy requirement of the population would be approximately 13 915 kJ km^{-2} day^{-1}. To place this in perspective, one can calculate the resources that must be consumed to provide that energy, assuming that the hagfishes are supported by feeding on fish flesh (cod) or shrimp. Cod flesh provides 6.36 kJ g^{-1} and shrimp flesh provides 3.53 kJ g^{-1} (Calloway and Carpenter, 1981). Thus each day, 2.2 kg of fish flesh or 3.9 kg of shrimp would provide enough energy to support the *E. deani* within a 1 km^2 area. On an annual basis, that is equivalent to 800 kg yr^{-1} km^{-2} of fish flesh, or 1400 kg yr^{-1} km^{-2} of shrimp flesh. For the annual density estimate over the range of 600–800 m, those values rise to 1500 kg yr^{-1} km^{-2} of fish or 2600 kg yr^{-1} km^{-2} of shrimp. The actual requirements are certainly higher, as this assumes (1) 100% energy capture from the diet, and (2) resting oxygen consumption. The energy capture rate has been estimated at 0.8 for a variety of fishes (Alexander, 1967), which would increase the basal estimate by 25%. Swimming, burrowing and feeding impose additional energy demands; data from Forster (1990) for aquarium-held *E. cirrhatus* indicate that oxygen consumption during periods of activity increases to 2–3 times resting levels.

Resting oxygen consumption for an average (136 g) *M. glutinosa* from the Gulf of Maine is roughly 5 mmol O_2 h^{-1} (Lesser *et al.*, 1996), or 2.39 kJ d^{-1}. At a density of 59 700 km^{-2}, each square kilometre occupied by *M. glutinosa* would require the energetic equivalent of 24 kg d^{-1} of fish flesh (8700 kg yr^{-1}), or 43 kg d^{-1} of shrimp (15 600 kg yr^{-1}). Because suitable habitat covers an estimated 5.4×10^4 km^2 within the Gulf of Maine, the impact of hagfishes on the benthic ecosystem may be considerable.

5.6 PREDATION ON HAGFISHES

Despite their cryptic habits, hagfishes are subject to predation by a variety of vertebrate and invertebrate predators. Table 5.2 summarizes reports of predation on hagfish eggs, juveniles and adults. Hagfishes and/or hagfish eggs have been identified in the stomach contents of cod (*Gadus*), spiny dogfish (*Squalus acanthias*), harbour seal (*Phoca vitulina*), harbour porpoise (*Phocoena phocoena*), the southern sea lion (*Otaria flavescens*), elephant seal (*Mirounga*), Peale's dolphin (*Lagenorhynchus australis*), and octopus (*Octopus magnificus*).

Hagfishes probably gain some protection from predation by their burrowing habits and by the slime they secrete. Hagfishes produce copious amounts of slime when feeding, when injured, or when encountering a potential threat. In addition to discouraging competitors, the slime may also interfere with the gill respiration of potential predators. (A hagfish left entrapped in its own slime will soon expire; the removal of secreted slime appears to be one important function of knotting behaviour.)

Intraspecific predation may also occur, as unfertilized eggs were reported from the gut of male hagfishes (*M. glutinosa*) by Holmgren

Table 5.2 Predation on hagfishes

Hagfish species	Predator	Remarks	Reference
M. glutinosa	Harbour porpoise, *Phocoena phocoena*	Frequency of occurrence in adults is 14.7%*	Smith and Read (1992) Recchia and Read (1989)
	Codfish, *Gadus callarias*	Egg predation	Muller (1875)
	Codfish, *Gadus callarias*; White hake, *Urophycis tenuis*; Halibut, *Hippoglossus hippoglossus*	Prey on small hagfish; eggs found in halibut stomachs	Scott and Scott (1988)
M. capensis	Octopus, *Octopus magnificus*	Comprise *c.* 17% of diet for octopus over 1 kg in weight	Villanueva (1993)
Myxine sp.	Peale's dolphin, *Lagenorhynchus australis*	Frequency of occurrence is 50–60%; eggs sometimes ingested	Schiavini and Goodall (1996)
	Southern sea lion, *Otaria flavescens*	Frequency of occurence is *c.* 10%; eggs sometimes ingested	Alonso *et al.* (in preparation)
Notomyxine (?)	Blue-eyed cormorant, *Phalacrocorax albiventer*	Frequency of occurrence below 3%	Gosztonyi and Kuba (1995)
E. stoutii	Harbour seal, *Phoca vitulina*	As percentage of total diet (wt): Feb.–April 5.55% May–July 2.69% Aug.–Oct. 2.17%	Oxman (1995)
	Harbour seal, *Phoca vitulina*	Fourth most common prey type, found in 24–30% of scat year-round	Hanson (1993)
Eptatretus sp.	Elephant seal, *Mirounga angustirostris*		Hacker (1986)

* *Frequency of occurrence* is the incidence of hagfish in stomachs containing food, expressed as a percentage.

(1946), Worthington (1905) and Shelton (1978a). However, the greatest threat to hagfish populations at present is human activity, either intentional (directed fisheries) or unintentional (habitat disruption).

5.7 HAGFISH–HUMAN INTERACTIONS

Hagfishes have a number of direct and indirect interactions with humans. Best known are the negative interactions between hagfishes and commercial fisheries. It is very unlikely that groundfishes or other vertebrate remains would form a substantial part of the hagfish diet without human assistance (Strahan, 1963). The extent of their direct impacts on the world's fisheries is difficult to assess, but they are certainly a nuisance in many areas. For example, Bigelow and Schroeder (1948) reported that *M. glutinosa* in the Gulf of Maine were damaging 'a large proportion of the fish caught on long lines, unless the latter are tended frequently'. They also noted that in the spring of 1913 hagfishes gutted 3–5% of all the haddock caught in gill nets around Jeffrey's Ledge. *E. stoutii, E. polytrema, E. burgeri* and *M. garmani* have all caused problems for commercial fishermen in other parts of the world (Strahan, 1963). However, the positive impacts of hagfishes on the environment, through substrate turnover and nutrient recycling, probably far outweigh their negative impact on local fisheries.

Hagfishes in many areas benefit indirectly from ongoing fisheries activities, by consuming discarded by-catch. The amount of by-catch released into the environment each year can be considerable. For example, the North Sea trawling fleet discards approximately 789 000 tonnes (metric tons) of offal and by-catch (vertebrate and invertebrate) each year. This is roughly equivalent to 22% of the commercial landings for the region (Garthe and Camphuysen, 1996). Seabirds consume an estimated 39% of the discards, but the rest (approximately 480 000 tonnes) is unavailable to them, primarily because it sinks too quickly. Although pelagic predators may consume a portion of the balance, the North Sea is relatively shallow and a significant portion of that 480 000 tonnes will undoubtedly reach the bottom and become accessible to hagfishes. The effects of this 'windfall' on hagfish populations in this area or other areas of intensive fishing effort have not been determined.

In addition, many hagfish species represent a commercial resource (*see Honma, this volume*). Although it has been demonstrated repeatedly that hagfish populations cannot withstand heavy fishing pressure, there have as yet been no studies to determine the short-term or long-term effects of a drastic reduction in the numbers of a potentially key species from the ecosystem.

Finally, hagfishes have the potential to play a useful role in our attempts to monitor the health of the marine environment. Because they are intimately associated with bottom sediments, hagfishes may be used as indicator species. Falkmer *et al.* (1977) found that *M. glutinosa* within the relatively enclosed Gullmar Fjord were apparently extremely susceptible to carcinogens (especially DDT and PCBs) in the environment. In 1972, prior to the banning of these compounds, the incidence of hepatomas and other neoplasms in adult hagfishes from the Gullmar Fjord approached 10%. The livers of these animals contained high concentrations of the carcinogens, presumably from the consumption of contaminated food. Over the next five years, the incidence of hepatomas decreased, reaching 0.66% in 1976. Although hagfish populations have not been sampled elsewhere to detect pollutants in the environment, this may be a promising area of research.

REFERENCES

Adam, H. (1960) Different types of body movement in the hagfish *Myxine glutinosa*. *Nature*, **188**(4750), 595–6.

Adam, H. and Strahan, R. (1963) Notes on the habitat, aquarium maintenance, and experimental use of hagfishes, in *The Biology of Myxine* (eds A. Brodal and R. Fänge), Oslo, Norway: Universitetsforlaget, pp. 33–41.

Alexander, R.M. (1967) *Functional Design in Fishes*, London, Hutchinson University Library, pp. 160.

Alonso, M.K., Crespo, S.N., Garcia, N.A. and Coscarella, M. (ms in preparation). *Feeding Habits of the Southern Sea Lion Otaria flavescens of Patagonia*. 27pp.

Andriyashev, A.P. (1965) A general review of the Antarctic fish fauna, in *Biogeography and Ecology in Antarctica*, vol. *Monographiae Biologicae*, The Hague: Dr W. Junk, Pubs. 491pp.

Bigelow, H.B. and Schroeder, W.C. (1948) Cyclostomes, in *Fishes of the Western North Atlantic*, Vol. 1 (ed. A.E. Parr), New Haven: Sears Foundation for Marine Research, Yale University.

Bigelow, H.B. and Schroeder, W.C. (1952) A new species of the cyclostome genus *Paramyxine* from the Gulf of Mexico. *Breviora*, **8**, 1–10.

Bigelow, H.B. and Schroeder, W.C. (1953) *Fishes of the Gulf of Maine*, Washington: US Government Printing Office.

Cailliet, G.M., McNulty, M. and Lewis, L. (1991) Habitat analysis of Pacific hagfish (*Eptatretus stouti*) in Monterey Bay, using the ROV Ventana, in *Western Groundfish Conference*. Union, WA.

Calloway, D.H. and Carpenter, K.O. (1981) *Nutrition and Health*, Philadelphia, PA: Saunders College Publishing, 341pp.

Cole, F.J. (1913) A monograph on the general morphology of the myxinoid fishes based on a study of *Myxine*, V. The anatomy of the gut and its appendages. *Transactions of the Royal Society of Edinburgh*, **49**, 293–344.

Cunningham, J.T. (1886/7) On the structure and development of the reproductive elements in *Myxine glutinosa*. *Quart. J. Micr. Sci.*, **27**, 49–76.

Davison, W., Baldwin, J., Davie, P.S., Forster, M.E. and Satchell, G.H. (1990) Exhausting exercise in the hagfish, *Eptatretus cirrhatus*: the anaerobic potential and the appearance of lactic acid in the blood. *Comp. Biochem. Physiol.*, **95A**, 585–9.

Dean, B. (1904) Notes on Japanese myxinoids. *J. Coll. Sci. Imp. Univ. Tokyo*, **19**, 1–24.

Falkmer, S., Marklund, S., Mathisson, P. and Rappe, C. (1977) Hepatomas and other neoplasms in the Atlantic hagfish (*Myxine glutinosa*): a histopathologic and chemical study. *Annals of the New York Academy of Sciences*, 342–55.

Fernholm, B. (1974) Diurnal variations in behaviour of the hagfish, *Eptatretus burgeri*. *Marine Biology*, **27**, 351–6.

Fernholm, B. (1982) *Eptatretus caribbeaus*: a new species of hagfish (Myxinidae) from the Caribbean. *Bulletin of Marine Science*, **32**, 434–8.

Fernholm, B. (1990) Myxinidae, in *Fishes of the Southern Ocean* (eds O. Gon and P.C. Hamstra), Capetown, So. Africa: JLB Smith Institute of Ichthyology, p. 77–8.

Fernholm, B. and Hubbs, C.L. (1981) Western Atlantic hagfishes of the genus *Eptatretus* (Myxinidae) with description of two new species. *Fishery Bulletin*, **79**, 69–83.

Forster, M.E. (1990) Confirmation of the low metabolic rates of hagfish. *Comp. Biochem. Physiol. A.*, **95A**, 113–16.

Foss, G. (1962) Some observations on the ecology of *Myxine glutinosa* L. *Sarsia*, **7**, 17–22.

Foss, G. (1968) Behaviour of *Myxine glutinosa* L. in natural habitat; investigation of the mud biotype by a suction technique. *Sarsia*, 31, 1–13.

Garthe, S., Camphuysen, C.J.K. and Furness, R.W. (1996) Amounts of discards by commercial fisheries and their significance as food for seabirds in the North Sea. *Marine Ecology Progress Series*, **136**, 1–11.

Goode, G.B. and Bean, T. (1895) *Deep-Sea and Pelagic Fishes of the World*, Washington, D.C.: Smithsonian Institution, US National Museum.

Gorbman, A., Kobayashi, H., Honma, Y. and Matsuyama, M. (1990) The hagfishery of Japan. *Fisheries*, **15**, 12–18.

Gosztonyi, A.E. and Kuba, L. (1995) Los Peces en la Dieta del Cormoran Real, *Phalacrocorax albiventer* en la Zona de Punta Loberia (Chubut, Argentina). In *VI Congreso Latinoamericano de Ciencias del Mar*. Mar del Plata, Argentina.

Gustafson, G. (1935) On the biology of *Myxine glutinosa* L. *Arkiv for Zoologi*, **28**, 1–8.

Hacker, E.S. (1986) Stomach content analysis of shoft-finned pilot whales and northern elephant seal in the southern California Bight: Report of the Southwest Fisheries Center, March 1986.

Hall-Arber, M. (1996) Workshop probes hagfish processing potential in US: fishery discards worry fishermen. In *Commercial Fisheries News*, Stonington, ME, March pp. 18B–19B.

Hanson, L.C. (1993) The foraging ecology of

harbour seals, *Phoca vitulina*, and California sea lions, *Zalopus californianus*, at the mouth of the Russian River, California. M.S. thesis, Sonoma State University, Rohnert Part, CA. 70pp.

Holmgren, N. (1946) On two embryos of *Myxine glutinosa*. *Acta Zoologica*, **27**, 1–90.

Howard, F.G. (1982) Of shrimps and sea anemones; of prawns and other things. *Scott. Fish. Bull.*, **47**, 39–40.

Johnson, E.W. (1994) Aspects of the biology of Pacific (*Eptatretus stouti*) and Black (*Eptatretus deani*) hagfishes from Monterey Bay, California. Masters thesis, School of Natural Sciences, Fresno, CA: California State University, 130pp.

Kabasawa, H. and Ooka-Souda, S. (1991) Circadian rhythms of locomotory activity in the hagfish and the effect of the light–dark cycle. *Bull. Jap. Soc. Sci. Fish.*, **57**, 1845–9.

Kench, J.E. (1989) Observations on the respiration of the South Atlantic hagfish *Eptatretus hexatrema* Mull. *Comp. Biochem. Physiol.*, **93A**, 877–92.

Lesser, M., Martini, F.H. and Heiser, J.B. (1996) Ecology of hagfish, *Myxine glutinosa* L., in the Gulf of Maine. I. Metabolic rates and energetics. *J. Exp. Mar. Biol. Ecol.*, **208**, 215–25.

Lindley, D. (1988) Bagging the Hag, in *Pacific Fishing*, pp. 55–61.

MacDonald, I.R., Reilly, J.F. II, Guinasso, N.L., Jr, Brooks, J.M., Carney, R.S., Bryant, W.A. and Bright, T.J. (1990) Chemosynthetic mussels at a brine-filled pockmark in the northern Gulf of Mexico. *Science*, **248**, 1096–9.

Martini, F.H. and Heiser, J.B. (1989) Field observations on the Atlantic hagfish, *Myxine glutinosa*, in the Gulf of Maine. *American Zoologist*, **29**, 38A.

Martini, F.H., Heiser, J.B. and Lesser, M.P. (1997) A population profile for hagfish, *Myxine glutinosa* L., in the Gulf of Maine: I. Morphometrics and Reproductive State. *Fishery Bulletin*, **95**(2), 312–21.

McInerney, J.E. and Evans, D.O. (1970) Habitat characteristics of the *Pacific hagfish*, Polistotrema stouti. J. *Fish. Res. Bd. Canada*, **27**, 966–8.

Müller, W. (1875) Ueber das urinogenitalsystem des Amphioxus und der Cyclostomen. *Jena Zeit. Naturwiss.*, **IX**, 94–129.

Munz, F.W. and Morris, R.W. (1965) Metabolic rate of the hagfish, *Eptatretus stoutii* (Lockington) 1878. *Comp. Biochem. Physiol.*, **16**, 1–6.

Nansen, F. (1887) A protandric hermaphrodite (*Myxine glutinosa* L.) amongst the vertebrates. *Bergen Mus. Aarsber.*, **7**, 1–34.

Newth, D.R. and Ross, D.M. (1955) On the reaction to light of *Myxine glutinosa* L., *J. Exp. Biol.*, **32**, 4–21.

Ooka-Souda, S., Kadota, T. and Kabasawa, H. (1991) In the hagfish visual information sets the circadian pacemaker via the pretectum. *Zool. Sci.*, **8**, 1037.

Oxman, D.S. (1995) Seasonal abundance, movements, and food habits of harbour seals (*Phoca vitulina*) in Elkhorn Slough, California. In *Department of Biology*. Stanislaus, California: California State University, 126pp.

Palmgren, A. (1927) Aquarium experiments with the hagfish (*Myxine glutinosa* L.). *Acta Zoologica*, **8**, 1–16.

Putnam, F.W. (1874) Notes on the genus Myxine. *Proc. Boston Soc. Nat. Hist.*, **16**, 127–35.

Recchia, C.R. and Read, A.J. (1989) Stomach contents of harbour porpoises *Phocoena phocoena* L. from the Bay of Fundy. *Can. J. Zool.*, **67**, 2140–6.

Richardson, L.R. (1958) A new genus and species of Myxinidae (Cyclostomata). *Transactions of the Royal Society of New Zealand*, **85**, 283–7.

Schiavini, A.C.M. and Goodall, R.N.P. (1996) A review of the food habits of Peale's dolphin, *Lagenorhynchus australis*. In *SC/48/SM47*: International Whaling Commission Scientific Committee.

Scott, W.B. and Scott, M.G. (1988) Atlantic fishes of Canada. *Can. Bull. fish. Aquat. Sci.*, **219**, 1–731.

Shelton, R.G.J. (1978a) Of hagfish, goats and sprats. *Scott. Fish. Bull.*, **44**, 47–50.

Shelton, R.G.J. (1978b) On the feeding of the hagfish *Myxine glutinosa* in the North Sea. J. *Mar. Biol. Ass.* UK, **58**, 81–6.

Shumway, S.E., Perkins, H.C., Schick, D.F. and Stickney, A.P. (1985) Synopsis of biological data on the pink shrimp, *Pandalus borealis* Kroyer, 1838. Washington, DC: US Department of Commerce.

Smith, C.R. (1985) Food for the deep sea: utilization, dispersal, and flux of nekton falls at the Santa Catalina Basin floor. *Deep Sea Research*, **32**, 417–42.

Smith, C.R. and Hessler, R.R. (1974) Respiration in benthopelagic fishes: in situ measurements at 1230 m. *Science*, **184**, 72–3.

Smith, R.J. and Read, A.J. (1992) Consumption of euphausiids by harbour porpoise (*Phocoena phocoena*) calves in the Bay of Fundy. *Can. J. Zool.*, **70**, 1629–32.

Strahan, R. (1962) Survival of the hag, *Paramyxine*

atami Dean, in diluted seawater. *Copeia*, **2**, 471–3.

Strahan, R. (1963) The behaviour of myxinoids. *Acta Zoologica*, **44**, 1–30.

Strahan, R. (1975) *Eptatretus longipinnis*, n.sp., a new hagfish (Family Eptatretidae) from South Australia, with a key to the 5–7 gilled Eptatretidae. *The Australian Zoologist*, **18**, 137–48.

Strahan, R. and Honma, Y. (1960) Notes on *Paramyxine atami* Dean (Family Myxinidae) and its fishery in Sado Strait, Sea of Japan. *Hong Kong Univ. Fish.* J., **3**, 27–35.

Sumich, J.L. (1992) Benthic communities, in *An Introduction to the Biology of Marine Life*, Dubuque, IA: Wm. C. Brown, 206pp.

Tambs-Lyche, H. (1969) Notes on the distribution and ecology of *Myxine glutinosa*, L. *FiskDir. Skr. Ser. HavUnders.*, **15**, 279–84.

Villanueva, R. (1993) Diet and mandibular growth of Octopus *mangnificus* (Cephalopoda). S. Afr. J. *Mar. Sci.*, **13**, 121–6.

Wakefield, W.W. (1990) Patterns in the distribution of demersal fishes on the upper-continental slope off central California, with studies on the role of ontogenetic vertical migration on particle flux. Doctoral dissertation, Scripps Institute of Oceanography, San Diego, CA.

Walvig, F. (1967) Experimental marking of hagfish (*Myxine glutinosa* L.). *Norwegian J. Zool.* (*Nytt. Mag. Zool.*), **15**, 35–9.

Wells, R.M.G., Forster, M.E., Davison, W., Taylor, H.H., Davie, P.S. and Satchell, G.H. (1986) Blood oxygen transport in the free-swimming hagfish, *Eptatretus cirrhatus. J. exp. Biol.*, **123**, 43–53.

Wisner, R.L. and McMillan, C.B. (1988) A new species of hagfish, genus *Eptatretus* (Cyclostomata, Myxinidae) from the Pacific Ocean near Valparaiso, Chile, with new data on *E. bischoffii* and *E. polytrema. Trans. San Diego Soc. of Nat. Hist.*, **21**, 227–44.

Wisner, R.L. and McMillan, C.B. (1995) Review of the new world hagfishes of the genus *Myxine* (Agnatha, Myxinidae) with descriptions of nine new species. *Fishery Bulletin*, **93**, 530–50.

Worthington, J. (1905) Contribution to our knowledge of the myxinoids. *The American Naturalist*, **39**, 625–63.

PART TWO

Development and Pathology

6
CHROMATIN DIMINUTION AND CHROMOSOME ELIMINATION IN HAGFISHES

Sei-ichi Kohno, Souichirou Kubota and Yasuharu Nakai

SUMMARY

Morphological analyses of mitotic and meiotic chromosomes were performed in hagfish species. The kinetochores in both types of chromosomes appeared as a three-layered structure along the surfaces of chromosomes (12–50% of the length of the chromosome) without any constriction. There are no similar reports at present in other vertebrates with long kinetochores.

The cytogenetic examination of hagfish species (*Eptatretus okinoseanus*, *E. burgeri*, *Paramyxine atami* and *Myxine garmani* from Japan, *E. stoutii* from Canada, *E. cirrhatus* from New Zealand, *P. sheni* from Taiwan, and *M. glutinosa* from Sweden) revealed differences in chromosome number between germ cells (spermatogonia) and somatic cells (liver, blood, gill and kidney). The differences in chromosome number between spermatogonia (48, 54, 54–62, 72, 80, 52, 48, 66–96, 44 and 16) and somatic cells (34, 34, 34, 34, 34, 36, 34, 34, 28 and 14) were 14, 20, 20–28, 38, 46, 16, 14, 32–62, 16 and 2 in *E. stoutii*, *E. okinoseanus* type A, *E. okinoseanus* type B, *E. cirrhatus* type A and *E. cirrhatus* type B, *E. burgeri*, *P. atami*, *P. sheni*, *M. glutinosa*, and *M. garmani*, respectively. The percentage of DNA decrease in presumptive somatic cells averaged 52.8% (*E. stoutii*), 44.2% (*E. okinoseanus* type A), 49.4–57.7% (*E. okinoseanus* type B), 48.7% (*E. cirrhatus* type A) and 54.6% (*E. cirrhatus* type B), 20.9% (*E. burgeri*), 40.0% (*P. atami*), 70.8–74.5% (*P. sheni*), 43.5% (*M. glutinosa*) and 29.8% (*M. garmani*). These results clearly indicate that chromosome elimination takes place during early cleavage in these eight species of Myxinidae, except in germ line cells.

C-banding of metaphase chromosome preparations of germ line and somatic cells revealed that the C-band-positive chromatin in the somatic cells had been almost completely eliminated. Two germ line-restricted DNA-families (174 and 85 bp long) in *E. okinoseanus* were isolated. They are highly and tandemly repeated independently. These two account for 19% of the total eliminated DNA in *E. okinoseanus* type A and are located on several C-band positive, small chromosomes that are limited to germ cells.

6.1 INTRODUCTION

In some invertebrates (Protostomia), such as nematodes and arthropods (Copepoda and Diptera), chromosome elimination occurs during the early stages of cleavage, in which a certain amount of chromatin is eliminated

The Biology of Hagfishes. Edited by Jørgen Mørup Jørgensen, Jens Peter Lomholt, Roy E. Weber and Hans Malte. Published in 1998 by Chapman & Hall, London. ISBN 0 412 78530 7.

from presumptive somatic cells (Boveri, 1887; reviewed by Tobler, 1986). In 'higher' animals (Deuterostomia), particularly vertebrates, there has been no evidence presented for such chromosome elimination. Consequently, chromosome elimination during cleavage had been considered a characteristic phenomenon limited to invertebrates.

Using earlier techniques, observations of features and numbers of chromosomes in hagfish species have been made and these are summarized in Table 6.1. Until the 1970s, the chromosomes of hagfishes were studied by using testis prepared by the paraffin section method, or by use of the squash method. After development of the air-dry method, investigators began to use somatic tissues (liver, kidney, gill and blood) together with testis. Thus now one can compare the numbers of chromosomes in spermatogonia with those in somatic cells in a specimen and/or a species.

Previous cytogenetic studies of the Japanese hagfish, *Eptatretus burgeri*, have shown that there is a diploid chromosome number of 48 in the testis (Nogusa, 1960) and 36 in the gills and other tissues (Kitada and Tagawa, 1975). The reported differences in the chromosome numbers has been attributed to the newly developed air-drying technique used for the latter study (Kitada and Tagawa, 1975). However, when this technique was used exclusively, different chromosome numbers between somatic cells and the germ cells have been observed consistently in every hagfish species examined (Kohno *et al.*, 1986; Nakai and Kohno, 1987; Nakai *et al.*, 1991, 1995; Kubota *et al.*, 1992, 1994). So, the phenomenon should be understood to be chromosome elimination during differentiation into somatic cells in the hagfish species. The chromosome elimination can be expected to take place in each hagfish species with some modification.

The following section describes the characteristics of the hagfish chromosomes, chromosome elimination in hagfish species, as well as the elimination of DNA from presumptive somatic cells.

6.2 MORPHOLOGICAL CHARACTERISTICS OF THE CHROMOSOMES IN HAGFISH SPECIES

In most vertebrates, the kinetochores are situated on the chromosomes in the region of the primary constriction, known as the centromere. But the chromosomes of hagfishes appear under light microscopy to have no primary constriction (Figure 6.1a). It is also noteworthy that in primary meiotic metaphase, some chromosomes formed so called associated dumbbell-shaped bivalent chromosomes (Figure 6.1b). These associated

Table 6.1 Previous cytogenetic studies of hagfishes

Reference	Species	Chromosome number	Tissue
Retzius (1890)	*Myxine glutinosa*	50	Testis
Schreiner and Schreiner (1904)	*Myxine glutinosa*	52	Testis
Makino (1951)	*Eptatretus burgeri*	48	Testis
Nogusa (1960)	*Eptatretus burgeri*	48	Testis
	Eptatretus okinoseanus	46	Testis
Taylor (1967)	*Eptatretus stoutii*	48	Testis
Nygren and Jahnke (1972)	*Myxine glutinosa*	17–112	Testis, liver, kidney
Kitada and Tagawa (1973)	*Eptatretus burgeri*	36	Gill, kidney, liver
	Paramyxine atami	36	Gill, kidney, liver

dumbbell-shaped chromosomes sometimes appeared to be separated into several blocks. Such chromosomes existed in primary spermatocytes in seven out of eight hagfishes examined.

Metaphases from germ line cells of *E. burgeri* were examined by electron microscopy. All surfaces of chromosomes to which microtubules are attached appear to be three-layered structures that occupy an extensive region along the long axis of the chromosome. However, this structure does not cover the entire poleward face of chromosomes, but is restricted to the central area. This structure appeared to be the kinetochore and was composed of an outer dense layer of 10–15 nm in thickness, an electron-translucent middle layer of 20–25 nm thickness and an inner dense layer of 10–15 nm thickness, usually associated with the chromatin (Figure 6.1c). Each layer of the kinetochore in *E. burgeri* was narrower than the analogous layers in other organisms that have a three-layered kinetochore. This kinetochore occupied 12–50% of the length of a chromosome and 12–45% of the width. Kinetochores during mitotic division were also examined. The kinetochores on some chromosomes were distinctly visible as three-layered structures similar to those observed in germ line cells.

All eukaryotic chromosomes are attached to spindle microtubules during mitosis, and most of the sites of such attachment can be identified by their unique morphology and staining characteristics under the light and electron microscope. In most eukaryotes, kinetochores can be identified as being either localized or diffuse. The kinetochore region can be easily identified as the primary constriction in those organisms that possess a localized kinetochore. By contrast, chromosomes with diffuse kinetochores have no primary constriction.

The metaphase chromosomes of hagfish species lack any visible constriction. For this reason, centromeres could not be detected under the light microscope. In the study by electron microscopy, the structure and position of the kinetochore of both mitotic and meiotic chromosomes in *E. burgeri* was clearly revealed (Shichiri *et al.*, submitted). The three-layered kinetochore was observed to cover about 50% (maximum value) of the length of a chromosome. The trilaminar type of kinetochore is highly conserved and can be widely observed from algae to mammals. However,

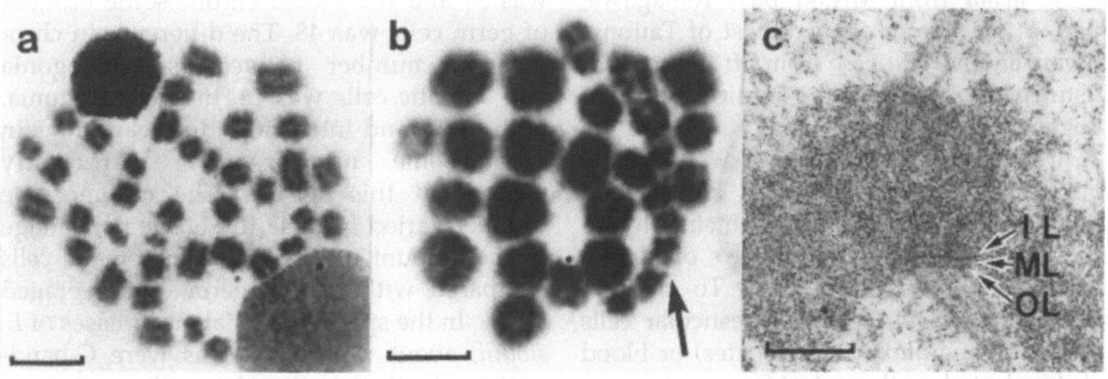

Figure 6.1 Mitotic and meiotic metaphases from *Eptatretus burgeri* as seen by light microscopy ((a) and (b)) and a chromosome by electron microscopy (c). The chromosomes have no primary constriction. Arrows indicate the associated dumbbell-shaped bivalent chromosomes (b) and the three-layered kinetochore structure (c). Typical three-layered kinetochore plates are composed of an outer electron-dense layer (OL), a middle electron-translucent layer (ML) and an inner electron-dense layer (IL). Bar represents 5μm((a) and (b)) or 0.5μm (c). (Courtesy Motoharu Shichiri, Toho University, Chiba, Japan.)

the ratio of the length of the kinetochore to that of the chromosome in *E. burgeri* is much higher than that in any other vertebrates studied to date. The kinetochore that appears over most of the chromosome's poleward face is called a diffuse kinetochore. In the case of chromosomes with diffuse kinetochores, holocentric chromosomes have only a single kinetochore on each poleward face. It is difficult to state whether the chromosomes of *E. burgeri* are holocentric because holocentric chromosomes usually have a kinetochore that extends over more than 75% of the chromosome length (Comings and Okada, 1972). Among vertebrates, however, the chromosomes of *E. burgeri* with their long kinetochores are unusual.

6.3 CHROMOSOME ELIMINATION AND CHROMATIN DIMINUTION IN HAGFISH SPECIES

Specimens of *Eptatretus stoutii* collected from the northeast Pacific Ocean off Bamfield, Canada, *E. okinoseanus*, *Paramyxine atami* and *Myxine garmani* from the sea of Kashima, off the coast of Ibaraki, Japan, *E. cirrhatus* from off the coast of Kaikoura, on the east coast of the South Island of New Zealand, *E. burgeri* and *P. atami* from Misaki Bay, Kanagawa, Japan, *P. sheni* from off the coast of Taitong, Taiwan, and *M. glutinosa* from off the coast of Kristineberg, Sweden, were studied by use of cytogenetic procedures.

Chromosome preparations were made from testis (germ cells), liver, kidney and blood (somatic cells) using colchicine, hypotonic treatment and routine air- or flame-drying technique (Ojima, 1983). To measure the amount of DNA per cell, testicular cells (spermatogonia and spermatocytes) or blood cells (nucleated erythrocytes) from each specimen and blood cells of *E. burgeri*, as controls, were fixed with 1:3 acetic acid–methanol, placed side by side on glass slides. The slides were air-dried, treated with RNase and stained with propidium iodide as described by Mazzini *et al.* (1980). The relative DNA content per cell was determined from erythrocytes of *E. burgeri* (about 0.86 times of that in a human diploid cell) and from the testicular cells (as germ cells) or erythrocytes (as somatic cells) of each specimen per slide, using a microscope based cytofluorometer (Ashihara *et al.*, 1986). In another procedure the air-dried slides were stained with Feulgen, and the relative DNA content per cell was measured by means of the two-wavelength method (Patau, 1952), using a microspectrophotometer.

Some metaphases were photographed and then destained with the fixative. After an overnight soak in absolute methanol, followed by drying in an incubator for 3–4 h at 60°C, the slides were processed further using a modified C-banding technique (Kuro-o and Kubota, submitted) or Sumner's barium/saline/Giemsa (BSG) technique (1970), to determine whether the chromatin was constitutive heterochromatin (positive for C-bands). Details of these methods have been described previously (Kohno *et al.*, 1986; Kubota *et al.*, 1992).

6.3.1 *Eptatretus stoutii*

The chromosome number of somatic cells was 34 and the modal chromosome number of germ cells was 48. The difference in chromosome number between spermatogonia and somatic cells was 14. In spermatogonia, both inter- and intra-individual variations in chromosome number were particularly evident in this species. The chromosome number varied from 47 to 61. The percentage of the amount of DNA (2C) in somatic cells compared with that in germ cells averaged 47.2%. In the spermatogonial metaphases of *E. stoutii*, about 8 chromosomes were C-band-positive in their entire chromatin, and these were observed as associated chromosomes in meiotic metaphases of the primary spermatocytes. In addition, C-band-positive chromatin was also observed at the terminal ends of about 27 chromosomes (partially C-band-positive chromosomes), while the other chromosomes were C-band-negative in spermatogonial metaphases. By contrast, C-band-positive chromatin was rarely observed in somatic metaphases (Kubota *et al.*, 1994; Nakai *et al.*, 1995).

6.3.2 *Eptatretus okinoseanus*

The chromosome number of somatic cells was 34 (Figures 6.2a and 6.2g) and the modal chromosome number of germ cells was 54 (Figures 6.2b and 6.2h). The difference in chromosome number between spermatogonia and somatic cells was 20. In this species, there are two types of individuals with respect to the quantity and distribution of C-band-positive chromatin in germ cells. Type A has less DNA and less C-band-positive chromatin than type B. The percentage of the amount of DNA (2C) in somatic cells compared with that in germ cells averaged 55.8% in type A and varied between 50.6 and 42.3% in type B. Both inter- and intra-individual variations in chromosome number were made clear and were particularly marked in type B specimens. The chromosome numbers in spermatogonia from type A specimens varied between 48 and 64 and the modal number was 54. That of the type B specimens varied between 52 and 65 and the modal number was also 54. Although the modal number of chromosomes in spermatogonia is the same (54) in both types, the percentages of cells with the modal number was rather low (42.4% in type A and 22.7% in type B).

In type A specimens, about 20 C-band-positive chromosomes were observed in spermatogonial metaphases (Figure 6.2e), and C-band-positive chromatin was found along almost the entire length of the associated dumbbell-shaped chromosomes in the meiotic metaphase of primary spermatocytes (Figure 6.2f). In germ cells of type B specimens, C-band-positive chromatin was observed at nearly all the terminal ends of the chromosomes, in addition to the C-band-positive chromosomes such as those found in the cells of type A (Figures 6.2k and 6.2l). C-band-positive chromatin was rarely observed in somatic metaphases from specimens of either type (Figures 6.2d and 6.2j). Furthermore, some intraindividual variability with respect to the quantity of C-band-positive chromatin in each type B specimen was apparent in the metaphases of germ cells (Nakai *et al.*, 1991; Kubota *et al.*, 1992; Kubota, unpublished data).

6.3.3 *Eptatretus cirrhatus*

In *E. cirrhatus*, type B and type A were cytogenetically distinguished, as the modal chromosome number in germ cells from type B was distinctly different from that of type A. Metaphases of spermatogonia had one large chromosome in *E. cirrhatus* type A and two in type B. The chromosome number of somatic cells in both types was 34 and the modal chromosome number of germ cells in type A was 72; that of type B was 80. The difference in chromosome number between spermatogonia and somatic cells was 38 and 46 in types A and B, respectively. In spermatogonia, both inter- and intra-individual variations in chromosome number were particularly evident in this species. The range of chromosome number was from 62 to 84. The percentage of the amount of DNA (2C) in somatic cells compared with that in germ cells averaged 51.3 and 45.4% in types A and B, respectively. In the spermatogonial metaphases of *E. cirrhatus*, 17–18 of entirely C-band-positive chromosomes in type A and 26–28 in type B were counted. The one large chromosome of *E. cirrhatus* type A was C-band-negative. On the other hand, one large chromosome was C-band-negative and the other was C-band-positive in *E. cirrhatus* type B. All of the large chromosomes were included among the eliminated chromosomes. The metaphases in somatic cells had no C-band-positive chromosomes in either type. The eliminated chromosomes in this species are about 38 in type A and 46 in type

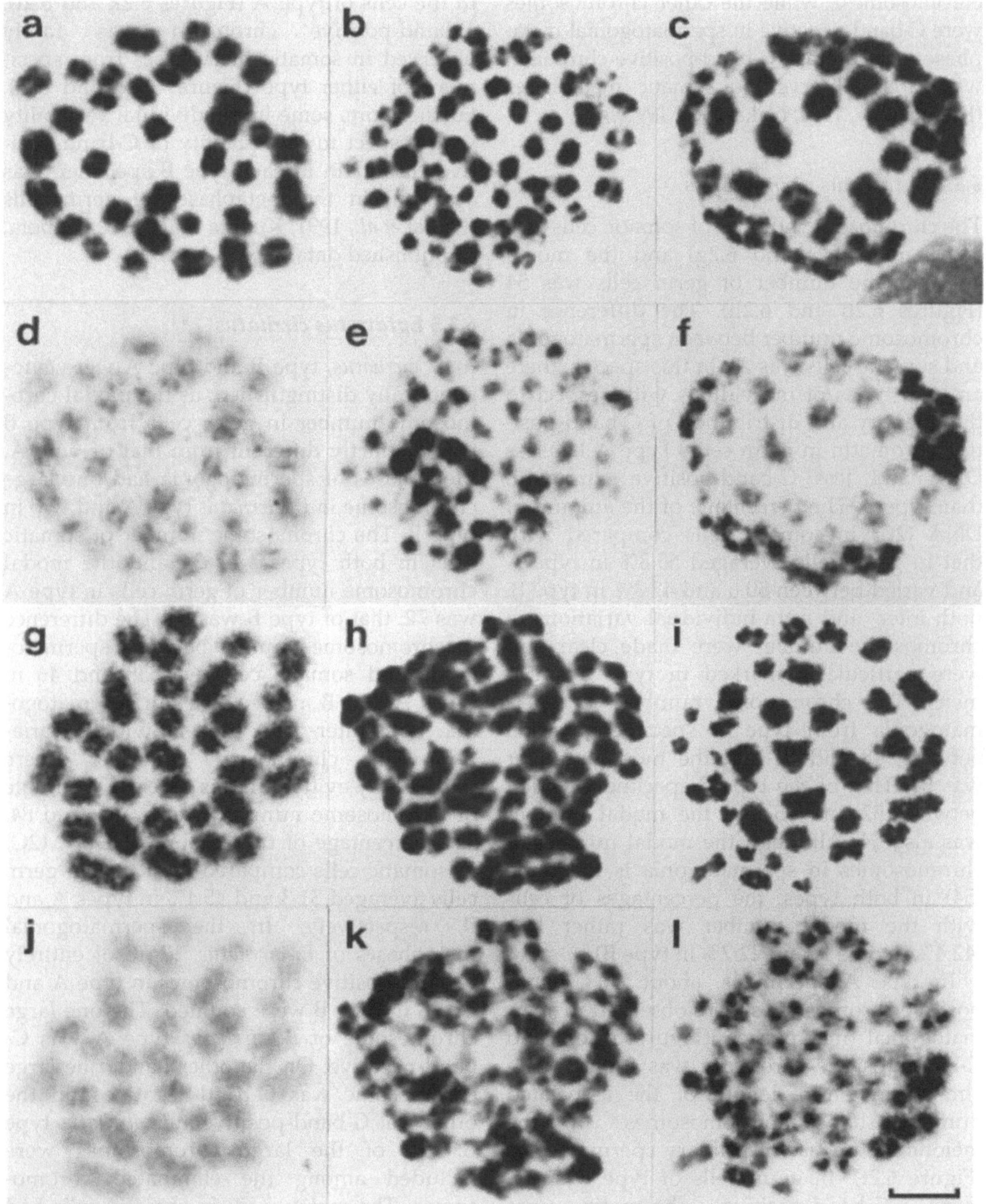

Figure 6.2 Metaphases from *Eptatretus okinoseanus* type A before and after the modified C-band treatment: somatic cell ((a) and (d)); spermatogonium ((b) and (e)); spermatocyte ((c) and (f)), and *Eptatretus okinseanus* type B before and after the modified C-band treatment: somatic cell ((g) and (j)); spermatogonium ((h) and (k)); spermatocyte ((i) and (l)). Bar represents 5μm.

B. This indicates that in *E. cirrhatus*, about 20 C-band-negative chromosomes were eliminated in addition to all of the C-band-positive chromosomes, during the process of differentiation of somatic cells (Nakai *et al.*, 1995).

6.3.4 *Eptatretus burgeri*

The chromosome number of somatic cells was 36 and the modal chromosome number of germ cells was 52. The difference in chromosome number between spermatogonia and somatic cells was 16. In spermatogonia, both inter- and intra-individual variations in chromosome number were also observed in this species. The range of chromosome number was from 46 to 54. The percentage of the amount of DNA (2C) in somatic cells compared with that in germ cells averaged 79.1%. In the spermatogonial metaphases of *E. burgeri*, about 16 chromosomes were entirely C-band-positive, and these were observed as associated chromosomes in meiotic metaphases of the primary spermatocytes. In contrast, C-band-positive chromatin was rarely observed in somatic metaphases (Kohno *et al.*, 1986; Nakai *et al.*, 1991).

6.3.5 *Paramyxine atami*

The chromosome number of somatic cells was 34 and that of the germ cells was 48. The difference in chromosome number between spermatogonia and somatic cells was 14. Both inter- and intra-individual variations in chromosome number were evident in germ cells of this species. The percentage of the amount of DNA (2C) in somatic cells compared with that in germ cells averaged 60.0%. In the spermatogonial metaphases of *P. atami*, about 14 chromosomes were entirely C-band-positive, and these were observed as associated chromosomes in meiotic metaphases of the primary spermatocytes. By contrast, C-band-positive chromatin was rarely observed in somatic metaphases (Nakai *et al.*, 1991).

6.3.6 *Paramyxine sheni*

The modal chromosome number of somatic cells was 34 and that of the germ cells in each specimen varied between 66 and 96. There were numerous micro-chromosomes in the germ cells. The difference in chromosome number between spermatogonia and somatic cells was between 32 and 62. Both inter- and intra-individual variations in chromosome number were particularly evident in germ cells of this species. The percentage of the amount of DNA (2C) in somatic cells compared with that in germ cells varied between 29.2 and 25.5%. In the spermatogonial metaphases of *P. sheni*, almost all microchromosomes were entirely C-band-positive. By contrast, C-band-positive chromatin was rarely observed in somatic metaphases.

6.3.7 *Myxine glutinosa*

The chromosome number of somatic cells was 28 and that of the germ cells was 44. The difference in chromosome number between spermatogonia and somatic cells was 16. There were no evident inter- and intra-individual variations in chromosome number, neither in somatic cells nor in germ cells in this species. The metaphases of neither somatic nor germ cells in *M. glutinosa* could be stained clearly by the C-banding method. Consequently, about 16 C-band-negative chromosomes were eliminated from the presumptive somatic cells, and they formed associated chromosomes in the primary spermatocytes of this species. The percentage of the amount of DNA (2C) in somatic cells compared with that in germ cells averaged 56.5%. For understanding the pattern of chromosome elimination in this species, more studies should be done, especially in C-banding (Nakai *et al.*, 1995).

6.3.8 *Myxine garmani*

The chromosome number of somatic cells was 14 and that of germ cells was mostly 16 chromosomes in spermatogonia, or eight bivalents in primary spermatocytes, respectively. The spermatogonia had a pair of remarkably large homologous chromosomes (marker chromosomes), which were about twice as long as the second largest chromosome. In the primary spermatocytes the marker chromosomes formed a large bivalent, but in about 36% (4/11) of the metaphases the marker chromosomes were unpaired. The somatic cells contained the same chromosomes as observed in the spermatogonia except for the marker chromosomes. The difference in chromosome number between spermatogonia and somatic cells is only 2 (one pair of the largest chromosome in the spermatogonia, the pair of the marker chromosome). The percentage of the amount of DNA (2C) in somatic cells compared with that in germ cells averaged 70.2%. While the C-banding patterns on the chromosomes of this species were somewhat ambiguous, C-band positive chromatin was observed in the middle part of the marker chromosomes and in some of the terminal positions of about six chromosomes in the spermatogonial metaphases. In the somatic metaphases, C-band positive chromatin was observed in the terminal positions of about six chromosomes (Nakai and Kohno, 1987; Nakai *et al.*, 1991).

6.3.9 Summary

The data mentioned above are summarized in Table 6.2. The differences between the modal chromosome numbers in somatic cells and those in the germ cells were observed in all the hagfishes examined. Thus chromosome elimination occurs in these eight species of hagfish, most likely during differentiation of the somatic cells. The elimination of chromosomes resulted in a loss of DNA equivalent to 52.8%, 44.2%, 49.4–57.7%, 48.7%, 54.6%, 20.9%, 40.0%, 70.8–74.5%, 43.5% and 29.8% in *E. stoutii*, *E. okinoseanus* types A and B, *E. cirrhatus* types A and B, *E. burgeri*, *P. atami*, *P. sheni*, *M. glutinosa* and *M. garmani*, respectively. Chromosome elimination in three genera and eight species of hagfish (five from northwest Pacific, one from northeast and south Pacific, and north Atlantic) was demonstrated. Therefore it appears probable that chromosome elimination may occur generally in the order Myxinida.

The chromosome elimination pattern in each hagfish species is different, but the patterns can be summarized according of the nature of the eliminated chromosomes or chromatin. A scheme for the chromosome elimination patterns is illustrated in Figure 6.3.

In *E. okinoseanus* type A, *E. burgeri* and *P. atami*, some chromosomes in germ-cell metaphases were C-band-positive over almost their entire length, but C-band-positive chromatin was rarely observed in somatic-cell metaphases, suggesting that the chromosomes that contained C-band-positive chromatin were eliminated selectively from the presumptive somatic cells.

In *E. stoutii* and *E. okinoseanus* type B, C-band-positive chromatin was observed at the terminal regions of nearly all chromosomes, in addition to the C-band-positive chromosomes seen in type A specimens. In contrast, C-band-positive chromatin was rarely observed in somatic-cell metaphases. Therefore, it seems very likely that, in these germ cells, excision of C-band-positive chromatin from the terminal regions of the chromosomes was followed by elimination of the C-band-positive fragments and of chromosomes that were C-band-positive over almost their entire length.

The elimination of C-band-positive chromatin observed in *M. garmani* differed from the two preceding patterns by the presence of C-band-positive chromatin in the middle of the largest chromosome pair in the germ cells only. In addition, C-band-positive chromatin was observed in the terminal regions of about six chromosomes in both somatic cells and germ cells. These observations suggest that only the C-band-positive chromatin in the

Table 6.2 Chromosome elimination in hagfishes

Species	Chromosome number (C-band)*			Percentage of DNA eliminated from presumptive somatic cells	Ratio of the amount of DNA†		References
	Somatic cells	Germ cells	Eliminated chromosomes		Somatic cells	Germ cells	
Eptatretus stoutii	34(-)	48(++)	14(++)	52.8%	0.94	1.99	Kubota *et al.* (1994); Nakai *et al.* (1995)
E. okinoseanus							Nakai *et al.* (1991); Kubota *et al.* (1992);
type A	34(-)	54(++)	20(++)	44.2%	0.90	1.65	Kubota (unpublished data)
type B	34(-)	54–62(++)‡	20–28(++)	49.4–57.7%	0.90	1.79–2.05	
E. cirrhatus							Nakai *et al.* (1995)
type A	34(-)	72(++)	38(++)	48.7%	0.79	1.52	
type B	34(-)	80(++)	46(++)	54.6%	0.75	1.68	
E. burgeri	36(-)	52(++)	16(++)	20.9%	1.00	1.26	Kohno *et al.* (1986)
Paramyxine atami	34(-)	48(++)	14(++)	40.0%	1.16	1.94	Nakai *et al.* (1991)
P. sheni	34(-)	66–96(++)§	32–62(++)	70.8–74.5%	0.85	2.94–3.37	Shichiri *et al.* (unpublished data)
Myxine glutinosa	28(-)	44(?)	16(?)	43.5%	1.44	2.55	Nakai *et al.* (1995)
M. garmani	14(+)	16(++)	2(++)	29.8%	1.54	2.19	Nakai and Kohno (1987)

* Marks in parenthesis indicate the presence of C-band positive chromatin in cells, as: -, absent; +, present; ++, present in large amounts; and ?, could not be obtained.
† Ratio of the amount of DNA per cell in each hagfish species to that in an erythrocyte of *E. burgeri*.
‡ Interindividual variations in modal chromosome number of germ cells were clearly evident in *E. okinoseanus* type B specimens.
§ The metaphases of germ cells in *P. sheni* have many micro-chromosomes.

largest chromosome pair was eliminated from the presumptive somatic cells of *M. garmani*, whereas C-band-positive chromatin was eliminated completely in the other six species of Eptatretidae.

The chromosome elimination patterns of *E. cirrhatus* and *E. stoutii* are different from those of *E. burgeri*, *E. okinoseanus*, and *P. atami*. The eliminated chromosomes or chromatin from presumptive somatic cells in the latter were thought to be almost all C-band-positive, because the number of eliminated chromosomes was almost the same as the number of wholly C-band-positive chromosomes. In the former, as the number of eliminated chromosomes was greater than that of wholly C-band-positive chromosomes, clearly some C-band-negative chromatin must have been eliminated in the form of some wholly or partially C-band-negative chromosomes.

The chromosome elimination patterns of *M. glutinosa* and *P. sheni* need further study. But it should be noted that the chromosomes in germ cells of *M. glutinosa* would not have C-band-positive chromatin. Therefore, the eliminated chromosomes would be C-band-negative.

All species of Eptatretidae examined showed inter- and intra-individual variation in the germ cell chromosome number. The variation in the number of chromosomes in *E. stoutii* was first described by Taylor (1967). B-chromosomes (supernumerary chromosomes),

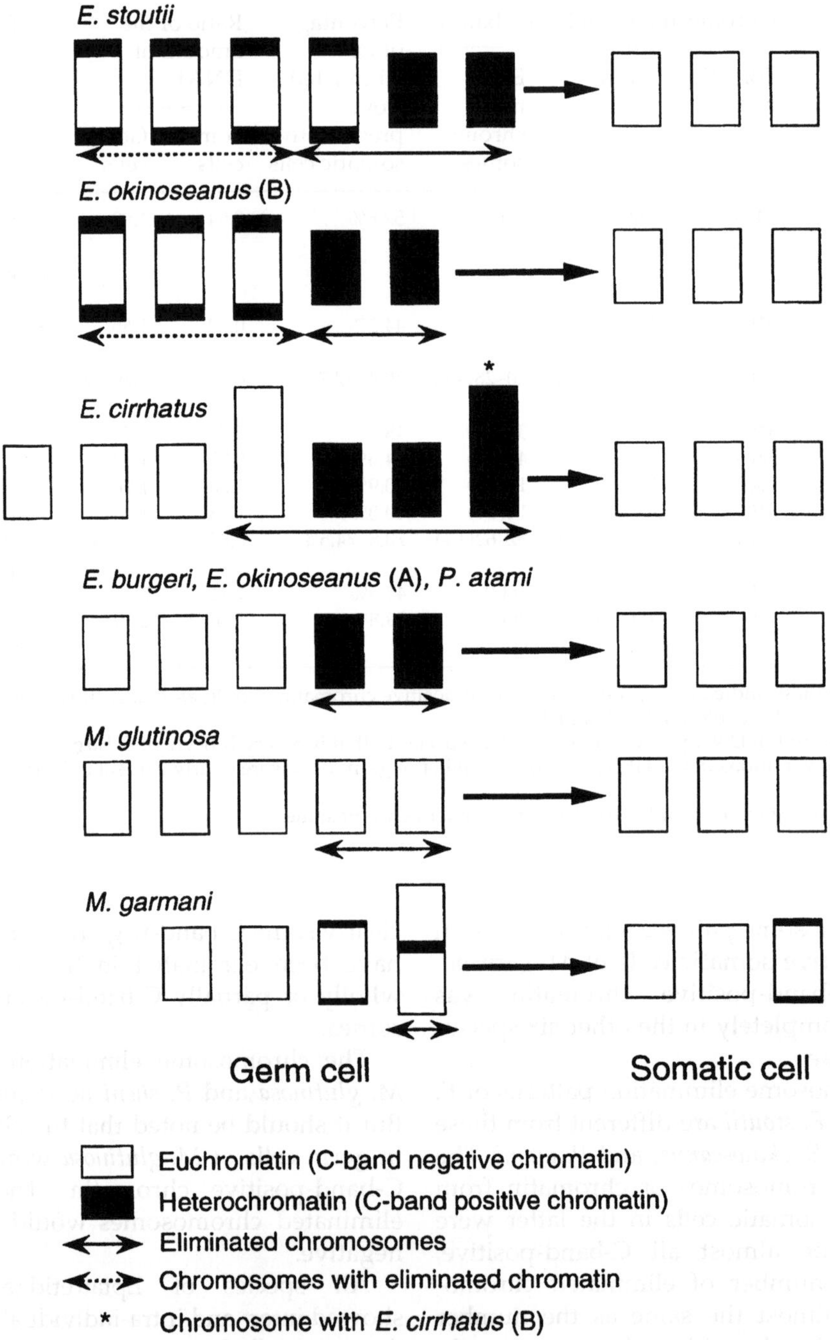

Figure 6.3 Schematic representation of patterns of chromosome elimination and chromatin diminution in hagfish species. Squares indicate chromosomes.

restricted to germ cells, which would be maintained in these species, were first described in *E. okinoseanus* (Kubota *et al.*, 1992). In all species of Eptatretidae studied, chromosome numbers and the amount of DNA in somatic cells are almost identical among the specimens. But in germ cells, the values of the two are inter- and intraspecifically variable. The cause is thought to be the existence of B-chromosome variation. B-chromosomes exist characteristically in the germ cells of the genera *Eptatretus* and *Paramyxine*.

E. cirrhatus, type B and type A were distinguished by their patterns of interindividual variation. In spermatogonia of *E. cirrhatus*, the modal chromosome number of type B was distinctly higher than that of type A. In addition, the amount of DNA in the germ cells of the single specimen of type B was higher than that of type A. Morphologically, type A and type B appeared identical. In a molecular genetic study (Kubota, unpublished data), the differences between type A and type B were not very distinct. So it seems that these two types are variants within one species. On the other hand, the large surplus chromosomes in type B were C-band-positive. The reason why the cells of type B have the surplus chromosomes is not known. It may be that the surplus chromosomes are also B-chromosomes. It is necessary to observe a greater number of specimens cytologically to make clear the extent of interindividual variation of B-chromosomes in *E. cirrhatus*.

Previous observations by Nygren and Jahnke (1972) show that there is variation in chromosome number in the somatic and germ cells of *M. glutinosa*, ranging from 17 to 116. Their results also show that there is a bimodal distribution of chromosome number, the mode of which in somatic cells is 28 and in germ cells 42–44. Thus, their values are exactly the same as those reported by Nakai *et al.* (1995). Unfortunately, they counted many incomplete metaphases from which a number of chromosomes had been lost, and they also counted two metaphases as one. In addition, they did not consider that the chromosome number in somatic cells was smaller than that in germ cells. As a result, they did not recognize chromosome elimination in this species.

In *Myxine*, the amount of DNA in *M. glutinosa* is slightly less than that of *M. garmani* in somatic cells. The chromosome number in *M. glutinosa* is more than twice that of *M. garmani* both in somatic and germ cells. Morphological studies of the number and distribution of the external gill opening (Adam and Strahan, 1963; Hardisty, 1979; Kuo *et al.*, 1994), and the fact that there is not a large amount of clear C-band-positive chromatin in germ cells, suggest that *M. glutinosa* is very similar to *M. garmani*. Considering that the chromosomal complex of *M. garmani* might have resulted from a cytogenetic dysploid decrease (Nakai and Kohno, 1987), it is likely that *M. glutinosa* could be the ancestor of *M. garmani*. Similar mechanisms of speciation have been proposed for Asian deer (Shi *et al.*, 1980), tropical fish (Carlton and Denton, 1974) and bats (Baker and Bickham, 1980; Ono and Obara, 1994). It would be of considerable interest to further study and clarify the cytogenetic relationships and phylogenetic affinities of Myxinidae.

6.4 GERM LINE-RESTRICTED, HIGHLY REPETITIVE DNA SEQUENCES IN HAGFISH SPECIES

The significance, as well as the cause and mechanism, of the chromosomal elimination process remain unknown. However, the elucidation of the genetic informational content of the eliminated chromosomes (DNA), using modern molecular biological techniques, might give us some clues to the nature and significance of chromosome elimination in hagfish species.

Except in the case of *M. garmani*, the eliminated chromosomes (or parts of chromosomes) are C-band positive, i.e. heterochromatin, so that none of the somatic cells retains

C-band positive chromatin. As a consequence, it could be expected that the eliminated materials are enriched in repetitive DNA sequences, and that the repetitive DNA sequences in eliminated materials are qualitatively different from the repetitive DNA sequences retained in the somatic genome. Molecular studies on *Ascaris lumbricoides* var. *suum* have provided new insights into the nature of chromatin and DNA elimination in nematodes (Tobler *et al.*, 1992). In this species, a large portion of germ line DNA is composed of a satellite DNA carrying about 106 copies of some 120 bp, AT-rich repeating units, whereas the somatic DNA contains at most 5000 copies of this unit. Thus, chromatin diminution removes more than 99.5% of the satellite DNA sequences from the presumptive somatic cells (Streeck *et al.*, 1982; Müller *et al.*, 1982). Recently, we have initiated an analysis of chromosome elimination in hagfish species at the molecular level by comparing germ lines and somatic genomes (Kubota *et al.*, 1993).

6.5 GERM LINE AND SOMATIC GENOMES SHOW DIFFERENT RESTRICTED ENZYME PATTERNS

To learn about the nature of the eliminated DNA sequences, we extracted and purified germ line and somatic DNA from members of three Japanese species, namely *E. okinoseanus* (type A and B), *E. burgeri* and *M. garmani*. Total DNAs of germ and somatic cells were digested with several kinds of restriction endonucleases, and digests were subjected to electrophoresis on agarose slab gels to compare the patterns of germ line DNA with those of somatic DNA. Two enzymes, *Bam* HI and *Dra* I, gave weak bands that were limited to the germ line DNA for both types of *E. okinoseanus*. Such differences in patterns between germ line DNA and somatic DNA could not be clearly detected after digestion with the other tested enzymes in any of the three species. Two weak bands, corresponding to fragments whose sizes were estimated to be about 90 and 180 bp, were generated by digestion with *Bam* HI of germ line DNA from both types of *E. okinoseanus*. One weak band, corresponding to a fragment whose size was estimated to be about 90 bp, was generated by digestion with *Dra* I of germ line DNA from both types of *E. okinoseanus*.

6.6 THE *BAM* HI AND *DRA* I FRAGMENTS ARE PARTS OF THE ELIMINATED DNA

To prove that the *Bam* HI fragments and the *Dra* I fragment are part of the eliminated material, these fragments were isolated from total *Bam* HI or *Dra* I digests of germ line DNA from type A of *E. okinoseanus*. These DNA fragments were labelled and hybridized to *Bam* HI- or *Dra* I-digested DNA that had been transferred from agarose gels to membrane filters. The *Bam* HI fragments hybridized mostly with germ line DNA digested with *Bam* HI from both types of E. okinoseanus. Restriction fragments of DNA corresponding to the *Bam* HI fragments were undetectable in the digests of somatic DNA from both types of *E. okinoseanus*, and in the digests of the germ line and somatic DNA from *E. burgeri* and *M. garmani*. These results indicate that the *Bam* HI-fragments and/or multiple copies of these fragments are present mostly in the germ line DNA of *E. okinoseanus*. That is to say, these DNA sequences are indeed part of the eliminated DNA in this species. In addition, the partial digestion of germ line DNA from both types led to the appearance of new ladder-like bands of regularly spaced fragments which corresponded to a monomer of 70–90 bp and related multimers. These bands were undetectable in the digests of somatic DNA. These results indicate that the *Bam* HI fragments are clustered in tandem arrays within the germ line genome of *E. okinoseanus*. After digestion with *Bam* HI of germ line DNA from type A of *E. okinoseanus* and agarose gel electrophoresis, it was estimated from densitometric tracings of a negative film that the *Bam* HI fragments account for about 2.6% of the total germ line genomic DNA in type A. If these fragments are assumed to consist of one repeating unit, these values imply that the respective sequences are repeated approximately 1.3×10^6 times in the diploid germ line genome in type A.

The *Dra* I-fragment also hybridized with a ladder-like pattern, mostly with germ line DNA digested with *Dra* I from both types of *E. okinoseanus*. Restriction fragments of DNA corresponding to the *Dra* I fragment were undetectable in the digests of somatic DNA from both types of *E. okinoseanus*, and in the genomic DNA from *E. burgeri* and *M. garmani*. This result indicates that the *Dra* I fragment, or multiples of it, is clustered in tandem arrays, and that it is present mainly in the germ line DNA of *E. okinoseanus*, as are also the *Bam* HI fragments. Moreover, the *Bam* HI-fragments, when used as probes, did not hybridize to the band that corresponded to the *Dra* I fragment. Thus, the *Bam* HI fragments and the *Dra* I fragment do not resemble one another in terms of nucleotide sequence. After digestion with *Dra* I of germ line DNA from type A of *E. okinoseanus* and agarose gel electrophoresis, it was estimated from densitometric tracings of a negative film that the *Dra* I-fragment accounted for about 0.4% of the total germ line genomic DNA. The pattern of hybridization of the *Dra* I-fragment to DNA that had been completely digested with *Dra* I was clearly ladder-like in the case of digests of germ line DNA from both types of *E. okinoseanus*. Densitometric scanning of autoradiographic hybridization signals obtained with a consensus sequence among cloned *Dra* I fragment (see below), as probe, revealed that the intensity of the band that corresponded to the *Dra* I fragment was equal to about 6.47% of the total intensity of bands generated from germ line DNA from type A of *E. okinoseanus*. From these values, it was estimated that the main components of the *Dra* I fragment represent about 6.15% of the germ line genomic DNA in type A of *E. okinoseanus* and correspond to approximately 6.2×106 copies per diploid genome.

6.7 THE *BAM* HI FRAGMENTS CONTAIN SEQUENCES FROM ONE DNA FAMILY (EEEO1)

To select the major components of the *Bam* HI fragments, these DNA fragments were isolated from total *Bam* HI digests of germ line DNA from type A of *E. okinoseanus* after gel electrophoresis and inserted into a plasmid vector. After transformation of the *Escherichia coli* host with the ligated vector, and screening by colony hybridization using the *Bam* HI fragments as probes, the recombinant plasmids of positive clones were isolated and their insert-DNAs were sequenced. All sequenced plasmid clones had from one to four *Bam* HI fragments as inserts. Their nucleotide sequences revealed that these fragments can be divided into two families in terms of length and sequence. The length of members from one family was about 79 bp, and that from another family was about 95 bp. DNA filter hybridization experiments using these plasmid clones as probes gave results that agreed with those obtained using the *Bam* HI fragments as probes. These results suggest the two DNA families are the major components of the *Bam* HI fragments.

Since in one sequenced clone the insert DNA had a sequence that encompassed both of these two sequences of the two DNA families, analysis with a restriction endonuclease, *Eco* RV, whose recognition site does not exist in the sequence of one family, but in that of the other, and subsequent DNA filter hybridization experiments, using sequences from the two DNA families were carried out. As a result, identical hybridization patterns that were limited to the germ line DNA for both types of *E. okinoseanus* were obtained. Two clear signals, corresponding to fragments whose sizes were estimated to be about 85 and 180 bp, were generated. This result indicated that the two DNA families were generated by digestion with *Bam* HI from one repeating unit.

The nucleotide sequences of the members in this family which comprise the major components of the *Bam* HI fragments, designated EEEo1 (for Eliminated Element of *E. okinoseanus* 1), differed by single-base substitutions and insertions or deletions, but the sequences were generally very similar to one another. The G+C content of the consensus sequence of this family (Figure 6.4) was 47.2% and the average divergence was 12.1 bp/174 bp (7.0%).

EEEo1 : 5′- *g* *GATCC*GGATTCGAATCCGCACCAAAATCGAATCACTTCTTCCTTGTCGCA CACCCAACGTGTACACCAAGTTTCGTCC*GGATCC*GTCCACAACTTTTTGA *GATATC*GTGCGATATGTCTGTTTAGGCCAATCTCTCGATGGTGAAGATTC GCCGAAACAAAATTATCAGATCC*G* *gatcc* -3′

(*Bam* HI, *Bam* HI, *Eco* RV, *Bam* HI sites marked; numbering 1–170)

EEEo2 : 5′- *ttt* AAAACGTCTTGTGGAAGTGATCTCTTCCTCTCTTTTTGGTGTCAGGAGTG GGAACTTTTCATTGCTCTTCACCAGAAGTGTT*TTT* *aaa* -3′

(*Dra* I sites marked; numbering 1–80)

Figure 6.4 Consensus nucleotide sequence of the major components of the *Bam* HI fragments, namely EEEo1 (top) and that of the *Dra* I fragment, namely EEEo2 (bottom). In EEEo1, one direct-repeat dyad is shown by one pair of open arrows below the consensus sequence. The sequence could be subdivided into four subrepeats as a hexanucleotide with some variations. Underlined sequences demonstrate the same motif of the subrepeats. Double underlined sequences demonstrate the other small direct repeats which were also found as heptanucleotide. One direct or inverted-repeat and one inverted-repeat dyads are shown by two pairs of solid arrows below the consensus sequence. In EEEo2, two inverted-repeat dyads are shown by two pairs of arrows below the consensus sequence. The nucleotide sequences of EEEo1 appear in the DDBJ, EMBL and Genbank Nucleotide Sequence Databases with the accession numbers D13915 (short *Bam* HI fragment as 79 bp) and D10254, D10255 and D10256 (long *Bam* HI fragment as 95 bp), respectively. The nucleotide sequence of EEEo2 appears in the DDBJ, EMBL and Genbank Nucleotide Sequence Databases with the accession number D12819.

The sequences had no continuous ORF (open reading frame) on either strand in any frame. Computer analysis of the sequences revealed the presence of one direct-repeat, one direct or inverted-repeat and one inverted-repeat dyad. The direct repeats, which were located between bases 168 and 22, and between bases 69 and 92 of the respective sequences, could be subdivided into four subrepeats, which all seem to be variants of the same sequence motif. The prototype sequence that could be deduced was a hexanucleotide with the sequence 5'GGATCC3'. The same motif was found, with some variations, between bases 29 and 33, and between bases 145 and 150 (underlined sequences in Figure 6.4). In addition, a similar small direct repeat as a heptanucleotide with the sequence 5'ACACCAA3' was also found with some variations (double underlined sequences in Figure 6.4). The sequence, which was located between bases 99 and 108, could be recognized as a direct or inverted-repeat dyad and the sequence which was located between bases 120 and 159, could be recognized as an inverted-repeat dyad.

6.8 THE *DRA* I FRAGMENT CONTAINS SEQUENCES FROM ONE DNA FAMILY (EEEO2)

To select the major components of the *Dra* I fragment, this DNA fragment isolated from germ line DNA in type A of *E. okinoseanus*, was also inserted into a plasmid vector and subsequent transformation, screening and sequencing were carried out. All sequenced plasmid clones had one *Dra* I fragment of about 85 bp long as an insert. Nucleotide sequences revealed that this fragment represents a single family in terms of length and sequence. This family is the major component of the *Dra* I fragment, because DNA filter hybridization experiments using the insert DNA of a plasmid clone from this family as probe gave results that were consistent with the results obtained using the *Dra* I fragment itself as probe. The nucleotide sequences of the members of this family, designated EEEo2, differed by single-base substitutions and insertions or deletions, but the sequences were generally very similar to one another. The G+C content of the consensus sequence was 41.2%. The average divergence was 1.13 bp/85 bp (1.3%). The sequences had no continuous ORF on either strand in any frame. Computer analysis of the sequences revealed the presence of two inverted-repeat dyads, which were located between bases 11 and 30, and between bases 72 and 14 of the respective sequences (Figure 6.4).

The sequences of EEEo1 and EEEo2 were used to search for homologous sequences in the EMBL nucleotide sequence database (Rel 23, 1990) and the LASL (GenBank) nucleotide sequence database (Rel 64, 1990), but no significant homologies were recognized.

6.9 EEEO1 AND EEEO2 ARE LOCATED ON GERM LINE-RESTRICTED CHROMOSOMES

To investigate the localization of EEEo1 and EEEo2, fluorescence *in situ* hybridization to metaphase chromosomes of somatic cells (cultured blood) and germ cells (testis) from both types of *E. okinoseanus* was carried out using the cloned DNA fragments, which belong to EEEo1 and EEEo2 as probes. Some metaphase plates from testes were analysed by a modified C-banding technique after hybridization.

Figures 6.5a and 6.5b show examples of metaphase chromosomes hybridized with EEEo1 as probe. Fluorescent signals were only observed in germ cells from both types of *E. okinoseanus*, and no signals were observed in somatic cells. In each metaphase from germ cells of both types, signals were concentrated on several germ line-restricted chromosomes which were C-band positive along almost their entire length and tended to form secondary associations during the primary meiotic metaphase of the spermatocyte.

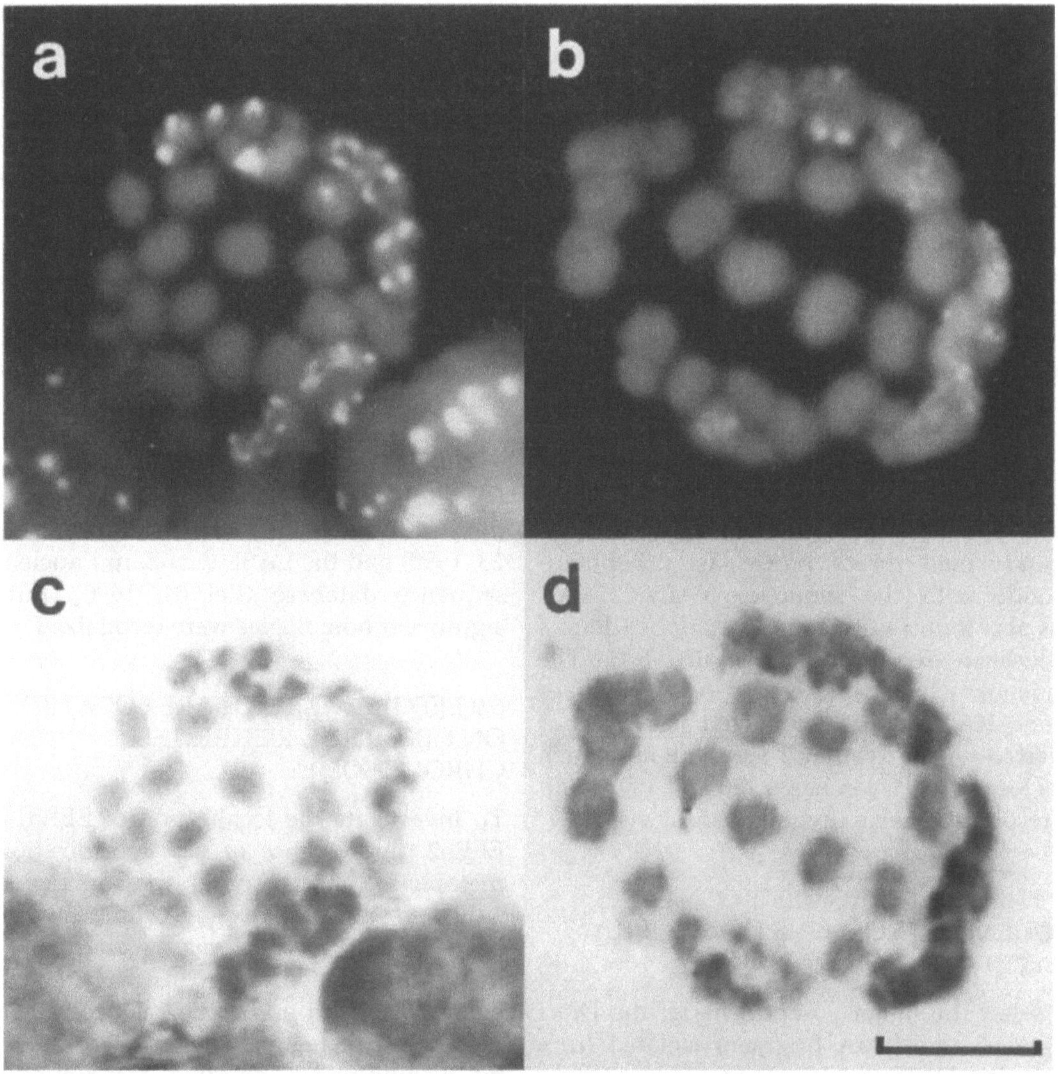

Figure 6.5 Metaphases of primary spermatocytes before ((a) and (b)) and after ((c) and (d)) modified C-banding treatment, from type A of E. *okinoseanus*. Each metaphase was hybridized with EEEo1 ((a) and (c)) and with EEEo2 ((b) and (d)), as probes. Signals of hybridization are detected as fluorescence of fluorescein isothiocyanate (FITC) and chromosomes are counterstained by propidium iodide ((a) and (b)). Bar represents 10μm.

As shown in Figures 6.5c and 6.5d, similar results were obtained when the EEEo2 was used as probe. Fluorescent signals were only observed on germ line-restricted, small chromosomes which were C-band positive. They did not appear on somatic cells. The intensity of fluorescent signals obtained with EEEo2 was clearly lower than that obtained with EEEo1. However, signals were numerous and dispersed on each chromosome as compared with those obtained with EEEo1.

6.10 PERSPECTIVE VIEW OF THE ELIMINATED ELEMENTS

The origin of repetitive DNA sequences within the genome undoubtedly involves some form of amplification (Smith, 1974, 1976, 1978). However, the actual processes of amplification and dispersal of a sequence, and its maintenance in the genome are less clear. Because of the method of selection, the DNA sequences, which are shown in Figure 6.4, are expected to be representative of the most abundant sequences in the two DNA families. According to the neutral theory of molecular evolution, nucleotide substitutions take place inherently in DNA as a result of point mutations that are followed by random genetic drift (Kimura, 1983), The extent of divergence in the respective sequences of the members in the two families appears to represent their ages. But EEEo2 has a very small value (1.3%) for the divergence by comparison with EEEo1 (7.0%). This observation implies that it is the difference pressure of molecular drive (Dover, 1982) that mediates concerted evolution (Arnheim, 1983) between the two DNA families, although the two families are distributed on identical chromosomes. Concerted evolution has been described as a process whereby a certain DNA sequence is amplified, homogenized throughout the genome, and distributed among both homologous and non-homologous chromosomes (Krystal *et al.*, 1978; Dover, 1982; Hamilton *et al.*, 1990, 1992). Several mechanisms, including unequal crossing over, gene conversion, sequence transposition and rolling circles, have been observed or postulated in certain genomes (Dover, 1982; Walsh, 1987), and these mechanisms may act together or independently during the process of concerted evolution. The concerted evolution of the two repetitive elements in *E. okinoseanus* may have been due to recombination on the secondary associations during the primary meiotic metaphase of the spermatocyte and the differences of intra-divergence of the two repetitive elements can be explained by their different patterns of chromosomal distribution. That is to say, DNA sequences which belong to EEEo1 appeared to be concentrated on a few limited regions of each C-band-positive chromosome. DNA sequences that belong to EEEo2 appeared to be dispersed along almost the entire length of each C-band-positive chromosome (Figure 6.5). The direct repeats which were located in EEEo1 could be subdivided into 4 hexanucleotide subrepeats, and the subrepeats with some variations were dispersed among EEEo1. In addition, similar small direct repeats as a heptanucleotide were also found with some variations. Such features in terms of nucleotide sequence may reflect saltatory replication (Britten and Kohne, 1968) in the evolution of this repetitive sequence. The scheme supported by Southern (1975), involves a series of steps in which repeats with periodic structure were generated successively from a small sequence.

The functional properties of the two germ line-restricted DNA elements from *E. okinoseanus* are still obscure. EEEo1 and EEEo2 appear to have no continuous ORFs. It may be that they have no biological functions as protein-coding elements in the germ line-restricted parts of the genome. Vogt (1990) hypothesized a potential genetic function of tandemly repeated sequence-blocks as an ability to develop and maintain a specific chromatin folding structure. From the present sequence analyses (Figure 6.4) it was demonstrated that the two repetitive sequences each have two inverted-repeat dyads. One possibility is that the regions of palindromic sequences function as binding sites for a protein that is necessary for chromosome elimination. Palindromes have frequently been found to play an important role in binding of proteins (Lee *et al.*, 1987; Halazonetis and Kandil, 1991). Besides, Bigot *et al.* (1990) suggested a functional role for palindromes in the amplification of repetitive DNA in two wasp species. Studies on the PSR chromosome of *Nasonia vitripennis* were also suggestive of a causal relationship between palindromic sequences and recombination events (Reed *et al.*, 1994). However, the random distribution of the point mutations in the respective sequences of the two families suggests that there is no strong selective pressure on any particular region within any of the sequences, and the diversity in the respective sequences of the two families appears to represent their concerted manner, as mentioned above.

In type A of *E. okinoseanus*, EEEo1 and EEEo2 account for about 2.6% and 6.2% of the total germ line genome, respectively. Thus, EEEo1 and EEEo2 account for only about 19% of the eliminated DNA in type A of this species. In recent molecular studies using *P. atami*, *E. cirrhatus*, *E. stoutii* and *M. glutinosa* in addition to the three Japanese hagfish species, EEEo2 seems to be more widely distributed over the germ line genomes of three hagfish species, *E. okinoseanus*, *E. cirrhatus* and *P. atami*. This evidence may provide a clue as to the origin of this shared repetitive element, EEEo2. Moreover, other eliminated elements which have highly repetitive sequences have been detected from some species, and analysed (Kubota, unpublished data).

6.11 CONCLUDING REMARKS

In general, one of the most obvious characteristics of eliminated chromatin (and/or chromosomes) is their heterochromatic nature (Hennig, 1986). Hence any assessment of the biological role of germ line-restricted parts of the genome inevitably leads to the question of the nature and biological function of heterochromatin. The possibility that heterochromatin may be 'junk' was raised by Hennig (1986) and Pardue and Hennig (1990). The concept that some genetic elements have no function other than their own replication or transmission – in other words, that they are 'junk', 'parasitic', or 'selfish' – has attracted a great deal of interest (Doolittle and Sapienza, 1980; Orgel and Crick, 1980). The significance at the molecular level of chromosome elimination in Japanese hagfishes is still a complete mystery. However, if these DNA elements are not 'junk' and if the programmed DNA rearrangement as elimination of heterochromatin does have any function, this function may be gene regulation of such phenomena as position effect, as hypothesized by Tobler *et al.* (1992), Goday and Pimpinelli (1993), Spradling (1993) and Zuckerkandl and Hennig (1995). Further studies of chromosome elimination in hagfishes are clearly necessary to address these issues.

ACKNOWLEDGEMENTS

We are grateful to Prof. J.-O. Strömberg and the staff of the Kristineberg Marine Biological Station, Dr S. Nilsson in University of Göteborg, Sweden, Dr W. Davison and Mr J. van Berkel of the Kaikoura Marine Station, University of Canterbury, New Zealand, Dr A. Gorbman, University of Washington, USA and the staff of the Bamfield Marine Station, Canada, Mr Y. Ono and coworkers at Ono Suisan Ltd, Onahama, Japan, the staff of the Misaki Marine Biological Station of University of Tokyo, Japan, Drs L. Liu and C. Kuo in Sun Yat-sen University, Taiwan, for supplying us with experimental materials. We are also grateful to Dr M. Kuro-o in Hirosaki University for his kind suggestions and helpful advice. We are also grateful to Mr M. Shichiri, Mr Y. Goto, Mr T. Ishibashi and Ms N. Sato for their providing data for this work. This work was supported in part by a Grant-in-Aid for Scientific Research from the Ministry of Education, Science and Culture, Japan (Nos. 63540508, 04640595 and 064633 for JSPS Research Fellow) and Scientific Exchange Program between Japan and Sweden in 1992.

REFERENCES

Adam, H. and Strahan, R. (1963) Systematics and distribution, in *The Biology of Myxine* (eds

A. Brodal and R. Fänge), Universitetsforlaget, Oslo, Norway, pp. 1–8.

Ashihara, T., Kamachi, M., Urata, Y., Kusuzaki, K., Takeshita, H. and Kagawa, K. (1986) Multiparametric analysis using autostage cytofluorometry. *Acta Histochemistry and Cytochemistry*, **19**, 51–9.

Arnheim, N. (1983) Concerted evolution of multigene families, in *Evolution of Genes and Proteins* (eds M. Nei and R.K. Koehn), Sinauer, Sunderland, Massachusetts, pp. 38–61.

Baker, R.J. and Bickham, J.W. (1980) Karyotypic evolution in bats: evidence of extensive and conservative chromosomal evolution in closely related taxa. *Systematic Zoology*, **29**, 239–53.

Bigot, Y., Hamelin, M.-H. and Periquet, G. (1990) Heterochromatin condensation and evolution of unique satellite-DNA families in two parasitic wasp species: *Diadromus pulchellus and Eupelmus vuilleti* (Hymenoptera). *Molecular Biology and Evolution*, **7**, 351–64.

Britten, R.J. and Kohne, D.E. (1968) Repeated sequences in DNA. *Science*, **161**, 529–40.

Boveri, T. (1887) Über Differenzierrung der Zellkerne während der Furchung des Eies von *Ascaris megalocephala*. *Anatomischer Anzeiger*, **2**, 688–93.

Carlton, M.S. and Denton, T.E. (1974) Chromosomes of the Chocorate Grami: a cytogenetic anomaly. *Science*, **185**, 616–19.

Comings, D.E. and Okada, T.A. (1972) Holocentric chromosomes in *Oncopeltus*: kinetochore plates are present in mitosis but absent in meiosis. *Chromosoma*, **37**, 177–92.

Doolittle, W.F. and Sapienza, C. (1980) Selfish genes, the phenotype paradigm and genome evolution. *Nature*, **284**, 601–03.

Dover, G. (1982) Molecular drive: a cohesive mode of species evolution. *Nature*, **299**, 111–17.

Goday, C. and Pimpinelli, S. (1993) The occurrence, role and evolution of chromatin diminution in Nematodes. *Parasitology Today*, **9**, 319–22.

Halazonetis, T.D. and Kandil, A.N. (1991) Determination of the c-MYC DNA-binding site. *Proceedings of the National Academy of Sciences ot the USA*, **88**, 6162–66.

Hamilton, M.J., Honeycutt, R.L. and Baker, R.J. (1990) Intragenomic movement, sequence amplification and concerted evolution in satellite DNA in harvest mice, *Reithrodontomys*: evidence from in situ hybridization. *Chromosoma*, **99**, 321–9.

Hamilton, M.J., Hong, G. and Wichman, H.A. (1992) Intragenomic movement and concerted evolution of satellite DNA in *Peromyscus*: evidence from in situ hybridization. *Cytogenetics and Cell Genetics*, **60**, 40–4.

Hardisty, M.W. (1979) *Biology of the Cyclostomes*, Chapman & Hall, London.

Hennig, W. (1986) Heterochromatin and germ line-restricted DNA, in *Germ Line-Soma Differentiation, Results and Problems in Cell Differentiation 13* (ed. W. Hennig), Springer, Berlin, Heidelberg, pp. 175–92.

Kimura, M. (1983) *The Neutral Theory of Molecular Evolution, Cambridge University Press, Cambridge.*

Kitada, J. and Tagawa, M. (1975) Somatic chromosomes of three species of Cyclostomata. *Chromosome Information Service*, 18, 10–12.

Krystal, M., Eustachio, P.B., Ruddle, F.H. and Arnheim, N. (1981) Human nucleolus organizers on homologous chromosomes can share the same ribosomal gene variants. *Proceedings of the National Academy of Sciences ot the USA*, **78**, 5744–8.

Kohno, S., Nakai, Y., Satoh, S., Yoshida, M. and Kobayashi, H. (1986) Chromosome elimination in Japanese hagfish, *Eptatretus burgeri* (Agnatha, Cyclostomata). *Cytogenetics and Cell Genetics*, **41**, 209–14.

Kubota, S., Nakai, Y., Kuro-o, M. and Kohno, S. (1992) Germ line-restricted supernumerary (B) chromosomes in *Eptatretus okinoseanus*. *Cytogenetics and Cell Genetics*, **60**, 224–8.

Kubota, S., Kuro-o, M., Mizuno, S. and Kohno, S. (1993) Germ line-restricted, highly repeated DNA sequences and their chromosomal localization in a Japanese hagfish (*Eptatretus okinoseanus*). *Chromosoma*, **102**, 163–73.

Kubota, S., Nakai, Y., Sato, N., Kuro-o, M. and Kohno, S. (1994) Chromosome elimination in northeast Pacific hagfish, *Eptatretus stoutii* (Cyclostomata, Agnatha). *Journal of Heredity*, **85**, 413–15.

Kuo, C., Huang, K. and Mok, H. (1994) Hagfishes of Taiwan (I): a taxonomic revision with description of four new *Paramyxine* species. *Zoological Studies*, **33**, 126–39.

Lee, W., Haslinger, A., Karin, M. and Tjian, R. (1987) Activation of transcription by two factors that bind promoter and enhancer sequences of human metallothionein gene and SV40. *Nature*, **325**, 368–72.

Makino, S. (1951) *Atlas of Chromosome Numbers in Animals*, Iowa State College Press, Ames.

Mazzini, G., Giordano, P., Montecucco, C.M. and

Riccardi, A. (1980) A rapid cytofluorometric method for quantitative DNA determination on fixed smears. *Histochemical Journal*, **12**, 153–68.

Müller, F., Walker, P., Aeby, P., Neuhaus, H., Felder, H., Back, E. and Tobler, H. (1982) Nucleotide sequence of satellite DNA contained in the eliminated genome of *Ascaris lumbricoides*. *Nucleic Acids Research*, **10**, 7493–510.

Nakai, Y., Kohno, S. (1987) Elimination of the largest chromosome pair during differentiation into somatic cells in Japanese hagfish, *Myxine garmani* (Cyclostomata, Agnatha). *Cytogenetics and Cell Genetics*, **45**, 80–3.

Nakai, Y., Kubota, S. and Kohno, S. (1991) Chromatin diminution and chromosome elimination in four Japanese hagfish species. *Cytogenetics and Cell Genetics*, **56**, 196–8.

Nakai, Y., Kubota, S., Goto, Y., Ishibashi, T., Davison, W. and Kohno, S. (1995) Chromosome elimination in three Baltic, south Pacific and north-east Pacific hagfish species. *Chromosome Research*, **3**, 321–30.

Nogusa, S. (1960) A comparative study of the chromosomes in fishes with particular considerations on taxonomy and evolution. *Memoirs of the Hyogo University of Agriculture*, **3**, 1–68.

Nygren, A. and Jahnke, M. (1972) Cytological studies in *Myxine glutinosa* (Cyclostomata) from the Gullmaren Fjord in Sweden. *Swedish Journal of Agriculture Research*, **2**, 83–8.

Ojima, Y. (1983) *Fish Cytogenetics (in Japanese)*, Suikohsha, Tokyo.

Ono, T. and Obara, Y. (1994) Karyotypes and Ag-NOR variations in Japanese vespertilionid bats (Mammalia: Chiroptera). *Zoological Science*, **11**, 473–84.

Orgel, L.E. and Crick, F.H.C. (1980) Selfish DNA: the ultimate parasite. *Nature*, **284**, 604–7.

Patau, K. (1952) Absorption microphotometry of irregular-shaped objects. *Chromosoma*, **5**, 341–62.

Pardue, M.L. and Hennig, W. (1990) Heterochromatin: junk or collectors item? *Chromosoma*, **100**, 3–7.

Reed, K.M., Beukeboom, L.W., Eickbush, D.G. and Werren, J.H. (1994) Junctions between repetitive DNAs on the PSR chromosomes of *Nasonia vitripennis*: association of palindromes with recombination. *Journal of Molecular Evolution*, **38**, 352–62.

Retzius, G. (1890) Über Zellenteilung bei *Myxine glutinosa*. *Biol Fören* (Stockholm), *Förhandl*, **2**(8), 80–91.

Schreiner, A. and Schreiner, K.E. (1904) Über die Entwicklung der mannlichen Geschleehtszellen von *Myxine glutinosa* (L.). I. Vermehrungsperiode, Reifungsperiode und Reifungsteilungen. *Archives de Biologie*, **21**, 183–355.

Shi, L., Ye, Y. and Duan, X. (1980) Comparative cytogenetic studies on the red muntjac, Chinese muntjac and their F1 hybrids. *Cytogenetics and Cell Genetics*, **26**, 22–7.

Smith, G.P. (1976) Evolution of repeated DNA sequences by unequal crossover. *Science*, **191**, 528–35.

Smith, G.P. (1978) What is the origin and evolution of repetitive DNAs? *Trends in Biochemical Science*, **3**, 34–6.

Southern, E.M. (1975) Long range periodicities in mouse satellite DNA. *Journal of Molecular Biology*, **94**, 51–69.

Spradling, A.C. (1993) Position effect variegation and genomic instability. *Cold Spring Harbor Symposia Quantitative Biology*, **58**, 585–96.

Streeck, R.E., Moritz, K.B. and Beer, K. (1982) Chromatin diminution in *Ascaris suum*: nucleotide sequence of the eliminated satellite DNA. *Nucleic Acids Research*, **10**, 3495–502.

Sumner, A.T. (1972) A simple technique for demonstrating centromeric heterochromatin. *Experimental Cell Research*, **75**, 304–06.

Taylor, K.M. (1967) The chromosomes of some lower chordates. *Chromosoma*, **21**, 181–8.

Tobler, H. (1986) The differentiation of germ and somatic cell lines in Nematodes, in *Germ Line-Soma Differentiation, Results and Problems in Cell Differentiation 13,* (ed. Hennig, W.), Springer, Berlin Heidelberg, pp. 1–69.

Tobler, H., Etter, A. and Müller, F. (1992) Chromatin diminution in nematode development. *Trends in Genetics*, **8**, 427–31.

Vogt, P. (1990) Potential genetic functions of tandem repeated DNA sequence blocks in the human genome are based on a highly conserved 'chromatin folding code'. *Human Genetics*, **84**, 301–36.

Walsh, J.B. (1987) Persistence of tandem arrays: Implications for satellite and simple-sequence DNAs. *Genetics*, **115**, 553–67.

Zuckerkandl, E. and Hennig, W. (1995) Tracking heterochromatin. *Chromosoma*, **104**, 75–83.

7
THE TUMOUR PATHOLOGY OF *MYXINE GLUTINOSA*

Sture Falkmer

SUMMARY

In really large specimens of the Atlantic hagfish, a whole spectrum of tumour nodules occurs in the liver; their incidence amounts (nowadays) to about 1–2%. Small whitish spots represent nodular hyperplasias; slightly larger ones benign adenomas; the really big nodules (0.5–1 cm in diameter) genuine adenocarcinomas. They can be of both hepatocellular and cholangiocellular types, sometimes in the same gross tumour nodule. There is no relationship to liver cirrhosis. So far, no metastatic lesions have been discovered.

Second in incidence come tumours of the islet organ. There, too, a broad spectrum of histopathologic pictures is present; several of the tumour-like lesions just represent non-neoplastic hamartomas; others are genuine adenomas or even outspoken adenocarcinomas. No signs of metastatic dissemination have been detected here, either.

7.1 INTRODUCTION

The Atlantic hagfish, *Myxine glutinosa*, occupies a pivotal position, not only in the pylogenetical evolution of the gastro-entero-pancreatic neuroendocrine system (cf. Thorndyke and Falkmer, this volume), but also in comparative oncology (Dawe, 1973, Falkmer *et al.*, 1976; Dawe *et al.*, 1981; Falkmer and Grimmelikhuijzen, 1981). The same statement also applies to other cyclostomes, both hagfishes and lampreys (Falkmer *et al.*, 1974; Hardisty, 1976). Of particular interest here is the fact that the Atlantic hagfishes belong to those aquatic bottom-feeder animals that are at the end of the nutritional chain; their incidence of liver cancer can serve as important 'danger signals' for carcinogenic pollution of our aquatic environment (Falkmer *et al.*, 1977; Falkmer and Grimmelikhuijzen, 1981). As a matter of fact, a dramatic fall during these last few decades in the incidence of primary liver carcinomas of those hagfishes caught inside the threshold at the mouth of the Gullmar Fjord (at the Kristineberg Marine Research Station at Fiskebäckskil) in Sweden—but not outside it—suggests that a pollution with chlorinated pesticides might have been an aetiologic/pathogenetic agent (Falkmer *et al.*, 1977; Falkmer and Grimmelikhuijzen, 1981).

7.2 GROSS ASPECTS

7.2.1 Lesions in the liver

When the viscera of the Atlantic hagfish of decent size (body lengths around 35–40 cm; body weights about 65–85 g) are dissected, it is not at all rare to find whitish tumour-like lesions on the surface of one or both of the two liver lobes (Figure 7.1a). Their size can vary from that of a barely visible whitish spot, about 1 mm in diameter, to that of a big nodule, measuring as much as 1 cm in width, occupying a large part of a liver lobe. Even

The Biology of Hagfishes. Edited by Jørgen Mørup Jørgensen, Jens Peter Lomholt, Roy E. Weber and Hans Malte. Published in 1998 by Chapman & Hall, London. ISBN 0 412 78530 7.

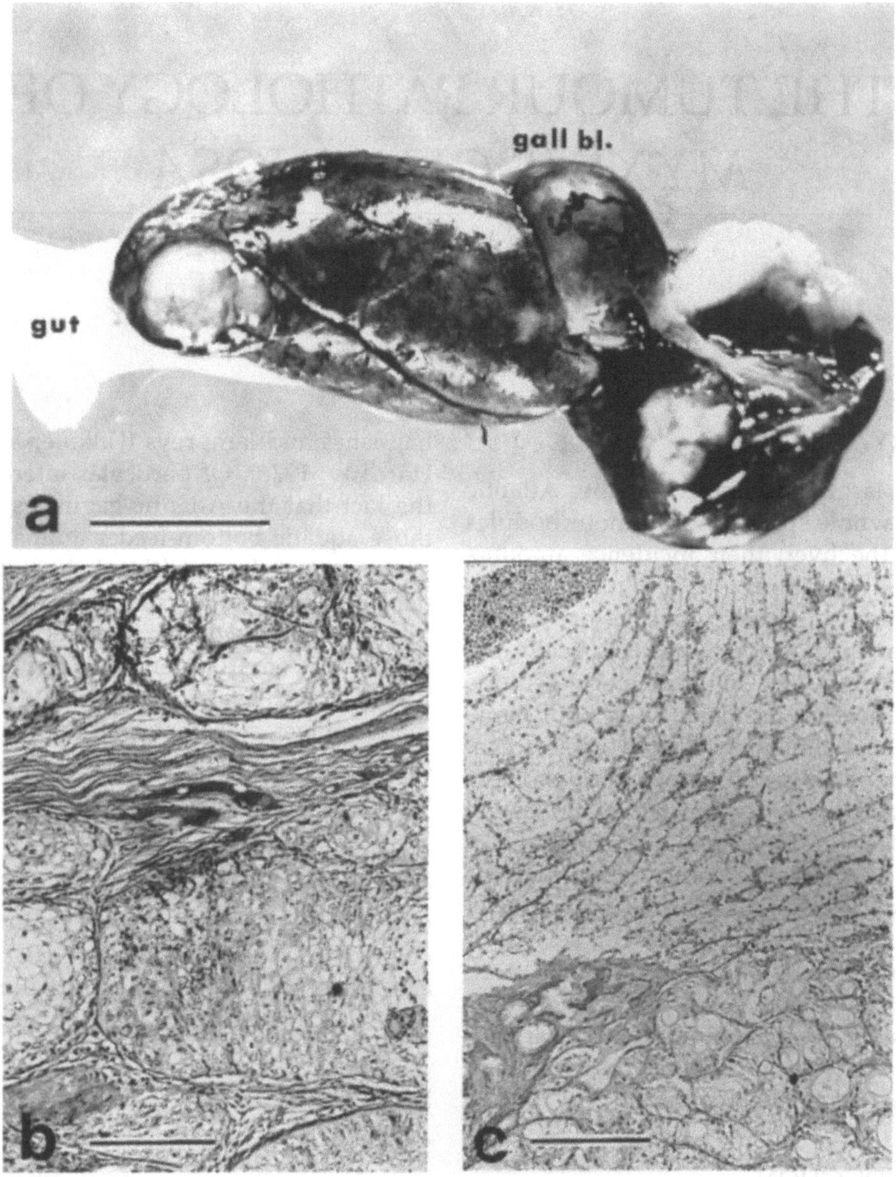

Figure 7.1 Slightly enlarged photograph (figure (a)) and medium-power photomicrographs (figures (b) and (c)) of hagfish liver carcinomas. The liver in figure (a) contains large, whitish tumour nodules, both in the caudal lobe (left) and two in the cranial lobe (right); in addition, there are a few minute whitish spots barely visible in the caudal lobe (scale bar 10 mm). The hisotopathologic picture of the tumour in figure (b) is that of rather highly differentiated, primary liver carcinoma of hepatocellular type with a marked predominance of clear cells, invading also the surrounding connective tissue stroma (scale bar 100 µm) (×200). The tumour shown in figure (c) displays histopathologically a primary neoplasm of the liver, consisting of both a hepatocellular carcinoma of clear-cell type (as in figure (b)) (upper 2/3 of the photomicrograph) and a cholangiocellular adenocarcinoma (lower 1/3 of the photomicrograph). Although the tumour nodule appeared homogenous, there was a rather distinct boundary between its two microscopic components (scale bar 200 µm) (×100).

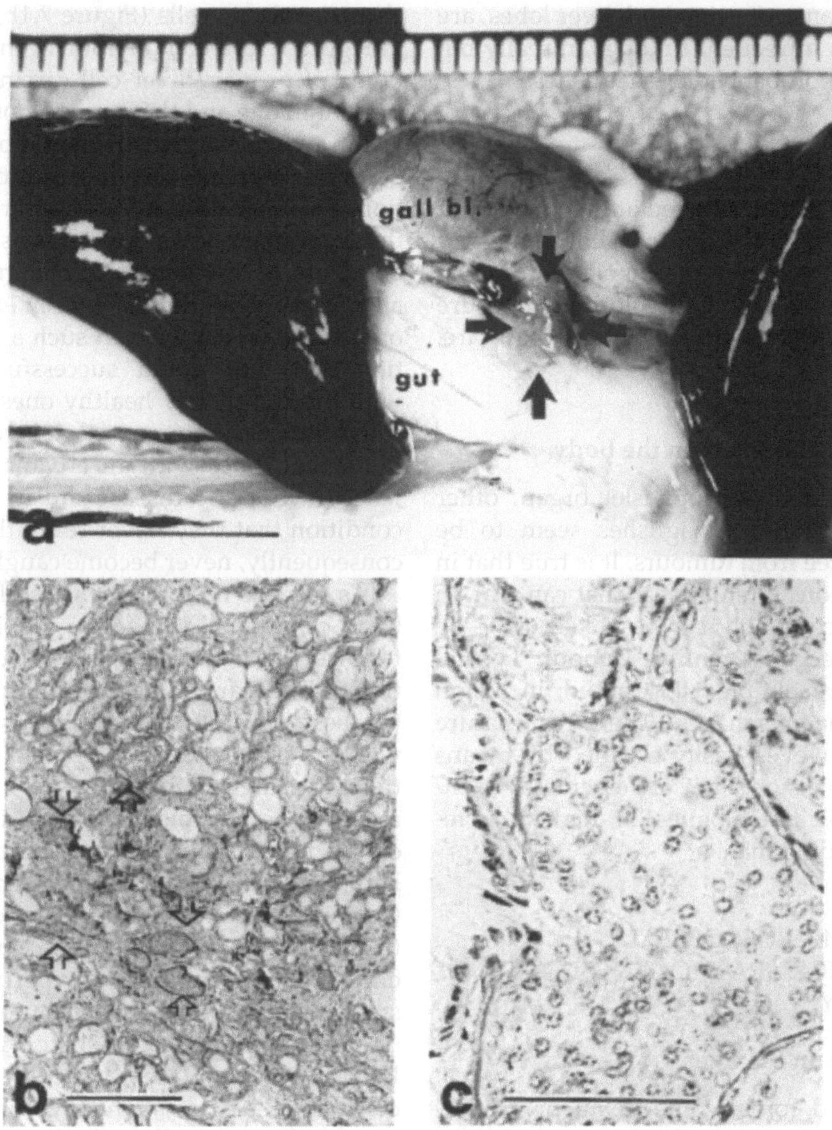

Figure 7.2 One slightly enlarged (mm scale in the top border) photograph (figure (a)), one low-power photomicrograph (figure (b)) and one photomicrograph of higher magnification (figure (c)) of tumours and tumour-like lesions of the hagfish islet organ. The islet organ (arrows) shown in figure (a) is enlarged and markedly discoloured. Normally, it is a small (1–2 mm in diameter) structure, appearing as distinctly whitish, swelling at the junction of the short bile duct in the gut (see Thorndyke and Falkmer in this volume). (Scale bar 10 mm.) The low-power microscopic appearance of this islet tumour (figure (b)) indicates a fairly high differentiation of its parenchyma, mainly consisting of densely packed cysts of normal appearance, filled with mucus and cell debris. There are, however, also small solid islets of parenchymal cells (arrows) among all these duct-like, partly cystic structures. A close-up of such an islet (figure (c)) displays a fairly normal appearance of the parenchyma; immuno-histochemically it was found to consist of 99% insulin cells. The lack of cellular atypia justifies a histopathological diagnosis of a hamartoma (i.e. a tumour-like malformation). (b) Scale bar 200 μm (×100); (c) scale bar 100 μm (×350).

multiple lesions, affecting both liver lobes, are by no means infrequent findings in really big – and, hence, probably old – hagfishes.

7.2.2 Lesions in the islet organ

Also the islet organ, situated at the junction of the bile duct with the gut (cf. Thorndyke and Falkmer, this volume), can display a tumour-like enlargement with discoloration (Figure 7.2a). Such lesions of the islet organ are, however, rare.

7.2.3 Lesions elsewhere in the body

Apart from the liver and islet organ, other tissues or organs of hagfishes seem to be essentially free from tumours. It is true that in the skin, lesions often appear that can – at an initial inspection only by the naked eye – be misinterpreted as a genuine tumour. Typical examples are simple epidermoid inclusion cysts and haemorrhages. Their real nature becomes, however, easily clarified by means of a simple incision. They are essentially identical to their homologues in higher vertebrates, including man.

7.3 LIGHT MICROSCOPICAL APPEARANCES, PATHOGENESIS AND AETIOLOGY

7.3.1 Lesions in the liver

As accounted for in some detail previously (Falkmer *et al.*, 1976), the histopathological appearance of the liver tumours is closely related to their size and gross aspects. Thus, all the small whitish spots and nodules with a diameter of about 3 mm, or less, are nothing other than nodular hyperplasias or completely benign hepatocellular adenomas. In contrast, tumour nodules, as large as about 1 cm in diameter, are almost always outspoken, invasive adenocarcinomas. They can be of either purely cholangiocellular type, or purely hepatocellular, often with distinctly clear neoplastic cells (Figure 7.1b). As shown in Figure 7.1c, a combination of the two major kinds of primary liver cell carcinoma can be present in one and the same tumour nodule.

Intravascular growth has not yet been observed, nor has any metastatic dissemination. This fact does not, however, imply that these primary liver carcinomas would be benign or of low degree of malignancy. After all, the hagfishes that have been examined are only those which were in such a good condition that they could successfully compete with the completely healthy ones for the bait – and then become trapped. It is reasonable to think that those with an advanced, disseminated, neoplastic disease are in such a bad condition that they never reach the bait, and, consequently, never become caught.

As regards the pathogenesis of the hagfish liver carcinomas, it seems obvious that it follows the same stepwise development as that of liver carcinomas in higher vertebrates; the best example is perhaps its similarities with the sequential development of lesions in the experimental induction of liver carcinomas in mice by means of chlorinated pesticides; here, striking homologies exist (cf. Falkmer *et al.*, 1977). The spectrum consists of (i) minute, focal, hyperplastic lesions ('spots'); (ii) somewhat larger (2–3 mm), but still non-encapsulated, nodular hyperplasias; (iii) small (3–4 mm) but now genuine adenomas of highly differentiated hepatocytes, encapsulated by a thin membrane of collagenous connective tissue, slightly compressing the most adjacent cells of the surrounding, normal liver parenchyma; (iv) larger (4–5 mm) adenomas of the same type but now with a slight cellular and nuclear polymorphism of the hepatocytes; (v) still larger (5–10 mm) lesions, consisting of outspoken adenocarcinomas (as described above), non-encapsulated, with early invasion of the surrounding normal liver parenchyma. The fact that, in one and the same hagfish, the whole spectrum of these five types of neoplastic and pre-neoplastic lesions can be found, gives additional support for the probability of such a working hypothesis. No relationship with liver cirrhosis has been found.

With regard to the aetiology, it has been discussed whether or not some kind of chemical carcinogenesis might be operating (Falkmer *et al.*, 1977; Falkmer and Grimmelikhuijzen, 1981). As already mentioned in the Introduction, the main support for such an idea is the fact that the Atlantic hagfishes belong to the scavenger feeders – they live burrowed down in the mud at the bottom of the deep seas – and, consequently, are at the end ('top') of the nutritional chain. The rather drastic decrease in the incidence of these liver neoplasms in those hagfishes belonging to the population in the Gullmar Fjord on the Swedish west coast in connection with the installation of sewage plants in the Gullmar Fjord, and a ban by law of the use of chlorinated pesticides in Sweden in the early 1970s, has also been used as a support for such an aetiology. Lastly, there are direct mass-spectrometric chemical assays of chlorinated pesticides of the livers of hagfishes, both with and without tumours (Falkmer *et al.* 1977); the results do not violate the working hypothesis (Falkmer and Grimmelikhuijzen, 1981).

7.3.2 Lesions in the islet organ

Also in the islet organ, the microscopic picture shows a broad spectrum of histopathological lesions (Falkmer *et al.*, 1976). Some of the grossly enlarged islet organs are so highly differentiated that they must be classified as hamartomas, i.e. non-neoplastic, tumour-like malformations (Figure 7.2b). Some of them are partly cystic. On the other end of the spectrum are clear-cut adenocarcinomas, usually highly differentiated. Still, small remnants of islet parenchyma with insulin- and somatostatin-immunoreactive parenchymal cells can be found (Figure 7.2c).

Any vascular invasions, or any signs of metastatic dissemination, have not been observed, so far. The reasons are probably the same as those given above for the liver neoplasms.

ACKNOWLEDGEMENTS

This review is based on works supported by the Swedish Medical Research Council (Project No. 102), the King Gustaf V Jubilee Foundation, Stockholm, and the Research Funds of the Faculty of Medicine at the Karolinska Institute and those of Astra/Hässle Co., Sweden.

REFERENCES

Dawe, C.J. (1973) Comparative neoplasia, in *Cancer Medicine* (eds J.D. Holland and E. Frei, III), Lea & Febiger, Philadelphia, pp. 193–240.

Dawe, C.J., Harshberger, J.C., Kondo, S. *et al.* (eds) (1981) *Phyletic Approaches to Cancer*, Japan Scientific Society Press, Tokyo.

Falkmer, S., Thomas, N.W. and Boquist, L. (1974) Endocrinology of the cyclostomata. *Chem. Zool.*, **8**, 195–257.

Falkmer, S., Emdin, S.O., Östberg, Y. *et al.* (1976) Tumor pathology of the hagfish, *Myxine glutinosa*, and the river lamprey, *Lampetra fluviatilis*. A light microscopical study with particular reference to the occurrence of primary liver carcinoma, islet-cell tumours, and epidermoid cysts of the skin. *Progr. Exp. Tumor Res.*, **20**, 217–50.

Falkmer, S., Marklund, S., Mattson, P.E. *et al.* (1977) Hepatomas and other neoplasms in the Atlantic hagfish (*Myxine glutinosa*). A histopathologic and chemical study. *Ann. N.Y. Acad. Sci.*, **298**, 342–55.

Falkmer, S. and Grimmelikhuijzen, C.J.P. (1981) Phyletic aspects of the tumor pathology of the liver and the gastro-entero-pancreatic (GEP) neuroendocrine system (carcinoid-islet-cell tumors). In *Phyletic Approaches to Cancer* (eds C.J. Dawe, J.C. Harshbarger, S. Kondo *et al.*), Japan Scientific Society Press, Tokyo, pp. 333–46.

Hardisty, M.W. (1976) Cysts and tumour-like lesions in the endocrine pancreas of the lamprey (*Lampetra fluviatilis*). J. Zool., **178**, 305–17.

PART THREE

The Integument and Associated Glands

8
HAGFISH SKIN AND SLIME GLANDS

Robert H. Spitzer and Elizabeth A. Koch

SUMMARY

The scaleless hagfish employs two modes of secretion, holocrine by the slime glands and to a lesser extent merocrine by the epidermis for release of its voluminous viscous exudate. In both cases, large organized entities of intermediate filaments (IFs) are requisite participants. In the epidermis, each small mucous cell contains an IF-rich, basket-like structure ('capsule') which serves to compartmentalize the mucin-rich secretory vesicles within the apical region for subsequent release. Each epidermal thread cell contains at least one large IF-biopolymer ('thread') which may affect both the physical properties of the epidermis and the epidermal exudate. By contrast, the single IF-rich thread biopolymer precisely localized in each gland thread cell, interacts synergistically with mucins from the gland mucous cells to loosely organize water into a viscous mass of slime (stage 1), which, after physical perturbation, forms even more massive IF-aggregates ('cables') accompanied by the release of water (stage 2). The massive gland thread exhibits a linearly aligned IF/microtubule motif and is destined for extracellular export to function in an aqueous environment. Comparisons of the hagfish thread IF-polypeptide sequences (γ and α, with high threonine-contents) reported herein show no preferred identity to any other type of sequenced IF including those from higher vertebrates, a cephalochordate and invertebrates. Inasmuch as several keratin traits exist, we have characterized γ and α as homologues of type I and II epidermal keratins, respectively.

8.1 INTRODUCTION

The hagfish is often described as an eel-like, primitive marine vertebrate with the notable ability to rapidly release a massive volume of slime when disturbed. The enormity and nature of this secretory product and its cellular origins in the epidermis and slime glands have elicited the interest not only of very early investigators, but of contemporary researchers alike. The present chapter, which is limited mainly to discussions of the epidermis and slime glands, extends Theodor Blackstad's 1963 review on 'The skin and the slime glands' that included classic references such as Schreiner (1916, 1918) as well as his own seminal 'pilot studies' utilizing electron microscopy. Other important aspects of skin are discussed in detail elsewhere in the present monograph including dermis and connective tissue, dermal capillaries and skin sensory organs. Among prescient observations noted by Blackstad were that the epidermis and slime glands contain several fibrous proteins that might provide insight into the general properties of proteins, and that the results derived from the limited number of histochemical and microscopic techniques then available, established a basis for continued work of general biological importance. It is now clear that the fibrous proteins correspond to intermediate filaments (IFs), some of

The Biology of Hagfishes. Edited by Jørgen Mørup Jørgensen, Jens Peter Lomholt, Roy E. Weber and Hans Malte. Published in 1998 by Chapman & Hall, London. ISBN 0 412 78530 7.

which are atypical, that are bundled in unusual and differing arrays. The hagfish provides a particularly useful animal to study merocrine secretion by the scaleless epidermis (Spitzer *et al.*, 1979) and holocrine secretion by the numerous slime glands (Downing *et al.*, 1981a, b). The results have functional and evolutionary consequences (Koch *et al.*, 1994, 1995). In recent years, a wide spectrum of techniques has been employed to reveal important and interrelated morphologic, metabolic and functional features of hagfish tissue, often indicative of the co-action of intermediate filaments and mucins. In addition, the first complete deduced amino acid sequences of two slime gland intermediate filament (IF) polypeptides were determined by DNA sequencing.

Finally, there are two aspects of hagfish integument which are of peripheral interest to the present review but should be mentioned. First, the oft-stated view that the hagfish is of no commercial value is now incorrect because of the emergence of the 'eelskin' industry whereby the hagfish integument is fabricated into leather-like items, primarily by the Koreans (Gorbman *et al.*, 1990; Honma, this volume). Second, the early speculation that genuine neoplasia would not be found in hagfish was invalidated by Falkmer *et al.* (1976) who found both epithelial and mesenchymal neoplasms in *M. glutinosa*, with the frequency of tumours in skin far less than in liver.

8.1.1 Choice of species and animal care

Eptatretus stoutii (Pacific hagfish) and *Myxine glutinosa* (Atlantic hagfish) represent the species most frequently studied, followed by *Eptatretus burgeri* and *Paramyxine atami*, hagfishes proximal to Japan, and *Eptatretus cirrhatus* (New Zealand hagfish). Currently the species of choice is related primarily to availability rather than to a particular virtue of a given species. Our experience is mainly with *E. stoutii* obtained from coastal waters of California and maintained in 30–50 gallon aquaria (16±1 °C) with circulating seawater ('Instant Ocean') as described elsewhere (Spitzer *et al.*, 1979; Koch *et al.*, 1993). Among the studies cited below, the immediate history or mode of animal husbandry was not always detailed, and the consequences are unclear. This review examines publications from a wide range of journals reflecting the usefulness of the hagfish, and also includes data from several 'pilot' studies in our laboratories, and notes some promising areas for future research. The recent work of G. Zaccone and coworkers (1995) on the epidermis of the sea lamprey (*Petromyzon marinus*) is notable in that it offers another primitive vertebrate for an experimental animal as well as providing a basis for comparison between two disparate cyclostomes. By contrast to the hagfishes, the lampreys are devoid of slime glands.

8.1.2 Some definitions, terminology and abbreviations

Unlike most species of advanced fish, the total secretory product released by the hagfish integument includes both merocrine components from the epidermis and holocrine substances from the slime glands. The term 'mucus' will be used to refer to the collective product from the several major cell types of the epidermis while 'slime' will be reserved for the collective product from the gland mucous cells (GMCs) and gland thread cells (GTCs), albeit these terms are sometimes used interchangeably in the literature. When the animal is disturbed, the holocrine mode produces the bulk of secretory product. Though some definitions of mucus and mucins are ambiguous, a consensus view is that mucins are a group of polydisperse highly glycosylated (mainly O-linked) glycoproteins secreted by, but not limited to goblet cells and mucuous glands (Devine and McKenzie, 1992; Sheehan *et al.*, 1991). For comparative purposes, as well as the marked disparity in nomenclature, see reviews on cyclosome and teleost epidermis (Whitear, 1986) and mucus (Whitear, 1986; Shephard, 1993). Some abbreviations include: TEM, transmission electron microscopy; SEM, scanning electron microscopy; IF, intermediate filament; MT, microtubule; IFAP, intermediate filament-associated protein.

8.2 EPIDERMIS

The thickness of the scaleless epidermis, as gleaned from numerous reports, varies from 75 to 125 µm and is a function of species, body site, mode of animal husbandry and nutritional state. In the middorsal-lateral region of adult *E. stoutii* (>30 cm in length) the epidermis is composed mainly of three distinct epithelial cell types which are arranged imprecisely into 6–8 cell layers (Figure 8.1A). Electron micrographs reveal that these three cell types are tightly knitted together by both interdigitating membranes and desmosomes, unlike cells within the slime glands. The small mucous cells (SMCs) are the most numerous, and account for approximately 91% of all epidermal cells. The conspicuous IF-rich epidermal thread cells (ETCs) and vesicular-rich large mucous cells (LMCs) represent about 7% and 1%, respectively, of total cells and appear to be dispersed irregularly throughout the epidermis. SMCs are relatively uniform in size with an average length and width of about 15 × 8 µm. Dimensions of ETCs and LMCs vary widely, but sizes of 35 × 20 µm and 60 × 35 µm, respectively, are common. Sensory cells and ionocytes are observed infrequently and are discussed elsewhere. Mucous cells ('Schleimzelle') are also found among the epidermal cells surrounding the sensory cells in the barbels of *Myxine* but neither their function nor relationship to the SMCs in the skin is known (Georgieva *et al.*, 1979). The anal gland of *Myxine glutinosa* consists of cell types similar to the epidermis, but functions remain speculative (Kosmath *et al.*, 1981).

The nomenclature employed herein for hagfish epidermal cells is consistent with that of Blackstad (1963) and others, e.g. Lethbridge and Potter (1982), Patzner *et al.* (1982). However, Whitear (1986), when addressing the disparate nomenclature applied to epidermal cells from various teleosts and cyclostomes, noted the probable homology of 'mucous' and 'goblet' cells and emphasized that 'epithelial' is appropriate for an SMC, and goblet for an LMC.

An extensive dermal capillary network was delineated by light and electron microscopy in *Myxine glutinosa, Paramyxine atami, Eptatretus stoutii* and in several adult species of lampreys (Potter *et al.*, 1995). The elegant SEM images of vascular corrosion casts obtained by Lametschwandtner *et al.* (1986) serve to illustrate the complexity of vascular patterns in the hagfish and contrast differences in *Myxine glutinosa* and *Eptatretus stoutii*, particularly around the slime glands. It was suggested that an extensive dermal capillary network in the hagfish is requisite for rapid dispersion of precursors used in biosynthesis of mucus. In support of this view, quantitative light-and-electron-microscopic autoradiography has provided insight into some interrelated metabolic-morphologic events that occur in the principal epidermal cell types during cell division, differentiation, biosynthesis and release of mucus by *E. stoutii* (Spitzer *et al.*, 1979). Corroborative results have been attained by radiolabelling epidermal tissue *in vitro* and also *in vivo* after injection of the radiolabelled compounds (15–500 µCi in 0.2–0.5 ml seawater) into the caudal sinus (subcutaneous sinus) of fish lightly anaesthetized with 3-aminobenzoic acid ethyl ester methanesulphonate salt (Sigma Chem., USA).

The distribution of radiolabelled nucleic acids, one hour after administration of [^{3}H]-thymidine *in vivo*, was determined to identify sites of DNA synthesis. In *E. stoutii* about 98% of radiolabelled nuclei were in cells localized in the basal third of the epidermis (Figure

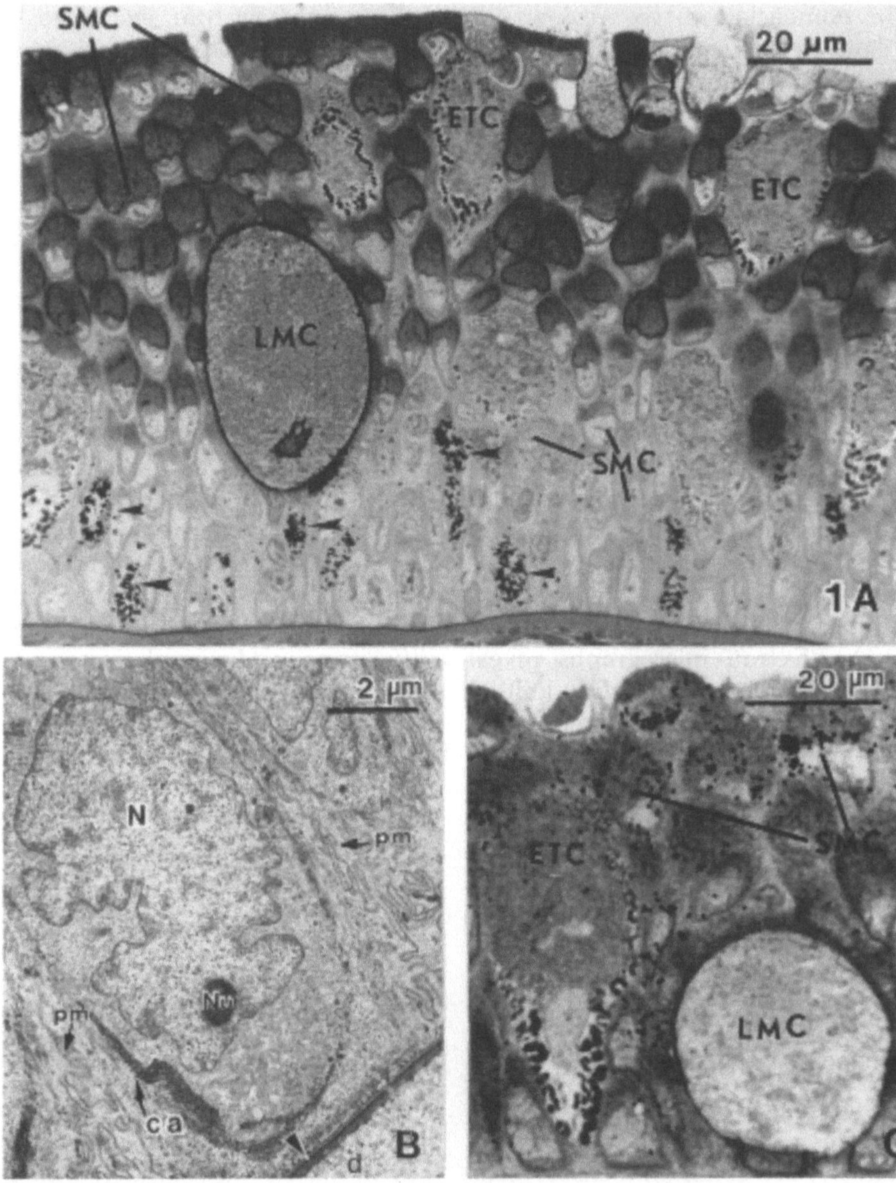

Figure 8.1 (A) Autoradiograph of cross-section through hagfish epidermis after 1 h incorporation of [^{3}H]-thymidine *in vivo*. Note three main cell types including large mucous cell (*LMC*), epidermal thread cell (*ETC*) and small mucous cell (*SMC*). Labelling (arrowheads) appears only in nuclei of cells in basal third of epidermis. Peripheral regions in ETCs appear darkly stained with the methylene blue-azure II-basic fuchsin reagent (MBBF) but do not represent radiolabelled areas. Several SMCs at epidermal surface have ruptured. ×960 (from Spitzer *et al.*, 1979, with permission of Springer-Verlag GmbH & Co. KG) (B) Electron micrograph of SMCs adjacent to basement membrane (arrowhead) with capsule *(ca)* already evident, nucleus *(N)*, nucleolus *(Nu)*, plasma membrane *(pm)* and dermis *(d)*. ×7200 (from Koch and Spitzer, unpublished). (C) Autoradiograph of epidermis after 45 min. incorporation of [^{3}H]-fucose *in vitro*. SMCs in distal third of epidermis show intense juxtanuclear labelling. ×1050 (from Spitzer *et al.*, 1979, with permission of Springer-Verlag GmbH & Co. KG).

8.1A). By contrast some, but not all, scaleless teleost species show labelling at all regions of the epidermis (Whitear, 1986; Spitzer *et al.*, 1979).

8.2.1 The small mucous cell (SMC): principal source of epidermal mucins

SMCs, the most abundant cell-type in hagfish epidermis, display a progressive and coordinated metabolic-morphologic view of mucigenesis as the cells differentiate from the basal epidermis to the epidermal surface (Spitzer *et al.*, 1979). Immature SMCs near the basement membrane (basal lamina) can be identified morphologically by the presence of a cytoplasmic structure called the capsule which is composed of densely packed intermediate filaments (Figure 8.1B). In addition, these immature SMCs contain a large nucleus and numerous mitochondria but little evidence of secretory vesicles. The radiolabelling pattern for SMCs in the basal third of the epidermis also reflects the expected host of biosynthetic activities for a mucous cell at early stages of differentiation prior to the terminal synthesis of mucins; relative incorporation of radiolabelled precursors is as follows: [^{3}H]-L-lysine>>[^{3}H]-D-glucosamine>>[^{3}H]-L-fucose.

In mid-epidermis, the capsule with a thickness of 0.2–0.4 μm appears cup-shaped in the peripheral cytoplasm where it surrounds the basally positioned nucleus. The capsule, clearly prominent only basally and laterally, becomes confluent in the apical region with loosely organized intermediate filaments which course around the relatively abundant mucin-containing secretory granules that are rapidly accumulating. The labelling intensity produced by lysine and glucosamine is sharply diminished when compared to cells in the basal layer, but fucose incorporation is elevated somewhat as a result of mucin synthesis. The relative order of radiolabelling now becomes [^{3}H]-D-glucosamine > [^{3}H]-L-lysine > [^{3}H]-L-fucose.

SMCs near the epidermal surface show a labelling pattern wherein [^{3}H]-L-lysine > [^{3}H]-L-fucose > [^{3}H]-D-glucosamine. However, the silver grain distribution is much more localized for fucose (e.g. Figure 8.1C) and glucosamine than for lysine, probably due to the incorporation of fucose and glucosamine during terminal synthesis of mucins. Quantitative electron microscopic autoradiography (Spitzer *et al.*, 1979) indicated that, after incorporation of [^{3}H]-L-fucose *in vivo* for 4.5 h, about 80% of the total label was localized juxtanuclearly in the region of the Golgi apparatus and 15% within the apical secretory vesicles.

Perhaps the capsule should be viewed as a specialized region of the cytoskeleton which not only provides mechanical strength to the SMC but also serves to compartmentalize secretory vesicles to the apical region (see electron micrographs, Spitzer *et al.*, 1979). Although the formation of a capsule by intermediate filaments is quite unusual, similar structures have been observed in epidermal cells from a few species of teleosts (Whitear, 1986). Bearing some similarity to the capsule is the cytoskeletal theca found in the intestinal goblet cells of some mammals. However, the theca represents a highly organized basket formed by both intermediate filaments and microtubules that is located distal to the nucleus where it serves to maintain the shape of the cell apex and segregate the secretory granules (Specian and Neutra, 1984).

The fine cytoarchitecture of the epidermal surface suggests a functional role for SMCs in mucous granule delivery in *M. glutinosa* (Patzner *et al.*, 1982). When the surface, consisting mainly of SMCs (Figure 8.2A), is viewed by SEM, a hexagonal honeycomb structure is evident. Within one SMC (one hexagon) is seen 100–200 combs, each with a diameter of about 0.42 μm (Figure 8.2A, C) representing a channel of 2–3 μm in length (Figure 8.2B) often containing mucous granules for delivery to the cell membrane. The

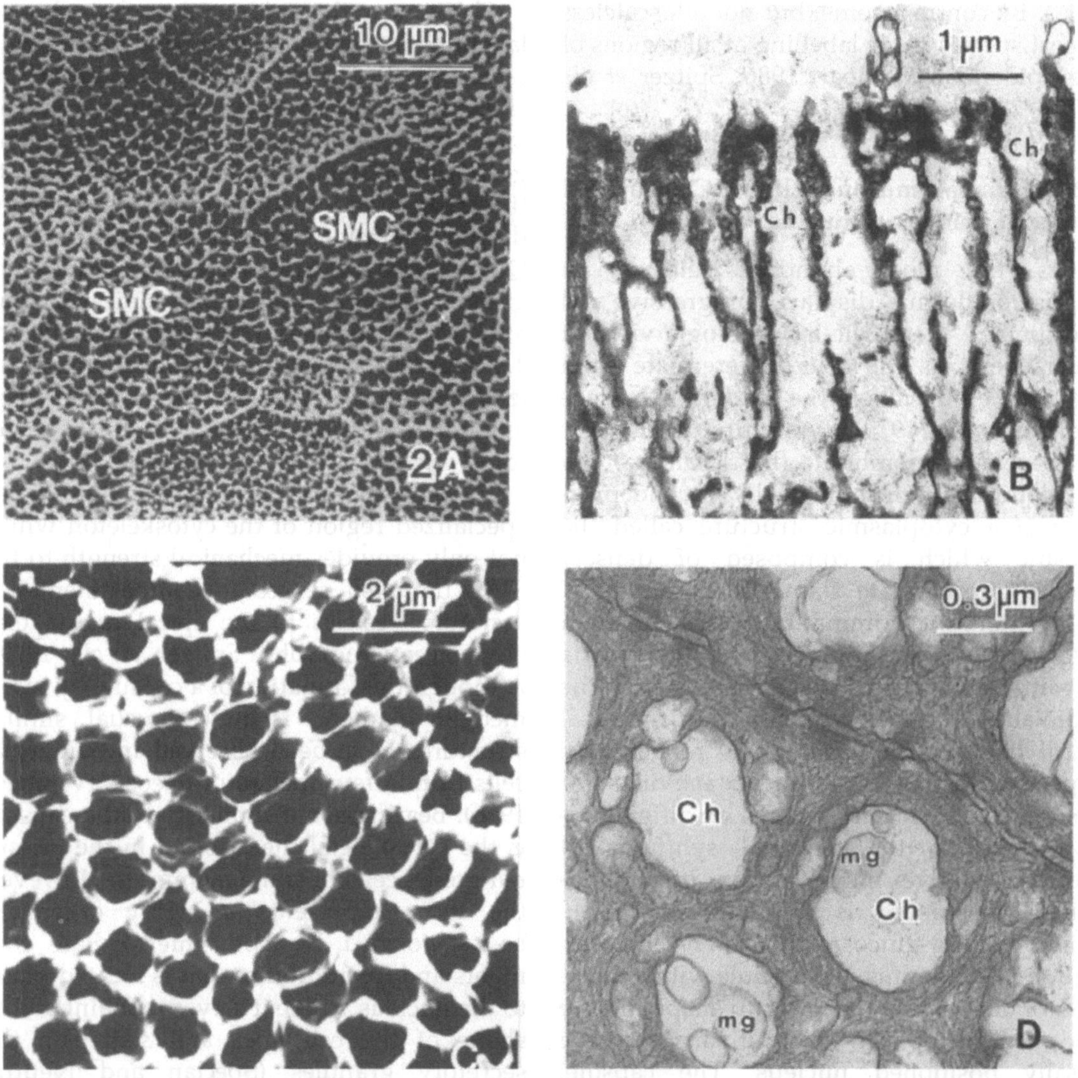

Figure 8.2 (A) Surface view of several small mucous cells *(SMC)*. (B) Longitudinal section through the apical channels of a small mucous cell. Few granules at the surface and within the channels. (C) Surface view of small mucous cells, microvilli around the channel openings. (D) Transverse section about 0.5 µm from the apical region, several desmosomes at the connection of two mucous cells, mucous granules *(mg)* within the channels *(Ch)*; tonofilaments and vesicles in the wall. (Modified from Patzner *et al.*, 1982, with permission from The Royal Swedish Academy of Sciences.)

cytoskeleton, in the form of IF arrays, circumvolutes and stabilizes the channels (Figure 8.2D) and perhaps is contiguous with the intertwined IFs comprising the capsule proximal to the perinuclear cytoplasm (Spitzer *et al.*, 1979).

8.2.2 The epidermal thread cell (ETC): a functional homologue of the gland thread cell (GTC)?

The unusual structural morphology of the elongated epidermal thread cell (Figures 8.1A, 8.3A, B) with its convoluted cytoplasmic

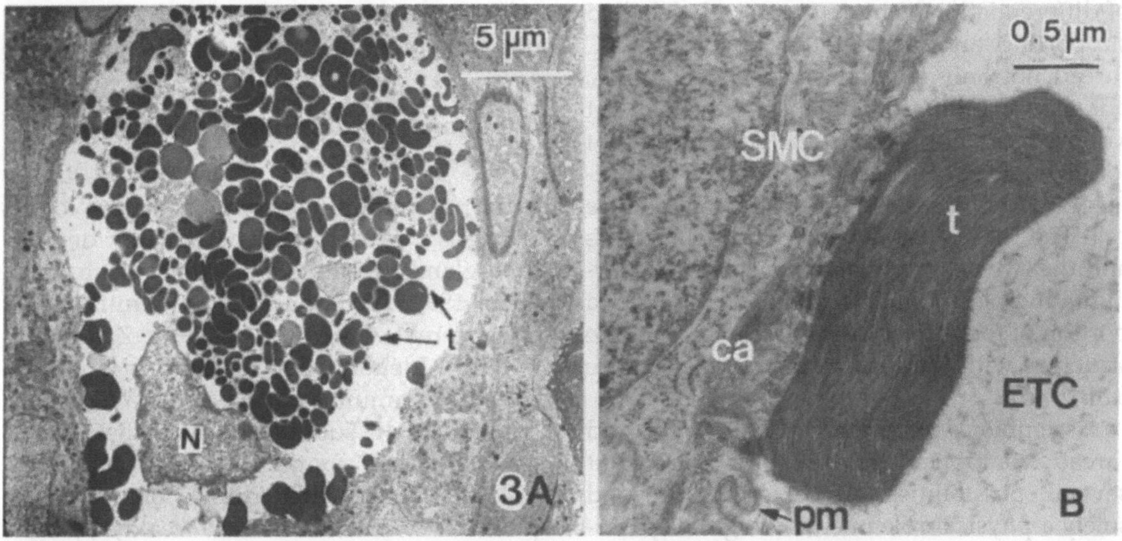

Figure 8.3 (A) Electronmicrograph of ETC with numerous sections of cytoplasmic thread *(t)*; nucleus *(N)*. ×3390. (B) Electronmicrograph of ETC and adjacent SMC at higher magnification than (A). Region of IF-rich cytoplasmic thread *(t)* is confluent with desmosomes. Note IF-rich capsule *(ca)* within SMC; plasma membrane *(pm)*. ×24 800. (Koch *et al.*, unpublished.)

secretory thread ('Sekretfaden') was described in detail via light microscopy by Schreiner (1916, 1918). Schreiner's observations were mainly concordant with the elegant electron microscopic images provided in 'pilot studies' by Blackstad in 1963. Electron micrographs revealed that a thread, with a cross-sectional diameter of about 1 μm, was actually composed of some 60–80 'lines' oriented in parallel along the long axis of the thread. Blackstad correctly concluded that the 10–12 nm 'lines' represented filaments that were the 'elementary building units' of the thread. We now view the thread as an IF biopolymer with keratin-like IFs bundled in parallel array (Figure 8.3B), bearing some similarity to the more concisely coiled single thread in a GTC (Figure 8.7A, B). Within the ETC, the thread appears relatively thick in the basal juxta-nuclear region of the cell, but distal to the nucleus, the thread shows great variation in thickness (Figure 8.3A). The length and precise coiling of the presumptive single thread is not established, and the amino acid sequence of the IF subunits is not yet known.

Unlike SMCs which differentiate progressively as they move across the epidermis (Figure 8.1A) and show distinct metabolic activities as indicated by radiolabelling patterns, ETCs at late stages of differentiation can be found at all levels of the epidermis including near-basal regions. Because the cells become very large, a single ETC may extend through 50% or more of the epidermis (Figure 8.1C). The pattern of radiolabelling in the ETCs, 24 h after administration of [^{3}H]-L-leucine via the subcutaneous sinus, is significantly different from that observed in gland thread cells (GTCs). A section through a single large ETC often reveals one to three localized sites of intense radiolabelling (hot spots) in the distal portion of the cell (unpublished data). These hot spots are surrounded by cross-sections of apparently mature thread because most regions of thread as well as the nucleus are unlabelled. By contrast, the thread in GTCs, even at relatively late stages of differentiation, is more extensively labelled supporting an interpretation that the GTC thread not only lengthens as the cell grows, but also continues to thicken. It is not yet known if the hot spots in the ETCs represent localized sites of thread synthesis or a secretory product produced by a relatively mature cell. The ETC is usually surrounded by SMCs and the two cell types are connected by interdigitating plasma membranes. Occasionally a portion of the thread forms a desmosome-like structure with the plasma membrane (Figure 8.3B).

Assignment of function to the ETC and thread component is yet premature, but several possibilities are emerging beyond solely a physical role in endowing the epidermis with requisite strength and resilience during mucigenesis by other cell types. Massive IF-rich biopolymers are found within large epidermal cell types of other fish species, albeit not thread-like (Whitear and Mittal, 1983; Whitear and Zaccone, 1984; Whitear, 1986, 1988; Henriksen and Matoltsy, 1968). These include club cells of teleosts and skein cells of lampreys. By analogy to the role in which the thread from hagfish GTCs co-functions with mucins to organize water in 'slime', the aforementioned IFs from ETCs might mimic this activity by interaction of mucins from goblet-like cells (LMCs, SMCs).

8.2.3 The large mucous cell (LMC): another source of epidermal mucins

Goblet-like, mature LMCs are the largest (35 × 60 μm) and the least numerous (1%) of all mucigenic cell types (Figure 8.1A). When first identifiable, the LMC is already a large, ovoidal cell in the basal third of the epidermis, but becoming more elongated when occupying mid-to-surface epidermal regions. With the methylene blue-azure II-basic fuchsin reagent (MBBF; Humphrey and Pittman, 1974), the nuclear and juxtanuclear sites are darkly stained (blue), while most of the substructure of the cytoplasm is weakly stained (magenta) and has a foamy appearance and represents a source of epidermal mucins.

8.2.4 Immunolocalization of keratins in cyclostome epidermis

Keratin or keratin-like components are found in hagfish epidermis, epidermally derived slime glands (Spitzer *et al.*, 1988; Koch *et al.*, 1991b, 1994, 1995), and the epidermis of the sea lamprey (Alarcòn *et al.*, 1994; Zaccone *et al.*, 1995). Identification of keratins is mainly limited to immunoassays, but the amino acid sequence of IF subunits of hagfish GTCs is established and found atypical (Koch *et al.*, 1994, 1995; see later discussion). Hagfish epidermis is immunostained weakly by antisera to mammalian keratins and often shows a relative staining of SMC > ETC > LMC (unpublished observation). Trypsinization, often effective to uncover keratin epitopes, is useful to enhance staining in hagfish tissue, probably by disrupting the tonofilament-rich thread in ETCs and the basket-like capsule in SMCs (Koch *et al.*, 1991b). It is notable that mammalian antisera induce somewhat more labelling in the epidermis of several teleosts, e.g. channel catfish (*Ictaluras punctatus*) and American eel (*Anguilla rostrata*) than in the hagfish (unpublished observation). Lack of comparative sequence data precludes a molecular or evolutionary interpretation at present. Keratin or other IF-sequences from the lamprey are not available at this time.

For the other group of cyclostomes, the scaleless sea lamprey (*Petromyzon marinus*) has proved to be a useful animal for immunolocalization of keratins in the epidermis of larval (ammocoete) and adult forms (Alarcòn *et al.*, 1994; Zaccone *et al.*, 1995). In ammocoete specimens (6–15 cm length), antibodies to bovine or human keratins (nos 7, 18, 19) elicited positive and cell-type specific staining among the three main types of epithelial cells: tonofilament-rich skein cells, granular cells and mucous cells (Alarcòn *et al.*, 1994). Subsequently, Zaccone *et al.* (1995)

utilized a panel of ten monoclonal antibodies to further define the cytokeratin distribution patterns in epidermis of juvenile and adult specimens including the horny teeth. All cell types and cell layers of the epidermis contained components which cross-reacted with one or more keratin antibodies and indicated general 8/18 keratin pair expression, high labelling of skein and granular cells by antisera to keratins 7, 18 and 19 and keratin labelling within the primary and secondary horny layers of the teeth. The results serve to indicate a complex cytokeratin expression pattern as in epidermis of teleost fish (Markl and Franke, 1988; Markl *et al.*, 1989) and evolutionary conservation of keratin epitopes among vertebrates. Further studies are required to compare structure–function relationships of keratins among cyclostomes, e.g. hagfish gland thread cells and lamprey skein cells. The earlier reports by Downing and Novales (1971a, b, c) on the fine structure of lamprey (*Ichthyomyzon unicuspis*) epidermis showed marked differences in cell types compared to hagfish and teleost species.

8.2.5 Identification of epidermal components

Very few compounds of note have been identified in Agnathan skin. Several acid mucopolysaccharides including hyaluronic acid, dermatan sulphate and dermatan polysulphate were found in *M. glutinosa,* and hyaluronic acid and dermatan sulphate were also found in *Lampetra fluviatilis* (Rahemtulla *et al.*, 1976). Naturally occurring haemagglutinins and several serum proteins were identified in the water-soluble fraction from the total exudate of *E. stoutii* (Spitzer *et al.*, 1976), but the cellular origin remains unknown.

8.3 THE SLIME GLANDS

8.3.1 Morphology

The ducted slime glands are responsible for most of the thick, viscous exudate released when the hagfish is disturbed. Blackstad in 1963 reviewed the important early investigations of the slime gland and slime products (Schaffer, 1925; Ferry, 1941; Newby, 1946) and also included results from his own pilot experiments. These exciting studies, mainly morphological, together with a popular account of the 'knot-tying' fish (Jensen, 1966) were a stimulus for subsequent investigations employing methodology not available in the 1960s. Recent results have provided a more incisive molecular view of the events of slime formation, its properties and functions, and a mode to study slime and mucus production.

Some 75–100 slime glands are located along each of the ventrolateral body walls, evenly and linearly dispersed from the head region to the cloaca. Each ovoid gland (1.5–3.5 mm, major axis) is encapsulated by peripheral connective tissue (Figure 8.4), surrounded by a thin *musculus decussatus* (Lametschwandtner *et al.*, 1986). The two main cell types, gland mucous cells (GMCs) and gland thread cells (GTCs) are distributed throughout the gland (Figure 8.4). Both types are large when fully developed; irregularly shaped GMCs (72–93 μm, major axis) and prolate ellipsoidal GTCs (~64 × 123 μm, minor and major axis). An extensive vascular network is interspersed between the capsule and skeletal muscle, where it forms a glandular sheath which exhibits a ball- or cage-like pattern (Figure 8.5) as judged by SEM examination of vascular corrosion casts (methylmethacrylate resin) of *Eptatretus* and *Myxine* (Lametschwandtner *et al.*, 1986). However, more capillary loops are seen directed to the interior of the glands of *Eptatretus*, providing a greater and presumably more efficient blood supply. The physiological advantage is not yet clear, but it was hypothesized that possible increased glandular activity might decrease recovery time following holocrine release from *Eptatretus*. The geometry of the vascular system may ensure a rapid supply of precursors for biosynthesis of slime components.

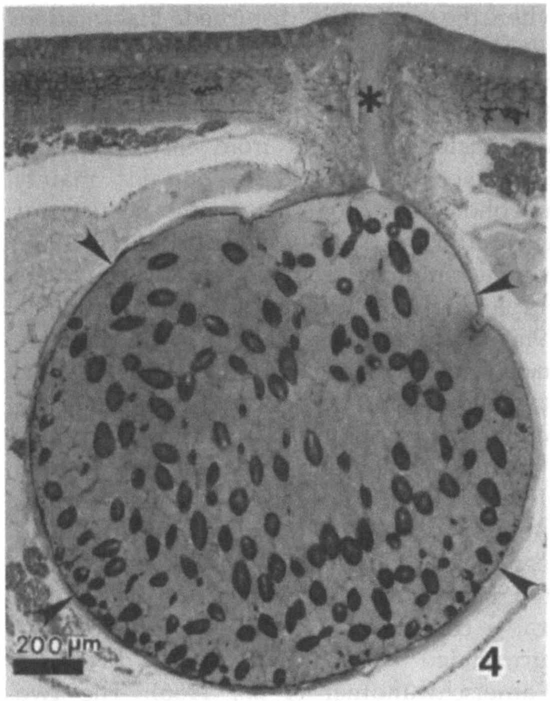

Figure 8.4 A methylene blue-basic fuchsin stained 1–2 µm plastic cross-section through a typical slime gland in *Eptatretus stoutii*. Each gland connects to the epidermal surface via a single pore (*). Within the gland are two types of cells, GTCs (dark staining) and GMCs (light staining). The largest thread cells are located deep within the gland; the smaller, less differentiated thread cells are found near the gland capsule (arrowheads). The mature cells are invariably elliptical in shape with a somewhat blunted end and a more pointed end and the nucleus is always located near the blunt end of the cell and the GTC-cytoplasm is filled with a highly coiled, single thread. Bar, 200 µm. (From Downing *et al.*, 1984; reproduced from the *J. Cell Biology* by copyright permission of the Rockefeller University Press.)

8.3.2 Slime production by holocrine secretion: a brief overview

During holocrine secretion, the gland cells are partially ruptured upon passage through the gland duct (~500 µm, length; 100 µm, width) onto the epidermal surface for contact with seawater. GTCs lose their plasma membrane and release a massive IF-rich thread biopolymer. GMCs are disassembled into constituent secretory vesicles (granules) which then disgorge mucins upon contact with water. Mucins and uncoiled threads interact in a synergistic manner during organization of water into a copious, viscous slime product (stage 1). Subsequent physical perturbation of the slime induces the threads to aggregate into 'cables' simultaneously retracting from the slime and expelling the organized water (stage 2). The experimental basis for this overview is described in the following sections.

8.3.3 Isolation of GMC vesicles and GTC thread requires a stabilization solution

When the anaesthetized hagfish is electrostimulated adjacent to the pores of the slime glands, a white exudate is rapidly expelled onto the epidermal surface (Figure 8.6A). The exudate from one or two glands can be removed and admixed with seawater (or distilled water) to quickly form a large extrudable mass of slime (Figure 8.6B). Although this sequence of events establishes the slime producing ability of gland exudate, the addition of water as a final step is counterproductive for the isolation, segregation and characterization of individual important products released from GMCs and GTCs. For this purpose, the exudate is instead collected in a 'stabilization solution,' of which several are useful (Salo *et al.*, 1983; Downing *et al.*, 1984; Spitzer *et al.*, 1988; Luchtel *et al.*, 1991).

Analyses of the stabilized exudate reveals mainly GTCs devoid of plasma membrane (Figures 8.6C, 8.7A) and intact, mucin-rich vesicles from disassembled GMCs (Figure 8.6D). The GTCs are easily segregated from the smaller vesicles by employing a series of graded nylon mesh filters (Salo *et al.*, 1983; Spitzer *et al.*, 1988). Of significance is the incisive observation by Luchtel *et al.* (1991) that GTCs and vesicles exhibit similar selective ionic permeability ('matched processes') which either serves to stabilize or rupture both entities – possibly indicative of a cofunction.

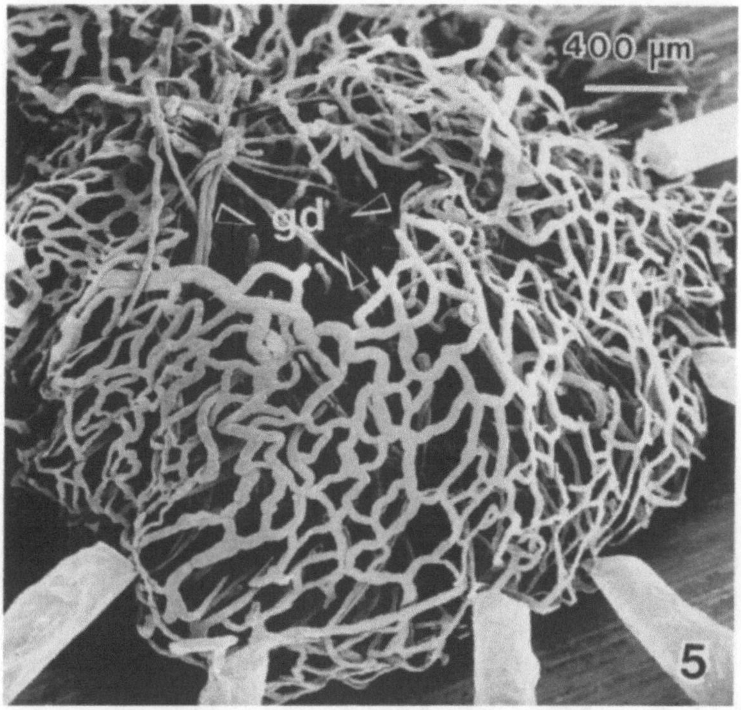

Figure 8.5 Ball-like vascular pattern surrounding a slime gland of *Eptatretus stoutii* as viewed by SEM of a corrosion cast. The peripheral vascular network gives rise to capillary loops extending towards the beginning of the glandular duct (*gd; arrowheads*). (From Lametschwandtner *et al.*, 1986, with permission from the Royal Academy of Sciences.)

8.3.4 GMCs provide mucin-rich vesicles: the mucins have a low content of carbohydrate but are rich in proline and threonine

The ovoidal vesicles (~7 µm, major axis) are each contained within a single membrane of 9–10 nm in width (for ultrastructural features, see Luchtel *et al.*, 1991). A turbidimetric analysis is useful to measure vesicle yield, i.e. turbidity at 350 nm vs dry weight of vesicular mucus per ml (Salo *et al.*, 1983).

Early work to characterize slime gland mucins included histochemical evaluations by Leppi (1966, 1967, 1968) who identified sulphomucins in *E. stoutii*, as did Lehtonen *et al.* (1966) in *M. glutinosa*. Subsequently, Salo *et al.* (1983) extracted and partially characterized mucins from vesicles 'stabilized' in several manners: 1.0 M ammonium sulphate in 1.75 M sucrose; 1.0 M ammonium sulphate; 1.0 M sodium citrate. Analysis of vesicular mucus solubilized by sulphitolysis indicated 77% protein, 12–18% carbohydrate, and 5% lipid. The value for carbohydrate content is very low and atypical of that found in mammalian mucins; e.g. ~75% (Gum *et al.*, 1991) and even higher (~85%) in a gel mucin from a teleost as suggested from a protein content of 15% (Lumsden and Ferguson, 1994). The most frequent amino acid residues in hagfish mucins are proline (27%), threonine (19%) and valine (14%). It is now evident that among mammalian apomucins, degenerate peptide motifs exist which have a variable number of tandem repeats often containing threonine, serine and proline domains for O-glycosylation (Wilson *et al.*, 1991). Structural determination of hagfish mucins may prove useful to understand the manner by which specific subdomains of the GMC-mucin and GTC-thread interact.

Figure 8.6 Collection of slime gland exudate after electrostimulation of anaesthetized hagfish. (A) The cellular contents of eight glands are being exuded onto the epidermal surface; the pores of non-stimulated glands (arrowheads) can be seen on the right. The exudates are readily removed with a spatula. When the exudate from 1–2 glands is stirred in seawater for about 15 s, a slime mass is produced (B). By contrast, when the exudate is collected in a stabilization buffer, the major entities can be segregated, and when viewed by SEM show (C) GTCs without plasma membranes and (D) secretory vesicles from GMCs. (A and B from Downing *et al.*, 1981b, with permission of *Science*, © AAAS; C from S.W. Downing, unpublished; D from Salo *et al.*, 1983, with permission of Marcel Dekker, Inc.)

8.3.5 The GTC differentiates to assemble a single IF-rich thread for extracellular export to function in seawater

After many decades of study by numerous investigators (Ferry, 1941; Strahan, 1959; Blackstad, 1963; Leppi, 1967, 1968; Terakado *et al.*, 1975; Fernholm, 1981; Downing *et al.*, 1981a, b; Koch *et al.*, 1994, 1995), the GTC can be summarily described as an 'IF machine' for the synthesis and assembly of atypical IFs into parallel arrays to form a single, large, tapered, coiled, thread biopolymer which is released efficiently by holocrine secretion during slime formation (Downing *et al.*, 1984; Koch *et al.*, 1994, 1995).

Many of these attributes can be inferred by light- and electron-microscopy. Coiling is already evident within small immature GTCs which are localized near the capsule (Downing *et al.*, 1984) and become intensely radiolabelled shortly after administration *in vivo* of [^{3}H]-L-leucine (data not shown). At a later developmental state, the relatively mature GTCs, now found more deeply in the gland, contain a larger thread that attains a highly coiled conformation (Figures 8.4, 8.6C, 8.7A). The nucleus, usually containing one nucleolus, resides in the blunter end of the ellipsoidal cell. As judged by TEM of cells *in situ*, the structural motif of the thread (exclusive of very mature GTCs) corresponds to IFs and a lesser number of MTs bundled together in parallel alignment (Figure 8.7C). This packaging motif bears striking similarity to the parallel alignment of neurofilaments and microtubules in spinal nerve roots of higher vertebrates (see TEM, in Eagles *et al.*, 1990). That the thread is tapered and coiled can be construed partly from a single section through the GTC which reveals thread sections (cross, oblique, longitudinal) of various diameters (Downing *et al.*, 1984). The precise coiling of the thread is best seen by SEM of cells isolated from gland exudate collected in stabilization buffer (Figure 8.7A, B). Calculations indicate that the thread length, in a GTC of average size (63 × 127 μm, minor and major axis) is about 22 cm, but may approach 60 cm in larger GTCs (Downing *et al.*, 1981a). The tapered thread approaches a maximum width of ~2.5 μm and can occupy about 70% of the cell volume. The presumptive origin of the thread, where the thread diameter is narrow, is near the nucleus in a mitochondrial-rich region. During thread growth, IFs and MTs are recruited in a manner to lengthen and thicken the biopolymer while maintaining a taper.

The role of MTs is not clear at present, but hypotheses centre on functions to provide temporary scaffolding for IF alignment, direction for coiling, binding of associated proteins, and ion transport. At late stages of differentiation, MTs disappear, integrity of individual IFs is lost and the thread becomes electron dense (Downing *et al.*, 1984). Terakado *et al.* (1975) reported that during thread maturation, two filaments smaller than IFs (1–3 nm, and 3–6 nm; cross sections) became detectable. The ultrastructural change accompanying thread maturation might be attributable to a post-assembly modification in one of the IF polypeptide chains ($\gamma \rightarrow$ ß) inducing partial disassembly of the IFs (Spitzer *et al.*, 1988; also subsequent discussion).

8.3.6 Mucins and threads organize water synergistically during formation of slime: stage 1

The hagfish slime gland has been viewed as a useful system for studying the biology of 'slime' or mucus (see 'Some definitions', p. 110) by providing gram quantities of interacting entities–mucous vesicles and fibrous proteins in the form of the threads (Downing *et al.*, 1981b). The results of one study, summarized in Figure 8.8, established an important role of the thread in modulation of the water-organizing ability of hagfish slime. The exudate collected in stabilization buffer (0.92 M sodium citrate, 0.14 M PIPES, 5% glycerol, 5.0 mM $MgCl_2$, 2.0 mM EGTA, 0.5 mM PMSF) had a ratio of vesicles to GTCs of ~500:1. After segregation by nylon mesh filters, stock suspensions of known concentrations (entities, ml^{-1}) were prepared to evaluate perturbations produced by hydration of these reformulated 'standard' slime samples.

Some of the principal events which occur following the incremental addition of water with stirring to a vesicle/GTC mixture (500:1), as monitored by light microscopy are summarized in Figure 8.8A–H. These include:

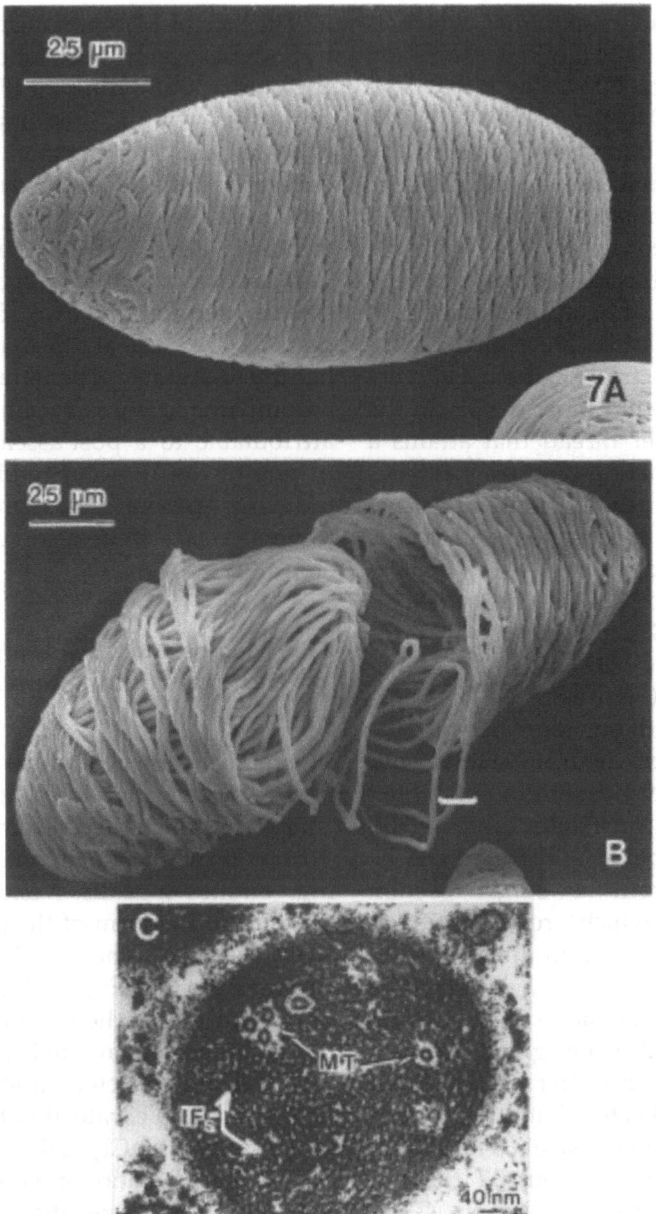

Figure 8.7 Conformation and ultrastructure of thread in hagfish gland thread cell (GTC). (A) Scanning electron micrograph of single coiled thread within cytoplasm of GTC. Plasma membrane removed during isolation of GTC from gland. (B) Partially uncoiled thread shows gross conformation of a thread which may reach a length of 60 cm and is compartmentalized within about 70% of the cytoplasm (Downing *et al.*, 1981a). Complete uncoiling of the thread would uncover nucleus at blunt end of cell. (C) Transmission electron micrograph of transverse section of single thread (~0.65 μm width, similar to half-bracket in (B)) showing parallel alignment of about 800 IFs and at least 7 MTs. Each IF consists of α and γ polypeptides (sequenced herein), and probable isoforms of γ. (From Koch *et al.*, 1994, with copyright permission of the Company of Biologists Limited.)

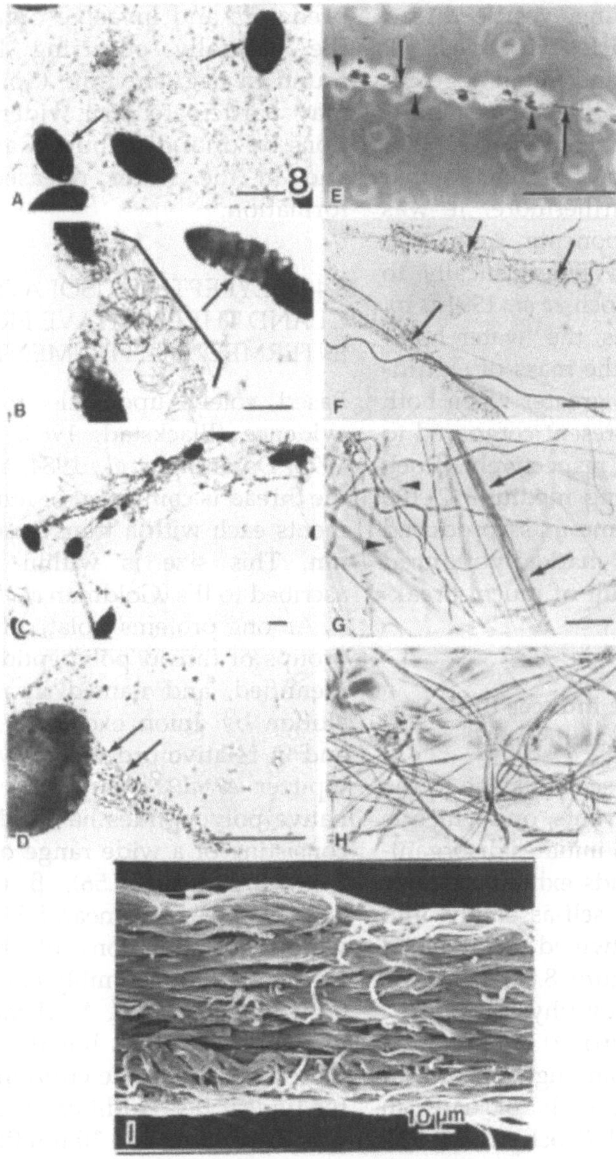

Figure 8.8 Major events during formation of slime *in vitro*. Light micrographs of GTCs and vesicles released during holocrine secretion and subsequent events leading to slime formation. (A) GMC-vesicles and GTCs (*arrows*) of exudate in stabilization buffer. (B–H) Increasing amounts of water admixed with vesicles and GTCs. (B) GTCs swelling (*arrow*) as single thread from each uncoils (*bracket*). (C) Threads beginning to tangle and aggregate with intact vesicles. (D) Vesicles aggregating with one another on thread (*arrows*). (E) Single vesicles (*arrowheads*) binding to thread (*arrows*). (F–H) Further uncoiling of threads (*arrows*), rupture of vesicles (*arrowheads*), and formation of viscous slime with various thread alignments. Unfixed and unstained samples. Bar: (A) 100 µm; (B) 100 µm; (C) 100 µm; (D) 100 µm; (E) 50 µm; (F) 100 µm; (G) 100 µm; (H) 100 µm (from Koch *et al.*, 1991b, with permission of Springer-Verlag GmbH and Co. KG). After stirring, large thread aggregates (cables) exhibit by SEM an overall twisted, parallel alignment of component threads (I) (Bar, 10 µm) (from Koch *et al.*, 1991a, with permission of Academic Press.)

progressive thread uncoiling, vesicle–thread interaction, rupture of vesicles with release of mucins (Salo *et al.*, 1983), and initial aggregation of threads. These related events (stage 1) result in a viscous product (Figure 8.8G, H) of high water content which is similar to that seen in Figure 8.6B. Furthermore, it was evident that these components from two gland cell types function synergistically to loosely organize water (Koch *et al.* 1991a, b). Under identical conditions, the 'water holding ability' as judged by the mass of extrudable slime is 99–165 times greater when both threads and vesicles are present compared to threads or vesicles alone, respectively (Koch *et al.*, 1991b). In a reducing medium (2–100 mM dithiothreitol), the slime mass produced by threads and vesicles is markedly reduced (29–95%), mainly as a result of mucin breakdown.

8.3.7 Perturbation of slime induces threads to aggregate and release water: stage 2

It is useful to designate a second stage to the post-holocrine secretory events, one which is confluent with stage 1 and initiates disorganization of water. The threads exhibit an early tendency for subsequent self-assembly into novel, aligned and intertwined cytoskeletal aggregates – 'cables' (Figure 8.8G–I). These entities can be produced by physical perturbation of the slime (e.g. stirring) but are often seen occurring naturally among the tufts of anchor filaments on the ends of eggs in aquaria-maintained hagfish (Koch *et al.*, 1993). Physically prepared cables are larger (15–40 cm length; 50–800 μm width) relative to those removed from eggs (0.1–10 cm length; 50–200 μm width) but exhibit a similar ultrastructure (Koch *et al.*, 1991a). While the cables may merely represent a useless end-product which is found among eggs in a quasi-natural habitat, some functions can be offered. The threads and cables may localize eggs to an appropriate spawning site (a possibility expressed also by Walvig, 1963), and deter predators and invasive organisms. Perhaps the naturally occurring haemagglutinins found in hagfish slime (Spitzer *et al.*, 1976) may participate in a wider range of functions by dual distribution among the cables and in the water released during cable formation.

8.4 POLYPEPTIDES ISOLATED FROM GLAND THREAD HAVE PROPERTIES OF INTERMEDIATE FILAMENT SUBUNITS

Based solely upon electron microscopical evidence (Blackstad, 1963; Terakado *et al.*, 1975; Downing *et al.*, 1984; and Figure 8.7C), the thread is composed of numerous long filaments each with a transverse section of 8–12 nm. This size is within the range now ascribed to IFs (Goldman *et al.*, 1990).

Among proteins isolated from GTCs, three groups of thread polypeptides (α, β, γ) were identified, and named in relative order of elution by anion exchange chromatography and in relative order of pI values: $\alpha > \beta > \gamma$ (Spitzer *et al.*, 1984, 1988). Each group of native polypeptides has an M_r of 63–66 kDa consisting of a wide range of isoforms: α (pI 7.26–7.87, mean 7.56), β (5.52–5.81, mean 5.67), γ (5.18–5.47, mean 5.31). Each group or various combinations of the three groups were tested for assembly *in vitro* into IF forms (Spitzer *et al.*, 1988). A 1:1 mixture of α and γ was most effective, but the assemblies were more linear than the curvilinear forms found for typical IFs. Addition of β to this mixture was disruptive, and 10 nm filaments were not formed by any of the three components alone. The results were indicative of a keratin-like intermediate filament, i.e. a heteropolymer formed from an acidic subunit (γ, type I) and a more basic subunit (α, type II). The existence and role of the β-category was enigmatic *vis-à-vis* keratin structure. Subsequent studies served to establish that β represents a post-translational form of γ that occurs during the late stages of thread development. Supportive of this view are observations that

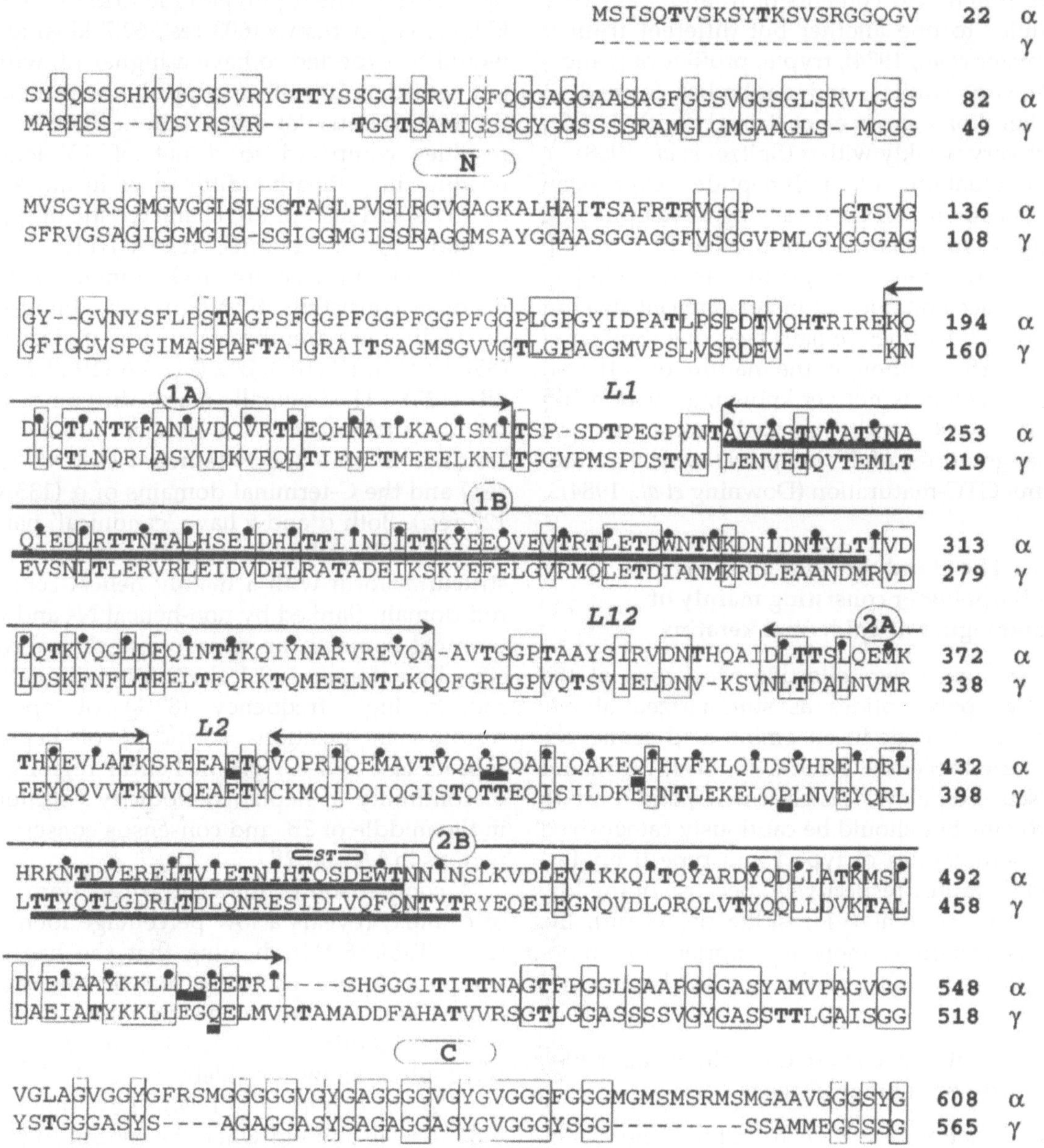

Figure 8.9 Sequence comparison of hagfish γ- and α-polypeptides. Identical sequences within boxes; subdomains of each central rod of 318 residues delineated by arrows; threonine residues in bold type; three threonine-rich (>20%) regions in central rod (non-linker segments) are underlined (▬). Residues not usually found in similar locations among most other IF types when aligned in a similar manner include: α:F (L2); GP, Q, DS (2B); γ: P, Q (2B), which are underlined (▬). Also atypical, but not underlined in central rod α-helical subdomains, are: α, five T–T; γ, two T–T. Alignment was by a modification of Bestfit (GCG) to achieve concordance with central homology profiles of Conway and Parry (1988). The nucleotide sequences in the GenBank™/EMBL Data Bank for α-chain and γ-chain have accession numbers of U11865 and U20546, respectively. (From Koch *et al.*, 1995, with permission of Elsevier Science B.V.)

the amino acid contents of β and γ are very similar to one another but different from α (Spitzer *et al.*, 1984), tryptic profiles of β and γ are very similar, and antibodies formed to either β or γ cross react strongly with β and γ but very weakly with α (Spitzer *et al.*, 1988).

Evaluation of polypeptide expression ratios among GTCs at varying developmental stages (sizes) is also supportive of a γ → β transformation. A progressive increase in β/γ ratios accompanies an increase in cell size but α/(β + γ) values remain near one (Spitzer *et al.*, 1988). Although the nature of γ-thread modification is not yet known, a relationship may exist with marked ultrastructural changes (loss of IF integrity) which accompany GTC-maturation (Downing *et al.*, 1984).

8.4.1 The gland thread is a massive IF-biopolymer consisting mainly of homologues of epidermal keratins

Based upon the collective properties of the native polypeptides as summarized above, and the two deduced amino acid sequences reported recently (Koch *et al.*, 1994, 1995), and discussed below, it is clear that γ and α are IF proteins but should be cautiously categorized as homologues of type I and type II epidermal keratins, respectively. Because of the low cysteine content (~1 residue per chain), the classification is more appropriate to homologues of type Ib and IIb – the soft keratins.

8.4.2 Central rods of α and γ chains have low identity with other IF-types

The isolation of mRNA from the slime glands requires special methodology to circumvent interference by the slime product (Gadbois *et al.*, 1988). Detailed procedures including sequencing are included in Koch *et al.* (1994, 1995). The complete amino acid sequences (Figure 8.9) are amenable to alignment and comparison (Table 8.1) with other IF subunits in the manner established by Conway and Parry (1988) and also by computer alignment by TFASTA. The α-polypeptide (643 res., 66.6 kDa) is larger than γ (603 res., 62.7 kDa) and would be expected to have a higher pI, without consideration of post-translational modifications, because of a net content of 16 basic residues compared to a net of 13 acidic residues in γ. Relative differences in molecular mass and also pI are characteristic of type II and type I keratins, respectively. Both chains contain a central rod domain of 318 residues consisting of subdomains (number of residues in parentheses) of similar size: 1A (35), L1 (13), 1B (101), L12 (21), 2A (19), L2 (8), 2B (121). The overall mass difference is mainly attributable to a greater number of residues in both the N-terminal (192 vs 158 res.) and the C-terminal domains of α (133 vs 127 res.). Both α and γ have 'canonical' parameters of IF subunits including a tripartite structural form with a mainly helical central rod domain flanked by non-helical N- and C-terminal domains, an appropriate central rod size, link domains (L1, L12, L2) within the rod, a high frequency (82%) of apolar residues in positions *a* and *d* of heptad repeats *(a–b–c–d–e–f–g)n* in helical regions, a discontinuity of heptad periodicity ('stutter') in the middle of 2B, and consensus conserved regions in 1A and 2B.

A comparison of the central rod domains of α and γ reveals a low percentage identity (23%, Table 8.1) indicating that the hagfish components are probably not of the same type. Chains of the same type usually show 50–95% identity while chains of different types often show rod identity of 30–35% or less (see Koch *et al.*, 1994). Low identity (<31%) is also found when the central rods of α or γ are compared to establish possible homology to numerous examples of the five types of vertebrate IF chains (data for γ in Table 8.1; for α see Koch *et al.*, 1994). Identity with a cephalochordate (*Branchiostoma lanceolatum*) was low (~29%) as were two invertebrates (~19%). Of particular significance was that γ (Table 8.1) and α showed no preferred homology to types Ia and b or IIa and b

Table 8.1 Comparisons of percentage identity in central rod domains of hagfish α and γ show low identity with α and other IF-types

IF chain	Subdomains of central rod*							Total (central rod)	TFASTA § Total (central rod) (mean % identity and range)
	1A	L1†	1B	L12†	2A	L2	2B		
Hagfish α 66.6	34	31	22	5	37	38	21	23	– –
Human Ib 50	43	0	29	19	37	38	27	29	
Mouse Ib 59	43	0	26	13	47	38	29	29	
Type Ib (mean of above)	(43)	(0)	(28)	(16)	(42)	(38)	(28)	(29)	29 (21–31)
Type Ia	46	6	30	31	26	25	30	31	
Human IIb 67	31	25	23	12	16	38	27	25	
Mouse IIb 67	31	25	21	18	11	50	27	25	
X. laevis IIb 55.7	40	10	25	18	26	25	29	27	
Goldfish IIb 58	40	10	26	12	11	38	28	26	
Type IIb (mean of above)	(36)	(18)	(24)	(15)	(16)	(38)	(28)	(26)	25 (22–27)
Type IIa	37	20	22	35	11	38	26	25	
Vimentin III 53.5	46	0	27	13	42	63	30	31	
GFAP III 50	40	0	27	6	42	50	30	29	
Type III (mean of above)	(43)	(0)	(27)	(10)	(42)	(57)	(30)	(30)	30 (29–30)
Torpedo californica III?	40	0	25	0	32	38	32	28	
Peripherin III	37	25	24	22	37	63	31	30	28
NFL IV 68	37	0	22	6	32	50	28	26	25 (18–29)
α-Internexin IV	43	30	25	13	37	50	30	30	28
X.laevis Lamin 3 type V‡	31	36	21	0	26	13	27	24	22 (19–25)
Branchiostoma lanceolatum	46	25	29	14	32	25	28	29	(28% vs Hagfish α)
Helix aspersa (mollusc)‡	40	18	12	0	11	25	24	20	
Aplysia californica (mollusc)	ND→						←ND	19	

*Alignment and number of residues in each subdomain similar to method of Conway and Parry (1988). Chains of the same type aligned and compared in this manner show a mean identity of 67.4% while a mean identity of 32.6% is evident among different types (Koch *et al.*, 1994). Comparisons among keratin subtypes (Ia vs Ib) show identity >50%; wool Ia vs human Ib, 58.5%; wool Ia vs mouse Ib, 51.8%. Origin of sequencescited in Koch *et al.* (1994, 1995).
†When linker regions differ in size, % identity was based on the number of identical residues per number of residues in the shorter linker.
‡The 42 residue (six heptad) insert within 1B of *Xenopus laevis* lamin 3 and the invertebrate *Helix aspersa* were excluded within the comparisons.
§Computer alignment with TFASTA indicates mean percent identity with range in parentheses, and includes: six type I with 326 residue overlap; eight type II with a 332 residue overlap; six type III with 336 residue overlap; peripherin, 329 residues; four type IV with 330 residue overlap; α-internexin of 327 residues; six type V with 329 residue overlap; *Aplysia californica* (invertebrate, IF protein A, 321 residue overlap. ND, not determined. (Modified from Koch *et al.*, 1994, 1995, with permission of Elsevier Science B.V.)

keratins, inasmuch as identities with types I–V were quite similar. Subdomain IB (101 res.) in α and γ was similar in size to that in the cephalochordate and non-neuronal cyloplasmic IF chains in higher vertebrates, hence devoid of the six heptad insert found in lamins and invertebrates. The collective comparative results appear concordant with the overall view of Riemer *et al.* (1992) that IF chains possibly arose from a lamin-like progenitor.

Among the deduced sequences, the strongest evidence for 'keratin-likeness' resides in the N- and C-terminal domains

Table 8.2 The central rods of α and γ subunits have a higher percentage of threonine than other types of IF-chains

	Subdomains of central rod†														Entire central rod	
	1A		L1		1B		L12		2A		L2		2B			
IF chains	T	S	T	S	T	S	T	S	T	S	T	S	T	S	T	S
Hagfish γ	9	3	15	15	10	3	5	10	11	0	13	0	11	3	10	4
Hagfish α	9	3	23	15	19	2	14	5	21	5	13	13	7	4	13	4
(mean)	9	3	19	15	15	3	10	8	16	3	13	7	9	4	12	4
Type Ib (mean)	3	4	0	0	6	4	3	3	3	3	0	0	6	10	5	6
Type Ia	3	6	0	9	5	7	6	0	11	0	0	0	6	8	5	6
Type IIb (mean)	4	6	25	8	4	5	4	15	0	4	3	16	4	5	4	6
Type IIa	3	0	0	10	3	6	6	18	0	5	0	25	5	3	4	3
Type III (mean)	0	2	6	6	3	3	0	3	11	5	6	0	3	5	3	4
Torpedo californica	0	3	0	13	1	2	0	6	5	11	13	0	7	6	4	5
Type IV	0	3	0	11	2	2	0	18	0	5	0	0	6	6	3	5
Type V	0	6	8	15	6	2	5	10	0	0	0	0	5	6	4	5
Cephalochordate *Branchiostoma lanceolatum*	3	9	13	0	2	3	9	13	11	0	0	0	5	4	4	4
Invertebrates																
Helix aspersa	0	3	0	17	5	4	0	7	0	11	13	0	3	7	3	6
A. lumbricoides	0	0	9	9	2	2	7	0	0	0	0	13	3	4	2	3

*Original sequences cited in Koch *et al.* (1994, 1995).
†Alignment of subdomains (Conway and Parry, 1988) similar to Table 8.1. Mean threonine (T%) and serine (S%) values for types I–III calculated from chains in Table 8.1. Individually determined were one chain each from types IV and V and *Torpedo californica* (probably type III), *Helix aspersa* (invertebrate, snail), type Ia (hard keratin, sheep wool), type IIa (hard keratin, sheep wool), *Ascaris lumbricoides* (invertebrate, muscle cells from nematode, protein B). Comparative chain analysis shows that threonine residues are not conservative substitutions for serine sites. The high frequency of T–T (five in α) is unusual, as are regions of very high threonine content (>19%) in coiled coil domains. (Modified from Koch *et al.*, 1995, with permission of Elsevier Science B.V.)

wherein both γ and α have a high combined content of glycine and serine residues, and contain tandem peptide repeats. Another keratin parameter is inferred by theoretical analyses of the entire chains to ascertain maximal interhelix ionic stabilization and probable organization of dimers (Koch *et al.*, 1995). A maximum of 12 ionic interactions (0.305 per heptad pair) is found when α and γ are aligned in parallel and in axial register. These values are within the range (0.23–0.77 per heptad) for other two stranded IF proteins (Conway and Parry, 1990). The fundamental building block of a keratin IF corresponds to a coiled coil heterodimer of a type I and type II chain aligned in the above manner.

8.4.3 Central rods of α and γ have high threonine content and exhibit residue substitutions in conserved sites

Further examination of the thread sequences (Figure 8.9) provides a basis to understand the low central rod identity with other IF types. The threonine content is notably high (α, 13%; γ, 10%) relative to threonine in the central rod of other types (2–5%, Table 8.2). Even higher threonine levels exist in two subdomains of α (1B, 19%; 2A, 21%) but are more evenly distributed in γ (1A, 9%; 1B, 10%; 2A, 11%; 2B, 11%). Both α and γ have regions within the helical subdomains which have threonine contents of 20–23% (Figure

8.9 and Table 8.2). Analyses indicate that Thr-residues do not represent conservative substitutions for serine, but mainly represent unique Thr-sites in the central rod (Koch *et al.*, 1995, 1996). The abundance of threonine residues along with some unexpected sequences in α and γ (Figure 8.9) serve collectively to negate the normal structural linear periodicity of ionic residues in some regions of the central rod and provide a basis for low percentage identity with keratins and other IF-chain types.

Specific roles for the threonine residues in the thread polypeptides are not yet clear. Fast Fourier transform analyses show no linear distribution period for this amino acid in α and γ and its high content does not appreciably alter the tripartite structure typical of IF chains as judged by secondary structure prediction algorithms (Koch *et al.*, 1995). Among coiled coil proteins such as IF dimers which exhibit a heptad repeat pattern (*a–b–c–d–e–f–g*)*n*, stabilization is partially achieved by apolar residues in *a* and *d*, and ionic residues often found in *e* and *g* (Parry, 1990). Significantly among the thread polypeptides, most of the threonine is located among outer *b*, *c* and *f* positions of the helical substructure. Threonine and serine are sites for phosphorylation/dephosphorylation which modulate IF assembly. This process may be of major importance because the hagfish thread subunits show a structural hierarchy greater than for any other known type of IF, e.g. dimer → IF → thread → thread aggregates (cables). These residues may interact with IFAPs and MTs to facilitate parallel alignment of IFs while the thread thickens, lengthens and coils while in the GTC. In addition, numerous aliphatic hydroxyl groups become available on the thread surface when it is expelled from the GTC and uncoils in seawater for interaction with the glycosylated, threonine-rich mucins during slime formation.

It is not clear if the α and γ subunits, each with an atypical central rod, represent primitive examples of keratins derived from a complex epithelium which emerged during the transitional period from exoskeleton to endoskeleton (see reference citations in Koch *et al.*, 1995). Perhaps for survival as a scavenger restricted mainly to a benthic environment, the slime gland was under selective pressure to yield gland products uniquely suitable for rapid deployment of large slime masses. Possibly similar IF chains with threonine-rich central rods might be found in other marine organisms which use specialized cells for specialized functions not yet recognized. The necessity for two modes of secretion, particularly the remarkable holocrine type of the slime glands, is probably not understandable from the behaviour of aquarium-maintained hagfishes, but will require more extensive observations on animals in the natural habitat, such as the *in situ* probes initiated by Martini and Heiser (1989). Even then, specific functions of fish mucus are often speculative and difficult to determine experimentally (e.g. Shephard, 1993). Perhaps similar roles may emerge for IF-rich skein cells/granule cells and mucigenic epidermal cells in the lamprey (Zaccone *et al.*, 1995) and IF-rich club cells and goblet cells in some teleosts (Whitear, 1986).

ACKNOWLEDGEMENT

We are most appreciative for the sustained support received from the Asthmatic Children's Aid (Chicago) and The Ceal and Dr Morris A. Kaplan Foundation. Much of the work herein, as documented in references, was done in collaboration with Drs S.W. Downing and W.L. Salo, both at the U. Minnesota (Duluth). At various times the work was supported by the National Institutes of Health and the National Science Foundation. We thank both Roland I. Saavedra and Ilene R. Brenner of our Institution, and W. Saavedra and S. Smith for their assistance in the preparation of this manuscript.

REFERENCES

Alarcòn, V.B., Filosa, M.F. and Youson, J.H. (1994) Keratin polypeptides in the epidermis of the larval (ammocoete) sea lamprey, *Petromyzon marinus L.*, show a cell type-specific immunolocalization. *Can. J. Zool.*, **72**, 190–4.

Blackstad, T.W. (1963) The skin and the slime glands. In *The Biology of Myxine* (eds A. Brodal and R. Fänge), Universtitsforlaget, Oslo, pp. 195–230.

Conway, J.F. and Parry, D.A.D. (1988) Intermediate filament structure: 3. Analysis of sequence homologies. *Int. J. Biol. Macromol.*, **10**, 79–98.

Conway, J.F. and Parry, D.A.D. (1990) Structural features in the heptad substructure and longer range repeats of two-stranded α-fibrous proteins. *Int. J. Biol. Macromol.*, **12**, 328–34.

Devine, P.L. and McKenzie, F.C. (1992) Mucins: structure, function and associations with malignancy. *Bioessays*, **14**, 619–25.

Downing, S.W. and Novales, R.R. (1971a) The fine structure of lamprey epidermis. I. Introduction and mucous cells. *J. Ultrastruct.* Res., **35**, 282–94.

Downing, S.W. and Novales, R.R. (1971b) The fine structure of lamprey epidermis. II. Club cells. *J. Ultrastruct.* Res., **35**, 295–303.

Downing, S.W. and Novales, R.R. (1971c) The fine structure of lamprey epidermis. III. Granular cells. *J. Ultrastruct.* Res., **35**, 304–13.

Downing, S.W., Salo, W.L., Spitzer, R.H. and Koch, E.A. (1981b) The hagfish slime gland: a model system for studying the biology of mucus. *Science*, **214**, 1143–5.

Downing, S.W., Spitzer, R.H., Koch, E.A. and Salo, W.L. (1984) The hagfish slime gland thread cell. I. A unique cellular system for the study of intermediate filaments and intermediate filament microtubule interactions. *J. Cell Biol.*, **98**, 653–69.

Downing, S.W., Spitzer, R.H., Salo, W.L., Downing, J.S., Saidel, L.J. and Koch, E.A. (1981a) Threads in the hagfish slime gland thread cells: organization, biochemical features and length. *Science*, **212**, 326–8.

Eagles, P.A.M., Pant, H.C. and Gainer, H. (1990) Neurofilaments, in *Cellular and Molecular Biology of Intermediate Filaments* (eds R.D. Goldman and P.M. Steinert), Plenum Press, New York, pp. 37–94.

Falkmer, S., Emdin, S.O., Östberg, Y. *et al.* (1976) Tumor pathology of the hagfish, *Myxine glutinosa*, and the river lamprey. *Lampetra fluviatilis. Prog. Exp. Tumor Res.*, **20**, 217–50.

Fernholm, B. (1981) Thread cells from the slime glands of hagfish (Myxinidae). *Acta Zoologica (Stockh.)*, **62**, 137–45.

Ferry, J.D. (1941) A fibrous protein from the slime of the hagfish. *J. Biol. Chem.*, **138**, 263–8.

Gadbois, D.M., Salo, W.L., Ann, D.K., Downing, S.W. and Carlson, D.M. (1988) The preparation of poly(A)$^+$mRNA from the hagfish slime gland. *Prep. Biochem.*, **18**, 67–76.

Georgieva, V., Patzner, R.A. and Adam, H. (1979) Transmissions-und rasterelektronen mikroskopische untersuchung an den sinnesknospen der tentakeln von Myxine glutinosa L. (Cyclostomata). *Zoologica Scripta*, **8**, 61–7.

Goldman, R.D., Zackroff, R.V. and Steinert, P.M. (1990) Intermediate filaments: an overview, in *Cellular and Molecular Biology of Intermediate Filaments* (eds R.D. Goldman and P.M. Steinert), Plenum Press, New York, pp. 3–17.

Gorbman, A., Kobayashi, H., Honma, Y. and Matsuyama, M. (1990) The hagfishery of Japan. *Fisheries*, **15**, 12–18.

Gum, J.R., Hicks, J.W., Lagace, R.E., Byrd, J.C., Toribara, N.W., Siddiki, B., Fearney, F.J., Lamport, D.T.A. and Kim, Y.S. (1991) Molecular cloning of rat intestinal mucin. Lack of conservation between mammalian species. *J. Biol. Chem.*, **266**, 22773–8.

Henriksen, R.C. and Matoltsy, A.G. (1968) The fine structure of teleost epidermis. III. Club cells and other cell types. *J. Ultrastruct. Res.*, **21**, 222–32.

Humphrey, C.D. and Pittman, F.E. (1974) A simple methylene blue-azure II-basic fuchsin stain for epoxy-embedded tissue sections. *Stain Technol.*, **49**, 9–14.

Jensen, E. (1966) The hagfish. *Scientific American*, **214**, 82–90.

Koch. E.A., Spitzer, R.H. and Pithawalla, R.B. (1991a) Structural forms and possible roles of aligned cytoskeletal biopolymers in hagfish (slime eel) mucus. *J. Struc. Biol.*, **106**, 205–10.

Koch, E.A., Spitzer, R.H., Pithawalla, R.B. and Downing, S.W. (1991b) Keratin-like components of gland thread cells modulate the properties of mucus from hagfish (*Eptatretus stoutii*). *Cell Tissue Res.*, **264**, 79–86.

Koch, E.A., Spitzer, R.H., Pithawalla, R.B., Castillos III, F.A. and Wilson, L.J. (1993) The hagfish egg at a late stage of oogenesis: structural and metabolic events at the micropylar region. *Tissue and Cell*, **25**, 259–73.

Koch, E.A., Spitzer, R.H., Pithawalla, R.B. and Parry, D.A.D. (1994) An unusual intermediate filament subunit from the cytoskeletal biopolymer released extracellularly into seawater by the primitive hagfish (*Eptatretus stoutii*). *J. Cell Sci.*, **111**, 3133–44.

Koch, E.A., Spitzer, R.H., Pithawalla, R.B., Castillos III, F.A. and Parry, D.A.D. (1995) Hagfish biopolymer: a type I/type II homologue of epidermal keratin intermediate filaments. *Int. J. Biol. Macromol.*, **17**, 283–92.

Kosmath, I., Patzner, R.A. and Adam, H. (1981) The cloaca of *Myxine glutinosa* (Cyclostomata): a scanning electron microscopical and histochemical investigation. *Z. Mikrosk.-Anat. Forsch., Leipzig*, **95**, 6, S. 936–942.

Lametschwandtner, A., Lametschwandtner, U. and Patzner, R.A. (1986) The different vascular patterns of slime glands in the hagfishes, *Myxine glutinosa* Linnaeus and *Eptatretus stoutii* Lockington. A scanning electron microscope study of vascular corrosion casts. *Acta. Zoologica (Stockh.)*, **67**, 243–8.

Lehtonen, A., Kärkkäinen, J. and Haahti, E. (1966) Carbohydrate components in the epithelial mucin of hagfish, *Myxine glutinosa*. *Acta Chem. Scand.*, **20**, 1456–62.

Leppi, T.J. (1966) Histochemical studies of carbohydrate-rich substances in the Pacific hagfish (*Polistrotrema stoutii.*). *Amer. Zool.*, **6**, 580.

Leppi, T.J. (1967) Histochemical studies on mucous cells in the skin and slime glands of hagfishes. *Anat. Rec.*, **157**, 278.

Leppi, T.J. (1968) Morphochemical analysis of mucous cells in the skin and slime glands of hagfish. *Histochem.*, **15**, 68–78.

Lethbridge, R.C. and Potter, I.C. (1982). The skin, in *The Biology of Lampreys*, Vol III (eds M.S.Hardisty and I.C. Potter), Academic Press, New York, pp. 377–448.

Luchtel, D.L., Martin, A.W. and Deyrup-Olsen, I. (1991) Ultrastructure and permeability characteristics of the membranes of mucous granules of the hagfish. *Tissue and Cell*, **23**, 939–48.

Lumsden, J.S. and Ferguson, H.W. (1994) Isolation and partial characterization of rainbow trout (*Oncorhynchus mykiss*) gill mucin. *Fish Physiol. and Biochem.*, **12**, 387–98.

Markl, J. and Franke, W.W. (1988) Localization of cytokeratins in tissues of the rainbow trout: fundamental differences in expression pattern between fish and higher vertebrates. *Differentiation*, **39**, 97–122.

Markl. J., Winter, S. and Franke, W.W. (1989) The catalog and the expression complexity of cytokeratins in a lower vertebrate: biochemical identification of cytokeratins in a teleost fish, the rainbow trout. *Eur. J. Cell Biol.*, **50**, 1–16.

Martini, H. and Heiser, J.B. (1989) Field observations of the Atlantic hagfish, *Myxine glutinosa* in the Gulf of Maine. *Amer. Zool.*, **29**, 38A.

Newby, W.W. (1946) The slime glands and thread cells of the hagfish, *Polistrotrema stoutii. J. Morph.*, **78**, 397–409.

Parry, D.A.D. (1990) Primary and secondary structure of IF protein chains and modes of molecular aggregation, in *Cellular and Molecular Biology of Intermediate Filaments* (eds R.D. Goldman and P.M. Steinert), Plenum Press, New York, pp. 175–204.

Patzner, R.A., Hanson, V. and Adam, H. (1982) Fine structure of the surface of small mucous cells in the epidermis of the hagfish *Myxine glutinosa* (cyclostomata). *Acta Zoologica (Stockh.)*, **63**, 183–6.

Potter, I.C., Welsch, U., Wright, G.M., Honma, Y. and Chiba, A. (1995) Light and electron microscope studies of the dermal capillaries in three species of hagfishes and three species of lampreys. *J. Zool. Lond.*, **235**, 677–88.

Rahemtulla, F., Höglund, N.-G. and Løvtrup, S. (1976) Acid mucopolysaccharides in the skin of some lower vertebrates (hagfish, lamprey and chimaera). *Comp. Biochem. Physiol.*, **53B**, 295–8.

Riemer, D., Dodemont, H. and Weber, K. (1992) Analysis of the cDNA and gene encoding a cytoplasmic intermediate filament (IF) protein from the cephalochordate *Branchiostoma lanceolatum*; implications for the evolution of the IF protein family. *Eur. J. Cell Biol.*, **58**, 128–35.

Salo, W.L., Downing, S.W., Lidinsky, W.A., Gallagher, W.H., Spitzer, R.H. and Koch, E.A. (1983) Fractionation of hagfish slime and gland secretions: partial characterization of the mucous vesicle fractions. *Prep. Biochem.*, **13**, 103–35.

Schaffer, J. (1925) Zur Kenntnis der Hautdrüsen bei den Säugetieren und bei Myxine. *Z. Anat. Entwickl.-Gesch.*, **76**, 320–37.

Schreiner, K.E. (1916) Zur Kenntnis der Zellgranula. Untersuchungen über den Feineren bau der Haut von *Myxine glutinosa*. I. Teil. Erste Hälfte. *Arch. Mikr. Anat. (Abt. I)*, **89**, 79–188.

Schreiner, K.E. (1918) Zur Kenntnis der Zellgranula. Untersuchungen über den

Feineren bau der Haut von *Myxine glutinosa*. I. Teil Zweite Hälfte. *Arch. Mikr. Anat. (Abt. I)*, **92**, 1–63.

Sheehan, J.K., Thornton, D.J., Sommerville, M. and Carlstedt, I. (1991) Mucin structure. The structure and heterogeneity of respiratory glycoproteins. *Ann. Rev. Respir. Dis.*, **144**, 54–59.

Shephard, K.L. (1993) Mucus on the epidermis of fish and its influence on drug delivery. *Advanced Drug Deliv. Revs.*, **11**, 403–17.

Specian, R.D. and Neutra, M.R. (1984) Cytoskeleton of intestinal goblet cells in rabbit and monkey. The theca. *Gastroenterology*, **87**, 1313–25.

Spitzer, R.H., Downing S.W., Koch, E.A. and Kaplan, M.A. (1976) Hemagglutinins in the mucus of Pacific hagfish (*Eptatretus stoutii.*). *Comp. Biochem. Physiol.*, **54B**, 409–11.

Spitzer, R.H., Downing, S.W. and Koch, E.A. (1979) Metabolic-morphologic events in the integument of the Pacific Hagfish (*Eptatretus stoutii*). *Cell Tissue Res.*, **197**, 235–55.

Spitzer, R.H., Downing, S.W., Koch, E.A., Salo, W.L. and Saidel, L.J. (1984) Hagfish slime gland thread cells. II. Isolation and characterization of intermediate filament components associated with the thread. *J. Cell Biol.*, **98**, 670–7.

Spitzer, R.H., Koch, E.A. and Downing, S.W. (1988) Maturation of hagfish gland thread cells: composition and characterization of intermediate filament polypeptides. *Cell Motility and the Cytoskeleton*, **11**, 31–45.

Strahan, R. (1959) Slime production in *Myxine glutinosa* Linnaeus. *Copeia*, **2**, 165–6.

Terakado, K., Ogawa, M., Hashimoto, Y. and Matsuzaki, H. (1975) Ultrastructure of the thread cells in the slime gland of Japanese hagfishes, *Paramyxine atami and Eptatretus burgeri. Cell Tissue Res.*, **159**, 311–23.

Walvig, R. (1963) The gonads and the formation of the sexual cells, in *The Biology of Myxine* (eds A. Brodal and R. Fänge), Universtitsforlaget, Oslo, pp. 530–80.

Whitear, M. (1986) The skin of fishes including cyclostomes, in *Biology of the Integument 2. Vertebrates* (eds J. Bereiter-Hahn, A.G. Matoltsy and K.S. Richards), Springer-Verlag, Berlin, pp. 8–38.

Whitear, M. (1988) Variations in the arrangement of tonofilaments in the epidermis of teleost fish. *Biology of the Cell*, **64**, 85–92.

Whitear, M. and Mittal, A.K. (1983) Fine structure of the club cells in the skin of ostariophysan fish. *Z. Mikrosk Anat. Forsch.*, **97**, 141–57.

Whitear, M. and Zaccone, G. (1984) Fine structure and histochemistry of club cells in the skin of three species of eel. *Z. Mikrosk Anat. Forsch.*, **98**, 481–501.

Wilson, I.B.H., Gavel, Y. and von Heijne, G. (1991) Amino acid distributions around O-linked glycosylation sites. *Biochem. J.*, **275**, 529–34.

Zaccone, G., Howie, A.J., Mauceri, A., Fasulo, S., Lo Cascio, P. and Youson, J.H. (1995) Distribution patterns of cytokeratins in epidermis and horny teeth of the adult sea lamprey (*Petromyzon marinus*). *Folia Histochem. Cytobiol.*, **33**, 69–75.

9
THE DERMIS

Ulrich Welsch, Simone Büchl and Rainer Erlinger

SUMMARY

In hagfishes the dermis is a distinct layer of the integument, composed of regularly arranged collagen fibres. It is located immediately below the epidermis and is bordered at its inner aspect by the dermal endothelium. The collagen forms up to 45 lamellae in which the collagen fibres cross each other at an angle of about 90–110°. Furthermore the dermis contains 12 nm thick microfibrils which are considered to be part of the elastic fibre system, and sulphated proteoglycans. The extracellular components are produced by active fibroblasts which usually occupy a space between the consecutive layers of collagen. The diameter of the collagen fibrils varies between 90 and 100 nm, their D-period measures about 42 nm. In addition to the fibroblasts the dermis contains melanocytes, nerve fibres and small blood vessels. The dermal endothelium is usually a 0.7 µm thick continuous layer of flat cells interconnected by desmosome-like contacts. A basal lamina accompanies the dermal endothelium both on its outer and inner aspect. The cell membrane of the dermal endothelial cells forms abundant tubular invaginations as occur in a very similar fashion in vascular endothelial cells.

9.1 INTRODUCTION

As in other fish groups (Krause, 1923) the dermis of hagfishes is a distinct layer of the integument between epidermis and subdermis (Figure 9.1a). It is composed of up to 45 layers of very densely packed collagen fibrils and thus differs significantly from the subdermis which is built up by a loose connective tissue with abundant adipose cells and blood vessels (Rabl, 1931). The outer border of the dermis is marked by the thick epidermal basal lamina (Figure 9.1b). The inner aspect of the dermis is marked by a continuous flat epithelium, the so-called dermal endothelium. This peculiar epithelium was originally described in teleosts by Whitear (1986) and also occurs in lampreys (Potter and Welsch, 1992). The subdermis is mainly composed of glycogen-rich adipocytes, which contain numerous individual lipid droplets, and abundant blood vessels.

In hagfishes the dermis does not contain any scales but forms a continuous tough envelope of the body. It is not only a protective coat but also serves as an indirect attachment site for the body musculature, since the collagen fibres of the myosepta often extend into the subdermis and dermis. However, this interconnection is not as intensive as in lampreys because of the vast subdermal blood sinus in hagfishes. The principal component of the dermis, the collagen fibrils, are non-elastic pull-resistant structures. The dermis thus gives shape to the hagfish body but due to its architecture also allows undulating and other, in part, extreme movements.

The integument of adult *Eptatretus burgeri*, *Myxine glutinosa* and *Paramyxine* (= *Eptatretus*) *atami* was studied. Skin of all body regions was analysed including nasal

The Biology of Hagfishes. Edited by Jørgen Mørup Jørgensen, Jens Peter Lomholt, Roy E. Weber and Hans Malte. Published in 1998 by Chapman & Hall, London. ISBN 0 412 78530 7.

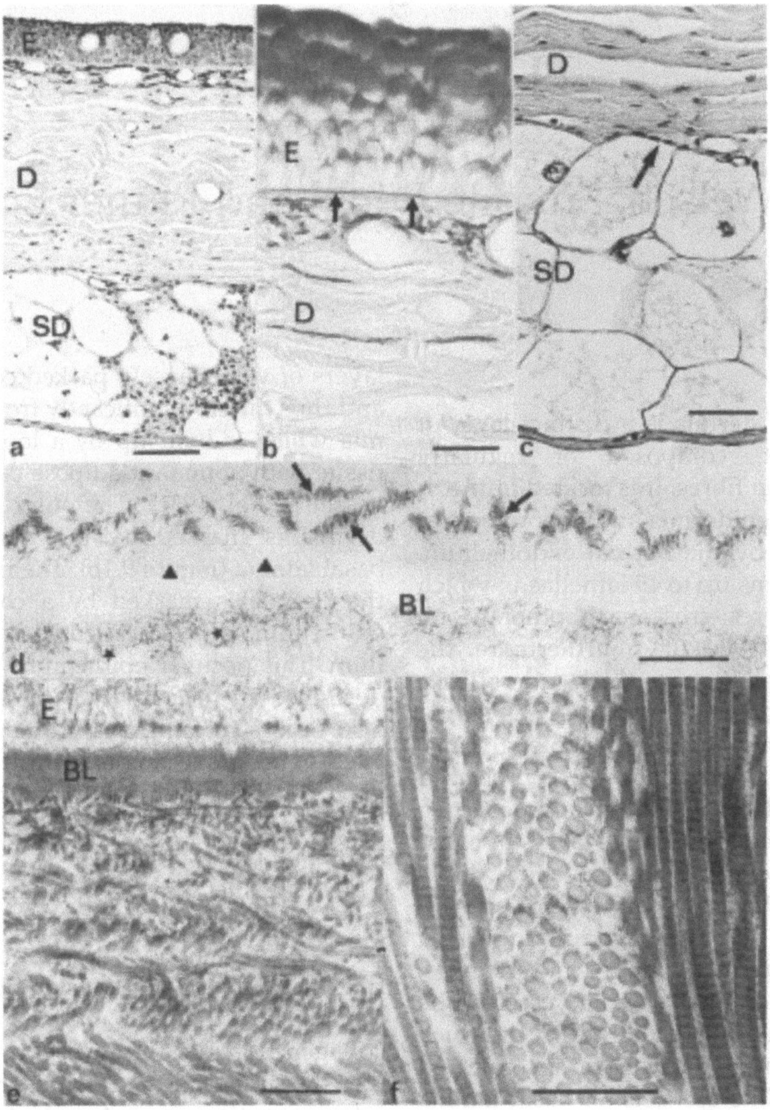

Figure 9.1 (a) *Eptatretus burgeri*, overview of the integument, E: epidermis, D: dermis, SD: subdermis; H.E. scale bar 100 µm. (b) *Eptatretus burgeri*, epidermis (E) and dermis (D) PAS-reaction, note strong reaction in small epidermal mucous cells and in the epidermal basal lamina (arrows); melanoctyes in the upper dermis are associated with dermal capillaries (oval profiles); scale bar 50 µm. (c) *Eptatretus burgeri*, abrupt transition between dermis (D) and subdermis (SD). The dermal endothelium is hardly recognizable (*arrow*), the subdermis is composed of large plurivacuolar adipocytes; H.E. scale bar 50 µm. (d) *Eptatretus burgeri*, proteoglycans associated with the epidermal basal lamina (BL) occur mainly in the *lamina rara* (*arrows*) and below (asterisks) the *lamina densa* (*triangles*), CMB 0.3 m MgCl; scale bar 250 nm. (e) *Eptatretus burgeri*, uppermost zone of the dermis immediately beneath epidermal basal lamina (BL), E: epidermis cell; scale bar 500 nm. (f) *P. atami*, three layers of collagen fibrils, in the middle layer with cross-sectioned, left and right with longitudinally sectioned fibrils; scale bar 500 nm.

and oral tentacles and tail fin. For technical details see Potter and Welsch (1992) and Erlinger *et al.* (1993).

9.2 FUNCTIONAL MORPHOLOGY OF THE HAGFISH DERMIS

The dermis, which measures in the truncal region of all species studied as a whole about 250 μm in thickness (Figure 9.1a), is composed of up to 45 layers of collagen. In the tentacles and the tail fin there are fewer layers. Within each layer the collagen fibres are roughly oriented in the same direction, whereas in the successive layers (= lamellae) the collagen fibres run in different directions (Figure 9.1a, f). The fibres of consecutive layers cross each other at an angle of about 90–110°. The deeper layers of the dermis are thicker (10 μm) than those at the outer periphery (4 μm). Within each layer the collagen is divided into closely packed bundles of fibrils which usually are running exactly in the same direction and separated by a narrow space of amorphous material which often separates the consecutive layers, too. In the space separating the layers, bundles of 12 nm thick microfibrils (Figure 9.3a) are also to be found which have been interpreted in the lamprey dermis by Potter and Welsch (1992) to represent oxytalan fibres, i.e. part of the elastic fibre system. Such bundles of oxytalan fibres can also run vertically through the dermis, as do single narrow bundles of collagen fibrils, which penetrate into the subdermis at the basis of the dermis. A thin layer of microfibrils usually covers the dermal side of the dermal endothelium. In *Myxine glutinosa* we found amorphous dense material among the microfibrils, suggesting the presence of elastic fibres (Schinko *et al.*, 1992). Immediately below the epidermis a 5–7 μm thick zone is to be found in which the collagen lacks the regular organization of the deeper zones (Figure 9.1e, see below). This brief description of the principal structure of the dermis applies to all areas of the hagfish body. A more detailed analysis of the orientation of the collagen fibres and fibrils may reveal a more complex pattern in the individual regions of the integument. In specialized areas as in the nasal tentacles the dermis is relatively thin and the general collagenous structure is less regular. The general structure of the hagfish dermis resembles closely that of the lamprey dermis (Wright and McBurney, 1992; Potter and Welsch, 1992).

As mentioned above, the uppermost zone of the dermis, immediately underneath the epidermis, is marked by irregularly arranged collagen fibrils of variable diameter (20–80 nm, Figure 9.1e). These collagen fibrils can be rather densely packed but can also be loosely arranged. In areas with few collagen fibrils abundant microfibrils occur. The collagen fibrils are separated by amorphous material containing proteoglycans and glycoproteins as indicated by positive PAS reaction and binding of various lectins. Lectins are carbohydrate binding proteins which have been widely used in histochemical studies (Spicer and Schulte, 1992). This zone also occurs in lampreys, in which Wright and McBurney (1992) have called it *pars fibroreticularis* of the basement membrane. It borders the epidermal basal lamina, which is about 350 nm thick and consists of a *lamina rara (=lucida)* (100 nm) and a broad *lamina densa* (250 nm). In the routine transmission electron microscope preparation densely packed fine-filamentous structures (Figure 9.2a, b) traverse the *lamina rara* and interconnect at regular intervals the basal plasma membrane of the basal epidermis cell and the *lamina densa,* such fine filamentous structures also occur in lampreys (Wright and McBurney, 1992). Positive staining with Cupromeronic Blue (CMB) reveals that these structures are composed of proteoglycans. Moreover CMB staining shows evenly distributed proteoglycans in a 100 nm thick layer underneath the lamina densa which in itself is free of CMB positive proteoglycans (Figure 9.1d). The lamina densa is composed of an unusually

Figure 9.2 (a–c) *Myxine glutinosa*. (a) Epidermal basal lamina, E: basal part of basal epidermis cell, (1) *lamina rara*, (2) *lamina densa*, arrow: anchoring fibril; scale bar 250 nm. (b) Basal lamina with interruptions (arrow); scale bar 500 nm. (c) Collagen fibril with wide spaced D-period of about 110 nm (arrow). N, nerve fibre; scale bar 500 nm.

homogeneous dense material and is perforated by single interruptions about 120–140 nm wide (Figure 9.2b). The basal aspect of the lamina densa is of irregular outline and interconnected with microfibrils or cross striated anchoring fibrils (Figure 9.2a) composed of collagen VII (Drenckhahn and Kugler, 1994).

At the ultrastructural level the individual collagen fibrils of the collagenous layers (= lamellae) are not homogenous with respect to their diameter, which varies between 90 nm and 100 nm (Figure 9.3a, b). In cross-sections the fibrils often have an electron translucent core. The D-period measures about 42 nm (Figure 9.3b). Such a relatively short D-period has also been observed in dermal collagen of lampreys by x-ray diffraction by Brodsky *et al.* (1994). Proteoglycans occur regularly at the surface of the collagen fibrils and interconnect neighbouring ones, keeping them apart at a constant distance (Figure 9.3c). The application of CMB in 0.3 m $MgCl_2$ (Scott, 1985) shows that the proteoglycans are sulphated. The needle-shaped proteoglycan-dye complex measures up to 30 nm in length (Figure 9.3c). In the vertical bundles of collagen fibrils, there is a greater variability of fibril diameters (40–100 nm) than in the horizontal layers. The bundles of microfibrils can be found within or at the periphery of the vertical bundles of collagen, which they usually accompany (Figure 9.3a).

Occasionally, especially close to basal laminae of Schwann cells or of the dermal endothelium, thick collagen fibrils (diameter about 220 nm) with a wide D-period of about 110 nm can be found (Figure 9.2c).

The following cell types occur in the hagfish dermis: fibroblasts and melanocytes, in addition to capillaries and nerve fibres, are a normal constituent of the dermis; occasionally neutrophil granulocytes are to be found outside blood vessels in the dermal connective tissue.

The fibroblasts are usually flattened cells with long extensions which occur predominantly between the consecutive layers of collagen. They can, however, also extend processes vertically or obliquely through a few layers. They are actively synthesizing cells with a nucleus rich in euchromatin and a big nucleolus and well-developed rough endoplasmic reticulum (RER). Also the other cell organelles are well developed. In the cellular periphery smooth vesicles and single tubular invaginations of the cell membrane occur. Single cells have been seen in our material which contain a fair number of lysosomes. It has not been determined whether such cells represent macrophages or fibroblasts in a particular physiological state in which degrading processes predominate. The latter assumption is supported by the observations of Wright and McBurney (1992) on the dermis of *Petromyzon marinus* during metamorphosis. The melanocytes occur in the outer parts of the dermis in the neighbourhood of the capillaries (Figure 9.1b). They have a pale, active nucleus and extend long processes. Their principal characteristic is oval melanosomes (Figure 9.3d). Their peripheral cytoplasm contains abundant small smooth vesicles and tubular profiles, which open up at the surface of the cell. Frequently melanocytes and RER-rich cells, presumably fibroblasts, are closely attached to each other. Nerve terminals with clear vesicles form contacts with melanocytes (Figure 9.4c) and possibly with the attached fibroblasts, too.

Nerve fibres are common in the outer parts of the dermis (Figure 9.4a) and occur usually in the neighbourhood of melanocytes or in the pericapillary space. They are particularly frequent in the nasal and oral tentacles. Regularly nerves running into the epidermis can be found. Often about 3–7 (occasionally up to 20) nerve fibres of variable diameters (about 100 nm up to 1 μm) are embedded in pockets of a Schwann cell which is surrounded by a basal lamina. The cell membrane of the Schwann cells – as in many other cells in hagfishes – forms abundant tubular invaginations. Often small groups of such small nerve fibres occur together. The individual nerve fibres contain neurofilaments and microtubules and can form varicosities with clear vesicles (Figure 9.4b) or dense core granules. Single varicosities have been found to contain 4–5 mitochondria and glycogen particles. The pericytes are possibly a target of such varicosities. Occasionally one Schwann cell encloses only one, relatively thick (diameter 1.15 μm) nerve fibre. Such Schwann cells are surrounded by an unusually prominent basal lamina (80 nm thick)

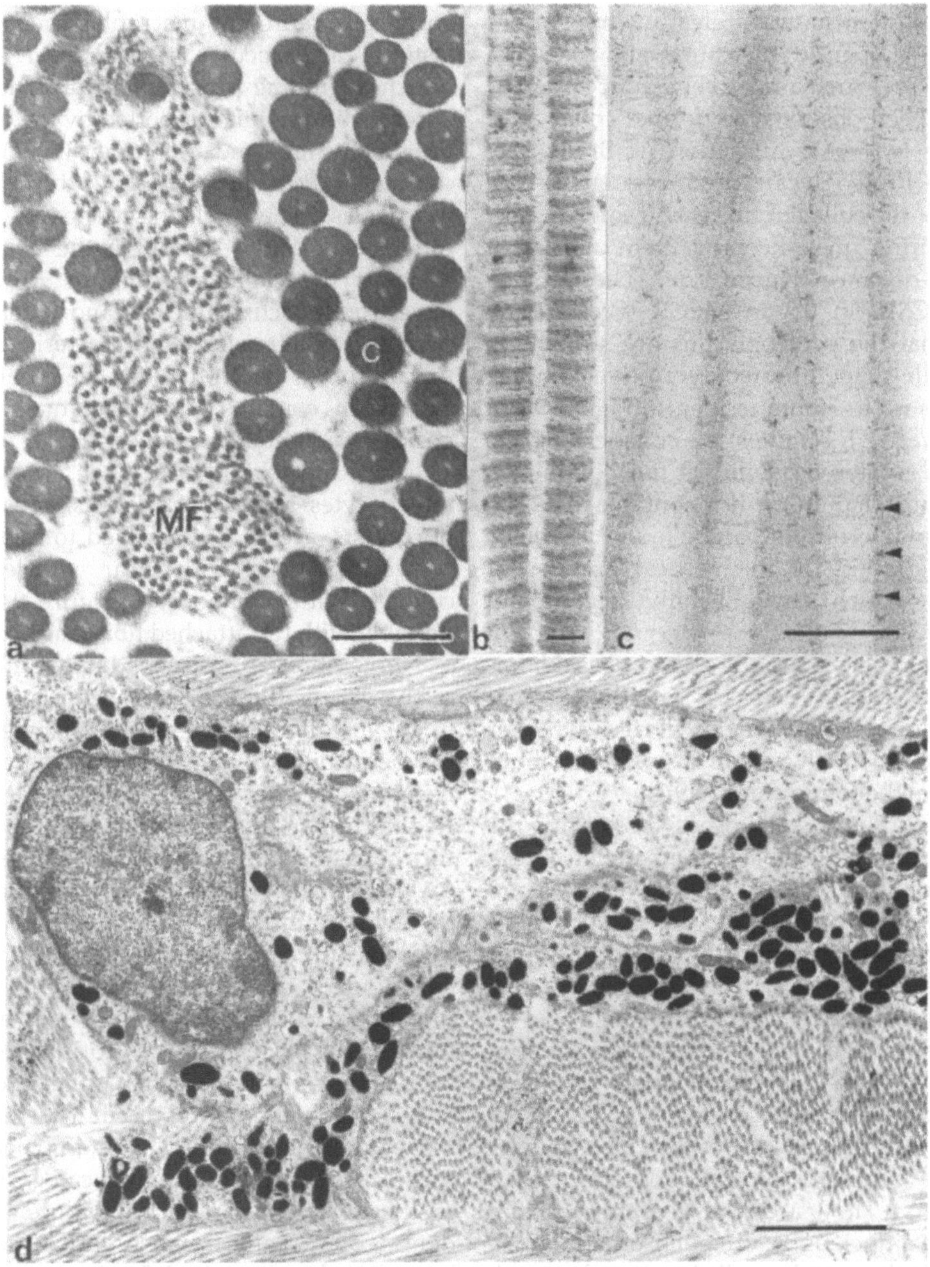

Figure 9.3 (a) *P. atami*, dermis, cross-sectioned collagen fibrils (C) often have an electron translucent core, MF: bundle of microfibrils; scale bar 200 nm. (b) *Eptatretus burgeri*, longitudinally cut collagen fibrils of the dermis; scale bar 50 nm. (c) *P. atami*, regularly spaced proteoglycans (arrowheads) at dermal collagen fibrils, CMB 0.3 m $MgCl_2$; scale bar 200 nm. (d) *P. atami*, melanocytes in the dermis; scale bar 3 μm.

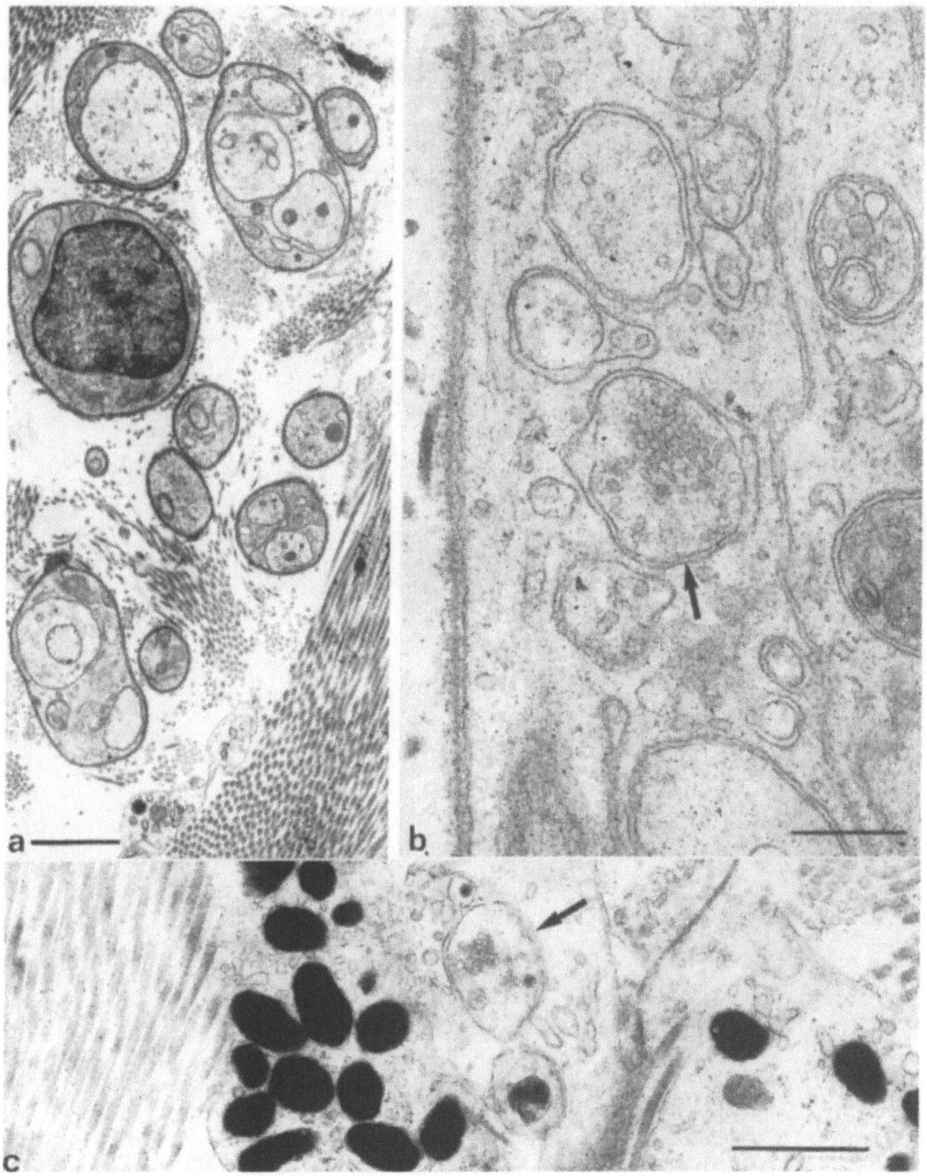

Figure 9.4 Nerves in the dermis of *P. atami.* (a) Small bundle of dermal nerves at the level of the capillaries; scale bar 2 μm. (b) Nerve fibres embedded in the cytoplasm of a Schwann cell which is surrounded by a basal lamina. Arrow: fibre with small clear vesicles; scale bar 500 nm. (c) Nerve fibres with clear vesicles (arrow) in close contact with a dermal melanocyte; scale bar 1 μm.

and are loosely ensheathed by processes of fibroblasts reminiscent of neural cells.

The dermal endothelium is a continuous flat epithelium accompanied by a typical basal lamina both on its inner and outer side (Figure 9.5a), an unusual and remarkable feature of unknown function and origin, demarcating the border between dermis and

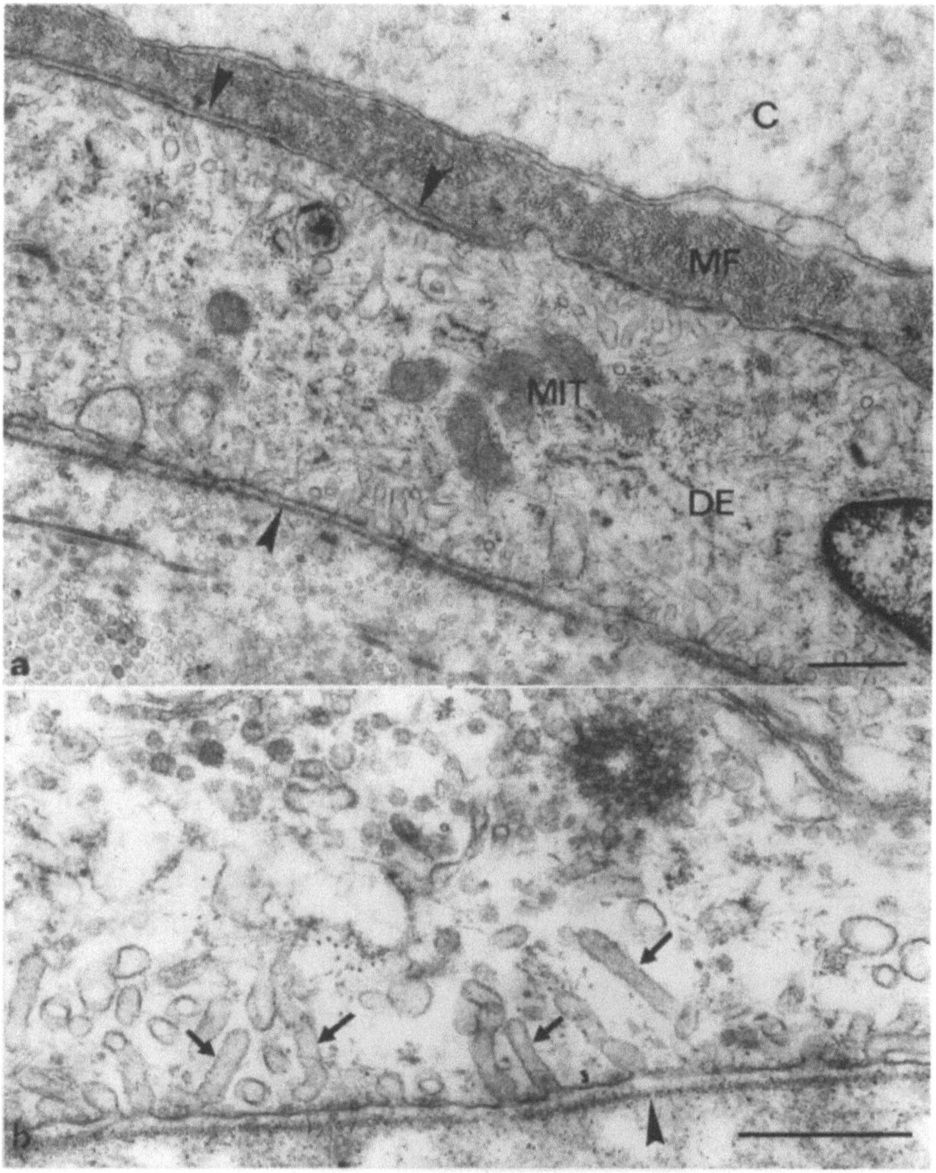

Figure 9.5. *P. atami*, dermal endothelium. (a) Dermal endothelial cell (DE) with basal lamina (arrowheads) at its dermal (above) and subdermal (below) side, MIT: mitochondria, MF: microfibrils, C: collagen; scale bar 500 nm. (b) Tubular invaginations (arrows) of the cell membrane. Arrowhead: basal lamina at subdermal side; scale bar 500 nm.

subdermis. Bundles of collagen fibrils, blood vessels and nerves pass through it. The endothelium measures about 0.7 μm in thickness, in the area of the nucleus the thickness increases to 2.5 μm. The individual epithelial cells are interconnected by interdigitations and desmosome-like contacts. Their flat nucleus is rich in euchromatin and contains a distinct nucleolus. Concentrated in the neighbourhood of the nucleus, a small Golgi apparatus, RER cisterns, mitochondria and lysosomal structures occur. All these organelles, however, can also be found in other more peripheral parts of the cell. The most characteristic feature of these cells are countless deep tubular invaginations of the cell membrane (Figure 9.5b). The entire cytoplasm is pervaded by numerous cytoplasmic intermediate filaments measuring 8 nm in diameter. This brief description refers to the dermal endothelium of the truncal region of the body. In the area of the nasal and oral tentacles it is less regular and can be two or three layers thick and interrupted by wide gaps. This dermal endothelium of hagfishes resembles vascular endothelial cells and may be engaged to some extent in transport processes. It presumably forms a barrier between dermis and subdermis, the physiological significance of which remains to be explored. With respect to its morphological origin its peculiar structure and localization suggest a comparison with the lateral myocoelomic epithelium as found in *Branchiostoma*. Possibly the dermal endothelium of hagfishes and other fish groups is a homologue of the myocoelomic epithelium of the cephalochordates. Holland and Holland (1990) have described this flat epithelium in *Branchiostoma* at the EM-level and found indications for secretory activities, which are not obvious in the hagfish dermal endothelium. However, such indications have been observed in the dermal endothelium of *Petromyzon* during metamorphosis (Wright and McBurney, 1992). On the other hand, the dermal endothelial cells also resemble flattened fibroblasts, which, however, lack a basal lamina. Such flat fibroblasts can form epithelium-like layers, e.g. at the border between *dura mater* and *arachnoidea* in the mammalian meninges (Andres, 1990) or around peripheral nerves (perineural sheath). Such perineural cells can even be associated with an irregular layer of extracellular material (*Myxine*, personal observations). Whitear (1986) and Wright and McBurney (1992) favour the idea that dermal endothelial cells correspond to fibroblasts. Also Schwann cells bear resemblance to the dermal endothelial cells. In the absence of experimental physiological and embryological work this layer remains somewhat enigmatic. It is absent in tetrapod vertebrates.

ACKNOWLEDGEMENTS

We are grateful to Professor Y. Honma and Professor A. Chiba (Niigata, Japan) for the generous gift of perfusion-fixed integuments of *Paramyxine atami* and *Eptatretus burgeri*, and to Profesor B. Strömberg (Kristineberg, Sweden) for providing fresh *Myxine glutinosa* for study.

REFERENCES

Andres, K.H. (1990) Über die Feinstruktur der Arachnoidea und Dura mater von Mammalia. *Zeitschrift für Zellforschung und mikroskopische Anatomie*, **79**, 272–295.

Brodsky, B., Belbruno, K.C., Hardt, T.A. and Eikenberry, E.F. (1994) Collagenfibril structures in lamprey. *Journal of Molecular Biology*, **243**, 38–47.

Drenckhahn, D. and Kugler, P. (1994) Bindegewebe, in *Benninghoff Anatomie*, Vol. 1, 15th edition (eds D. Drenckhahn and W. Zenker), Urban & Schwarzenberg.

Erlinger, R., Welsch, U. and Scott, J.E. (1993) Ultrastructural and biochemical observations on proteoglycans and collagen in the mutable connective tissue of the feather star *Antedon bifida* (Echinodermata, Crinoidea). *Journal of Anatomy*, **183**, 1–11.

Holland, N.D. and Holland, L.Z. (1990) Fine structure of the mesothelia and extracellular materials in the coelomic fluids of the fin boxes, myocoels and skelerocoels of a lancelet, *Brachiostoma floridae* (Cephalochordata = Acrania). *Acta Zoologica*, **71**, 193–250.

Krause, R. (1923) *Mikroskopische Anatomie der Wirbeltiere*, IV. Teleostier, Plagiostomen, Zyklostomen und Leptokardier, de Gruyter, Berlin.

Potter, I.C. and Welsch, U. (1992) Arrangement,

histochemistry and fine structure of the connective tissue architecture of lampreys. *Journal of Zoology, London*, **226**, 1–30.

Rabl, H. (1931) Integument. 1. Integument der Anamnier, in *Handbuch der vergleichenden Anatomie der Wirbeltiere* (eds L. Bolk, E. Göppert, E. Kallius, and W. Lubosch), Vol. 1, Urban & Schwarzenberg, Berlin and Wien, pp. 271–374.

Schinko, I., Potter, I.C., Welsch, U. and Debbage, P. (1992) Structure and development of the notochord elastica externa and nearby components of the elastic fibre system of agnathans. *Acta Zoologica*, **73**, 57–66.

Scott, J.E. (1985) Proteoglycan histochemistry – a valuable tool for connective tissue biochemists. *Collagen Relations Research*, **5**, 765–74.

Spicer, S.S. and Schulte, B.A. (1992) Diversity of cell glycoconjugates shown histochemically: a perspective. *Journal of Histochemistry and Cytochemistry*, **40**, 1–38.

Welsch, U. and Potter, I.C. (1994) Variability in the presence of elastic fibre-like structures in the ventral aorta of Agnathans (hagfishes and lampreys). *Acta Zoologica*, **75** (4), 323–7.

Whitear, M. (1986) The skin of fishes including cyclostomes. dermis, in *Biology of the Integcument 2. Vertebrates* (eds J. Bereiter-Hahn, A.G. Maltolsy and K.S. Richards, Springer Verlag, Berlin, pp. 39–64.

Wright, G.M. and McBurney, K.M. (1992) Changes in the ventral dermis and development of iridiophores in the anadromous sea lamprey, *Petromyzon marinus*, during metamorphosis: an ultrastructural study. *Histology Histopathology*, **7**, 237–50.

PART FOUR
Supporting Tissues

10
THE NOTOCHORD

Ulrich Welsch, Akira Chiba and Yoshiharu Honma

SUMMARY

The notochord of hagfishes is composed of a central rod of epithelial cells which are surrounded by a complex sheath; there are no traces of any vertebral structures. The notochord epithelial cells are interconnected by abundant desmosomes which are intracellularly associated with a well-developed system of intermediate filaments (diameter 10–12 nm). The vast majority of the notochord epithelial cells is characterized by a 60 µm wide vacuole containing heteromorphic material which reacts positively with several histochemical tests for carbohydrates. The peripheral notochord cells are rich in RER and other cell organelles, suggesting active protein synthesis. In these cells the initial steps of vacuole formation can be found. The notochord sheath consists of two main layers, the inner and the outer sheath, which are separated by a distinct *elastica externa*, a layer of electron dense extracellular material. The acellular PAS-positive and 50 µm thick inner sheath is composed of densely packed 8–11 nm tubular microfibrils, the chemical composition of which has not yet been determined. The outer 20–35 µm thick sheath consists of typical 30–50 nm thick cross-striated collagen fibrils, tubular microfibrils, elastic-like fibres, proteoglycans, fibroblasts and single blood capillaries.

10.1 INTRODUCTION

The notochord is the primary axial skeleton of the vertebrates. It is a firm but flexible rod, which runs originally – as in hagfishes – undivided from the pituitary region to the tip of the tail. With the exception of hagfishes, lampreys and some bony fishes, in all vertebrates the notochord is to various degrees replaced or supplemented by the vertebrae; in lampreys and some teleosts (e.g. sturgeons) vertebral elements are confined to dorsal or dorsal and ventral elements (Remane, 1936). It, however, never disappears completely, in the embryo of all vertebrates it has an important inductive role and even in adult mammals often traces of the notochord can be detected in the intervertebral discs.

In contrast to the heterogeneous morphology of the notochord in invertebrate chordates (Welsch and Storch, 1969; Flood, 1972), the microscopic anatomy of the vertebrate notochord is rather uniform (Remane, 1936). The principal component is a central rod of epithelial cells the vast majority of which is marked by a large intracellular vacuole. Only the peripheral 1–2 layers of cells, the so-called notochord 'epithelium' (a somewhat misleading term, since all notochordal cells are epithelial cells), usually lack this large vacuole and have a typical cytoplasm of epithelial cells with numerous organelles and a large nucleus. Close examination reveals that a few of these notochordal epithelial cells already posses a small vacuole and that a series of cells, beginning with avacuolar cells and ending with large vacuolated cells, can be recognized in the periphery of notochord. The rod of notochord epithelial cells is surrounded by a notochord sheath, which in

The Biology of Hagfishes. Edited by Jørgen Mørup Jørgensen, Jens Peter Lomholt, Roy E. Weber and Hans Malte. Published in 1998 by Chapman & Hall, London. ISBN 0 412 78530 7.

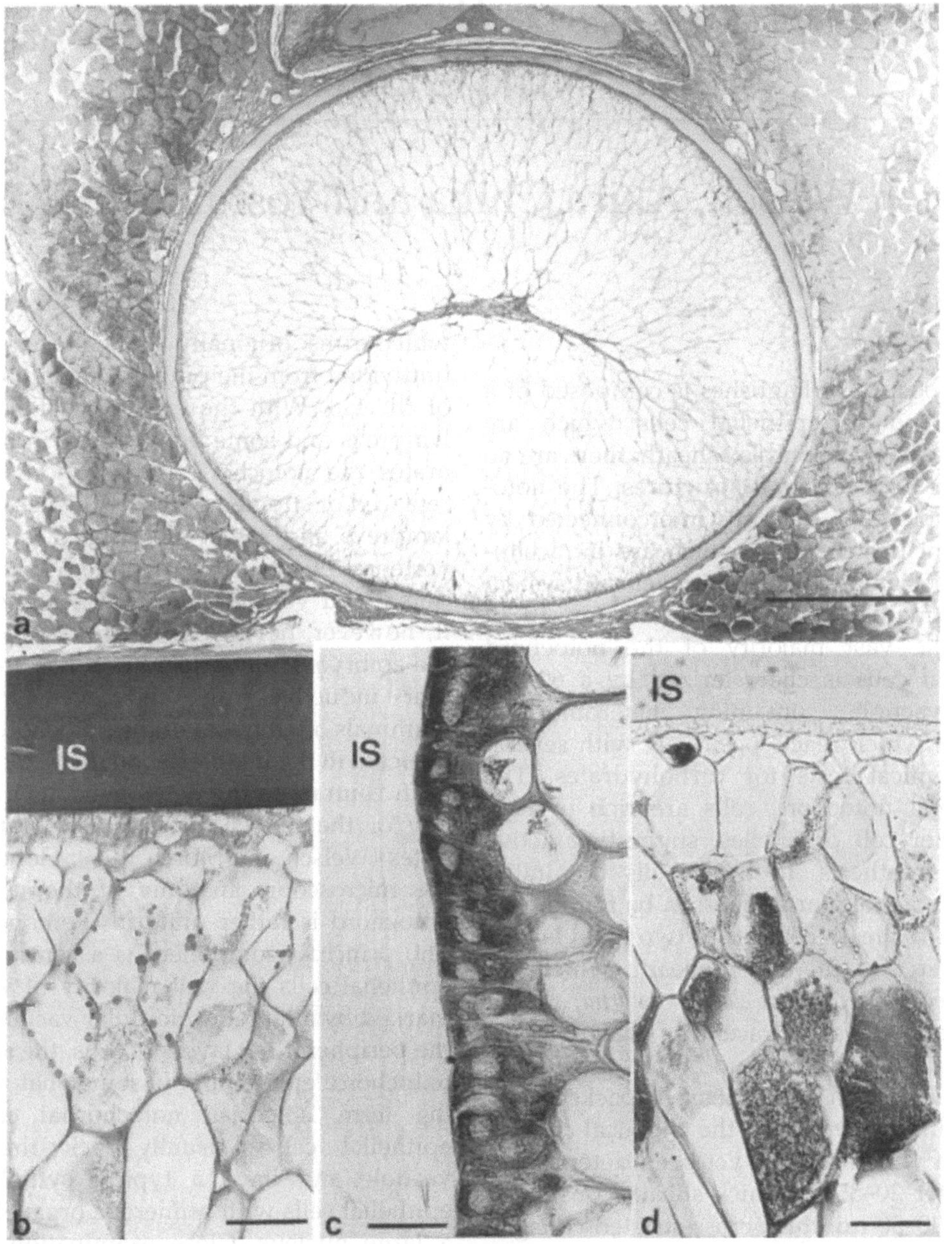

Figure 10.1 Histologic sections of the notochord of *Eptatretus burgeri*. (a) Cross-section through the entire notochord, note crescent-shaped condensation of cells in the centre; scale bar 1 mm. (b) PAS reaction, note strong reaction in the inner notochord sheath (IS) and of globular bodies in the peripheral vacuolated cells; scale bar 50 µm. (c) Strong binding of the lectin Con A in the notochord epithelium; scale bar 25 µm. (d) Binding of lectin UEA I to the contents of the vacuoles of many vacuolar cells; scale bar 50 µm.

hagfishes and various other fish groups is composed of inner acellular and outer cellular parts. Immediately outside the notochord epithelium, a peculiar thin layer of extracellular material has been termed *elastica interna*, while another thin layer of extracellular material separating the inner and outer part of the notochord sheath has been termed *elastica externa* (Krause, 1923; Remane, 1936). The detailed examinations carried out for this communication have shown that the structure of the hagfish notochord falls fully into the above outlined pattern of vertebrate notochord structure (Figures 10.1a, 10.2a). The literature on the hagfish notochord is scanty; Flood (1969, 1972) reported briefly on the fine structural characters of the notochord epithelium of *Myxine glutinosa* and Schinko *et al.* (1992) gave observations on the connective tissue of the notochord sheath of different species of hagfishes and lampreys.

It is the aim of this contribution to give a survey on the morphology (light microscopy, transmission and scanning electron microscopy) of the hagfish notochord based on the analysis of different species from the Atlantic and Pacific. Functional aspects as derived from ultrastructural observations will be included. For this contribution freshly fixed notochord material of adult *Myxine glutinosa, Paramyxine* (= *Eptatretus*) *atami* and *Eptatretus burgeri* were studied. For techniques see Potter and Welsch (1992) and Scott (1985). Attempts to characterize keratins in the sections with different antibodies to human keratins failed, except for KL 1, a pankeratin marker. Attempts to demonstrate collagen types I, III and IV and laminin with laminin antibodies were not successful in our hands.

10.2 NOTOCHORD SHEATH

In all hagfishes studied the notochord sheath is composed of special connective tissue and consists of an inner and an outer part, the structure of which differs distinctly and which are sharply separated by a distinct layer of homogenous extracellular material, traditionally termed *elastica externa*; a thin *elastica interna* is less distinct and occurs between inner sheath and notochord cells. Only in the periphery of the external sheath are small blood vessels found, the inner part of the sheath and the notochord itself are avascular, pointing to specialized metabolic conditions for the vacuolated notochordal epithelial cells.

The external sheath measures about 20–35 µm in thickness and is composed of densely packed collagen fibres and single fibroblasts (Figures 10.1a, 10.3a). The collagen fibres are predominantly arranged in a circular or spiral manner, with single fibres running longitudinally. The collagen fibres consist of very densely packed relatively thin collagen fibrils measuring 30–50 nm in diameter and having a relatively short periodicity (D-period) of about 40 nm. Staining with Cupromeronic Blue (CMB) (Scott, 1985) reveals that the collagen fibrils are surrounded and interconnected by proteoglycans with strongly sulphated glycosaminoglycans (Figure 10.3b). Among the collagen fibrils single bundles of microfibrils and single elastic-like fibres (bundles of microfibrils with a core of homogenous material) running mainly in a longitudinal direction are to be found. For a recent discussion on the morphology of the elastic fibre system in the Agnatha see Wright (1984) and Schinko *et al.* (1992). The fibroblasts are elongated active cells with well-developed rough endoplasmic reticulum (RER) and euchromatin-rich nuclei with distinct nucleoli. The infrequent blood capillaries are lined by a continuous typical flat endothelium.

The *elastica externa* measures about 2–4 µm in thickness and consists of highly electron dense material (Figure 10.3a) which stains positively with aldehyde fuchsin and Verhoeff's stain for elastic fibres. It does not stain specifically, however, with orcein and resorcin–fuchsin, classical stains for elastic

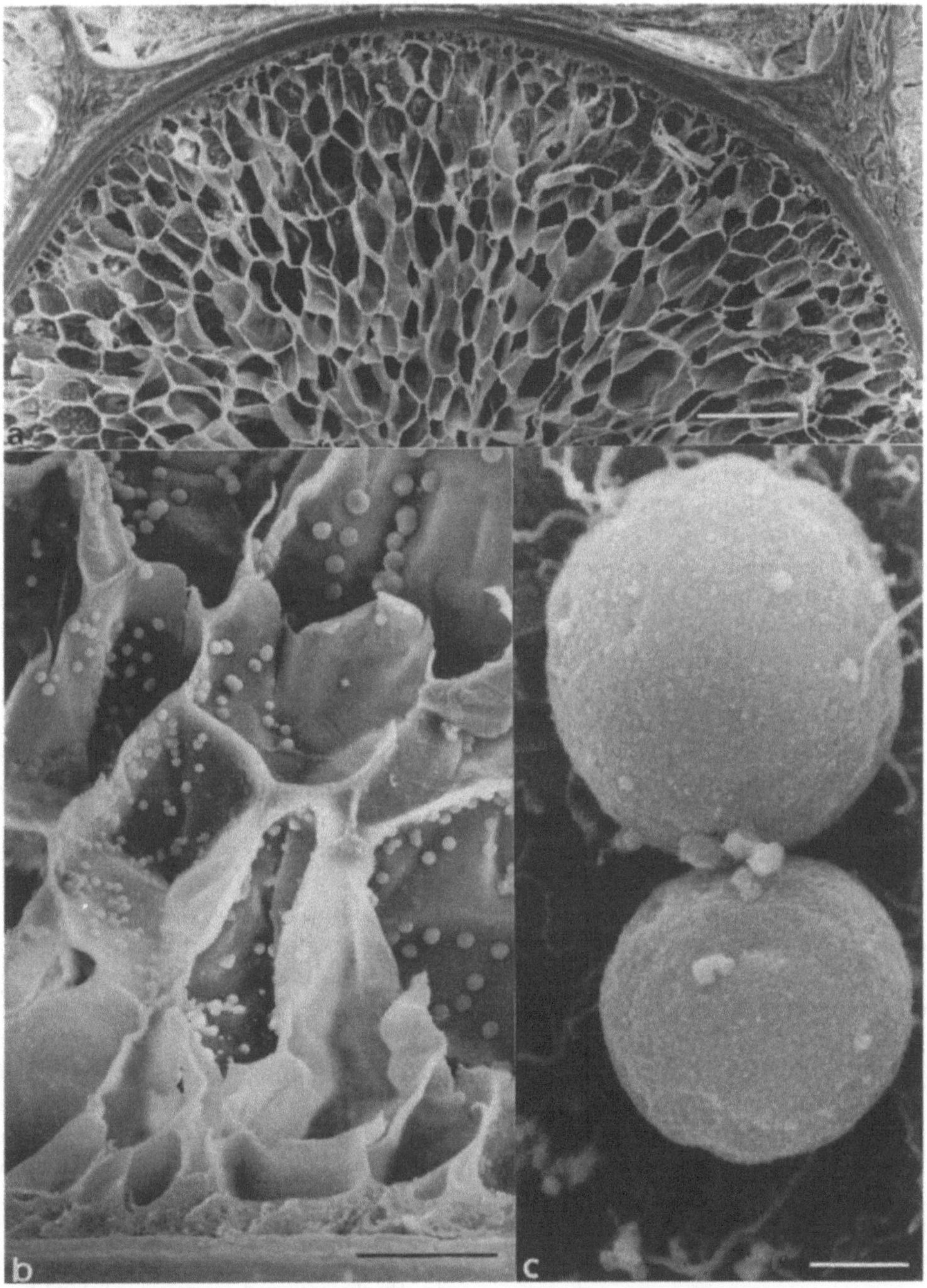

Figure 10.2 Scanning electron micrographs of the notochord of *Eptatretus burgeri*. (a) Overview of the dorsal part of the notochord; scale bar 100 μm. (b) Periphery of the notochord with non-vacuolated and vacuolated cells, note globular deposits at the vacuolar wall; scale bar 20 μm. (c) High power of two globular deposits and thread-like structures, representing slender microvilli; scale bar 0.5 µm.

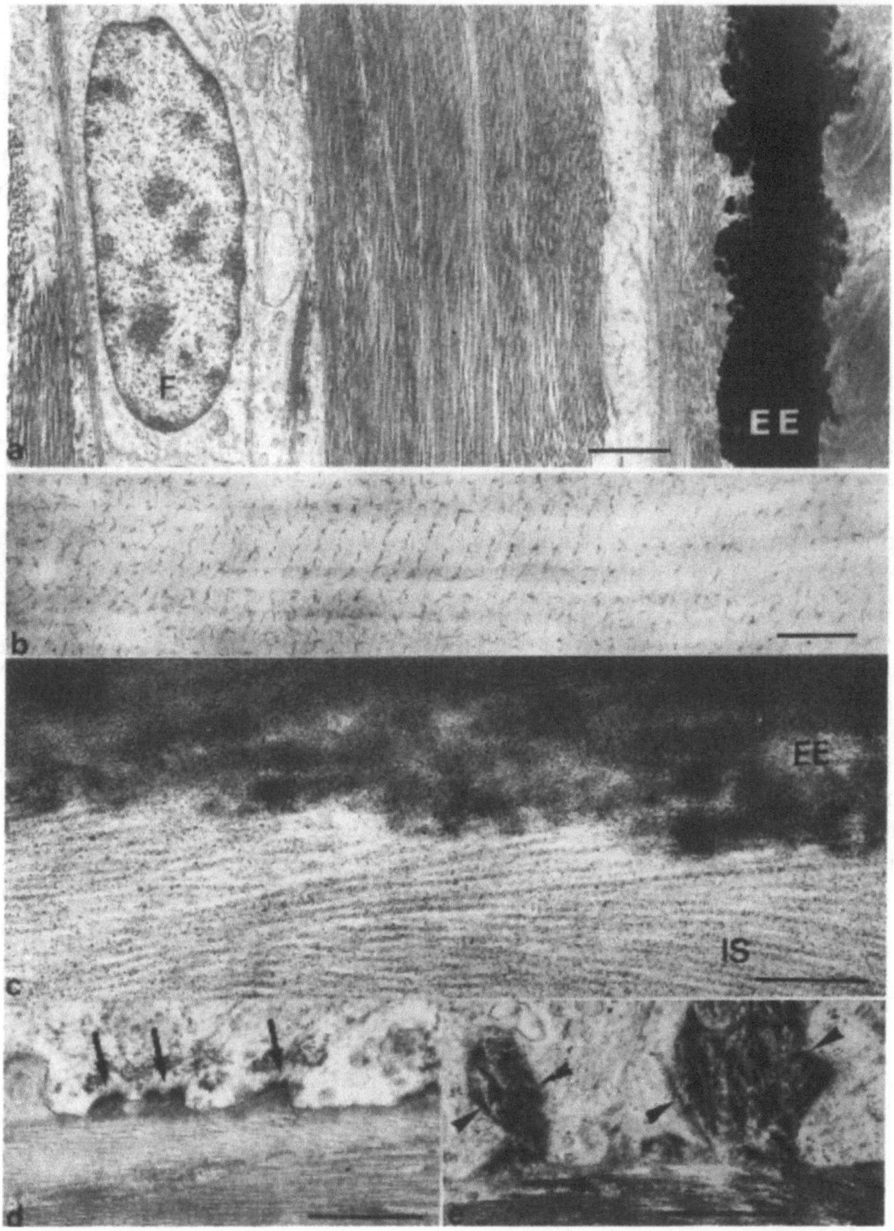

Figure 10.3 Notochord sheath. (a) *P. atami*, outer sheath with fibroblast (F) and densely packed collagen fibrils; EE: *elastica externa*; scale bar 2 µm. (b) *Eptatretus burgeri*, proteoglycans (thin, needle-like precipitates) attached to collagen fibrils, CMB with 0.3 m MgCl; scale bar 200 nm. (c) Elastica externa (EE) and inner sheath (IS) built up by densely packed thin tubular fibrils; scale bar 200 nm (d) *P. atami*, basal cell membrane of notochord epithelial cells with hemidesmosomes (arrows); opposite condensations in the zone of the basal lamina; scale bar 0.5 µm. (e) *P. atami*, knob-like protrusions of extracellular material (arrowheads) extending from the zone of the basal lamina; scale bar 1 µm.

fibres in mammals. With Masson's trichrome and Heidenhain's azan it stains red, with Goldner's trichrome it stains red-brown. The high density of the material of the *elastica externa* resembles that of hagfish elastic fibres (Schinko *et al.*, 1992) but differs from the low electron density of elastic fibres in higher vertebrates. It is still an open question whether this structure has, indeed, elastic properties, since corresponding biochemical or biomechanical studies are lacking. Its inner and outer border are of irregular outline (Figure 10.3a). Embedded in its outer half, single collagen fibrils can be observed. Several observations suggest that the *elastica externa* grows at its outer aspect of apposition of the amorphous component of elastic-like fibres (Schinko *et al.*, 1992). At its inner aspect the thin fibrils of the inner sheath (see below) penetrate into the dense material of the *elastica externa* (Figure 10.3c) which, on the other hand, may send irregularly shaped extensions deeply into the inner sheath. Thus, there is a possibility that the *elastica externa* also grows at its inner edge. In any case there is an intimate structural relationship of the *elastica externa* with the fibrils bordering its inner and outer edge.

The inner sheath is about 50 μm thick and free of cells. In the light microscope it stains homogeneously red with haematoxylin and eosin, with the trichrome stains used (Masson's trichrome, Heidenhain's azan, Goldner's trichrome) it stains with both dominant colours, i.e. some areas stain blue and others red (Masson, Heidenhain) or green and red-brown (Goldner). A modified Weigert technique for the demonstration of oxytalan fibres (Montes, 1992) stains specifically the inner sheath in *Myxine*, *Paramyxine*, *Eptatretus* and lampreys. Oxytalan fibres correspond to bundles of microfibrils. Histochemical tests reveal a strong periodic acid-Schiff (PAS) reaction, weak binding of several lectins (proteins with a high affinity for carbohydrates) and a generally weak to medium intensive alcian blue stain (both at pH 1 and pH 2.5). In semithin sections it becomes clear that the inner sheath is of a complex internal structure composed of fibrous material forming 2–3 main layers running predominantly in circular or spiral arrangement and each being composed of densely packed bundles of variable diameters.

At the electron microscopical level the inner sheath consists of a multitude of densely packed 8–11 nm microfibrils (Figure 10.3c). These fibrils have a very faint cross-striation and cross-sectioned have an electron translucent core, and can thus be called 'tubular'. Their principal orientation is circular or oblique to the long axis of the notochord: single bundles run in other directions. A detailed analysis of the complex and tight arrangement of these fibrils is lacking. The fibrils are accompanied by proteoglycans (Figure 10.4c), as revealed by CMB. The CMB precipitates measure about 20–25 nm in length, are unbranched and usually run in parallel with the fibrils. Locally narrow cleft-like spaces between the fibrils are filled with abundant CMB precipitates, which appear to correspond to the more strongly stained areas in the alcian blue preparations. Locally homogeneous dense material indistinguishable from that of the *elastica externa* can be found in the inner sheath and also stains with aldehyde-fuchsin, Verhoeff and Masson's trichrome. An interpretation of the principal constituent of the inner sheath, the microfibrils, is still difficult. The inner sheath of lamprey notochords has been interpreted on the basis of biochemical and morphological data to be a collagenous structure (Potter and Welsch, 1992). On the other hand, the fibrils in question closely resemble the microfibrils, e.g., in the ventral aorta of various Agnatha (Wright, 1984; Welsch and Potter, 1994). The microfibrils of Agnatha seem to correspond to the elastic microfibrils in elastic tissues of mammals, in which they are composed of glycoproteins, which could explain the strong PAS reaction. Proteoglycans are associated

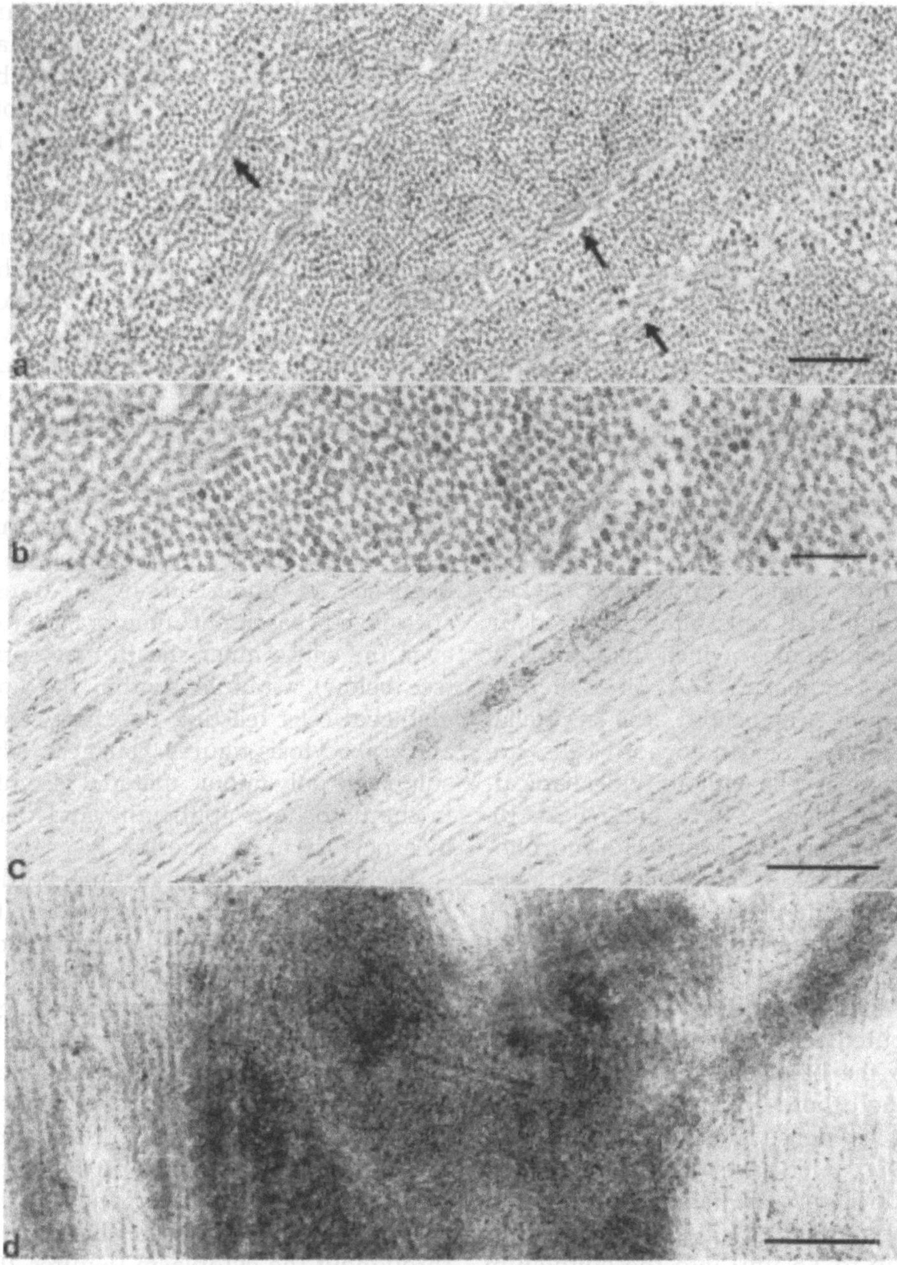

Figure 10.4 Inner sheath of the notochord. (a, b) *P. atami* longitudinal section, most of the 8–11 nm tubular microfibrils are sectioned transversely indicating a circular orientation; regularly, however, also bundles of such fibrils occur running at different angles (arrows) to these circular fibrils; (a) scale bar 200 nm, (b) scale bar 100 nm. (c) *Eptatretus burgeri*, needle-shaped proteoglycans in parallel to the thin fibrils, CMB with 0.3 m MgCl; scale bar 200 nm. (d) *Myxine glutinosa* diffusely arranged dense material in the inner sheath, deposited among the thin fibrils; scale bar 200 nm.

both with collagen fibrils and microfibrils (Erlinger, 1995). The histological trichrome stains cannot decide on the presence of collagenous material in the inner sheath but it is known that also collagen fibrils, e.g. in mammalian cartilage, can be very thin. From a functional point of view (see below) a predominantly elastic inner sheath would not appear to be advantageous to the animal. An additional aspect is the presence of collagen type II in the notochord sheath of lampreys (Brodsky *et al.*, 1994); this type of collagen is typical for cartilage and in the hagfish notochord may be located in the outer sheath.

10.3 *ELASTICA INTERNA*, NOTOCHORD EPITHELIUM AND NOTOCHORD CELLS

The notochord consists of densely packed vacuolated epithelial cells interconnected by countless desmosomes which had already been seen by Studnicka (1897). Since there is no lumen in the notochord, the epithelial cells, also simply called notochord cells, are usually polygonal cells without typical apical-basal polarity (Figures 10.1a and 10.2a, b). Only the outermost cell layer has a morphological basis resting on the extracellular layer of the inner notochord sheath.

In adult hagfishes no typical basal lamina occurs outside the notochord epithelium. Instead, an 80–100 nm thick layer of irregularly distributed dense material occurs closely attached to the basal cell membrane of the peripheral notochord epithelial cells (without intervening less dense layer corresponding to a *lamina rara*). This layer can be relatively dense and thick opposite hemidesmosomes. It extends numerous knob-like projections (diameter up to 1.5 μm) into the epithelium (Figures 10.3d, e) and contains parts composed of polyanions, as revealed by the presence of small needle-shaped precipitates after CMB staining. Presumably these precipitates visualize sulphated proteoglycans. This layer appears to correspond to the *elastica interna* of the light microscopical literature. It seems noteworthy that in young lamprey larvae the notochord epithelium is underlain by a typical basal lamina which in older animals becomes partly replaced by dense material (Potter and Welsch, 1992), similar to that found in adult hagfishes; this material also contains proteoglycans in lampreys (Welsch *et al.*, 1991). In smaller, about 15 cm long, specimens of *Myxine glutinosa* this layer is more homogeneous, less dense, and resembles more closely a *lamina densa* of a basal lamina. However, as in other hagfishes, the *lamina lucida* cannot be discerned. The presence of this dense material suggests a strong structural connection between notochord epithelium and notochord sheath. The *lamina densa* of the hagfish epidermis is also unusually dense and thick. We assume that the notochord epithelium of adult hagfishes is underlain by a transformed specialized basement membrane, to which even the entire inner sheath may correspond (see below). Granules within the notochord epithelial cells reacting positively with CMB and exocytosis-figures suggest that the sulphated polyanionic material is synthesized in the notochord epithelium and discharged by exocytosis to the exterior.

The outermost layer of notochord epithelial cells, the so-called notochord epithelium, consists of one layer of prismatic cells (Figures 10.2c and 10.5a) which are characterized by a relatively strong binding of the lectin concanavalin A, a mannose-specific lectin (Figure 10.1c). Their large nuclei are rich in euchromatin and contain one or two prominent nucleoli (Figures 10.5a and 10.6a); heterochromatin occurs in single small clumps. Among the organelles the RER predominates, it occurs in numerous cisterns throughout the cytoplasm (Figures 10.5a and 10.6b). Often the cisterns are dilated and contain fine particular matter of medium density. Free ribosomes are common, too. The Golgi apparatus is well developed, giving rise to dense granules. Mitochondria are scattered throughout the cytoplasm. Regularly granular inclusions occur (Figures 10.5a and 10.6a),

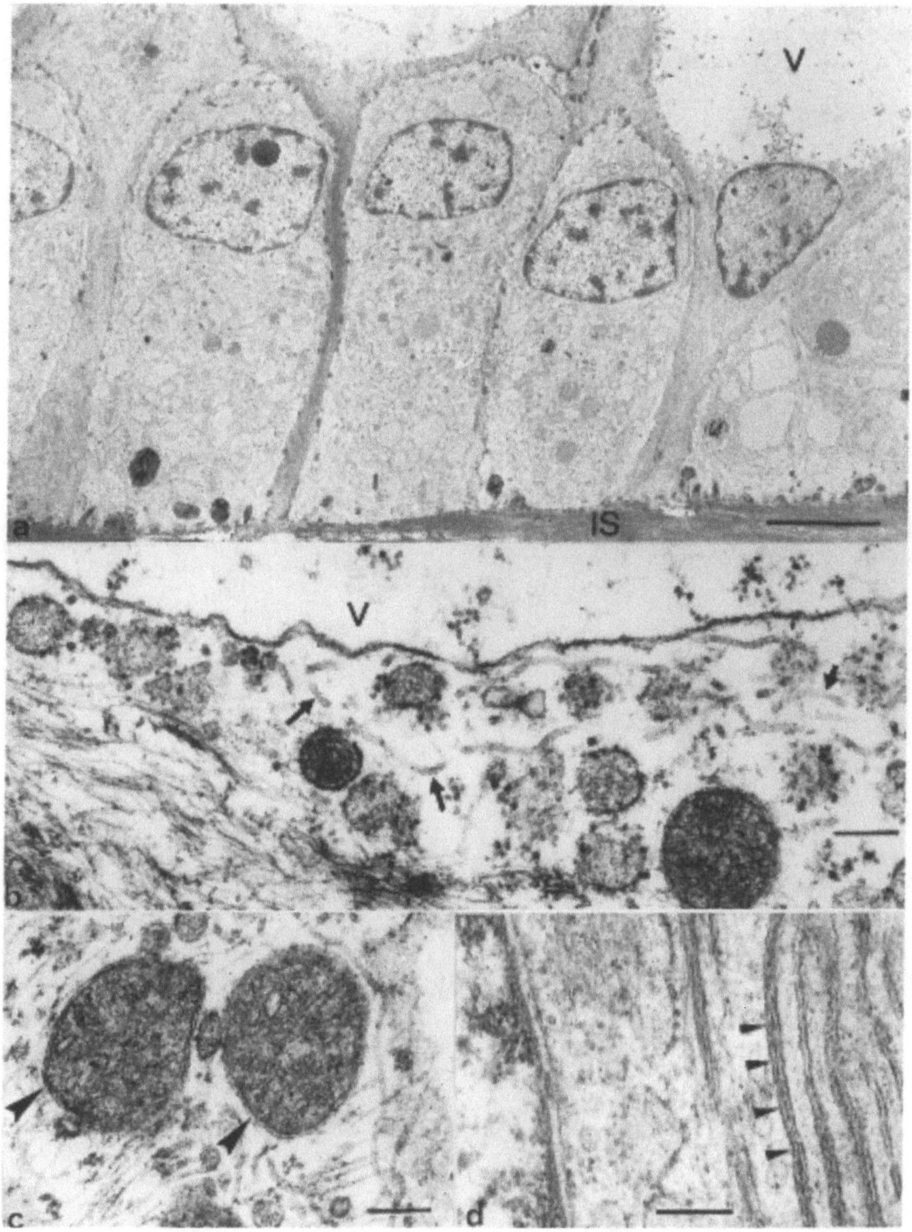

Figure 10.5 Fine structural details of the peripheral notochord cells of *P. atami* (a–c) and *Eptatretus burgeri* (d). (a) Low power electron micrograph of notochord epithelial cells; V, vacuole; IS, inner sheath; note abundant RER and single cytoplasmic granules; scale bar 5 μm. (b) Periphery of a newly formed vacuole (V), note small membrane bound cytoplasmic granules; arrows, microtubules; scale bar 0.2 μm. (c) Mitochondria (arrowheads) near the vacuole; scale bar 0.2 μm. (d) Interdigitating lateral cell protrusions of notochord epithelial cells, the membranes of neighbouring cells are often united by special junctions (arrowheads) which appear to be tight junctions; scale bar 0.2 μm.

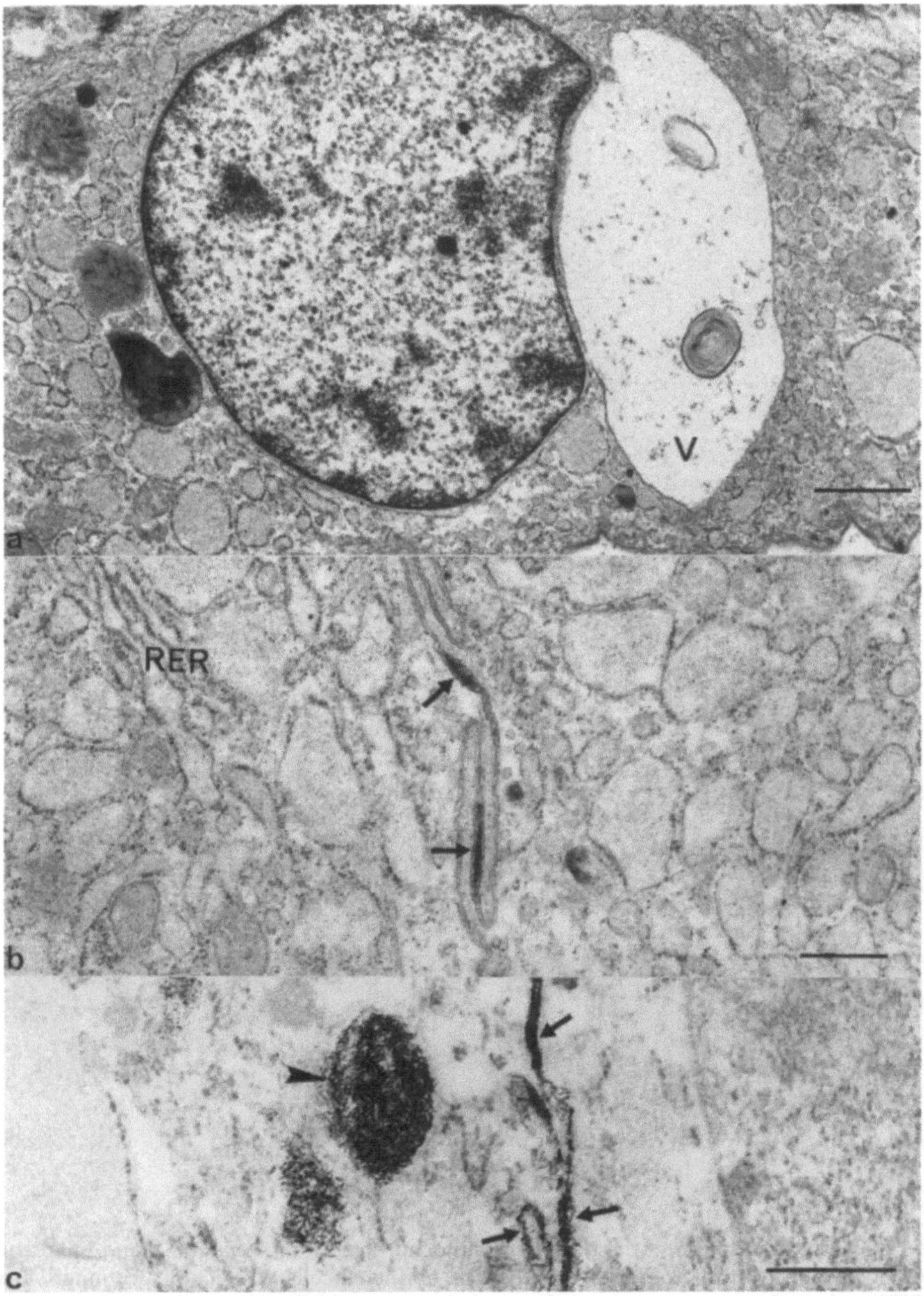

Figure 10.6 Ultrastructural details of the peripheral notochordal cells of *P. atami* (a,b) and *Eptatretus burgeri* (c). (a) Newly formed vacuole near nucleus; scale bar 1 µm. (b) Abundant RER in two neighbouring notochordal epithelial cells; arrows: dense material in the intercellular space; scale bar 0.5 µm. (c) Dense bodies (arrowhead) and intercellular space with negatively charged material, visualized by CMB in the presence of 0.3 m $MgCl_2$; scale bar 400 nm.

many of which contain material resembling that of the subepithelial *elastica interna* and also contain polyanionic components as visualized with CMB (Figure 10.6c). In the light microscope there are PAS and alcian blue positive granules of variable dimensions. Furthermore, the cytoplasm is pervaded by 9–10 nm filaments. Lipid droplets and glycogen particles have been found occasionally. Basally the notochord epithelial cells form narrow invaginations and hemidesmosomes (Figure 10.3d). Laterally and apically neighbouring cells form numerous interdigitations and are connected by numerous desmosomes and also by specialized junctions (Figure 10.5d) since the membranes of neighbouring cells apparently fuse locally. The intercellular space is usually narrow and measures about 18–20 nm. It usually contains negatively charged material (Figure 10.6c) possibly representing membrane bound mucosubstances. Locally the intercellular space is widened and contains dense fine-particular matter (Figure 10.6b), composed again partly of polyanionic material as revealed by CMB.

A membrane-bound vacuole of variable dimensions occurs within single cells of the outermost layer of the notochord epithelium (Figure 10.6a). The vacuole usually contains some loosely distributed fine filamentous and occasionally also lamellar material or single small particles. This vacuole obviously is the precursor of the large vacuole of the epithelial cells in the centre of the notochord. In such cells with a vacuole the cytoplasm assumes a partly different organization compared with the non-vacuolar cells, especially in the perivacuolar cytoplasm. Here, numerous mitochondria, intermediate filaments and microtubules occur (Figure 10.5b, c), the latter are concentrated at that side of the vacuole facing the notochord sheath. In addition, small membrane delimited granular inclusions occurred close to the membrane (Figure 10.5b). This study could not reveal the very first steps of formation of the vacuole within the cytoplasm.

The typical notochord epithelial cells with their large vacuole (diameter about 60 μm) form by gradual transformation from the peripheral epithelial cells. Fully developed they contain a large vacuole, which in well-fixed material is membrane-bound, and a narrow rim of cytoplasm with densely packed 10–12 nm filaments (Figure 10.7a, b, c) and single dense mitochondria. The vacuolar membrane has the structure of a biological unit membrane. The nucleus of these cells is flat but not inactive as demonstrated by its still well-developed euchromatin. The cells are interconnected by desmosomes (Figure 10.7c, d). Other types of intercellular junctions have not been observed in thin-sectioned and freeze-fractured material. The plasma membrane forms abundant small tubular invaginations (Figure 10.7c), which are associated with anionic material as shown by CMB staining. The membrane of the vacuole frequently extends very slender long processes into the lumen of the vacuole, somewhat resembling very thin microvilli (Figure 10.7a). The periphery of the vacuole often contains dense material (Figures 10.2b, c, and 10.7a, b) which is attached to the vacuolar membrane or distributed throughout the vacuole. The morphology of this material is variable. It often forms irregular and variably sized small bodies or a layer or cushion-shaped clumps (Figures 10.2b and 10.7a). Occasionally it appears to be dissociated or forms cord-like aggregates. Not infrequently it is attached to the long thin processes (Figure 10.7b). The positive PAS reaction (Figure 10.1b), as well as binding of several lectins (wheat germ agglutinin, *Ulex europeaus* agglutinin I, *Ricinus communis* agglutinin I, *Helix pomatia* agglutinin, concanavalin A) (Figure 10.1d), indicates that the intravacuolar content comprises abundant complex carbohydrates, presumably as components of glycoproteins. It is striking that the positive PAS reaction and the lectin binding do not show up evenly in all vacuoles but predominantly occur in the

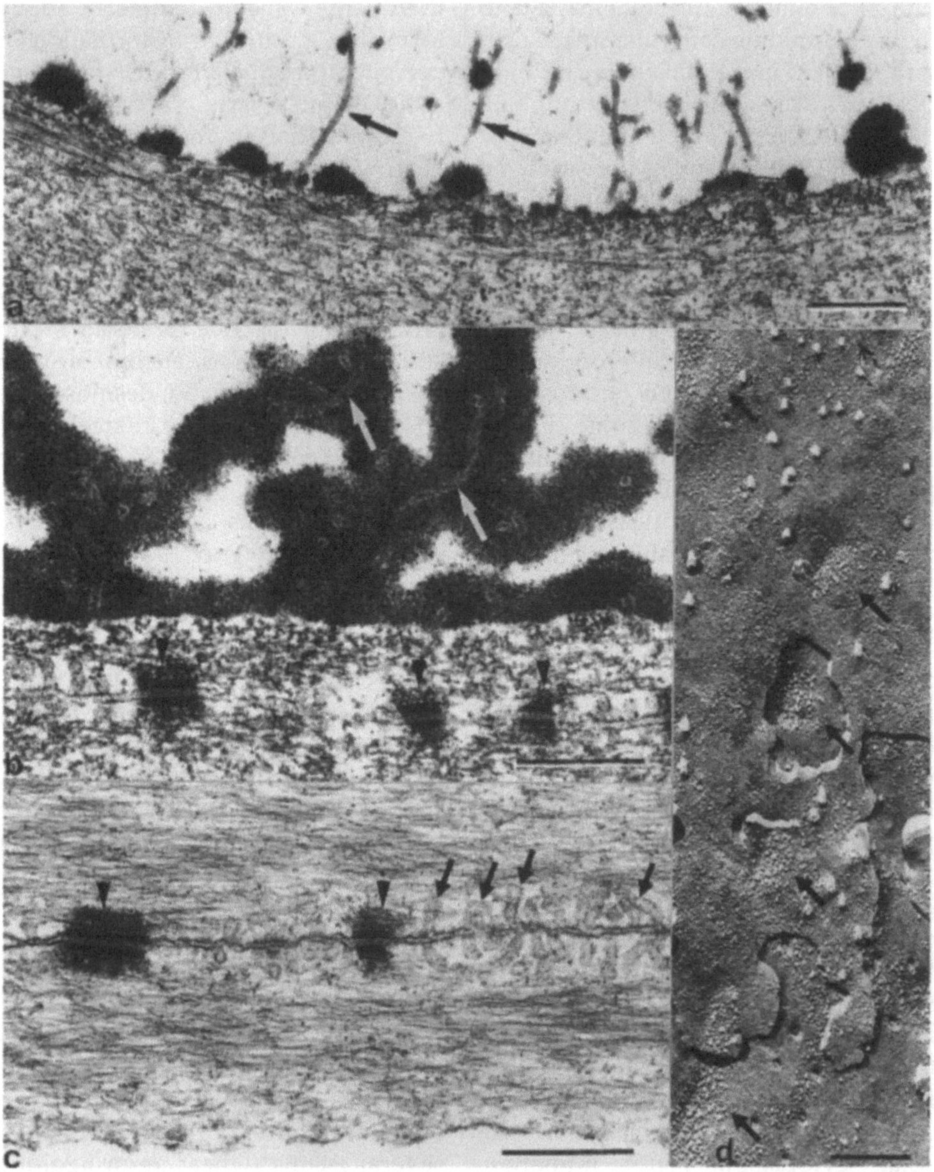

Figure 10.7 *Paramyxine atami*, cytoplasmic details of vacuolated notochordal cells. (a) Periphery of the vacuole with slender microvilli and dense globular depositions at the vacuolar membrane; scale bar 0.5 µm. (b) Dense material attached to the microvilli (arrows) extending into the vacuole; scale bar 0.5µm. (c) Tubular invaginations (arrows) of the cell membrane into the cells, which are interconnected by desmosomes (arrowheads) and which contain abundant intermediate filaments; scale bar 0.5 µm. (d) Freeze fracture preparation of the cell membranes of neighbouring notochord cells; arrows: aggregations of membrane particles marking the presence of desmosomes; scale bar 0.25 µm.

vacuoles of the more peripheral cells (Figure 10.1d). We suspect that this staining pattern is related to penetration of the fixation fluid, but cannot exclude changes in the contents of the vacuoles from periphery to centre.

In adult animals the centre of the notochord is marked by cells which are compressed to flat sac-like structures, forming a condensation termed the notochordal string (Figure 10.1a; Krause, 1923; Flood, 1972).

A number of differences have been observed regularly in the notochord tissues of *Myxine* if compared with those of *P. atami* and *Eptatretus*, to which the above description mainly refers. In the inner sheath the dense homogeneous material is more common than in *P. atami* and *Eptatretus* and often extends in numerous broad irregularly shaped strands from the *elastica externa* deep into the inner sheath suggesting that the *elastica externa* and the inner sheath indeed form a functional unit. The microscopic structure of the notochord is less regular than in *P. atami* and *Eptatretus*. Often cells with abundant intermediate filaments and a relatively large vacuole approach the *elastica interna*. Usually peripheral notochordal epithelial cells without vacuole and with well-developed organelles (mainly RER) occur only in small groups or individually.

10.4 CONCLUSIONS

The cells of the notochordal epithelium are active protein synthesizing cells. Their abundant RER was observed in *M. glutinosa* by Flood (1969, 1972). In *P. atami* and *Eptatretus* the RER is by far the dominating organelle and free ribosomes are also common in these cells. Also the large, pale nucleus with its prominent nucleoli supports the assumption of active protein synthesis. Generally speaking, one product of synthesis obviously is the abundant intermediate filaments, characterizing in particular the large vacuolated cells. However, intracellular structural proteins are usually produced by free ribosomes, which are also common in the peripheral notochord epithelial cells. The abundant RER cisterns suggest production of export proteins, and it appears plausible that it is the inner cell-free part of the notochord sheath which is built up by these cells. In addition, the layer of subepithelial material (*elastica interna*) is presumably produced by these cells, as well as components of the contents of the large vacuoles, in the periphery of which small membrane bound granules occur in the stage of vacuole growth.

The question as to which cells produce the acellular, strongly PAS-positive inner part of the notochord sheath has been discussed for more than 100 years, since the notochord sheath can be considered to be a 'matrix' for the future vertebrae (Koelliker, 1872; v. Ebner, 1895). The inner part of the sheath is composed of thin microfibrils, which differ markedly from the typical collagen fibrils of the outer part of the sheath, and of negatively charged material clearly visualized by CMB staining, the latter in all probability representing glycosaminoglycans as components of proteoglycans (Scott, 1985). While there is no doubt that the outer sheath is produced by fibroblasts – as suggested above – the inner sheath may be a product of the notochord epithelial cells and thus may correspond possibly to a special basement membrane-like structure. The positive PAS reaction of the inner sheath is reminiscent of basement membranes which in general show a strong PAS reactivity due predominantly to the presence of laminin, a glycoprotein, which, however, we could not detect immunohistochemically in the notochord sheath of hagfishes with antibodies to mammalian laminin. The strong PAS reaction will therefore indicate the presence of other glycoproteins and, indeed, of microfibrils which, in mammals, are composed of fibrillin, a glycoprotein. The ordinary collagen of basal laminae (type IV collagen) is non-fibrillar and thus differs from the fibrillar structure of this unique inner notochordal sheath found in lampreys and hagfishes. In conclusion, morphological methods apparently cannot yet decide with certainty the origin and nature of the microfibrils of the inner sheath.

The inner notochordal epithelial cells with their large vacuole are highly specialized cells and by no means degenerate elements. Flood (1969, 1972) had already observed in *M. glutinosa* that the vacuole is lined by a membrane and that the cell membrane forms abundant tubular invaginations. This is confirmed by this study on the notochord of *Myxine*, *P. atami* and *Eptatretus*. The numerous mitochondria in the periphery of small and medium-sized vacuoles may indicate active transport processes into the vacuole. The abundant microtubules near small vacuoles may indicate transport processes too, but may also be the expression of structural needs. The existence of the vacuolar membrane also indicates that the composition of the vacuolar fluid is controlled. Remarkable are the unusually thin and long processes reminiscent of special microvilli which extend into the vacuole.

It is the general assumption (Remane 1936; Flood 1972) that the vacuolar fluid exerts a turgor rendering structural strength to the notochord and one wonders how this turgor is produced. As, for example, exemplified in a paper on the lamprey notochord by Schwarz (1961), it is often believed that glycogen is the principal component of the vacuole which binds water. However, glycogen particles are normally deposited free in the cytoplasm and not in vacuoles. In hagfish notochordal cells we have no evidence that glycogen is a component of the vacuoles. The morphology of the dense material in the vacuoles differs considerably from that of glycogen, which also in hagfishes does not occur in membrane-delimited compartments but free in the cytoplasm (of muscle cells and of adipocytes). It can be assumed that the hagfish notochordal cells do contain a glycoproteinaceous material with complex carbohydrate components in their vacuoles. The complexity of the carbohydrates is suggested by binding of different lectins. In the peripheral vacuolated cells the carbohydrate-containing material has often been found in the form of large globular bodies or as an inner lining of the vacuole also covering the peculiar slender microvillous-like processes. This localization may represent an artefact of fixation; *in vivo* the material may be evenly distributed in the vacuole in a less condensed form. This is also suggested by the heteromorphic appearance of the contents of the vacuoles in the more central parts of the notochord. It remains to be analysed whether there is a gradient of changing metabolic processes from the periphery of the notochord to its centre, which may influence the vacuolar contents.

The abundant intracellular filaments are interpreted to represent keratin filaments, although most antibodies to the mammalian keratins did not react in these cells. Their function is presumably to render rigidity to the cellular periphery. The cells of the notochord act mechanically as a whole, since they are interconnected by abundant desmosomes.

Questions remain as to the biomechanical properties of the notochord, which have not yet been studied in hagfishes but appear to be different from Osteichthyes, e.g. the sturgeon (Long, 1995). The notochord allows extensive coiling in *Eptatretus* species or even formation of 'knots' in *Myxine*. The absence of vertebrae may be advantageous in the formation of such postures. The *elastica externa* and the inner sheath may be the elastic components which could be stretched during coiling. The complex arrangement of the different types of extracellular fibrils in the entire sheath may be altered during movements or coiling. The inner rod of the vacuolar notochord cells obviously allows bending with either compression or stretching of the cells. Without doubt the body musculature is the active component also in extreme movements as, e.g., coiling, since holding animals cautiously in the hand one can observe a strangely flaccid posture, with both ends of the animal hanging down vertically from the hand. The sinus system in the skin may facilitate the more extreme movements.

ACKNOWLEDGEMENTS

We are grateful to Professor W. Buchheim, Federal Dairy Research Institute, Kiel, Germany, for the preparation of several freeze fracture replicas.

REFERENCES

Brodsky, B., Belbruno, K.C., Hardt, T.A. and Eikenberry, E.F. (1994) Collagen fibril structure in lamprey. *Journal of Molecular Biology*, **243**, 38–47.

v. Ebner, J. (1895) Über den feineren Bau der Chorda dorsalis der Cyclostomen (Vorl. Mitt.) Sitzgsber. *Akad. Wiss. Wien, Math.-naturwiss. K1. III*, **104**, 7–16.

Erlinger, R. (1995) Glycosaminoglycans in porcine lung. An ultrastructural study using Cupromeronic blue. *Cell and Tissue Research*, **281**, 473–83.

Floor, R. (1969) Fine structure of the notochord in *Myxine glutinosa*. *Journal of Ultrastructure Research*, **29**, 573–4.

Flood, R. (1972) The notochord of *Myxine glutinosa* L. related to that of other chordates, Acta Regiae societatis scientiarum et literarum Gothoburgensis. *Zoologica*, 8. *Myxine glutinosa*. Reports from a Symposium in Göteborg, R. Fänge (ed.).

Koelliker, A. (1872); Kritische Bemerkungen zur Geschichte der Untersuchungen über die Scheiden der Chorda dorsalis. *Würzburger Verhandlungen N.F.*, **3**, 335–9.

Krause, R. (1923) *Mikroskopische Anatomie der Wirbeltiere*, IV. *Teleostier, Plagiostomen, Zyklostomen und Leptokardier*, de Gruyter, Berlin.

Long, J.H. (1995) Morphology, mechanics and locomotion – the relation between the notochord and swimming motions in sturgeon. *Environmental Biology, Fishes*, **44**, 199–211.

Montes, G.S. (1992) A modified Weigert's technique for the demonstration of oxytalan fibres. *Journal of the Brazilian Association for the Advancement of Science*, **44**, 224–233,

Potter, I.C. and Welsch, U. (1992) Arrangment, histochemistry and fine structure of the connective tissue architecture of lampreys. *Journal of Zoology, London*, **226**, 1–30.

Remane, A. (1936) Skelettsystem, I. Wirbelsäule und ihre Abkommlinge, in *Handbuch der vergleichenden Anatomie der Wirbeltiere*, Vol. 4 (eds L. Bolk, E. Göppert, E. Kallius and W. Lubosch) Urban & Schwarzenberg, Berlin, pp. 1–206.

Schinko, I., Potter, I.C., Welsch, U., Debagge, P. (1992) Structure and development of the Notochord 'Elastica Externa' and nearby components of the elastic fibre system of Agnathans. *Acta Zoologica*, **73** (1), 57–66.

Schwarz, W. (1961) Elektronenmikroskopische Untersuchungen an den Chordazellen von *Petromyzon*. *Zeitschrift fur Zellforschung und mikroskopische Anatomie*, **55**, 597–609.

Scott, J.E. (1985) Proteoglycan histochemistry – a valuable tool for connective tissue biochemists. *Collagen Relation Research*, **5**, 765–74.

Studnicka, F.K. (1897) Über die Histologie und Histogenese des Knorpels der Cyclostomen. Quoted after Remane (1936).

Welsch, U. and Potter, I.C. (1994) Variability in the presence of elastic fibre-like structures in the ventral aorta of Agnathans (hagfishes and lampreys). *Acta Zoologica*, **75** (4), 323–7.

Welsch, U. and Storch, V. (1969) Zur Feinstruktur der Chorda dorsalis niederer und höherer Chordaten. *Zoologischer Anzeiger*, **33**, 160–8.

Welsch, U., Erlinger, R. and Potter, I.C. (1991) Proteoglycans in the notochord sheath of lampreys. *Acta Histochemica*, **91**, 59–65.

Wright, G.M. (1984) Structure of the conus arteriosus and the ventral aorta in sea lamprey, *Petromyzon marinus* and the Atlantic hagfish, *Myxine glutinosa*: microfibrils, a major component. *Canadian Journal of Zoology*, **62**, 2445–56.

11
HAGFISH CARTILAGE

Glenda M. Wright, Fred W. Keeley and M. Edwin DeMont

SUMMARY

Morphological, biochemical, molecular biological and biomechanical analysis of cartilages from the Atlantic hagfish, *Myxine glutinosa*, reveal they are unusual tissues. At least three different cartilages designated Types 1, 2 and 3 have been identified.

Type 1 appears superficially to resemble other vertebrate cartilages; however, ultrastructural analysis reveals that the extracellular matrix (ECM) is non-collagenous. Type 2 bears no morphological resemblance to any known vertebrate cartilage, Type 3 is similar in appearance to Type 1; however, biochemical analysis of Types 1 and 3 cartilages reveal that each is composed primarily of a cyanogen bromide (CNBr) insoluble protein of unique composition. Myxinin, the principle structural protein of Type 1 cartilage, is similar but not identical to lamprin, the main structural protein of lamprey annular cartilage. The major structural protein of Type 3 cartilage has a composition very different from myxinin and as yet has not been named. The CNBr insoluble material from Type 2 cartilage is a minor component with a composition different from myxinin, Type 3 cartilage and lamprin.

Molecular biological studies suggest no homology between hagfish and lamprey cartilages. Comparative biomechanical studies indicate that the modulus of elasticity of hagfish cartilage is about one order of magnitude higher than lamprey and bovine cartilages.

The morphological and biochemical differences between hagfish and lamprey cartilages support the concept that these two agnathans followed long independent evolutionary histories.

11.1 INTRODUCTION

Except for the notochord, the skeletal system of the hagfish is completely cartilaginous and almost all the cartilages in the head are continuous with one another. Cole (1905) defined the skeleton of *Myxine* as consisting of two tissue types: cartilage and pseudocartilage and further described each tissue as either hard or soft. Pseudocartilage is a term adopted by Cole (1905) to describe skeletal tissue in *Myxine* that had a superficial resemblance to true cartilage but differed in morphology and microchemical reactions. Based on his studies of 'its minute structure and microchemical reactions', Cole (1905) compared the hard cartilage of *Myxine* directly with hyaline cartilage of other animals. More recent studies of the ultrastructural morphology and biochemistry of the lingual cartilages of the Atlantic hagfish, *Myxine glutinosa*, have shown that they are composed of two different types of cartilage, designated Type 1 and 2, and that Type 1 cartilage is not a collagen-based hyaline-like cartilage but represents an unusual non-collagenous cartilage (Wright *et al.*, 1984).

The Biology of Hagfishes. Edited by Jørgen Mørup Jørgensen, Jens Peter Lomholt, Roy E. Weber and Hans Malte. Published in 1998 by Chapman & Hall, London. ISBN 0 412 78530 7.

The presence of non-collagen-based cartilages in hagfishes is intriguing. Hagfishes synthesize collagens and have an abundance of collagen composing their soft connective tissues yet they produce and utilize unique structural proteins which form a non-collagen-based cartilaginous skeleton. Presumably these structural proteins of hagfish cartilages impart unique properties which the collagens that are typical structural proteins of other vertebrate cartilages cannot. This chapter presents an overview of the current information on the morphological, biochemical, molecular biological and biomechanical characteristics of hagfish cartilages.

11.2 LINGUAL CARTILAGES

11.2.1 Gross morphology

The lingual cartilages, originally termed basal plate cartilages by Cole (1905), form the large complex tongue skeleton located in the floor of the mouth. The tongue skeleton is composed of three regions: the anterior, middle and posterior segments (Figure 11.1a). The anterior lingual cartilages consist of a single central bar and two lateral bars. The lateral bars are situated on either side of the central bar except at the rostral-most end where the lateral bars lie dorsal to the edges of the central bar. The lateral labial cartilage fuses with each lateral bar rostrally. The lateral bars are separated from the central bar and middle segment by connective tissue. The central bar bifurcates caudally and is continuous with the wide middle segment which consists of one piece of cartilage. The first branchial arch cartilage fuses with the middle segment at the caudal lateral aspects. The elements of the anterior and middle segments are all hard cartilages and are designated as Type 1 cartilage (Wright *et al.*, 1984). Immediately caudal to the middle lingual and attached by collagenous connective tissue is the posterior segment. The posterior lingual is a long, flexible, rod-like piece of hard pseudocartilage (Cole, 1905) that tapers to a point caudally. A deep trough extends along most of the dorsal surface. The posterior lingual cartilage is designated as Type 2 cartilage (Wright *et al.*, 1984).

11.2.2 Light microscopy

Anterior and middle lingual cartilages (Type 1 cartilage)

All elements forming the anterior and middle lingual segments appear superficially, at the light microscopic level, to resemble hyaline cartilage. Chondrocytes of variable shape and size are surrounded by a dense extracellular matrix (Figure 11.1b) which stains positively with Verhoeff's elastin stain, periodic acid-Schiff reagent and alcian blue. Although Verhoeff's stain is considered to be specific for elastin, its mode of action in unknown (Sage and Gray, 1979). Most chondrocytes are large rounded or polygonal-shaped cells occurring singly or in groups of two or four. Chondrocytes immediately adjacent to the perichondrium are smaller and more elongated to oval shape. The territorial extracellular matrix immediately surrounding each single or group of chondrocytes is arranged in the form of concentric rings (Figure 11.1b). Interterritorial matrix forms a branched network of fibrous material which, in toluidine blue stained, 1 μm thick resin sections, stained moderately between individual chondrocytes and cell groups. Adjacent to each piece of cartilage is a dense connective tissue perichondrium which contains collagen fibres, fibroblasts and blood vessels. Dispersed throughout the perichondrium and extending into the surrounding connective tissue are clusters of very large, spherical-shaped cells subdivided by collagen fibres.

Posterior lingual cartilage (Type 2 cartilage)

The pseudocartilage forming the posterior lingual cartilage is very similar, at least at the

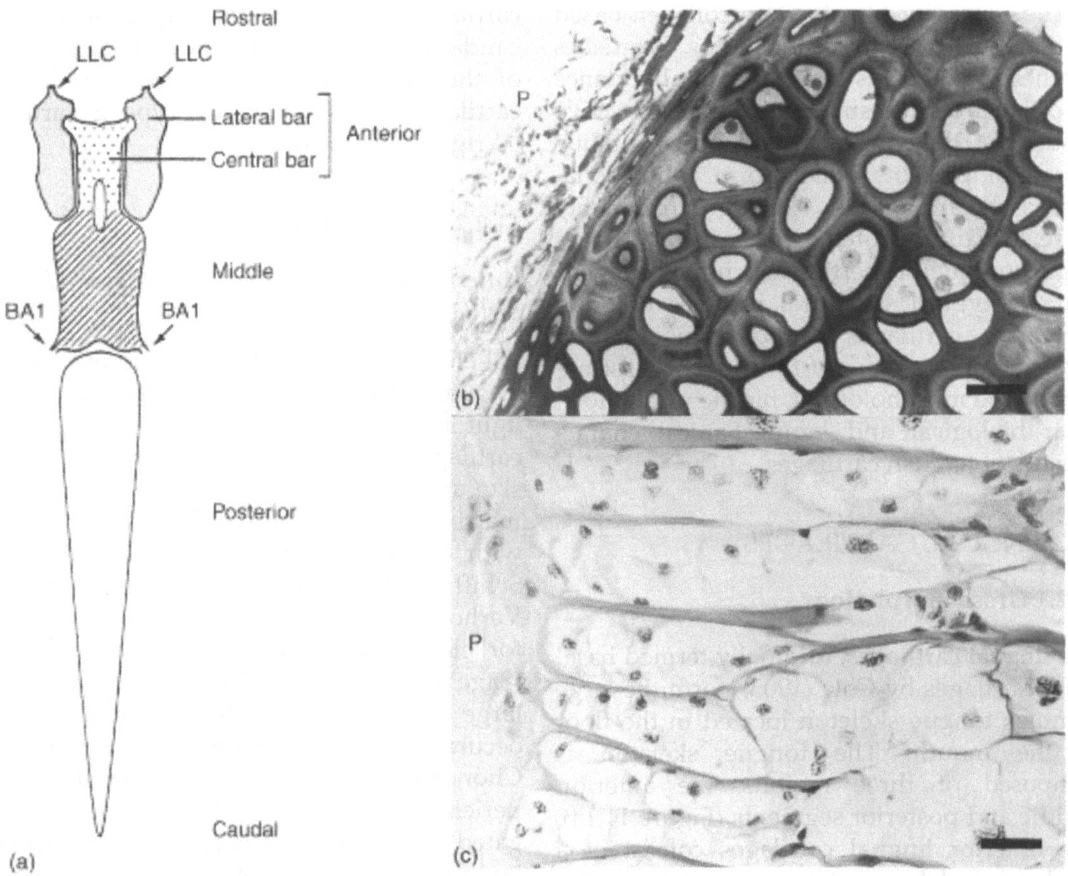

Figure 11.1 (a) Schematic diagram of the lingual cartilages of *Myxine glutinosa*. Ventral view. BA1: first branchial arch, LLC: lateral labial cartilage. (b) Light micrograph of a portion of anterior lingual cartilage. Chondrocytes adjacent to the perichondrium (P) are smaller than those more central. Extracellular matrix material is arranged in concentric rings (arrows) around chondrocytes. Bar: 30 µm. (c) Light micrograph of a portion of posterior lingual cartilage (Type 2 cartilage) showing the dense collagenous perichondrium (P) and septae (S) between cords of large spherical cells and nests of smaller cells. Bar: 30 µm. (b) and (c) reproduced from Wright, G.M., Keeley, F.W., Youson, J.H. and Babineau, D.L. (1984) Cartilage in the Atlantic hagfish, *Myxine glutinosa*. *Am. J. Anat.*, **169**, 407–24. Copyright 1984 Alan R. Liss, Inc. Reprinted by permission of John Wiley and Sons, Inc.

light microscopic level, to a type of invertebrate cartilage found in the odontophore of the marine snail, *Busycon canaliculatum*, and parts of the endoskeletal complex of the marine worm, *Eudistylia polymorpha* (Person and Philpott, 1967). It is surrounded by a thick perichondrium consisting of dense regular collagenous connective tissue. Thick collagenous trabeculae (septae) extend inward from the perichondrium segregating large cells into cords, columns or islets (Figure 11.1c). Nests of smaller cells are found at the inntermost part of the perichondrium or more centrally at regions where trabeculae anastomose. The posterior lingual cartilage does not stain with Verhoeff's elastin stain but does show positive staining with periodic acid-Schiff and alcian blue.

11.2.3 Ultrastructural morphology

Anterior and middle lingual cartilages (Type 1 cartilage)

Active chondrocytes with numerous organelles are situated near the periphery of the cartilage (Figure 11.2a). They contain an elongated or oval nucleus, extensive rough endoplasmic reticulum and well-developed juxtanuclear Golgi apparatus. The plasma membrane is directly apposed to the extracellular matrix. Many *caveolae intracellulares* (50–65 nm in diameter) are found within the cytoplasm just beneath the plasma membrane or in continuum with the plasma membrane. The large Golgi apparatus consists of several sets of saccules, small vesicles and larger secretory vacuoles. Dilated *trans* Golgi saccules and secretory vacuoles frequently contain dense material in the form of globules, straight or curved rods and rings (Figure 11.2b). The material is similar in appearance to that which constitutes the fibrils in the extracellular matrix. Inactive chondrocytes are very large cells in which the rough endoplasmic reticulum and Golgi apparatus are inconspicuous components.

Territorial extracellular matrix consists of electron dense globular material, most of which has fused together to form beaded fibrils 70–120 nm in diameter (Figure 11.2c) that are arranged in concentric lamellae around each chondrocyte (Figure 11.2a). These concentric lamellae of matrix fibrils in the Type 1 cartilage are reminiscent of the concentric lamellar pattern of rod-like elements in the matrix of the gill cartilage of the horseshoe crab, *Limulus polyphemus* (Person and Philpott, 1969). Interterritorial extracellular matrix consists of branched fibrils, 50–80 nm in diameter (Figure 11.2d) that do not have the beaded appearance of the territorial fibrils and appear quite similar to those found throughout the matrix of adult lamprey cartilage (Wright and Youson, 1983). Very little collagen is observed within the extracellular matrix of hagfish Type 1 cartilage.

Between the territorial and interterritorial maxtrix fibrils is a fine flocculent material containing small particles, 10 nm in diameter. Presumably, as in the extracellular matrix of other vertebrate cartilages, these dense granules represent monomer portions of proteoglycan aggregates containing chondroitin sulphate. Mathews (1967, 1975) has shown that most of the glycosaminoglycans of hagfish cartilage proteoglycans have an unusual chemical structure: they are a type of chondroitin sulphate that contains galactosamine sulphated at both C4 and C6. This unusual glycosaminoglycan separates the proteoglycans of hagfish cartilage from those of lamprey and other vertebrate cartilages (whose proteoglycans contain chondroitin sulphates with galactosamine sulphated at either C4 or C6 but not both) and possibly reflects analogy to some invertebrate cartilages, since this unusual chondroitin sulphate is the major glycosaminoglycan found in squid cartilage (Mathews, 1975).

Posterior lingual cartilage (Type 2 cartilage)

Large spherical cells of the posterior lingual cartilage have an irregular sometimes convoluted outline. The cytoplasm to nucleus ratio is high and nuclei are most commonly eccentrically positioned. The cytoplasm exhibits a paucity of organelles but is characterized by a network of intermediate filaments, 8.5–10 nm in diameter, which are found concentrated in the central cytoplasm (Figure 11.3a). Numerous smooth tubulovesicular structures are located immediately beneath and in continuum with the cell membrane. Autophagic vacuoles and lipid droplets are common inclusions in most cells.

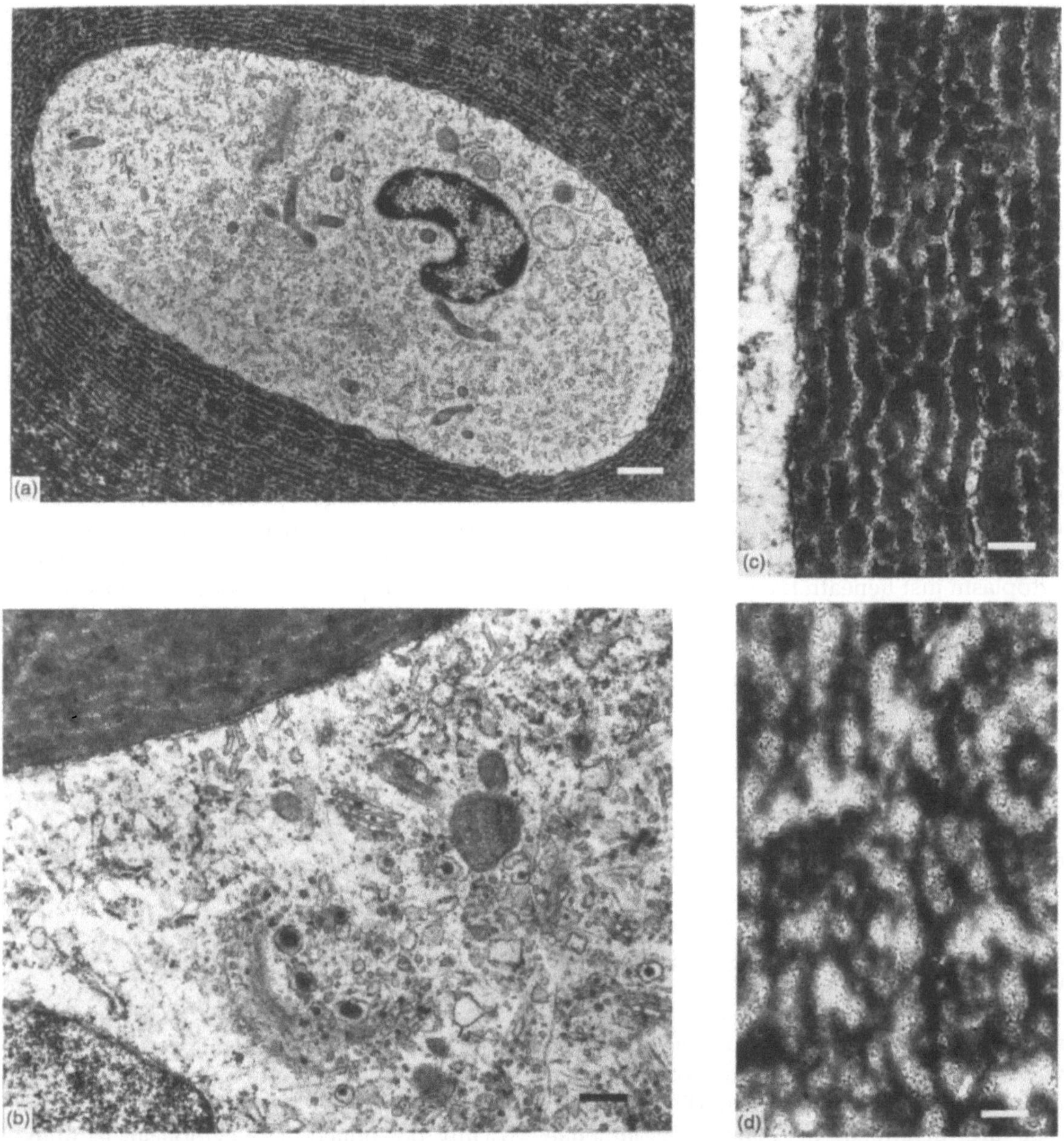

Figure 11.2 (a) Electron micrograph of a peripherally located chondrocyte from Type 1 cartilage (anterior lingual) surrounded by matrix fibrils which are arranged in concentric lamellae in the territorial extracellular matrix. Bar: 830 nm. (b) Golgi region in a peripheral chondrocyte from Type 1 cartilage containing dense globular material. Bar: 435 nm. (c) Higher magnification of the beaded matrix fibrils and interfibrillar material forming the territorial matrix of Type 1 cartilage. Bar: 235 nm. (d) Branched fibrils in the interterritorial matrix of Type 1 cartilage. Bar: 210 nm. (a)–(d) reproduced from Wright, G.M., Keeley, F.W., Youson, J.H. and Babineau, D.L. (1984) Cartilage in the Atlantic hagfish, *Myxine glutinosa*. *Am. J. Anat.*, **169**, 407–24. Copyright 1984 Alan R. Liss, Inc. Reprinted by permission of John Wiley and Sons, Inc.

The smaller nest cells contain more densely packed organelles than the large spherical cells (Figure 11.3b). The Golgi apparatus is prominent and consists of several sets of saccules with many associated vesicles. Rough endoplasmic reticulum is widely distributed throughout the cytoplasm and usually displays dilated cisternae. No autophagic vacuoles are apparent in the nest cells.

The ultrastructural features of small nest and large cells and their close association with each other suggests that the nest cells may give rise to the larger cells. Ultrastructural features of the nest cells reflect their active involvement in the synthesis of the extracellular matrix. Larger cells appear not to be involved actively in the synthesis of the extracellular matrix. The arrangement of the large cells and extracellular matrix combine to make the Type 2 cartilage into a flexible supporting structure much like the notochord. Type 2 cartilage may also provide a cushion-like protection to the tendon of the clavatus muscle that lies within the dorsal trough of the cartilage.

The composition of the extracellular matrix of the posterior lingual cartilage varied with position. At the periphery and in the trabeculae are collagen fibrils 50–70 nm in diameter. Most of the large cells are separated by an extracellular space of uniform dimension containing an extracellular matrix consisting most commonly of a single band of basal lamina-like dense fibrillar material (Figure 11.3c).

Perichondrium

The perichondrium surrounding all elements of the anterior, middle and posterior lingual segments contains fibroblasts between collagen fibres and tubular microfibrils (11–13 nm in diameter). Although no elastin has been identified in hagfish or lamprey (Sage and Gray, 1981; Sage, 1983), elastic-like fibres consisting of dense patches of amorphous material associated with bundles of microfibrils are frequently seen within the perichondrium (Figure 11.3d, e). The perichondrium immediately adjacent to the matrix of the anterior and middle lingual cartilages consists of amorphous material similar to that associated with the microfibrils. Collagen fibrils frequently penetrate this material.

11.2.4 Biochemistry

Samples of Type 1 (anterior and middle lingual) and Type 2 (posterior lingual) cartilage from *Myxine glutinosa* have been extracted with hot alkali (0.1 M NaOH) or treated with cyanogen bromide (CNBr) as described previously for lamprey cartilage (Wright *et al.*, 1983). The amino acid compositions of the insoluble residues after such treatments have been determined (Wright *et al.*, 1984).

Unlike lamprey annular cartilage, none of the hagfish lingual cartilages is resistant to extraction with hot 0.1 M NaOH and all are solubilized by the process. However, as with lamprey cartilage, treatment with CNBr effectively removes collagens and most other proteins, leaving an almost completely intact skeleton composed of insoluble residual cartilage material that contains no methionine (Figure 11.4a). The amino acid composition of the insoluble residues after CNBr treatment of the Type 1 cartilage and the Type 2 cartilage are given in Table 11.1 along with the amino acid composition of lamprin (the main structural protein of lamprey annular cartilage) for comparison.

The amino acid composition of the CNBr residues of Type 1 and Type 2 cartilage samples are very different. The fine structure of the residual material after CNBr treatment of the Type 1 cartilage consists only of matrix fibrils and these fibrils comprise the major protein component of the cartilage. The principle structural matrix protein of Type 1 cartilage is called myxinin (Wright *et al.*, 1984).

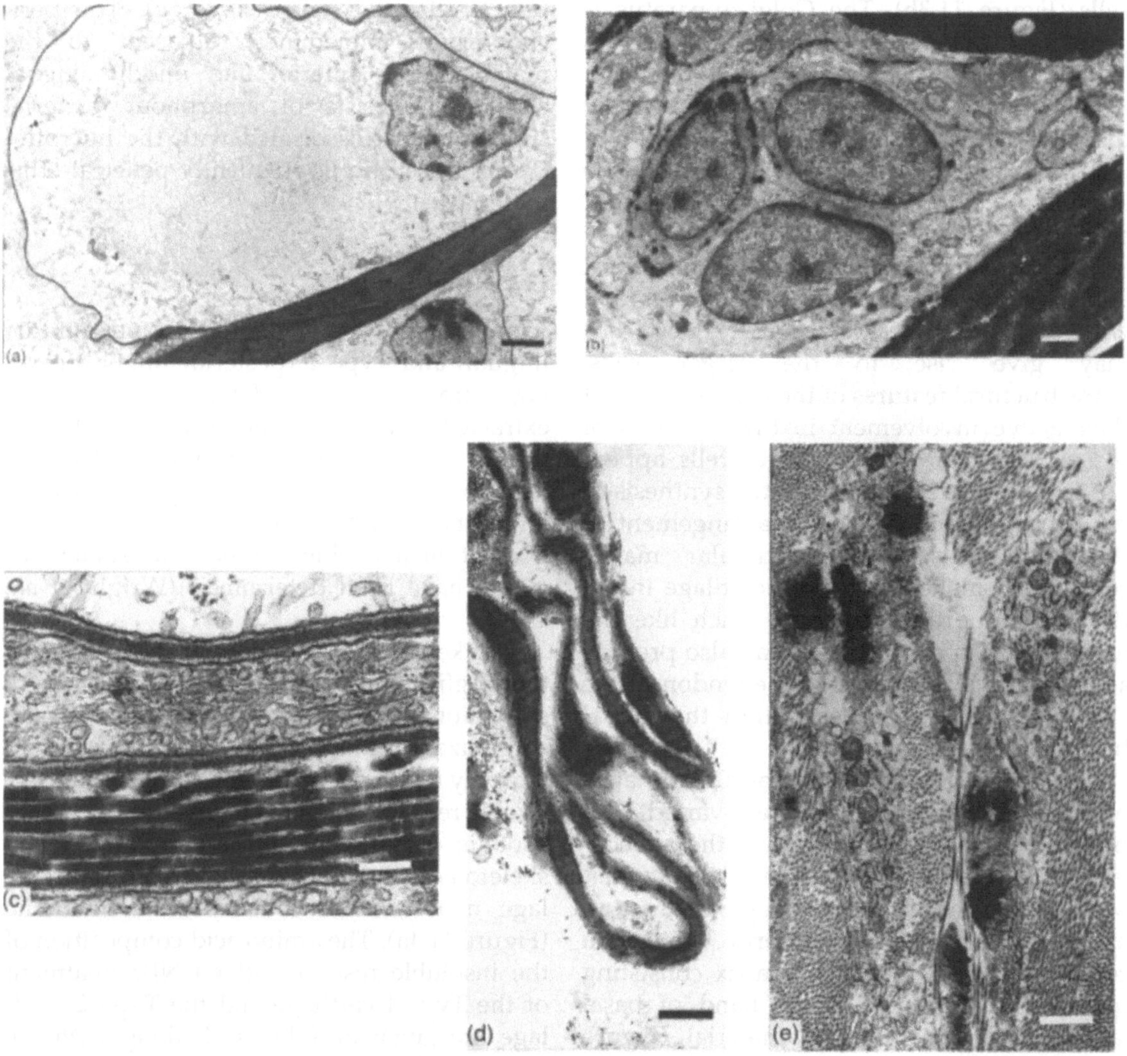

Figure 11.3 (a) Large spherical cells from the posterior lingual cartilage bounded by a collagenous trabecula (CT) and thin basal lamina-like extracellular material (arrows). Bar: 2.2 μm. (b) Nest cells between collagenous septae in Type 2 cartilage. Bar: 1.6 μm. (c) Basal lamina-like material (arrows) and collagen fibrils (CF) form the extracellular matrix of Type 2 cartilage. Bar: 250 nm. (d) Wavy elastic-like fibres in the perichondrium. Bar: 1.3 μm. (e) Portion of the perichondrium showing transverse sections of elastic-like fibres containing dense amorphous material (asterisk) surrounded by microfibrils (mf). Bar: 91 nm. (a)–(e) reproduced from Wright, G.M., Keeley, F.W., Youson, J.H. and Babineau, D.L. (1984) Cartilage in the Atlantic hagfish, *Myxine glutinosa*. *Am. J. Anat.*, 169, 407–24. Copyright 1984 Alan R. Liss, Inc. Reprinted by permission of John Wiley and Sons, Inc.

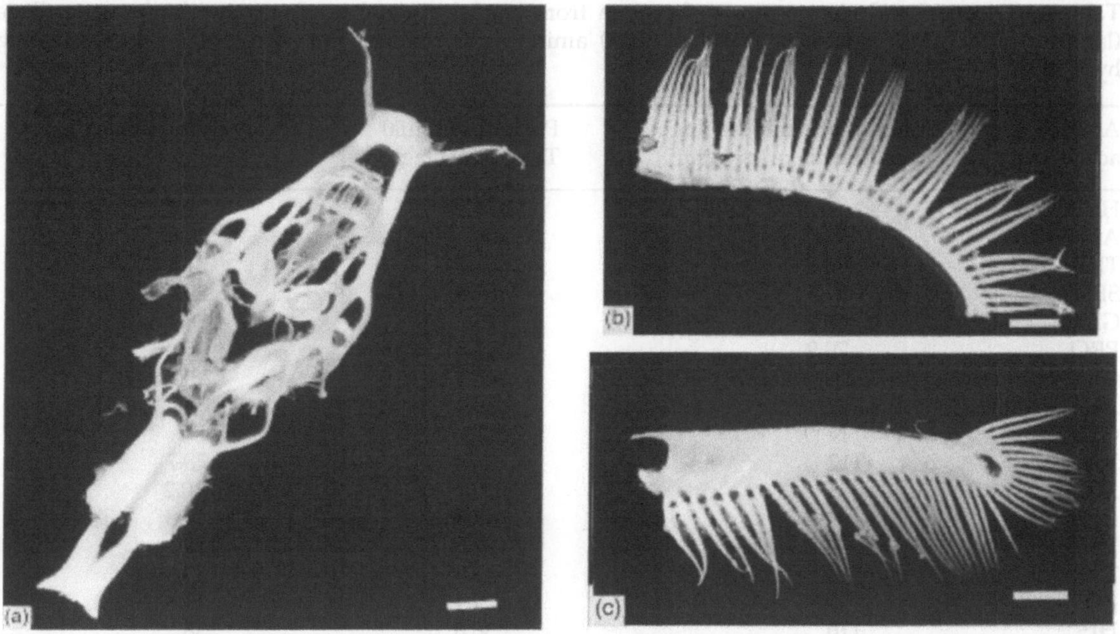

Figure 11.4 (a) Dorsal view of the CNBr isolated head skeleton. A portion of the anterior and middle lingual cartilages (*) have been displaced dorsally for better viewing. Nasal capsule (NC). Bar: 2.4 mm. (b and c) Lateral views of CNBr isolated tail skeleton: (b) medical dorsal bar (MDB) and fin rays; (c) medial ventral bar (MVB) and fin rays. Bar: 2.4 mm.

Although myxinin has similarities in amino acid composition to elastin and collagen (high glycine, high neutral amino acid and low acidic amino acid content) its composition is clearly not that of elastin or collagen, nor can this composition be accounted for by the intermixing of quantities of collagen and/or elastin with other proteins. Myxinin does not contain significant amounts of hydroxylysine, an amino acid characteristic of collagen, or desmosines, the cross-linking amino acids characteristic of elastin.

Even though there are some similarities in the amino acid compositions of lamprin and myxinin, particularly with respect to high content of glycine, valine and leucine, their compositions are clearly not identical. Myxinin has a tyrosine content and a tyrosine/phenylalanine ratio which is even higher than that seen in lamprin. Myxinin also has a larger proportion of the basic amino acids than lamprin and the large amount of histidine in myxinin poses a particularly striking difference between the two proteins.

Only strands of basal lamina-like material are present in the Type 2 cartilage after treatment with CNBr. This residue after CNBr treatment is only a minor component of the Type 2 cartilage. The amino acid composition of the CNBr insoluble residue is similar neither to that of myxinin nor to that of lamprin.

11.2.5 Molecular biology

Preliminary molecular biological studies of hagfish lingual cartilage show no homology between lamprey and hagfish cartilages. Northern analysis of total RNA isolated from lingual Type 1 cartilage probed with full-length lamprin cDNA (Robson *et al.*, 1993)

Table 11.1 Amino acid compositions of protein from hagfish lingual cartilages and lamprey cartilage (lamprin) after CNBr treatment (residues/1000 amino acids): values not corrected for losses during hydrolysis

Amino acids	Anterior and middle lingual Type I cartilage (myxinin)*	Posterior lingual Type 2 cartilage*	Lamprey annular cartilage†
HYP	0	4.2	1.4
ASX	39.8	108	20.8
THR	40.3	61.7	31.1
SER	27.5	61.8	27.4
GLX	45.2	148	38.1
PRO	70.9	24.8	95.6
GLY	231	113	282
ALA	80	59.4	156
CYS	N.D.	N.D.	N.D.
VAL	112	67	78.7
MET	0	0	<0.2
ILU	8.6	47.9	19.8
LEU	74.3	87.8	119
TYR	94.8	29.3	53.3
PHE	14.9	29.4	13.8
HIS	110	36.6	38
HYP	0	0	<0.2
LYS	15.5	67.1	7.3
ARG	35.8	54.5	17.4
IDES	0	0	<0.2
DES	0	0	<0.2

* From Wright *et al.* (1984).
†From Wright *et al.* (1983).
N.D. not determined

and an end-labelled oligonucleotide made to two of the tandem repeats of peptide sequence GGLGY of lamprin (Robson *et al.*, 1993) demonstrate no hybridization to the hagfish lingual cartilage.

Western blotting of SDS (sodium dodecylsulphate) soluble extracts of hagfish cartilages using antibodies to lamprin indicates no cross-reactivity, adding further weight to the fact that hagfish cartilage matrix proteins differ from lamprin.

11.2.6 Mechanical studies

Comparative mechanical studies between hagfish anterior lingual (myxinin-based) cartilage, lamprey annular (lamprin-based) cartilage and bovine elastic cartilage have been completed and preliminary results have been analysed. The experiments measured both the flow-independent (intrinsic) physical properties of the extracellular matrix and the swelling kinetics. The interactions between the proteoglycans and the structural proteins can be examined using the swelling kinetics. Essentially, the experiments modified the internal balance of forces in the extracellular matrix by applying an external load and changing the ambient counter-ion concentration (Akizuki *et al.*, 1986, 1987).

Preliminary results indicate that the modulus of elasticity of hagfish lingual cartilage is about one order of magnitude higher than the other two cartilages. The intrinsic moduli of both the hagfish Type 1 cartilage and lamprey annular cartilage are independent of the external counter-ion concentration. This implies that the interactions between the proteoglycans and the myxinin are different than those found in mammalian articular cartilage, where they play an important role in determining the intrinsic physical properties. The intrinsic modulus of the bovine elastic cartilage is dependent on the external counter-ion concentration.

11.3 OTHER HAGFISH CARTILAGES

Samples of nasal capsule cartilage and fin ray cartilage from *Myxine glutinosa* have recently been extracted with CNBr (Figure 11.4b, c). Preliminary analysis indicates that these cartilages are each composed primarily of a CNBr insoluble protein. Amino acid compositions of the CNBr residues of nasal capsule and fin ray cartilages are similar to each other but very different from myxinin, Type 2 cartilage and lamprin. The cartilages of the nasal capsule and fin rays represent yet another type of hagfish cartilage which is non-collagen-based and is designated as Type 3.

The insoluble CNBr residue of Type 3 cartilage has a high glycine and valine content similar to lamprin. However, Type 3 cartilage is clearly not identical to lamprin, having only half the levels of valine and histidine present in lamprin. Type 3 cartilage is unlike myxinin, having only one quarter of the histidine content and one half of the tyrosine content present in myxinin. The very low amounts of valine and leucine in Type 3 cartilage are particularly striking differences between Type 3 cartilage and myxinin and lamprin.

Type 3 cartilage is similar in appearance, at least at the light microscopic level, to type Type 1 cartilage of the anterior and middle lingual cartilages. Analysis of the ultrastructural morphology of the cells and the extracellular matrix of this new hagfish cartilage is presently underway.

11.4 CONCLUDING REMARKS

Three types of cartilage have been identified in hagfish based on morphological, biochemical, molecular biological and biomechanical analysis. The morphological and biochemical differences between hagfish and lamprey cartilages support the concept that these two agnathans followed long independent evolutionary histories.

REFERENCES

Akizuki, S., Mow, V.C., Muller, F., Pita, J.C., Howell, D.S. and Manicourt, D.H. (1986) Tensile properties of human knee joint cartilage: I. Influence of ionic conditions, weight bearing, and fibrillation on the tensile modulus. *Journal of Orthopedic Research*, **4**, 379–92.

Akizuki, S., Mow, V.C., Muller, F., Pita, J.C. and Howell, D.S. (1987) Tensile properties of human knee joint cartilage: II. Correlations between weight bearing and tissue pathology and the kinetics of swelling. *Journal of Orthopedic Research*, **5**, 173–86.

Cole, F.J. (1905) A monograph on the general morphology of the myxinoid fishes, based on a study of *Myxine*. Part 1. The anatomy of the skeleton. *Transactions of the Royal Society of Edinburgh*, **XLI**, Part III (No. 30) 749–91.

Mathews, M.B. (1967) Macromolecular evolution of connective tissue. *Biological Review*, **42**, 499–551.

Mathews, M.B. (1975) *Connective Tissue, Macromolecular Structure and Evolution*, Springer-Verlag, New York.

Person, P. and Philpott, D.E. (1967) On the occurrence and biologic significance of cartilage tissues in invertebrates. *Clinical Orthopedic Related Research*, **53**, 185–212.

Person, P. and Philpott, D.E. (1969) The biology of cartilage. I. Invertebrate cartilages: Limulus gill cartilage. *Journal of Morphology*, **128**, 67–94.

Robson, P., Wright, G.M., Sitarz, E., Maiti, A., Rawat, M., Youson, J.H. and Keeley, F.W. (1993) Characterization of lamprin, an unusual matrix protein from lamprey cartilage. *Journal of Biological Chemistry*, **268**, 1440–7.

Sage, H. and Gray, W.R. (1979) Studies on the evolution of elastin, I. Phylogenetic distribution. *Comparative Biochemistry and Physiology*, **64B**, 313–27.

Sage, H. and Gray, W.R. (1981) Studies on the evolution of elastin. III. The ancestral protein. *Comparative Biochemistry and Physiolology*, **68B**, 473–80.

Sage H. (1983) The evolution of elastin: correlation of functional properties with protein structure and phylogenetic distribution. *Comparative Biochemistry and Physiology*, **78B**, 373–80.

Wright, G.M. and Youson, J.H. (1983) Ultrastructure of cartilage from young adult sea lamprey, *Petromyzon marinus L.*: A new type of vertebrate cartilage. *American Journal of Anatomy*, **167**, 59–70.

Wright, G.M., Keeley, F.W. and Youson, J.H. (1983) Lamprin: a new vertebrte protein comprising the major structural protein of adult lamprey cartilage. *Experientia*, **39**, 495–7.

Wright, G.M., Keeley, F.W., Youson, J.H. and Babineau, D.L. (1984) Cartilage in the Atlantic hagfish, *Myxine glutinosa*. *American Journal of Anatomy*, **169**, 407–24.

PART FIVE
The Muscular System

12
THE SKELETAL MUSCLE FIBRE TYPES OF *MYXINE GLUTINOSA*

Per R. Flood

SUMMARY

In *Myxine glutinosa* three distinct skeletal muscle fibre types are present in the parietal muscle. The *red, slow* or *tonic fibres* are thin, sarcoplasma- and lipid-rich, highly vascularized, oxidative and multiply innervated. They respond to electrical stimulation by distributed junctional potentials, graded, slow contractions, and show little tendency for fatigue upon repetitive stimulation. The *white, fast* or *phasic fibres* are thick, sarcoplasma-poor, almost non-vascularized, anaerobic and focally innervated. They respond to electrical stimulation by a propagated end plate potential and a twitch contraction, and are highly susceptible to fatigue. A third, *intermediate* fibre type is intermediary between the abovementioned types in several morphological and histochemical traits and is best classified as a fatigue-resistant twitch fibre type in functional terms. A fourth fibre type is the *red, slow craniovelaris muscle fibre*. This differs from the red parietalis fibre in overall diameter, diameter of myofibrils, content of lactate dehydrogenase isoforms, lack of lipid droplets and by a denser innervation by distributed nerve terminals containing more dense core synaptic vesicles. A variety of white fibres, found in the longitudinalis linguae muscle, can work under extremely high lactate concentrations and show an intracellular buffering capacity among the highest recorded in any animal.

12.1 INTRODUCTION AND HISTORY

A resurgence in comparative myology research took place in the late 1940s and early 1950s after the discovery of a functional diversification of skeletal muscle fibres of vertebrates into fast and slow fibre types (e.g. Kuffler and Gerard, 1947; Katz, 1949; Kuffler and Vaughan Williams, 1953). Krüger (1952) maintained that the fast fibres corresponded to fibres with tightly packed thin myofibrils (his *Fibrillenstruktur* fibres) and the slow fibres to fibres with thicker and irregularly outlined myofibrils separated by more sarcoplasm (his *Felderstruktur* fibres). Krüger's view was criticized because certain mammalian skeletal muscle fibres revealed 'field'-like myofibrils in cross-section and failed to show slow fibre characteristics in electrophysiological experiments (Kuffler and Vaughan Williams, 1953). However, for lower vertebrates (e.g. amphibia) his hypothesis was supported, among others, by Gray (1958), Hess (1960, 1963) and Peachey and Huxley (1962). This new insight resulted in a renewed research interest in the since long-known structural distinctions between skeletal muscle fibre types in numerous other lower vertebrates including *Myxine* (Maurer, 1894; Cole, 1907; Marcus, 1925). The numerous reports on *Myxine* skeletal muscles in the 1960s and 1970s may be seen in this perspective.

The state of the art was thoroughly reviewed by Jansen and Andersen (1963) in

The Biology of Hagfishes. Edited by Jørgen Mørup Jørgensen, Jens Peter Lomholt, Roy E. Weber and Hans Malte. Published in 1998 by Chapman & Hall, London. ISBN 0 412 78530 7.

the previous book on the biology of *Myxine* (Brodal and Fänge, 1963). Later a briefer update was published by Hardisty (1979). Since the 1963 review our knowledge of *Myxine* muscles has advanced greatly, particularly concerning ultrastructure, vascularization, histochemistry and metabolism of distinct fibre types. It will be a primary aim of the present communication to review this new knowledge.

The presence of a distinct intermediary type of muscle fibre in addition to the classical red (slow) and white (fast) fibre types in the parietal muscles of the Atlantic hagfish (Flood and Mathisen, 1962) represents a central issue in some of these papers. The present communication offers an opportunity to report some unpublished data on the electrophysiological characterization of this third fibre type.

In recent years the research interest in cyclostome muscles has shifted much towards lampreys where an *in vitro* spinal cord–myotome preparation has been used successfully to provide data on the spinal motoneurons that govern the somatic muscles (Teräväinen and Rovainen, 1971; Wallén *et al.*, 1985; Brodin and Grillner, 1986a, b). Since it is still uncertain to what extent these data are relevant for *Myxine* they will not be discussed any further here.

12.2 GROSS ANATOMY OF THE MUSCLES

To my knowledge no new information on the gross anatomy of the skeletal muscles of *Myxine* has appeared in any recent publication. Those interested in this topic are therefore referred to the older literature (Grenacher, 1867; Fürbriger, 1875; Cole, 1907; Nishi, 1938; Luther, 1938; Marinelli and Strenger, 1956) and to the summary of this as presented by Jansen and Andersen (1963).

12.3 FIBRE TYPE NOMENCLATURE

The many terms used to designate distinct skeletal muscle fibre types have generally been related to the phenomenon under investigation and may seem confusing for readers beyond the field of myology. In *Myxine*, as for other animals, combined morphological, histochemical and electrophysiological studies have made considerable contributions to relate these terms to each other and to resolve this confused state of nomenclature (Table 12.1).

Table 12.1 Nomenclature used for skeletal muscle fibre types in the Atlantic hagfish *Myxine glutinosa*

Red fibres	Intermediary fibres	White fibres	Author and year
Parietal		Central	Maurer 1894
Plasmic		Aplasmic	Cole 1907
Peripheren	Übergangsformen	Zentralen	Marcus 1925
A		B	Bone 1963
II		I	Jansen *et al.*, 1963
Slow		Fast	Jansen and Andersen 1960, 1963; Andersen *et al.*, 1963; Alnæs *et al.*, 1964; Nicolaysen 1966
Slow	Intermediary	Fast	Flood and Mathisen 1962
Red	Intermediary	White	Flood 1965, 1973, 1979; Mellgren and Mathisen 1966; Dahl and Nicolaysen 1971; Korneliussen 1973a, b; Korneliussen and Nicolaysen 1973, 1975
		Twitch	Brautaset and Nicolaysen 1968
(Red)	(Pink)	(White)	Baldwin *et al.*, 1991 for *Eptatretus cirrhatus*

A complicating factor is that, whereas most of the cranial skeletal muscles of *Myxine* are homogeneous with respect to the fibre types they contain (Cole, 1907; Nicolaysen, 1964, 1966), there exist pronounced differences between distinct muscles (e.g. Korneliussen and Nicolaysen, 1973). Both red and white fibres may therefore represent fibres with a wide range of metabolic and functional properties, and it is only when we see these fibres in mixed populations, as in the parietal and oblique abdominal muscle of *Myxine*, that we introduce the term 'intermediary' as well (Flood and Mathisen, 1962). The presence of such intermediary fibres was also noted by Marcus (1925). He interpreted them as transitional forms (Übergänge) and postulated that the peripheral (red) fibres differentiated into central (white) fibres.

12.4 HISTOLOGY OF THE MUSCLES AND DISTRIBUTION OF THEIR FIBRE TYPES

The cranial skeletal muscles of *Myxine* are homogeneous with respect to the fibre types they contain (Cole, 1907; Nicolaysen, 1964, 1966; Mellgren and Mathisen, 1966; Dahl and Nicolaysen, 1971; Korneliussen and Nicolaysen, 1973). The large *m. longitudinalis linguae* consists exclusively of white fibres whereas several smaller muscles of the velar apparatus (*mm. craniovelaris anterior, -superior, -inferior* and *posterior*, and *m. spinovelaris* as described by Marinelli and Strenger 1956) contain red fibres only. In comparison, the paraxial *m. parietalis* and two abdominal muscles, *mm. obliquus et rectus abdominis*, all contain three distinct fibre types (Marcus, 1925; Flood and Mathisen, 1962; Jansen and Andersen, 1963).

Each (ipsilateral) muscle segment of the parietal muscle of an adult *Myxine glutinosa* (30 to 40 cm long) contains some 2800 muscle fibres, distributed into some 35 compartments (Flood, unpubl. observation on frozen sections stained by Sudan Black B, Figure 12.1). Andersen *et al.* (1963) arrived at 1935 fibres in a comparable hemisegment peeled apart under the dissecting microscope. Each compartment contains 3 to 5 layers of muscle fibre where all the red fibres are located in a single ventral layer and in 1 to 3 lateral layers. The white fibres are located in 2 to 4 dorsal layers. Intermediary fibres are mostly concentrated in the lateral region of each compartment; here often as a complete layer between the lateral red fibres and the deeper white fibres. Further, some intermediary fibres are interposed between the red fibres in the ventral layer and less frequently between the white fibres in the dorsal layers of each compartment (Figure 12.2) (Flood and Mathisen, 1962).

A well-defined, but thin connective tissue lamina covers the ventral and lateral surfaces of each compartment. The dorsal compartment surface, however, seems to lack such a layer and the compartment is easily separated from the next above at this level. The connective tissue component of each compartment accordingly is best developed next to the red muscle fibres where also most of the blood capillaries are found. Since these blood capillaries also penetrate to the deeper side of the red fibres the connective tissue lamina is difficult to detach from these fibres. However, on the dorsal surface of each compartment the white fibres are easily separated from the rest of the compartment and from each other due to their exceedingly delicate basement membrane (Flood, unpubl. obs.).

The proportion of white (type I or type A) to red (type II or type B) fibres in the parietal muscle was reported to be 3:1 by Andersen *et al.* (1963) and 10:1 by Bone (1963). Some of this difference might be due to a reduced number of red fibres in the anterior and posterior ends of the body as reported by Cole (1907). However, a more likely explanation is perhaps that the reported fibre ratios refer to paraffin-embedded material where the distinction between the fibre types is less prominent than in frozen sections stained for lipids. In material processed according to the latter technique, Flood (1979) reported red

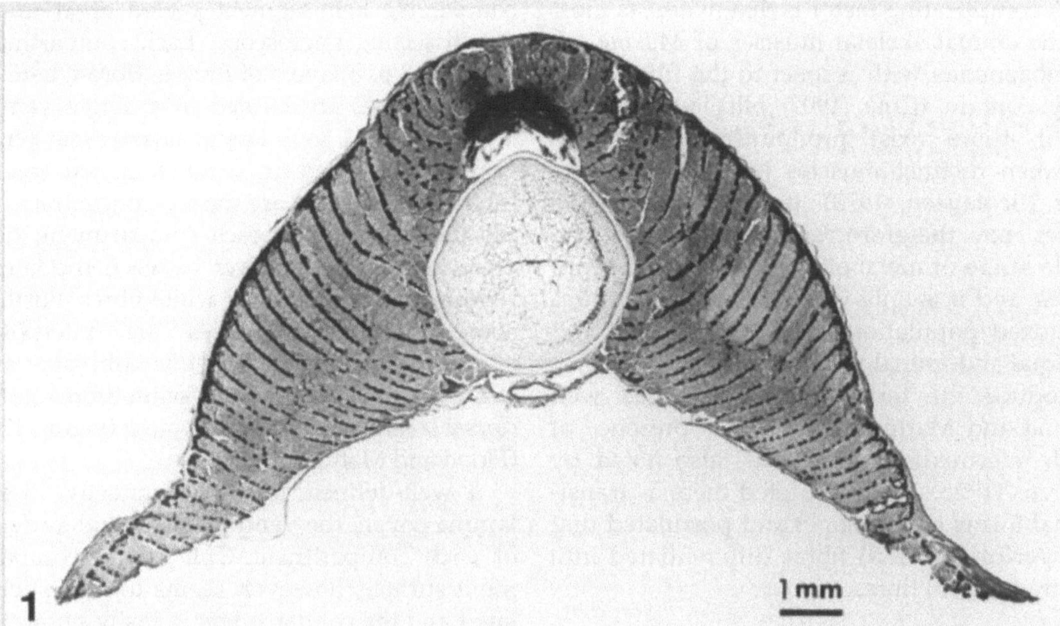

Figure 12.1 Transverse frozen section through midbody region of an eviscerated *Myxine glutinosa* stained by Sudan Black B. Note the compartmental organization of the parietal muscle.

fibres to constitute 36%, white fibres 49%, and intermediary fibres the remaining 15% of the fibre population in a typical midbody segment of the parietal muscle. Even if the intermediary fibres are added to the white fibre population the white to red fibre ratio will be more like 2:1. Since the diameter, and accordingly the volume, of the red fibres is strikingly different from that of the white fibres (Flood, 1979), the comparable volume fractions of distinct fibre types within the parietal muscle is more like: 70% white, 13% intermediate and 17% red fibres. In 84 species of marine fish Geer-Walker and Pull (1975) found the red muscle fibres never to constitute more than 25% of the myotomal musculature. In most species they rather constituted <10%. Bone (1978) reported the caudal region of *Scyliorhinus canicula* to have 18% of the myotome as red fibres.

The temperature to which the animals are acclimated may influence the proportion of distinct fibre types to some extent. A shift towards more oxidative metabolism and higher proportions of red and pink (intermediate) fibres has been found in teleost fish acclimated to a colder than usual environment for 3 months (Johnston and Lucking, 1978; Johnston and Maitland, 1980). Likewise starvation for prolonged periods increase the proportion of oxidative fibres (Johnston, 1982).

Although red and intermediary fibres are present throughout the mediolateral and dorsoventral diameter of the midbody segments, their concentration seems to be higher near the lateral surface of the trunk at the level of the notochord (Andersen *et al.*, 1963). Here, they are ideally situated for optimal effect on the undulatory movements of the body during swimming.

12.5 STRUCTURAL CHARACTERISTICS OF THE FIBRE TYPES

The existence of two distinct skeletal muscle fibre types in *Myxine* was established by the

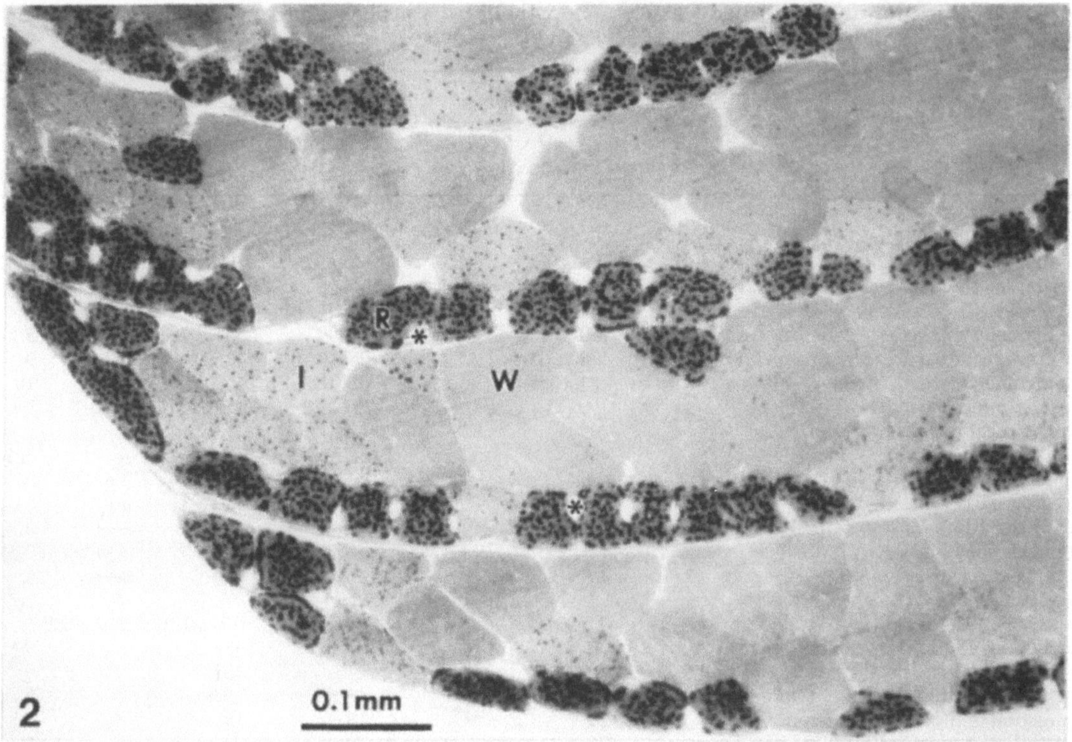

Figure 12.2 Transverse, 1μm thick section of perfusion fixed and plastic embedded parietal muscle from *Myxine glutinosa* stained by paraphenylendiamine (see Korneliussen 1972a). Note the distribution of the red (R), intermediate (I) and white (W) muscle fibre types and blood capillaries (*asterisks*) within the compartments of the muscle.

early works of Maurer (1894) and Cole (1907). See also Jansen *et al.* (1963) and Jansen and Andersen (1963). The presence of an additional third fibre type in the parietal muscle was documented in frozen sections stained for lipids by Flood and Mathisen (1962) and in ultrathin sections prepared for transmission electron microscopy by Flood (1965).

The finer structure of the three fibre types of the parietal muscle was described more thoroughly by Korneliussen (1972, 1973a, b) and Korneliussen and Nicolaysen (1973, 1975). These authors also described a fourth (red/slow) fibre type in the craniovelar muscle, and claimed that these in several respects were radically different from the red/slow fibres of the parietal muscle. The same authors also studied the fibres of *m. longitudinalis linguae* in detail, but concluded that these were quite similar to the white (fast) fibres of the parietal muscle and accordingly should not be considered as a fifth fibre type. However, since histochemical differences are present between the white fibres of *m. parietalis* and those of *m. longitudinalis linguae* (Mellgren and Mathisen, 1966), it is probably best to treat them separately in a review like this. In Table 12.2 the available quantitative and semiquantitative data on these five fibre types are compiled.

In general terms the red fibres of the parietal muscle are characterized by myofibrils separated by large areas of sarcoplasm containing numerous large mitochondria,

Table 12.2 Structural and histochemical characteristics of distinct skeletal muscle fibre types in the Atlantic hagfish *Myxine glutinosa*

Parameter	White fibres		Intermediate fibres	Red fibres		Ref.
	Parietal muscle	M. long. linguae	Parietal m.	Parietal muscle	M. cranio velaris	
Fibres						
Proportion of fibre in muscle in %	36	100	15	49	100	3, 4, 9
Fibre diameter in μm	80–120	78–112		35–50	20–28	2, 10
Fibre diameter in native state in μm	58–159			26–71		6
Fibre diameter in plastic sections in μm	63±10	53±7	49±6	26±2		6
Fibre diameter in plastic sections in μm	130		95	80	25	5
Fibre cross sectional area in frozen sections in μm^2	4700±1000		2700±800	1600±400		9
Equivalent circle diam. in μm	77.4±8.5		58.6±8.8	45.1±5.8		10
Fibre circumference in μm	250±26		200±33	147±22		9
Histochemical profile of fibres*						
Lipid droplets	(+)	–	+	+++	–	3, 1
Glycogen	++	+++	+++	+++	+++	3
Cytochrome oxidase	(+)	(+)	+	+++	++	3
Succinic dehydrogenase	(+)	(+)	+	+++	+++	3, 4
Lactate dehydrogenase (w/PMS)*	++	+++	++	+++	+++	3, 7
Actomyosin ATPase	++	++	+	+++	+++	4
Fine structure of fibres						
Cross-sect. area of single myofibril in μm^2	0.58±0.06		0.62±0.32	0.79±0.10	0.17±0.08	6
Vol. fraction of lipid droplets in fibre in %	0.1		6	38		9
No. of mitochondria per cross-sect. of fibre	102		245	381		9
Vol. fraction of mitochondria in fibre in %	0.7		4	15.5		9
Vol. fraction of SR and glycogen in %	23		18	18		9
Vol. fraction of myofibrils in fibre in %	76		70	28		9
Vol. fraction of nuclei in %	0.7		2	0.5		9
Transverse and sarcotubular system						
Mean no. of triads (T-tubules) per sect. sarcomer	1.64		1.22	1.11	1.04	6
Surface to T-tubule membrane area ratio	0.109		0.073	0.084	0.122	6
Myosatellite (MS) cells						
MS-nuclei as % of total 'myo' nucl.	2.8±1.6		11.1±2.3	22.6±2.0		8
MS-profiles per cross-sectioned fibre	0.23		0.41	0.60		8
Nucl. MS-profiles as % of all MS profiles	27±11		27±5	30±1		8

* Semiquantitative evaluation: +++ = strong reaction, ++ = moderate reaction, + = weak but evident reaction, (+) or – = trace or no reaction, PMS = phenazine methosulphate.

1. Flood and Mathisen (1962) 2. Jansen *et al.* (1963) 3. Mellgren and Mathisen (1966)
4. Dahl and Nicolaysen (1971) 5. Korneliussen and Nicolaysen (1973) 6. Korneliussen and Nicolaysen (1975)
7. Dahl and Korneliussen (1976) 8. Sandset and Korneliussen (1978) 9. Flood, unpubl. obs.
10. Nicolaysen (1964)

glycogen granules and lipid droplets. Likewise, the white fibres of the parietal muscle are characterized by tightly packed myofibrils, little sarcoplasm, glycogen granules and lipid droplets. The mitochondria of these fibres are extremely few and small. These characteristics are little more than a modern cell biological transcription of the old designations; plasmic and aplasmic fibres as defined by Cole (1907). The intermediary fibres of the parietal muscle are characterized by a content of mitochondria and lipid droplets intermediary to that of red and white fibres. Further, whereas the myofibrillar packing resembles that of the white fibres, the concentration of glycogen granules tends to be higher than in both red and white fibres (Figures 12.3 to 12.5).

The appearance of the myofibrils in both transverse (Figure 12.3a) and longitudinal

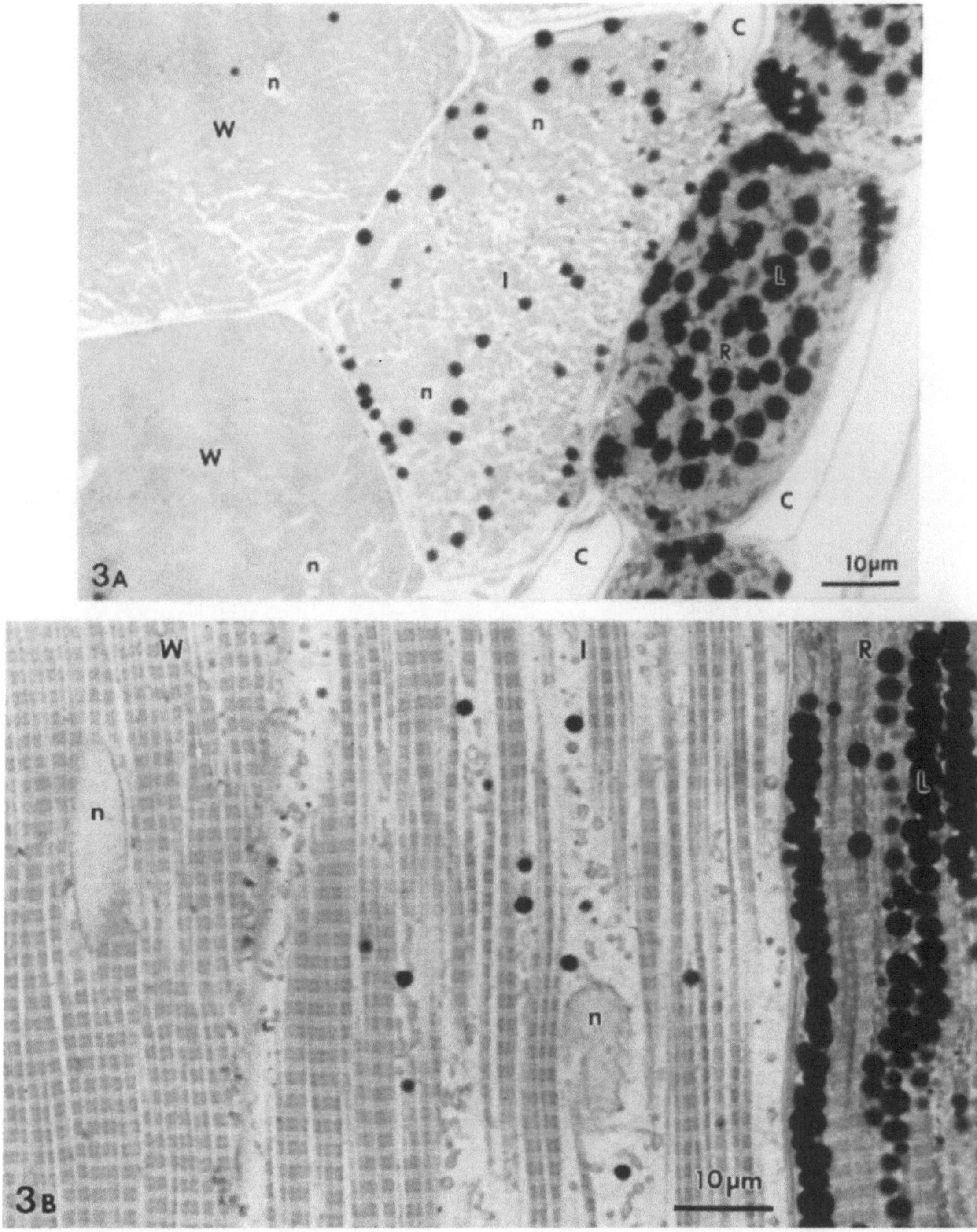

Figure 12.3 (a) Transverse section as in Figure 12.2 at high light microscopical magnification to reveal cytological details of the three fibre types. C = capillaries, L = lipid droplets, n = centrally located nuclei. Otherwise same labelling as in Figure 12.2. (b) Longitudinal section of the same material as in Figures 12.2 and 12.3a. Same staining and labelling.

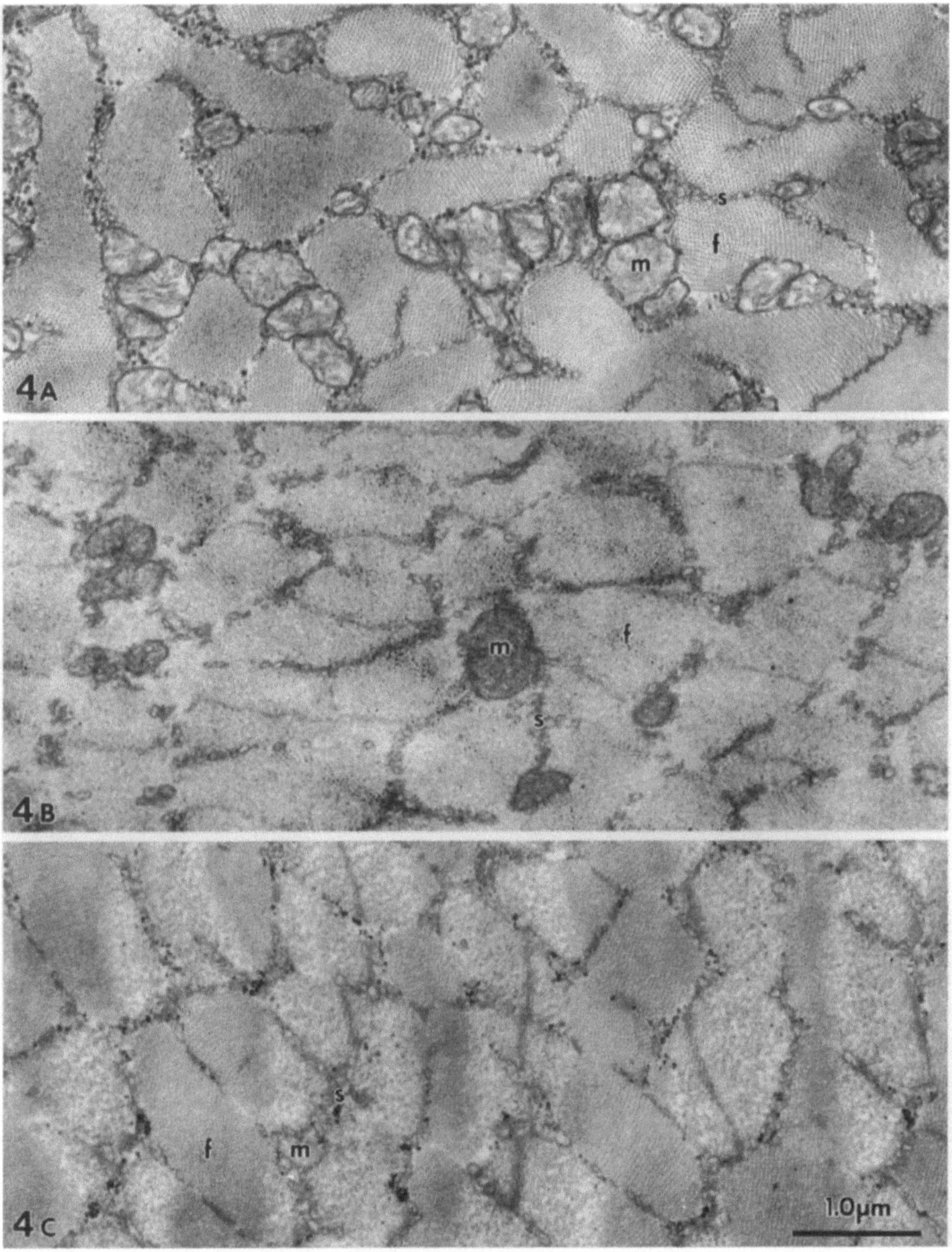

Figure 12.4 Transmission electron micrograph of cross-sectioned (a) red, (b) intermediate, and (c) white muscle fibres from the parietal muscle of *Myxine glutinosa*. f = myofibrils, m = mitochondria, s = sarcotubular elements (T-tubules and sarcoplasmic reticulum).

Figure 12.5 Transmission electron micrographs of longitudinally sectioned (a) red, (b) intermediate and (c) white muscle fibres from the parietal muscle of *Myxine glutinosa*. *Arrowheads* = triads of Transverse tubules and sarcoplasmic reticulum. M and Z identify the banding pattern of the myofibrils.

(Figure 12.3b) section is rather similar in these three fibre types of the parietal muscle. Accordingly, the general fibre type designations *Felderstruktur* and *Fibrillenstruktur*, as used by Krüger (1952), seem inappropriate for *Myxine*. In longitudinal sections no difference could be noted in the sarcomere length or myofibrillar banding pattern in the three fibre types (Korneliussen and Nicolaysen, 1973). In particular the M-band, which is absent in red or slow fibres from frog (Peachey and Huxley, 1962), is present in the red fibres of the parietal muscle (Figure 12.5a).

In an attempt to give a quantitative expression of the fine structural features of the three fibre types of the parietal muscle I have analysed a couple of high power light and low power transmission electron micrographs from each fibre type by standard morphometrical techniques (see Weibel and Elias, 1967). Although these results are too few to be statistically evaluated they hopefully give a rough indication of the distinctive features of the three fibre types (Table 12.2). For example the red fibres seem to have more of their volume occupied by lipid droplets than of myofibrils, further that their myofibrillar volume is only twice as large as their mitochondrial volume. In comparison the white fibres have more than ³/₄ of their volume occupied by myofibrils whereas lipid droplets and mitochondria occupy <1% each. As expected the intermediary fibres, for the same parameters, came out with values intermediary between those of the red and white fibres. The difference in the amount of intermyofibrillar spaces between intermediary and white fibres, so evident in frozen sections and material fixed by perfusion of aldehydes (Figure 12.2), was not reflected in the morphometrical data. The spaces occupied by the sarcotubular system (T-tubuli and sarcoplasmic reticulum) and glycogen appeared to be larger in the white than both the red and intermediary fibres. Unequal shrinkage of the distinct fibre types during embedding for electron microscopy may be an explanation for this failure. In fact, the intermediary fibre type rarely fixed well and always gave poor electron contrast and crumbled internal cytomembranes by direct osmication (see Flood, 1967).

The presence of three fibre types of structural characteristics compatible with those described above has been reported in the axial trunk muscle of *Branchiostoma lanceolatum* (Flood, 1968), lamprey (Lie, 1973, 1974), sharks (Kryvi, 1877), sturgeons (Kryvi *et al.*, 1980), and salamander (Totland, 1976a, b). In the dogfish and several teleosts more than three fibre types have been identified on structural and histochemical grounds (Bone, 1966, 1978; Johnston *et al.*, 1975; Johnston, 1981a, b).

The red muscle fibres of the craniovelaris muscle of *Myxine* are much thinner than the red fibres of the parietal muscle and they contain fewer and smaller lipid droplets than this fibre type. Further, glycogen is abundant and the mitochondria tend to be larger than in the red fibres of the parietal muscle. The myofibrils of the craniovelar red muscle fibres are much thinner than those of the parietal red fibres and they lack a prominent M-line in the middle of the myosin filaments or A-bands.

As stated above, the white fibres of the longitudinalis linguae muscle are indistinguishable from the white fibres of the parietal muscle by all structural criteria reported so far.

12.6 TRANSVERSE TUBULES AND SARCOPLASMIC RETICULUM

In all the five fibre types mentioned above the transverse tubular system is located at the A–I junction of the myofibrils (Korneliussen and Nicolaysen, 1973, 1975). This is notably different from the situation in numerous other lower chordates and vertebrates, including *Petromyzon* (Jasper, 1967; Teräväinen, 1971), where these tubules are located at the Z-disk level of the myofibrils. In the cephalochordate *Branchiostoma lanceolatum* the cytological units

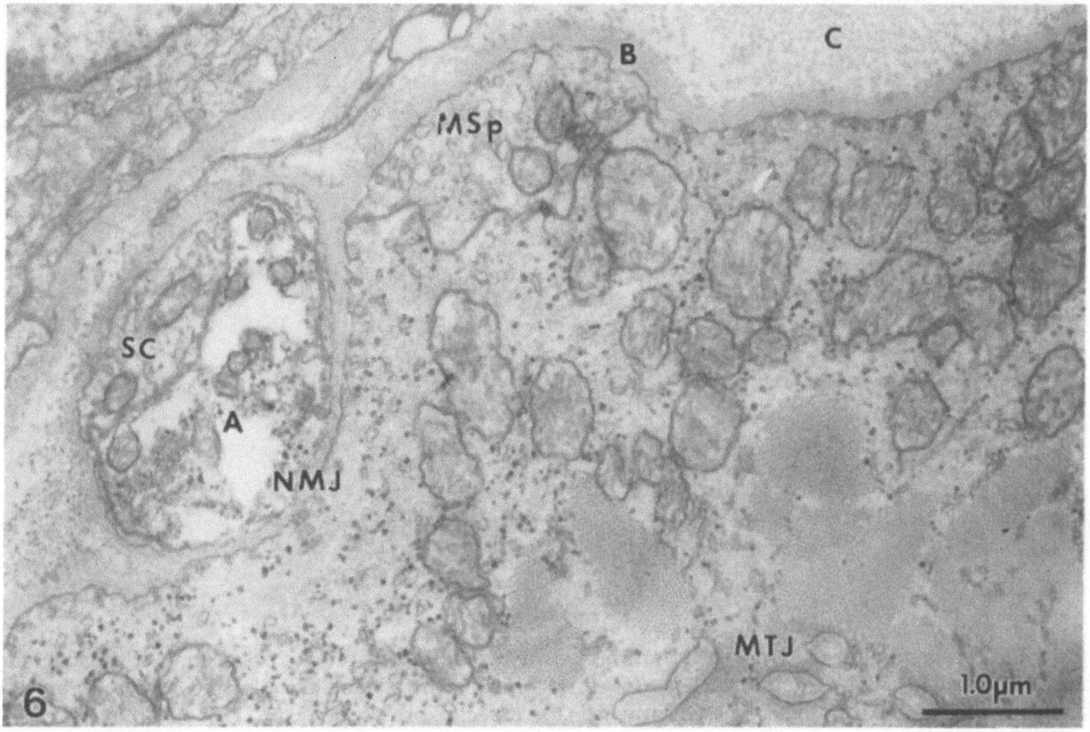

Figure 12.6 Transmission electron micrograph of cross-sectioned red muscle fibre of *Myxine glutinosa* including a neuromuscular junction (NMJ) with axon (A) and Schwann cell cytoplasm (SC), a myosatellite cytoplasmic profile (MSp), and some tube-like plasma membrane infoldings of the myotendinous junction (MTJ). B = basement lamina, C = collagen fibrils.

of skeletal muscle are organized as thin lamellae without T-tubules. However, even here the sarcoplasmic reticulum vesicles are gathered at the Z-disk level (Flood, 1977). As discussed by Korneliussen and Nicolaysen (1973, 1975), it is difficult to give this remarkable fact a phylogenetic explanation.

In the parietal muscle the average number of triads (maximally 4) adjacent to each longitudinally sectioned sarcomere (Figure 12.5), is higher in white than in red parietal fibres and lowest in the red craniovelar fibres. In the intermediary fibres the average number of triads per sarcomere was closer to that of red parietal fibres than that of white fibres (Table 12.2). Based on these and other data the membrane area of the T-system tubules was calculated to be 3–4 times the surface membrane area in a typical white parietal fibre (80 µm fibre diameter) and 1–2 times the surface area in a typical red parietal fibre (60 µm fibre diameter) (Korneliussen and Nicolaysen, 1975).

Quantitative aspects of terminal cisterns and longitudinal elements of the sarcoplasmic reticulum, beyond what may be implied from the triad counts reported above, have not been reported. In my own material these elements seem developed to a larger extent in the intermediary than in the red and white fibres of the parietal muscle (Figure 12.5).

12.7 MYOSATELLITE CELLS

Myosatellite (MS) cells are defined as cells lodged between the plasma membrane and

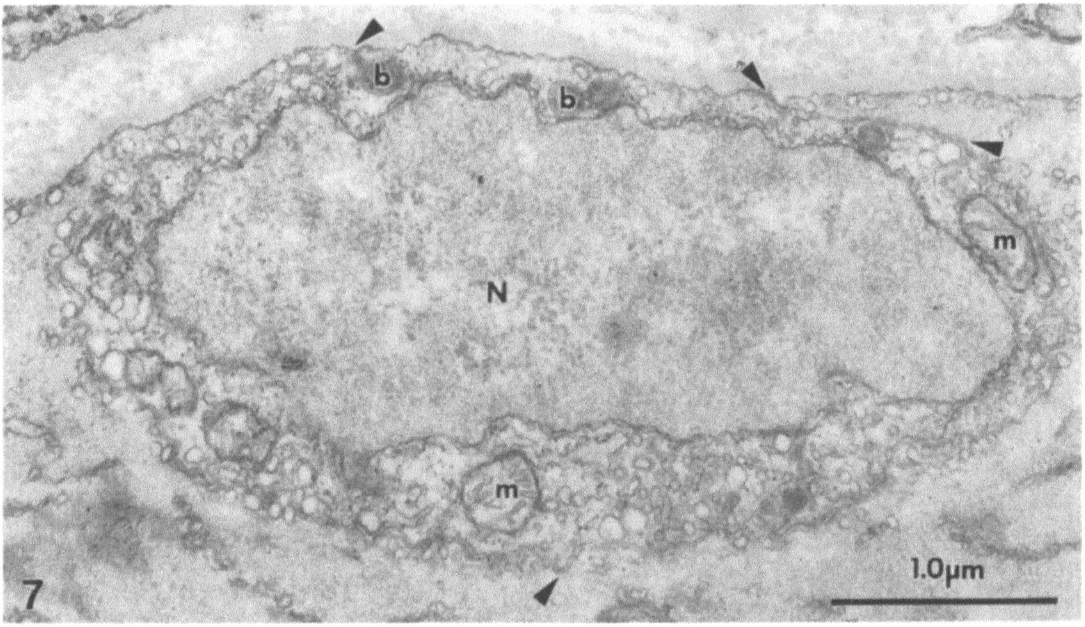

Figure 12.7 Transmission electron micrograph of a myosatellite cell on a cross-sectioned intermediary fibre of *Myxine glutinosa*. Note its nucleus (N), mitochondria (m), dense cytoplasmic bodies (b) and the plasma membrane separation (*arrowheads*) towards the muscle fibre proper.

the basal lamina of skeletal muscle fibres (Mauro, 1961). Rather than bulging out on the fibre surface they occupy a trench in the muscle fibre proper. In *Myxine* this cell type has been briefly described by Flood (1964) and more extensively studied by Sandset and Korneliussen (1978). The nuclei in the myosatellite cells are generally more heterochromatic than the myonuclei and this difference may enable their light microscopical identification. However, in general, MS cells are identified by transmission electron microscopy by their plasma membrane separation from the myoplasm proper. For *Myxine* in particular, their identification depends on detailed examination of the plasma membrane contours, since the MS cell membrane interdigitates extensively with the underlying sarcolemma (Figures 12.6 and 12.7). Further, the incidence of such profiles per cross-sectioned muscle fibre is rather low; only one of four sectioned parietal white fibres and six of ten craniovelar red fibres will show such a profile (Table 12.2). The incidence of MS cells of intermediary fibres of the parietal muscle was not given as a percentage by Sandset and Korneliussen (1978). It was, however, mentioned to resemble that of parietal white fibres. The MS cells tend to be elongated along the muscle fibres since purely cytoplasmic profiles are seen 3 to 4 times more frequently than nucleated profiles.

The cytoplasm of the satellite cells usually contains many free ribosomes and vesicles, and few larger, membrane bound organelles. However, in some cells prominent Golgi stacks and electron dense bodies resembling secretory granules, may be found near the nucleus (Figure 12.7). A few mitochondria, some smooth and granulated elements of the endoplasmic reticulum contribute to the general impression of these cells as being rather undifferentiated. This cell type is now generally accepted to represent dormant myoblasts.

12.8 MYOTENDINOUS JUNCTIONS

The myotendinous junctions in the parietal muscle of *Myxine* were described by Jansen *et al.* (1963) and Korneliussen (1973b) and subjected to a more detailed electron microscopical analysis by Korneliussen (1973c). The white and intermediary fibres end bluntly at the myocommata in front and behind the fibres, and the end which possess the neuromuscular end plate is flattened and dilated beyond the cross-sectional dimension elsewhere on the fibre. The red fibres, on the other hand, have pointed ends that never reach the myocommata proper and are attached to these by short tendons.

In the electron microscope tube-like, sometimes branching, invaginations of the muscle fibre plasma membrane were seen to extend for several microns into the fibre ends. These pockets were coated by a prominent basement lamina and filled by collagen fibrils extending from the myocomma. I-filaments of the contractile apparatus of the fibre interior attached to a thickened cytoplasmic leaflet of the plasma membrane and the collagen fibrils in the extracellular pockets attached to the lamina densa of the basal lamina (Figure 12.6). The otherwise electron lucent space between the lamina densa of the basal lamina and the extracellular leaflet of the plasma membrane was most conspicuously traversed by 6 nm thick electron dense lines with a very regular, 15 nm, periodic spacing. These lines ran perpendicular to the long axis of the tube-like pockets and probably represent proteins involved in the mechanical force transfer from the contractile apparatus to the tendons (cf. Nakao, 1975; Trotter *et al.*, 1981, 1983).

Myomuscular junctions between cross-striated muscle fibres surrounding the gill sacs of *Myxine glutinosa* have also been studied in the electron microscope (Bartels, 1986). Sarcolemmal differentiations, identical with those described for the myotendinous junctions of the parietal muscle by Korneliussen (1973c), are here present at end-to-end and side-to-side junctions between the muscle fibres. However, since these junctions are not associated with tendons they may be described as symmetrical myomuscular junctions.

12.9 HISTOCHEMISTRY

Mellgren and Mathisen (1966) described the distribution of oxydative enzymes, glycogen and lipid in several muscles of *Myxine* (Table 12.2). They reported high activity of cytochrome oxidase and succinate dehydrogenases in *m. craniovelaris,* heart muscle, and in red fibres of *m. parietalis*. In *m. cordis caudalis* there was a somewhat weaker reaction and the intermediary fibres of *M. parietalis* showed a still lower, but easily detectable activity. In the white fibres of the latter muscle and in *m. longitudinalis linguae* only traces of reaction product were observed. In the presence of phenazine methosulphate, NAD- and NADP-linked isocitrate dehydrogenases were active in red fibres of all muscles examined, and particularly in red fibres of the *m. craniovelaris*. No activity was observed in white fibres. NAD- and NADH-linked maleate dehydrogenases, in the presence of phenazine methosulphate, were much less reactive in the red fibres. Further, the white and intermediate fibres also showed some activity of this enzyme.

'Menadione-linked' α-glycerophosphate dehydrogenase could not be detected in any fibre type from *Myxine* whereas a positive reaction was present in control sections of rat muscles. NAD-linked α-glycerophosphate dehydrogenase and lactate dehydrogenase, in the absence of phenazine methosulphate, produced a picture identical with that observed for the Krebs cycle dehydrogenases and cytochrome oxydase as mentioned above. In the presence of phenazine methosulphate the picture changed drastically: an enhanced reaction of intermediary fibres and a strong activity of white fibres then became apparent. Further, provided that cyanide was included in the incubation medium (Mathisen and Mellgren, 1965), the staining of red fibres was enhanced.

The distribution of succinic dehydrogenase and lactate dehydrogenase in *Myxine* muscles, as described above, was verified by Dahl and Korneliussen (1976). These authors also visualized the histochemical and electrophoretic distribution of lactate dehydrogenase isoenzymes in the parietal and craniovelaris muscles. As might be expected, the muscular isoenzyme was not liable to product inhibition by 8 mM pyruvate to the incubation medium. On the other hand, the addition of 1–4 M urea had a pronounced inhibitory effect on the histochemical staining intensity for lactate dehydrogenase, especially in the red fibres of the parietalis and craniovelaris muscles.

According to Mellgren and Mathiesen (1966) all red fibres contained glycogen, but those of *m. craniovelaris*, heart muscle and *m. parietalis* were richer in glycogen than those of *m. cordis caudalis*. The amount of glycogen in white fibres often was somewhat smaller than in the red fibres. *Musculus longitudinalis linguae*, as a rule, contained more glycogen than the white fibres of the *m. parietalis* at the corresponding level. Intermediate fibres were sometimes intermediately stained, but often more intensely coloured than the other fibre types.

In sections stained with Fat Red 7B and Sudan Black B a great quantity of fat droplets was observed in the red fibres of *m. parietalis*. Intermediate fibres contained similar but generally fewer and smaller droplets. These lipid droplets stained equally well with the two dyes and were extracted with acetone. In *m. craniovelaris* and *m. cordis caudalis* only small dense granules were observed. These were only stained by Sudan Black B and some of this staining persisted after acetone extraction. Comparable sudanophilic granules, probably of mitochondrial or sarcoplasmic reticulum origin, were also present in red and intermediate fibres of *m. parietalis*.

Mellgren and Mathisen (1966) conclude that the red muscle fibres and heart muscle cells contain large quantities of most enzymes studied as well as glycogen. In addition, the red fibres of *m. parietalis* contain large quantities of storage lipid. These fibres are therefore equipped to provide ATP anaerobically as well as aerobically. This may represent a specialization permitting the important muscles of circulation and locomotion to function during long periods of anoxia, e.g. when the animal burrows in the mud or feeds in the interior of a fish carcass. Hagfishes have been observed to remain completely submerged in the mud for more than one hour (Strahan, 1963). Apart from permitting life for long periods at low oxygen tension, this equipment would, under aerobic conditions, render the red muscle fibres well suited for sustained work.

White fibres contain practically no cytochrome oxydase and Krebs cycle dehydrogenases, except for some activity of NAD-dependent malic dehydrogenase. However, these fibres are approximately as rich in lactate dehydrogenase and NAD-linked α-glycerophosphate dehydrogenase as the red fibres. Accordingly, the energy source of the white fibres seems to depend almost exclusively upon anaerobical metabolism. Compared to the aerobic metabolism of the red fibres, such anaerobic metabolism provides less ATP and energy. This, combined with their poor vascularization (which may lead to accumulation of intermediary metabolites, acidification of the sarcoplasm and blocking of the contraction mechanism), is consistent with the finding that white fibres of *m. parietalis* are easily fatigued (Andersen *et al.*, 1963). Intermediate fibres generally showed an intermediate activity of enzymes and content of glycogen and lipid.

Dahl and Nicolaysen (1971) found the fast twitch white fibres of *m. longitudinalis linguae* and *m. parietalis* to show relatively high activity of Ca^{++}-activated myosin ATPase, whereas the intermediary fibres of the latter muscle showed low activity. More surprisingly, the slow non-twitch, superficial red fibres of *m. parietalis* and the fibres of *m. craniovelaris* showed an ATPase activity even higher than that of the white fibres.

Since the activity of Ca^{++}-activated myosin ATPase is generally supposed to parallel the speed of contraction (Bárány, 1967; Guth and Samaha, 1969) a high activity in white fibres seems reasonable. According to Mellgren and Mathisen (1966) the low activity of the intermediate fibres was 'difficult to comment upon since little is known about their physiological properties. Theoretically, they ought to be slow twitch.' The high ATPase activity in the red fibres, which physiologically are known to be slow non-twitch, producing small junction potentials only (Andersen *et al.*, 1963; Nicolaysen, 1964, 1966), could not be interpreted in accordance with the mentioned hypothesis of speed of contraction.

Acetylcholine esterases have been histochemically detected in neuromuscular junctions of *Myxine* (Jansen *et al.*, 1963; Jansen and Andersen, 1963; Nicolaysen, 1964). For more details see the section on innervation below.

12.10 CHEMISTRY

Little information is available in the literature on this topic. Only the properties of one enzyme present in the muscles seem to have been examined so far. This is the lactate dehydrogenase (LDH, EC 1.1.1.27) which has been studied in *Myxine glutinosa* by Dahl and Korneliussen (1976), Sidell and Beland (1980), and more extensively in other hagfishes like *Eptatretus stoutii* by Wilson *et al.* (1964, 1967), Arnheim *et al.* (1967), Ohno *et al.* (1967), Makert *et al.* (1975) and *Eptatretus cirrhatus* by Baldwin *et al.* (1989). This specific interest for the LDH enzyme is based primarily on its use as a marker for gene evolution throughout the vertebrate line (see Markert *et al.*, 1975). All LDH isoforms, from bacterial to human origin, are tetramers of molecular weight close to 140 000 Da. Three distinct subunits of the enzyme are coded for by three genes in most vertebrates and since these three subunits have a strong tendency to form homo- and hetero-tetramers a total of nine isoforms of the enzyme may be defined. Current theory postulates that the three distinct subunits came into existence by duplication of a single ancestral LDH gene. This is reflected in the fact that in hagfishes only two genes, and in the lampreys only one LDH gene are expressed. In other words, hagfishes have a central position near the important evolutionary events leading to the emergence of vertebrates from more primitive animals (see Baldwin *et al.*, 1987, 1988 and 1989).

Dahl and Korneliussen (1976) describe the distribution of LDH isoforms in distinct skeletal muscle fibre types and heart muscle of *Myxine*. They found only two isoenzymes and were unable to detect or produce heterotetramers of them *in vivo*, respectively *in vitro*. The cathodal isoform is present in all three fibre types of the parietal muscle and the white fibres of the *m. longitudinalis linguae*. The anodal isoenzyme, on the other hand, was present in heart muscle and red fibres of the *m. craniovelaris* and, in a much lower concentration, in the red parietal fibres.

A functional and evolutionary evaluation of these two isoforms of LDH was given by Sidell and Beland (1980). Based on substrate specificity and susceptibility to inhibition by pyruvate at high concentration, they proposed distinct functions for the two LDH isoforms found in *Myxine*. In general terms they described the heart isoform (H_4 or B_4) (the anodal isoform of Dahl and Korneliussen, 1976) as being more susceptible to pyruvate inhibition and suitable for aerobic dehydrogenation of lactate, and the skeletal muscle isoform (M_4 or A_4) (the cathodal isoform of Dahl and Korneliussen, 1976) as being more tolerant to high pyruvate concentrations, and suitable for reduction of pyruvate to lactate under conditions of limited oxygen supply. However, in agreement with the habit of *Myxine* to feed by invading coelomic cavities of other dead or dying fishes, and thus to be exposed to extreme hypoxic conditions, they found the heart isoform of LDH to be less geared towards aerobic functions than the comparable isoforms of higher vertebrates. Later Baldwin and Lake (1987) and Baldwin *et al.*, (1987, 1988, 1989) have ascribed these unusual features of the *Myxine* LDH H_4 isozyme to metabolic adaptation rather than to conservation of evolutionary ancient properties.

The intracellular buffering capacity and activity of several enzymes involved in anaerobic glycolysis in the dental retractor muscles (including *m. longitudinalis linguae*) and the parietal muscle of the New Zealand hagfish (*Eptatretus cirrhatus*) were studied by Baldwin (1989) and Baldwin *et al.* (1991). The buffering capacity of the white dental retractor muscles was found to be extremely high, comparable to that of the skipjack tuna (Castellini and Somero, 1981; Dickson and Somero, 1987), and three times higher than the myotomal muscle. Further, although the activity of glycolytic enzymes was moderate in the same muscles, the lactate concentration within them increased linearly during bursts of continuous feeding, for periods from 1.5 to 6.45 min, reaching a maximum of 132 μmol lactate g^{-1} wet weight muscle. This lactate concentration is among the highest ever reported for normal muscle work by unrestrained animals and it outperforms the skipjack tuna (Dickson and Somero, 1987) by quite a margin.

12.11 INNERVATION

As convincingly documented by Jansen and Andersen (1960), Bone (1963) and Jansen *et al.* (1963) the white parietal fibres are innervated by 3–5 μm thick axons ending in typical *'en plaque'* motor end plates localized at or very close to the end of the muscle fibres where these insert into the connective tissue septum at the anterior or posterior side of the segment. Convincing electrophysiological evidence presented by Andersen *et al.* (1963) and Jansen and Andersen (1963) indicates that there is only one end plate per white, fast fibre. In another white muscle of *Myxine* (*m. protractor dentium profundus*) Bone (1963) found *'en plaque'* endings localized to the middle of the fibres. In the white, fast fibres of the *m. longitudinalis linguae* Nicolaysen (1964, 1966) also identified one end plate per fibre.

In contrast to such localized or focal innervation, the red parietal fibres are innervated by axons, only 1–2 μm in diameter, running along much of the length of the muscle fibre with numerous, presumably synaptic, expansions distributed along their course. Quite often two such varicose nerve fibres, one from the myoseptum in front and one from the myoseptum behind the muscle fibre, are found along one and the same red muscle fibre. For the red (slow) muscle fibres of *m. spinovelaris* and *m. craniovelaris*, Nicolaysen (1964, 1966) found an average of >5 nerve fibres around each muscle fibre. These nerve fibres run along the entire muscle fibres and have synaptic swellings distributed approximately every 10 μm throughout.

Korneliussen (1973a, b, c) identified one terminal end plate on intermediary fibres of the parietal muscle. These end plates were rather similar to those of the white fibres within the same muscle, and could only be distinguished from these based on the cytological features of the muscle fibres.

Korneliussen (1973b, c) also gave important ultrastructural details on the nerve terminals, their Schwann cells, and the postsynaptic elements in the three muscle fibre types of the parietal muscle and the red craniovelar fibres.

The preterminal regions of all motor axons are invested in one or two layers of Schwann cell cytoplasm as real myelin does not exist in this animal (Peters, 1963). In synaptic areas on all fibre types the cytoplasmic leaflets of the Schwann cells are retracted and the axonal membrane is exposed directly to the muscle membrane across a synaptic cleft about 45 to 60 nm wide. The Schwann cells, however, insulate most of the motor nerve fibres completely from the extracellular tissue space. This is particularly evident for the terminal end plates on white and intermediary muscle fibres where the axon terminals are deeply embedded in niches of the muscle fibre and completely covered by plug-like Schwann cells. Extensive folding of the post-synaptic membrane, as typically found in neuromuscular end plates of higher animals, is absent or rare in *Myxine* (Figures 12.6 and 12.7a–b).

The most distinctive ultrastructural feature reported by Korneliussen (1973a, b) for the nerve terminals to distinct muscle fibre types was their content of dense core synaptic vesicles (Figure 12.7a), expressed as percentage of total synaptic vesicle population. Nerve terminals to white and intermediary fibres contained only 1–2% of such dense core vesicles, those to parietal red fibres an average of 9%, and those of craniovelaris red fibres an average of 15% dense core vesicles. Dense core vesicles may store serotonin, and a formaldehyde-induced fluorescence, probably caused by serotonin, was described by Korneliussen (1973a) in structures interpreted as terminal axons along the red craniovelar muscle fibres. However, a green background autofluorescence, also present without formaldehyde treatment, made the results difficult to interpret.

The 'empty'-looking synaptic vesicles were classified as either round or flattened and subjected to a thorough study with fixatives of distinct tonicity by Korneliussen (1973b). His conclusion was that the tonicity of the fixative influenced the morphology of the vesicles to a great extent. Any sign of dual innervation of white fibres, as documented for white muscle fibres of dogfish (*Scyllorhinus caniculus*) by Bone (1972) and ammocoete larvae of lamprey (*Lampetra fluviatilis*) by Kashapova and Sakharov (1976), was not found.

Differences in the size of 'empty'-looking synaptic vesicles in nerve terminals to red and white fibres are obvious for the lancelet (*Branchiostoma lanceolatum*) (Flood, 1966) and the sturgeon (Kashapova and Sakharov, 1977). A difference has also been observed in the dogfish (Bone, personal communication), frog, garter snake and chicken. For the latter three species the observations were made by the present author on electron micrographs kindly supplied by A. Hess (see Hess, 1960, 1963) many years ago. A functional interpretation of this difference in size of the synaptic vesicles remains entirely speculative. Distinct transmitter substances or distinct quantum sizes of the same transmitter may be involved.

Histochemical tests identify acetylcholine esterase in the neuromuscular junctions of white fibres in the parietal muscle (Jansen *et al.*, 1963; Jansen and Anderson, 1963) and *m. longitudinalis linguae* (Nicolaysen, 1964). For the neuromuscular junctions of red fibres in the parietal muscle the histochemical reaction is much weaker but still positive for this enzyme (Mathisen in Jansen and Andersen, 1963). Electrophysiological evidence, based on the use of d-tubocurarine and eserine, favours the view that acetylcholine acts as the transmitter substance in the neuromuscular junctions to both white and red muscle fibres of the parietal muscle (Jansen and Andersen, 1963). The abundance of dense core vesicles and the probable presence of a monoamine-like formaldehyde-induced fluorescence in the neuromuscular junctions to red fibres in the parietal and craniovelaris muscles may indicate that other neuro-transmitters or neuro-modulators are involved (cf. Korneliussen, 1973a).

12.12 ELECTROPHYSIOLOGY

As convincingly documented by Jansen and Andersen (1963) and Andersen *et al.* (1963), two functionally distinct fibre types were identified by basic electrophysiological parameters and mechanical responses in the parietal muscle of the Atlantic hagfish (Table 12.3). Later, numerous details have been added on (1) the spatial, amplitude and interval distribution of spontaneous neuromuscular junctional potentials and the electrical cable properties of the two fibre types (Alnæs *et al.*, 1964), (2) the effect of chloride ions on the surface membrane and the T-tubule membrane of twitch (white) fibres (Brautaset and Nicolaysen, 1968), (3) the spread of the action potential through the T-tubule system of twitch (white) parietal fibres (Nicolaysen 1976a) and (4) the spread of junctional potentials through the T-tubule system of slow (red) parietal fibres (Nicolaysen, 1976b). Beyond this, the electrophysiology of slow red fibres in the spinovelaris and craniovelaris posterior muscles and the fast white fibres in *m. longitudinalis linguae* were described in a masters thesis in Norwegian (Nicolaysen, 1964) and in a brief English abstract thereof (Nicolaysen, 1966). Some data from these publications are compiled in Table 12.3.

In none of the above-mentioned studies was any attempt made to distinguish the intermediary fibres from the other fibre types of the parietal muscle. Alnæs *et al.* (1964, p. 253) rather believed the intermediary fibre type to be excluded from their material.

In my own experiments, performed during a visit to the Institute of Neurophysiology in Oslo during 1975, the resting membrane potential recorded from the intermediary fibres tended to be slightly lower than those of the white fibres (Table 12.3). Further, the intermediary fibres could not be distinguished from the white fibres either on shape and distribution of miniature end plate potentials or on size and shape of propagated and overshooting end plate potentials.

Both the intermediary and white fibre types responded to single stimuli by twitch contractions of comparable tension rise and decline phases (Figures 12.9a–b and 12.11). As might be expected, both the rise and decline phases lasted longer in my experiments than in the previously published experiments due to the temperature difference. Similarly, the graded response of the red fibres (Figure 12.9c) had a much slower time course in my experiments than in experiments performed at room temperature as reported by others (Table 12.3).

By repetitive stimulation at 1 Hz the white fibres were susceptible to prominent fatigue (tension reduced to less than 10% of maximum) within 5 to 10 min (Figure 12.10) (cf. Andersen *et al.*, 1963; Jansen and Andersen, 1963). The intermediary fibres, however, maintained the amplitude of their tension above 50% of maximum tension for more than 20 min at this stimulus frequency (Figure 12.11).

Based on these preliminary observations the white muscle fibres of *Myxine* could conveniently be considered as 'fatiguable fast fibres' and the intermediary fibres as 'fatigue-resistant fast fibres'. Comparable distinctions were established for different motor units in the anterior tibial and soleus muscles of the rat by Edström and Kugelberg (1968) and Kugelberg and Edström (1968) and in cat gastrocnemius muscle by Burke *et al.* (1971).

12.13 VASCULAR SUPPLY

The rich supply of blood capillaries to the parietal red fibres and poor supply to the central white fibres within the parietal muscle were noted by Jansen *et al.* (1963) and several later authors (Figure 12.2). This difference was given a more thorough and quantitative evaluation by Flood (1979) (Table 12.4). In this paper several different techniques to express capillary distribution within a mixed muscle were explored and some new numerical and

Table 12.3 Functional characteristics of distinct skeletal muscle fibre types in the Atlantic hagfish *Myxine glutinosa*

Parameter	White (fast) fibres		Intermediate fibres	Red (slow) fibres		Refs.
	Parietal muscle	M. long. linguae	Parietal muscle	Parietal muscle	M. cranio velaris	
Electrical characteristics						
Resting memb. pot. at 18–20°C	74.6 mV (0.3 SE)	70.4 mV		46 mV (0.5 SE)	35.3 mV	1, 4
	66 mV (1.7 SE)			40 mV (2.0 SE)		3
Resting memb. pot. at 8–10°C	65.40 mV (±3.62 SD) (n = 16)		59.21 mV (±4.24 SD) (n = 19)	36.82 mV (±8.80 SD) (n = 76)		2
Membr. capacitance	4.0 μFcm^{-2}			4.0 μFcm^{-2}		3
Membrane resistance	17000 Ωcm^{-2}			5000 Ωcm^{-2}		3
Response to el. stim.	Overshooting action potential			Gradual depolarization		1, 4
Origin of MEEPs	Focal	Focal	Focal	Distributed	Distributed	2, 3, 4
Distance between synaptic sites				<0.1 mm	10μm	1, 4
Spread of membrane potential changes along T-tubule system	Active			Passive		5, 6
Mechanical characteristics of single contractions						
Time to peak tension at 18–20°C	60–80 ms	60 ms		120 ms	125	1, 4
Time to peak tension at 7–11°C	226 ms (±71 SD) (n = 8)		257 ms (±84 SD) (n = 6)	478 ms (±159 SD) (n = 8)		2
Total duration at 18–20°C	150 ms		>400 ms			1
Total duration at 7–11°C	~400 ms		~400 ms	~2500 ms		2
Response to repetitive stimulation						
Tetanic fusion frequency	50 Hz	50 Hz		35 Hz	35 Hz	1, 4
Fatiguability*	<10 min		>30 min	>30 min		2
Vascularization of fibres						
No. of caps. around each fibre	0.76±0.99		1.95±1.04	0.19±0.86		7
Vasc. surface/volume of fibre in μm^2/μm^{-3}	0.0002		0.011	0.029		7

1. Andersen *et al.* (1963) (SE = Standard error of the mean)
2. Flood (unpubl. res. from 1975) (SD = Standard deviation of population). (Number of observations in parentheses on line below mean and SD)
3. Alnæs *et al.* (1964) (SE = Standard error of the mean).
4. Nicolaysen (1964, 1966)
5. Nicolaysen (1976a)
6. Nicolaysen (1976b)
7. Flood (1979)

* = time before tension has fallen to <50% of maximum tension at a stimulus frequency of 1 Hz (see Fig. 10B)

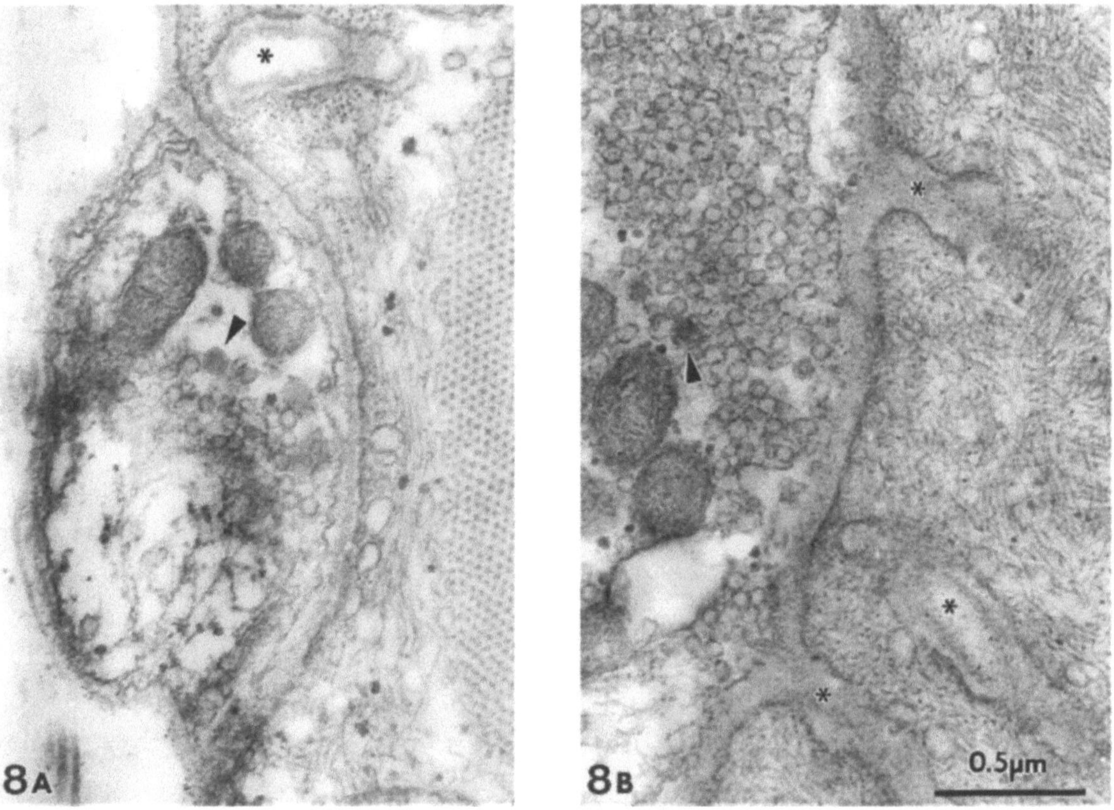

Figure 12.8 Transmission electron micrographs of neuromuscular junctions on (a) a red muscle fibre and (b) a white muscle fibre of *Myxine glutinosa*. Note the presence of dense core vesicles (*arrowheads*), the apparent equal size of the 'empty'-looking synaptic vesicles and the presence of postjunctional plasma membrane folds (*asterisks*) in both junctions.

graphical ways to express the capillary density relative to distinct fibre types were presented (Figure 12.8a–b). Of these perhaps the most functionally informative way to express the capillary density relative to fibre types was the 'mean vascularized surface area per unit volume of fibre'. For red fibres this was $0.029 \mu m^2/\mu m^3$, for intermediate fibres $0.011 \mu m^2/\mu m^3$, and for white fibres $0.0002 \mu m^2/\mu m^3$. These numbers indicate that the intermediate fibres are 50 times better supplied by capillaries than the white fibres and that there is an almost 150 times difference in the supply of blood to the red and white fibres.

This extremely specific distribution of blood capillaries to distinct muscle fibre types is in good agreement with the ultrastructure and enzymatic profiles of the three fibre types as previously worked out by electron microscopy (Flood, 1965; Korneliussen and Nicolaysen, 1973), respectively by histochemistry (Mellgren and Mathisen, 1966). Comparable data exist for some elasmobranchiomorph fishes (Totland *et al.*, 1981), for sturgeon (Kryvi *et al.*, 1980) and axolotl (Totland, 1984) (Table 12.3). From the values presented in this table it is evident that the red and intermediary fibres of *Myxine* are among the better vascularized fibres and the

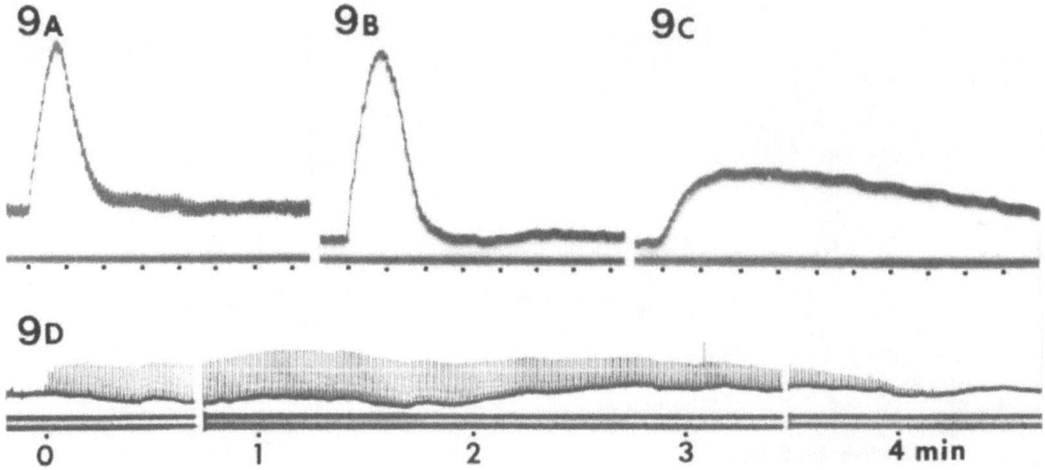

Figure 12.9 Oscilloscope tracks of the mechanical response of a single (a) white, (b) intermediate and (c) red parietal muscle fibre of *Myxine glutinosa* to single electrical stimuli. Time marks every 200 ms. (d) Oscilloscope track of the mechanical response of a single white parietal muscle fibre of *Myxine glutinosa* to repetitive electrical stimulation at 1 Hz. Note the initial facilitation and the later progressive reduction of tension for each contraction.

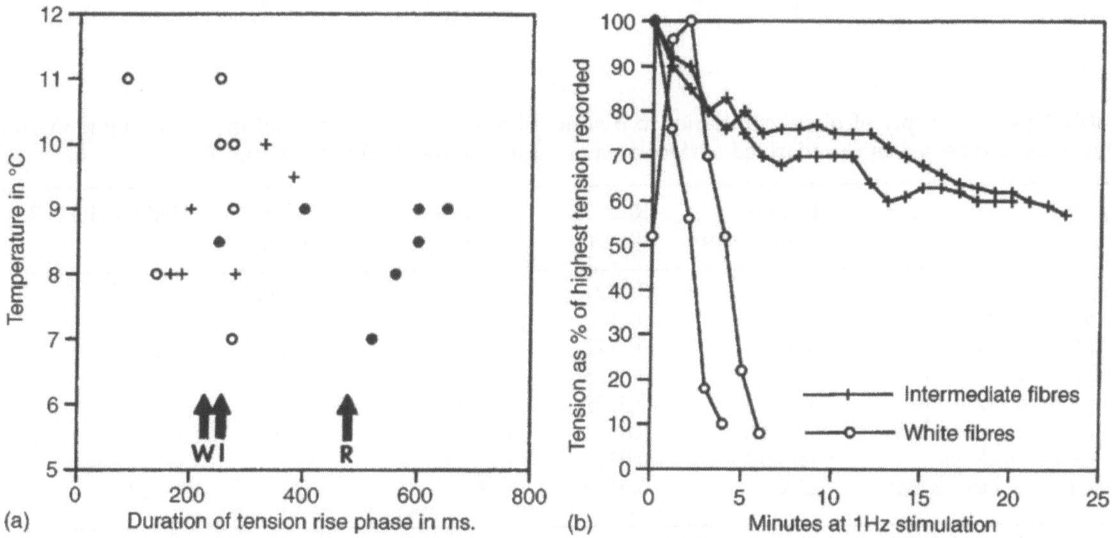

Figure 12.10 (a) Two dimensional plot of duration of the tension rising phase of individual red (•), intermediate (+) and white (o) parietal muscle fibres of *Myxine glutinosa* relative to the temperature at which the recording was made. *Arrows* and *letters* indicate the mean duration of the rising phase of the plotted values for red (R), intermediate (I) and white (W) fibres respectively. (b) Two-dimensional plot of the tension of repetitively stimulated (at 1 Hz) individual white (o) and intermediate (+) parietal muscle fibres of *Myxine glutinosa*, expressed as percentage of their maximal tension, relative to duration of the stimulation period.

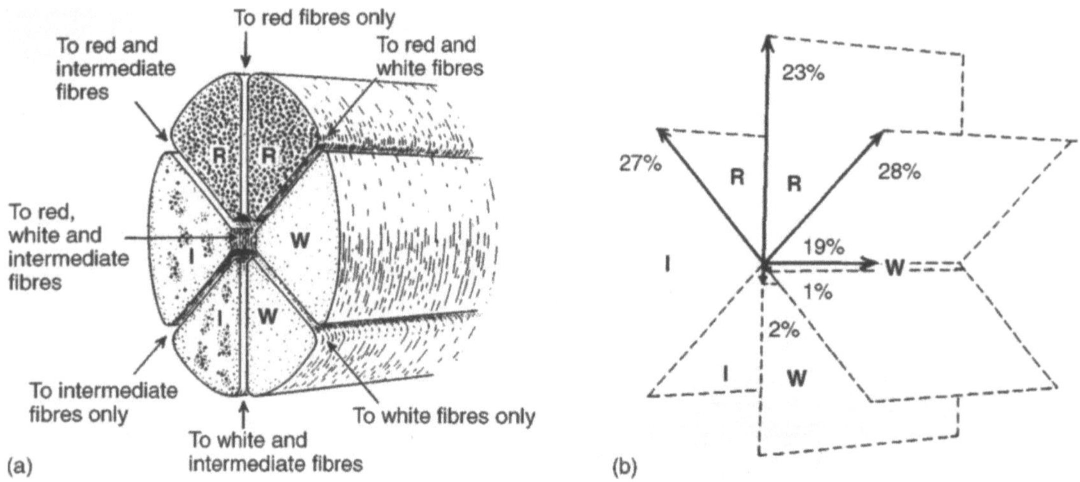

Figure 12.11 Graphical representation of capillary distribution relative to possible constellations of neighbouring muscle fibre types in a mixed muscle containing three fibre types. (a) Schematic presentation of possible fibre constellations around a capillary and (b) vector diagram of the incidence of each of seven possible capillary-fibre type constellations encountered in the parietal muscle of *Myxine glutinosa*, plotted as percentages and arrows of comparable length. (From Flood, 1979, with permission.)

Table 12.4 The supply of blood capillaries to distinct fibre types in the paraxial muscle of some marine chordates. Expressed as vascularized surface area per unit volume of fibre ($\mu m^2/\mu m^3$)

Species	Habitat temperature	Red fibres	Intermediate fibres	White fibres	Red/white ratio	Ref.
Myxine glutinosa	6–7°C	0.029	0.011	0.0002	145	1
Galeus melastomus	6–7°C	0.0169	0.0021	0.0005	33.8	2
Etmopterus spinax	6–7°C	0.0177	0.0032	0.0001	177	2
Chimaera monstrosa	6–7°C	0.0065	0.0022	0.0005	13	2
Scyliorhinus canicula I	6–7°C	0.0128	0.0030	0.0022	5.8	2
Scyliorhinus canicula II	12°C(?)	0.0332	0.0098	0.0049	6.8	2
Acipenser stellatus	18–20°C	0.031	0.0099	0.0028	11.7	3
Ambystoma mexicanum	25°C	0.0159	0.0068	0.0007	22.7	4

1. Flood (1979) 2. Totland *et al.* (1980) 3. Kryvi *et al.* (1981) 4. Totland (1984)

white fibres among the poorest vascularized ones, even when the possible effect of a variable habitat temperature (see Johnston and Lucking, 1978; Johnston and Maitland, 1980) is disregarded. A comparable distribution of capillaries relative to fibre types has also been found in the New Zealand hagfish (Baldwin *et al.*, 1989), in lampreys (Lie, 1974; Czopek, 1980) and in several teleosts (Mosse, 1978, 1979).

12.14 FUNCTION

As discussed in detail by Bone (1978) and Johnson (1981), it is now generally agreed that the superficial slow (red) fibres of fishes are utilized for sustained slow-speed swimming or cruising and the deeper fast (white) fibres for bursts of higher speed. Although this has never been proved for *Myxine*, the structural organization and histochemical profiles of the red and white fibre types of *Myxine*, as well as their huge difference in vascularization, is in good agreement with this hypothesis. However, it should be remembered that the behaviour of *Myxine* in its natural habitat is poorly known and that it may spend hours buried in the mud or feeding inside fish carcass under rather extreme hypoxic conditions (Foss, 1963, 1968; Strahan, 1963). Vigorous spontaneous swimming has never been reported for *Myxine* in its natural habitat, but the animal may need all its muscle power during feeding, when it needs to tear loose incompletely cut food bites, by the 'knotting' movement described by Strahan (1963).

In *Myxine* (and lampreys) the special arrangement of red and white muscle fibres within numerous compartments of each segment results in a unique intermingling of the two fibre types, not seen in higher vertebrates (Flood, 1973; Flood *et al.*, 1977). In all the cyclostomes one or two layers of highly vascularized red fibres will be interposed between every 2–5 layers of white fibre. The conditions for 'cooperative metabolism', between red and white muscle fibre types (see Bilinski, 1974) therefore seem much more favourable in *Myxine* than in most higher animals, as for example the teleosts where red muscle fibres are almost completely segregated from the white ones. The most likely compounds involved in such a hypothetical cooperative metabolism would be lactate produced by pyruvate reduction in the almost completely anaerobic white fibres, the diffusion of this lactate through the intercellular space towards the red muscle fibres, and the aerobic dehydrogenation of this lactate to pyruvate within these fibres. The high capacity for lactate production of hagfish white fibres (Baldwin *et al.*, 1991) and the ample amounts of lactate dehydrogenase of the more 'aerobic' heart isoform (Sidell and Beland, 1980) in the red muscle fibres (Dahl and Korneliussen, 1976) may support such a hypothesis.

Several details related to the intermediary fibres of the parietal muscle may support a hypothesis of higher fatigue resistance than in the white fibres. First of all, the intermediary fibres contain more mitochondria and oxidative enzymes, and are better vascularized. Thus they appear to be suited for aerobic metabolism which will not result in the accumulation of intermediary metabolites and acidification of the sarcoplasm. Secondly, they contain higher energy reserves in the form of stored lipid droplets than the white fibres. Thirdly, they are almost without exception located next to the red fibres and the blood vessels, which both may assist in removing metabolites from the vicinity of intermediary fibres, thus preventing accumulation of such metabolites within their sarcoplasm. Finally, electrophysiological data reported in this review indicate that they are far more resistant to fatigue than the white fibres when stimulated repetitively. The intermediary fibres may be recruited at intermediary levels of muscular activity, before the white fibres are recruited, in a scheme comparable to that found in rat (Edström and Kugelberg, 1968) and cat skeletal muscles (Burke *et al.*, 1971).

The red fibres of the craniovelaris muscle are active (1) to provide water currents past the olfactory organs and through the branchial sacs, thus serving both feeding as well as respiration (Johansen and Strahan, 1963), (2) to serve as a venous heart by its pumping action on an adjacent blood sinus ('the cardinal heart') which is furnished with valves (Johansen, 1963). These muscles contract rhythmically at a frequency varying between 10 and 100 per minute, depending on temperature and activity of the animal. These contractions are not myogenic as in heart muscle since they stop immediately upon removal of the brain of the animal (Nicolaysen, 1964). Based on the ultrastructure and histochemical profiles of these fibres they seem well suited for continuous activity.

The white fibres of *m. longitudinalis linguae* is an important muscle for the retraction of the dental plates and the chewing act of *Myxine*. Its character as an almost completely anaerobic muscle is in good agreement with the short bursts of feeding activity observed for *Myxine* (Strahan, 1963). However, in another hagfish (*Eptatretus cirrhatus*) it has been shown that the white fibres of the dental retractor muscles (including *m. longitudinalis linguae*) have an exceptionally high buffering capacity, and that this capacity probably allows the animal to feed for prolonged periods under rather anoxic conditions (Davidson *et al.*, 1990; Baldwin *et al.*, 1991). The same may well be true for *Myxine*.

12.15 FIBRE TYPES IN OTHER MYXINOIDS

Maurer (1894) based his description of the parietal (red) and central (white) muscle fibre types on sections from *Myxine australis*. According to Adam and Strahan (1963), this is now considered for inclusion in the species *Myxine glutinosa*.

The skeletal muscle fibre types of the New Zealand hagfish *Eptatretus cirrhatus* were examined by histochemical and ultrastructural techniques by Baldwin *et al.* (1991). They found the parietal (myotome) muscle to possess three fibre types in a compartmental arrangement and supplied by blood capillaries in a similar way to that previously described for *Myxine glutinosa*. Further, they found only white fibres, with a mean diameter larger than the equivalent fibres of the myotome, in *m. longitudinalis linguae* and *m. tubularis*. Baldwin *et al.* (1991) also give important data on the intracellular buffering capacity, on the concentration of lactate, glycogen and glucose, and on the activity of phosphorylase, phosphofructokinase, pyruvate kinase, lactate dehydrogenase, hexokinase, 3-hydroxyacyl CoA dehydrogenase, maleate dehydrogenase and citrate synthase in the two dental retractor muscles and in the mixed myotomal muscle.

12.16 ONTOGENY AND GROWTH

The mystery about the reproduction of *Myxine glutinosa* remains as unsolved today (see Gorbman, this volume) as it has always been (Walvig, 1963). Foss (1963, 1968), who made a strong effort to obtain larval material of the Atlantic hagfish, was unsuccessful. Accordingly, no information on the development of skeletal muscle fibre types exists for this animal. Even for other cyclostomes, where embryonal material is easier to obtain, no clear explanation exists for the origin of distinct fibre types in relation to the subdivisions of the early somite or to the compartmental subdivisions of the adult parietal muscle.

The view of Maurer (1894), that only the medial wall of the somite (the myotome) gives origin to paraxial skeletal muscle fibres, has now been generally accepted and expressed in many textbooks. However, this may represent an oversimplification. Based on experiments on larval *Acipenser stellatus*, incorporation of tritiated thymidine and autoradiography, rapidly dividing, undifferentiated cells, lateral to the primitive red and white fibres and derived from the lateral (dermatome) layer of the somite, have been found. At specific developmental stages these cells migrate medially past the red fibres to contribute to the white and intermediate fibre population there (Flood and Kryvi, 1982; Flood *et al.*, 1987). Therefore, in spite of the excellent works of Waterman (1969), Nag and Nursall (1972), Raamsdonk *et al.* (1974, 1978) and Nakao (1977) there should be good reason to reinvestigate, by modern techniques, the origin and differentiation of distinct paraxial skeletal muscle fibre types, particularly in the lower chordate groups, including the cyclostomes, where the compartmental organization of the muscle fibre types offers unique interpretational advantages.

More is known about the (postlarval) growth of *Myxine* muscle fibres. Marcus (1925) studied three specimens of *Myxine* ranging between 9 and 40 cm in length. (The length of the third animal was not reported.) He noted a pronounced increase in the dorsoventral diameter of each compartment within the segments of the parietal muscle and stated this diameter to be 1:2:3 in his three animals. Further, Marcus (1925) noted that the thickness of the ventral sheet of peripheral (red) fibres increased more than the thickness of the remaining central (white) fibres of the compartment. In the 9 cm long specimen this red fibre sheet made up 1:8 of the compartment thickness. In his medium-sized specimen this ratio was 1:6 and in his 40 cm long specimen 1:4. Marcus (1925) also noted transitional forms (*alle Übergänge*) between the vacuolated peripheral (red) fibres and the thick central (white) fibres. He concluded that the peripheral (red) fibres were the forerunners of the central (white) fibres. However, he also observed that the number of muscle fibres in essence did not increase, since in young as well as grown animals an average of 4 to 5 rows of fibre are present within each compartment. The 'transitional forms' mentioned above are now interpreted as the intermediary fibre type.

Korneliussen and Nicolaysen (1975) presented convincing data on a more than doubling of the cross-sectional diameter of all fibre types of the parietal muscle in animals between 19 and 38 cm body length. This increase took place without a concomitant increase in the cross-sectional area and diameter of the myofibrils or in the distribution and size of the T-tubules. A comparable increase in the length of the muscle fibres to keep pace with the increasing length of the animal seems likely. This lengthwise growth evidently takes place without any change in sarcomere length since this varied regionally between 2.1 and 2.6 μm without any significant difference between fibre types or between animals of different size. In other words, the growth of the animal is unlikely to be associated with any pronounced increase in the number of fibres within each segment or number of segments within the animal.

12.17 PHYLOGENY

The paraxial muscle of all swimming chordates represents a unique material for phylogenetic comparison of fibre type differentiations and arrangements within the tissue. Since this muscle, in all such animals, is primarily used to produce relatively slow undulatory movements to move the animal through the dense water, anatomical differences caused by distinct functional requirements are kept at a minimum. Maybe this conserved function of the parietal muscle contributes to its retained segmental arrangement in all aquatic chordates from cephalochordates to reptiles. Further, at least from cyclostomes to amphibia (Flood, 1973; Flood *et al.*, 1977), but probably even up to apes and mammals (Grenacher, 1867; Schneider, 1879; and Maurer, 1899, 1906) there is also a definite subdivision of each muscle segment into numerous compartments stacked on top of each other in a ventrodorsal direction, each with a very specific and repetitive distribution of distinct fibre types. A clear phylogenetic trend of red and intermediary fibres to shift from an almost unpolarized mediolateral distribution towards a lateral accumulation was demonstrated by Flood (1973) and Flood *et al.* (1977).

It is still rather unclear how these compartments and their contained fibre types relate to subdivisions of the somite during early ontogenesis (see above).

ACKNOWLEDGEMENTS

I wish to express my gratitude to Prof. Jan Jansen (†) and Prof. Th. Blackstad who introduced me as a young student of medicine to *Myxine*, and to basic biological research, back in 1961. Drs Q. Bone and G.K. Totland most kindly read my manuscript and gave many constructive comments incorporated in the final version of this paper. Mrs Teresa Cieplinska and photographers Ragnar Jensen and Jan Reidar Lothe are acknowledged for expert technical assistance. This work was supported by The Norwegian Research Council.

REFERENCES

Adam, H. and Strahan, R. (1963) Systematics and geographical distribution of myxinoids, in *The Biology of Myxine* (eds A. Brodal and R. Fänge), Universitetsforlaget, Oslo, pp. 1–8.

Alnæs, E., Jansen, J.K.S. and Rudjord, T. (1964) Spontaneous junctional activity of fast and slow parietal muscle fibres of the Atlantic hagfish. *Acta Physiologica Scandinavica*, **60**, 240–55.

Andersen, P., Jansen, J.K.S. and Løyning, Y. (1963) Slow and fast muscle fibres in the Atlantic hagfish. *Acta Physiologica Scandinavica*, **57**, 167–79.

Arnheim, N., Cocks, G.T. and Wilson, A.C. (1967) Molecular size of hagfish muscle lactate dehydrogenase. *Science*, **157**, 568–9.

Bárány, M. (1967) ATPase activity of myosin correlated with speed of muscle shortening. *Journal of General Physiology*, **50 Suppl.**, 197–218.

Baldwin, J. (1989) Intracellular pH buffering of hagfish muscle. *Australian Comparative Physiology*, Abs., **6**, 1.

Baldwin, J., Davison, W. and Foster, M.E. (1989) Properties of the muscle and heart lactate dehydrogenase of the New Zealand hagfish, Eptatretus cirrhatus: functional and evolutionary implications. *Journal of Experimental Zoology*, **250**, 135–9.

Baldwin, J., Davison, W. and Foster, M.E. (1991) Anaerobic glycolysis in the dental plate retractor muscles of the New Zealand hagfish *Eptatretus cirrhatus* during feeding. *Journal of Experimental Zoology*, **260**, 295–301.

Baldwin, J. and Lake, P.S. (1987) Lactate dehydrogenase homopolymer of hagfish heart and the single lactate dehydrogenase of lampreys display greater immunochemical similarity to LDHC4 than to LDHB4 of teleost fish. *Journal of Experimental Zoology*, **242**, 99–102.

Baldwin, J., Lake, P.S. and Moon, T.W. (1987) Immunochemical evidence that the single lactate dehydrogenase of lampreys is more similar to LDHB4 than to LDHA4 of hagfish. *Journal of Experimental Zoology*, **241**, 1–8.

Baldwin, J., Mortimer, K. and Patak, A. (1988) Do ascidians possess the ancestral subunit type of vertebrate lactate dehydrogenase? *Journal of Experimental Zoology*, **246**, 109–14.

Bartels, H. (1986) Myomuscular junctions in the gill-sac muscle of the Atlantic hagfish, Myxine glutionosa: analogy with myotendinous junctions. *Cell and Tissue Research*, **246**, 223–35.

Bilinski, E. (1974) Biochemical aspects of fish swimming, in *Biophysical Perspectives in Marine Biology*, Vol. 1 (eds D.C. Malins and J.R. Sargent), Academic Press, New York, pp. 239–88.

Bone, Q. (1963) Some observations upon the peripheral nervous system of the hagfish, *Myxine glutinosa. Journal of the Marine Biological Association U.K.*, **43**, 31–47.

Bone, Q. (1972) The dogfish neuromuscular junction: dual innervation of vertebrate striated muscle fibres? *Journal of Cell Science*, **10**, 657–65.

Bone, Q. (1978) Locomotor muscle, in *Fish Physiology*, Vol. 7, Academic Press, London, pp. 361–424.

Brautaset, N.J. and Nicolaysen, K. (1968) A possible difference in the effect of chloride on the surface membrane and the T-tubuli membrane of the hagfish twitch membrane. *Acta Physiologica Scandinavica*, **74**, 2A–3A (Abstract).

Brodin, L. and Grillner, S. (1986a) Effects of magnesium on fictive locomotion induced by activation of N-Methyl-D-Aspartate (NMDA) receptors in the lamprey spinal cord in vitro. *Brain Research*, **380**, 244–52.

Brodin, L. and Grillner, S. (1986b) Tonic inhibition of a new type of spinal interneurone during fictive locomotion in the lamprey. *Acta Physiologica Scandinavica*, **128**, 327–9.

Brodal, A. and Fänge, R. (eds) (1963) *The Biology of Myxine*, Universitetsforlaget, Oslo.

Burke, R.E., Levine, D.N., Zajac, III, F.E., Tsairis, P. and Engel, W.K. (1971) Mammalian motor units: physiological-histochemical correlation in three types in cat gastrocnemius. *Science*, **174**, 709–12.

Castellini, M.A. and Somero, G.N. (1981) Buffering capacity of vertebrate muscle: correlations with potentials for anaerobic function. Journal of Comparative Physiology B, **143**, 191–8.

Cole, J.F. (1907) A monograph on the general morphology of the myxinoid fishes, based on a study of *Myxine.* II. The anatomy of the muscles. *Transactions of the Royal Society, Edinburgh*, **45**, 683–757. 4pl.

Czopek, J. (1980) Vascularization of the skeletal muscles in the river lamprey (*Lampetra fluviatilis L.*). *Zoologica Poloniae*, **27**, 577–86. 2pl.

Dahl, H.A. and Korneliussen, H. (1976) Lactate dehydrogenase isoenzymes in different types of muscle fibres in the atlantic hagfish (*Myxine glutinosa* L.). *Comparative Biochemistry and Physiology B*, **55B**, 381–5.

Dahl, H.A. and Nicolaysen, K. (1971) Actomyosin ATPase activity in Atlantic hagfish muscles. *Histochemie*, **28**, 205–10.

Davison, W., Baldwin, J., Davie, P.S., Foster, M.E. and Satchell, G.H. (1990) Exhaustive exercise in the hagfish, *Eptatretus cirrhatus*: the anaerobic potential and the appearance of lactic acid in the blood. *Comparative Biochemistry and Physiology* A, **95**, 585–9.

Dickson, K.A. and Somero, G.N. (1987) Partial characterization of the buffering components of the red and white myotomal muscle of marine teleosts, with special reference to scombrid fishes. *Physiological Zoology*, **60**, 699–706.

Edström, L. and Kugelberg, E. (1968) Histochemical composition, distribution of fibres and fatiguability of single motor units. *Journal of Neurology, Neurosurgery and Psychiatry*, **31**, 424–33.

Flood, P.R. (1964) Myo-satellite cells in *Myxine* and Axolotl. *Proc. 3rd European Conference on Electron Microscopy, Prague*, pp. 575–6.

Flood, P.R. (1965) Skeletal muscle fibre types in *Amphioxus lanceolatus* and *Myxine glutinosa. Journal of Ultrastructure Research*, **12**, 238 (Abstract).

Flood, P.R. (1966) A peculiar mode of muscular innervation in amphioxus. Light and electron microscopic studies of the so-called ventral roots. *Journal of Comparative Neurology*, **126**, 181–218.

Flood, P.R. (1967) The effect of different fixatives on mitochrondrial ultrastructure in amphioxus muscle. *Journal of Ultrastructure Research*, **18**, 228 (Abstract).

Flood, P.R. (1968) Structure of the segmental trunk muscle in Amphioxus. With notes on the course and 'endings' of the so-called ventral root fibres. *Zeitschrift für Zellforschung*, **84**, 398–416.

Flood, P.R. (1973) The skeletal muscle fibre types of *Myxine glutinosa* L related to those of other chordates. *Acta Regiaa Societatis Scientiarum et Litterarum Gothenburgensis, Zoologica*, **8**, 17–20.

Flood, P.R. (1977) The sarcoplasmic reticulum and associated plasma membrane of trunk muscle lamellae in *Branchiostoma lanceolatum* (Pallas). A transmission and scanning electron microscopic study including freeze-fractures, direct replicas and x-ray microanalysis of calcium oxalate deposits. *Cell and Tissue Research*, **181**, 169–96.

Flood, P.R. (1979) The vascular supply of three fibre types in the parietal trunk muscle of the Atlantic hagfish (*Myxine glutinosa L*). A light microscopic quantitative analysis and an evaluation of various methods to express capillary density relative to fibre types. *Microvascular Research*, **17**, 55–70.

Flood, P.R. and Mathisen, J.S. (1962) A third type of muscle fibre in the parietal muscle of the Atlantic hagfish *Myxine glutinosa* (L.)? *Zeitschrift für Zellforschung*, **58**, 638–40.

Flood, P.R., Kryvi, H. and Totland, G.K. (1977) Onto-phylogenetic aspects of muscle fibre types in the segmental trunk muscle of lower chordates. *Folia Morphologica*, **25**, 64–7, 2 Pl.

Flood, P.R. and Kryvi, H. (1982) The origin and differentiation of red and white muscle fibres in the sturgeon (*Acipenser stellatus*). *Journal of Muscle Research and Cell Motility*, **3**, 125–6.

Flood, P.R., Gulyaev, D. and Kryvi, H. (1987) Origin and differentiation of muscle fibre types in the trunk of the sturgeon, *Acipenser stellatus* Pallas. *Sarsia*, **72**, 343–4.

Foss, G. (1963) Behaviour of *Myxine glutinosa L.* in its natural surroundings, in *The Biology of Myxine* (eds A. Brodal and R. Fänge), Universitetsforlaget, Oslo, pp. 42–9.

Foss, G. (1968) Behaviour of *Myxine glutinosa* in its natural habitat. Investigation of the mud biotope by suction technique, *Sarsia*, **31**, 1–13.

Fürbriger, P. (1875) Untersuchungen zur vergleichended Anatomie der Muskulatur des Kopfskelets der Cyclostomen. *Jena Zeitschrift für Medizin und Naturwissenschaft*, **9**, 1–93.

Geer-Walker, M. and Pull, G.A. (1975) A survey of red and white muscle in marine fish. *Journal of Fish Biology*, **7**, 295–300.

Gray, E.G. (1958) The structures of fast and slow muscle fibres in the frog. *Journal of Anatomy*, **92**, 559–63.

Grenacher, H. (1867) Beiträge zur häheren Kenntnis der Muskulatur der Cyclostomen und

Leptocardier. *Zeitschrift für wissenschaftliche Zoologie*, **17**, 577–97.

Guppy, M., Hulbert, W.C. and Hochachka, P.W. (1979) Metabolic sources of heat and power in tuna muscles. II: Enzyme and metabolite profiles. *Journal of Experimental Biology*, **82**, 303–20.

Guth, L. and Samaha, F.J. (1969) Quantitative differences between actomyosin ATPase of slow and fast mammalian muscle. *Experimental Neurology*, **25**, 138–52.

Hardisty, M.W. (1979) *Biology of the Cyclostomes*, Chapman & Hall, London. 428pp.

Hess, A. (1960) The structure of extrafusal muscle fibres in the frog and their innervation studied by the cholinesterase technique. *American Journal of Anatomy*, **107**, 129–52.

Hess, A. (1963) Two kinds of extrafusal muscle fibres and their nerve endings in the garter snake. *American Journal of Anatomy*, **113**, 347–64.

Jansen, J. jr. and Andersen, P. (1960) Observations on slow and fast muscle fibres in the hagfish (*Myxine glutinosa*). *Acta Physiologica Scandinavica*, **50**, Suppl. 175, 76–7.

Jansen, J.K.S. and Andersen, P. (1963) Anatomy and physiology of the skeletal muscles, in *The Biology of Myxine* (eds A. Brodal and R. Fänge), Universitetsforlaget, Oslo, pp. 161–94.

Jansen, J., Andersen, P. and Jansen, J.K.S. (1963) On the structure and innervation of the parietal muscle of the hagfish (*Myxine glutinosa*). *Acta Morphologica Neerlando-Scandinavica*, **5**, 329–38.

Jasper, D. (1967) Body muscles of the lamprey. *Journal of Cell Biology*, **32**, 219–27.

Johansen, K. (1963) The cardiovascular system of *Myxine glutinosa* L, in *The Biology of Myxine* (eds A. Brodal and R. Fänge), Universitetsforlaget, Oslo, pp. 289–316.

Johansen, K. and Strahan, R. (1963) The respiratory system of *Myxine glutinosa L*, in *The Biology of Myxine* (eds A. Brodal and R. Fänge), Universitetsforlaget, Oslo, pp. 352–71.

Johnston, I.A. (1981a) Structure and function of fish muscles. *Symposiums of the Zoological Society, London*, **48**, 71–113.

Johnston, I.A. (1981b) Specialization in fish muscle, in *Development and Specialization of Muscle* (ed. D.F. Goldspink), Cambridge Univ. Press, Cambridge, pp. 123–48.

Johnston, I.A. (1982) Physiology of muscle in hatchery raised fish. *Comparative Biochemistry and Physiology*, **73B**, 105–24.

Johnston, I.A. and Lucking, M. (1978) Temperature induced variation in the distribution of different types of muscle fibre in the goldfish (*Carassus auratus*). *Journal of Comparative Physiology*, **124**, 111–16.

Johnston, I.A. and Maitland, B. (1980) Temperature acclimation in crucian carp: a morphometric analysis of muscle fibre ultrastructure. *Journal of Fish Biology*, **17**, 113–25.

Katz, B. (1949) The efferent regulation of the muscle spindle of the frog. *Journal of Experimental Biology*, **26**, 210–17.

Kashapova, L.A. and Sakharov, D.A. (1976) Dual innervation of fast fibres in trunk muscles of larval lamprey. *Dokl. Akad. Nauk SSSR*, **231**, 1495–6.

Kashapova, L.A. and Sakharov, D.A. (1977) Sturgeon neuromuscular junctions: synaptic vesicles in the red and white muscles are of different size. *Dokl. Akad. Nauk SSSR*, **235**, 1423–4.

Korneliussen, H. (1972) Identification of muscle fibre types in 'semithin' sections stained with p-phenylene-diamine. *Histochemie*, **32**, 95–8.

Korneliussen, H. (1973a) Dense-core vesicles in motor nerve terminals. Monoaminergic innervation of slow non-twitch muscle fibres in the Atlantic hagfish (*Myxine glutinosa L.*). *Zeitschrift für Zellforschung*, **140**, 425–32.

Korneliussen, H. (1973b) Ultrastructure of motor nerve terminals on different types of muscle fibres in the atlantic hagfish (*Myxine glutinosa*, L.). Occurrence of round and elongated profiles of synaptic vesicles and dense-core vesicles. *Zeitschrift für Zellforschung*, **147**, 87–105.

Korneliussen, H. (1973c) Ultrastructure of myotendinous junctions in Myxine and rat. Specializations between the plasma membrane and the lamina densa. *Zeitschrift für Anatomie und Entwicklungs-Geschichte*, **142**, 91–101.

Korneliussen, H. and Nicolaysen, K. (1973) Ultrastructure of four types of striated muscle fibres in the Atlantic hagfish (*Myxine glutinosa* L.). *Zeitschrift für Zellforschung*, **143**, 273–90.

Korneliussen, H. and Nicolaysen, K. (1975) Distribution and dimension of the T-system in different muscle fibre types in the Atlantic hagfish (*Myxine glutionosa* L.). *Cell and Tissue Research*, **157**, 1–16.

Krüger, P. (1952) *Tetanus und Tonus der quergestreiften Skelettmuskeln der Wirbeltiere und des Menschen*, Akademische Verlagsgesellschaft, Leipzig.

Kryvi, H. (1977) Ultrastructure of the different

fibre types in axial muscles of the sharks *Etmopterus spinax* and Galeus melastomus. *Cell and Tissue Research*, **184**, 287–300.

Kryvi, H., Flood, P.R. and Gulyaev, D. (1980) The ultrastructure and vascular supply of the different fibre types in the axial muscle of the sturgeon *Acipenser stellatus* Pallas. *Cell and Tissue Research*, **212**, 117–26.

Kuffler, S.W. and Gerhard, R.W. (1947) The small-nerve motor system to skeletal muscle. *Journal of Neurophysiology*, **10**, 383–94.

Kuffler, S.W. and Vaughan Williams, E.M. (1953) Properties of the 'slow' skeletal muscle fibres of the frog. *Journal of Physiology* (*Lond.*), **121**, 318–40.

Kugelberg, E. and Edström, L. (1968) Differential histochemical effects of muscle contractions on phosphorylase and glycogen in various types of fibres: relation to fatigue. *Journal of Neurology, Neurosurgery and Psychiatry*, **31**, 415–23.

Lie, H.R. (1973) An intermediate muscle fibre type in the river lamprey, Lampetra fluviatilis L. *Acta Regiae Societatis Scientiarum et Litterarum Gothenburgensis, Zoologica*, **8**, 21–2.

Lie, H.R. (1974) A quantitative identification of three muscle fibre types in the body muscles of *Lampetra fluviatilis*, and their relation to blood capillaries. *Cell and Tissue Research*, **154**, 109–19.

Luther, A. (1938) Die Visceralmuskulatur der Acranier, Cyclostomen und Fische, in *Handbuch der vergleichenden Anatomie der Wirbeltiere*, Vol. 5 (eds L. Bolk, E. Göppert, E. Kallius and W. Lubosch), Urban & Schwarzenberg, Berlin, pp. 468–542.

Marcus, H. (1925) Über zweierlei Muskelnfasern in der Rumpfmuskulatur von Cyclostomen. *Zeitschrift für Anatomie und Entwicklungs-Geschichte*, **76**, 578–91.

Marinelli, W. and Strenger, A. (1956) *Vergleichende Anatomie und Morphologie der Wirbeltiere*, Vol. 1, F. Deutiche, Wien, pp. 81–172.

Markert, C.L., Shaklee, J.B. and Whitt, G.S. (1975) Evolution of a gene. *Science*, **189**, 102–14.

Mathisen, J.S. and Mellgren, S.I. (1965) Some observations concerning the role of phenazine methosulphate in histochemical dehydrogenase methods. *Journal of Histochemistry and Cytochemistry*, **13**, 408–9.

Maurer, F. (1894) Die Elemente der Rumpfmuskulatur bei Cyclostomen und höeren Wirbelthieren. Ein Beitrag zur Phylogenie der quergestreiften Muskelfaser. *Morphologische Jahrbuch*, **21**, 473–619.

Maurer, F. (1899) Die Rumpfmuskulatur der Wirbeltiere und die Phylogenese der Muskelfaser, in *Ergebnisse der Anatomie und Entwicklungs-Geschichte*, Vol. 9 (eds Merkel und Bonnet), pp. 691–819.

Maurer, F. (1906) Die Entwicklung des Muskelsystems und der elektrischen Organe. In *Handbuch der vergleichenden und experimentellen Entwicklungslehre der Wirbeltiere*, Vol. 3 (ed. O. Hertwig), pp. 1–80.

Mauro, A. (1961) Satellite cells of skeletal muscle fibres. *Journal of Biophys. Biochemistry and Cytology*, **9**, 493–5.

Mellgren, S.I. and Mathisen, J.S. (1966) Oxidative enzymes, glycogen and lipid in striated muscle. A histochemical study in the Atlantic hagfish (*Myxine glutinosa* L.). *Zeitschrift für Zellforschung*, **71**, 169–88.

Mosse, P.R.L. (1978) The distribution of capillaries in the somatic musculature of two vertebrate types with particular reference to teleost fish. *Cell and Tissue Research*, **187**, 281–303.

Mosse, P.R.L. (1979) Capillary distribution and metabolic histochemistry of the lateral propulsive musculature of pelagic teleost fish. *Cell and Tissue Research*, **203**, 141–60.

Nag, A.C. and Nursall, J.R. (1972) Histogenesis of white and red muscle fibres of trunk muscles of a fish *Salmo gairdneri*. *Cytobios*, **6**, 227–46.

Nakao, T. (1975) Fine structure of the myotendinous junction and 'terminal coupling' in the skeletal muscle of the lamprey, *Lampetra japonica*. *Anatomical Record*, **182**, 321–38.

Nakao, T. (1976) An electron microscopic study of the neuromuscular junction in the myotomes of the larval lamprey, *Lampetra japonica*. *Journal of Comparative Neurology*, **165**, 1–16.

Nakao, T. (1977) Electron microscopic studies on the myotomes of larval lamprey, *Lampetra japonica*. *Anatomical Record*, **187**, 383–404.

Nicolaysen, K. (1964) Hurtige og langsomme muskelfibre hos slimålen. En histologisk og fysiologisk studie av en 'hvit' muskel, *M. longitudinalis linguae*, og av en 'rød' muskel, *m. cranio-velaris posterior*. Masters thesis in zoophysiology, University of Oslo, Olso. 55pp.

Nicolaysen, K. (1966) On the functional properties of the fast and slow cranial muscles of the Atlantic hagfish. *Acta Physiologica Scandinavica*, 68, Suppl. 277, **142** (Abstract).

Nicolaysen, K. (1976a) The spread of the action potential in the T-system in hagfish twitch muscle fibres. *Acta Physiologica Scandinavica*, **96**, 29–49.

Nicolaysen, `K. (1976a) Spread of the junctional potential in the T-system in hagfish slow muscle fibres. *Acta Physiologica Scandinavica*, **96**, 50–57.

Nishi, S. (1938) Muskeln des Rumpfes, in *Handbuch der vergleichenden Anatomie der Wirbeltiere*, Vol. 5 (eds L. Bolk, E. Göppert, E. Kallius and W. Lubosch), Urban & Schwarzenberg, Berlin, pp. 351–446.

Ohno, S., Klein, J., Destree, A. and Morrison, M. (1967) Genetic control of lactate dehydrogenase formation in the hagfish *Eptatretus stoutii*. *Science*, **156**, 96–8.

Peters, A. (1963) The peripheral nervous system, in *The Biology of Myxine* (eds A. Brodal and R. Fänge), Universitetsforlaget, Oslo, pp. 92–123.

Peters, A. and Mackay, B. (1964) The structure and innervation of the myotomes of the lamprey. *Journal of Anatomy*, **95**, 575–85.

Peachey, L.D. and Huxley, A.F. (1962) Structural identification of twitch and slow striated muscle fibres of the frog. *Journal of Cell Biology*, **13**, 177–80.

Raamsdonk, W. van, Stelt, A. van der, Diegerbach, P.C., Berg, W. van de, Bruyn, H. de, Dijk, J. van and Mijsen, P. (1974) Differentiation of the musculature of the teleost *Brachydanio rerio*. *Zeitschrift für Anatomie und Entwicklungs-Geschichte*, **145**, 321–42.

Raamsdonk, W. van, Pool, C.W. and TeKronnie, G. (1978) Differentiation of muscle fibre types in the teleost *Brachydanio rerio*. *Anatomy and Embryology*, **153**, 137–55.

Sakharov, D.A. and Kashapova, L.A. (1979) The primitive pattern of the vertebrate body muscle innervation: Ultrastructural evidence for two synaptic transmitters. *Comparative Biochemistry and Physiolgy*, **62A**, 771–6.

Sandset, P.M. and Korneliussen, H. (1978) Myosatellite cells associated with different muscle fibre types in the Atlantic hagfish (*Myxine glutinosa* L.). *Cell and Tissue Research*, **195**, 17–27.

Schneider, A. (1879) *Beiträge zur vergleichenden Anatomie und Entwicklungsgeschichte der Wirbeltiere*, Berlin.

Sidell, B.D. and Beland, K.F. (1980) Lactate dehydrogenase of atlantic hagfish: physiological and evolutionary implications of a primitive heart isoenzyme. *Science*, **207**, 769–770.

Strahan, R. (1963) The behaviour of *Myxine* and other myxinoids, in *The Biology of Myxine* (eds A. Brodal and R. Fänge), Universitetsforlaget, Oslo, pp. 22–32.

Teräväinen, H. (1971) Anatomical and physiological studies on muscles of lamprey. *Journal of Neurophysiology*, **34**, 954–73.

Teräväinen, H. and Rovainen, C.M. (1971) Fast and slow motoneurons to body muscle of the sea lamprey. *Journal of Neurophysiology*, **34**, 990–8.

Totland, G.K. (1976a) Histological and histochemical studies of the segmental muscle in the Axolotl *Ambystoma mexicanum* Shaw (Amphibia: Urodela) *Norwegian Journal of Zoology*, **24**, 79–90.

Totland, G.K. (1976b) Three muscle fibre types in the axial muscle of Axolotl (*Ambystoma mexicanum* Shaw). A quantitative light and electron microscopic study. *Cell and Tissue Research*, **168**, 65–78.

Totland, G.K. (1984) Capillary distribution in the lateral muscle of Axolotl (*Ambystoma mexicanum* Shaw) (Amphibia, Urodela). *Acta Zoologica (Stockh.)*, **65**, 221–5.

Totland, G.K., Kryvi, H., Bone, Q. and Flood, P.R. (1981) Vascularization of the lateral muscle of some elasmobranchiomorph fishes. *Journal of Fish Biology*, **18**, 223–34.

Trotter, J.A., Corbett, K. and Avner, B.P. (1981) Structure and function of the murine muscle-tendon junction. *Anatomical Record*, **201**, 293–302.

Trotter, J.A., Eberhard, S. and Samora, A. (1983) Structural domains of the muscle-tendon junction. 1. The internal lamina and the connecting domain. *Anatomical Record*, **207**, 573–91.

Wallén, P., Grillner, S., Feldman, J.L. and Bergelt, S. (1985) Dorsal and ventral myotome motoneurons and their input during fictive locomotion in lamprey. *Journal of Neuroscience*, **5**, 654–61.

Walvik, F. (1963) The gonads and the formation of the sexual cells, in *The Biology of Myxine* (eds A. Brodal and R. Fänge), Universitetsforlaget, Oslo, pp. 530–80.

Waterman, R.E. (1969) Development of the lateral musculature in the teleost, *Brachydanio rerio*: a fine structural study. *American Journal of Anatomy*, **125**, 457–94.

Weibel, E.R. and Elias, H. (1967) *Quantitative Methods in Morphology*, Springer Verlag, Berlin.

Wilson, A.C., Kitto, G.B. and Kaplan, N.O. (1967) Enzymatic identification of fish products. *Science*, **157**, 82–3.

Wilson, A.C., Kaplan, N.O., Levine, L., Pesce, A., Reichlin, M. and Allison, W.S. (1964) Evolution of lactate dehydrogenase. *Federal Proceedings*, **23**, 1258–66.

PART SIX

The Respiratory System

13

THE GILLS OF HAGFISHES[1]

Helmut Bartels

SUMMARY

The gills of hagfishes form lens-shaped pouches, rather than holobranches that characterize those of lampreys and gnathostome fishes. Their internal surface is enlarged by radial folds, which extend between the medial and lateral walls of the pouch and contain the branchial microcirculation. Blood and water flow through the gills are arranged in a countercurrent system, thereby facilitating gas exchange.

The circulation of the hagfish gills consists of an arterio-arterial and an arteriovenous component. The arterio-arterial circulation connects the afferent and efferent branchial arteries. The arteriovenous circulation comprises an intrabranchial sinusoid system, which is supplied by the arterio-arterial circulation through arteriovenous anastomoses and connected to the peribranchial sinus through gaps in the muscle layer of the gill pouch.

The majority of the epithelial surface is formed by mucus-secreting pavement cells. In the lateral half of the gill pouch, mitochondria-rich (MR) cells are intercalated between the pavement cells, which are characterized by an extensive amplification of the basolateral cell membrane and, in freeze-fracture replicas, by assemblies of linear arrays of particles and fibrils in the apical cell membrane. Since the MR cells show strong histochemical reactions for both Na^+/K^+-ATPase and carbonic anhydrase, they may be engaged in ion transport and/or acid-base regulation.

13.1 THE GILLS OF HAGFISHES ARE UNIQUE AMONG VERTEBRATE GILLS

The branchial region of agnathans (jawless vertebrates) differs in several aspects from that of gnathostome (jawed) fishes. Significant differences are also present between the gills of hagfishes and lampreys, the only two extant representatives of the agnathan stage of vertebrate evolution, in that the gills of lampreys resemble more closely those of gnathostome fishes than those of hagfishes (Mallatt, 1984). Rauther (1935) already noted that the term 'Marsipobranchii', which was then commonly used for these two agnathan groups, suggests a larger conformity than is indeed existing. He added that, strictly speaking, this term is only applicable to myxinoids.

The main differences between the gills of hagfishes and lampreys are: (1) the presence of true gill pouches in hagfishes (*bursae branchiales*, Marinelli and Strenger, 1956) instead of a holobranch formed by two hemibranches as found in lampreys (Mallatt, 1984); (2) the extent of the branchial skeleton; (3) the structure of the internal gill surface; and (4) the organization of the branchial circulation.

13.1.1 The branchial skeleton

The branchial skeleton of hagfishes is represented only by small cartilaginous plates in

[1] This paper is dedicated to Prof. Dr. h.c. Andreas Oksche on the occasion of his 70th birthday.

The Biology of Hagfishes. Edited by Jørgen Mørup Jørgensen, Jens Peter Lomholt, Roy E. Weber and Hans Malte. Published in 1998 by Chapman & Hall, London. ISBN 0 412 78530 7.

the connective tissue surrounding the external gill pores. They may assist in keeping the external gill pore open and thereby reduce the resistance to the outflowing water (Marinelli and Strenger, 1956).

Although the branchial skeleton of hagfishes is less developed than that of lampreys, which possess a continuous system of cartilagenous plates, the gill basket, it is located superficially beneath the body surface in both of these groups (Bourne, 1892; Marinelli and Strenger, 1954, 1956). Thus, the gills and their muscles are situated internally (medially) to the branchial skeleton in agnathans, a situation contrasting with that typically found in gnathostomous fishes.

13.1.2 The gill pouches and ducts

The number of gill pouches and the arrangement of their efferent ducts differ in the two groups (families) of hagfishes (Myxinidae and Eptatretidae) with the smaller variability among Myxinidae (Adam and Strahan, 1963). In this group, usually six (e.g. *Myxine glutinosa*) and rarely five (e.g. *Myxine circifrons*) or seven lens-shaped gill pouches are located on each side of the pharynx to which each pouch is connected by an afferent (internal) branchial duct. The efferent (external) branchial ducts on either side converge and lead to a single external gill pore (*porus branchialis externus*). On the left side of the body, a pharyngocutaneous duct is located posterior to the last gill pouch. It lacks a gill pouch and connects the posterior part of the pharynx directly with the left external gill pore. The only known exception to this rule is found in *Notomyxine tridentiger*, where the pharyngocutaneous duct has a separate opening, so that there are two openings on the left side of the branchial region (Adam and Strahan, 1963).

In Eptatretidae, there are large interspecies variations in the number of pairs of gill pouches, ranging from 5 (*Eptatretus profundum*) to 14 (*Eptatretus polytrema*) (Adam and Strahan, 1963). Although the numbers are fairly constant within a species, considerable intraspecies variations are described for *Eptatretus stoutii*, which in most cases have 12 pairs of gills with a range from 10 to 15 (Worthington, 1905; Adam and Strahan, 1963). Since the efferent branchial ducts of Eptatretidae usually open separately on the body surface, the number of external gill pores is the key taxonomic character by which Myxinidae and Eptatretidae can be distinguished (Adam and Strahan, 1963; Hardisty, 1979). However, as in Myxinidae, the pharyngocutaneous duct and the efferent duct of the last gill of Eptatretidae usually have a common external pore (Worthington, 1905; Hardisty, 1979). In about 10% of *Eptatretus burgeri*, the pharnygocutaneous duct has a separate opening (Adam and Strahan, 1963). In addition, differences in the shape of the gill pouches were observed in *Eptatretus stoutii*, where only the first pouch is disk-shaped, while the following are bilobed (Mallatt and Paulsen, 1986).

The internal structure of the gill pouches is very similar in both groups of hagfishes (Hofbauer, 1934; Rauther, 1935; Pohla *et al.*, 1977; Bartels, 1985; Mallat and Paulsen, 1986; Elger, 1987). However, the typical arrangement of filaments and lamellae found in the gills of lampreys and jawed fishes is not present. Instead, the internal surface of the hagfish gill is enlarged by several folds, which are attached with their base to the wall of the pouch. These folds vary considerably in height and are oriented in parallel with the axis that connects the medial and the lateral pole of the pouch. The free margin of only a few (about five or six) large folds borders the central water channel which lies in the axis of the gill pouch (Hofbauer, 1934). The spaces between these large folds ('first-order folds', Rauther, 1935) are filled with smaller folds which gradually decrease in height ('second- to sixth-order folds', Rauther, 1935). As already noted 150 years ago by Johannes Müller (1845), the surface of the folds is enlarged by smaller transverse folds, called secondary and tertiary folds by Hofbauer (1934). The transverse folds are best developed in the central part of the pouch, while they are absent from those regions of the folds close to the medial and lateral wall of the pouch, which have a smooth surface (Rauther, 1935; Pohla *et al.*, 1977; Mallatt and Paulsen, 1986).

Mallatt and Paulsen (1986) concluded that the folds in the gills of hagfishes correspond to the filaments in the gills of lampreys and jawed fishes. This view is based on the observations that (i) the lateral and medial parts of the folds contain blood vessels (see 13.1.3), which correspond to those on the afferent and efferent sides of the filament of the gnathostome-type of gills, respectively, and (ii) that the mitochondria-rich cells concentrate in the epithelium on the lateral (afferent) part of the fold (see 13.2.2) as do the chloride cells, their putative counterparts in teleosts (see 'Possible functions', p. 216), on the afferent side of the filament. Striking similarities are also present between the structure of the transverse folds in the hagfish gills and that of the lamellae in the gills of lampreys and jawed fishes. Since the lamellae are responsible for gas exchange, the transverse folds were also termed 'respiratory lamellae' (Bartels and Decker, 1985; Mallatt and Paulsen, 1986) and 'lamellar' part of the gill fold (Elger, 1987). Mallatt and Paulsen (1986) observed that the respiratory lamellae are apparently more developed in *Myxine glutinosa* than in *Eptatretus stoutii* and suggested that the resulting smaller surface area available for gas exchange in an individual gill pouch may be compensated by the larger number of gill pouches in the latter species.

The outer wall of the pouch is formed by striated muscle fibres. Hofbauer (1934) distinguished an inner layer of obliquely running and crossing bundles of muscle fibres (*constrictor branchialis internus*) and an outer layer of circular muscle fibres (*constrictor branchialis externus*). The circular muscle layer is best developed where the internal and external branchial ducts enter and leave the gill pouch, respectively.

13.1.3 The gill circulation

The organization of the branchial circulation in hagfishes is also unique among vertebrates. Each gill pouch is supplied by a branch of the ventral aorta, the afferent gill artery, which approaches the gill from the lateral side and runs parallel with the external branchial duct. This arrangement differs from that in lampreys and jawed fishes, where the afferent gill artery enters the gills from the medial side. As with lampreys and gnathostome fishes, the gills of hagfishes have an arterio-arterial and an arteriovenous circulation (Laurent, 1984). For a detailed ultrastructural description of the branchial circulation, the reader is referred to the studies by Mallatt and Paulsen (1986) and Elger (1987).

Arterio-arterial circulation

The afferent gill artery forms the afferent ring artery ('Ringkanal', Hofbauer, 1934, Pohla *et al.*, 1977), which surrounds the efferent branchial duct, where it leaves the gill pouch. Afferent radial arteries, which have a meridional course and anastomose with each other, rise at regular intervals from the afferent ring artery and form the base of the gill folds. They supply a *corpus cavernosum* in the unbranched lateral part of the folds, which represents a plexus of sinusoidal vessels (Tomonaga *et al.*, 1975; Mallatt and Paulsen, 1986; Elger, 1987). The afferent radial arteries and the *corpora cavernosa* are lined by a high endothelium, which is capable of phagocytosis and is surrounded by one or two layers of smooth muscle cells (Tomonaga *et al.*, 1975; Elger, 1987). Branches of the *corpora cavernosa* form the core of the secondary folds in the lateral part of the gill pouch.

The core of the small transverse folds in the central part of the gill pouch is formed by a continuous lacunar blood space lined by pillar cells (Jakobshagen, 1920; Hughes and Morgan, 1973). This blood space is crossed by posts (pillars), consisting of bundles of collagen fibrils and microfibrils that are enclosed by (Elger, 1987) or lie between the pillar cells (Mallatt and Paulsen, 1986). The pillar cells contain numerous microfilaments (Elger, 1987), which, in teleost fishes, have been shown to be contractile (Bettex-Galland and Hughes, 1973; Smith and Chamley-Campbell, 1981). Typical endothelial cells line only the marginal channels, i.e. blood spaces below the free margins of the folds. They contain endothelial cell specific organelles (Weibel-Palade bodies), which are absent in pillar cells (Mallatt and Paulsen, 1986; Elger, 1987).

The blood is drained from the respiratory portion of the folds by efferent *corpora cavernosa* and efferent radial arteries in the unbranched medial part of the gill fold. These efferent vessels are less ramified than the corresponding afferent vessels and lined by a squamous endothelium. The efferent radial arteries lead into the efferent ring artery, which surrounds the internal branchial duct and is connected to the paired dorsal aorta by two efferent gill arteries (Cole, 1912). Thus, water and blood flow through the hagfish gills in a countercurrent system which facilitates gas exchange (see also Malte and Lomholt, this volume).

Arteriovenous circulation

The gills of both groups of agnathans are surrounded by an extended peribranchial sinus, which communicates with sinusoidal spaces within the gills. In hagfishes, the intrabranchial sinuses are located between the gill epithelium and the vessels of the arterio-arterial circulation and between the gill epithelium and the striated muscle fibres, which form the wall of the gill pouch. Such a sinus is absent in the respiratory lamellae of the folds. The sinus wall consists only of a thin endothelium. Erythrocytes are occasionally present in the sinus lumen ('red lymphatics', Cole, 1925; Lomholt and Franko-Dossar, this volume). These intrabranchial sinuses are connected with the afferent and efferent radial arteries and the *corpora cavernosa* through arterio-venous anastomoses. They communicate with the peribranchial sinus through gaps in the muscular wall of the gill pouch. The peribranchial sinus is drained by the cardinal veins.

13.2 THE GILL EPITHELIUM

Although there have been no studies which have compared directly the gill epithelium of Myxinidae and Eptatretidae, a comparison of the results of ultrastructural studies of two representative species, *Myxine glutinosa* and *Eptatretus stoutii*, reveals no obvious differences between the two families (Bartels, 1984, 1985, 1988; Mallatt and Paulsen, 1986; Elger, 1987).

The gill epithelium of hagfishes shows regional differences in height and cellular composition. The wall of the gill pouch and the unbranched folds are lined by a multilayered columnar epithelium in the lateral part of the gill and by a two-layered squamous epithelium in its medial part. The epithelium which covers the pillar cell-lined lacunar blood space of the transverse folds ('respiratory lamellae') is almost two-layered squamous and thinnest (c. 1 µm) above the marginal channels. Studies performed in teleost fishes suggest that the blood flow is not homogeneous through the lamellae but, through the contractile activity of the pillar cells, directed towards the marginal channel, where it is assumed to be greater than in the remaining lamella (Hughes, 1984). Provided that this situation applies also to the gills of hagfishes and since the diffusion distance between water and blood in these animals is also shortest (c. 1.6 µm) above the marginal channels, they are likewise, at least at rest, probably the preferential sites of gas exchange.

The gill epithelium consists of four cell types. Its surface is formed by pavement cells and mitochondria-rich cells, while basal cells and intermediate cells are found in the deeper layers of the epithelium. In addition, granulocytes and dense-core granular (lymphoid) cells, which have emigrated from the blood, frequently occur in the epithelium. The existence of mucous cells, which had been identified in the gill epithelium of *Myxine glutinosa* by light microscopy in paraffin sections (Hofbauer, 1934; Morris, 1965) and by scanning electron microscopy (Pohla *et al.*, 1977), has not been confirmed by thin section electron microscopy of gill material from either Atlantic or Pacific hagfishes (Bartels, 1985; Mallatt and Paulsen, 1986; Elger, 1987). However, intraepithelial cysts, which contain mucus, occur occasionally in the gills of individual *Myxine glutinosa* and may have been mistaken for mucous cells in paraffin sections. These cysts are lined by a squamous epithelium and store the mucus extracellularly instead of in cytoplasmic secretory granules as would be the case for mucous cells.

The epithelium rests on a *basal lamina* with a compact, amorphous *lamina densa*, which increases in thickness from the lateral to the medial side of the gill.

13.2.1 Pavement cells

The pavement cells, which form by far the largest part of the gill surface, secrete mucus (Figure 13.1). These are almost squamous cells which, mainly in the multilayered columnar epithelium of the lateral part of the gill, send thin cytoplasmic processes towards the basal lamina. However, contacts between pavement cells and the basal lamina were not observed. The apical surface area of the pavement cells is enlarged by short microvilli, which contain an axial bundle of thin (5 nm) filaments (Figure 13.1b, c) or by a system of low microplicae (Pohla *et al.*, 1977). The lateral surface area of these cells is amplified by thin processes, which interdigitate with those of neighbouring pavement cells (Figure 13.1e).

Mucus synthesis and secretion

Light microscopy of Toluidine blue-stained, 1 µm thick sections of the hagfish gills shows that a narrow border of metachromatic staining is present at the epithelial surface (Figure 13.1a), corresponding to the 'Randsaum' (cuticular border) described by Hofbauer (1934). This metachromatic staining is confined to the pavement cells and frequently focally interrupted in the lateral part of the gills, whenever a mitochondria-rich cell is in contact with the environment (Figure 13.1a). It also produces a positive reaction after staining with PAS and the cationic dyes Alcian blue at pH 1.0 and pH 2.5, and Safranin O.

The ultrastructural investigation of the apical cytoplasm of the pavement cells shows that numerous tightly packed ovoid or elliptical mucous granules lie immediately beneath the cell surface with their long axis oriented perpendicular to that surface (Figure 13.1b, c, d). Since granules are occasionally caught in the process of discharging their content, they are secretory. Their content appears finely granular and electron-lucent after conventional fixation in aldehydes and postfixation with buffered OsO_4, but homogeneous and electron-dense after postfixation with reduced OsO_4 (Figure 13.1c). Staining with uranyl acetate *en bloc* produces very fine filaments within the granules, while, after the addition of the cationic dye Safranin O to the fixative, filaments are visible which are 5 nm thick and up to 150 nm long (Figure 13.1d). The glycocalyx of the apical membrane shows similar reactions to these staining procedures as the content of the granules (Figure 13.1d). In contrast, addition of the cationic dye Ruthenium red to the fixative produces only a strong electron-dense reaction of the glycocalyx but not of the content of the granules. The failure of Ruthenium red to stain the granular content is apparently caused by the inability

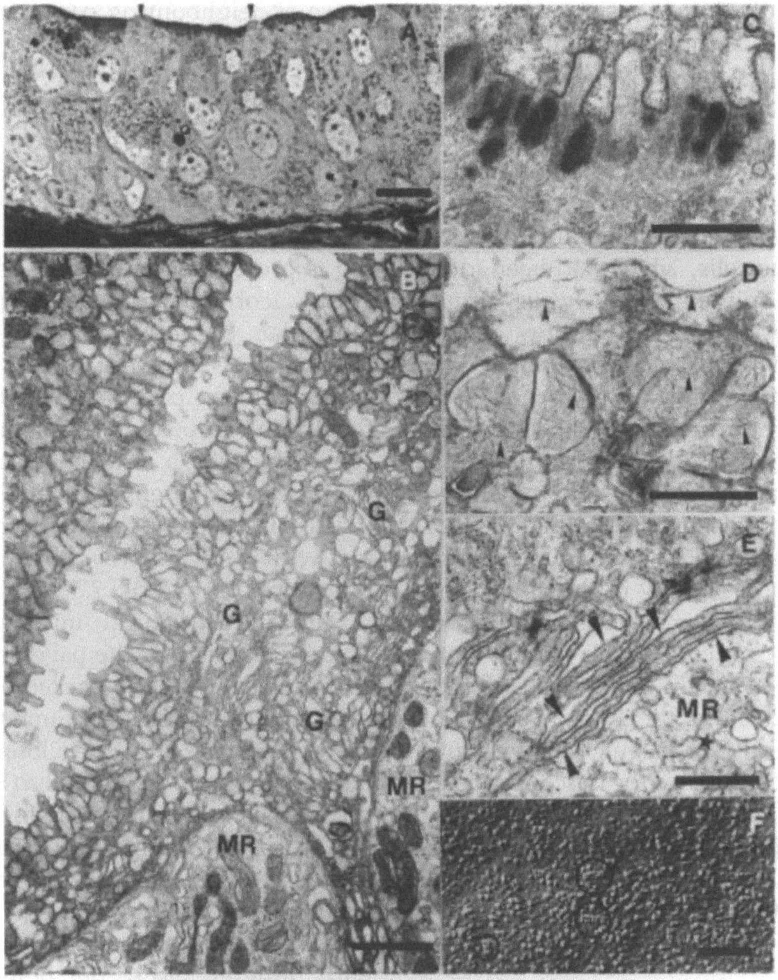

Figure 13.1 Pavement cells. (a) Light micrograph of a one-micron thick Toluidin-blue stained section showing the multilayered columnar epithelium covering a branch of an afferent radial artery. The cuticular border of the epithelium (black in this micrograph due to its metachromatic staining) is focally interrupted, where mitochondria-rich cells are in contact with the epithelial surface (arrows); bar = 20 μm. (b) Ultrastructure of pavement cells on a respiratory lamella in the lateral part of the gill after conventional fixation in aldehydes and buffered OsO_4. The apical cytoplasm contains tightly packed secretory granules with an electron-lucent content. G, Golgi apparatus; MR, mitochondria-rich cell; bar = 1 μm. (c) After postfixation with reduced OsO_4, the content of the secretory granules appears homogeneously electron-dense, whereas fine electron-dense granules are visible in the glycocalyx; bar = 0.5 μm. (d) Addition of the cationic dye Safranin O produces filaments (arrows) within the granules and the glycocalyx; bar = 0.5 μm. (e) The surface area of the lateral pavement cell membrane is enlarged by thin, interdigitating processes (arrows), while that of the mitochondria-rich cell (MR) is amplified by a system of membrane tubules (asterisk); bar = 0.5 μm. (f) Freeze-fracture replica of a lateral pavement cell membrane showing orthogonal arrays of particles (encircled) in addition to globular particles; bar = 0.1 μm.

of this dye to permeate the intact cell and granule membranes.

From these ultrastructural findings, it is evident that the metachromatic staining at the epithelial surface is caused by the glycocalyx and the secretory granules stored directly beneath the apical membrane of the pavement cell. The histochemical demonstration of anionic sites by cationic dyes (Alcian blue, pH 1.0; Safranin O) within the granules and the glycocalyx strongly indicates that the pavement cells secrete sulphated glycoconjugates. In addition, the presence of several large Golgi complexes (Figure 13.1b) in the pavement cell combined with the presence of only a few *cisternae* of the rough endoplasmic reticulum in these cells further supports the view that (sulphated) carbohydrates rather than proteins are the predominant component(s) of the secretions of pavement cells.

The synthesis and secretion of mucus by pavement cells is also a characteristic feature of these cells in lamprey gills (Youson and Freeman, 1976; Bartels, 1989). Mucus could help protect the epithelial surface from toxic effects of pollutants and infectious microorganisms in the water. Such protection would be particularly important in the respiratory lamellae, where the water–blood barrier is thinnest. The microridges, which are present at the surface of some pavement cells, may facilitate the adhesion of the mucus to the epithelial surface (Andrews, 1976; Sperry and Wassersug, 1976).

The content of the intra-epithelial mucous cysts shows histochemical reactions identical to those described for the cuticular border at the epithelial surface. Although the squamous cells, which line the cysts, usually lack secretory granules characteristic of pavement cells, it is suggested that the cysts are formed by pavement cells. The question remains open at present, as to whether these cysts are normal constituents of the gill epithelium or whether they represent pathological changes. The factors which cause the formation of these cysts are also unknown.

Membrane structure

Freeze-fracture replicas of pavement cells show that the particle populations of the apical and lateral membrane are different. The apical membrane is characterized only by globular particles (diameter 9–10 nm), which occur mainly on the P-(protoplasmic) face of the split membrane, whereas the lateral membrane contains orthogonal arrays of particles (OAPs) in addition to scattered globular particles (diameter of 6–9 nm) (Figure 13.1f). Each of the OAPs consists of 4–16 tightly packed square particles (length 5–6 nm) and occupies a mean area of 340 nm^2. They appear on the P-face, while complementary arrays of pits are present on the E-face. The density of the OAPs varies considerably between individual pavement cells between 0 and 55μm^{-2} lateral membrane (Bartels, 1984).

Similar OAPs have been identified in numerous vertebrate cell membranes, particularly those of glial cells (astrocytes), lens fibres, skeletal muscle fibres and a variety of epithelia, where they are always restricted to the (baso-) lateral membrane (reviews by Bartels and Miragall, 1986; Wolburg, 1994). Since several plasma membranes that have been shown by immunocytochemistry to contain the mercurial-insensitive water channel are characterized by OAPs in freeze-fracture replicas, it has been proposed that the OAPs contain proteins related to water channels (Verbavatz *et al.*, 1994; Brown *et al.*, 1995).

13.2.2 Mitochondria-rich cells

The mitochondria-rich (MR) cells are the largest cells in the gill epithelium and are identified in Toluidine blue-stained thick sections by the granular appearance of their cytoplasm (Figure 13.1a). These cells are predominantly located in the lateral part of the gill, decrease in number from the lateral to the medial side of the respiratory lamellae and are absent from the epithelium covering the efferent *corpora cavernosa* and the medial wall of the gill pouch. The MR cells are always singly intercalated between the pavement cells and, even in the depth of the epithelium, remain separated from one another by thin pavement cell processes (Figure 13.2). They are linked to pavement cells at the epithelial surface by junctional complexes (Figure 13.2) and, in deeper layers of the epithelium, to pavement, intermediate and basal cells by desmosomes. A continuous layer of basal cells separates the MR cells from the *basal lamina* (Figure 13.2). The MR

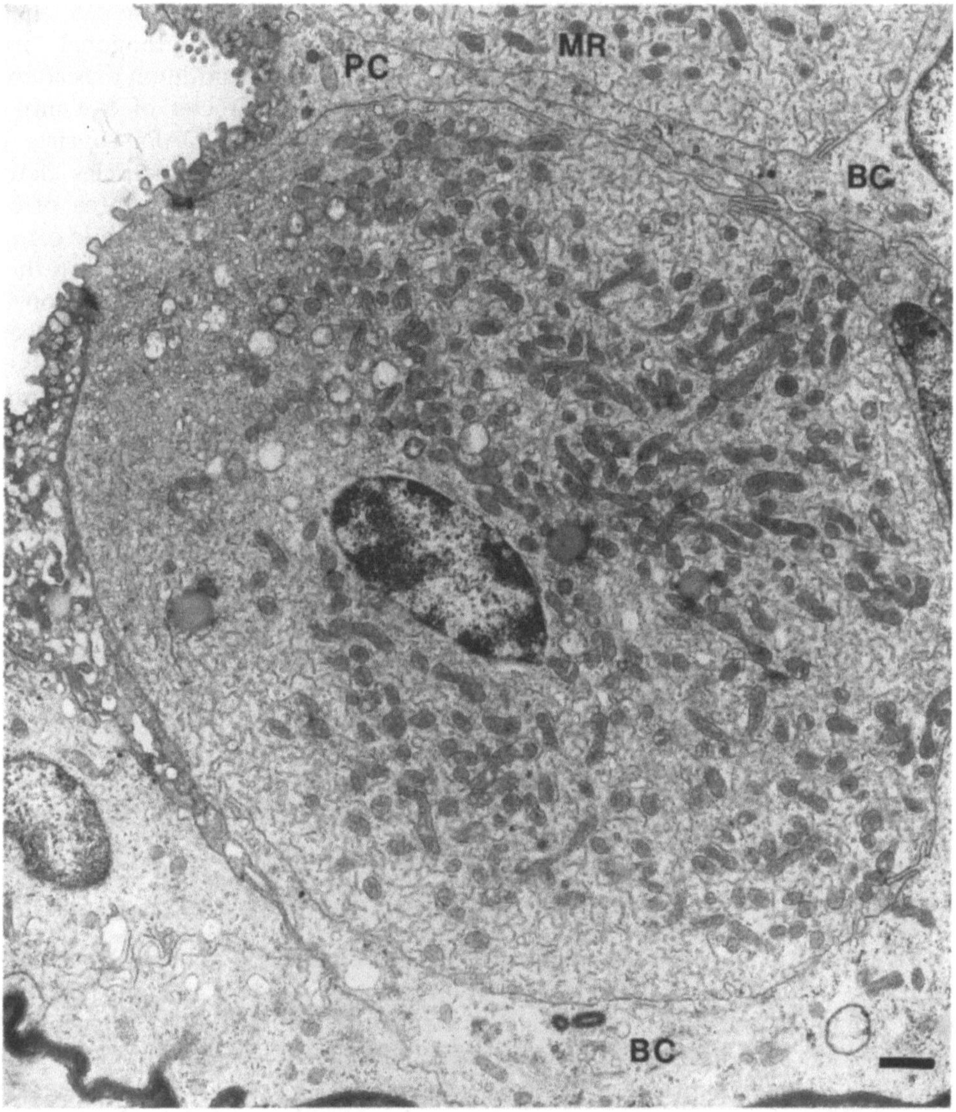

Figure 13.2 A flask-shaped mitochondria-rich (MR) cell in the epithelium of a respiratory lamella is separated from the basal lamina by a basal cell (BC) and from another MR cell by thin processes of a pavement cell and a basal cell, respectively. Note the absence of the tubular system in the apical cytoplasm of the MR cell; bar = 1μm.

cells are identical to the 'plasmareichen Zellen' (Hofbauer, 1934) and the acidophilic cells (Rauther, 1935) described in early light microscopical studies of the hagfish gills. They were called 'ionocytes' by Mallatt and Paulsen (1986) and Elger (1987).

The shape of the MR cells depends on their location. They are tall columnar in the multilayered columnar epithelium covering the lateral part of the gill pouch and the corresponding unbranched portions of the folds, and flask-shaped or spherical in the epithelium of the lamellar part of the folds (Figure 13.2). In the latter case, the cytoplasm of the MR cells is mainly covered by thin processes of adjacent pavement cells and lies in the medial layer of the epithelium. Most of the flask-shaped MR cells are in contact with the environment only by a short narrow neck (Figure 13.2), whereas others have an extended apical surface. While the apical surface area is usually enlarged by short microvilli (Figures 13.2 and 13.3a), which sometimes branch, it is rather smooth in some cells, with a few dome-like protrusions. The microvilli contain an axial bundle of thin (5 nm) filaments.

The striking ultrastructural feature of the MR cells, besides their large number of mitochondria, is the presence of an extensive system of membranous tubules with a diameter of 60–100 nm (Figures 13.2 and 13.3c, d, e). This tubular system passes through the entire cytoplasm except the apical cell pole (Figure 13.2). Its membrane is continuous with the lateral cell membrane and thus represents an intracellular amplification of the latter (Figure 13.3c, e). As a consequence, the lumen of the tubular system is part of the intra-epithelial extracellular space. The numerous mitochondria lie in a close spatial relationship to the tubular system. In many MR cells, they are concentrated in one part of the cell, while in the remaining cytoplasm, although filled with membranous tubules, mitochondria are absent. In addition, small bundles of intermediate filaments are frequently found in the immediate vicinity of the tubular system.

The MR cells usually possess several Golgi complexes lying in a supranuclear position, whereas the *cisternae* of the rough endoplasmic reticulum are sparse. The most apical region of the cytoplasm, which is devoid of the tubular system, contains numerous spherical vesicles with an electron lucent content (Figure 13.2). These vesicles differ from the mucous granules of the pavement cells by their shape, size, arrangement and histochemical properties. Their content is neither electron dense after postfixation with reduced OsO_4 solutions nor does it produce filament-like reaction products after treatment with Safranin O. Glycogen particles and lipid droplets are scattered throughout the cytoplasm.

Membrane structure

As with many epithelial cells, freeze-fracture replicas show that the intramembrane particles of the apical and basolateral membranes of the MR cell in the hagfish gill epithelium differ in size, shape and arrangement (Bartels, 1985).

Apical membrane

The striking character of the apical membrane of the MR cell in freeze-fracture replicas is the presence on the P-face of assemblies of linear arrays of particles and fibrils, as well as scattered globular particles (Figure 13.3a). The linear arrays are composed of globular (diameter 8–9 nm) and rod-shaped particles (length 16–20 nm) and short fibrils and are arranged almost in parallel to each other within the assemblies. The rod-shaped particles apparently consist of two globular subunits, the fibrils of the arrays probably comprising two or more rod-shaped particles that are tightly attached end-to-end. The distances between the arrays of an assembly measure 10–15 nm. The E-face contains assemblies of grooves that are complementary to the assemblies of arrays of particles and fibrils (Figure 13.3b).

The size of the assemblies is correlated

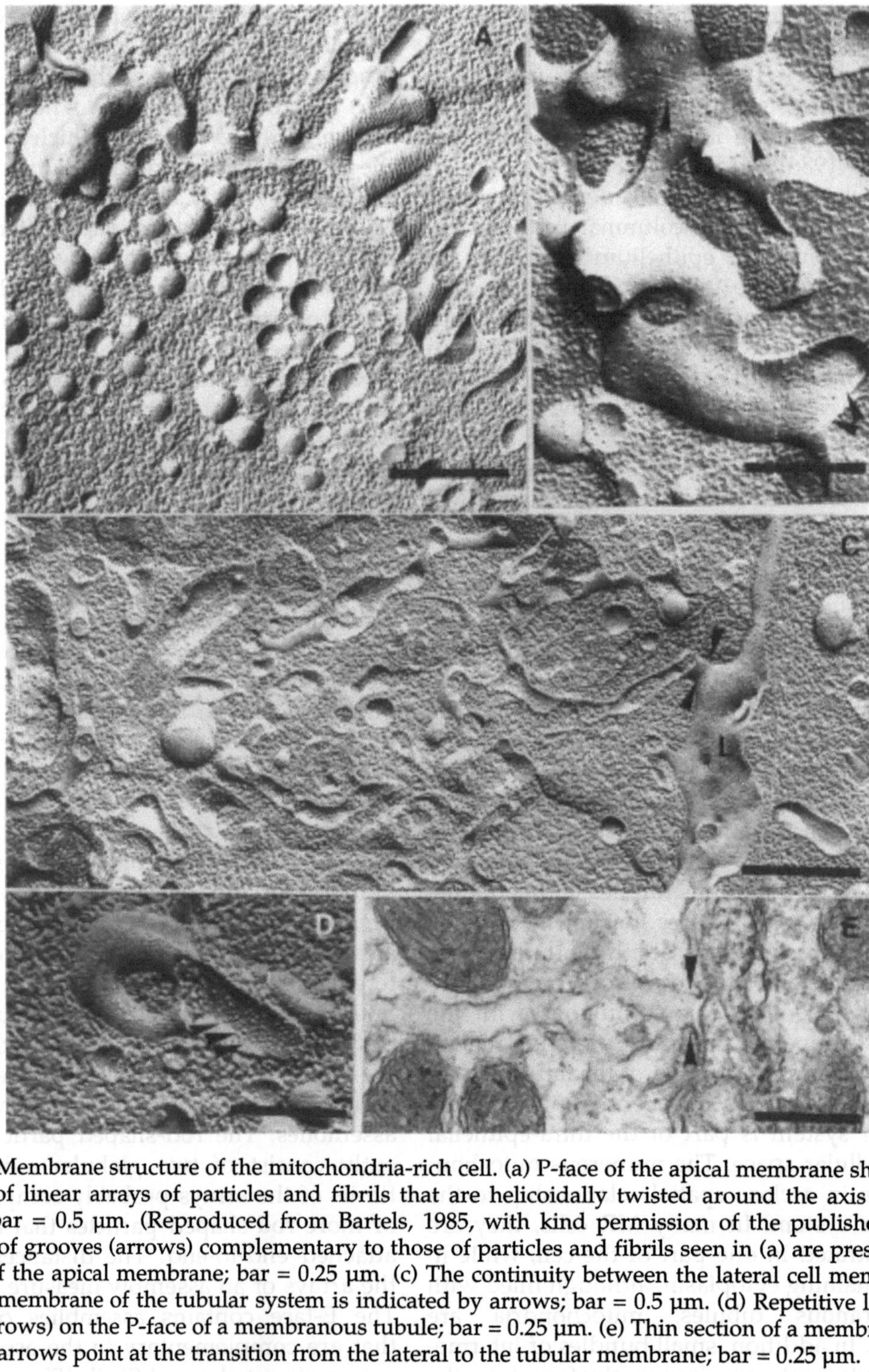

Figure 13.3 Membrane structure of the mitochondria-rich cell. (a) P-face of the apical membrane showing assemblies of linear arrays of particles and fibrils that are helicoidally twisted around the axis of the microvilli; bar = 0.5 µm. (Reproduced from Bartels, 1985, with kind permission of the publisher.) (b) Assemblies of grooves (arrows) complementary to those of particles and fibrils seen in (a) are present on the E-face of the apical membrane; bar = 0.25 µm. (c) The continuity between the lateral cell membrane (L) and the membrane of the tubular system is indicated by arrows; bar = 0.5 µm. (d) Repetitive lines of particles (arrows) on the P-face of a membranous tubule; bar = 0.25 µm. (e) Thin section of a membranous tubule. The arrows point at the transition from the lateral to the tubular membrane; bar = 0.25 µm.

with the shape of the apical surface of the cell (Bartels, 1985). The largest assemblies, which consist of up to 20 arrays of particles and fibrils, occur in the membrane of the longest microvilli (up to 0.7 μm).The arrays of these assemblies, which measure more than 150 μm in length and are helicoidally twisted around the microvillar axis with a pitch of about 45°, are almost entirely composed of fibrils arranged in parallel (Figure 13.3a). In contrast, the assemblies of those cells, which, on their surface, are characterized by small dome-like protrusions instead of microvilli, are not restricted to these protrusions but occur also in membrane areas between the protrusions. Their arrays are smaller and less regularly arranged than those of the microvilli. Between these two extremes of surface structure and assembly size, transitions characterized by short stub-like microvilli and medium-sized assemblies are present.

The correlation between surface structure and the size of the assemblies suggests that these assemblies have a specific function, which, however, has not yet been determined. The only vertebrate cell membranes which possess similar arrays of globular and rod-shaped particles are the brush-border membranes of a small population of epithelial cells in the primate rectum and rabbit caecum and of the so-called 'cup cells' in the small intestine of various mammals (Neutra, 1979; Madara, 1982; Gebert and Bartels, 1995). The specific function of these intestinal cells that is related to the linear arrays of particles also remains to be determined.

Rod-shaped particles, similar to those which are constituents of the linear arrays in the apical membrane of the MR cells of the hagfish gill epithelium, are present in cell membranes containing a H^+-ATPase. These membranes include the apical membrane of proton-secreting (α-type) and the basolateral membrane of bicarbonate-secreting (β-type) MR cells in various epithelia (e.g. the intercalated cell in the renal collecting duct, the MR cells in the toad and turtle urinary bladders, the flask cell in the amphibian skin) and the ruffled border membrane of osteoclasts (Humbert *et al.*, 1975; Brown *et al.*, 1978; Stetson and Steinmetz, 1985, 1986; Brown, 1989; Akisaka *et al.*, 1990). It has been proposed that the rod-shaped particles either represent the transmembrane part of the proton pump or are at least intimately associated with this pump (Brown *et al.*, 1987).

Several differences exist, however, between the rod-shaped particles of the arrays in the MR cells of the hagfish gill epithelium and these particles in those membranes known to be capable of electrogenic proton transport. (1) The rod-shaped particles in the latter membranes are usually scattered and not aligned in linear arrays or assemblies with a characteristic two-dimensional arrangement. (2) A coat of regularly spaced studs, to which antibodies directed against the cytoplasmic tail of the H^+-ATPase bind, is only visible on the cytoplasmic side of membranes which possess scattered rod-shaped particles (Brown *et al.*, 1987). (3) Membranes of cytoplasmic vesicles, which are inserted into the cell membrane by exocytosis on physiological demand regularly contain rod-shaped particles (and H^+-ATPase) in MR cells characterized by scattered intramembrane particles (Stetson and Steinmetz, 1985, 1986; Brown, 1989), but only occasionally in MR cells of the hagfish gills.

Basolateral membrane

The basolateral membrane of the MR cells and the membrane of the tubular system which represents a vast intracellular amplification of the former are characterized in freeze-fracture replicas by particles with a diameter of 7–8 nm (Figure 13.3c, d). These particles usually occur on the P-face of the split membrane, while the E-face appears almost smooth (Figure 13.3d). They are tightly packed, but do not show a characteristic two-dimensional arrangement in the superficial cell membrane (Figure 13.3c), whereas an orderly arrangement of the particles is observed in the membrane of the tubular system, particularly in straight tubules. In these locations, the particles are arranged in circular or slightly helicoidal, periodical rows (Figure 13.3d). The distances between the centres of adjacent rows are 10–12 nm. Discrete furrows, complementary to the circular rows of particles, are present on the E-face of straight tubules. In thin sections cut tangentially to the membrane of the tubular system, periodical striations can be seen at distances of c. 10 nm, particularly after staining with uranyl acetate *en bloc*. They correspond to the rows of particles seen in freeze-fracture replicas (Bartels, 1985).

Membranous tubules which represent an amplification of the basolateral cell membrane are the prominent feature of the chloride cells in the gill epithelium of teleost fishes and lampreys (Philpott, 1966; Karnaky, 1986; Nakao, 1974; Peek and Youson, 1979). These membranes harbour most of the Ouabain-sensitive Na^+/K^+-ATPase (Karnaky *et al.*, 1976b; Hootman and Philpott, 1979). Acclimation of euryhaline teleosts to increased external salinities causes increases in the activity of Na^+/K^+-ATPase in gill homogenates and isolated chloride cells and in the number of ouabain-binding sites in the tubular membranes and also an extension of the tubular system (Kamiya, 1972; Sargent *et al.*, 1975; Karnaky *et al.*, 1976a, b). Periodical structures, similar to those described above for the tubular system of the hagfish MR cells, have also been reported for these membranes in the chloride cells of teleosts and lampreys (Ritch and Philpott, 1969; Dendy *et al.*, 1973; Nakao, 1974; Sardet *et al.*, 1979; Hatae and Benedetti, 1982). It has been suggested that the orderly arranged particles in the membranes of the tubular system are the sites of active ion transport and thus represent Na^+ pumps (Sardet *et al.*, 1979; Hatae and Benedetti, 1982).

Enzyme histochemical investigations of Na^+/K^+-ATPase activity in the gills of the Pacific hagfish *Eptatretus stoutii* have shown that, at the light microscopical level, the cytoplasm of the MR cells stains almost homogeneously for this enzyme, indicating that it is localized in the membranes of the tubular system (Mallatt *et al.*, 1987). However, biochemical studies showed that the Na^+/K^+-ATPase activity in hagfish gill homogenates is in the order of that of teleosts in freshwater and significantly below that of marine teleosts (Mallatt *et al.*, 1987).

Possible functions of the mitochondria-rich cells

Since the MR cells in the hagfish gill epithelium have not yet been studied under controlled experimental conditions, speculation on the function of these cells has to be based only on their morphological characteristics. These cells possess characteristics which have not been observed together in any other vertebrate epithelial cell type so far. On the one hand, they have a vastly amplified basolateral membrane, which forms a tubular system similar to that of teleost and lamprey chloride cells and provides the space for a large number of Na^+ pumps. On the other hand, the apical membrane contains rod-shaped particles similar to those of MR cell membranes that are involved in electrogenic proton transport, although in a different arrangement.

In contrast to the situation in the gill epithelium of teleosts and lampreys in seawater, where the chloride cells form small (teleosts) or large (lampreys) groups and are linked to one another by leaky junctions (Sardet *et al.*, 1979; Ernst *et al.*, 1980; Bartels and Potter, 1991; Bartels *et al.*, 1996), the MR cells of the strictly marine hagfishes always occur singly between pavement cells. This situation rather resembles that of chloride cells in the gills of freshwater teleosts and also of the MR cells characterized by scattered rod-shaped particles in urinary epithelia and amphibian skin (Rosen, 1972; Karnaky, 1986; Katz and Gabbay, 1988). In addition, as the

MR cells in the frog skin and toad urinary bladder (Rick *et al.*, 1978; Wade, 1978), those in the hagfish gill epithelium are probably also electrically isolated from the remaining epithelial cells, since gap junctions were not observed between the MR cells and their adjacent cells (Bartels, 1988). Finally, the almost exclusive localization of Na^+/K^+-ATPase and carbonic anhydrase activity in the hagfish gill epithelium to the MR cells (Mallatt *et al.*, 1987) parallels the situation of Na^+/K^+-ATPase in chloride cells of teleost gills on the one side (Karnaky, 1986) and of carbonic anhydrase in MR cells of urinary epithelia and amphibian skin on the other (Rosen, 1972; Rosen and Friedley, 1973; Steinmetz, 1986).

Since, in contrast to lampreys and teleost fishes, hagfishes are unable to regulate efficiently the osmolality of their internal milieu independent of that of the environment by means of the Na^+ and Cl^- concentrations, they are generally considered to be osmoconformers (Morris, 1965; Cholette *et al.*, 1970; Robertson, 1974; Lutz, 1975). However, the morphological and enzyme histochemical data reviewed above suggest that the MR cells in their gill epithelium are engaged in ion transport and/or acid-base regulation. It is thus tempting to speculate that these cells are responsible for the Na^+/H^+ and Cl^-/HCO_3^- exchanges postulated for the hagfish gill epithelium (Evans, 1984).

Intriguingly, the electroneutral, but independent uptakes of Na^+ and Cl^- in exchange for H^+ and HCO_3^-, respectively, are the osmoregulatory mechanisms used by animals in freshwater (Krogh, 1939; Kirschner, 1983). The Na^+ concentration in the hagfish serum has repeatedly been reported to be higher than in the environment by from 10 to 19% (Robertson, 1954; Bellamy and Chester-Jones, 1961; McFarland and Munz, 1965; Morris, 1965). This situation is indeed similar, although much less significant, to that of vertebrates in freshwater. Since the hagfish kidneys produce a nearly isotonic urine and do not significantly reabsorb Na^+ in the archinephric duct (McFarland and Munz, 1964; Morris 1965; Alt *et al.*, 1980; Fels *et al.*, 1989), they are unlikely to be responsible for the higher Na^+ concentration in serum than seawater and thus not an effector organ of 'Na^+ regulation' in these agnathans.

In conclusion, since the MR cells in the gill epithelium are the only cells with the pronounced ultrastructural characteristics of ion-transporting cells (Oschman and Berridge, 1972) that have so far been found in various epithelia of hagfishes, including those of the skin, intestine, gallbladder and kidney, they are the favourite candidate for the uptake of Na^+ from the marine environment. This uptake may occur in exchange for protons, as proposed by Evans (1984), which, in the MR cells, are generated by the activity of carbonic anhydrase (Mallatt *et al.*, 1987). It remains to be established if the protons are indeed secreted by the activity of an H^+-ATPase, as suggested by the presence of rod-shaped particles in the apical membrane (see above), and that Na^+ then enters the cell through a Na^+ channel, or if the exchange occurs through an antiport. Independent from the model operating in the apical membrane of the MR cells, the Na^+ pumps that are most probably localized in the membrane of the tubular system (Mallatt *et al.*, 1987) help to keep the intracellular Na^+ concentration low and thereby facilitate the entry of this ion through the apical membrane. Finally, the higher Na^+ concentrations in the serum of individual hagfishes than in their environment (see above) may result from acidotic conditions and the need of these animals to secrete excess H^+.

If this hypothesis proves to be correct, a couple of problems remain to be resolved. (1) When H^+, generated by the carbonic anhydrase activity, leaves the MR cell through the apical membrane, the remaining HCO_3^- should leave it through the basolateral membrane. In most H^+-secreting vertebrate cells, this is achieved by an exchange for Cl^-

(Drenckhahn *et al.*, 1985; Schuster *et al.*, 1986). Since a band 3 anion exchanger, which mediates this exchange in cell membranes of higher vertebrates, is not operating in the membrane of agnathan erythrocytes (Ohnishi and Asai, 1985; Ellory *et al.*, 1987; Tufts and Boutilier, 1989; Peters and Gros, this volume), it is not expected that this protein is expressed in the basolateral membrane of the MR cells. (2) In addition to the Na^+/H^+ exchange, Evans (1984) also observed a Cl^-/HCO_3^- exchange, although only during the winter season. Whereas, in urinary epithelia (collecting duct, toad and turtle urinary bladder), this task is performed by the β-type of MR cells which are characterized by rod-shaped particles and H^+-ATPase in the basolateral membrane and an anion exchanger in the apical membrane (Stetson and Steinmetz, 1985; Schuster *et al.*, 1986), there is no morphological evidence for such a subtype of MR cells in the hagfish gill epithelium.

13.2.3 Basal and intermediate cells

The basal cells form a continuous layer of cells on the basal lamina to which they are connected by hemidesmosomes. Their content of organelles (mitochondria, ribosomes, rough endoplasmic reticulum, Golgi apparatus, lysosomes) and cytoplasmic inclusions is inconspicuous (Figure 13.2) with the exception of sporadically occurring myelin bodies. The basal cells contain bundles of intermediate filaments and are connected to each other by desmosomes. Since basal cells are the only cells in the gill epithelium that were observed during mitosis, they are considered to be the proliferating cell pool from which the other cell types develop.

Intermediate cells are present in those regions of the epithelium which consist of more than two layers. They lie above the basal cells but are not in contact with the epithelial surface. Intermediate cells resemble basal cells in their supply with organelles. Ovoid granules with a fine granular content, that are similar to the secretory (mucous) granules of pavement cells, are occasionally present in the peripheral cytoplasm. These cells are thus intermediate stages in the development from basal to pavement cells.

13.2.4 Dense-core granule cells

In the middle layers of the multilayered epithelium, single cells differ from the surrounding cells by an electron-lucent cytoplasm. These cells contain a few round granules with an electron-dense core. They have a large nucleus which is deeply indented, but few organelles. Some of these cells possess long processes which contain a few axial microtubules but no organelles.

The function and origin of these cells is not entirely clear. Since they are not connected by desmosomes to their neighbouring cells and morphologically resemble the lymphocyte-like or lymphoid cells (Tomonaga *et al.*, 1973; Mattison and Fänge, 1977), they could have invaded from the blood. On the other hand, the electron-lucent cytoplasm and dense-core granules of these cells are reminiscent of the neuroepithelial cells in the gill epithelium of gnathostome fishes (Dunel-Erb *et al.*, 1982). In contrast to the neuroepithelial cells, the cells with dense-core granules in the hagfish gill epithelium are not restricted to a defined region of the gill and do not have a relationship to intra- or subepithelial nerve fibres.

13.2.5 Granulocytes

Cells characterized by numerous electron-dense granules frequently occur in the middle layers of the multilayered epithelium and in the epithelium of the respiratory lamellae. The granules are elliptical or rod-like and measure up to 400 nm in length and up to 200 nm in thickness. The nucleus is indented and appears lobulated in some sections. Its envelope possesses only a few nuclear pores which are shown by freeze-fracture replicas to be arranged in clusters. The Golgi apparatus lies in a close spatial relation to the nucleus. Ribosomes, *cisternae* of the rough endoplasmic reticulum and small mitochondria are regularly present, although in small numbers.

Identical cells are also present in the lumen of blood vessels and in the connective tissue (Mattison and Fänge, 1977). They are thus considered to be granulocytes that have emigrated from the blood.

ACKNOWLEDGEMENTS

I am greatly indebted to Prof. Hilmar Stolte, Hannover Medical School, for generously providing the hagfishes and to Prof. Ian C. Potter, Murdoch University, for critically reading the text. Gratitude is also expressed to Ursula Fazekas, Gudrun Voss-Wermbter, Dipl.Ing., and Hans Heidrich for excellent technical assistance.

REFERENCES

Adam, H. and Strahan, R. (1963) Systematics and geographical distribution in myxinoids. In *The Biology of Myxine* (eds A. Brodal and R. Fänge), Universitetsforlaget, Oslo, pp. 3–8.

Akisaka, T., Yoshida, H., Kogaya, Y. *et al.* (1990) Membrane modifications in chick osteoclats revealed by freeze fracture replicas. *American Journal of Anatomy*, **188**, 381–92.

Alt, J.M., Stolte, H., Eisenbach, G.M. *et al.* (1980) Renal electrolyte and fluid excretion in the Atlantic hagfish *Myxine glutinosa. Journal of Experimental Biology*, **91**, 323–30.

Andrews, P.M. (1976) Microplicae: characteristic ridge-like folds of the plasmalemma. *Journal of Cell Biology*, **68**, 420–9.

Bartels, H. (1984) Orthogonal arrays of particles in the gill epithelium of the Atlantic hagfish, *Myxine glutinosa. Cell and Tissue Research*, **238**, 657–9.

Bartels, H. (1985) Assemblies of linear arrays of particles in the apical plasma membrane of mitochondria-rich cells in the gill epithelium of the Atlantic hagfish (*Myxine glutinosa*). *Anatomical Record*, **211**, 229–38.

Bartels, H. (1988) Intercellular junctions in the gill epithelium of the Atlantic hagfish, *Myxine glutinosa. Cell and Tissue Research*, **254**, 573–83.

Bartels, H. (1989) Freeze-fracture study of the pavement cell in the lamprey gill epithelium. Analogy of membrane structure with the granular cell in the amphibian urinary bladder. *Biology of the Cell*, **66**, 165–71.

Bartels, H. and Decker, B. (1985) Communicating junctions between pillar cells in the gills of the Atlantic hagfish, *Myxine glutinosa. Experientia*, **41**, 1039–40.

Bartels, H. and Miragall, F. (1986) Orthogonal arrays of particles in the plasma membrane of pneumocytes. *Journal of Submicroscopic Cytology*, **18**, 637–46.

Bartels, H. and Potter, I.C. (1991) Structural changes in the zonulae occludentes of the chloride cells of young adult lampreys following acclimation to seawater. *Cell and Tissue Research*, **265**, 447–57.

Bartels, H., Moldenhauer, A. and Potter, I.C. (1996) Changes in the apical surface of chloride cells following acclimation of lampreys to seawater. *American Journal of Physiology*, **270**, R125–R133.

Bellamy, D. and Chester Jones, I. (1961) Studies on *Myxina glutinosa*. I. The chemical composition of the tissues. *Comparative Biochemistry and Physiology*, **3**, 175–83.

Berridge, M.J. and Oschman, J.L. (1972) *Transporting Epithelia*, Academic Press, New York, London.

Bettex-Galand, M. and Hughes, G.M. (1973) Contractile filamentous material in the pillar cells of fish gills. *Journal of Cell Science*, **13**, 359–70.

Bourne, R.H. (1892) On the presence of a branchial basket in *Myxine glutinosa. Proceedings of the Zoological Society (London)* **1892**, 706–8.

Brown, D. (1989) Membrane recycling and epithelial cell function. *American Journal of Physiology*, **256**, F1–F12.

Brown, D., Ilic, V. and Orci, L. (1978) Rod-shaped particles in the plasma membrane of the mitochondria-rich cell of amphibian epidermis. *Anatomical Record*, **192**, 269–76.

Brown, D., Gluck, S. and Hartwig, J. (1987) Structure of the novel membrane-coating material in proton-secreting epithelial cells and identification as an H^+ ATPase. *Journal of Cell Biology*, **105**, 1637–48.

Brown, D., Katsura, T., Kawashima, M. *et al.* (1995) Cellular distribution of the aquaporins: a family of water channel proteins. *Histochemistry and Cell Biology*, **104**, 1–9.

Cholette, C., Gagnon, A. and Germain, P. (1970) Isosmotic adaption in *Myxine glutinosa* L. I. Variations of some parameters and role of the amino acid pool of the muscle cells. *Comparative Biochemistry and Physiology*, **33**, 333–46.

Cole, F.J. (1912) A monograph on the general morphology of the myxinoid fishes, based on a study of *Myxine*. Part IV. On some peculiarities of the afferent and efferent branchial arteries of *Myxine*. *Transactions of the Royal Society, Edinburgh*, **48**, 215–30.

Cole, F.J. (1925) A monograph on the general morphology the myxinoid fishes, based on a study of *Myxine*. Part VI. The morphology of the vascular system. *Transactions of the Royal Society, Edinburgh*, **54**, 309–42.

Dendy, L.A., Philpott, C.W. and Deter, R.L. (1973) Localization of Na^+, K^+-ATPase and other enzymes in teleost pseudobranch. II. Morphological characterization of intact pseudobranch, subcellular fractions and plasma membrane substructure. *Journal of Cell Biology*, **57**, 689–703.

Drenckhahn, D., Schlüter, K., Allen, D.P. *et al.* (1985) Colocalization of band 3 with ankyrin and spectrin at the basal membrane of intercalated cells in the rat kidney. *Science*, **230**, 1287–9.

Dunel-Erb, S., Bailly, Y., Laurent, P. (1982) Neuroepithelial cells in fish gill primary lamellae. *Journal Applied Physiology*, **53**, 1342–53.

Elger, M. (1987) The branchial circulation and the gill epithelia in the Atlantic hagfish, *Myxine glutinosa* L. *Anatomy and Embryology*, **175**, 489–504.

Ellory, J.C., Wolowyk, M.W. and Young, J.D. (1987) Hagfish (*Eptatretus stoutii)* erythrocytes show minimal chloride transport activity. *Journal of Experimental Biology*, **129**, 377–83.

Ernst, S.A., Dodson, W.B., Karnaky, K.J. Jr (1980) Structural diversity of occluding junctions in the low-resistance chloride-secreting opercular epithelium of seawater-adapted killifish (Fundulus heteroclitus). *Journal of Cell Biology*, **87**, 488–97.

Evans, D.H. (1984) Gill Na^+/H^+ and Cl^-/HCO_3-exchange systems evolved before the vertebrates entered fresh water. *Journal of Experimental Biology*, **113**, 465–9.

Fels, L., Raguse-Degener, G., Stolte, H. (1989) The Archinephron of *Myxine glutinosa* L. (Cyclostoma), in *Structure and Function of the Kidney* (ed. R.K.H. Kinne), *Comparative Physiology*, Vol. 1, Karger, Basel, pp. 73–152.

Flöge, J., Stolte, H. and Kinne, R. (1984) Presence of a sodium-dependent D-glucose transport system in the kidney of the Atlantic hagfish (*Myxine glutinosa*). *Journal of Comparative Physiology*, **B154**, 355–64.

Gebert, A. and Bartels, H. (1995) Linear arrays of intramembranous particles characterize a subpopulation of epithelial cells in the rabbit caecum. *Journal of Submicroscopic Cytology and Pathology*, **27**, 125–7.

Hardisty, M.W. (1979) *Biology of Cyclostomes*, Chapman & Hall, London.

Hatae, T. and Benedetti, E.L. (1982) Mosaic structure in the plasma membrane: spiral arrays of subunits in the cytoplasmic tubules of lamprey chloride cells. *Journal of Cell Science*, **56**, 441–52.

Hofbauer, M. (1934) Anatomischer und histologischer Bau der Kiemensäcke von *Myxine glutinosa*. *Biologia Generalis*, **12**, 330–48.

Hootman, S.R. and Philpott, C.W. (1979) Ultracytochemical localization of Na+, K+-activated ATPase in chloride cells from the gills of a euryhaline teleost. *Anatomical Record*, **193**, 99–129.

Hughes, G.M. (1984) General anatomy of the gills. In *Fish Physiology*, Vol. 10, Part A. (eds W.S. Hoar and D.J. Randall), Academic Press, Orlando, pp. 1–72.

Hughes, G.M. and Morgan, M. (1973) The structure of fish gills in relation to their respiratory function. *Biological Review*, **48**, 419–75.

Humbert, F., Pricam, C., Perrelet, A. *et al.* (1975) Specific plasma membrane differentiations in the cells of the kidney collecting tubule. *Journal of Ultrastructure Research*, **52**, 13–20.

Jakobshagen, E. (1920) Die Homologie der Wirbeltierkiemen. *Jenaische Zeitschrift für Naturwissenschaften*, **57**, 87–142.

Kamiya, M. (1972) Sodium-potassium-activated adenosine-triphosphatase in isolated chloride cells from eel gills. *comparative Biochemistry and physiology*, **43B**, 611–17.

Karnaky, K.J. Jr (1986) Structure and function of the chloride cell of *Fundulus heteroclitus* and other teleosts. *American Zoologist*, **26**, 209–24.

Karnaky, K.J. Jr, Ernst, S.A. and Philpott, C.W. (1976a) Teleost chloride cell. I. Response of pupfish *Cyprinodon variegatus* gill Na, K-ATPase and chloride cell fine structure to various high salinity environments. *Journal of Cell Biology*, **70**, 144–56.

Karnaky, K.J. Jr, Kinter, L.B., Kinter, W.B. *et al.*

(1976b) Teleost chloride cell. II. Auto-radiographic localization of gill Na, K-ATPase in killifish *Fundulus heteroclitus* adapted to low and high salinity environments. *Journal of Cell Biology*, **70**, 157–77.

Katz, U. and Gabbay, S. (1988) Mitochondria-rich cells and carbonic anhydrase content of toad skin epithelium. *Cell and Tissue Research*, **251**, 425–31.

Kirschner, L.B. (1983) Sodium chloride absorption across the body surface: frog skins and other epithelia. *American Journal of Physiology*, **244**, R429–R443.

Krogh, A. (1939) *Osmotic Regulation in Aquatic Animals*, Cambridge University Press, London.

Laurent, P. (1984) Gill internal morphology, in *Fish Physiology*, Vol. 10, Part A (eds W.S. Hoar and D.J. Randall), Academic Press, Orlando, pp. 73–183.

Lewis, S.V. and Potter, I.C. (1982) A light and electron microscope study of the gills of larval lampreys (*Geotria australis*) with particular reference to the water–blood pathway. Journal of Zoology London, **198**, 157–76.

Lutz, P. (1975) Adaptive and evolutionary aspects of the ionic content of fishes. *Copeia*, **1975**, 369–73.

Madara, J.L. (1982) Cup cells: structure and distribution of a unique class of epithelial cells in guinea pig, rabbit and monkey small intestine. *Gastroenterology*, **83**, 981–94.

Mallatt, J. (1984) Early vertebrate evolution: pharyngeal structure and the origin of gnathostomes. *Journal of Zoology London*, **204**, 169–83.

Mallatt, J. and Paulsen, C. (1986) Gill ultrastructure of the Pacific hagfish *Eptatretus stoutii*. *American Journal of Anatomy*, **177**, 243–69.

Mallatt, J., Conley, D.M. and Ridgway, R.L. (1987) Why do hagfish have gill 'chloride cells' when they need not regulate plasma NaCl concentration? *Canadian Journal of Zoology*, **65**, 1956–65.

Marinelli, W. and Strenger, A. (1954) *Vergleichende Anatomie und Morphologie der Wirbeltiere*, Vol. 1: *Lampetra fluviatilis* L, Deuticke, Wien.

Marinelli, W. and Strenger, A. (1956) *Vergleichende Anatomie und Morphologie der Wirbeltiere*, Vol. 2: *Myxine glutinosa* L, Deuticke, Wien.

Mattison, A.G.M. and Fänge, R. (1973) Light- and electron microscopic observations on the blood cells of the Atlantic hagfish, *Myxine glutinosa (L.)*. *Acta Zoologica*, **58**, 205–21.

McFarland, W. and Munz, F.W. (1965) Regulation of body weight and serum composition by hagfish in various media. *Comparative Biochemistry and Physiology*, **14**, 383–98.

Morris, R. (1965) Studies on salt and water balance in Myxine glutinosa (L.). *Journal of Experimental Biology*, **42**, 359–71.

Müller, J. (1845) Untersuchungen über die Eingeweide der Fische. Schluß der *Vergleichenden Anatomie der Myxinoiden*. Abhandlungen der Akademie der Wissenschaften Berlin.

Munz, F.W. and McFarland, W.N. (1964) Regulatory functions of a primitive vertebrate kidney. *Comparative Biochemistry and Physiology*, **13**, 381–400.

Nakao, T. (1974) Fine structure of the agranular cytoplasmic tubules in the lamprey chloride cells. *Anatomical Record*, **178**, 49–62.

Neutra, M. (1979) Linear arrays of intramembrane particles on microvilli in primate large intestine. *Anatomical Record*, **193**, 367–82.

Ohnishi, S.T. and Asai, H. (1985) Lamprey erythrocytes lack glycoproteins and anion transport. *Comparative Biochemistry and Physiology*, **81B**, 405–7.

Peek, W.D. and Youson, J.H. (1979) Ultrastructure of chloride cells in young adults of the anadromous sea lamprey, *Petromyzon marinus* L., in fresh water and during adaptation to sea water. *Journal of Morphology*, **160**, 143–63.

Philpott, C.W. (1966) The use of horseradish peroxidase to demonstrate functional continuity between the plasmalemma and the unique tubular system of the chloride cell. *Journal of Cell Biology*, **31**, 86.

Pohla, H., Lametschwandtner, A. and Adam, H. (1977) Die Vaskularisation der Kiemen von *Myxine glutinosa* L. (Cyclostomata). *Zoologica Scripta*, **6**, 331–41.

Rauther, M. (1935) Zur Kenntnis der Myxinoiden-Kiemen. *Gegenbaurs Morphologisches Jahrbuch*, **75**, 613–33.

Rick, R., Dörge, A., v. Arnim, E. *et al.* (1978) Electron microprobe analysis of frog skin epithelium: evidence for a syncytial sodium transport compartment. *Journal of Membrane Biology*, **39**, 313–31.

Ritch, R. and Philpott, C.W. (1969) Repeating particles associated with an electrolyte-transport membrane. *Experimental Cell Research*, **55**, 17–24.

Robertson, J.D. (1954) The chemical composition of the blood of some aquatic chordates, including members of the Tunicata, Cyclostomata and

Osteichthyes. *Journal of Experimental Biology*, **31**, 424–42.

Robertson, J.D. (1974) Osmotic and ionic regulation in cyclostomes, in *Chemical Zoology*, Vol. 8 (eds M. Florkin and B.T. Scheer), Academic Press, New York, pp. 149–93.

Rosen, S. (1972) Localization of carbonic anhydrase activity in turtle and toad urinary bladder mucosa. *Journal of Histochemistry Cytochemistry*, **20**, 696–702.

Rosen, S. and Friedley, N.J. (1973) Carbonic anhydrase activity in Rana pipiens skin: biochemical and histochemical analysis. *Histochemistry*, **36**, 1–4.

Sardet, C., Pisam, M. and Maetz, J. (1979) The surface epithelium of teleostean fish gills. Cellular and junctional adaptations of the chloride cell in relation to salt adaptation. *Journal of Cell Biology*, **80**, 96–117.

Sargent, J.R., Thomson, A.J. and Bornancin, M. (1975) Activities and localization of succinic dehydrogenase and Na^+/K^+-activated adenosine triphosphatase in the gills of fresh water and sea water eels (*Anguilla anguilla*). *Comparative Biochemistry and Physiology*, **51B**, 75–9.

Schuster, V.L., Bonsib, S.M. and Jennings, M.L. (1986) Two types of collecting duct mitochondria-rich (intercalated) cells: lectin and Band 3 cytochemistry. *American Journal of Physiology*, **251**, C347–C355.

Smith, D.G. and Chamley-Champbell, J. (1981) Localization of smooth-muscle myosin in branchial pillar cells of snapper (*Chrysophis auratus*) by immunofluorescence histochemistry. *Journal of Experimental Zoology*, **215**, 121–4.

Sperry, D.G. and Wassersug, R.J. (1976) A proposed function of microridges on epithelial cell. *Anatomical Record*, **185**, 253–8.

Steinmetz, P.R. (1986) Cellular organization of urinary acidification. *American Journal of Physiology*, **251**, F173–F187.

Stetson, D.L. and Steinmetz, P.R. (1985) α and β types of carbonic anhydrase-rich cells in turtle bladder. *American Journal of Physiology*, **249**, F553–F565.

Stetson, D.L. and Steinmetz, P.R. (1986) Correlation between apical intramembrane particles and H^+ secretion rates during CO_2 stimulation in turtle bladder. *Pflügers Archiv*, **407**, S80–S84.

Tomonaga, S., Hirokane, T. and Awaya, K. (1973) Lymphoid cells in the hagfish. *Zoological Magazine Tokyo*, **82**, 133–5.

Tomonaga, S., Sakai, K., Tashiro, J. *et al.* (1975) High-walled endothelium in the gills of the hagfish. *Zoological Magazine Tokyo*, **84**, 151–5.

Tufts, B.L. and Boutilier, R.G. (1989) The absence of rapid chloride/bicarbonate exchange in lamprey erythrocytes: implications for CO_2 transport and ion distributions between plasma and erythrocytes in the blood of *Petromyzon marinus*. *Journal of Experimental Biology*, **144**, 565–76.

Verbavatz, J.-M., Van Hoek, A.N., Ma, T. *et al.* (1994) A 28 kDa sarcolemmal antigen in kidney principal cell basolateral membranes: relationship to orthogonal arrays and MIP26. *Journal of Cell Science*, **107**, 1083–94.

Wade, J.B. (1978) Membrane structural specializations of the toad urinary bladder revealed by the freeze-fracture technique. III. Location, structure and vasopressin dependence of intramembrane particle arrays. *Journal of Membrane Biology*, **40**, 281–96.

Worthington, J. (1905) Contribution to our knowledge of myxinoids. *American Naturalist*, **39**, 625–63.

Youson, J.H., Freeman, P.A. (1976) Morphology of the gills of larval and parasitic adult sea lamprey, *Petromyzon marinus* L. *Journal of Morphology*, **149**, 73–104.

14
VENTILATION AND GAS EXCHANGE

Hans Malte and Jens Peter Lomholt

SUMMARY

The respiratory system of hagfishes consists of 6–14 pairs of lens-shaped or bilobed gill pouches each receiving a ventilatory water flow from a short afferent gill duct given off from the pharynx. Depending on species, efferent gill ducts either open separately to the exterior or unite to form a single opening on each side of the animal. Each gill pouch has a number of internal folds carrying successive orders of lamellae. Blood in the lamellae flows countercurrent to the water flowing in between the lamellae. The ventilatory water flow is generated by the velum situated in the anterior part of the pharynx. The mechanism of velar pumping is discussed based on ultrasonic scanning images of velar movements. The simultaneous efficiences of oxygenation of the arterial blood and deoxygenation of the ventilated water can only be explained by countercurrent exchange in a well-matched exchanger. A well-developed dermal capillary network secures an adequate supply of oxygen to the skin when the animal has burrowed into sediment. *Myxine* is very anoxia-tolerant and may retract into the sediment and stop ventilating for hours.

14.1 INTRODUCTION

The respiratory system and the mechanism of ventilation of hagfishes are unique compared to those of cartilaginous and bony fishes. The morphology of the respiratory system is well described (Cole, 1905, 1908; Marinelli and Strenger, 1956; Strahan, 1958; Johansen and Strahan, 1963). Water is taken in through the nasopharyngeal duct and not through the mouth. The duct leads to an expanded part of the anterior pharynx, the velar chamber which houses the velum, the structure responsible for creating the ventilatory water current. In cross-section the velum has a shape somewhat like an inverted T attached in the dorsal midline of the velar chamber. The openings of the afferent gill ducts are located in the pharynx behind the velar chamber. Short, narrow afferent gill ducts lead to the lens-shaped gill pouches. The morphology of the efferent gill ducts varies between different groups of hagfishes (Fernholm, this volume). In *Eptatretus* short efferent ducts lead to separate external gill openings, whereas in *Myxine* the efferent ducts discharge through common external openings located behind the gill pouches. A short pharyngocutaneous duct leads directly to the surface on the left side of the animal.

14.2 MECHANISM OF VENTILATION

On the basis of anatomical studies Cole (1908) proposed that the velum was the structure responsible for the ventilatory water current and not a valve to close the nasopharyngeal duct as had been previously believed. In his classical textbook Goodrich (1930) did not mention the velum but maintained that the water current was produced by contractions of

The Biology of Hagfishes. Edited by Jørgen Mørup Jørgensen, Jens Peter Lomholt, Roy E. Weber and Hans Malte. Published in 1998 by Chapman & Hall, London. ISBN 0 412 78530 7.

the muscular coat surrounding the gill pouches. He also believed that, while the animal was feeding, water could be pumped in and out via the external branchial openings. This has, however, never been observed in live animals. Gustafson (1935), working at the Kristineberg Zoological Station, Sweden, observed the pumping movements of the velum in dissected, live specimens of *Myxine glutinosa*. He could not detect any contractions of the gill pouches or gill ducts and thus concluded that ventilation was exclusively the result of the activity of the velum.

Strahan (1958) gave a detailed description of the cartilaginous skeleton of the velum and the muscles causing its movements (Figure 14.1a, b). Contraction of the posterior craniovelar muscle pulls the lateral bar towards the midline. Because the transverse bars resist this movement the lateral bar is rotated with the effect that the external bar comes closer to the midline than the internal bar. The end result is that the velar scroll is curled while at the same time moving ventrally. The ventral and dorsal anterior craniovelar muscles and the spinovelar muscle are antagonistic and

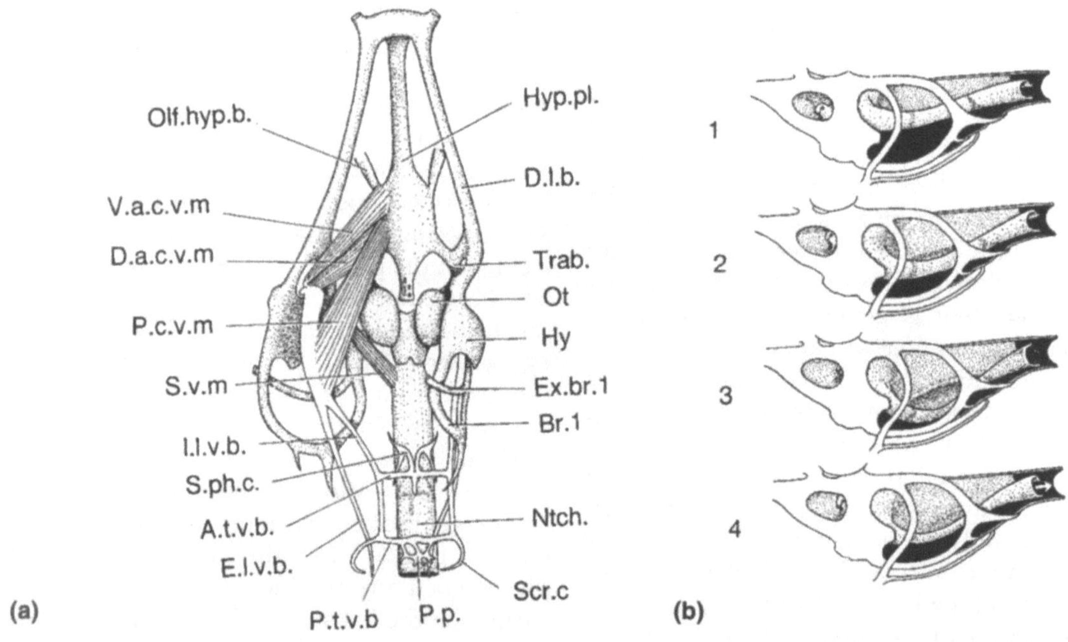

Figure 14.1 (a) Ventral view of the skeleton in the velar region. The right side has been pulled outwards to display the velar muscles. A.t.v.b., anterior transverse bar; Br.1, first branchial arch; D.a.c.v.m., dorsal anterior craniovelar muscle; D.l.b., dorsal longitudinal bar (cut); E.l.v.b., external lateral velar bar; Ex.br.1, first extrabranchial arch (cut); Hy., hyoid arch; Hyp.pl., hypophysial plate (cut); I.l.v.b., internal lateral velar bar; Ntch., notochord; Olf.hyp.b., bar connecting olfactory capsule to hypophysial plate (cut); Ot., otic capsule; P.c.v.m., posterior craniovelar muscle; P.p., posterior process of posterior transverse velar bar; P.t.v.b., posterior transverse velar bar; Scr.c., scroll cartilage; S.ph.c. suprapharyngeal cartilage; S.v.m., spinovelar muscle; Trab., trabeculae; V.a.c.v.m., ventral anterior craniovelar muscle. (b) Four successive stages in the velar cycle. The velar chamber is shown from left side with the left pharyngeal wall removed. 1, resting; 2, velar scroll beginning to unroll; 3, velar scroll maximally unrolled; 4, velar scroll beginning to move dorsally. The arrow shows the direction of water flow. (From Strahan, 1958, with permission.)

their contraction unrolls the velar scroll bringing its margin close to the lateral wall of the velar chamber. At the same time the velar scroll is moved dorsally and water above the scroll is forced posteriorly towards the gills, while at the same time water is drawn in below the velum through the nasopharyngeal duct. Strahan (1958) agreed with Gustafson (1935) that velar pumping is exclusively responsible for creating the water current.

In contrast to the studies discussed above Johansen and Hol (1960) concluded that contraction of the gill pouches in combination with activity of muscular sphincters in the afferent and efferent gill ducts forms an integral part of the mechanism of ventilation in *Myxine glutinosa*. They used x-ray analysis of intact, unanaesthetized, albeit restrained animals in combination with a water soluble contrast medium. When another contrast medium consisting of a suspension of particles was used, the animals reacted by 'sneezing' or 'coughing'. This involved contractions of the gill pouches as well as the entire pharynx and material could be expelled both via the nasal duct and via the pharyngocutaneous duct.

We have recently used ultrasonic scanning to visualize velar movements in *Myxine glutinosa* (Jørgensen, Lomholt, Malte and Nielsen, unpublished observations). This technique permits observations on unanaesthetized, unrestrained and completely undisturbed animals. The pattern of movements described by Strahan (1958) was confirmed with the velar scrolls curled during downstroke and extended during upstroke. Two points of interest appear from the recordings (Figure 14.2). Firstly, the scrolls are extended to such a degree that their margins touch the walls of the velar chamber during the upstroke. This means that during the upstroke the velar chamber may be effectively divided in two and the velum may act in the manner of a piston pump. In some cases frequencies of velar movements as low as 4–6 per minute were recorded. It seems unlikely that such slow velar movements could result in any flow of water unless the velum does work in the manner of a piston pump. Secondly, the ultrasonic scanner revealed the downstroke to be the more rapid part of the cycle, whereas the upstroke lasted longer. This pattern is opposite to what was found by Strahan (1958). The difference is likely to be caused by the fact that during the ultrasonic recordings the velum was working against the naturally occurring resistance to water flow, whereas the observations of Strahan (1958) involved opening the velar chamber. From a functional point of view it appears logical that the upstroke with extended scrolls should be the slower one, since this is the phase where water is moved and the pump is doing work. During the rapid downstroke with the scrolls curled the velum will move against a much lower resistance.

14.3 GILL STRUCTURE AND GAS EXCHANGE

Considerable attention has been paid to the structure of the gill pouches and their blood supply (Pohla *et al.*, 1977; Elgar, 1987; Mallatt and Paulsen, 1986; Bartels, this volume). In the following only a brief account of the arterio-arterial circulation is given. Each gill pouch receives an afferent branch of the ventral aorta. The afferent gill arteries join the afferent circular arteries which encircle the efferent gill ducts at the point where these emerge from the gill pouches. From the afferent circular arteries a number of afferent radial arteries are given off. These run in the wall of the gill pouch towards the equator. From the system of radial arteries a number of gill folds inside the gill pouch are supplied with blood via the afferent cavernous tissue. The gill folds protrude from the internal surface of the gill pouch and extend towards the central axis (the axis pointing from the afferent to the efferent gill duct). Relatively few gill folds reach the central axis and the spaces between these large folds are filled

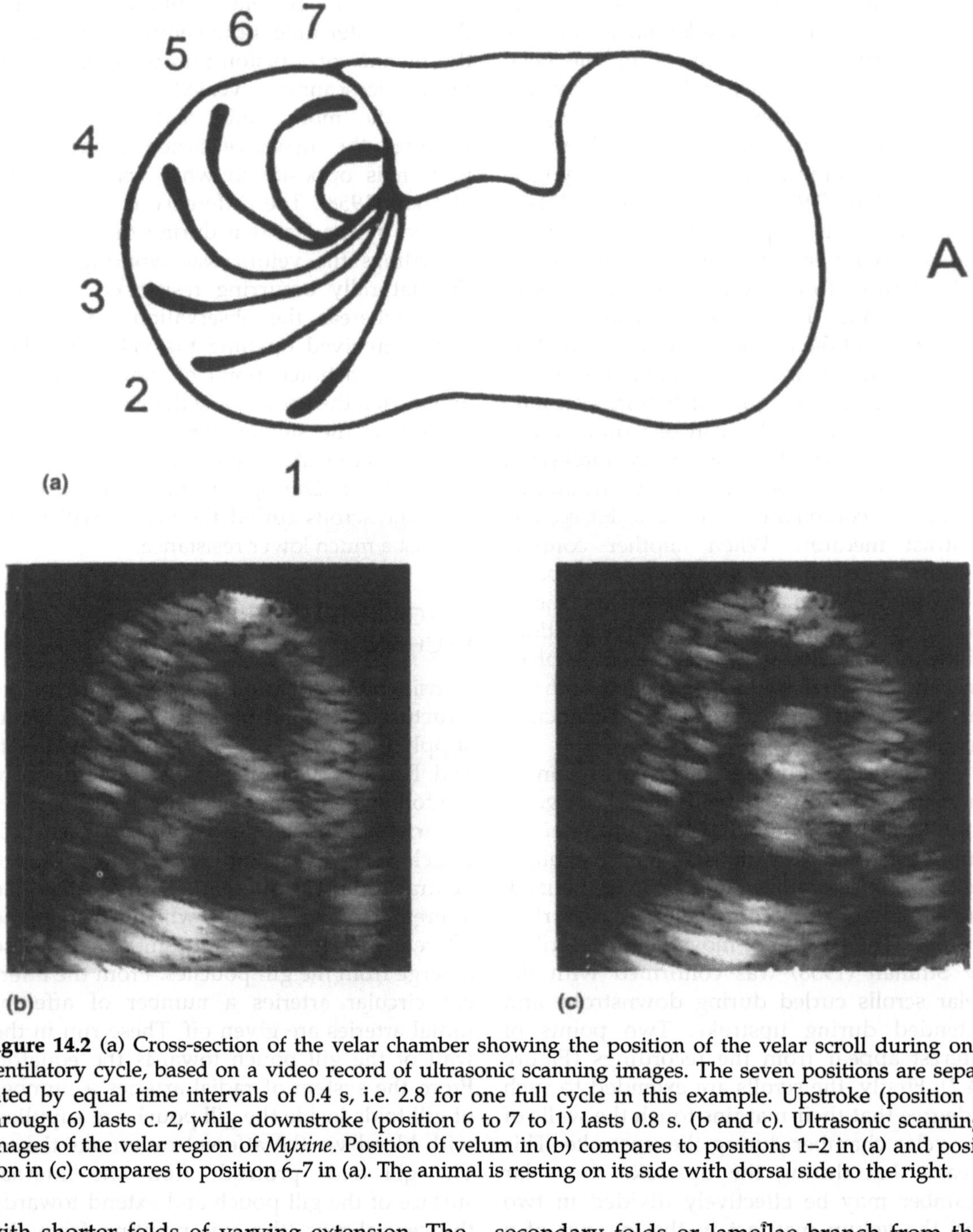

Figure 14.2 (a) Cross-section of the velar chamber showing the position of the velar scroll during one ventilatory cycle, based on a video record of ultrasonic scanning images. The seven positions are separated by equal time intervals of 0.4 s, i.e. 2.8 for one full cycle in this example. Upstroke (position 1 through 6) lasts c. 2, while downstroke (position 6 to 7 to 1) lasts 0.8 s. (b and c). Ultrasonic scanning images of the velar region of *Myxine*. Position of velum in (b) compares to positions 1–2 in (a) and position in (c) compares to position 6–7 in (a). The animal is resting on its side with dorsal side to the right.

with shorter folds of varying extension. The primary gill folds remain undivided in the afferent and efferent ends, the so-called afferent and efferent zones. In the middle part, secondary folds or lamellae branch from the primary folds and these again may give rise to higher order lamellae. All the lamellae are oriented parallel to the central axis of the gill pouch. This middle zone is the respiratory zone where gas exchange takes place. The blood which runs countercurrent to the waterflow is drained *via* the efferent cavernous tissue to the efferent radial arteries which join the efferent circular artery, encircling the afferent gill duct. The efferent circular artery gives off two efferent branchial arteries which discharge into the common carotid arteries (aortic roots). The arrangement of the blood supply to the gill pouches and the arrangement of the gill folds in *Myxine* are shown in Figure 14.3.

From cross-sections of the gill pouch published by Pohla *et al.* (1977) it appears that the arrangement of gill folds and lamellae of successive orders in the respiratory zone is such that water comes in intimate contact with the blood over the entire gill surface with little possibility for water shunts. The regions afferent and efferent to the respiratory zone, where the primary gill folds remain undivided, are much less densely packed with lamellar tissue and serve as manifolds distributing waterflow over the entire surface of the respiratory zone. There are no quantitative, morphometric data available relating to gill surface area and/or water to blood diffusion distances. Estimation of diffusion distance from high magnification cross-sections from Mallatt and Paulsen

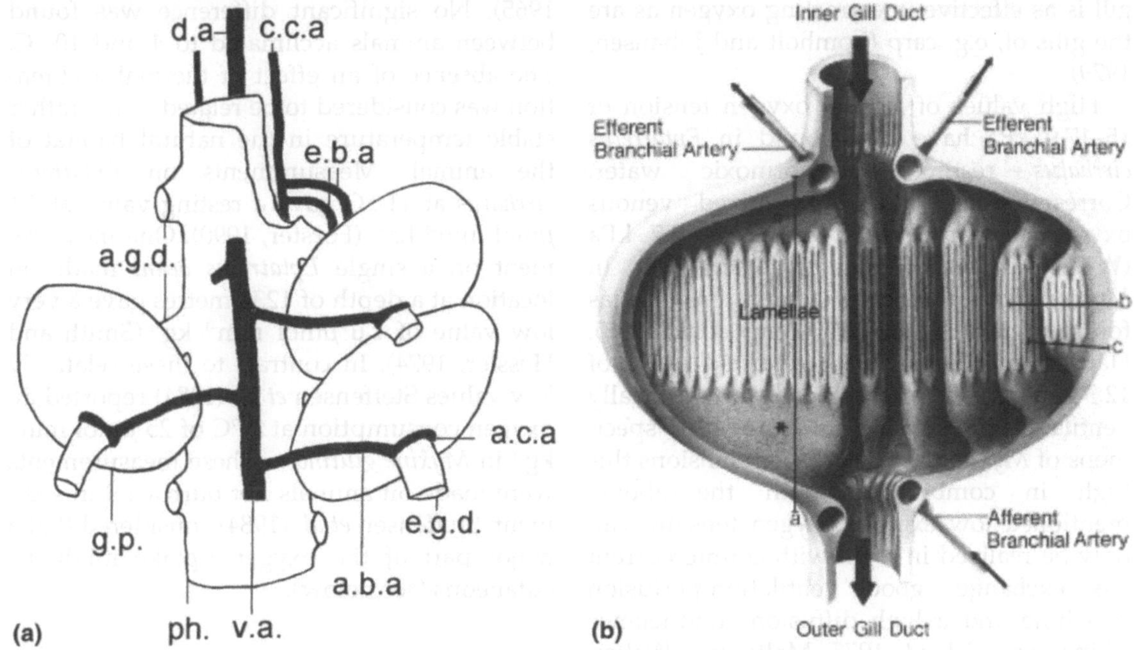

Figure 14.3 Blood supply to the gill pouch of *Myxine glutinosa*. (a) Schematic view of blood vessels and gill pouches relative to the pharynx. d.a., dorsal aorta; c.c.a., common carotid artery; e.b.a., efferent branchial arteries; a.b.a., afferent branchial artery; v.a., ventral aorta; a.c.a., afferent circular artery; a.g.d., afferent gill duct; e.g.d., efferent gill duct; g.p., gill pouch; ph., pharynx. (b) Schematic drawing of gill pouch cut longitudinally. To the left a large gill fold extending between the two poles of the gill pouch and its central axis is seen. To the right several smaller gill folds which do not extend all the way to the central axis are seen. Each gill fold consists of an afferent and an efferent unbranched region (marked by asterisks) and a central region, the respiratory zone, covered by lamellae. The blood flow inside the lamellae is counter-current to the water flow (thick arrows) through the gill pouch. (Figure 14.3b from Elger, 1987, with permission.)

(1986) yield values in the range 1–5 µm, which is comparable to, or even shorter than, that in many teleost fishes (Hughes, 1984).

No reports exist in the literature on the ability of the hagfish gill to extract oxygen from the respiratory water current. Preliminary measurements of our own indicate that oxygen extraction may be as high as 75–80% when the inspired water is near air saturation. Values of expired oxygen tension were quite variable, but this variation may be the result of difficulties in obtaining exhaled water not contaminated with ambient water of high oxygen tension. The fact that with an inspired oxygen tension of about 20 kPa (150 mmHg) exhaled oxygen tensions as low as 4.7 kPa could be recorded shows that the hagfish gill is as effective in extracting oxygen as are the gills of, e.g. carp (Lomholt and Johansen, 1979).

High values of arterial oxygen tension of 15–17.9 kPa have been found in *Eptatretus cirrhatus* resting in normoxic water. Corresponding values for mixed venous oxygen tension were from 2.3 to 2.7 kPa (Wells *et al.*, 1986; Forster *et al.*, 1992). In *Myxine glutinosa* arterial oxygen tension was found to be 10.8–11.2 kPa (Perry *et al.*, 1993). Hanson and Sidell (1983) reported a P_{O_2} of 12.3 kPa in blood arterialized by an artificially ventilated gill pouch of anaesthetized specimens of *Myxine*. Arterial oxygen tensions this high in combination with the above-mentioned low expired oxygen tensions can only be realized in a gill with countercurrent gas exchange, good ventilation-perfusion matching and a high diffusion conductance (Piiper and Scheid, 1975; Malte and Weber, 1985).

14.4 VENTILATION

Only one study is available on the magnitude of gill ventilation in hagfish. Steffensen *et al.* (1984) used an electromagnetic flowmeter mounted on a funnel placed over the snout of specimens of *Myxine glutinosa* buried in sediment. Measurements were made at 7, 15 and 20 °C. At these temperatures average ventilation volumes of 23, 45 and 120 ml min^{-1} kg^{-1} were found. Although the temperatures of 15 and 20 °C may seem high compared to conditions in the natural habitat, the results indicate a considerable capacity to increase ventilation above the level at the more natural temperature of 7 °C.

14.5 OXYGEN UPTAKE

Few data exist on the metabolic rate of hagfishes. Oxygen uptake in *Eptatretus stoutii* was found to be about 3.0–7.4 µmol min^{-1} kg^{-1} (4–10 ml kg^{-1} h^{-1}) at 5 °C rising to about 7.4–14 µmol min^{-1} kg^{-1} at 15 °C (Munz and Morris, 1965). No significant difference was found between animals acclimated to 4 and 10 °C. The absence of an effect of thermal acclimation was considered to be related to the rather stable temperature in the natural habitat of the animal. Measurements on *Eptatretus cirrhatus* at 11 °C gave a resting value of 5.1 µmol min^{-1} kg^{-1} (Forster, 1990). One measurement on a single *Eptatretus deani* made on location at a depth of 1230 metres gave a very low value of 1.6 µmol min^{-1} kg^{-1} (Smith and Hessler, 1974). In contrast to these relatively low values Steffensen *et al.* (1984) reported an oxygen consumption at 7 °C of 25 µmol min^{-1} kg^{-1} in *Myxine glutinosa*. These measurements were made on animals not burrowed in sediment. Steffensen *et al.* (1984) considered that a major part of the oxygen uptake might be cutaneous (see below).

14.6 CUTANEOUS GAS EXCHANGE

Hagfishes have a well-developed capillary network beneath the epidermis (Hans and Tabencka, 1938; Lametschwandtner *et al.*, 1989; Welsch and Potter, this volume). The existence of these skin capillaries was taken as an indication that the skin might act as an important accessory gas exchange organ. Although the thickness of the epidermis was reported to be 70–100 μm, the implications of this barrier for oxygen transfer from water to blood was not considered. Steffensen *et al.* (1984) estimated that a very large fraction of the oxygen uptake may be cutaneous. Their estimate was based on measurements of total oxygen uptake, gill ventilation and an assumed value for gill oxygen extraction. As discussed above their value for oxygen uptake is considerably higher than what has otherwise been reported and their assumed value of 25% for gill oxygen extraction is likely to be too low. Consequently the estimate of the relative importance of cutaneous oxygen uptake is probably too high.

Several factors seem to argue against the idea that the cutaneous capillary network has an important function in oxygen uptake. The capillaries are perfused with arterial blood of a high oxygen tension and haemoglobin-oxygen saturation and consequently a low capacitance coefficient for oxygen. Furthermore the thickness of the barrier separating blood and water is at least an order of magnitude larger than what is found in effective gas exchangers such as fish gills (Hughes, 1984). Finally the boundary layer of water adjacent to the skin surface may impose a significant, additional resistance to diffusion.

In order for the subepidermal capillaries to play a role in oxygen uptake, there must be a net diffusion of oxygen across the epidermis, which will depend on its thickness, the oxygen tension gradient and Krogh's constant of diffusion. The conditions for diffusion across the skin are complicated by the fact that the epidermis is not only a diffusive barrier but the tissue itself consumes oxygen. In Figure 14.4a we have modelled the oxygen tension profile across the epidermis for different values of tissue oxygen consumption, m (see Appendix for details). The epidermal tissue has been assumed to be homogenous with respect to oxygen consumption. On the inside is assumed a sheet of blood of oxygen tension P_i and on the outside water of oxygen tension P_o. If no oxygen is consumed by the tissue (m = 0), oxygen tension drops along a

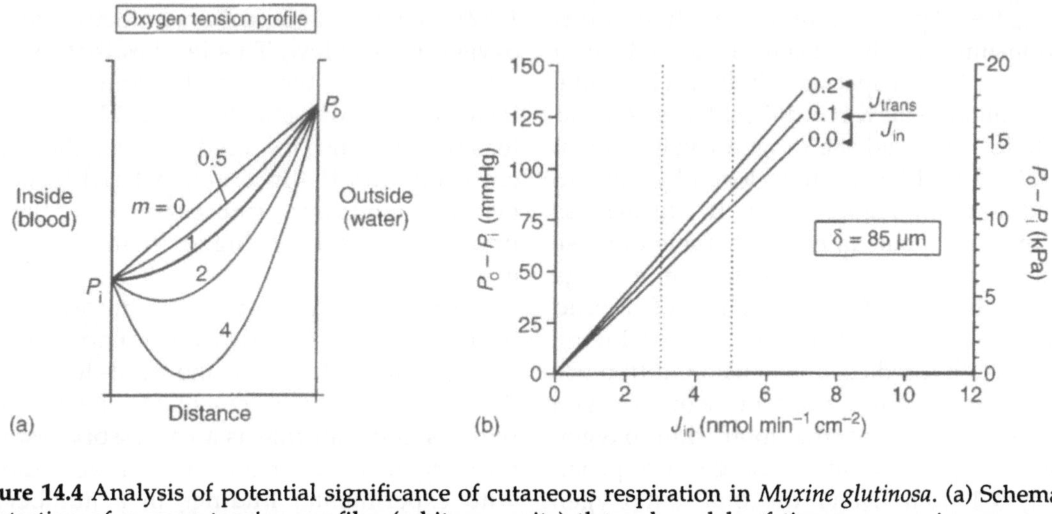

Figure 14.4 Analysis of potential significance of cutaneous respiration in *Myxine glutinosa*. (a) Schematic illustration of oxygen tension profiles (arbitrary units) through a slab of tissue consuming oxygen at different rates (m) and with the tension held constant at both interfaces. (b) The driving oxygen tension difference across the skin necessary to obtain three different levels of the transcutaneous to total oxygen flux ratio, J_{trans}/J_{in}, plotted as a function of the total cutaneous oxygen uptake, J_{in}. The two vertical dotted lines enclose a range of values for total cutaneous oxygen uptake normally found in fishes. The value of K used was that found for eel skin by Kirch and Nonnotte (1977): 4.34×10^{-6} nmol s^{-1} cm^{-1} mmHg^{-1}.

straight line. For increasing values of m, the gradient becomes steeper near the outer surface and less steep near the inner surface. Consequently more oxygen will be taken up at the outer surface and less will move into the blood at the inner surface. For $m = 1$ (arbitrary units) the gradient becomes zero at the inner surface and no oxygen will be taken up by the blood. When m increases further the oxygen tension will have its lowest value inside the epidermis and oxygen will be taken up both from the blood and from the water. From the oxygen tension profile the total flux of oxygen into the skin from the outside, J_{in}, as well as the flux out of the skin to the inside J_{out}, can be calculated (= $K{\cdot}[dP/dx]_o$ and $K{\cdot}[dP/dx]_i$ respectively; dimension, e.g., nmol min^{-1} cm^{-2}). The flux out of the skin to the inside, J_{out}, is equal to the transcutaneous flux, J_{trans}, which represents the part of the skin oxygen uptake available to other tissues. Figure 14.4b is a plot of the tension difference across the skin, $P_o - P_i$, required to obtain three different levels of transcutaneous oxygen flux relative to total oxygen flux, J_{trans}/J_{in}, as a function of the total oxygen flux, J_{in} (see Appendix for a derivation of this relationship). J_{in} has been measured in a number of teleost species (Kirch and Nonotte, 1977; Nonotte and Kirch, 1978; Steffensen and Lomholt, 1985) and the range of values most typically found has been enclosed by the two dotted lines in Figure 14.4b. If *Myxine* has skin oxygen uptake rates comparable to those of most fishes, and if we assume an average water to blood diffusion distance of around 85 μm (Welsch and Potter, this volume), Figure 14.4b predicts the tension difference required to obtain a transcutaneous oxygen flux of say 20% of the total skin oxygen uptake (= J_{in}) to be in the range of approximately 8–13 kPa. Even if the total skin oxygen uptake is as high as 30% of the oxygen uptake of the animal at rest, only 6% of this oxygen is diffusing through the skin. A transcutaneous oxygen flux large enough to contribute significantly to the total oxygen uptake of the animal would, of course, require an even larger tension difference. The oxygen tension at the water-skin interface, P_o, referred to in the above calculations, is not synonymous with the oxygen tension in the ambient water but is lower due to the inevitable existence of a boundary layer over the surface of a body immersed in a fluid medium. Based on the boundary layers around simple cylindrical objects (Rosenhead, 1988; Holman, 1989) it can be estimated that the oxygen tension difference across the boundary layer at water velocities > 10 cm s^{-1} is likely to be in the range 0.53–5.3 kPa. The exact value depends mainly on whether water flow is perpendicular (small difference) or parallel (largest difference) to the long axis of the cylinder. Although it is difficult to obtain a closer estimate, it appears that the tension drop across the boundary layer should not be neglected.

Furthermore, the oxygen tension on the inside, P_i, is not identical to the arterial oxygen tension obtained by oxygenation of the blood in the gill. Since the haemoglobin-oxygen saturation is close to 100% in the arterial blood (Wells *et al.*, 1986; Forster *et al.*, 1992), the blood capacitance coefficient for oxygen is very low. This implies that oxygen diffusing into the blood would rapidly increase the oxygen tension. The oxygen tension difference required between the ambient water and the incoming arterial blood to get a significant transcutaneous oxygen uptake is thus even higher than the above estimates of $P_o - P_i$.

Despite its simplifying assumptions we believe that the analysis presented above strongly indicates that the well-developed, dermal capillary network of *Myxine* does not play a significant role as an accessory respiratory organ. This conclusion is in accordance with experimental findings in a number of fishes where the *in vivo* oxygen uptake by the skin was shown to make up a large fraction, often around 25%, of the total oxygen uptake at rest. However, little or none of this oxygen crossed the skin to enter the blood but was consumed by the epidermal tissue itself (Kirsch and Nonnotte, 1977; Nonnotte and Kirsch, 1978; Steffensen and Lomholt, 1985).

In muddy sediments where *Myxine* is found, oxygen is typically present only in the top 5–6 mm (Barker-Jørgensen and Revsbech, 1989). Thus, when *Myxine* is burrowed in the sediment, its skin will be exposed to an anoxic environment (see below) and the oxygen demand of the skin itself cannot be met by oxygen diffusing from the ambient water but must be supplied by the dermal capillaries. This extreme situation places a higher demand on the blood supply to the skin of *Myxine* than in most fishes. It also places stringent demands on the dimensioning of the capillary bed and on the control of blood flow, since a too generous supply of oxygen would be lost by diffusion into the anoxic sediment. We propose that the well-developed dermal capillary network of *Myxine* is an adaptation to secure oxygen supply to the skin while burrowed in the sediment. This is in agreement with the recent finding of better developed dermal capillary networks in the two burrowing species of hagfish *Myxine glutinosa* and *Paramyxine* (= *Eptatretus) atami* than in *Eptatretus stoutii* which spends less time burrowed (Potter *et al.*, 1995).

14.7 RESPIRATORY ASPECTS OF BURROWING

Among the better known species of hagfish, *Eptatretus stoutii* and *E. cirrhatus* remain on the surface of the bottom for much of their time whereas *Myxine glutinosa* and *Paramyxine atami* spend the larger part of their life burrowed in the sediment. This behaviour of *M. glutinosa* has been described by Gustafson (1935), Strahan (1963a, b) and Martini (this volume).

Our own observations of *Myxine* in aquaria with a bottom layer of mud largely conform to the observations of previous authors. Certain points pertinent to respiratory function will be discussed in the following. The sediment was a black, completely anoxic mud. In spite of a constant oxygen tension of the water close to air saturation only the top few millimetres were oxidized and consequently showed a lighter, brownish colour.

The burrowed animals took up an almost horizontal position only 3–5 cm beneath the surface. After an animal had penetrated into the mud it would forcefully blow water out of the nasal duct ('cough'). This often resulted in the creation of a depression in the sediment. The snout protruded into the bottom of this depression but was kept below the general surface. At the same time water was ejected from the exhalant openings and the pharyngocutaneous duct. A spurt of mud could be seen to appear above the approximate position of the exhalant openings. This apparently caused a loosening of the sediment above the exhalant openings and during normal ventilation the expired water kept coming up at this spot, which was often made conspicuous because black sediment washed up from deeper layers formed a small heap on the otherwise brownish surface. When an animal was made to inhale dyed water the dye also appeared at this localized spot.

In some cases an animal burrowed along the glass wall of the aquarium so the fish could be observed within its burrow. When dyed water was injected into the burrow little or no movement of water could be observed. It thus appears that normally there is practically no renewal of water in the burrow and expired water finds the shortest possible route to the surface. Even after an animal had remained in the same position for a day or so, the sediment inside the burrow still looked black. This indicates that the hagfish does not bring significant amounts of oxygen into its burrow. Furthermore, the burrows are not permanent. Whenever the animal burrows itself it will do so in a new place and thus expose its skin to 'fresh' anoxic sediment. In contrast to this it was striking to observe the permanent burrows of the polychaete *Nereïs diversicolor*, which happened to inhabit the same sediment. This animal actively ventilates its burrows, which were lined by a conspicuous layer of light brown, oxidized sediment.

From time to time a burrowed animal would retract its snout in such a way that the sediment collapsed and covered the snout. Often an animal could stay in this position for many hours with no signs of ventilation. As the environment of the burrow is anoxic, the animal must rely on oxygen stores and eventually anaerobic metabolism during these periods of breathhold. Before resuming ventilation the animal clears away the mud covering the snout by a series of vigorous coughs.

The unified exhalant openings located far behind the snout have been considered an adaptation related to the burrowing habit. The observations given above show that the system ensures that expired water exits as a unified current removed from where inspired water is taken in, thus avoiding rebreathing of exhaled water.

14.8 HYPOXIA AND ANOXIA

Although there are no systematic studies of the minimal environmental oxygen tensions required for survival in hagfish, *Myxine* is known to be very hypoxia tolerant. Strahan (1958) reported that specimens which had their nasopharyngeal duct blocked lived for several days with one specimen living for almost two weeks. Since *Myxine* is not known to ventilate with its mouth open the gills must have remained unventilated throughout this period. However, due to the special situation with a high oxygen tension in the water combined with a very low oxygen tension in the arterial blood supplying the skin, transcutaneous oxygen uptake in this case may have contributed significantly to the total energy metabolism of the organism. Apparently *Myxine* can retreat into the anoxic sediment for hours (see above), an observation indicating that the potential for anaerobic metabolism is well developed. At least for the heart, which in *Myxine* tolerates complete anoxia, the metabolic basis for this anoxia-tolerance is well established (Hanson and Sidell, 1983). Despite its anoxia tolerance *Myxine* responds to hypoxia with an increase in ventilation frequency as do cartilaginous and teleost fishes. In *Myxine,* however, this increase in frequency can be much more dramatic (up to ten-fold) than in fishes since there is no possibility for increasing stroke volume (Jørgensen, Nielsen, Lomholt and Malte, unpublished results).

Cardiac output is maintained even in severe hypoxia (water oxygen tensions 1.5–2.2 kPa, Axelsson *et al.*, 1990) and *Myxine* is able to narrow the inspired to arterial oxygen tension difference considerably under these conditions (Perry *et al.*, 1993). The adjustments responsible for these observations may be a result of the catecholamines that are released during hypoxia (Perry *et al.*, 1993; Axelsson *et al.*, 1990). In *Eptatretus cirrhatus* the ability to defend the arterial oxygen tension in environmental hypoxia seems much less developed. Thus, at a water oxygen tension of 5.3 kPa the arterial oxygen tension was only 0.6 kPa and the arterio-venous oxygen concentration difference was correspondingly low (Forster *et al.*, 1992). Even though this was partly compensated by an increased cardiac output the resulting oxygen consumption was only between one-fourth and one-third that obtained under normoxic conditions (Forster *et al.*, 1992). Apparently this reduced aerobic metabolism was not accompanied by a metabolic acidosis since arterial and venous pH remained constant.

14.9 EXERCISE

Inducing *Eptatretus cirrhatus* to swim steadily in a flume at 20 cm s^{-1} had no effect on its arterial oxygen tension which remained above 12 kPa. The venous oxygen tension, however, decreased from 2.3 kPa (haemoglobin-oxygen saturation 58%) to 0.47 kPa (haemoglobin-oxygen saturation $< 10\%$) (Wells *et al.*, 1986). Thus *Eptatretus* exploits its venous reserve in the same way as fishes and mammals do. It appears that the 20 cm s^{-1}, which corresponds to about 0.5 body lengths s^{-1} is the approximate aerobic swimming capacity of this species (Davison *et al.*, 1990). Compared to teleost fishes this is a very low aerobic swimming capacity (Beamish, 1978). When forced to swim to exhaustion at higher velocities, lactate and protons rapidly accumulated in the blood of the central circulation as well as in that of the subcutaneous sinus. There was no indication of a preferential accumulation in the subcutaneous sinus of either protons or lactate (Davison *et al.*, 1990), but the myotomal lactate concentrations were found to be very high.

14.10 APPENDIX: OXYGEN TENSION PROFILES THROUGH THE SKIN

The skin is modelled as a uniform sheet of tissue of thickness δ and characterized by a constant, volume-specific oxygen consumption, m (e.g. μmol min^{-1} cm^{-3}). The oxygen tension profile through the tissue is obtained from the diffusion equation:

$$\frac{\partial P}{\partial t} = K\frac{\partial^2 P}{\partial x^2} - m \quad (1)$$

where K is Krogh's diffusion constant. In steady state this reduces to

$$K\frac{d^2P}{dx^2} - m = 0 \quad (2)$$

With the boundary conditions $P(0) = P_i$ and $P(\delta) = P_o$ the solution is:

$$P(x) = P_i + \left[\frac{P_o - P_i}{\delta} - \frac{m\delta}{2K}\right] x + \frac{m}{2K}x^2 \quad (3)$$

The flux of oxygen into the sheet at $x = \delta$ is given by:

$$J_{in} = K\left[\frac{dP}{dx}\right]_{x=\delta} = \frac{K}{\delta}(P_o - P_i) + \frac{m\delta}{2} \quad (4)$$

The flux of oxygen out of the sheet at $x = 0$, which is equal to the transcutaneous oxygen flux, J_{trans} is given by:

$$J_{out} = J_{trans} = K\left[\frac{dP}{dx}\right]_{x=0} = \frac{K}{\delta}(P_o - P_i) - \frac{m\delta}{2} \quad (5)$$

The ratio between the transcutaneous and the total oxygen flux is given by:

$$\frac{J_{trans}}{J_{in}} = \frac{\frac{K}{\delta}(P_0 - P_i) - \frac{m\delta}{2}}{J_{in}} \quad (6)$$

Solving for $m\delta/2$ in eq.(4) and inserting this into eq. (6) yields:

$$\frac{J_{trans}}{J_{in}} = \frac{\frac{K}{\delta}(P_o - P_i) - J_{in} + \frac{K}{\delta}(P_o - P_i)}{J_{in}}$$

$$= \frac{2K}{J_{in}\delta}(P_o - P_i) - 1 \quad (7)$$

which can be rearranged to the relation between $P_o - P_i$ and J_{in} plotted in Figure 14.4b:

$$P_o - P_i = \frac{J_{in}\delta}{2K}\left(1 + \frac{J_{trans}}{J_{in}}\right) \quad (8)$$

REFERENCES

Axelsson, M., Farrell, A.P. and Nilsson, S. (1990) Effects of hypoxia and drugs on the cardiovascular dynamics of the atlantic hagfish *Myxine glutinosa*. *J. Exp. Biol.*, **151**, 297–316.

Barker-Jørgensen, B. and Revsbech, N.P. (1989) Oxygen uptake, bacterial distribution, and carbon-nitrogen-sulfur cycling in sediments from the Baltic Sea–North Sea transition. *Ophelia*, **31**(1), 29–49.

Beamish, F.W.H. (1978) Swimming capacity, in *Fish Physiology*, Vol. VII (eds W.S. Hoar and D.J. Randall), Academic Press, New York, pp. 101–85.

Cole, F.J. (1905) A monograph on the general morphology of the Myxinoid fishes based on a study of *Myxine*. Part I. The anatomy of the skeleton. *Trans. Roy. Soc. Edinburgh*, **41**, 749–88.

Cole, F.J. (1908) A monograph on the general morphology of the Myxinoid fishes based on a study of *Myxine*. Part II. The anatomy of the muscles. *Trans. Roy. Soc. Edinburgh*, **45**, 683–757.

Davison, W., Baldwin, J., Davie, P.S., Forster, M.E. and Satchell, G.H. (1990) Exhausting exercise in the hagfish, *Eptatretus cirrhatus*: the anaerobic potential and the appearance of lactic acid in the blood. *Comp. Biochem. Physiol.*, **95A**, 585–9.

Elger, M. (1987) The branchial circulation and the gill epithelia in the Atlantic hagfish, *Myxine glutinosa L. Anat. Embryol.*, **175**, 489–504.

Forster, M.E. (1990) Confirmation of the low metabolic rate of hagfish. *Comp. Biochem. Physiol.*, **96A**, 113–16.

Forster, M.E., Davison, W., Axelsson, M., Farrel, A.P. (1992) Cardiovascular responses to hypoxia in the hagfish, *Eptatretus cirrhatus*. *Respir. Physiol.*, **88**, 373–86.

Goodrich, E.S. (1958) *Studies on the Structure and Development of Vertebrates*, ch. IX, p. 494, Dover Publications, New York. (Orig. edn 1930.)

Gustafson, G. (1935). On the biology of *Myxine glutinosa*. *Arkiv Zool.*, **28**, 1–8.

Hans, M. and Tabenca, Z. (1938) Über die blutgefässe der haut von *Myxine glutinosa L*. *Bull. Int. Acad. Pol. Sci. Ser. BII*, **1–3**, 69–77.

Hanson, C.A. and Sidell, B.D. (1983) Atlantic hagfish cardiac muscle: metabolic basis of tolerance to anoxia. *Amer. J. Physiol.*, **244**, 356–62.

Holman, J.P. (1989) *Heat Transfer*, McGraw-Hill, Singapore.

Hughes, G.M. (1984) General anatomy of the gills, in *Fish Physiology*, Vol. X, part A (eds W.S. Hoar and D.J. Randall) Academic Press, New York, pp. 1–72.

Johansen, K. and Hol, R. (1960) A cineradiographic study of respiration in *Myxine glutinosa*. *J. Exp. Biol.*, **37**, 474–80.

Johansen, K. and Strahan, R. (1963) The respiratory system of *Myxine glutinosa* L., in *The Biology of Myxine* (eds A. Brodal and R. Fänge), Universitetsforlaget, Oslo, pp. 352–71.

Kirch, R. and Nonotte, G. (1977) Cutaneous respiration in three freshwater teleosts. *Respir. Physiol.*, **29**, 339–54.

Lametschwandtner, A., Weiger, T., Lametschwandtner, U., Georgieva-Hanson, V., Patzner, R.A. and Adam, H. (1989) The vascularization of the skin of the Atlantic hagfish, *Myxine glutinosa*, as revealed by scanning electron microscopy of vascular corrosion casts. *Scanning Microscopy*, **3**, 305–14.

Lomholt, J.P. and Johansen, K. (1979) Hypoxia acclimation in carp – how it affects O_2 uptake, ventilation and O_2 extraction from water. *Physiol. Zool.*, **52**, 38–49.

Mallatt, J. and Paulsen, C. (1986) Gill ultrastructure of the Pacific hagfish *Eptatretus stoutii*. *Amer. J. Anat.*, **177**, 243–69.

Malte, H. and Weber, R.E. (1985) A mathematical model for gas exchange in the fish gill based on non-linear blood gas equilibrium curves. *Respir. Physiol.*, **62**, 359–74.

Marinelli, W. and Strenger, A. (1956) *Vergleichende Anatomie und Morphologie der Wirbeltiere. II. Myxine glutinosa* (L.). Franz Deuticke, Wien.

Munz, F.W. and Morris, R.W. (1965) Metabolic rate of the hagfish, *Eptatretus stoutii*. *Comp. Biochem. Physiol.*, **16**, 1–6.

Nonnotte, G. and Kirch, R. (1978) Cutaneous respiration in seven sea-water teleosts. *Respir. Physiol.*, **35**, 111–18.

Perry, S.F., Fritsche, R. and Thomas, S. (1993) Storage and release of catecholamines from the chromaffin tissue of the Atlantic hagfish *Myxine glutinosa*. *J. Exp. Biol.*, **183**, 165–84.

Piiper, J. and Scheid, P. (1975) Gas transport efficacy of gills, lungs and skin: theory and experimental data. *Respir. Physiol.*, **23**, 209–21.

Pohla, H., Lametschwandtner, A. and Adam, H. (1977) Die Vaskularisation der Kiemen von *Myxine glutinosa L.* (Cyclostomata). *Zoologica Sripta*, **6**, 331–41.

Potter, I.C., Welsch, U., Wright, G.M., Honma, Y. and Chiba, A. (1995). Light and electron microscope studies of the dermal capillaries in three species of hagfishes and three species of lampreys. *J. Zool.*, **235**, 677–88.

Rosenhead, L. (ed.) *Laminar Boundary Layers*, Dover Publications, New York.

Smith, K.L. and Hessler, R.R. (1974) Respiration of benthopelagic fishes: *in situ* measurements at 1230 metres. *Science*, **184**, 72–3.

Steffensen, J.F., Johansen, K., Sindberg, C.D., Sørensen, J.H. and Møller, J.L. (1984) Ventilation and oxygen consumption in the hagfish, *Myxine glutinosa L. J. Exp. Mar. Biol. Ecol.*, **84**, 173–8.

Steffensen, J.F. and Lomholt, J.P. (1985) Cutaneous oxygen uptake and its relation to skin blood perfusion and ambient salinity in the plaice, *Pleuronectes platessa*. *Comp. Biochem. Physiol.*, **81A**, 373–5.

Strahan, R. (1958) The velum and respiratory current of *Myxine*. *Acta Zool.*, **39**, 227–40.

Strahan, R. (1963a). The behaviour of myxinoids. *Acta Zool., Stockh.*, **44**, 73–102.

Strahan, R. (1963b). The behaviour of *Myxine* and other myxinoids, in *The Biology of Myxine* (eds A. Brodal and R. Fänge), Universitetsforlaget, Oslo, pp. 22–32.

Wells, R.M.G., Forster, M.E., Davison, W., Taylor, H.H., Davie, P.S. and Satchell, G.H. (1986) Blood oxygen transport in the free-swimming hagfish, *Eptatretus cirrhatus*. *J. Exp. Biol.*, **123**, 43–53.

PART SEVEN

The Cardiovascular System

15
CARDIOVASCULAR FUNCTION IN HAGFISHES

Malcolm E. Forster

SUMMARY

Perhaps 30% or more of the high total blood volume of the hagfish is contained within a venous sinus system and may circulate at a lower rate than the central blood volume. Caudal and cardinal 'hearts' return subcutaneous sinus blood to the central circulation. The presence of a portal heart supplying the liver has also been associated with low venous blood pressures. The systemic heart of the hagfish generates the lowest arterial pressure of any vertebrate animal, but cardiac output is comparable with that of many types of fish.

The gills of hagfishes are efficient countercurrent extractors of oxygen. As water duct pressures generated by the velum are low, it may be that arterial blood pressures must also remain low. High blood pressures might also force too much blood into the sinus system.

The hypoxia tolerance of heart muscle, the prolonged action potential, and the relative sparsity of contractile material in the myocardium can be related to the low power output of the systemic heart. That it is functionally aneural, contrasts with the potentially sophisticated control of the rest of the circulation and may not represent a 'primitive' condition.

15.1 INTRODUCTION

In his 1963 chapter on the cardiovascular system of *Myxine glutinosa*, Kjell Johansen presented a detailed account of the morphology of the blood system. The chapter also gave the first values for blood pressures in living hagfish, from animals which were partially anaesthetized and/or restrained on a board. Since that time blood pressures and flows have been measured in unrestrained and exercising hagfish and these studies have reported higher dorsal aortic pressures (Wells *et al.*, 1986; Forster *et al.*, 1988; Axelsson *et al.*, 1990). Nevertheless, it remains true that hagfishes have the lowest arterial blood pressures recorded within the Class Vertebrata (Johansen, 1963; Satchell, 1986; Forster *et al.*, 1991), a phenomenon which has implications for a number of physiological systems, including kidney function (Riegel, this volume). In this chapter I will concentrate on the physiology of the hagfish cardiovascular system and possible control mechanisms. Readers are also referred to the final chapter of Satchell's (1991) book for a comparison of the hagfish cardiovascular system with those of true 'fishes'.

15.2 THE ANATOMY OF THE BLOOD SYSTEM

The blood system of hagfishes is partitioned (Figure 15.1), and the venous sinuses can contain a significant portion of the total blood volume. In resting animals more than 30% of

The Biology of Hagfishes. Edited by Jørgen Mørup Jørgensen, Jens Peter Lomholt, Roy E. Weber and Hans Malte. Published in 1998 by Chapman & Hall, London. ISBN 0 412 78530 7.

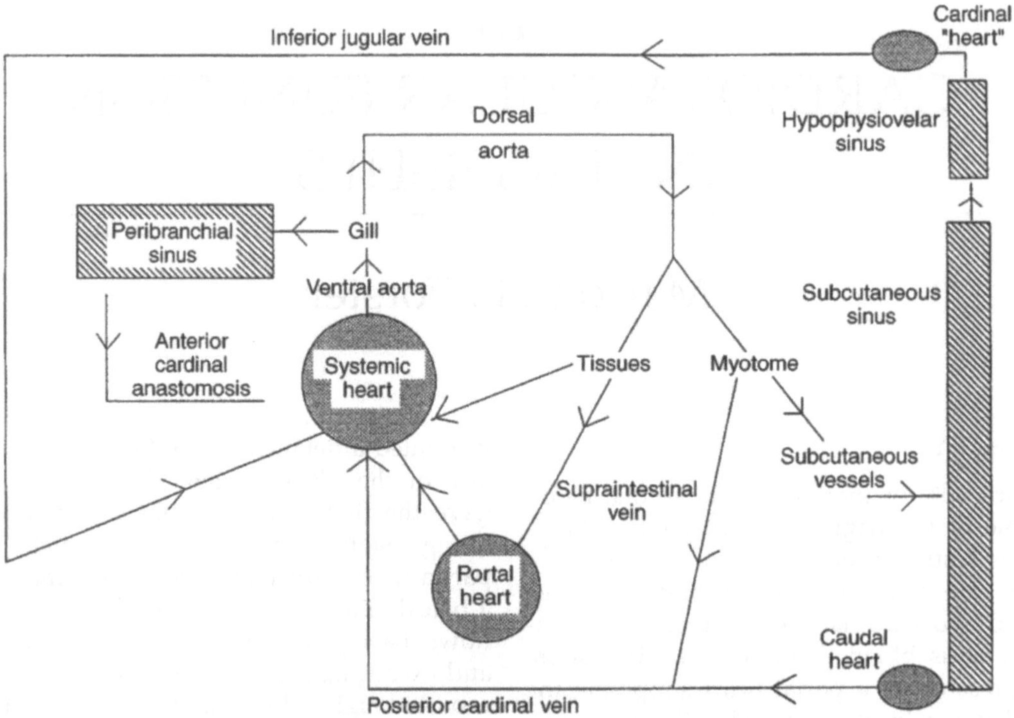

Figure 15.1 An oversimplified representation of the sinus systems of hagfishes and their relationship to the circulatory pumps. (Based on Cole, 1925, and redrawn from Forster, 1997.)

the total blood volume has been estimated to be contained within the sinus system (Forster *et al.*, 1989). While this value could be exaggerated by the animals' immobility when confined in boxes, it appears to be characteristic of hagfishes that they have long periods of inactivity (Strahan, 1963). Blood enters the peribranchial sinuses through papillae on the branchial arteries (Cole, 1912; Pohla *et al.*, 1977; Elger, 1987) and returns to the central circulation via the anterior cardinal anastomosis. The caudal and cardinal hearts return blood from the subcutaneous sinus (SCS) to the posterior cardinal and inferior jugular veins, respectively (Satchell, 1984). Partitioning and the slow turnover time of the SCS blood (Forster *et al.*, 1989) complicate measurement of blood volume in hagfishes, since it takes an extremely long time for equilibration of the marker with the total blood volume. As at least certain capillaries of hagfishes are reported to be more leaky than those of other vertebrates (Casely-Smith and Casely-Smith, 1975) and low MW proteins can be lost into the extracellular space (Forster *et al.*, 1989) we must treat the high estimates of plasma volumes with caution. The variable haematocrit of blood at different parts of the circulation adds to the difficulties of estimating total blood volume. Until a more sophisticated methodology is employed we must accept the high values for blood volume estimated by McCarthy and Conte (1966) and Forster *et al.* (1989), 187 and 177 ml kg^{-1} respectively.

The high blood volume of hagfishes implies a high turnover time for the total circulation. Extremely low venous oxygen partial pressures (P_{O_2}) have been measured in swimming hagfishes, which contrasts with the high P_{O_2} of postbranchial blood (Wells *et al.*, 1986). Although the oxygen-carrying capacity of hagfish blood is low, the highly efficient extraction of oxygen from the blood by metabolically active tissues suggests a relatively long residence time at the tissues. However, hagfishes have a low metabolic rate at rest (Munz and Morris, 1965; Kench, 1989; Forster, 1990), and a high rate of oxygen delivery may not be necessary when the animal is inactive. Also, as a consequence of partitioning, a smaller pool of blood may be circulating at a higher rate than the blood volume as a whole, and this might supply the oxygen requirements of some red muscle, for instance.

15.3 CARDIOVASCULAR FUNCTION

15.3.1 Pressures and flows in the circulation

The hagfish circulation has been characterized as a low pressure, moderate flow system (Forster *et al.*, 1991). As Table 15.1 indicates, arterial pressures of hagfishes are the lowest recorded in 'fishes' and in the Vertebrata as a whole (Johansen, 1972). Operating with the lowest afterload (systolic ventral aortic pressure) the ventricular myocytes of hagfishes need to exert less force to propel the blood around the body than do myocytes in other vertebrates and this will require less energy. The ventricle of the hagfish is of the spongy, trabeculated type (Leak, 1969). Compact myocardium occurs in the hearts of some teleost and elasmobranch fishes, where it has been associated with the ability to generate high arterial pressures (Farrell, 1991).

15.3.2 The composition of hagfish blood vessels

Hagfish arteries lack elastin (Sage and Gray, 1979, 1980) but the *tunica media* of the ventral aorta contains large numbers of densely packed microfibrils which are thought to be extensible (Wright, 1984), and presumably function to make the vessel compliant (Davison *et al.*, 1995). Both the structure of

Table 15.1 Ventral (VAP) and dorsal (DAP) aortic blood pressures, cardiac outputs and branchial vascular resistances in various 'fishes'

Species	Condition	VAP (kPa)	DAP (kPa)	Cardiac output (ml min^{-1} kg^{-1})	Branchial resistance (Pa min kg ml^{-1})	Reference
Myxine glutinosa (hagfish)	Resting	1.04	0.77	8.7	31	Axelsson *et al.* (1990)
Eptatretus cirrhatus (hagfish)	Resting	1.44	1.07			Forster *et al.* (1988)
	Swimming	1.55	1.00			
Eptatretus cirrhatus (hagfish)	Normoxia	1.6	1.3	15.8	20.3	Forster *et al.* (1992)
	Hypoxia	2.2	1.4	22.2	32	
Entosphenus tridentata (lamprey)		2.5–4.3				Johansen *et al.* (1973)
Scyliorhinus canicula (dogfish)	Resting	5.1	3.9	32.1	37	Short *et al.* (1979)
	Hypoxia	4.3	3.2	35.7	29	
Gadus morhua (Atlantic cod)	Resting	4.9	3.2	17.3	99	Axelsson and Nilsson (1986)
	Swimming	6.2	4.0	25.4	97	
	Hypoxia	6.3	5.0	19.2	67	Fritsche and Nilsson (1989)

their walls and the stress–strain inflation curves of lamprey and hagfish aortas are similar to those of other vertebrates, although with lower elastic moduli, reflecting the lower ventral aortic pressures (Davison *et al.*, 1995). The distensibility of the ventral aorta reduces the pulse pressure experienced by the gills and generates continuing blood flow to the gills during ventricular diastole. The compliance of the ventral aorta of *Eptatretus cirrhatus* is nicely tuned to peak ventricular pressures (Forster, 1989). Interestingly, 'elastic' tissue is absent from the ventral aorta of *M. glutinosa* (Welsch and Potter, 1994) and Axelsson *et al.* (1990) observed zero diastolic flow in the ventral aorta of this species.

15.3.3 Blood propulsors

(a) The systemic heart and portal heart

These two hearts are considered together as both are composed of cardiac muscle and they have many features in common. The portal heart is a structure unique to hagfish (Fänge *et al.*, 1963). It pumps blood derived from the gut through the liver and forward to the systemic heart. Its presence has been linked to the need to maintain flow through the liver where otherwise the venous blood has little or no kinetic energy (Satchell, 1991) but it is unclear why hagfishes should need such a pump when lampreys, elasmobranchs and teleosts have similarly low venous pressures (Satchell, 1992). We do not know if the vascular resistance of the hagfish liver is unusually high. Davison (1995) reports finding a large specimen of *E. cirrhatus* that lacked a portal heart. The rates of beating of the systemic and portal hearts are not synchronous (Johansen, 1960; Davie *et al.*, 1987).

The long conduction times of the action potential of the hagfish systemic heart are remarkable (Arlock, 1975; Davie *et al.*, 1987) (Figure 15.2). Thus the 'Pt' wave associated with atrial repolarization occurs before the QRS wave of the ventricle and is not obscured by it, as it is in other vertebrates (Satchell, 1991). The long conduction time of the action potential of the heart of *M. glutinosa* is associated with the absence of fast-conducting tracts (Satchell, 1986) and the coupling of excitation and contraction may be slowed by the absence of transverse tubules in *Myxine* myocytes (Helle and Lönning, 1973). In mammals the ventricle is filled to 70% of its end-diastolic volume by passive flow from the central veins (Guyton, 1986). The long *P–Q* interval of fish hearts has been associated with low venous pressures and the requirement that atrial contraction alone fills the ventricle (Farrell and Jones, 1992). Most teleost hearts beat with a frequency of 20–60 beats per min (Farrell, 1984), with the exception of the tunas, which are a special case (Farrell *et al.*, 1992). The highest heart rate recorded from the systemic heart of hagfishes *in vivo* is the mean of 27 b.p.m. for hypoxic *E. cirrhatus* at 17°C. (Forster *et al.*, 1992).

In hagfishes the pressure generated by the ventricle is lower than in other vertebrates and in a highly compartmentalized circulation with multiple resistances in series the force generated by the ventricle must be almost fully dissipated before the blood returns to the sinus venosus. How then, is the heart of a hagfish filled? In elasmobranch and most teleost fishes the presence of a rigid pericardium allows them to use aspirational force to fill the sinus venosus and atrium (Farrell, 1991) and lampreys have a rigid pericardium (Hardisty, 1979). Recent measurements confirm that venous pressures are indeed low, but positive in resting hagfishes (c. 40 Pa, Johnsson *et al.*, 1996). Satchell (1986) recorded diastolic pressures in the atrium of *M. glutinosa* of 20 Pa. Therefore, filling should continue as long as central venous pressures remain slightly higher. Perhaps this is the significance of the low heart rates of hagfishes; allowing prolonged filling times to offset the small pressure gradient? Although hagfishes lack a rigid pericardium of the sort found in lampreys and many fishes the

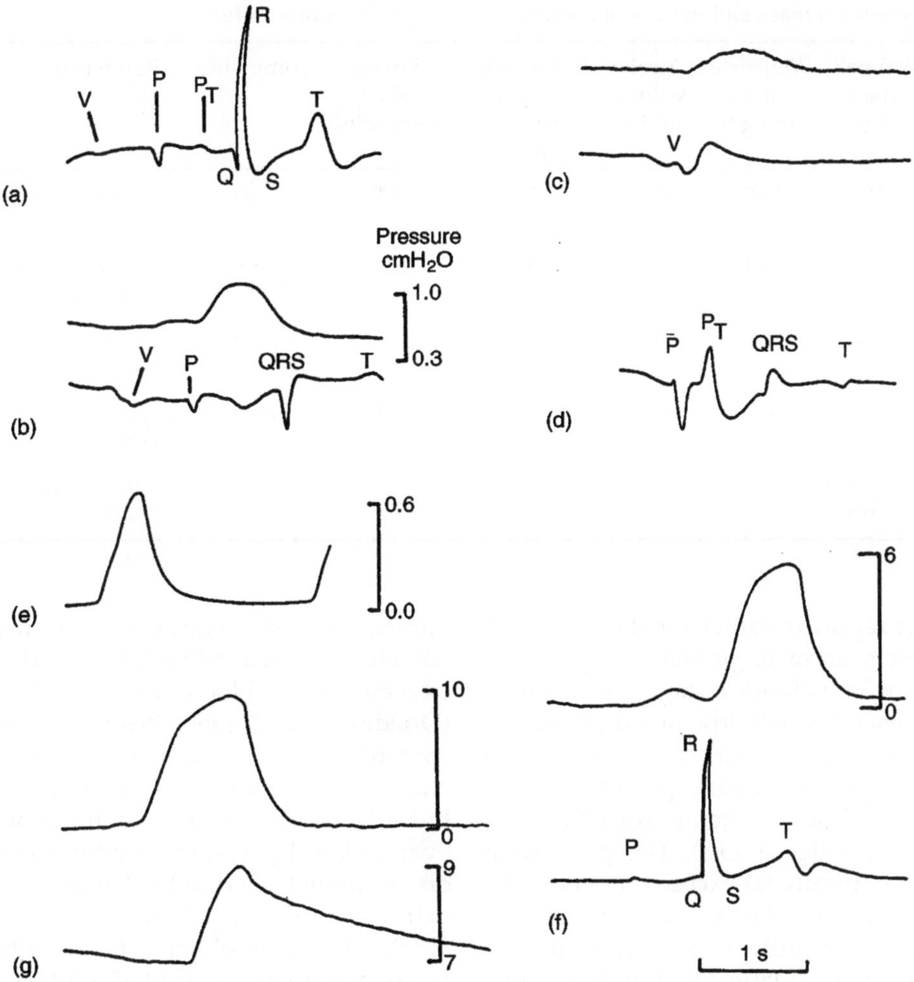

Figure 15.2 Electrocardiograms of *Eptatretus cirrhatus*, and above them in some cases, intravascular pressures. (a) The whole heart; (b) the sinus venosus; (c) the sinus, plus mechanogram (upper trace); (d) atrium; (e) atrial pressure; (f) ventricle; (g) pressures in ventricle (above) and ventral aorta. (From Davie *et al.*, 1987; reproduced with permission of the University of Chicago Press.)

systemic and portal hearts of hagfishes are surrounded by connective tissue. A recent study of portal heart function *in situ* demonstrated a fall in its output once surrounding connective was cut (Johnsson *et al.*, 1996).

Cardiac output in hagfishes is comparable with that of many elasmobranch or teleost fishes with similar relatively inactive lifestyles (Table 15.2). Body movements associated with swimming may compromise cardiac function *in vivo* and prevent the living animals from achieving the maximum values for cardiac output determined from studies on isolated, perfused hearts. *In situ* preparations are now favoured over those in which the heart is removed from the pericardium (Farrell and Jones, 1992). The *in situ* preparation recently used by Johnsson and Axelsson (1996) gives

Table 15.2 Ventricle mass and indices of maximal cardiac performance in fish

Species	Animal's mass (g)	Ventricle mass ($g\ kg^{-1}$)	Stroke volume ($ml\ kg^{-1}$)	Cardiac output ($ml\ kg^{-1}\ min^{-1}$)	Power ($mW\ g^{-1}$ ventricle)	Comments	Reference
Eptatretus cirrhatus	973	1.007	0.68	21.6	0.37	*in vitro*	Forster (1989)
Myxine glutinosa	48	0.917	1.68	31.7	0.59	*in situ*	Johnsson and Axelsson (1996)
Squalus acanthias	1920	0.67	0.87	34.2	1.76	*in vitro*	Davie and Farrell (1991)
Hemitripterus americanus	670–1400		0.64	30.9	3.13	*in vivo*	Axelsson *et al.* (1989)
Ophiodon elongatus	3500–6500		0.70	15.2	3.13	*in vivo*	Farrell (1981, 1982)

the highest reported values for the output of the systemic heart of *M. glutinosa*.

While stroke volume is high, the hagfish ventricle generates only low blood pressures. As a consequence the heart of *E. cirrhatus* can fuel 70% of its maximum power through anaerobic metabolism when perfused *in vitro* (Forster, 1991) (Figure 15.3). The proportion fuelled anaerobically far exceeds that possible in other vertebrates, but we must remember that the export of protons and lactate may be maximized under *in vitro* conditions and thus lead to an overestimate of anaerobic capabilities. Davison *et al.* (1990) and Satchell (1991) have suggested that the high blood volume of hagfishes might provide a large 'sink' for anaerobic end-products, and thus facilitate the removal of lactic acid from an anaerobically functioning heart. However, the low buffering capacity of hagfish blood (Wells *et al.*, 1986) will offset this potential means of maintaining a favourable proton gradient from the myocytes into the blood.

Extremely low venous oxygen tensions have been measured in *E. cirrhatus* (Wells *et al.*, 1986) and the heart lacks an arterial (coronary) blood supply and therefore must operate in a hypoxic environment. High intracellular concentrations of myoglobin have been associated with the need to extract oxygen from blood at low P_{O_2} values (Driedzic and Gesser, 1994). Hagfish hearts contain myoglobin, showing a marked colour change on treatment with hydroxylamine, but afterwards continue to function *in vitro* even at low P_{O_2} values (Forster, unpublished observations). Certainly, hagfish hearts are extremely tolerant of hypoxia (Hansen and Sidell, 1983; Axelsson *et al.*, 1990). The isolated systemic heart of *M. glutinosa* continued to beat with a normoxic rate and force when made anoxic and when treated with sodium cyanide and sodium azide (Hansen and Sidell, 1983; Sidell, 1983). It failed when treated with iodoacetic acid, a compound which blocks glycolysis. This demonstrates the importance of carbohydrate metabolism to its function. Glycolysis can continue to generate ATP in the absence of oxygen (Driedzic and Gesser, 1994). There are numerous glycogen particles in the hagfish ventricle (Helle and Lönning, 1973) – which is typical of animals that are periodically exposed to hypoxia, such as turtles, muskrats and seals (Hochachka and Somero, 1984; McKean, 1984). This also implies that hagfish

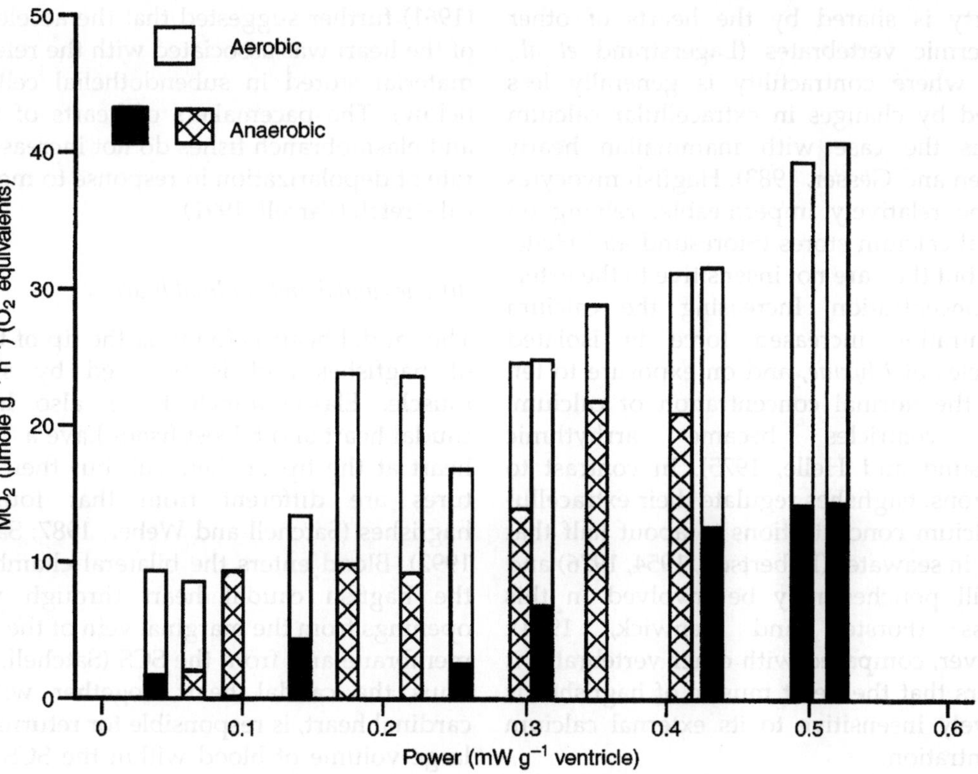

Figure 15.3 The sum of aerobic and anaerobic components of cardiac energy supply, expressed as oxygen equivalents, for hearts of *Eptatretus cirrhatus* perfused *in vitro*. The anaerobic component is shown for hearts perfused at low (P_{IO_2} 1.6–2.7 kPa, hatched) and high (P_{IO_2} 3.9–5.9 kPa, solid) P_{O_2}s values. For the purposes of calculating the anaerobic contribution to ATP production, the export of 6 moles of lactate was considered to be equivalent to the uptake of 1 mole of oxygen (see Forster, 1991). It can be seen that under *in vitro* conditions, a power output of *c.* 0.3 mW g^{-1} might be fuelled anaerobically. All hearts released lactic acid. (Data modified from Forster, 1991.)

hearts will experience hypoxia *in vivo* – see below.

As determined from its oxygen usage and lactate production rates, the efficiency of the hagfish heart is apparently similar to the hearts of teleost fishes (Forster, 1991). Though the anaerobic capability of hagfish cardiac muscle appears remarkable, we should remember that it is made possible by the low power output of the heart. The relatively low heart rates and the need to develop the lowest arterial pressures of any vertebrate animal reduce its energy requirements. The high power requirements of the hearts of small endothermic animals seem to require the use of lipid as a fuel, which precludes anaerobic metabolism (Driedzic *et al.*, 1987).

Upon reperfusion following periods of zero flow in their coronary arteries the hearts of mammals are damaged by the transient rise in cytosolic calcium (Tani and Neely, 1990; Daniels and Duncan, 1993). Poupa *et al.* (1984) first reported that the hagfish ventricle did not show this 'calcium paradox'. The property is shared by the hearts of other ectothermic vertebrates (Lagerstrand *et al.*, 1983), where contractility is generally less affected by changes in extracellular calcium than is the case with mammalian hearts (Nielsen and Gesser, 1983). Hagfish myocytes may be relatively impermeable, relying on internal calcium stores (Storesund and Helle, 1975), but they are not insensitive to the external concentration. Increasing the calcium concentration increased force in isolated ventricles of *Myxine*, and on exposure to ten times the normal concentration of calcium, some ventricles became arrhythmic (Storesund and Helle, 1975). In contrast to other ions, hagfishes regulate their extracellular calcium concentrations to about half that found in seawater (Robertson, 1954, 1976) and the gill pouches may be involved in this process (Forster and Fenwick, 1994). However, compared with other vertebrates it appears that the heart muscle of hagfishes is relatively insensitive to its external calcium concentration.

Vertebrate hearts show a Frank–Starling effect, where increased end-diastolic volumes result in a proportionate increase in the volume ejected on systole (Sarnoff and Mitchell, 1962; Farrell, 1991) and the hagfish heart conforms to this rule (Chapman *et al.*, 1963; Forster, 1989; Johnsson and Axelsson, 1996). Jensen (1961) first described the pressure-sensitivity of the pacemaker in the systemic heart of *E. stoutii* and the heart of *M. glutinosa* was shown subsequently to have a similar mechanism (Bloom *et al.*, 1963). These early observations have been discounted because the hearts were not performing at physiological work loads (Forster *et al.*, 1991). However, Axelsson *et al.* (1990) provided confirmatory evidence for the pressure-sensitive pacemaker in *Myxine in vivo*. Johnsson *et al.* (1996) demonstrated a similar effect in the portal heart of *E. cirrhatus* and Johnsson and Axelsson (1996) have recently quantified the effect in the systemic and portal hearts of *M. glutinosa*, using *in situ* preparations. Jensen (1961) further suggested that the acceleration of the heart was associated with the release of material stored in subendothelial cells (see below). The pacemakers of hearts of teleost and elasmobranch fishes do not increase their rate of depolarization in response to mechanical stretch (Farrell, 1991).

(b) The caudal and cardinal hearts

The caudal heart is found at the tip of the tail of hagfishes and is powered by skeletal muscle. Elasmobranch fishes also have a caudal heart and teleost fishes have a 'lymph' heart at the tip of their tail, but their structures are different from that found in hagfishes (Satchell and Weber, 1987; Satchell, 1992). Blood enters the bilateral chambers of the hagfish caudal heart through valved openings from the marginal vein of the tail fin membrane and from the SCS (Satchell, 1984). Thus the caudal heart, together with the cardinal heart, is responsible for returning the large volume of blood within the SCS to the central circulation. In contrast to the systemic and portal hearts, the caudal heart is neurogenic (Jensen, 1965). The propulsor muscles of the two sides of the heart contract alternately and are active intermittently (Greene, 1902). Typically the caudal heart starts to beat shortly after the animal stops swimming, and at this time beat frequency is maximal (Satchell, 1984). Raised blood pressures in swimming hagfishes may increase blood movement through the arterial papillae into the PBS and SCS, with the consequent need for the return of that blood to the central circulation. However, we have no knowledge of receptors monitoring central blood volume in hagfishes and their possible relationship to the spinal nerves controlling caudal heart activity. When the hagfish swims, activity of the caudal heart muscles may be unnecessary, as Satchell (1984) has observed movement of blood from the SCS to the chambers of the caudal heart in the absence of caudal heart muscle activity. Pressures generated by the swimming muscles may be sufficient to force blood through the caudal heart.

According to Kampmeier (1969) the cardinal heart of hagfishes should be designated as a 'propulsor' rather than a 'heart', since unlike the animal's caudal heart it is powered by extrinsic muscles. Satchell (1991) who supported this distinction also notes that the term 'cardinal heart' is firmly established in the literature. The pumping action of the cardinal heart depends upon the movement of the velum (Cole, 1926; Johansen, 1963; Satchell, 1991). Study of its activity is made difficult by its proximity to the brain. Administration of anaesthetics which are necessary for analgaesia also abolish contractions of the velar muscles.

15.3.4 The gill pouch

The hagfish gill pouch with its highly branched, radially symmetrical blood vessels and countercurrent seawater flow, functions as an efficient exchanger (Mallatt and Paulsen, 1986; Elger, 1987; Bartels, this volume; Malte and Lomholt, this volume). Despite blood to water distance in the finest hagfish gill lamellae being a relatively high 5 µm (Mallat and Paulsen, 1986), the very high P_{O_2} found in the dorsal aortic blood attest to the gas exchange efficiency of the structures (Wells *et al.*, 1986). Whereas the anatomy of blood vessels supplying the gill pouches and their functional relationship with the systemic heart are very similar in hagfishes, lampreys, and in elasmobranch and teleost fishes, the methods of gill ventilation in lampreys and hagfishes are very different from that in the other two groups. Hagfish gill pouches are irrigated through the action of a velum, a structure powered by a limited amount of muscle, compared to the buccal and opercular pumps of teleosts, for example. Tidal ventilation of the gill pouches of lampreys is also effected by a mechanism which differs from that in hagfishes. Extrinsic muscles compress the branchial basket and force water out of the pouches (Randall, 1972). Perhaps as a consequence of the velum's limited muscularity, velar contractions in hagfishes have been recorded at relatively high frequencies (50–90 b.p.m.: Johansen and Strahan, 1963; Steffensen *et al.*, 1984; Forster *et al.*, 1988), though recent measurements in resting *M. glutinosa* record lower rates (Malte and Lomholt, this volume). Ventilation frequencies of c. 16–90 b.p.m. are reported in teleost and elasmobranch fishes (Shelton, 1970). Steffensen *et al.* (1984) measured inflow to the nostril of *M. glutinosa* and estimated ventilatory flow to be only 15 to 25% of that reported for teleost fishes. Thus ventilatory stroke volume appears to be low in hagfishes compared to teleosts, though this might be expected as the latter have much higher metabolic rates.

Unpublished measurements by M.E. Forster and W. Davison of pressures in afferent water ducts and arteries of isolated, perfused gill pouches of *E. cirrhatus* reveal how pressures in the blood vessels respond to changes in water duct pressure and *vice versa* (Figure 15.4). The smooth muscle of the water ducts contracts in response to acetylycholine and the ducts show spontaneous muscular activity. If hydrostatic pressures are necessarily low in water ducts, as a consequence of their irrigation by a velar pump, then this might explain why blood pressures must also be low in hagfishes, given a relatively thin barrier between the blood and water (Forster, 1997). A high pulse pressure in the branchial arteries might otherwise prevent seawater flow through the gill pouch. Equally, the dramatic increase in branchial vascular resistance during hypoxia in *E. cirrhatus*, described below, could be due to a sustained contraction of either the smooth muscle of the water ducts, or the striated muscle of the gill pouch itself (Johansen, 1960). Rhythmic contractions were observed in isolated, perfused gill pouches.

Hol and Johansen (1960) described muscular activity in the walls of gill pouches, and its effect on the movement of contrast medium through the branchial blood vessels of *M.*

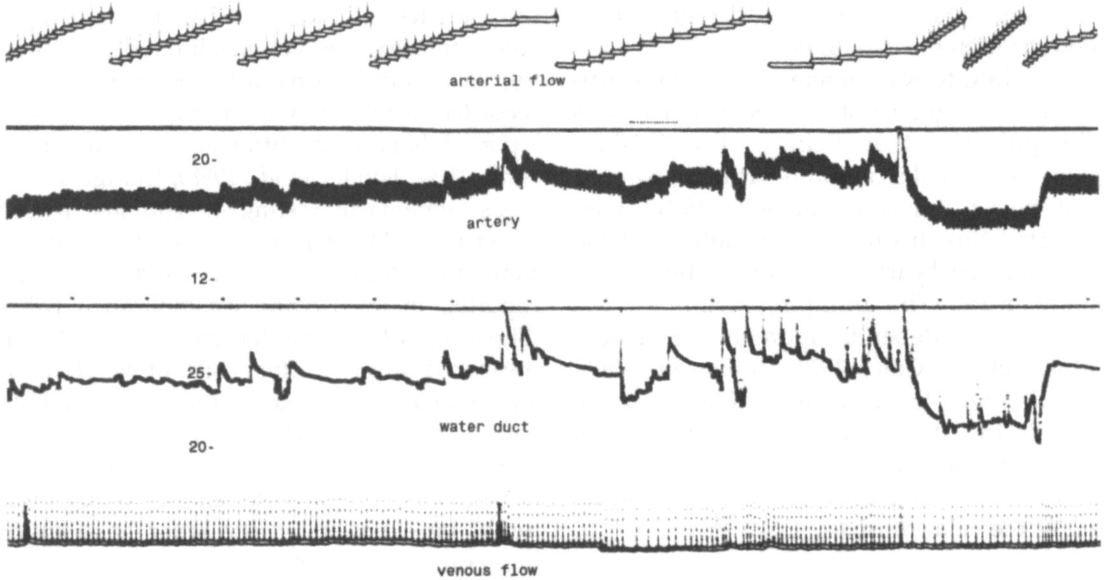

Figure 15.4 Spontaneous pressure changes (in cm H_2O) measured simultaneously in the afferent branchial artery and afferent water duct of a hypoxic, perfused gill pouch of *Eptatretus cirrhatus*. The branchial arteries and water ducts were cannulated close to their junction with the gill pouch which was held in a sealed chamber. The venous outflow dripped from a tube extending from the chamber. Shown at the top and bottom are the drops of perfusate passing respectively from the arterial and venous outflows of the gill pouch. The two pressures change synchronously and the distribution of the outflow from the afferent branchial artery is altered as pressures change. (M.E. Forster and W. Davison, unpublished observations.)

glutinosa, but the periodic cessation of branchial blood flow they observed could have been due to the abnormally low blood pressures in their experimental animals. We should note that the delicate balance that is suggested between blood vascular and water duct pressures should not be affected by myotomal contractions when the animal swims. The gill pouches are internalized and pressures should rise and fall equally in both blood and water channels. Indeed, in swimming, hagfish Pa_{O_2} is maintained at high levels (Wells *et al.*, 1986).

15.3.5 Variations in haematocrit

Thanks to 'plasma skimming' at the papillae where blood enters the sinus system, the haematocrit of SCS blood is less than that of the central circulation in resting animals (Johansen *et al.*, 1962; Forster *et al.*, 1989). When a hagfish struggles the haematocrit of SCS blood approaches that in the central circulation, suggesting a significant redistribution of the blood volume from the central circulation to the SCS (Johansen, 1963) and potentially imposing a greater energy demand on the caudal and cardinal hearts. The haematocrit of blood in the central circulation of *E. cirrhatus* is suboptimal for oxygen transport (Crowell and Smith, 1967; Wells and Forster, 1989). Although red cells are deformable, their passage through small blood vessels will involve frictional energy losses. Operating with a low proportion of circulating red blood cells will reduce blood viscosity and hence the energy requirement of the systemic and portal hearts. Wells and Forster (1989) also suggest that the plasma skimming that occurs at points where blood enters the sinus system will further reduce the work required of the caudal and cardinal hearts. As blood velocity must approach zero in the venous sinuses and viscosity increases at low shear rates, the energy saving may be significant.

15.4 CONTROL OF THE CARDIOVASCULAR SYSTEM

The role of the autonomic nervous system, the nature of putative neuromuscular transmitters and factors influencing the release of catecholamines are described elsewhere in this volume (Nilsson and Holmgren; Bernier and Perry). We have little knowledge of the transducers involved in cardiorespiratory control systems and their relationship to the central nervous system in hagfishes. From studies in which drugs have been administered to living animals and tested on organ systems *in vitro*, we can describe probable effector mechanisms of cardiovascular control, but as yet can only speculate on how these mechanisms operate *in vivo*.

15.4.1 Control of the heart

In both the systemic and portal hearts of hagfishes there are catecholamine stores, within specialized granular cells bordering the luminal surface (Bloom *et al.*, 1961; Euler and Fänge, 1961; Fänge *et al.*, 1963; Perry *et al.*, 1993). The catecholamines seem to be associated with the granular matrix of cisternae which are situated immediately underneath the sarcolemma (Leak, 1969; Helle and Lönning, 1973). Enzymes for catecholamine synthesis are present within the heart (Jönsson, 1983; Reid *et al.*, 1995). Perhaps because of these endogenous stores, the hagfish heart is remarkably insensitive to the application of endogenous catecholamines. Catecholamines acclerate cardiac frequency of both the systemic and portal hearts, but the increases are modest (Forster *et al.*, 1992; Johnsson *et al.*, 1996; Johnsson and Axelsson, 1996). The sinus venosus acts as the pacemaker of the systemic heart (Davie *et al.*, 1987). In the absence of the sinus, the systemic heart of *E. cirrhatus* perfused *in vitro* was unresponsive to adrenaline, and heart rate was remarkably constant over a period of 2 to 3 hours' perfusion (Forster, 1989). With its supra-intestinal vein pacemaker present, an *in situ* portal heart preparation from *E. cirrhatus* showed a modest increase in frequency when exposed to adrenaline (Johnsson *et al.*, 1996). The β-adrenergic blocking drugs sotalol and propranolol produced a pronounced fall in heart rate when injected into *M. glutinosa* (Axelsson *et al.*, 1990) and *E. cirrhatus* (Forster *et al.*, 1992). The α-adrenergic blocking drug, thymoxamine, also slowed or stopped the isolated heart of *E. cirrhatus* and diminished the force of contraction in atrial strips (Table 15.3). These observations reinforce the view that endogenous catecholamines are important in the control of cardiac function in hagfishes. Axelsson *et al.* (1990) have suggested that the release of endogenous catecholamines may 'kick-start' the heart at times of critically low venous return. This could be important in an animal which is unable to generate subambient pressures in its pericardium.

Although nerves supplying the systemic and portal hearts of hagfishes have been described (Hirsch *et al.*, 1964; Yamauchi, 1980), hagfish hearts have been described as being 'functionally aneural' (Augustinsson *et al.*, 1956; Jensen, 1965). Recently it has been discovered that acetylcholine slows the systemic and portal heart of *M. glutinosa*, but only at concentrations exceeding 10^{-6} M (Johnsson and Axelsson, 1996). In earlier work hagfish hearts were unresponsive to acetylcholine even at considerably higher concentrations (Jensen, 1958; Östlund, 1954).

Table 15.3 Effects of adrenergic blocking drugs on atrial strips from the hagfish *Eptatretus cirrhatus*. The α-adrenergic blocking drug thymoxamine and β-blocking drug metoprolol added to a concentration of 1 × 10^{-4} M. * and ** – significantly different from control value at 500 s post-treatment, $P < 0.05$, $P < 0.01$

	Thymoxamine		Metoprolol	
	Rate (b.p.m.)	Force (arbitrary)	Rate (b.p.m.)	Force (arbitrary)
Control	23.3 + 2.5	17.1 + 0.8	20.2 + 0.9	14.1 + 0.6
Treated	19.0 + 1.1*	13.3 + 1.1*	19.1 + 1.4	11.4 + 0.8**
n	6	6	7	7

15.4.2 Control of the vasculature

The blood vessels of the isolated hagfish gill pouch showed vasoactivity when exposed to a number of drugs. As with teleost fishes, there are both arterio-arterial and arterio-venous efferent pathways from the hagfish gill pouch, the latter emptying into the peri-branchial sinus (Figure 15.1). The β-adrenergic agonist isoprenaline lowered gill resistance in a dose-dependent manner. The naturally occurring catecholamine adrenaline had a pressor effect at low concentrations, which reversed to a depressor effect at high concentrations (Sundin *et al.*, 1994). These results indicate the presence of both α- and β-adrenergic receptors in the hagfish vasculature, the former activating the pressor effect. Despite their different effects on afferent perfusion pressure, both catecholamines induced a dose-dependent change in the pattern of outflow, with the perfusate being directed to the arterio-arterial route (Figure 15.5). This suggests that catecholamines are acting at at least two sites in the hagfish vasculature, with α-adrenergic receptors being situated upstream from the separation of the arterio-arterial and arteriovenous routes.

In vivo, adrenaline injection did not change branchial resistance in *M. glutinosa* (Axelsson *et al.*, 1990), possibly because the α- and β-adrenergic effects cancelled each other out. Adrenaline did increase cardiac output, through rises in both beat frequency and stroke volume. It also reduced systemic vascular resistance. With an isolated, perfused gill preparation from *M. glutinosa*, Axelsson *et al.* (1990) observed a dose-dependent vasoconstriction with adrenaline and vasodilation with noradrenaline. There appears to be a difference between *E. cirrhatus* and *M. glutinosa*, for afferent and efferent branchial arteries of the former constrict in response to both noradrenaline and adrenaline at what might be expected to be physiological concentrations (Figure 15.6; Forster, unpublished results). It is not clear if the very different responses to noradrenaline of the two species can be related to the different preparations used. That used by Axelsson *et al.* (1990) was the more complex, an *in situ* perfusion of the entire branchial apparatus. Despite this anomaly, the dramatic effect of the β-adrenergic blocking drug propranolol, which induced a large increase in branchial resistance when injected into *E. cirrhatus in vivo*, further supports the concept of a sustained adrenergic tonus in the vasculature of the hagfish gill (Forster *et al.*, 1992). As noted above, we do not know whether this tonus is supplied by catecholamines released upstream by the hearts.

Sundin *et al.* (1994) tested two other potentially vasoactive agents on the gill pouch preparation. Both 5-hydroxytryptamine and cholecystokinin increased flow to the arterio-arterial route, though the former lowered vascular resistance and the latter raised it. Acetycholine vasoconstricted the perfused gills of *M. glutinosa* in a dose-dependent manner (Axelsson *et al.*, 1990). Thus a number

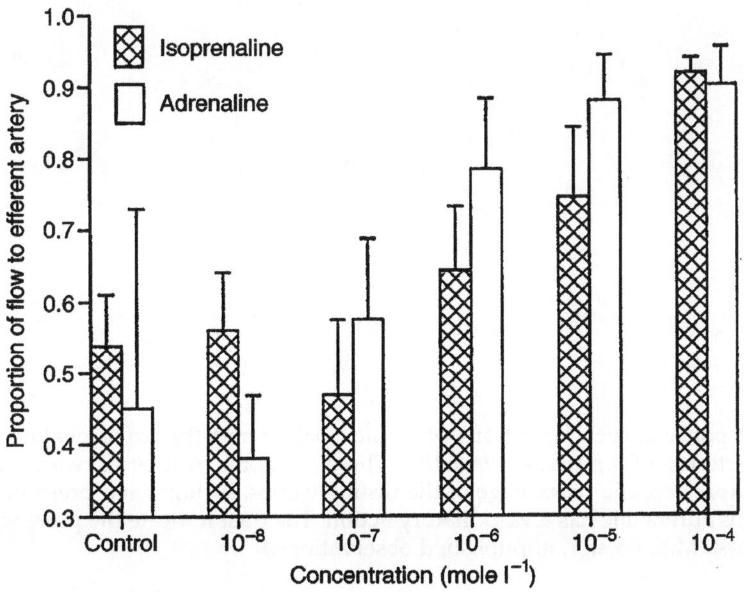

Figure 15.5 Changes in the proportion of the perfusate passing to the efferent artery rather than leaving via the 'venous' route in isolated hagfish gills exposed to different concentrations of adrenaline and the β-adrenergic agonist isoprenaline. At concentrations of 1×10^{-7} M and below, adrenaline vasoconstricted the gill and isoprenaline caused vasodilation. (From Sundin *et al.*, 1994, with permission.)

of agents may affect the vascular resistance and pattern of blood flow through the hagfish gill pouch.

Preconstricted rings of ventral aorta of *M. glutinosa* dilated in response to natriuretic peptide (Evans, 1991; Evans *et al.*, 1993) and there are NP binding sites in this tissue (Kloas *et al.*, 1988). The presence of natriuretic peptides (NPs) in the heart, brain and plasma of *M. glutinosa* has been demonstrated by immunohistochemical methods (Evans *et al.*, 1989; Donald *et al.*, 1992). There are also both atrial natriuretic peptide and C-type natriuretic peptide receptors on the epithelia of gill lamellae in *M. glutinosa,* where the absence of smooth muscle suggests a possible role of the peptides in ionic regulation rather than regulating blood flow (Toop *et al.*, 1995). If blood volume is regulated in hagfishes and NPs have a role, the regulatory system must be able to cope with the partitioning of blood between the sinus systems and the central circulation (Forster, 1996). Satchell (1986) has recorded his own observations and those of Chapman *et al.* (1963) and Alt *et al.* (1981) that small injections of saline caused a rise in blood pressure in living animals that could last for several hours. These observations are not consistent with a rapid, fine control of blood volume in hagfish, but indicate that if blood pressures are to be maintained within narrow limits then central blood volume must also be controlled. There is a need for further studies on the effects of NPs in hagfishes, but the data we do have suggest that the actions of this group of compounds were established early in vertebrate evolution (Evans *et al.*, 1989). It appears that hagfishes resemble marine invertebrates in their ability to compensate for volume loading on transfer to low salinity (McFarland and Munz, 1958) although they are unable to compensate for

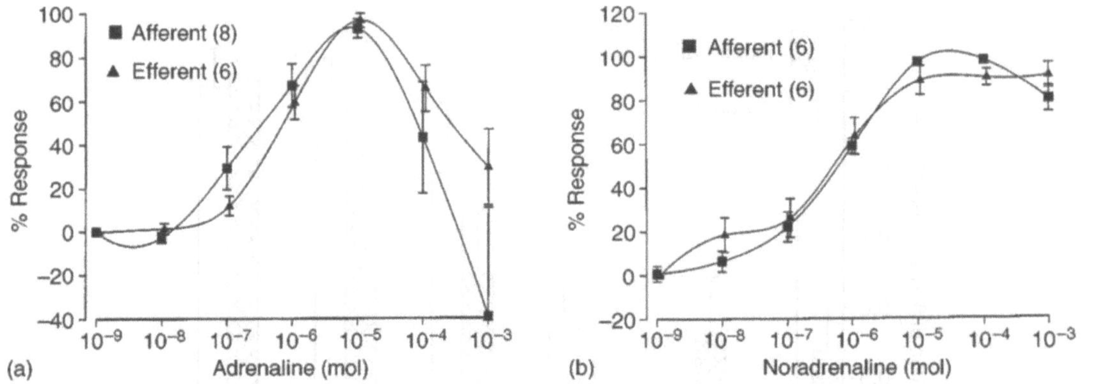

Figure 15.6 Dose–response curves for actions of (a) adrenaline and (b) noradrenaline on afferent and efferent branchial arteries of *Eptatretus cirrhatus*. The vasomotor responses were recorded with a myograph and are expressed as a percentage of the resting values. At high, and presumably non-physiological, concentrations adrenaline has a vasodilatory action. The data indicate the presence of both α- and β-adrenergic receptors. (M.E. Forster, unpublished observations.)

volume decrease (Toop and Evans, 1993). Potentially, volume loading in dilute media is a means of investigating the induction of NP release.

Eicosanoids may also play a role in blood pressure regulation in hagfishes, as prostaglandins E_1, E_2 and A_2 and thromboxane B_2 lowered dorsal aortic blood pressure in *M. glutinosa*, where administration of prostaglandin $F_{2\alpha}$ raised blood pressure (Wales, 1988). Adenosine injection also reduced blood pressure in *M. glutinosa in vivo*, at a time when it increased cardiac output, which would increase pressure *per se* (Axelsson *et al.*, 1990). Where it has an effect, adenosine generally vasoconstricts arteries in teleost fishes (Nilsson and Holmgren, 1992), But it vasodilated the coronary arteries of the mako shark (Farrell and Davie, 1991) as it does in mammals (Berne *et al.*, 1983).

The evolutionary line that led to hagfishes may have diverged from that which led to gnathostome fishes more than 500 Mya, but it appears that the chemical agents involved in cardiovascular regulation are remarkably similar in all vertebrate animals. In this respect the Vertebrata seems to be a remarkably conservative group. That so many vasoactive agents influence vascular resistance in hagfishes (Table 15.4) also suggests that even though blood pressures are unusually low, they are regulated.

15.5 EXERCISE

Eptatretids have been observed and filmed swimming in the wild (Jensen, 1965; Martini, this volume) and experimental evidence suggests that aerobic metabolism may fuel much of this activity. When forced to swim for 15 min at relatively low swimming speeds (*c.* 0.4 body length s^{-1}) the postbranchial blood of *E. cirrhatus* was almost fully saturated with oxygen. The tissues extracted 42% of the oxygen at rest. When swimming, blood returning to the heart had been depleted of a further 40% of its oxygen, being only about 10% saturated (Wells *et al.*, 1986). Despite the very low P_{O_2} of venous blood, heart rates increased by 11% in swimming hagfishes, and rose a further 7% once exercise ceased (Forster *et al.*, 1988). Cardiac output was not

Table 15.4 Drugs with demonstrated actions on hagfish blood vessels

Drug	Arteries where tested
Acetylcholine/carbachol	Afferent/efferent branchials[1,2,3]
Catecholamines	Afferent/efferent branchials[1,2,4]
5-Hydroxytryptamine	Afferent/efferent branchials[4]
Adenosine	Perfused gills[1]
Cholecystokinin	Afferent/efferent branchials[4]
Prostaglandins	Dorsal aortic pressure changed[5]
Natriuretic peptides	Ventral aorta[6]

1. Axelsson *et al.* (1990); 2. Reite (1969); 3. Glover and Forster (unpublished); 4. Sundin *et al.* (1994); 5. Wales (1988); 6. Evans (1991)

measured, but indications are that it did not fall, as ventral and dorsal aortic pressures rose in swimming animals. An increased venous return when swimming is likely to increase cardiac output, through the Starling effect (Chapman *et al.*, 1963; Satchell, 1986; Forster, 1989; Johnsson and Axelsson, 1996). However, ventral aortic blood pressure rose by only 7.5% (Table 15.1) compared to rises of from 21.2 to 34.5% in teleosts (Bushnell *et al.*, 1992). As noted above, patterns of blood and seawater flow through the gill pouches are extremely sensitive to changes in pressures and it may be that there are physiological constraints restricting the blood pressure rise.

15.6 HYPOXIA TOLERANCE

It has been assumed that hagfishes are physiologically and biochemically adapted to withstand hypoxia (Hansen and Sidell, 1983) and that the group retains a well-developed anaerobic capacity which evolved in the early vertebrates and facilitated an active lifestyle (Ruben and Bennett, 1980). However, *M. glutinosa* and *E. cirrhatus* responded differently to hypoxia, perhaps reflecting their very different habitats. *M. glutinosa* lives in burrows in hypoxic mud whereas *E. cirrhatus* is found coiled in rocky substrates, where presumably it does not encounter hypoxic water. The white myotomal and tongue muscles of *E. cirrhatus* do have a well-developed capacity for anaerobic glycolysis (Baldwin *et al.*, 1991), but in the case of the tongue muscle this is poorly vascularized and there is a presumed limitation of oxygen supply.

When *E. cirrhatus* were allowed to deplete the oxygen content of the water in a box in which they were held, they made vigorous attempts to escape the container once the P_{O_2} fell below 5 kPa (Forster, 1990). Some teleost fishes can withstand much lower inspired oxygen tensions (Ott *et al.*, 1980). In a different series of experiments when the P_{O_2} of inspired water was reduced to 5.3 kPa, oxygen consumption fell to 29% of the normoxic value, despite a 40% increase in cardiac output (Forster *et al.*, 1992). At a P_{O_2} of 5.3 kPa branchial vascular resistance was increased by 59% and arterial blood was only 15% saturated, despite *E. cirrhatus* haemoglobin having a P_{50} of only 1.64 kPa (Wells *et al.*, 1986). This indicates that the diffusive resistance of the gills had increased (Forster *et al.*, 1992). This impaired uptake of oxygen across the gill contrasted with normoxia where the Po_2 of dorsal aortic blood was 86% of that of the inspired water, which indicated that the hagfish gill pouch can be a most efficient countercurrent exchanger. It is hard to ascribe an adaptive significance to the increased diffusive resistance of the gill when *E. cirrhatus* is made hypoxic, other than as a mechanism that may minimize *loss* of oxygen from the blood when the respiratory water supply is interrupted. It is possible that the nostril is occluded and ventilation is prevented at times when the animal envelopes itself in mucus, or when feeding within marine carcases (Steffensen *et al.*, 1984). At such times the animal might have to utilize anaerobic metabolism to maintain vital functions. As noted above, injection of the β-adrenergic blocking drug propranolol increased branchial vascular resistance and reduced oxygen uptake, suggesting that α-adrenergic receptors might be involved in the vasoconstrictor response to hypoxia (Forster *et al.*, 1992). Perry *et al.* (1993) found that hypoxia elevated catecholamine concentrations in the blood of *M. glutinosa*, but the subendothelial cells of the hearts did not release catecholamines directly when exposed to hypoxic saline.

In contrast to *E. cirrhatus*, *M. glutinosa* withstood a severe hypoxia, down to an inspired P_{O_2} of 1.5–2.2 kPa. Ventral and dorsal aortic pressures rose by up to 25%, and cardiac output was maintained, demonstrating that the heart can continue to function effectively in hypoxic water. The experimenters did not observe the struggling behaviour observed in *E. cirrhatus* (Axelsson *et al.*, 1990; Forster, 1990). While the heart of *E. cirrhatus* shares this hypoxic tolerance with *M. glutinosa* (Forster, 1991), it appears that *E. cirrhatus* is less likely to exploit that capability voluntarily. The intolerance of *E. cirrhatus* to hypoxia is surprising, given the animal's low metabolic rate.

In summary, although hagfishes of the families Eptatretidae and Myxindae share potentially similar effector mechanisms for cardiovascular control, there are differences between species in their responses to hypoxia. It may be possible to exploit these differences to investigate the regulatory activity of cardiorespiratory control systems. We have limited knowledge of such processes at present.

15.7 'PRIMITIVE' FEATURES OF THE HAGFISH BLOOD SYSTEM: A REASSESSMENT

As the fossil record can provide only limited information on the evolution of the blood system in vertebrate animals and tells us nothing of cardiovascular control mechanisms, it is natural to look to the two groups of living agnathan vertebrates, the hagfishes and lampreys, for clues to the condition in the 'ancestral vertebrate'. A number of features of the hagfish cardiovascular system have been described as 'primitive' (Table 15.5). However, recent findings cast doubt on such a designation.

The possession of large venous blood sinuses, a property shared with elasmobranch fishes, has been correlated with getting blood back to the heart with as little resistance as possible, where the venous blood possesses little kinetic energy. Such an arrangement allows forces developed by the swimming muscles to assist in the venous return

Table 15.5 Features of the cardiovascular system of hagfish that have been described as 'primitive'

1. Possession of large venous sinuses.
2. A high blood volume.
3. Low blood pressures.
4. The heart:
   (a) long action potential
   (b) sparse myofibrils
   (c) hypoxia tolerance
   (d) relative insensitivity to high calcium concentrations
   (e) cardiac control mechanisms – functionally aneural.

(Satchell, 1991) and as such may be energetically efficient.

The marked contrast between the 18% of the body mass that is occupied by the blood volume of a hagfish and the 2–3% blood volume of a teleost fish is considerably diminished with the finding that the secondary circulation of the teleost fish is double the size of the primary volume (Steffensen and Lomholt, 1992). This was not considered when Thorson (1961) suggested that possession of a low blood volume was an advanced feature in fishes. Further, some teleosts, such as icefishes, also have high primary blood volumes (Acierno *et al.*, 1995).

As has been described, the low arterial blood pressures in hagfishes can probably be explained as a consequence of their reliance on a velar pump for ventilation and the need for matching hydrostatic pressures across the exchange surfaces of the gill. In this regard the relative lack of myofibrils in the heart can be seen as a necessary correlate of blood pressure regulation. The need to generate only low arterial blood pressures also explains the well-documented hypoxia tolerance of the hagfish heart. The relative insensitivity of the hagfish heart to external calcium concentrations, as compared to mammals, is a feature shared with the hearts of other ectothermic vertebrates.

The absence of neural control and a relative insensitivity of hagfish hearts to acetylcholine and adrenaline have been described as primitive characteristics. But other sites in the cardiovascular system are sensitive to these agents, and, as described above, we have considerable evidence for a potentially sophisticated control of the vasculature. This must be balanced against the lack of extrinsic control of the heart. As other parts of the blood system possess receptors for these vasoactive agents, their absence from the heart cannot be described as a primitive feature. In most 'fishes' it is the venous return which is the primary influence on cardiac output, through the Frank–Starling effect (Farrell, 1991). Given the additional effect of the volume of the venous return on the sinus venosus pacemaker in hagfishes (Jensen, 1961) and the potential to release endogenous catecholamines, it may be that extrinsic control of cardiac function in hagfishes is redundant and has disappeared in their evolutionary history.

The hagfish 'design' is clearly a successful one that has survived with perhaps little modification for about 300 million years. The characteristics which differentiate the hagfish cardiovascular system from those of other living vetebrates should perhaps be described as 'specialized' rather than 'primitive'.

REFERENCES

Acierno, R., Macdonald, J.A., Agnisola, C. and Tota, B. (1995) Blood volume in the hemoglobinless antarctic teleost *Chionodraco hamatus* (Lönnberg). *Journal of Experimental Zoology*, **272**, 407–9.

Alt, J.M., Stolte, H., Eisenbach, G.M. and Walvig, F. (1981) Renal electrolyte and fluid excretion in the Atlantic hagfish *Myxine glutinosa*. *Journal of Experimental Biology*, **91**, 323–30.

Arlock, P. (1975) Electrical activity and mechanical response in the systemic heart and the portal vein heart of *Myxine glutinosa*. *Comparative Biochemistry and Physiology*, **51A**, 521–2.

Augustinsson, K.B., Fänge, R., Johnels, A. and Östlund, E. (1956) Histological, physiological and biochemical studies on the heart of two cyclostomes, hagfish (*Myxine*) and lamprey (*Lampetra*). *Journal of Physiology*, **131**, 257–76.

Axelsson, M. and Nilsson, S. (1986) Blood pressure control during exercise in the Atlantic cod, *Gadus morhua*. *Journal of Experimental Biology*, **126**, 225–36.

Axelsson, M., Driedzic, W.R., Farrell, A.P. and Nilsson, S. (1989) Regulation of cardiac output and gut blood flow in the sea raven, *Hemitripterus americanus*. *Fish Physiology and Biochemistry*, **6**, 315–26.

Axelsson, M., Farrell, A.P. and Nilsson, S. (1990) Effects of hypoxia and drugs on the cardiovascular dynamics of the Atlantic hagfish *Myxine glutinosa*. *Journal of Experimental Biology*, **151**, 297–16.

Baldwin, J., Davison, W. and Forster, M.E. (1991)

Anaerobic glycolysis in the dental plate retractor muscles of the New Zealand hagfish *Eptatretus cirrhatus* during feeding. *Journal of Experimental Zoology*, **260**, 295–301.

Berne, R.M., Winn, H.R., Knapp, R.M., Ely, S.W and Rubio, R. (1983) Blood flow regulation by adenosine in heart, brain and skeletal muscle, in *Regulatory Function of Adenosine* (eds R.M. Berne, T.W. Tall and R. Rubio), Martinus Nijhoff, The Hague, pp. 293–317.

Bloom, G., Östlund, E., Euler, U.S.v., Lishajko, F., Ritzen, M. and Adams-Ray, J. (1961) Studies on catecholamine-containing granules of specific cells in cyclostome hearts. *Acta Physiologica Scandinavica*, **53**, suppl. 185, 1–34.

Bloom, G., Östlund, E. and Fänge, R. (1963) Functional aspects of cyclostome hearts in relation to recent structural findings. In The *Biology of Myxine* (eds A. Brodal and R. Fänge), Universitetsforlaget, Oslo, pp. 317–39.

Bushnell, P.G., Jones, D.R. and Farrell, A.P. (1992) The arterial system, in *Fish Physiology*, Vol. *XIIA* (eds W.S. Hoar, D.J. Randall and A.P. Farrell), Academic Press, San Diego, pp. 89–139.

Casely-Smith, J.R. and Casely-Smith, J.R. (1975) The fine structure of the blood capillaries of some endocrine glands of the hagfish *Eptatretus stoutii*: implications for the evolution of blood and lymph vessels. *Revue Suisse de Zoologie*, **82**, 35–40.

Chapman, C.B., Jensen, D. and Wildenthal, K. (1963) On circulatory control mechanisms in the Pacific hagfish. *Circulation Research*, **12**, 427–40.

Cole, F.J. (1912) A monograph on the general morphology of the Myxinoid fishes, based on a study of *Myxine*. Part IV. On some peculiarities of the afferent and efferent branchial arteries of Myxine. *Transactions of the Royal Society Edinburgh*, **49**, 215–30.

Cole, F.J. (1925) A monograph on the general morphology of the Myxinoid fishes, based on a study of *Myxine*. Part VI. The morphology of the vascular system. *Transactions of the Royal Society Edinburgh*, **54**, 309–42.

Crowell, J.W. and Smith, E.E. (1967) Determinant of the optimal hematocrit. *Journal of Applied Physiology*, **22**, 501–4.

Daniels, S. and Duncan, C.J. (1993) The effect on creatine kinase release of altered conditions in the Ca^{2+} paradox. *Comparative Biochemistry and Physiology*, **106A**, 557–60.

Davie, P.S. and Farrell, A.P. (1991) Cardiac performance of an isolated heart preparation from the dogfish (*Squalus acanthias*): the effects of hypoxia and coronary artery perfusion. *Canadian Journal of Zoology*, **69**, 1822–8.

Davie, P.S., Forster, M.E., Davison, W. and Satchell, G.H. (1987) Cardiac function in the New Zealand hagfish, *Eptatretus cirrhatus*. *Physiological Zoology*, **60**, 233–40.

Davison, I.G., Wright, G.M and DeMont, M.E. (1995) The structure and physical properties of invertebrate and primitive vertebrate arteries. *Journal of Experimental Biology*, **198**, 2185–96.

Davison, W. (1995) What is the function of the hagfish portal heart? *New Zealand Natural Sciences*, **22**, 95–8.

Davison, W., Baldwin, J., Davie, P.S., Forster, M.E. and Satchell, G.H. (1990) Exhausting exercise in the hagfish, *Eptatretus cirrhatus*: the anaerobic potential and the appearance of lactic acid in the blood. *Comparative Biochemistry and Physiology*, **95A**, 585–9.

Donald, J.A., Vomachka, A.J. and Evans, D.H. (1992) Immunohistochemical localisation of natriuretic peptides in the brain and hearts of the spiny dogfish *Squalus acanthias* and the Atlantic hagfish *Myxine glutinosa*. *Cell and Tissue Research*, **270**, 535–45.

Driedzic, W.R. and Gesser, H. (1994) Energy metabolism and contractility in ectothermic vertebrate hearts: hypoxia, acidosis, and low temperature. *Physiological Reviews*, **74**, 221–58.

Driedzic, W.R., Sidell, B.D., Stowe, D. and Branscombe, R. (1987) Matching of vertebrate cardiac energy demand to energy metabolism. *American Journal of Physiology*, **252**, R930–37.

Elger, M. (1987) The branchial circulation and the gill epithelia in the Atlantic hagfish, *Myxine glutinosa* L. *Anatomy and Embryology*, **175**, 489–504.

Euler, U.S.v. and Fänge, R. (1961) Catecholamines in nerves and organs of *Myxine glutinosa*, *Squalus acanthias* and *Gadus morhua*. *Journal of General and Comparative Endocrinology*, **1**, 191–4.

Evans, D.H. (1991) Rat atriopeptin dilates vascular smooth muscle of the ventral aorta from the shark (*Squalus acanthias*) and the hagfish (*Myxine glutinosa*). *Journal of Experimental Biology*, **157**, 551–6.

Evans, D.H., Chipouras, E. and Payne, J.A. (1989) Immunoreactive atriopeptin in plasma of fishes: its potential role in gill haemodynamics. *American Journal of Physiology*, **257**, R939–45.

Evans, D.H., Donald, J.A. and Stidham, J.D. (1993) C-type natriuretic peptides are not particularly potent dilators of hagfish (*Myxine glutinosa*) vascular smooth muscle. *Bulletin of the Mt Desert Island Biological Laboratory*, **32**, 106.

Fänge, R., Bloom, G. and Östlund, E. (1963) The portal vein heart of myxinoids. In *The Biology of Myxine* (eds A. Brodal and R. Fänge), Universitetsforlaget, Oslo, pp. 340–51.

Farrell, A.P. (1981) Cardiovascular changes in the lingcod (*Ophiodon elongatus*) following adrenergic and cholinergic drug infusions. *Journal of Experimental Biology*, **91**, 293–305.

Farrell, A.P. (1982) Cardiovascular changes in the unanaesthetised lingcod (*Ophiodon elongatus*) during short-term, progressive hypoxia and spontaneous activity. *Canadian Journal of Zoology*, **60**, 933–41.

Farrell, A.P. (1984) A review of cardiac performance in the teleost heart: intrinsic and humoral regulation. *Canadian Journal of Zoology*, **62**, 523–36.

Farrell, A.P. (1991) From hagfish to tuna: a perspective on cardiac function in fish. *Physiological Zoology*, **64**, 1137–64.

Farrell, A.P. and Davie, P.S. (1991) Coronary artery reactivity in the mako shark, *Isurus oxyrinchus*. *Canadian Journal of Zoology*, **69**, 375–9.

Farrell, A.P., Davie, P.S., Franklin, C.E., Johansen, J.A. and Brill, R.W. (1992) Cardiac physiology in tunas. I. *In vitro* perfused heart preparations from yellowfin and skipjack tunas. *Canadian Journal of Zoology*, **70**, 1200–10.

Farrell, A.P. and Jones, D.R. (1992) *The heart*. In *Fish Physiology*, Vol. XIIA (eds W.S. Hoar, D.J. Randall and A.P. Farrell), Academic Press, San Diego, pp. 1–88.

Forster, M.E. (1989) Performance of the heart of the hagfish, *Eptatretus cirrhatus*. *Fish Physiology and Biochemistry*, **6**, 327–31.

Forster, M.E. (1990) Confirmation of the low metabolic rate of hagfish. *Comparative Biochemistry and Physiology*, **96A**, 113–16.

Forster, M.E. (1991) Myocardial oxygen consumption and lactate release by the hypoxic hagfish heart. *Journal of Experimental Biology*, **156**, 583–90.

Forster, M.E. (1997) The blood sinus system of hagfish: its significance in a low pressure circulation. *Comparative Biochemistry and Physiology*, **116A**, 239–44.

Forster, M.E., Axelsson, M., Farrell, A.P. and Nilsson, S. (1991) Cardiac function and circulation in hagfishes. *Canadian Journal of Zoology*, **69**, 1985–92.

Forster, M.E., Davie, P.S., Davison, W., Satchell, G.H. and Wells R.M.G. (1988) Blood pressures and heart rates in swimming hagfish. *Comparative Biochemistry and Physiology*, **89A**, 247–50.

Forster, M.E., Davison, W., Satchell, G.H. and Taylor, H.H. (1989) The subcutaneous sinus of the hagfish, *Eptatretus cirrhatus*: its relation to the central circulating blood volume. *Comparative Biochemistry and Physiology*, **93A**, 607–12.

Forster, M.E., Davison, W., Axelsson, M. and Farrell, A.P. (1992) Cardiovascular responses to hypoxia in the hagfish, *Eptatretus cirrhatus*. *Respiration Physiology*, **88**, 273–86.

Forster, M.E. and Fenwick, J.C. (1994) Stimulation of calcium efflux from the hagfish, *Eptatretus cirrhatus*, gill pouch by an extract of corpuscules of Stannius from an eel (*Anguilla dieffenbachii*): Teleostei. *General and Comparative Endocrinology*, **94**, 92–103.

Fritsche, R. and Nilsson, S. (1989) Cardiovascular responses to hypoxia in the Atlantic cod, *Gadus morhua*. *Experimental Biology*, **48**, 153–60.

Greene, C.W. (1902) Contributions to the physiology of the California hagfish, *Polistotrema stoutii*. I. The anatomy and physiology of the caudal heart. *American Journal of Physiology*, **3**, 366–82.

Guyton, A.C. (1986) *Textbook of Medical Physiology* (7th edn). W.B. Saunders, Philadelphia.

Hansen, C.A. and Sidell, B.D. (1983) Atlantic hagfish cardiac muscle: metabolic basis of tolerance to anoxia. *American Journal of Physiology*, **244**, R356–62.

Hardisty, M.W. (1979) *Biology of the Cyclostomes*, Chapman & Hall, London.

Helle, K.B. and Lönning, S. (1973) Sarcoplasmic reticulum in the portal vein heart and ventricle of the cyclostome *Myxine glutinosa* (L.). *Journal of Molecular and Cellular Cardiology*, **5**, 433–9.

Hirsch, E.F., Jellinek, M. and Cooper, T. (1964) Innervation of the hagfish heart. *Circulation Research*, **14**, 212–17.

Hochachka, P.W. and Somero, G.N. (1984) *Biochemical Adaptation*, Princeton University Press, Princeton.

Hol, R. and Johansen, K. (1960) A cineradiographic study of the central circulation in the hagfish, *Myxine glutinosa* L. *Journal of Experimental Biology*, **37**, 469–73.

Jensen, D. (1958) Some observations on cardiac automatism in certain animals. Journal of General Physiology, **42**, 289–302.

Jensen, D. (1961) Cardioregulation in an aneural heart. *Comparative Biochemistry and Physiology*, **2**, 181–201

Jensen, D. (1965) The aneural heart of the hagfish. *Annals of the New York Academy of Sciences*, **127**, 443–58.

Johansen, K. (1960) Circulation in the hagfish, *Myxine glutinosa* L. *Biological Bulletin, Woods Hole*, **118**, 289–95.

Johansen, K. (1963) The cardiovascular system of *Myxine glutinosa*. In *The Biology of Myxine* (eds A. Brodal and R. Fänge), Universitetsforlaget, Oslo, pp 289–316.

Johansen, K. (1972) Heart and circulation in gill, skin and lung breathing. *Respiration Physiology*, **14**, 193–210.

Johansen, K. and Hol, R. (1960) A cineradiographic study of respiration in *Myxine glutinosa*. *Journal of Experimental Biology*, **37**, 474–80.

Johansen, K. and Strahan, R. (1963) The respiratory system of *Myxine glutinosa*. In *The Biology of Myxine* (eds A. Brodal and R. Fänge), Universitetsforlaget, Oslo, pp. 353–71.

Johansen, K., Fänge, R. and Johannessen, M.W. (1962) Relations between blood, sinus fluid and lymph in *Myxine glutinosa* L. *Comparative Biochemistry and Physiology*, **7**, 23–8.

Johansen, K., Lenfant, F. and Hansen, D. (1973) Gas exchange in the lamprey, *Entosphenus tridentata*. *Comparative Biochemistry and Physiology*, **44A**, 107–19.

Johnsson, M. and Axelsson, M. (1996) Control of the systemic heart and the portal heart of *Myxine glutinosa*. *Journal of Experimental Biology*, **199**, 1429–34.

Johnsson, M., Axelsson, M., Davison, W., Forster, M.E. and Nilsson, S. (1996) Effects of preload and afterload on the performance of the *in situ*, perfused portal heart of the New Zealand hagfish, *Eptatretus cirrhatus*. *Journal of Experimental Biology*, **199**, 401–5.

Jönsson, A-C. (1983) Catecholamine formation *in vitro* in the systemic and portal hearts of the Atlantic hagfish, *Myxine glutinosa*. *Molecular Physiology*, **3**, 297–304.

Kampmeier, O.F. (1969) *Evolution and Comparative Morphology of the Lymphatic System*, Charles C. Thomas, Illinois.

Kench, J.E. (1989) Observations on the respiration of the South African hagfish, *Eptatretus hexatrema* Mull. *Comparative Biochemistry and Physiology*, **93A**, 877–92.

Kloas, W., Flugge, G., Fuchs, E. and Stolte, H. (1988) Binding sites for atrial natriuretic peptide in the kidney and aorta of the hagfish (*Myxine glutinosa*). *Comparative Biochemistry and Physiology*, **91A**, 685–8.

Lagerstrand, G., Mattisson, A. and Poupa, O. (1983) Studies on the calcium paradox in cardiac muscle strips in poikilotherms. *Comparative Biochemistry and Physiology*, **76A**, 601–13.

Leak, L.V. (1969) Electron microscopy of cardiac tissue in a primitive vertebrate *Myxine glutinosa*. *Journal of Morphology*, **128**, 131–58.

Mallatt, J. and Paulsen, C. (1986) Gill ultrastructure of the Pacific hagfish *Eptatretus stoutii*. *American Journal of Anatomy*, **177**, 243–69.

McCarthy, J.E. and Conte, F.P. (1966) Determination of the volume of vascular and extravascular fluid in the Pacific hagfish, *Eptatretus stoutii* (Lockington). *American Zoologist*, **6**, 605.

McFarland, W.N. and Munz, F.W. (1958) A reexamination of the osmotic properties of the Pacific hagfish *Polistotrema stoutii*. *Biological Bulletin, Woods Hole*, **114**, 348–56.

McKean, T.A. (1984) Response of isolated muskrat and guinea pig hearts to hypoxia. *Physiological Zoology*, **57**, 557–62.

Munz, F.W. and Morris, R.W. (1965) Metabolic rate of the hagfish, *Eptatretus stoutii* (Lockington) 1878. *Comparative Biochemistry and Physiology*, **16**, 1–6.

Nielsen, K.E. and Gesser, H. (1983) Effects of [Ca^{2+}] on contractility in the anoxic cardiac muscle of mammal and fish. *Life Sciences*, **32**, 1437–42.

Nilsson, S. and Holmgren, S. (1992) Cardiovascular control by purines, 5-hydroxytryptamine, and neuropeptides, in *Fish Physiology*, Vol. XIIB (eds W.S. Hoar, D.J. Randall and A.P. Farrell), Academic Press, San Diego, pp. 301–41.

Östlund, E. (1954) The distribution of catecholamines in lower animals and their effect on the heart. *Acta Physiologica Scandinavica*, **31** suppl. **112**, 1–67.

Ott, M., Heisler, N. and Ultsch, G.R. (1980) A reevaluation of the relationship between temperature and critical oxygen tension in freshwater fishes. *Comparative Biochemistry and Physiology*, **67A**, 337–40.

Perry, S.F., Fritsche, R. and Thomas, S. (1993) Storage and release of catecholamines from the chromaffin tissue of the Atlantic hagfish *Myxine glutinosa*. *Journal of Experimental Biology*, **183**, 165–84.

Pohla, H., Lametschwandtner, A. and Adam, H.

(1977) Die Vaskularisation der Kiemen von *Myxine glutinosa* L. (Cyclostomata). *Zoologica Scripta*, **6**, 331–41.

Poupa, O., Ask, J.A. and Helle, K.B. (1984) Absence of calcium paradox in the cardiac ventricle of the Atlantic hagfish (*Myxine glutinosa*). *Comparative Biochemistry and Physiology*, **78A**, 181–3.

Randall, D.J. (1972) Respiration, in *The Biology of Lampreys*, Vol. 2.. (eds M.W. Hardisty and I.C. Potter), Academic Press, London, pp. 287–316.

Reid, S.G. Fritsche, R. and Jonsson, A.C. (1995) Immunohistochemical localization of bioactive peptides and amines associated with the chromaffin tissue of five species of fish. *Cell and Tissue Research*, **280**, 499–512.

Reite, O.B. (1969) The evolution of vascular smooth muscle responses to histamine and 5-hydroxytryptamine. I. Occurrence of stimulatory actions in fish. *Acta Physiologica Scandinavica*, **75**, 221–39.

Robertson, J.D. (1954) The chemical composition of the blood of some aquatic chordates, including members of the Tunicata, Cyclostomata and Osteichthyes. *Journal of Experimental Biology*, **31**, 424–42.

Robertson, J.D. (1976) Chemical composition of the body fluids and the muscle of the hagfish *Myxine glutinosa* and the rabbit-fish *Chimaera monstrosa*. *Journal of Zoology*, London, **178**, 261–77.

Ruben, J.A. and Bennett, A.F. (1980) Antiquity of the vertebrate pattern of activity metabolism and its possible relation to vertebrate origins. *Nature, London*, **286**, 886–8.

Sage, H. and Gray, W.R. (1970) Studies on the evolution of elastin. I. Phylogenetic distribution. *Comparative Biochemistry and Physiology*, **64B**, 313–27.

Sage, H. and Gray, W.R. (1980) Studies on the evolution of elastin. II. Histology. *Comparative Biochemistry and Physiology*, **66B**, 13–22.

Sarnoff, S.J. and Mitchell, J.H. (1962) The control of the function of the heart. In *Handbook of Physiology, Sect. 2, Circulation, vol 1* American Physiological Society, Washington. pp. 409–532.

Satchell, G.H. (1984) On the caudal heart of *Myxine* (Myxinoidea: Cyclostomata). *Acta Zoologica (Stockholm)*, **65**, 125–33.

Satchell, G.H. (1986) Cardiac function in the hagfish, *Myxine* (Myxinoidea: Cylostomata). *Acta Zoologica (Stockholm)*, **67**, 115–22.

Satchell, G.H. (1991) *Physiology and Form of Fish Circulation*, Cambridge University Press, Cambridge.

Satchell, G.H. (1992) The venous system, in *Fish Physiology*, Vol. XIIA (eds W.S. Hoar, D.J. Randall and A.P. Farrell), Academic Press, San Diego, pp. 141–83.

Satchell, G.H. and Weber, L.J. (1987) The caudal heart of the carpet shark *Cephaloscyllium isabella*. *Physiological Zoology*, **60**, 692–8.

Shelton, G. (1970) The regulation of breathing, in *Fish Physiology*, Vol. IV (eds W.S. Hoar and D.J. Randall), Academic Press, New York, pp. 293–359.

Short, S., Taylor, E.W. and Butler, P.J. (1979) The effectiveness of oxygen transfer during normoxia and hypoxia in the dogfish (*Scyliorhinus canicula* L.) before and after cardiac vagotomy. *Journal of Comparative Physiology*, **132**, 289–95.

Sidell, B.D. (1983) Cardiac metabolism in the Myxinidae: physiological and phylogenetic considerations. *Comparative Biochemistry and Physiology*, **76A**, 495–505.

Steffensen, J.F., Johansen, K., Sindberg, C.D., Sørensen, J.H. and Møller, J.L. (1984) Ventilation and oxygen consumption in the hagfish, *Myxine glutinosa* L. *Journal of Experimental Marine Biology and Ecology*, **84**, 173–8.

Storesund, A. and Helle, K.B. (1975) Practolol, caffeine and calcium in the regulation of mechanical activity of the cardiac ventricle in *Myxine glutinosa*. *Comparative Biochemistry and Physiology*, **52C**, 17–22.

Strahan, R. (1963) The behaviour of myxinoids. *Acta Zoologica*, **44**, 73–102.

Sundin, L., Axelsson, M., Nilsson, S., Davison, W. and Forster, M.E. (1994) Evidence of regulatory mechanisms for the distribution of blood between the arterial and the venous compartments in the hagfish gill pouch. *Journal of Experimental Biology*, **190**, 281–6.

Tani, M. and Neely, J.R. (1990) Vascular washout reduces Ca^{2+} overload and improves function of reperfused ischemic hearts. *American Journal of Physiology*, **258**, H354–61.

Thorson, T.B. (1961) The partitioning of body water in Osteichthyes: phylogenetic and ecological implications in aquatic vertebrates. *Biological Bulletin, Woods Hole*, **120**, 238–54.

Toop, T. and Evans, D.H. (1993) Whole animal volume regulation in the Atlantic hagfish, *Myxine glutinosa*, exposed to 85% and 115% sea water. *Bulletin of the Mt Desert Island Biological*

Laboratory, **32**, 98–9.

Toop, T., Donald, J.A. and Evans, D.H. (1995) Localisation and characteristics of natriuretic peptide receptors in the gills of the Atlantic hagfish *Myxine glutinosa* (Agnatha). *Journal of Experimental Biology*, **198**, 117–26.

Wales, N.A.M. (1988) Hormone studies in Myxine glutinosa: effects of the eicosanoids arachidonic acid, prostaglandin E_1, E_2, A_2, $F_{2\alpha}$, thromboxane B_2 and of indomethacin on plasma cortisol, blood pressure, urine flow and electrolyte balance. Journal of Comparative Physiology, **158**, 621–6.

Wells, R.M.G. and Forster, M.E. (1989) Dependence of blood viscosity on haematocrit and shear rate in a primitive vertebrate. *Journal of Experimental Biology*, **145**, 483–7.

Wells, R.M.G., Forster, M.E., Davison, W., Taylor, H.H., Davie, P.S. and Satchell, G.H. (1986) Blood oxygen transport in the free-swimming hagfish, *Eptatretus cirrhatus*. *Journal of Experimental Biology*, **123**, 43–53.

Welsch, U. and Potter, I.C. (1994) Variability in the presence of elastic fibre-like structures in the ventral aorta of agnathans (hagfishes and lampreys). *Acta Zoologica (Stockholm)*, **75**, 232–327.

Wright, G.M. (1984) Structure of the conus arteriosus and ventral aorta in the sea lamprey, *Petromyzon marinus*, and the Atlantic hagfish, *Myxine glutinosa*: microfibrils, a major component, *Canadian Journal of Zoology*, **62**, 2445–56.

Yamauchi, A. (1980) Fine structure of the fish heart, in *Hearts and Heart-like Organs* (ed. G.H. Bourne), Academic Press, New York, pp. 119–48.

16

THE SINUS SYSTEM OF HAGFISHES – LYMPHATIC OR SECONDARY CIRCULATORY SYSTEM?

Jens Peter Lomholt and Frida Franko-Dossar

SUMMARY

Apart from the vascular system of arteries, capillaries and veins, hagfish possess an extensive system of vessels and sinuses originally believed to be lymphatic. The occurrence of red cells in parts of the system was considered to be an artefact. Later, connections were found between certain sinuses and arteries. These sinuses were termed 'red lymphatics'. Other parts of the system were termed 'white' lymphatics and believed to be true lymphatics. New observations show that even 'white' lymphatics communicate with the arterial system. It is concluded that the system is comparable to the secondary vascular system of teleosts. A new system of fine channels surrounded by outgrowths from mesothelial cells of the serosal surface of the intestine is described. These channels communicate with secondary vessels of the intestine. Finally characteristic crypts in the serosal surface of the intestine are described.

16.1 INTRODUCTION

Since the early nineteenth century hagfishes have been known to possess an extensive and complicated system of vessels and sinuses in addition to the ordinary vascular system of arteries, capillaries and veins. This system was considered a lymphatic system because it appeared not to be connected to the arterial system (Müller, 1841). As early as 1824 Anders Retzius noticed the presence of red blood cells in the subcutaneous sinus of *Myxine*. Ever since the nature and functional role of the system of sinuses and vessels containing blood of a low but variable hematocrit has been controversial.

Some authors (Müller, 1841; Grodzinski, 1926, 1932; Hoyer, 1934 quoted in Kampmeier, 1969) believed the system to be lymphatic and considered the occurrence of red blood cells to be an artefact. By analogy with occasional, partly pathological, venolymphatic connections in man, it was thought that rough handling could cause rupture of delicate separations between veins and lymphatics. Extravasation caused by hauling the animals up from deep water was also suggested (Retzius, 1890; Jackson, 1901; Grodzinski, 1926). Why this should happen is not obvious, since intra- and extravascular changes in pressure will be the same during the ascent from depth. This argument is also weakened by the fact that some species of hagfishes do in fact live in rather shallow waters (Martini, this volume).

Others saw the hagfish as representing a primitive stage in evolution where there was not yet a clear distinction between the venous and the lymphatic systems. Hence the sinus system was spoken of as a mixed 'venolymphatic' system (Cole, 1912, 1925; Allen, 1913).

The Biology of Hagfishes. Edited by Jørgen Mørup Jørgensen, Jens Peter Lomholt, Roy E. Weber and Hans Malte. Published in 1998 by Chapman & Hall, London. ISBN 0 412 78530 7.

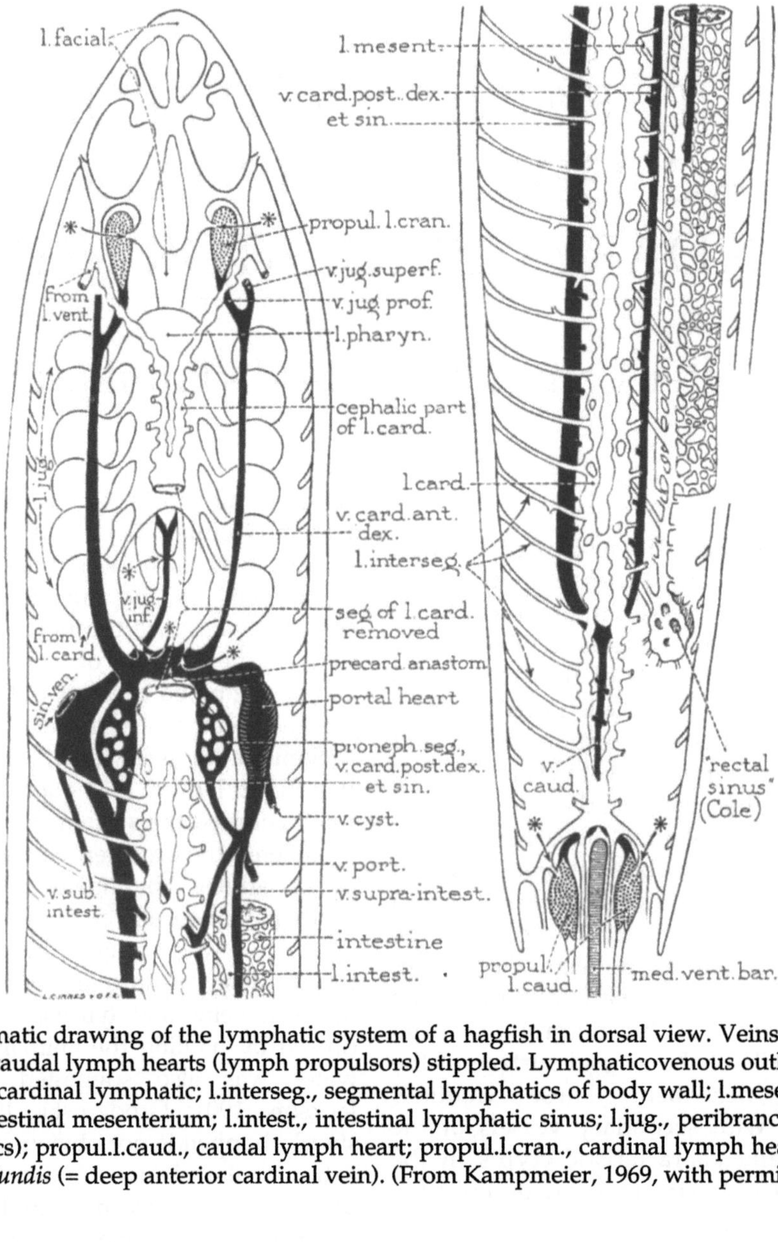

Figure 16.1 Schematic drawing of the lymphatic system of a hagfish in dorsal view. Veins black. Cranial (= cardinal) and caudal lymph hearts (lymph propulsors) stippled. Lymphaticovenous outlets marked by asterisks. l.card., cardinal lymphatic; l.interseg., segmental lymphatics of body wall; l.mesent., segmental lymphatics of intestinal mesenterium; l.intest., intestinal lymphatic sinus; l.jug., peribranchial sinuses (= jugular lymphatics); propul.l.caud., caudal lymph heart; propul.l.cran., cardinal lymph heart; v.jug.prof., *vena jugularis profundis* (= deep anterior cardinal vein). (From Kampmeier, 1969, with permission.)

16.2 MORPHOLOGY OF THE SINUS SYSTEM

The older literature on the morphology of the sinus system has been summarized by Kampmeier (1969). Figure 16.1 gives a schematic representation of the lymphatic system of a 'generalized' hagfish. A pair of cardinal lymphatics runs along the aorta and cardinal veins for nearly the entire length of the body. The cardinal lymphatics give rise to segmental lymphatics in the body wall and in the intestinal mesenterium. The latter communicate with the intestinal sinus and a network of intestinal lymphatics.

The subcutaneous sinus surrounds the body almost completely. It communicates with the segmental lymphatics of the body wall through a series of subcutaneous anastomoses. The head and branchial region are the site of a complicated system of sinuses. The peribranchial sinuses are associated with the gill pouches. Other sinuses not shown in the figure are the hypophysiovelar, the lingual, the dental, the ventral and the suboesophageal sinuses. A detailed description of these sinuses is given by Cole (1925).

The sinus system communicates with the venous system through the caudal and cardinal (or cranial) lymph hearts ('lymph propulsors') and through several valved openings between the sinus system and veins in the anterior region (Figure 16.1). The cardinal lymph hearts propel blood into the veins called deep jugular veins (*vena jugularis profundis* in Figure 16.1) by Kampmeier (1969) or deep anterior cardinal veins by Cole (1925). They communicate directly with the hypophysiovelar sinus and via this with the subcutaneous sinus. Their inflow and outflow openings are guarded by valves. Their pumping actions has not been directly observed, but was inferred from anatomical observations to be caused by the muscles, which produce the respiratory movements of the velum (Cole, 1925; Malte and Lomholt, this volume).

The caudal lymph hearts are situated in the tip of the tail on either side of a median plate of cartilage (Figure 16.2a). Each chamber is covered by a sheet of striated muscle innervated from the spinal chord. The pumping mechanism is well known from observations on live animals (Satchell, 1984, 1991, 1992). Contraction of the muscle on one side causes the cartilage to bend in such a way that the chamber on the same side is enlarged whereas the chamber on the opposite side is compressed (Figure 16.2b).

16.3 'RED' AND 'WHITE' LYMPHATICS

Cole (1912) presented an explanation for the regular occurrence of red blood cells in certain parts of the sinus system. He discovered hollow vascular papillae on the afferent and efferent branchial arteries and on the common carotid arteries (aortic roots). The papillae formed narrow communications between the arteries and the peribranchial sinuses. Cole suspected similar connections to occur in other parts of the body, but did not find them. Grodzinski (1932) saw Cole's papillae, but doubted if they were in open communication with the sinuses. More recently other connections between the gill vasculature and sinuses associated with the gills have been found by Pohla *et al.* (1977) and Elger (1987).

Cole (1912) considered the possibility that the papillae were vestigial structures. Thus the hagfish could represent a primitive situation where the sinus system was not yet completely separated from the arterial system.

Cole (1925) did not, however, find red cells in all parts of the sinus system. He thus spoke of 'red lymphatics' and 'white lymphatics'. The red lymphatics comprised the subcutaneous sinus as well as the peribranchial and other sinuses of the head and gill region. The cardinal, segmental and intestinal lymphatics were designated as white lymphatics, and considered true lymphatics originating from blind ending terminal lymphatics in the tissues.

Johansen (1963) discussed how the low and variable haematocrit in the sinuses of the red lymphatic system could be explained by plasma skimming taking place as blood passes through the papillae of Cole. Plasma from the subcutaneous sinus was found to have the same protein concentration and composition as plasma from the vascular system proper. On the contrary, plasma from the intestinal sinus was found to have a lower protein concentration. It was concluded that

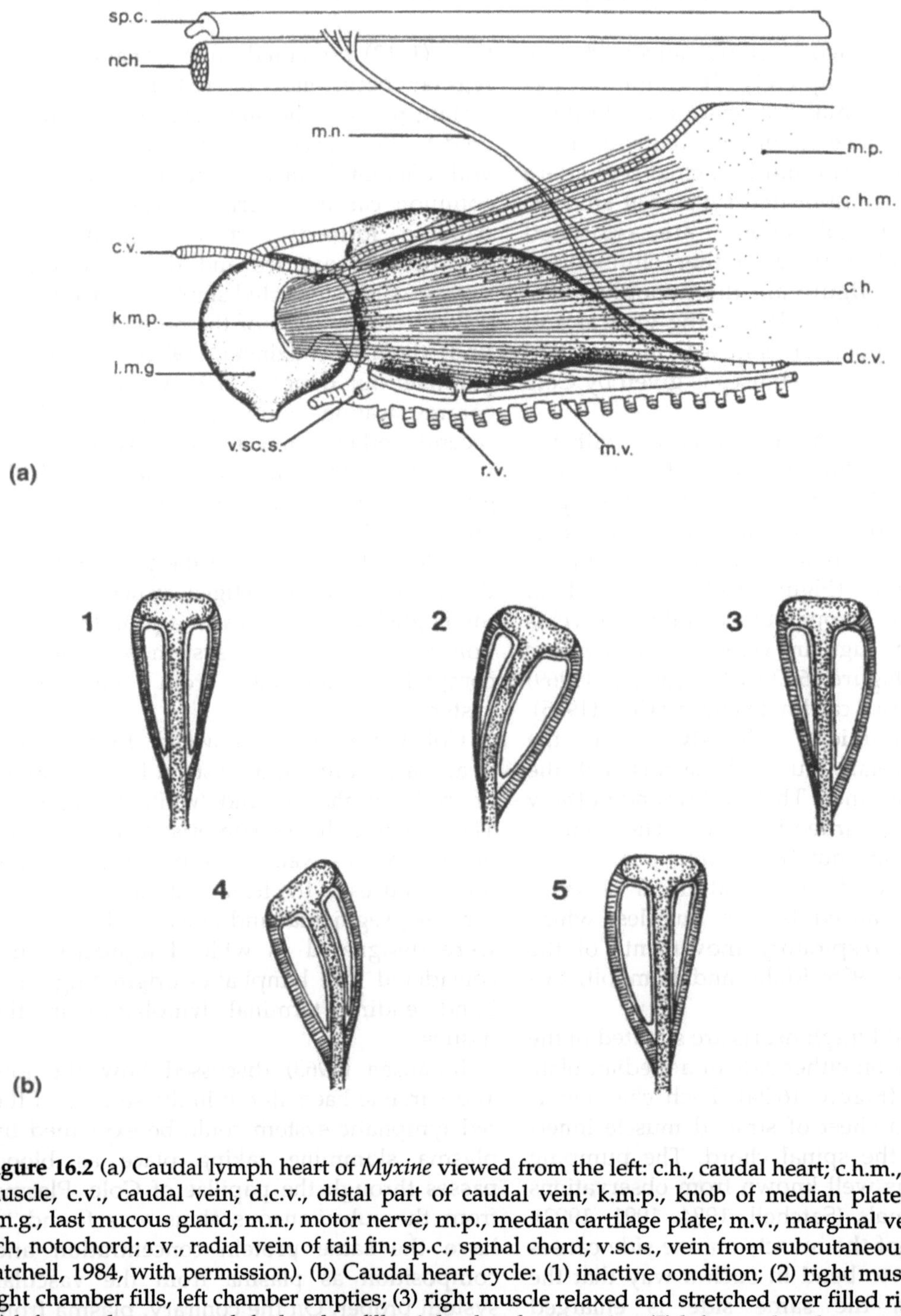

Figure 16.2 (a) Caudal lymph heart of *Myxine* viewed from the left: c.h., caudal heart; c.h.m., caudal heart muscle, c.v., caudal vein; d.c.v., distal part of caudal vein; k.m.p., knob of median plate of cartilage; l.m.g., last mucous gland; m.n., motor nerve; m.p., median cartilage plate; m.v., marginal vein of tail fin; nch, notochord; r.v., radial vein of tail fin; sp.c., spinal chord; v.sc.s., vein from subcutaneous sinus (from Satchell, 1984, with permission). (b) Caudal heart cycle: (1) inactive condition; (2) right muscle contracts, right chamber fills, left chamber empties; (3) right muscle relaxed and stretched over filled right chamber; (4) left muscle contracts, left chamber fills, right chamber empties; (5) left muscle relaxes and is stretched over left chamber as median cartilage returns to midline. Cartilage stippled. Muscle cross-striated. Anterior end up. (Redrawn from Satchell, 1984.)

the white lymphatics were true lymphatic vessels containing lymph, which had originated through a process of ultrafiltration, whereas the red lymphatics carried true blood, which had undergone a variable degree of plasma skimming (Johansen *et al.*, 1962; Johansen, 1963).

16.4 THE SECONDARY VASCULAR SYSTEM OF TELEOSTS

The concept of a lymphatic vessel system in teleost fish was challenged with the discovery of the so-called secondary vascular system (Vogel, 1981a, b; Vogel and Claviez, 1981; Vogel, 1985a, b). Vogel showed that vessels, which had previously been considered to be lymphatics, originate through open communications with the arterial system. These communications take the form of clusters of tiny vessels branching from various arteries. They immediately coalesce to form secondary arteries running parallel to the artery from which they originate. Consequently Vogel termed these communications interarterial anastomoses. Secondary arteries feed secondary capillaries located in the skin, fin membranes, mucus membranes of the mouth and peritoneum of the fish. They are in turn drained by secondary veins, which discharge into the primary venous circulation close to the heart and in the tail, in some cases via a caudal 'lymph' heart. No evidence for the existence of blind ending terminal lymphatics was found. A review of the secondary vascular system was given by Steffensen and Lomholt (1992).

16.5 CONNECTIONS BETWEEN ARTERIES AND 'WHITE' LYMPHATICS IN *MYXINE*

It is tempting to compare the vascular papillae of Cole with the interarterial anastomoses in teleosts described by Vogel. They both have a lumen just large enough to permit the passage of red blood cells. *In vivo* observations have shown plasma skimming to take place as blood enters the secondary circulation in a teleost (Steffensen *et al.*, 1986). Satchell (1992) suggested that the hagfish sinus system may be a forerunner of the secondary circulatory system in teleosts.

Prompted by the lack of evidence for true lymphatics in teleosts a re-examination of the system of 'white' lymphatics in *Myxine glutinosa* was undertaken (Franko-Dossar and Lomholt, 1997). The main emphasis in this study was on the vasculature of the intestinal tract. The findings are summarized in Figure 16.3. The intestine is supplied through segmental arteries (7) branching from the dorsal aorta (1). Along each segmental artery two segmental lymphatics (8) run, which communicate proximally with the cardinal lymphatics (3). After traversing the mesenterium (6), the segmental lymphatics communicate with the longitudinal intestinal lymphatic sinus (10) as well as with the network of lymphatic vessels (14), which extends over the entire external (serosal) surface of the intestine.

As soon as they reach the intestine, the arteries give off branches (13) to the ovaries as well as along the intestinal sinus. As the arteries penetrate further, small branches approach the serosal surface, whereas larger branches move into the deeper layers of the intestinal wall and supply the dense capillary network (18) of the mucosa. The mucosal capillaries are drained by a dense network of veins (16) which communicate with the intestinal portal vein (11) via segmental veins (17).

In the intestinal mesentery anastomoses (9) between the artery and segmental lymphatics are frequently seen. They resemble the interarterial anastomoses of teleosts (Vogel and Claviez, 1981; Vogel, 1981a, b) in that they connect a large artery with large secondary vessels. All over the serosal surface many very small arterial branches (21) were found to communicate with the network of lymphatics. Similar connections (20) were found along the intestinal sinus. No evidence

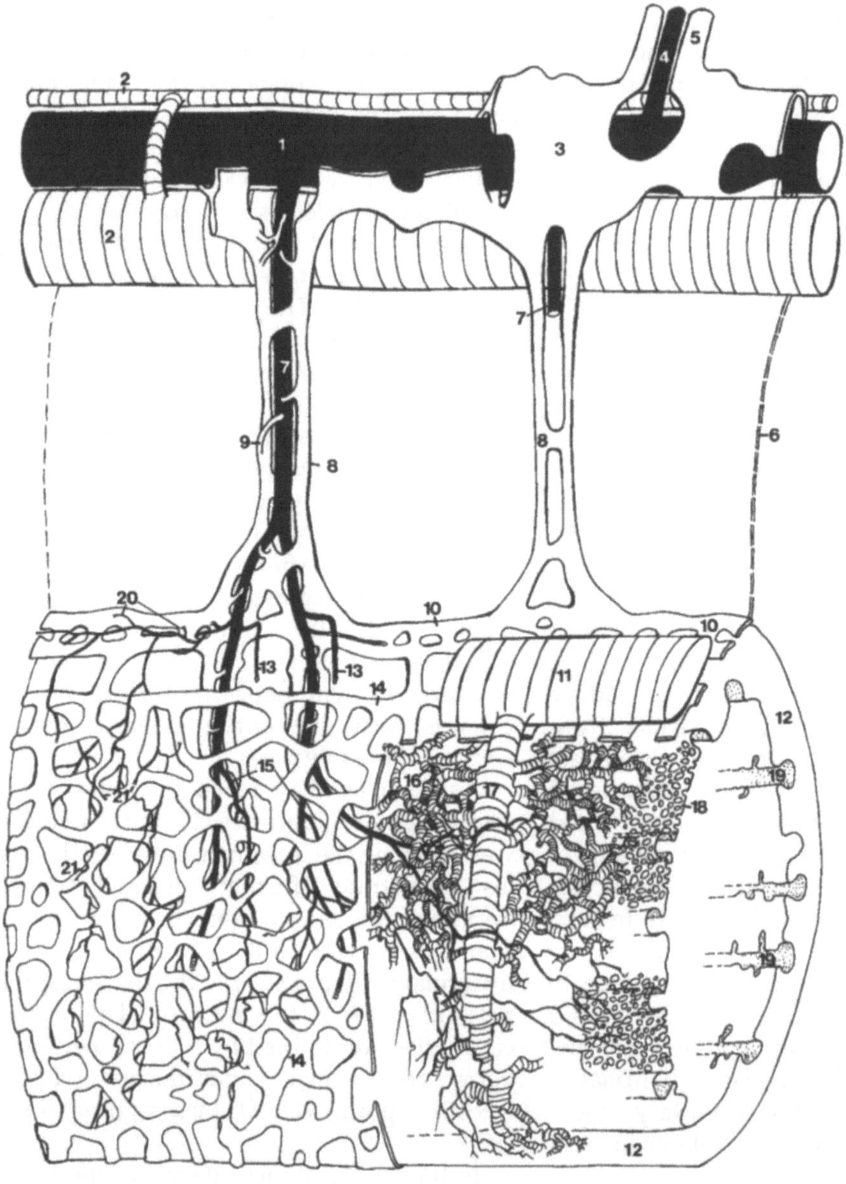

Figure 16.3 Schematic drawing of the intestinal vasculature of *Myxine glutinosa* (from Franko-Dossar and Lomholt, in press, with permission). (1) Dorsal aorta; (2) cardinal veins; (3) cardinal lymphatic sinus; (4) segmental artery of body wall; (5) segmental lymphatic of body wall; (6) intestinal mesentery; (7) segmental artery in mesentery; (8) segmental lymphatics in mesentery; (9) anastomosis between artery and lymphatic; (10) intestinal lymph sinus; (11) portal vein; (12) intestinal wall; (13) cut artery to ovary; (14) large lymphatic network of intestinal serosa; (15) arterial branches of intestinal wall; (16) venous plexus of intestinal wall; (17) segmental intestinal vein; (18) capillary network of intestinal mucosa; (19) longitudinal mucosal folds; (20) small arteries opening into intestinal lymph sinus; (21) small arteries opening into large lymphatic network.

was found for lymphatics in the deeper layers of the intestinal wall. This is in accordance with the findings of Cole (1925) and Fänge (1973), whereas Grodzinski (1932) claimed that lymphatics penetrate through the intestinal wall towards the mucosa. Being unaware of connections between arteries and the lymphatic network, Grodzinski may have mistaken arteries filled from the lymphatics for small lymphatic vessels.

Concerning the rest of the system of 'white' lymphatics the following observations of a more preliminary nature were made. Dye injected into segmental arteries of the body wall musculature was observed to return via the paired segmental lymphatics. The dye first appeared at the peripheral end of the segmental lymphatics and moved towards the centre and into the cardinal lymphatics. Although a few anastomoses were seen between segmental arteries and lymphatics, this observation suggests that deeper in the musculature many more must exist, which remain to be precisely located. No lymphatic vessels are present along the segmental veins.

Immediately after leaving the dorsal aorta, the segmental artery gives off a dorsal branch which eventually passes through the dorsal midline into the skin overlying the subcutaneous sinus. This artery and its branches are located superficially facing the lumen of the subcutaneous sinus. When the sinus is opened in live, anaesthetized animals, the artery and its branches – but not the corresponding veins – are seen to be followed by paired lymphatic vessels. Injection of the plastic methyl methacrylate into the arteries revealed many anastomoses between the arteries and their accompanying lymphatic vessels. In addition the injection material was sometimes seen to penetrate into the subcutaneous sinus. This suggests the presence of connections like the subcutaneous anastomoses between the distal ends of segmental lymphatics of the body wall and the subcutaneous sinus (see also Figure 31.7). Finally the arteries penetrated the subdermal and dermal connective tissue to give rise to the dense capillary network beneath the epidermis described by Hans and Tabencka (1938) and Lametschwandtner *et al.* (1989). See also Welsch and Potter, this volume.

In spite of the inspection of many casts of the dorsal aorta, no traces of anastomoses between the aorta and the cardinal lymphatics were found.

In accordance with previous investigators (Cole, 1925; Johansen *et al.*, 1962) we have found that the fluid in the 'white' lymphatics often appears colourless. On close inspection we have, however, always observed a few red cells in these vessels. In some cases we have found the cardinal and segmental lymphatics of freshly opened animals to appear pink from a high content of red cells. With the demonstration of numerous anastomoses with the arterial system this should no longer be considered an artefact.

The conclusion to be drawn from the observations described above is that many anastomoses exist between the arterial system and all parts of the system of 'white lymphatics' with the exception of the cardinal lymphatics. Consequently the notion that the 'white' lymphatics are true lymphatics originating from blind ending vessels in the tissue can no longer be upheld. Instead it appears that the hagfish has a secondary circulatory system resembling that of teleosts. There are, however, major differences between the secondary circulatory systems of the two groups, which will be discussed below.

16.6 A NOVEL SYSTEM OF FINE CHANNELS IN THE SEROSAL MESOTHELIUM OF THE INTESTINE

The network of lymphatics of the serosal surface of the intestine has been described before (Cole, 1925; Grodzinski, 1932; Fänge, 1973), although the connections with the arterial system were not noticed. In addition to this network, plastic injections revealed a finer network of channels, which has not been described before. Electron microscopical sections of the serosa gave the rather surprising result that these fine channels are enclosed within the serosal mesothelium. Figure 16.4a shows how mesothelial cells, resting on the basal lamina, give off thin flaps, which make contact and surround the channels of this fine network. The structure of these mesothelial cells is somewhat like that of the endothelial pillar cells of fish gills, which line the blood spaces of the secondary gill lamellae. Light microscopical sections revealed that this fine network is densely distributed over the entire intestinal surface. Plastic injections through either the cardinal lymphatics or the dorsal aorta showed that the fine network can be filled from the coarse lymphatic network. In some cases small arterial branches injected with plastic appeared to have many small openings in the wall, which looked similar to those found in the arterial branches in the skin. Whether these openings form secondary vessels parallel to the artery as in the skin could not, however, be determined. In a few cases minute arterial branches were found to communicate directly with the fine network of channels. A similar system of fine channels was found in the mesothelium covering the intestinal mesenterium.

Figure 16.4b presents our present understanding of the structure of the intestinal serosa and its various vessels. Superficially located small arteries (7) communicate (8) with the previously described lymphatic network (5) embedded in the *tunica serosa* (1). The mesothelium is shown in black in cross-section and encloses a fine network of channels (4) which covers the entire surface. Scattered over the serosal surface of the intestine groups of deep infoldings or crypts (12) were found. The mesothelial cells in and around the crypts are thinly covered by *microvilli*. Around the crypts are canals in the tissue of unknown affiliation. Whether the lumen of the crypts communicates with the secondary vessel system of the intestine has not been established. Along the margins of the vessels of the large lymphatic network outgrowths from endothelial cells enclose a system of small vessels (6) which probably communicate with the network surrounded by the mesothelial cells.

16.7 COMPARISON BETWEEN THE HAGFISH SINUS SYSTEM AND THE TELEOST SECONDARY VASCULAR SYSTEM

The finding in *Myxine* of numerous anastomoses connecting arteries to vessels until now believed to be lymphatics, invites a comparison with the secondary vascular system of teleosts. When attempting such a comparison it should be kept in mind that the hagfishes and the teleosts are, after all, only remotely related.

In those parts of the secondary vascular system of teleosts, which are reasonably well known, a clear separation exists between secondary arteries, capillaries and veins (Vogel, 1985a; Steffensen *et al.*, 1986). In the moderately developed fin membrane surrounding the hagfish tail the sinus is divided into channels running parallel to the fin rays. The radial channels communicate with a vessel running along the base of the tail fin (Cole, 1925). These structures may correspond to the venous and capillary part of the secondary vessel system in the anal fin of the glass catfish (Steffensen *et al.*, 1986), but there are no secondary arteries. Apart from this nothing in the way of secondary capillary networks has been found in *Myxine*. Instead the entire body is surrounded by the subcutaneous sinus. In a comparison with the teleost system, the subcutaneous sinus would represent the capillary and venous part of the secondary circulation of the skin as suggested by Satchell (1991).

Since the various segmental lymphatics of the hagfish have been shown to anastomose with the arterial system, these vessels should be comparable to the secondary arteries of

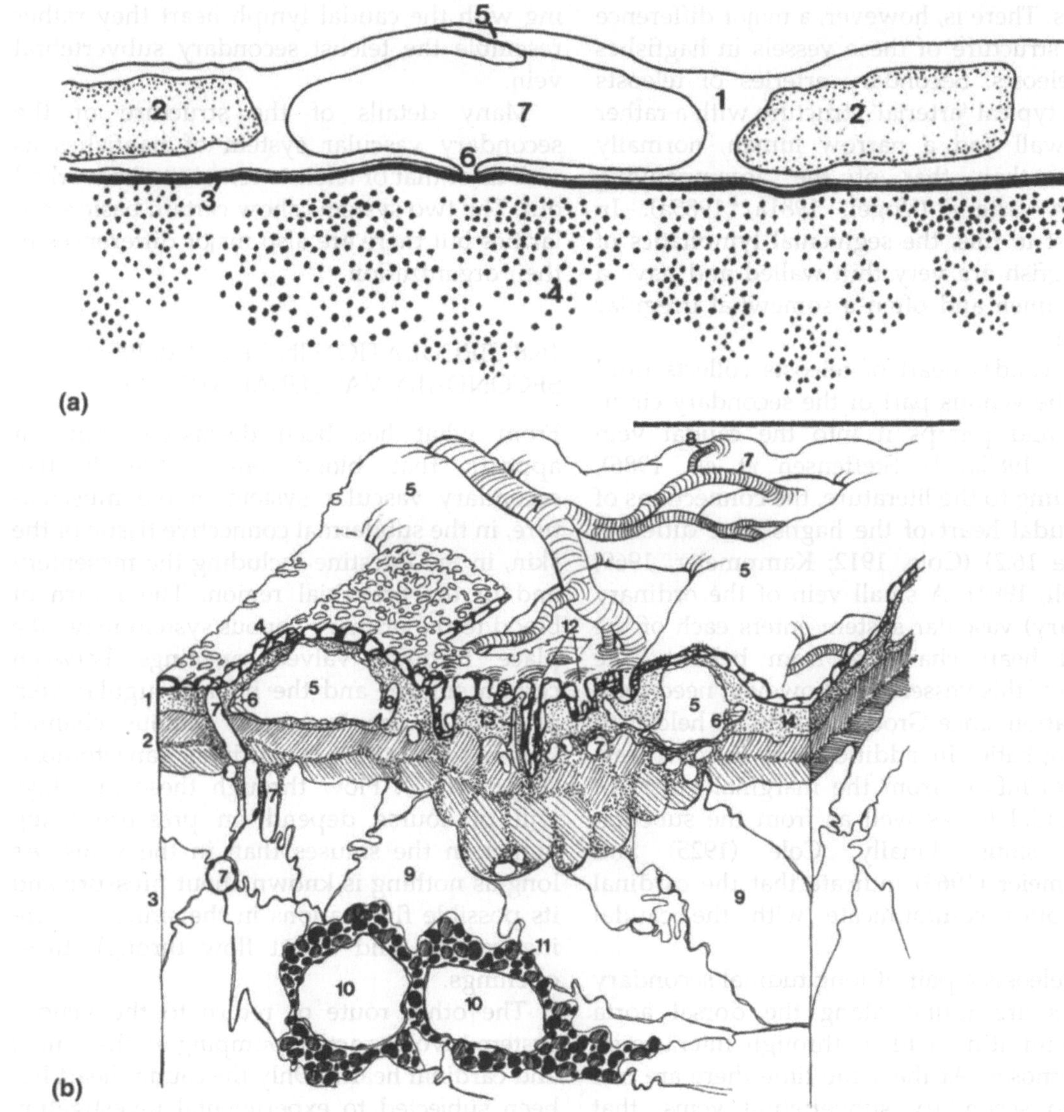

Figure 16.4 (a) Cross-section of a channel enclosed by mesothelial cells of the intestinal serosa. Based on an electron micrograph. Scale bar 5 microns. (1) Mesothelium of intestinal serosa; (2) nucleus of mesothelial cell; (3) basal lamina; (4) collagen; (5) contact between two outgrowths from neighbouring mesothelial cells; (6) a more open contact between mesothelial cells; (7) lumen of channel surrounded by mesothelial cells. (b) Block diagram of the outer part of the intestinal wall (from Franko-Dossar and Lomholt, in press, with permission). (1) *Tunica serosa*; (2) *tunica muscularis*; (3) *submucosa*; (4) fine network of channels within mesothelium; (5) large lymphatic network; (6) fine channels surrounded by outgrowths from endothelium of large lymphatic network; (7) superficial artery; (8) anastomosis between artery and large lymphatic network; (9) fatty tissue of *submucosa*; (10) vein; (11) lymphoid tissue; (12) cluster of infoldings of *serosa* (crypts); (13) tissue channels of unknown affiliation around crypts; (14) connective tissue of *tunica serosa*.

teleosts. There is, however, a major difference in the structure of these vessels in hagfishes and teleosts. Secondary arteries of teleosts have a typical 'arterial' structure with a rather thick wall and a narrow lumen, normally smaller than that of the accompanying primary artery (Vogel, 1981a, 1985a). In contrast to this, the segmental lymphatics of the hagfish are very thin walled and have a wide lumen and often a somewhat irregular outline.

The caudal heart of teleosts collects fluid from the venous part of the secondary circulation and pumps it into the caudal vein (Vogel, 1985a, b; Steffensen *et al.*, 1986). According to the literature, the connections of the caudal heart of the hagfish are different (Figure 16.2) (Cole, 1912; Kampmeier, 1969; Satchell, 1984). A small vein of the ordinary (primary) vascular system enters each of the caudal heart chambers from behind. The nature of this vessel may, however, need reinvestigation since Grodzinski (1932) held it to be lymphatic. In addition, the caudal heart receives inflow from the marginal vessel of the caudal fin as well as from the subcutaneous sinus. Finally, Cole (1925) and Kampmeier (1969) indicate that the cardinal lymphatics communicate with the caudal heart.

In teleosts a pair of longitudinal secondary arteries are found along the dorsal aorta communicating with it through interarterial anastomoses. At the same time there are one or two secondary subvertebral veins, that discharge into the caudal heart (Vogel, 1985a, b). In the hagfish there is only one set of longitudinal channels, the cardinal lymphatics in the terminology of Kampmeier (1969). The cardinal lymphatics communicate with the segmental lymphatics which, as suggested above, may compare to secondary arteries. In that respect the cardinal lymphatics resemble the longitudinal secondary arteries of teleosts. However, as mentioned previously, no anastomoses with the dorsal aorta have been found, and in communicating with the caudal lymph heart they rather resemble the teleost secondary subvertebral vein.

Many details of the structure of the secondary vascular system of hagfishes as well as of that of teleosts remain to be worked out. The two systems show certain basic similarities but there are also major differences in their organization.

16.8 CIRCULATION IN THE HAGFISH SECONDARY VASCULAR SYSTEM

From what has been discussed above it appears that blood enters the hagfish secondary vascular system in the musculature, in the subdermal connective tissue of the skin, in the intestine including the mesentery and in the branchial region. The return of blood to the primary venous system may take place through valved openings between certain sinuses and the inferior jugular vein as well as the transverse venous channel known as the precardinal anastomosis (Figure 16.1). Flow through these openings will of course depend on pressure being higher in the sinuses than in the veins. As long as nothing is known about pressure and its possible fluctuations in the sinuses, nothing can be said about flow through these openings.

The other route of return to the venous system involves active pumping by the caudal and cardinal hearts. Only the caudal heart has been subjected to experimental investigation (Satchell, 1984). Its pumping activity was found to be quite variable. In animals resting motionless on the bottom of an aquarium it often stopped beating, whereas the highest rate of beating was seen immediately after bouts of swimming activity. This suggests that more blood enters the sinus system from the arterial system during activity.

Satchell (1984) recorded pressure pulses in the caudal vein and in the subcutaneous sinus corresponding to the activity of the caudal heart. The pressure fluctuations were small, a few mm of water, and their exact magnitude was dependent on the location of the tip of the catheter used to record them. In some cases a negative pressure was recorded in the subcutaneous sinus. A negative pressure is clearly evidence of the activity of the caudal heart (and possibly the cardinal heart). Since the skin covering the subcutaneous sinus is flaccid, a negative pressure could hardly occur unless the lymph hearts drain the sinus of fluid almost completely.

The volume and turnover of blood of the subcutaneous sinus was studied by Forster *et al.* (1989) in *Eptatretus cirrhatus*. Tracer injected into the subcutaneous sinus and left to mix for a period short enough to exclude any appreciable mixing into the rest of the vascular system yielded a volume of the sinus of about 55 ml kg^{-1} body weight. Based on the distribution of marked red cells the volume of the primary vascular system was found to be 122 ml kg^{-1} body weight. It may not be correct when the authors add these two figures to obtain a total blood volume of 177 ml kg^{-1}, since other parts of the secondary vascular system may have a significant volume.

Forster *et al.* (1989) also determined the distribution space of radioactive serum albumin. After injection of albumin into the central circulation it took 20–24 h before the specific activity of plasma from the subcutaneous sinus was equal to that of plasma from the central circulation. This indicates a low rate of exchange between the primary and secondary vessel system and resembles what was found in rainbow trout (Steffensen and Lomholt, 1992). The albumin space was found to be 285 ml kg^{-1}, which is considerably larger than the calculated value for total blood volume. It was concluded that the tracer must have escaped into the interstitial space and that the albumin space thus represented the entire extracellular volume.

The assumption that albumin will distribute easily to the interstitial space was based on the findings of Hargens *et al.* (1974) who reported a high permeability of fish capillaries to protein. The high capillary permeability was not demonstrated directly but deducted from the presence of equal concentrations of protein in plasma and interstitial fluid. At that time the classical lymphatic function of removing protein from the interstitium was taken for granted in fish. If this function of protein removal is not operative because terminal lymphatics do not exist in fish, equal protein concentrations in plasma and interstitium do not necessarily indicate a high protein permeability of the capillary wall. Casley-Smith and Casley-Smith (1975) reported capillaries from hagfish thyroid and pituitary to show many open junctions and they saw this as a general, primitive trait. On the contrary Bundgaard and Cserr (1981) examined brain capillaries and found very few if any open junctions. They also found a blood-brain barrier comparable to that in mammals. This shows that at least one capillary bed of hagfish, that of the brain, is not very permeable to protein.

In accordance with earlier work by Johansen *et al.* (1962), Forster *et al.* (1989) found haematocrit to be low in the subcutaneous sinus compared to the central circulation (4.3% vs 13.5%) and this difference was maintained during moderate swimming activity.

Sundin *et al.* (1994) used a perfused gill preparation from *Eptatretus cirrhatus* to investigate the control of the distribution of blood flow between the arterio-arterial and the arteriovenous pathways through the gills. The arteriovenous pathway may be considered the counterpart of the secondary circulation in the gills. They concluded that the partition of blood flow between these pathways is under the control of circulating catecholamines. Whether similar mechanisms control the vascular resistance of the anastomoses, and hence the flow between the arteries and the secondary vessels in the rest of the body, is not known.

16.9 POSSIBLE FUNCTIONS OF THE HAGFISH SECONDARY VASCULAR SYSTEM

Very little physiological work has been carried out to throw light on functional aspects of the secondary circulation of teleosts, and even less is known about the sinus system of hagfishes.

As pointed out for teleosts (Steffensen *et al.*, 1986; Steffensen and Lomholt, 1992) the low haematocrit of the blood of the secondary system rules out a significant role in blood gas transport. There is evidence in trout that the secondary system plays a role in the exchange across the skin of acid-base relevant ions (Ishimatsu *et al.*, 1992, 1995). Since the subcutaneous sinus and the secondary vessels along the arteries of the skin are separated from the surface by a substantial layer of subdermal, dermal and epidermal tissue it seems impossible that they take part in any significant exchanges with the ambient water.

The discovery of the system of fine channels enclosed within the mesothelium of the intestinal surface, as well as the deep infoldings or crypts of this surface, invites speculations concerning fluid exchange of the peritoneal cavity. The crypts may be reminiscent of the peritoneal stomata of mammals, which are believed to form a link between the lymphatic system and the peritoneal cavity (Wang, 1975; Leak and Rahil, 1978; Negrini *et al.*, 1991).

The most conspicuous part of the secondary vascular system of hagfishes is the large subcutaneous sinus, the functional significance of which is by no means clear. It has a potential volume much larger than the volume of fluid which it actually contains. An extremely simplified description of the circulatory system of hagfishes would be to say that it consists of a high pressure primary circulation connected via narrow anastomoses to a low pressure, flaccid subcutaneous sinus. Could it be that the subcutaneous sinus acts as a kind of volume buffer in an animal with a limited potential for volume regulation? A change in total blood volume could be accommodated, so to speak, in the flaccid subcutaneous sinus with little effect on its pressure. Consequently the volume and pressure of the primary circulation could be kept constant in spite of a change in total blood volume.

16.10 EVOLUTIONARY IMPLICATIONS

Until the discovery of the secondary circulatory system of teleosts it was assumed that all vertebrates possess lymphatic vessels in the mammalian sense of the term, although parts of the system may be poorly differentiated from the venous system in cyclostomes and some elasmobranch fishes (Kampmeier, 1969). In contrast Vogel (1985a) concluded that true lymphatics do not exist in teleost fishes. He found the first unequivocal evidence for true terminal lymphatics in lungfishes (Vogel, 1988). Although interarterial anastomoses have not been demonstrated, Satchell (1992) considers the cutaneous veins and caudal sinus of sharks to be parts of a secondary circulatory system. The demonstration in the hagfish – the most primitive among extant vertebrates – of anastomoses between arteries and vessels believed to be lymphatics, supports the conclusion that no truly aquatic vertebrates have a lymphatic vessel system in the strict sense of the term. Instead they have a secondary circulatory system, the functional significance of which is as yet largely unknown. It is doubtful whether this secondary vascular system may be considered an evolutionary forerunner of the lymphatic system of terrestrial vertebrates (Steffensen and Lomholt, 1992). A true lymphatic system appears to have evolved with the evolution of terrestrial vertebrates, possibly as a consequence of a selection pressure resulting from hydrostatic forces acting on the vasculature of terrestrial animals but not on that of aquatic animals.

ACKNOWLEDGEMENT

The work of the authors has been supported by a grant from Elisabeth and Knud Petersens Foundation.

REFERENCES

Allen, W.F. (1913) Studies on the development of the venolymphatics in the tail region of *Pollistotrema* (*Bdellostoma*) *stoutii*. *Quart. J. Micr. Sci.*, **59**, 309–60.

Bundgaard, M. and Cserr, H.F. (1981) Impermeability of hagfish cerebral capillaries to radio-labelled polyethylene glycols and to microperoxidase. *Brain Research*, **206**, 71–81.

Casley-Smith, J.R. and Casley-Smith, Judith R. (1975) The fine structure of the blood capillaries of some endocrine glands of the hagfish, *Eptatretus stoutii*: implications for the evolution of blood and lymph vessels. *Revue Suisse Zool.*, **82**, 35–40.

Cole, F.J. (1912) A monograph of the general morphology of the myxinoid fishes, based on a study of *Myxine*. IV: On some peculiarities of the afferent and efferent branchial arteries of *Myxine*. *Trans. Roy. Soc. Edin.*, **48**, 215–30.

Cole, F.J. (1925) A monograph of the general morphology of the Myxinoid fishes based on a study of *Myxine*. VI: The morphology of the vascular system. *Trans. Roy. Soc. Edin.*, **54**, 309–42.

Elger, M. (1987) The branchial circulation and the gill epithelia in the Atlantic hagfish, *Myxine glutinosa* L. *Anat. Embryol.*, **175**, 489–504.

Forster, M.E., Davison, W., Satchell, G.H. and Taylor, H.H. (1989) The subcutaneous sinus of the hagfish, *Eptatretus cirrhatus* and its relation to the central circulating blood volume. *Comp. Biochem. Physiol.*, **93A**, 607–12.

Franko-Dossar, F. and Lomholt, J.P. (1997) Anastomoses between arteries and 'white' lymphatics in the Atlantic hagfish, *Myxine glutinosa*. (In press.)

Fänge, R. (1973) The lymphatic system of *Myxine*. *Acta Regiae Societatis Scientiarum et Litterarum Gothenburgensis, Zoologica*, **8**, 57–64.

Grodzinski, Z. (1962) Über das Blutgefässsystem von *Myxine glutinosa* L. *Bull. Int. Acad. Pol. Sci. Lett., Cl. Sci. Math. Nat.*, **B38**, 123–57.

Grodzinski, Z. (1932) Bemerkungungen über das Lymphgefässsystem von *Myxine glutinosa* L. *Bull. Int. Acad. Pol. Sci. Lett., Cl. Sci. Math. Nat.*, **B44**, 221–36.

Hans, M. and Tabencka, Z. (1938) Über die Blutgefässe der Haut von *Myxine glutinosa* L. *Bull. Int. Acad. Pol. Sci., Ser. BII Zool.*, 69–77.

Hargens, A.R., Millard, R.W. and Johansen, K. (1974) High capillary permeability in fishes. *Comp. Biochem. Physiol.*, **48A**, 675–80.

Hoyer, H. (1934) Das Lymphgefässsystem der Virbeltiere vom Standpunkte der vergleichende Anatomie. *Mem. Acad. Pol. Sci.* (*Cracow*), **A(1)**, 1–205.

Ishimatsu, A., Iwama, G.K., Bentley, G.K. and Heisler, N. (1992) Contribution of the secondary circulatory system to acid-base regulation during hypercapnia in rainbow trout (*Oncorrhynchus mykiss*). *J. Exp. Biol.*, **170**, 43–56.

Ishimatsu, A., Iwama, G.K. and Heisler, N. (1995) Physiological roles of the secondary circulatory system in fish, in *Advances in Comparative and Environmental Physiology*, Vol 21 (ed. N. Heisler), Springer, Berlin, pp. 215–36.

Jackson, C.M. (1901) An investigation of the vascular system of *Bdellostoma dombeyi*. *J. Cincinnati Soc. Nat. Hist.*, **20**, 13–47.

Johansen, K. (1963) The cardiovascular system of *Myxine glutinosa* L., in *The Biology of Myxine* (eds A. Brodal and R. Fänge), Universitetsforlaget, Oslo, pp. 289–316.

Johansen, K., Fänge, R. and Johannesen, M.W. (1962) Relations between blood, sinus fluid and lymph in *Myxine glutinosa* L. *Comp. Biochem. Physiol.*, **7**, 23–8.

Kampmeier, O.F. (1969) *Evolution and Comparative Morphology of the Lymphatic System*, Charles C. Thomas, Springfield, Illinois, pp. 198–210.

Lametschwandtner, A., Weiger, T., Lametschwandtner, U., Gerorgieva-Hanson, V., Patzner, R.A. and Adam, H. (1989) The vascularization of the skin of the Atlantic hagfish, *Myxine glutinosa*, as revealed by scanning electron microscopy of vascular corrosion casts. *Scanning Microscopy*, **3**, 305–14.

Leak, L.V. and Rahil, K. (1978) Permeability of the diaphragmatic mesothelium: the ultrastructural basis for 'Stomata'. *Amer. J. Anat.*, **151**, 557–94.

Müller, J. (1841) Vergleichende Anatomie der Myxinoiden. Dritte Fortsetzung. Über das Gefässsystem. *Abhandlungen der königlichen Akademie der Wissenschaften zu Berlin*, aus dem Jahre **1839**, 175–303.

Negrini, D., Mukenge, S., Del Fabro, M., Gonano, C. and Miserocchi, G. (1991) Distribution of diaphragmatic lymphatic stomata. *J. Appl. Physiol.*, **70**, 1544–9.

Pohla, H., Lametschwandtner, A. and Adam, H. (1977) Die Vaskularisation der kiemen von *Myxine glutinosa* L. (Cyclostomata). *Zool. Scripta*, **6**, 331–41.

Retzius, A. (1824) Ytterligare bidrag til anatomien af *Myxine glutinosa. Kongl. Vetenskaps Akademiens Handlinger*, Stockholm, 408–31.

Retzius, G. (1890) Eine sogenanntes Caudalherz bei *Myxine glutinosa. Biologische Untersuchungen* (Neue Folge), Stockholm, **1**, 94–6.

Satchell, G.H. (1984) On the caudal heart of *Myxine. Acta Zool. (Stockholm)*, **65**, 125–33.

Satchell, G.H. (1991) *Physiology and Form of Fish Circulation*, Cambridge University Press, Cambridge.

Satchell, G.H. (1992) The venous system, in *Fish Physiology*, Vol XII, Part A, *The Cardiovascular System* (eds W.S. Hoar, D.J. Randall and A.P. Farrell), Academic Press, San Diego, pp. 141–83.

Steffensen, J.F., Lomholt, J.P. and Vogel, W.O.P. (1986) In *vivo* observations of a specialized microvasculature, the primary and secondary vessels in fish. *Acta Zool. (Stockholm)*, **67**, 193–200.

Steffensen, J.F. and Lomholt, J.P. (1992) The secondary vascular system, in *Fish Physiology*, Vol XII, Part A, *The Cardiovascular System* (eds W.S. Hoar, D.J. Randall and A.P. Farrell), Academic Press, San Diego, pp. 185–217.

Sundin, L., Axelson, M., Nilsson, S., Davison, W. and Forster, M.E. (1994) Evidence of regulatory mechanisms for the distribution of blood between the arterial and the venous compartments in the hagfish gill pouch. *J. Exp. Biol.*, **190**, 281–6.

Vogel, W.O.P. (1981a) Struktur und Organisationsprinzip im Gefässsystem der Knochenfische. *Gegenbaurs. Morph. Jahrb*, **127**, 772–84.

Vogel, W.O.P. (1981b) Das Lymphgefässsystem der Knochenfische – Eine Fehlinterpretation? *Vehr. Anat. Ges. Antwerpen*, **75**, 733–5.

Vogel, W.O.P. (1985a) Systemic vascular anastomosis, primary and secondary vessels in fish, and the phylogeny of lymphatics, in *Cardiovascular Shunts*, Benzon Symposium no. 21 (eds K. Johansen and W.W. Burggren), Munksgaard, Copenhagen, pp. 143–51.

Vogel, W.O.P. (1985b) The caudal heart of fish: not a lymph heart. *Acta Anat.*, **121**, 41–5.

Vogel, W.O.P. (1988) Structure and evolution of lymphatics. Abstract #509, *Proc. 2nd Int. Congr. Comp. Physiol. Biochem*, Baton Rouge, Louisiana.

Vogel, W.O.P. and Claviez, M. (1981) Vascular specialization in fish, but no evidence for lymphatics. *Z. Naturforsch.*, **36c**, 490–2.

Wang, N..S (1975) The preformed stomatas connecting the pleural cavity and the lymphatics in the parietal pleura. *Amer. Rev. Respir. Dis.*, **111**, 12–20.

17
DERMAL CAPILLARIES

Ulrich Welsch and Ian C. Potter

SUMMARY

Hagfishes possess a very well developed dermal capillary network, with most of the capillaries lying 4–10 μm below the epidermis, which is c. 80 μm in depth. The capillaries are lined by a continuous single layer of endothelial cells (thickness 120–170 nm). Flange-like extensions from adjacent endothelial cells form complex interdigitations, thereby facilitating intercellular contacts through simple desmosome-like specializations. The basal lamina of the endothelial cells, which comprises a very conspicuous *lamina lucida* and *lamina densa*, is contiguous in places with that of pericytes. The pericapillary space surrounding the endothelium (diam. 0.5–0.9 μm) contains irregularly arranged collagen fibrils and oxytalan fibres. The ultrastructural characteristics of the dermal capillaries of hagfishes, and particularly of their endothelial cells, differ in several respects from those of lampreys, the only other extant agnathan group. The dermal capillary network is best developed in hagfishes that burrow. This suggests that, when the skin of hagfishes is surrounded by an anoxic environment, this network could play an important role in supplying, to the metabolically demanding epidermal cells, oxygen derived from gill respiration. The network would also supply the epidermal cells with precursors of the copious amounts of mucus they produce.

17.1 INTRODUCTION

The hagfishes (Myxiniformes), together with the lampreys (Petromyzontiformes), are the sole survivors of the agnathan (jawless) stage in vertebrate evolution (Hardisty, 1979). Unlike extinct agnathans, such as the cephalaspids, heterostracans and anaspids, the hagfishes and lampreys do not possess a dermal skeleton consisting of minute scales and/or bony plates (Forey and Janvier, 1993). Furthermore, their skin is not covered by the denticles or scales that are typically found in that region in gnathostome (jawed) fishes (Whitear, 1986). Since a dermal skeleton is lacking, the epidermis that forms the outer layer of the skin of living agnathans is in direct contact with the aquatic environment in which these jawless fish live (Lethbridge and Potter, 1981; Whitear, 1986).

The dermis underlying the epidermis varies in thickness, both between hagfishes and lampreys and amongst the life cycle stages and species of the latter group (Lethbridge and Potter, 1980; Potter *et al.*, 1995; Welsch *et al.*, this volume). Furthermore, capillaries are present in the dermis of presumably all hagfishes and the adults of several lamprey species (Lametschwandtner *et al.*, 1989; Potter *et al.*, 1995). Capillaries are not present in the thin dermis of larval lampreys (Potter *et al.*, 1995).

Since the epidermal cells are metabolically active, producing copious amounts of mucus (Blackstad, 1963; Hardisty, 1979; Spitzer and Koch, this volume), they are obviously dependent on both a good supply of oxygen and the precursors of the substances they secrete. Steffensen *et al.* (1984) estimated that about 80% of the oxygen taken up by the hagfish *Myxine glutinosa* can occur across the skin.

The Biology of Hagfishes. Edited by Jørgen Mørup Jørgensen, Jens Peter Lomholt, Roy E. Weber and Hans Malte. Published in 1998 by Chapman & Hall, London. ISBN 0 412 78530 7.

Burggren *et al.* (1985) went on to suggest that cutaneous exchange in hagfishes would be facilitated by the presence of a 'very thin epidermis and dense sub-epidermal (= dermal) capillary network' in this group. However, Malte and Lomholt, in a chapter in this book, provide evidence which indicates that the skin does not act as an important accessory gas exchange organ in hagfishes.

This chapter has focused on describing the histochemistry and ultrastructure of the dermal capillary network of hagfishes. Comparisons are then made with the corresponding network found in lampreys. Finally, the possible functional significance of the dermal capillary network in transporting oxygen, other molecules and various ions is discussed.

17.2 STRUCTURE OF DERMAL CAPILLARIES IN HAGFISHES

In this chapter, all dermal blood vessels are termed capillaries, irrespective of their diameter. Previous light and electron microscopical studies have shown that the hagfishes *Myxine glutinosa*, *Paramyxine* (= *Eptatretus*) *atami* and *Eptatretus stoutii* each possess an elaborate network of dermal capillaries (Rabl, 1931; Hans and Tabencka, 1938; Lametschwandtner *et al.*, 1989; Potter *et al.*, 1995). Our current work shows that this also applies to *E. burgeri* (Figure 17.1a). The elaborate meshwork of capillaries within and immediately beneath the dermis of *M. glutinosa* has been described by Lametschwandtner *et al.* (1989) on the basis of vascular corrosion casts. In each species of hagfish, the capillary network is best developed in the outer third of the dermis and has a similar structure and arrangement in the dorsal, lateral and ventral parts of the anterior, middle and posterior regions of the trunk. However, the degree to which this network is developed varies among species, the possible significance of which is discussed later. The capillary networks in the oral and nasal tentacles of *Myxine* are particularly well developed, comprising wide and frequently anastomosing vessels.

The majority of the capillaries are found in a zone 4 to 10 μm beneath the basal lamina of the epidermis (Figure 17.1a, b, c). Since the depth of the epidermis is c. 80 μm, there is a barrier of about 84 to 90 μm between the body surface and blood space in the integument. The capillaries are usually loosely surrounded by the long cytoplasmic extensions of melanocytes, which are also found in this zone and are particularly prominent in *E. stoutii* and *E. burgeri*.

Although the capillaries are located mainly in the outer region of the dermis, they can also be seen, at intervals, running vertically or at an angle through the dermis (Figure 17.1a, b). In cross-sections, the capillaries are usually oval in outline (Figure 17.1d), the long diameter running parallel to the surface of the epidermis. The capillaries in the outer part of the dermis vary in diameter from about 7 to 50μm, a range which encompasses the mean diameter recorded for these capillaries in *M. glutinosa* by Lametschwandtner *et al.* (1989). Here and elsewhere in the dermis, the capillaries are lined by a continuous endothelium, which consists of a single layer of cells, whose thickness ranges from 120 to 700 nm (Figure 17.1d). The endothelial cells are flattened and each contains a nucleus rich in euchromatin, but with clumps of heterochromatin towards the periphery (Figure 17.2a), and a prominent nucleolus. The capillary wall binds, with equal intensity, the lectins HPA (*Helix pomatia* agglutinin), PNA (peanut agglutinin) and UEA I (*Ulex europaeus* agglutinin I). Lectins are proteins, often isolated from plant seeds, which selectively bind to carbohydrates and can thus be used as tools to demonstrate the presence of sugars in tissue sections. Different lectins often have different sugar specificities; HPA binds to *N*-acetylgalactosamine, PNA to galactose, UEA I to fucose (Schumacher *et al.*, 1996). Although it could not be determined with absolute

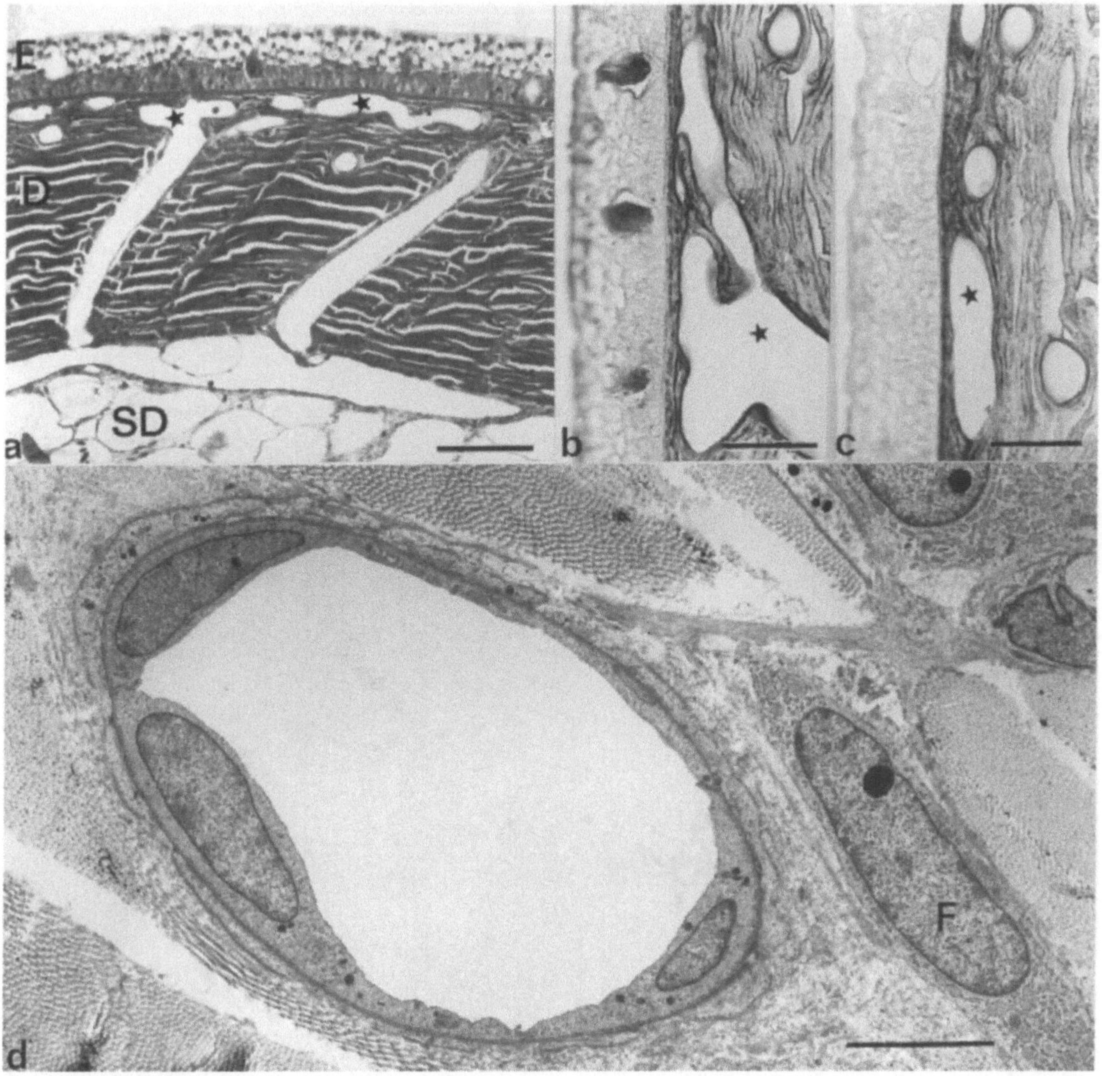

Figure 17.1 Dermal capillaries of *Eptatretus burgeri*. (a) Section of the integument of the dorsal region of the trunk. E, epidermis; D, dermis; SD, subdermis. Two capillaries run obliquely through the dermis, while the majority of the capillaries (asterisks) lie immediately beneath the epidermis; scale bar 100 µm. (b) WGA binding in the pericapillary space. Asterisk denotes capillary lumen; scale bar 50 µm. (c) RCA I binding in the pericapillary space and in the uppermost zone of the dermis. Asterisk denotes capillary lumen; scale bar 50 µm. (d) Low power electron micrograph of a typical dermal capillary. F, fibroblast; scale bar 5 µm.

certainty whether these lectins bind to the thin endothelial cytoplasm or subendothelial structures, it is noteworthy that, in the endothelia of mammals, such binding occurs in the cytoplasm (Holthofer *et al.*, 1982). The dermal capillaries of all hagfishes studied also stain with the PAS-reaction and WGA (wheat germ agglutinin) which binds to *N*-acetylglucosamine (Figure 17.1b) and furthermore react with RCA I (*Ricinus communis* agglutinin

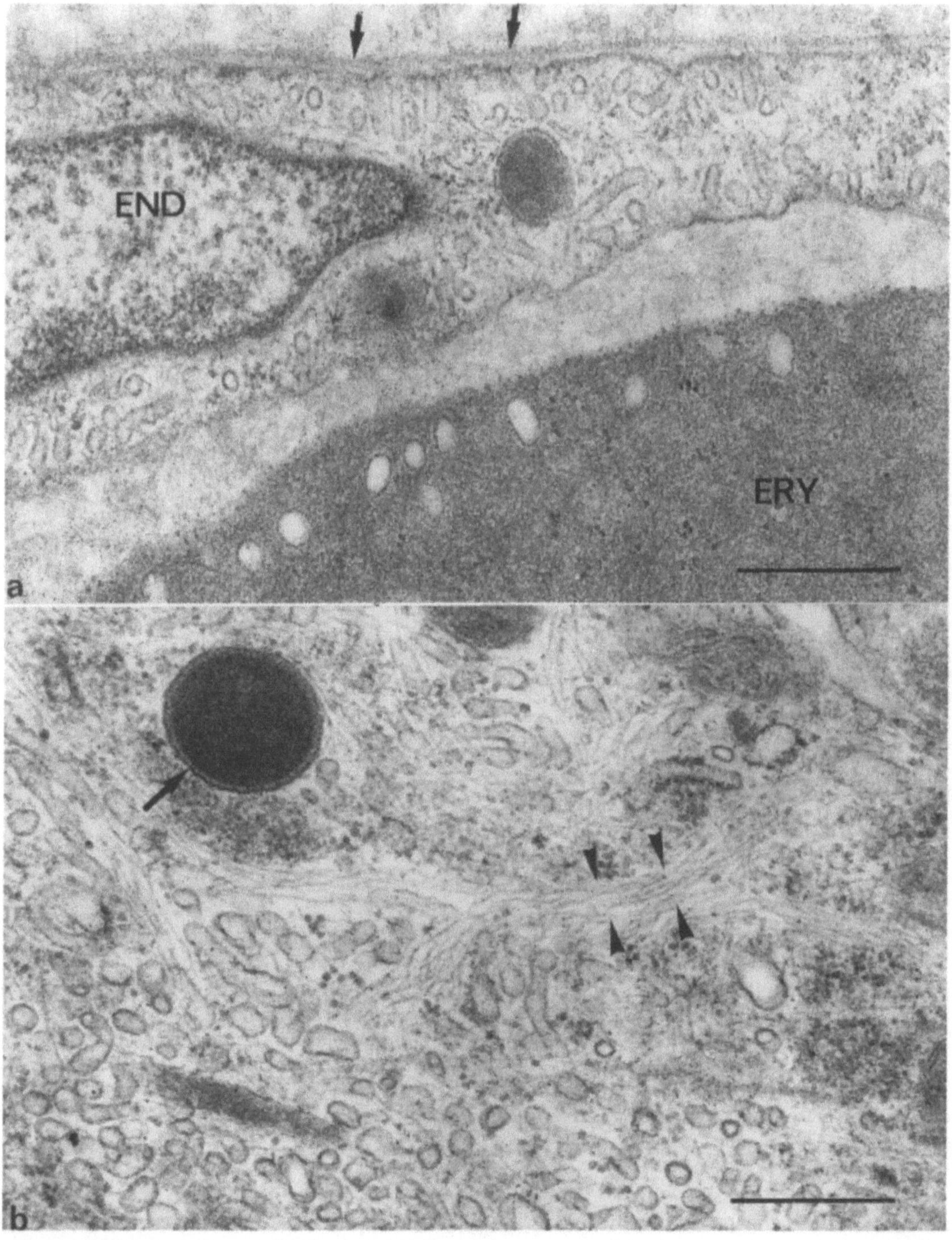

Figure 17.2 Ultrastructural details of the endothelial cells of the dermal capillaries of *Paramyxine atami*. (a) Endothelial cell (END), with typical smooth tubular invaginations of the cell membrane. Arrows denote basal lamina. ERY: erythrocyte; scale bar 0.5 μm. (b) Bundles of intermediate filaments (arrow heads) traverse the cytoplasm. Arrow denotes secretory granule; scale bar 0.5 μm.

I) which binds to galactose (Figure 17.1c). We assume that, as in mammals, the strong reaction with PAS denotes the presence of glycoproteins in the basement membrane and pericapillary space, as do WGA and RCA I. This assumption is supported by the fact that the positive reaction zone around the capillary lumen is relatively broad.

The cytoplasm of the endothelial cells contains an elaborate system of tubular invaginations, which have a diameter of c. 85 to 95 nm and open to the apical and basal regions of the plasma membrane (Figure 17.2a, b). Studies on the endothelium of the cerebral capillaries of *Myxine glutinosa* have shown that these tubular structures do not provide a direct communication between the capillary lumen and pericapillary space, but are blind-ending structures which are functionally associated with smooth-surfaced cisternae and may be involved in the regulation of the cytosolic Ca^{++}-concentration (Bundgaard, 1987). The endothelial cells also contain a small or medium-sized Golgi apparatus located near two centrioles, and moderate numbers of mitochondria, some small cisterns of the rough endoplasmic reticulum, smooth-surfaced cisternae, glycogen particles, single lipid droplets, dense lysosomal bodies and single peroxisomes with a crystalloid core and roundish secretory granules (Figure 17.2b). Microtubules, with a diameter of c. 30 nm and small bundles of intermediate filaments, with a diameter of 6 to 8 nm (Figure 17.2b), are found throughout the cytoplasm. The plasma membrane contains coated pits, which presumably eventually become transformed into the coated vesicles found within the cytoplasm. The basal plasma membrane forms local hemidesmosomal structures, that are associated with intracellular intermediate filaments. Flange-like extensions from adjacent cells form complex interdigitations, thereby facilitating intercellular contact through simple, small desmosome-like specializations. Although the width of the intercellular space is typically between 16 and 25 nm, the neighbouring lateral cell membranes are in places only about 4–5 nm apart. Morphologically the latter areas are characterized by thickened cell membranes and amorphous material in the intercellular space which may represent specialized intercellular contacts.

The endothelial cells possess a well-developed basal lamina (Figures 17.2a and 17.4c), which comprises a very conspicuous *lamina lucida* (c. 20–30 nm thick) and *lamina densa* (c. 30–40 nm thick) and is rich in proteoglycans (Figure 17.3a, b). This basal lamina is contiguous in places with that of the pericytes that are regularly associated with the endothelium. The nucleus and cytoplasm of the pericytes closely resemble those of the endothelial cells (Figures 17.1d, 17.2a, 17.4a, b, c). The pericytes also contain abundant smooth tubular invaginations of the plasma membrane (Figure 17.4c) and small bundles of thin filaments (diameter about 4 nm), among which thicker filaments occur. The presence of these filaments suggests that actin and myosin are present, which would give contractile properties to these cells (Figure 17.4d), as has been observed in mammals (Simms, 1986; Nehls and Drenckhahn, 1991). Long processes extend from the pericytes along the outer aspect of the endothelium. Pericytes are surrounded by a basal lamina of variable thickness, which is discontinuous on the surface adjacent to the endothelium. Simple hemidesmosomes are present along this surface (Figure 17.4c). However, the pericyte processes are often separated from the endothelium by a narrow space containing individual collagen fibrils, indicating that the pericyte processes can extend for a short distance into the pericapillary connective tissue.

The pericapillary space surrounding the endothelium, which is usually only 0.5 to 0.9μm in width, contains irregularly arranged collagen fibrils (diam. 40 to 120 nm, Figure 17.3b) and single bundles of tubular microfibrils (diam. c. 12 nm), that constitute oxytalan

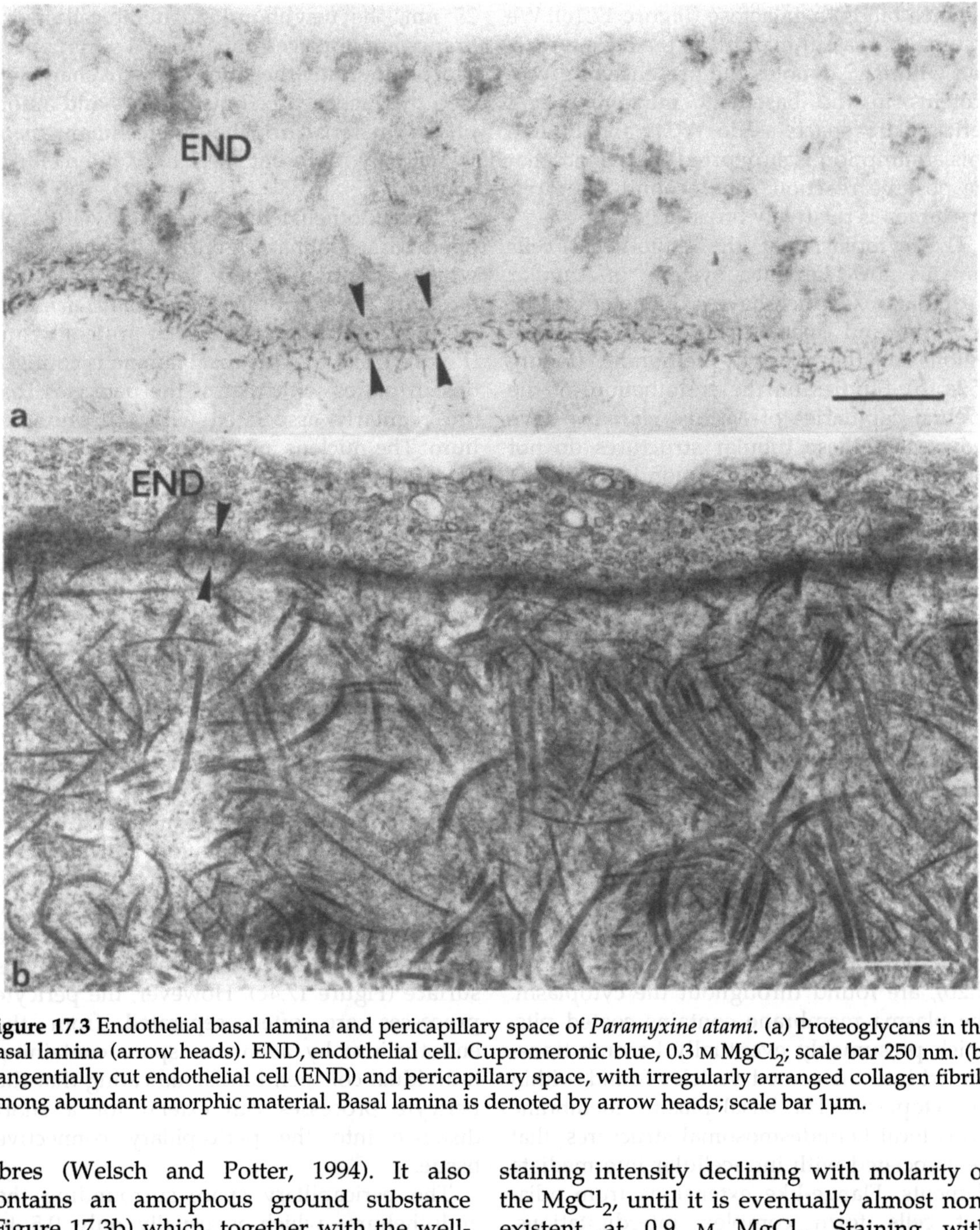

Figure 17.3 Endothelial basal lamina and pericapillary space of *Paramyxine atami*. (a) Proteoglycans in the basal lamina (arrow heads). END, endothelial cell. Cupromeronic blue, 0.3 M $MgCl_2$; scale bar 250 nm. (b) Tangentially cut endothelial cell (END) and pericapillary space, with irregularly arranged collagen fibrils among abundant amorphic material. Basal lamina is denoted by arrow heads; scale bar 1µm.

fibres (Welsch and Potter, 1994). It also contains an amorphous ground substance (Figure 17.3b) which, together with the well-developed basal lamina, appear to be responsible for the strong reaction to PAS-stain, as well as that of at least some of the lectins (see above). Alcian blue also stains this space, the staining intensity declining with molarity of the $MgCl_2$, until it is eventually almost non-existent at 0.9 M $MgCl_2$. Staining with Cupromeronic blue (CMB) in the presence of 0.3 M $MgCl_2$ resulted in the production of needle-shaped precipitates which occurred at regular intervals on collagen fibrils. This

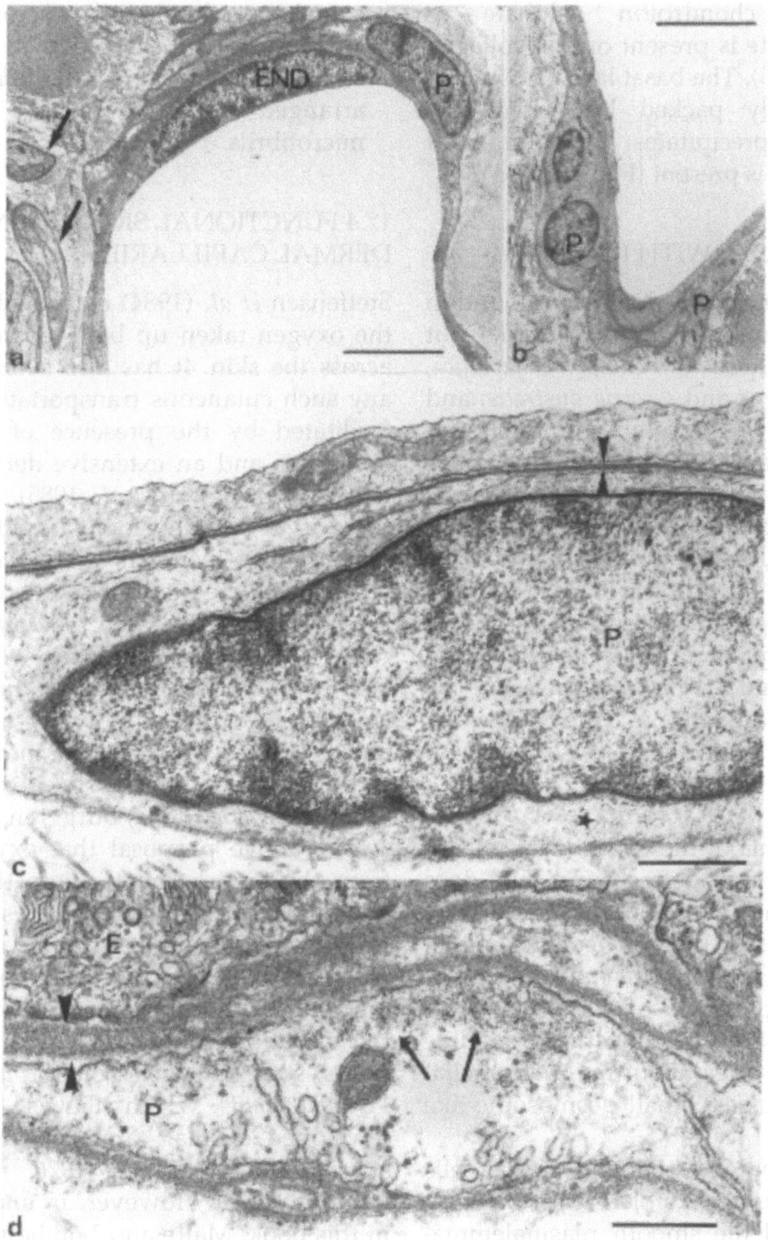

Figure 17.4 Pericytes of dermal capillaries of *Paramyxine atami*. (a) Pericyte (P) with long cytoplasmic processes attached to the endothelium (END). Arrows denote nerves in the pericapillary space; scale bar 5 μm. (b) Pericytes (P) apposing the endothelium of a dermal capillary; scale bar 5 μm. (c) Note the prominent basal lamina of the endothelial cell (arrow heads), the nearby adjacent hemidesmosome-like portions of the basal lamina of the pericyte nearest the endothelial cell and the continuous and better defined pericyte basal lamina on the surface farthest away from the endothelial cell. Asterisks denote the presence of abundant smooth vesicles and tubules; scale bar 1 μm. (d) *Myxine glutinosa*, detail of a pericyte (P) with a bundle of contractile filaments (arrows). E: endothelial cell. Arrow heads: common basal laminae; scale bar 400 nm.

indicates that chondroitin sulphate or dermatan sulphate is present on the collagen fibrils (Scott, 1985). The basal lamina contains relatively densely packed but irregularly arranged CMB precipitates, indicating that heparin sulphate is present (Figure 17.3a).

17.3 COMPARISONS WITH LAMPREYS

A dermal capillary network was not found in adults of the lamprey *Lampetra fluviatilis*, but is present in those of *Lampetra japonica*, *Petromyzon marinus* and *Geotria australis*, and was particularly well developed in the last of these species (Krause, 1923; Rabl, 1931; Potter *et al.*, 1995). Capillaries are not present in the dermis of ammocoetes of *Petromyzon marinus*, *Lampetra japonica*, *Geotria australis* (Lethbridge and Potter, 1980, 1981; Potter *et al.*, 1995) and *Lampetra planeri* (personal observations).

The dermal capillaries of adults of *L. japonica*, *P. marinus* and *G. australis* (Potter *et al.*, 1995) differ markedly in the following respects from those just described for hagfishes.

1. The basal lamina of the endothelial cells of the capillaries is ill defined and thus does not possess the thick and discrete clearly structured basal lamina, comprising a distinct *lamina lucida* and *lamina densa*, that is found in hagfishes.
2. The interdigitations between the flange-like extensions of neighbouring endothelial cells do not possess small desmosome-like structures.
3. The cell membrane of the endothelial cells contains numerous uniform oval or round infoldings and the smooth plasmalemmal vesicles that are characteristic of vertebrate endothelia. In contrast, the corresponding cell membrane of hagfishes contains numerous smooth tubular invaginations.
4. Pericytes were never observed in association with the endothelia in lampreys, as they are in hagfishes.
5. The pericapillary space of lampreys contains predominantly microfibrils, which are abundant, whereas that of hagfishes contains numerous and often irregularly arranged collagen fibrils, as well as microfibrils.

17.4 FUNCTIONAL SIGNIFICANCE OF DERMAL CAPILLARIES

Steffensen *et al.* (1984) estimated that 80% of the oxygen taken up by hagfishes can occur across the skin. It has also been argued that any such cutaneous transportation would be facilitated by the presence of both a thin epidermis and an extensive dermal capillary network (Burggren *et al.*, 1985). Furthermore, there is a well-developed capillary network immediately beneath the dermis, which would facilitate the transport to other parts of the body of any oxygen that entered via this route. Cutaneous respiration would presumably be of value to hagfishes during feeding, since at such times they may not always be able to draw a water current into the pharynx through their anteriorly located nostril (Steffensen *et al.*, 1984; Burggren *et al.*, 1985).

The above proposal that oxygen diffuses from the environment across the epidermis and into the dermal capillaries and is then transported to other regions of the body, where it makes a significant contribution to metabolism, would provide a parallel with the situation found in certain amphibious fishes, such as species of *Boleophthalmus* and *Periophthalmus*, which likewise have a well-developed dermal capillary network in the skin (Tamura *et al.*, 1976; Yokoya and Tamura, 1992). However, in another chapter in this book, Malte and Lomholt question the validity of the estimate by Steffensen *et al.* (1984) that approximately 80% of the oxygen used by hagfishes is derived by diffusion across the skin. They have also pointed out that 'the capillaries are perfused with arterial blood of a high oxygen tension and haemoglobin-oxygen saturation and consequently a low capacitance for oxygen. Furthermore, the thickness of the barrier separating the blood and water is at least an order of magnitude larger than what is found in effective gas exchangers such as fish gills.' It is argued that each of these features suggests that the uptake of oxygen across the skin and into the dermal capillary network is likely to be minimal and would thus not make a major contribution to the oxygen that is used overall by the animal. In the context of the water–blood barrier, it would certainly appear relevant that the distance between the water at the epidermal surface and underlying capillaries is far greater in hagfishes, i.e. 84–90 μm, than in mudskippers (2–4 μm), a group which can derive a large amount of oxygen by diffusion across the skin in both water and air (Tamura *et al.*, 1976; Yokoya and Tamura, 1992).

Malte and Lomholt do not deny that the very well developed dermal capillaries in hagfishes play an important role. Indeed, they propose that, particularly when hagfishes are buried in highly anoxic substrates, the dermal capillaries play a crucial role in supplying oxygen to the cells of the epidermis which are metabolically active. It therefore appears relevant that the capillary network is best developed in species such as *M. glutinosa* and *P. atami* (Potter *et al.*, 1995), which tend to burrow (Hardisty, 1979). In the case of hagfishes, substantial amounts of oxygen will be required by the numerous cells in this layer which produce substantial amounts of mucus. The amount of oxygen consumed by the skin has been measured in several teleost fishes and, at least in these cases, has been shown to be high (Nonnotte and Kirsch, 1978; Nonnotte, 1981). The oxygen in the dermal capillaries of hagfishes is presumed by Malte and Lomholt to have been extracted in the gills from water that entered the pharyngeal chamber via the anteriorly located nostril which lies close to the substrate surface.

The ability of ammocoetes and adults of the lamprey *G. australis* to survive out of water for at least 4 days and 36 hours, respectively, at 15°C, implies that these two life cycle stages are able to respire cutaneously (Potter *et al.*, 1996a, b). However, the trunk and tail only account for about 13% of the oxygen taken up by the adults of this species at 15°C (Potter *et al.*, in press), which implies that the dermal capillary network in these regions does not play a major role in facilitating the transport of oxygen from the environment to the various organs within the body for metabolism. Furthermore, only about 18% of the oxygen taken up by adults of the lamprey *Lampetra fluviatilis* in water occurs across the skin (Korolewa, 1964). Thus, the oxygen taken up by the skin of adult lampreys may be used mainly or even solely for metabolically demanding processes in the skin, such as the transport of ions and the production and secretion of mucus (Feder and Burggren, 1985). There is some indirect evidence that cutaneous respiration might play a greater role in ammocoetes than adult lampreys (Potter *et al.*, 1996b). If this is the case, the transfer of oxygen from the environment to the subdermal capillary space would be facilitated by the relatively thin dermis and relatively large surface area to body size ratio that is found in larval lampreys.

It is relevant that the epidermis of hagfishes contains numerous mucous cells (Blackstad, 1963). Since the production and secretion of mucus is a metabolically demanding process (Feder and Burggren, 1985), the presence of capillaries in the dermis just below the epidermis could help ensure that the mucous cells are well supplied with oxygen. A dermal capillary network could also play a crucial role in transporting to the epidermis the precursors required for the production of large amounts of mucus by the mucous cells. The dermal capillaries may also play a role in transporting ions to the epidermis prior to their excretion.

The results of the reactions of the pericapillary space in hagfishes to alcian blue with varying molarities of $MgCl_2$ indicate that this region outside the endothelial cells contains carboxylated and sulphated glycosaminoglycans (Potter *et al.*, 1995). The former is presumably hyaluronic acid, while the latter is chondroitin or dermatan sulphate (Scott and Dorling, 1965). These negatively charged glycosaminoglycans bind water and thus establish a water-rich space that would facilitate metabolic exchange. The functional significance of the glycoproteins, which the PAS reaction and presumably also the binding by several lectins have shown to be abundant in this space, remains to be explored.

ACKNOWLEDGEMENTS

We are grateful to Professor Y. Honma and Professor A. Chiba for the gift of excellently fixed material of *E. burgeri* and *P. atami*.

REFERENCES

Blackstad, J.W. (1963) The skin and slime glands, in *The Biology of Myxine* (eds A. Brodal and R. Fänge), Universititsforlaget, Oslo, pp. 195–230.

Burggren, W., Johansen, K. and McMahon, B. (1985) Respiration in phyletically ancient fishes, in *Evolutionary Biology of Primitive Fishes* (eds R.E. Foreman, A. Gorbman, J.M. Dodd and R. Olsson), Plenum Press, New York, pp. 217–52.

Bundgaard, M. (1987) Tubular invaginations in cerebral endothelium and their relation to smooth surfaced cisternae in hagfish (*Myxine glutinosa*). *Cell and Tissue Research*, **249**, 359–65.

Feder, M.E. and Burggren, W.W. (1985) Cutaneous gas exchange in vertebrates; design, patterns, control and implications. *Biological Reviews*, **60**, 1–45.

Forey, P. and Janvier, P. (1993) Agnathans and the origin of jawed vertebrates. *Nature, London*, **361**, 129–34.

Hans, M. and Tabencka, Z. (1938) Über die Blutgefäße der Haut von *Myxine glutinosa*. *Bulletin International de L'Academic des Sciences et des Lettres Cracovie* (B), **1938**, 69–77.

Hardisty, M.W. (1979) *Biology of the Cyclostomes*, Chapman & Hall, London.

Holthofer, H., Virtanen, I., Karinienn, A.I., Hormia, M., Linder, E. and Miettinen, A. (1982) *Ulex europaeus* I, Lectin as a marker for vascular endothelium in human tissues. *Laboratory Investigations*, **14**, 929–1002.

Korolewa, N.W. (1964) Water respiration of lamprey and its survival in a moist atmosphere. *Izvestiya Vsesoyuznogo Nauchno-Issledovatel'sklogo Instituta Ozernogoi Rechnogo Rybnogo Khozyaistva*, **58**, 186–90 (in Russian).

Krause, R. (1923) *Mikroskopische Anatomie der Wirbeltiere*, IV. *Teleostier, Plagiostomen, Zyklostomen und Leptokardier*, de Gruyter, Berlin.

Lametschwandtner, A., Weiger, T., Lametschwandtner, U., Georgieva-Hanson, V., Patzner, R.A. and Adam, H. (1989) The vascuarlization of the skin of the Atlantic hagfish, *Myxine glutinosa* L., as revealed by scanning electron microscopy of vascular corrosion casts. *Scanning Microscopy*, **3**, 305–14.

Lethbridge, R.C. and Potter, I.C. (1980) Quantitative studies on the skin of the paired species of lampreys, *Lampetra fluviatilis* and *Lampetra planeri*. *Journal of Morphology*, **164**, 39–46.

Lethbridge, R.C. and Potter, I.C (1981) The skin, in *The Biology of Lampreys* Vol. 3 (eds M.W. Hardisty and I.C. Potter), Academic Press, London, pp. 377–448.

Nehls, V. and Drenckhahn, D. (1991) Heterogeneity of microvascular pericytes for smooth muscle type alpha actin. *Journal of Cell Biology*, **113**, 147–54.

Nonnotte, G. (1981) Cutaneous respiration in six freshwater teleosts. *Comparative Biochemistry and Physiology*, **70A**, 541–3.

Nonnotte, G. and Kirsch, R. (1978) Cutaneous respiration in seven seawater teleosts. *Respiration Physiology*, **35**, 111–18.

Potter, I.C., Macey, D.J. and Roberts, A.R. (1996a) Oxygen consumption by adults of the southern hemisphere lamprey *Geotria australis* in air. *Journal of Experimental Zoology*, **276**, 254–61.

Potter, I.C., Macey, D.J. and Roberts, A.R. (1997) Oxygen uptake and carbon dioxide excretion by the branchial and postbranchial regions of adults of the lamprey *Geotria australis* in air. *Journal of Experimental Zoology* (in press).

Potter, I.C., Macey, D.J., Roberts, A.R. and Withers, P.C. (1996b) Oxygen consumption by ammocoetes of the lamprey *Geotria australis* in air. *Journal of Comparative Physiology*, **166B**, 331–6.

Potter, I.C., Welsch, U., Wright, G.M., Honma, Y. and Chiba, A. (1995) Light and electron microscope studies of the dermal capillaries in three species of hagfishes and three species of lampreys. *Journal of Zoology London*, **235**, 677–88.

Rabl, H. (1931) Integument. 1. Integument der Anamnier, in *Handbuch der vergleichenden Anatomie der Wirbeltiere*, Vol. 1 (eds L. Bolk, E. Göppert, E. Kallius and W. Lubosch), Urban & Schwarzenberg, Berlin and Wien, pp. 271–374.

Schumacher, U., v. Armansperg, N., Kreipe, H. and Welsch, U. (1996) Lectin binding and uptake in human (myelo)monocytic cell lines: HL60 and U937. *Ultrastructural Pathology*, **20**, 463–71.

Scott, J.E. (1985) Proteoglycan histochemistry – a valuable tool for connective tissue biochemists. *Collagen and Related Research*, **5**, 765–74.

Scott, J.E. and Dorling, I. (1965) Differential staining of acid glycosaminoglycans (mucopolysaccharides) by alcian blue in salt solutions. *Histochemie*, **5**, 221–3.

Simms, D.E. (1986) The pericyte – a review. *Tissue and Cell*, **18**, 153–72.

Steffensen, J.F., Johansen, K., Sindberg, C.D., Sorensen, J.H. and Mooler, J.L. (1984) Ventilation and oxygen consumption in the hagfish, *Myxine glutinosa*. *Journal of Experimental Marine Biology and Ecology*, **84**, 173–8.

Tamura, S.O., Morii, H. and Yuzuriha, M. (1976) Respiration of the amphibious fishes, *Periophthalmus cantonensis* and *Boleophthalmus chinensis* in water and on land. *Journal of Experimental Biology*, **65**, 97–107.

Welsch, U. and Potter, I.C. (1994) Variability in the presence of elastic fibre-like structures in the ventral aorta of agnathans (hagfishes and lampreys). *Acta Zoologica Stockholm*, **75**(4), 323–7.

Whitear, M. (1986) The skin of fishes including cyclostomes. Dermis, in *Biology of the integument*, Vol. 2: *Vertebrates* (eds J. Bereiter-Hahn, A.G. Maltolsy and K.S. Richards), Springer Verlag, Berlin, pp. 39–64.

Yokoya, S. and Tamura, O.S. (1992) Fine structure of the skin of the amphibious fishes, *Boleophthalmus pectinirostris* and *Periophthalmus cantonensis*, with special reference to the location of blood vessels. *Journal of Morphology*, **214**, 287–97.

PART EIGHT

The Blood and Immune System

18

HAGFISH BLOOD CELLS AND THEIR FORMATION

Ragnar Fänge

SUMMARY

Hagfishes have a large volume of extracellular fluid. The plasma is iso-osmotic to seawater and contains organic components in similar concentrations as in other vertebrates. However, the composition of the plasma proteins is poorly known. Suspended free cells in the blood are erythrocytes and leukocytes. Prominent among the latter are granulocytes, spindle cells, lymphocyte-like cells and immature blood cells (blast cells). Spindle cells and lymphocyte-like cells may to a limited extent resemble lymphocytes but also seem to be progenitors of other blood cells. Erythropoiesis goes on mainly in the circulating blood while granulopoiesis occurs extravascularly around portal vein branches in the intestinal submucosa. The blood coagulates by similar mechanisms as in other vertebrates, but the coagulation factors are insufficiently known. Some kind of thrombocytes are involved.

18.1 COMPOSITION AND VOLUME OF PLASMA AND LYMPH

The blood volume of hagfishes is remarkably large (17.7%) exceeding that in most fishes. About 30% of the blood is contained in a subcutaneous blood sinus (Fänge, 1985; Forster *et al.*, 1989). In addition to blood vessels and blood sinuses lymph-filled vessels or spaces are widely distributed (Johansen *et al.*, 1962; Lomholt and Franko-Dossar, this volume). Hagfishes hold almost the same concentrations of sodium and chloride in the blood plasma as in seawater but calcium and magnesium are physiologically regulated (Robertson, 1976). This special kind of osmotic balance should be kept in mind when preparing physiological salines or cell culture fluids suitable for experimental work on hagfish tissues. Thus, according to my own deliberation a 3.3% sodium chloride solution would be about iso-osmotic to hagfish blood plasma. The concentration of protein in the plasma of newly caught *Myxine glutinosa* was estimated to be 4.1 g per 100 ml, about equal to that in several marine fishes (Larsson *et al.*, 1976). Separate fractions of plasma proteins have been recognized by electrophoresis but were not further analysed (Johansen *et al.*, 1962). Transferrin and a copper protein similar to ceruloplasmin are among the few plasma proteins of hagfishes which have been investigated to some extent (Aasa, 1972; Aisen *et al.*, 1972). A few further data are known on prothrombin in the blood (Banfield *et al.*, 1994), and a complement-like protein (Fujii *et al.*, 1992), but on the whole investigations on the composition and functions of plasma proteins in hagfishes are still an open field for new discoveries. Glucose, lactate and a few other organic compounds in plasma show concentration levels comparable to those in various fishes (Davison *et al.*, 1990).

The Biology of Hagfishes. Edited by Jørgen Mørup Jørgensen, Jens Peter Lomholt, Roy E. Weber and Hans Malte. Published in 1998 by Chapman & Hall, London. ISBN 0 412 78530 7.

18.2 THE BLOOD CELLS

According to Larsson *et al.* (1976), newly caught *Myxine glutinosa* showed an average hematocrit of 19.1%, which is of the same magnitude as in slowly moving marine teleosts. The haemoglobin concentration of the blood was 4.1 g per 100 ml and the mean corpuscle haemoglobin concentration 22.5%. Wells *et al.* (1986) recorded a haematocrit of only 8.5% in dorsal aortic blood of *Eptatretus cirrhatus*. However, haematocrits of blood from different vascular regions vary considerably due to plasma skimming effects (Johansen *et al.*, 1962; Forster *et al.*, 1989).

Previously, Müller (1845) recognized different types of cells in the blood of *Myxine glutinosa*. About the same subject Jordan (1938) stated: 'Considering the relatively low evolutionary level of the hagfish, the variety of its blood cells is remarkable. This blood contains the principal types of cells found in mammalian blood. The additional fact that the blood cells of *Myxine* are relatively large renders this form of exceptional value.' The relatively large size of blood cells and tissue cells in hagfishes may be related to the cellular content of DNA, which exceeds that in lampreys and many teleosts (Fänge, 1985). Light microscopic studies show that in addition to red cells (erythrocytes) there exist several types of white cells (leucocytes) and numerous not fully differentiated cells (blast cells, haematoblasts) (Jordan, 1938; Holmgren, 1950; Linna *et al.*, 1975). Since about 1970 electron microscopy and various haematological methods have supplemented light microscopic studies of the blood cells (Sekhon and Maxwell 1970; Tomonaga *et al.*, 1973a, b, c; Mattisson and Fänge, 1977; Tanaka *et al.*, 1981). Gilbertson *et al.* (1986), by flow cytometric analysis of hagfish blood cells, distinguished two distinct leucocyte populations analogous to mammalian granulocytes/ monocytes and small lymphocytes. Various types of blood cells are described below.

18.2.1 Erythrocytes

The red cells (Figure 18.1d, left) are oval discs with an estimated length of 15–35 µm (*Myxine glutinosa*: Jordan and Speidel, 1930; Mattisson and Fänge, 1977). The cytoplasm is almost homogeneous but may carry small mitochondria and granular matter. In mature red cells not only the cytoplasm but the nucleus too contains haemoglobin (Tomonaga *et al.*, 1973b). Electron microscopy (Sekhon and Maxwell, 1970) shows marginal bundles of microtubules along the cell margins and micropinocytotic vesicles. The latter may represent sites of uptake of iron-containing ferritin from the plasma (*Eptatretus burgeri*: Tomonaga *et al.*, 1973b; *Myxine glutinosa*: Mattison and Fänge, 1977). A considerable fraction of the erythrocytes are immature cells (erythroblasts). Late erythroblasts or proerythrocytes have acquired red cell-like contours and show signs of haemoglobin synthesis (polychromatic staining), and the cytoplasm contains abundant polyribosomes (Sekhon and Maxwell, 1970). Erythroblasts are often in mitosis showing chromosomes (Figure 18.1d, right).

Early erythroblasts are probably featured as small round lymphocyte-like cells with a basophilic cytoplasm. At growing and differentiating, the nucleus and the cytoplasm of the erythoblasts increase in size and the cells pass through repeated mitotic divisions. Late erythroblasts or proerythrocytes in *Myxine* have red cell-like contours and show signs of haemoglobin synthesis (polychromatic staining). Sekhon and Maxwell (1970) reported abundant polyribosomes in the cytoplasm of proerythrocytes. Strik and Fänge (1974) observed a transient red fluorescence in fresh unstained preparations in erythrocytes of *Myxine*, especially strong in immature cells (erythroblasts). The fluorescence was probably due to free porphyrins formed during synthesis of haemoglobin. A red fluorescence at 630 nm in erythrocytes of *Eptatretus stoutii* has been

A

B

C

D

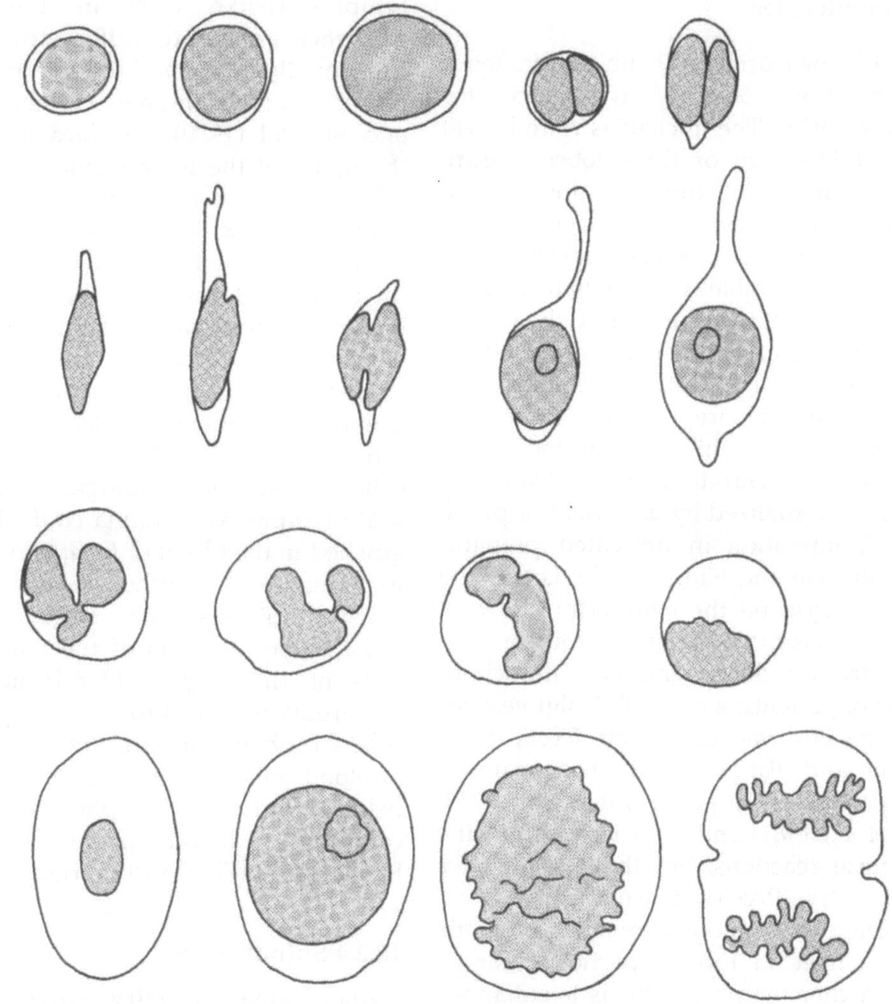

Figure 18.1 Contours of cells and nuclei in May–Grünwald–Giemsa stained blood smear of *Myxine glutinos*. Light microscope, high magnification, combination of camera lucida drawings. (A) *To the left*, lymphocyte-like cells of different size; *to the right*, two binucleate lymphocyte-like cells (or spindle cells). The cell with a large round nucleus may be characterized as a young lymphocyte-like cell (blast cell). (B) Spindle cells. *To the left*, three aged cells; *to the right*, two young spindle cells. (C) Granulocytes with different nuclear shape. (D) Erythrocytes. *To the left*, a mature cell; *to the right*, young cells (large nucleus, stages of division). May–Grünwald–Giemsa stained blood smear, camera lucida drawings.

noted (Gilbertson *et al.*, 1986). The haemoglobin of hagfish erythrocytes is mainly monomeric with a molecular weight of about 17 000 (Svedberg and Hedenius, 1934). Although the erythrocytes contain carbonic anhydrase (Carlsson *et al.*, 1980), the capacity of carbon dioxide transport of the blood is low (Tufts and Boutilier, 1990a, b). The red cells contain variable amounts of ATP and high concentrations of ADP (Bartlett, 1982). They have remarkably low capacities of volume regulation and anion exchange (Dohn and Malte, and Peters and Gros, this volume).

18.2.2 Granulocytes

These cells, measuring 7–9 µm in diameter, constitute about 50% of the leucocytes (Linthicum, 1975). The nucleus is round, oval or divided into two or three lobes (Figure 18.1c). A fraction of the granulocytes are immature promyelocytes with a central non-segmented nucleus. Morphologically the granulocytes resemble mammalian neutrophils (polymorphonucleates), but they have also been characterized as heterophilic leucocytes (Östberg *et al.*, 1976). The minute cytoplasmic granules are neither distinctly eosinophilic or basophilic by routine stains. The granules concentrate neutral red supravitally and are visualized by dark field or phase contrast illumination in untreated preparations (Mattisson and Fänge, 1977). Under the electron microscope the cytoplasm shows a rough endoplasmic reticulum, ovoid or rod-shaped electron dense granules, mitochondria, microfilaments, a centriol, Golgi vesicles and phagocytic vacuoles (Linthicum, 1975; Mattisson and Fänge, 1977). The granules give no peroxidase response (Johansson, 1973; Tanaka *et al.*, 1981) and no acid and alkaline phosphatase reactions, but the granulocytes show a positive PAS response (Östberg *et al.*, 1976). They are spontaneously phagoytic taking up injected foreign particles (Fänge, 1973). Whether the phagocytosis is enhanced by opsonins or similar factors in unknown.

18.2.3 Lymphocyte-like cells and lymphoid hemoblasts (blast cells)

The blood contains numerous lymphocyte-like (lymphoid) cells, sometimes considered as lymphocytes (Holmgren, 1950; Tomonaga *et al.*, 1973a, c) (Figure 18.1a, left). Linthicum (1975) stated that small lymphocytes (5–6 µm in diameter) comprise approximately 15% of the buffy coat of *Eptatretus stoutii*. Tanaka *et al.* (1981) observed 'small cells with lymphoid appearance' but not real lymphocytes in the blood of *Eptatretus burgeri*. The lymphocyte-like cells in the blood of hagfishes appear to differ structurally and functionally from the immunologically active lymphocytes of higher vertebrates. Jordan and Speidel (1930) and Jordan (1933, 1938) thought that the lymphocyte-like (lymphoid) cells in hagfishes are immature cells, haemoblasts, with ability of further development. These authors distinguished large, medium-sized and small haemoblasts in the blood of *Myxine glutinosa*. Similar views were expressed by Good and Finstad (1964) who characterized small round cells in the California hagfish, *Eptatretus stoutii*, as lymphoid haematoblasts and stated that the cells differed morphologically from vertebrate lymphocytes. Linna *et al.* (1975) distinguished in the blood of *Eptatretus stoutii* small mononuclear lymphocyte-like cells and 'haemocytoblasts', i.e. large mononuclear cells. Large numbers of the lymphocyte-like cells of the hagfish blood may be early erythroblasts (erythrocytic stem cells) but others probably are precursors of other types of blood cells, such as spindle cells (see 18.4.1 below). Binucleate lymphocyte-like cells are common in blood smears (Mattisson and Fänge, 1977) (Figure 18.1a, right).

18.2.4 Spindle cells

Jordan and Speidel (1930) used the term spindle cell to designate oval or fusiform blood cells which are very abundant and characteristic in hagfishes (Figure 18.1b). The same term has also been applied to thrombocytes in teleostean fishes and other non-mammalian vertebrates, but the resemblance with those cells is uncertain. Spindle cells usually constitute about 50% of the leucocytes of hagfish blood. Nevertheless, in an electron microscopic investigation Linthicum (1975) did not recognize any spindle cells probably because they were categorized as lymphocytes. The spindle cells are oval or fusiform with tapering ends measuring 6–10 µm in breadth and 8–20 µm in length, but they vary in shape, size and abundance. The nucleus is oval, often with a deep groove or cleft (Tomonaga *et al.*, 1973c). In stained blood smears the cytoplasm of the spindle cells appears weakly basophilic and non-granulated. The electron microscope shows free ribosomes, a few small mitochondria, Golgi complex, centriole and microtubules in the cytoplasm (Mattisson and Fänge, 1977), and according to Tomonaga *et al.* (1973b) spindle cells of *Eptatretus burgeri* contain minute electron-dense granules (0.2 μm). Young spindle cells possess a round nucleus and a large volume of basophilic cytoplasm (Figure 18.1b, right). Intermediate forms between spindle cells and lymphocyte-like cells are frequently seen. According to Jordan and Speidel (1930) complete understanding of the spindle cell must await further comparative investigations, and especially a study of their behaviour in fresh blood. The statement is still appropriate.

18.2.5 Are spindle cells haematoblasts, or primitive lymphocytes?

The nature of the hagfish spindle cells, in comparison with the types of blood cells generally found in vertebrates, is not well understood. Jordan and Speidel (1930) noted that the spindle cells vary between fusiform and spheroidal shape, and Smith (1969) reported that increase of temperature causes spindle cells of *Myxine* to assume a round lymphocyte-like shape. Even chemical agents which destroy the assemblage of microtubles provoke transformation of spindle cells into lymphocyte-like cells (Fänge *et al.*, 1974). Fänge and Edström (1973) noted that spindle cells take up uridine *in vitro*, indicating an intense RNA synthesis. The lectin, phytohaemagglutinin (PHA) caused appearance of cells with features of late erythroblasts (proerythrocytes) in a suspension of spindle cells and lymphocyte-like cells of *Myxine glutinosa* (Fänge and Zapata, 1985) and has been used to increase the frequency of mitoses among cultivated blood cells of *Eptatretus burgeri* (Kohno *et al.*, 1986). The observations may indicate that spindle cells and lymphocyte-like cells are able to transform mutually into each other, and when influenced by special stimuli (for instance PHA) these cells have a capacity of further growth and differentiation. Probably at least part of the spindle cells/lymphocyte-like cells serve as a pool of stem cells/blast cells giving rise to erythrocytes in accordance to Jordan's (1938) lymphoid haemoblast theory.

Gilbertson *et al.* (1986) observed that more than 70% of a population of small leucocytes (probably mainly spindle cells and lymphocyte-like cells) were stained with a rabbit antiserum directed against hagfish immunoglobin. Regardless of the nature of the alleged immunoglobin, the observation indicates that hagfish spindle cells or lymphocyte-like cells may have properties in common with vertebrate lymphocytes active in immune reactions. However, in hagfishes the humoral immune system seems to be represented by a simplified complement system instead of immunoglobulins (Fujii *et al.*, 1992; Raison and dos Remedios, this volume). More information on the non-granulated leucocytes of hagfishes may give clues for increased understanding of the evolution of the vertebrate lymphocytic system.

18.2.6 Thrombocytes

Jordan and Speidel (1930) and Jordan (1938) considered certain small, variable cells in the blood of *Myxine* as thrombocytes, which constituted less than 10% of the leucocytes. Similar cells were observed by Tomonaga *et al.* (1973b) and by Mattisson and Fänge (1977). On the role played by some kind of thrombocytes in blood coagulation, see section 18.5 below.

18.2.7 Monocytes and macrophages

Cells morphologically resembling mammalian monocytes are unusual in hagfish blood

(Jordan, 1938). Linthicum (1975) described ultrastructural features of presumed monocytes. Tanaka *et al.* (1981) noticed rare blood cells in *Eptatretus burgeri* which showed positive acid phosphatase reactions and were regarded as lysosome-containing macrophages.

18.2.8 Degenerate blood cells

Blood smears regularly contain numerous destroyed cells which are difficult to identify as to the original cell type (Holmgren, 1950; personal observations). Many of these cells may be degenerate erythrocytes.

18.3 HAEMATOPOIETIC TISSUES

Hagfish blood cells orginate both in haematopoietic (lymphomeloid) tissues and in the circulating blood. Haematopoietic cell aggregations have been found in the intestine, the pronephros (head kidney), and more diffusely in the connective tissue of a few other organs.

18.3.1 The intestinal myeloid tissue

Haematopoietic cell masses associated with the hepatic portal vein system form an extensive disperse tissue in the submucosa of the intestine (Figure 18.2). This 'perivascular lymphoid organ' (Schreiner, 1898), or 'perivenous myeloid tissue', is also called a primitive or disperse spleen or has been compared to mammalian bone marrow (Jordan, 1933 and 1938; Tomonaga *et al.*, 1973c). The intestinal wall is supplied by a series of mesenteric arteries (Grodzinski, 1926; Tanaka *et al.*, 1981) from which smaller arteries run to the capillary network at the base of the inner epithelium. Capillaries from the network connect with plexiform veins or sinusoids. These join the supraintestinal vein (hepatic portal vein), which widens anteriorly to form the portal heart at the surface of the intestine to the right of the branchial heart. The venous sinusoids are covered by haematopoietic cell masses, and both are embedded between large fat cells occupying the bulk of the intestinal submucosa. Jordan (1938) equalized the capillaries supplying the venous sinusoids with penicillar arterioles of the vertebrate spleen. However, Holmgren (1950) noticed a few isolated erythropoietic foci and believed that only these represented spleen-line tissue, while Tanaka *et al.* (1981) emphasized that the haematopoietic tissue of the hagfish intestinal submucosa differs fundamentally from a vertebrate spleen in its vascular supply.

The predominating cells of the haematopoietic cell islands of the submucosa are granulocytes in different stages of development. Mature granulocytes are observed migrating from the haematopoietic cell masses through an endothelial barrier into the venous lumen (Tomonaga *et al.*, 1973c). The tissue further contains lymphocyte-like cells and reticulum cells. A great deal of the lymphocyte-like elements probably are stem cells of granulocytes (granuloblasts), and other undifferentiated cells may be precursors of other types of blood cells. No incorporation of ^{59}Fe into erythroblasts in the intestinal submucosa was observed by Tomonaga *et al.* (1973c), which may indicate that late phases of erythropoiesis involving haemoglobin synthesis are not taking place in the intestine. As the submucosal myeloid tissue contains both granulocytes and lymphocyte-like cells, it may be termed lymphogranulopoietic.

18.3.2 The central mass of the pronephros

In most vertebrates the pronephros (head kidney) is a vestigial embryonic structure usually with no obvious function. Only in a few freshwater forms (urodeles, dipnoans, lampreys) it probably works as a water-eliminating organ during a brief period of the larval development (Christensen, 1964). Hagfishes are unique in that a modified pronephros persists in the adults. It consists of ciliated tubules, which on each side of the body connect the anterior (pericardial) part of

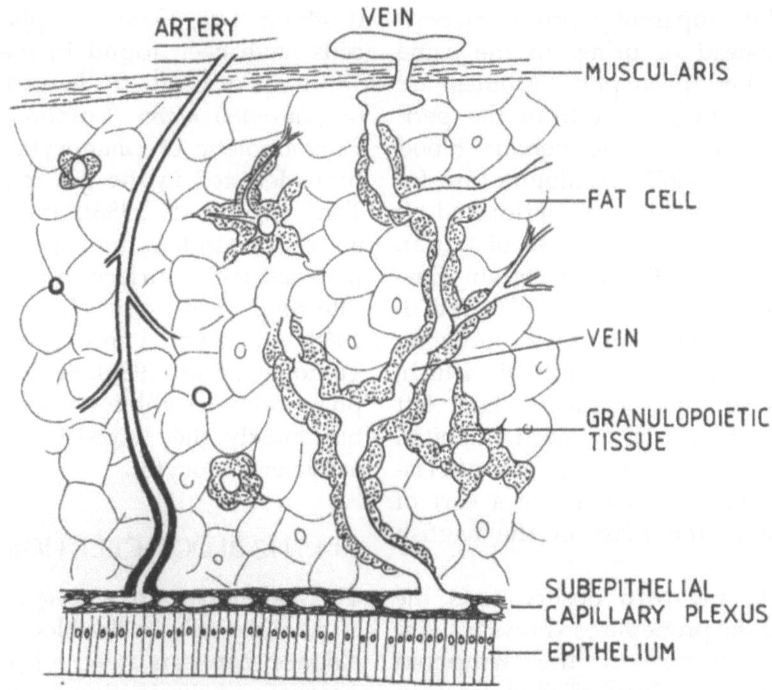

Figure 18.2 Diagram of section through the intestine of *Myxine glutinosa* showing the principal mode of vascular supply of the intestinal layers, and venous sinusoids surrounded by sheaths of granulopoietic tissue.

the body cavity with the lumen of a pronephric vein. The pronephric tubules, instead of forming a duct, penetrate the wall of the vein and transform into an amorphous 'central mass' with a lymphoid or haematopoietic appearance inside the vein (Price, 1910). The histology of the central mass is complex and variable. Holmgren (1950) examined hundreds of sections through the pronephros of hagfishes of different ages. He found the central mass to consist of a fibrous collagenous network, the meshes of which 'were either empty or contained cells of different characters, small cells, cells of medium and large size and giant cells. Many of the cells are lymphocytes and phagocytes.' The central mass was further stated to harbour a confusing array of degenerating lymphocytes and erythrocytes, alveolar vacuoles, 'megacaryocytes', stages of haematoblasts, etc., but such a complex cell composition was present in only about 25% of the animals. The pronephros seemed to pass through cycles of increased lymphoid (or haematopoietic) activity during the life of the hagfish (Holmgren, 1950). The finding of cells with the morphological features of plasma cells in the central mass of the pronephros of *Myxine* contributes to the cellular complexity of the central mass (Zapata *et al.*, 1984).

Compared to the intestinal myeloid tissue, the central mass of the pronephros is a small structure which may not play a great role quantitatively during blood formation. However, the pronephros and its central mass are of considerable theoretical interest. A remarkable transport of fluid is taking place in the pronephros region. An arterial glomus (glomerulus), which is associated with the pronephros, is not connected with any urinary duct but apparently produces peritoneal fluid instead of urine. In the same region the ciliated pronephric tubules are continuously pumping fluid from the peritoneal cavity back into the venous blood (Price, 1910; Fänge, 1973). Analogue transfer of body fluid into the blood is performed by ciliated nephrostomes in the kidney of anuran amphibians (Morris, 1981) and certain elasmobranch fishes (Schneider, 1897). The return of fluid into the venous blood may counteract loss of extracellular water in animals which, ontogenetically or phylogenetically, have left freshwater for a terrestrial or marine habitat. Lymphoid tissue at the nephrostomes of elasmobranchs (Borcea, 1905) may be a sort of parallel to the central mass of the hagfish pronephros.

According to my own observations the central mass of the pronephros shows a slight structural resemblance to the lymphoid nodules of the pericardium of chondrostean fishes (Fänge, 1986).

The central mass may play a role as a lymphoid structure guarding passages through the pronephric tubules between the peritoneal cavity and the blood. The abundant phagocytes in the pronephric region may hinder noxious particles in the peritoneal cavity from entering the blood (Fänge, 1973).

18.3.3 Other tissues

Leucocytes are found in various tissues besides the intestinal submucosa and the pronephros. These may be migrating leucocytes and it is not known if any haematopoiesis is taking place at the sites in question. Adam (1963) noted a subepithelial haemolymphatic tissue in the wall of the nasopharyngeal duct, and aggregations of lymphocyte-like cells in the velum or its muscle (Cole, 1925) have been likened to a primitive thymus (Riviere *et al.*, 1975). Granulocytes and lymphocyte-like cells are present within the intestinal epithelium (Östberg *et al.*, 1976), and phagocytic Kupffer cells have been found in the lining of liver sinusoids of *Eptatretus burgeri* and *Paramyxine* (=*Eptatretus*) *atami* (Tomonaga *et al.*, 1986). Haemopoietic or phagocytic cells have also been detected in the gill region (Cole, 1913; Tomonaga *et al.*, 1986), and phagocytic cells are very abundant in the peritoneal cavity, for instance at the surface of the gonads and the intestinal mesentery (Fänge and Gidholm, 1968; Thoenes and Hildemann, 1970; Tomonaga *et al.*, 1986). Some of the peritoneal phagocytes resemble monocytes structurally, but mostly they consist of large and very active granulocytes.

18.4 THE BLOOD CELL HOMEOSTASIS

The maintenance of a suitable number of free circulating cells in the blood depends on the balance between new formation (haematopoiesis) and elimination of blood cells.

18.4.1 Haematopoiesis

In mammals, and in vertebrates generally, blood cells are supposed to originate by proliferation and differentiation of pluripotential stem cells of a lymphocyte-like and undifferentiated appearance. When influenced by hormones like erythropoietin or other factors, stem cells transform into blast cells which proliferate mitotically and differentiate into mature blood cells (Hoffbrand and Pettit, 1980). Jordan (1933) suggested that the formation of erythrocytes goes on intravascularly, the synthesis of haemoglobin being facilitated by 'a sluggish or relatively stagnant blood current' and a 'relatively high carbon dioxide concentration', conditions which may be met with in the venous sinusoids of the hagfish intestine. The development of granulocytes, on the other hand, proceeds extravascularly in the hagfish intestinal submucosa.

The late phases of erythropoiesis undoubtedly take place in the circulating blood, but new formation of red cells has also been supposed to occur in the liver, in the walls of gill vessels, in the central mass of the pronephros and in the intestine (Cole, 1913; Holmgren, 1950). According to Jordan and Speidel (1930) free reticular cells in the intestinal submucosa may form haematoblasts. Some of these migrate into venous blood where they proliferate and differentiate into erythrocytes, spindle cells and thrombocytes.

The blast cells (or stem cells) of the blood with the capacity to give rise to erythrocytes or other blood cells probably belong to the categories of lymphocyte-like cells and spindle cells. However, the stem cells of different lines of blood cells still await to be structurally identified with certainty.

The enzyme δALA-D (delta-amino levulinic acid dehydrase) is one of the enzymes acting in the synthesis of porphyrins and needed at the formation of haemoglobin and other haeme compounds. Blood and blood cells of *Myxine glutinosa* showed relatively low activities of δALA-D (Olsson, 1973). Perhaps synthesis of porphyrins is going on only at a special haematoblast stage representing only a minor fraction of the total erythrocyte population.

The granulocytes of the blood arise from precursors (stem cells, granuloblasts) in the submucosal myeloid cells islands. After maturing extravascularly the granulocytes migrate into blood of the venous sinusoids in the intestinal submucosa (Tanaka *et al.*, 1981). The physiological regulation of haematopoiesis in hagfishes has not been investigated. Thus it is not known if some kind of erythropoietin initiates the formation of red cells. The fact that erythropoiesis is largely taking place in the circulating blood invites investigations of cytological details during growth and differentiation of erythrocytes.

18.4.2 Destruction and elimination of blood cells

Probably erythrocytes and other blood cells are worn out and eliminated at the same rate as they are formed. Effete erythrocytes may be destroyed by phagocytes such as Kupffer cells of the liver (Tomonaga *et al.*, 1986), and Holmgren (1950) observed phagocytosis of erythrocytes in the central mass of the pronephros. Tanaka *et al.* (1981) described cells in the intestinal submucosa with vesicles containing electron-dense material. Nothing was stated on the origin of this material, which might have emanated from a breakdown of red cells. A rich formation of bile-containing biliverdin (Sakai *et al.*, 1989) shows that the final breakdown of haemoglobin is taking place in the hagfish liver. Migrating leucocytes in the gut epithelium (Östberg *et al.*, 1976) may partly represent cells leaving the body via the gut lumen.

18.5 BLOOD COAGULATION

The blood of *Myxine* coagulates at a relatively slow rate if exposed to seawater, and the mass of cells and fibrin threads formed undergoes typical clot retraction (Figure 18.3a). The clotting process seems to be initiated by disintegrating cells, but the type of cells serving as thrombocytes has not been identified with certainty (Figure 18.3, right). Initiation of clotting in blood of non-mammalian vertebrates by spindle-shaped thrombocytes has been reported repeatedly, but few recent microscopic studies exist. *Myxine* blood is well suited for such studies. Clotting and clot retraction are blocked by heparin (Figure 18.3b) (Fänge and Gidholm, 1973). The hagfish blood contains a mammalian-like fibrinogen and a prothrombin with an amino acid sequence showing resemblance to that of chicken and mammalian prothrombins (Banfield *et al.*, 1994).

18.6 REMARKS ON HAGFISH PATHOLOGY

Although hagfishes may lack a vertebrate type of humoral immune system, lymphocyte-like cells circulate in the blood and

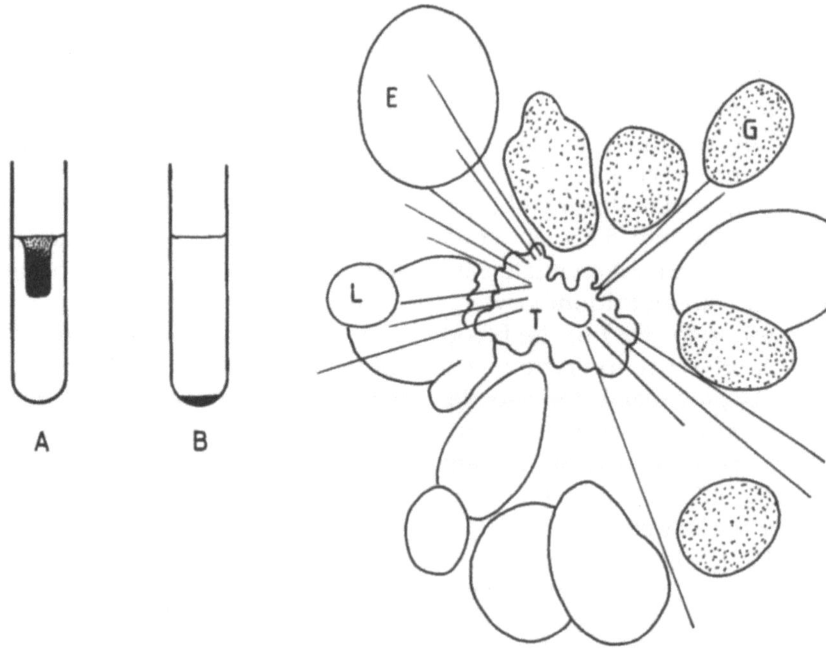

Figure 18.3 Blood coagulation in *Myxine glutinosa*. **To the left**: (A) Test tube containing a mixture of blood and iso-osmotic sodium chloride solution. The erythrocytes get trapped in a fibrin clot at the surface of the fluid. (B) As in (A) but with addition of heparin. Because no fibrin is formed erythrocytes settle on the bottom of the tube. **To the right**: cells and fibrin threads centring around a 'thrombocyte' (T). E, erythrocyte, G, granulocyte, L, lymphocyte-like cell. Light microscopy of a fresh sample of coagulating blood, high magnification, camera lucida drawing.

migrate through the tissues. It is hard to avoid the thought that some of the ubiquitous lymphocyte-like cells may possess not yet understood defence functions, but future research hopefully will clear up the question about the functions of those cells.

The body cavity communicates with the exterior through abdominal pores, but parasites from the exterior seldom or never penetrate into the peritoneal space, which probably is well protected by phagocytes associated with the pronephros, etc. Any foreign particles in the peritoneal cavity are immediately attacked by granulocytes, larger and phagocytically more active than those of the blood (Fänge, 1973). Whether phagocytosis in hagfishes may be further enhanced by opsonins or similar factors is an open question.

Parasites are rarely seen in hagfishes. In a *Bdellostoma* (*Eptatretus*) species nematodes have been found in the subdermal blood sinus (Ayers, 1933), and trypanosomes may appear in the blood of *Heptatretus* (=*Eptatretus*?) (Laird, 1952). Monogeneans infesting the skin of some species of hagfishes (Malmberg and Fernholm, 1991) are ectoparasites which do not penetrate into the hosts.

Tumours of unknown aetiology, primarily of the liver and the skin and sometimes of cancerous character, are fairly common in *Myxine glutinosa* (Falkmer *et al.*, 1976; Falkmer, this volume).

REFERENCES

Aasa, R. (1972) Studies of hagfish transferrin by

electron paramagnetic resonance (EPR) spectroscopy. *Acta Regiae Societatis Scientarum et Litterarum Gothoburgensis, Zoologica*, **8**, 46.

Adam, H. (1963) Structure and histochemistry of the alimentary canal, in *The Biology of Myxine* (eds A. Brodal and R. Fänge), Scandinavian University Books, Oslo, pp. 256–88.

Aisen, P., Leibman, A. and Sin, C.-L. (1972) Molecular weight and subunit structure of hagfish transferrin. *Biochemistry*, **11**, 3461–4.

Ayers, H. (1933) A nematode parasite in myxinoids. *Science*, **78**, 459.

Banfield, D.K., Irwin, D.M., Walz, D.A. and MacGillivray, R.T.A. (1994) Evolution of prothrombin: isolation and characterization of the cDNAs encoding chicken and hagfish prothrombin. *Journal of Molecular Evolution*, **38**, 177–87.

Bartlett, G.R. (1982) Phosphates in red cells of a hagfish and a lamprey. *Comparative Biochemistry and Physiology* (A), **73**, 141–5.

Borcea, I. (1905) Recherches sur le système urogenital des elasmobranches. *Archives de Zoologie Experimentales et Génerales (4 sér.)*, **5**, 199–484.

Carlsson, U., Kjellström, B. and Antonsson, B. (1980) Purification and properties of cyclostome carbonic anhydrase from erythrocytes of hagfish. *Biochimica et Biophysica Acta*, **612**, 160–70.

Christensen, A.K. (1964) The structure of the functional pronephros in larvae of *Ambystoma opacum* as studied by light and electron microscopy. *American Journal of Anatomy*, **115**, 257–78.

Cole, F.J. (1913) A monograph on the general morphology of the myxinoid fishes, etc., Part V. The anatomy of the gut and its appendages. *Transactions of the Royal Society, Edinburgh*, **49**, 293–344.

Cole, F.J. (1925) Ibidem, Part VI. The morphology of the vascular system. *Transactions of the Royal Society, Edinburgh*, **54**, 309–42.

Davison, W., Baldwin, J., Davie, P.S., Forster, M.E. and Satchell, G.H. (1990) Exhausting exercise in the hagfish, *Eptatretus cirrhatus:* the anaerobic potential and the appearance of lactic acid in the blood. *Comparative Biochemistry and Physiology* (A), **95**, 585–90.

Falkmer, S., Embdin, S.O., Östberg, Y., Mattisson, A., Johansson-Sjöbeck, M.-L. and Fänge, R. (1976) Tumor pathology of the hagfish, *Myxine glutinosa*, and the river lamprey, *Lampetra fluviatilis. Progress in Experimental Tumor Research*, **20**, 217–50.

Fänge, R. (1973) The lymphatic system of *Myxine. Acta Regiae Societatis Scientarum et Litterarum Gothoburgensis, Zoologica*, **8**, 57–64.

Fänge, R. (1985) Regulation of blood and body fluids in primitive fish groups, in *Evolutionary Biology of Primitive Fishes* (eds R.E. Foreman, A. Gorbman, J.M. Dodd and R. Olsson), Plenum Press, New York and London, pp. 253–73.

Fänge, R. (1986) Lymphoid organs in sturgeons (Acipenseridae). *Veterinary Immunology and Immunopathology*, **12**, 153–61.

Fänge, R. and Edström, A. (1973) Incorporation of thymidine, uridine and leucine into blood cells of *Myxine glutinosa. Acta Regiae Societatis Scientarum et Litterarum Gothoburgensis, Zoologica*, **8**, 49–50.

Fänge, R. and Gidholm, L. (1968) A macrophage system in *Myxine glutinosa* L. *Naturwissenschaften*, **55**, **44**.

Fänge, R. and Gidholm, L. (1973) Blood coagulation. *Acta Regia Societatis Scientarum et Litterarum Gothoburgensis, Zoologica*, **8**, 51–52.

Fänge, R., Johansson-Sjöbeck, M.-L. and Kanje, M. (1974) Transformation of spindle cells into lymphocyte-like cells in the blood from *Myxine glutinosa. Acta Physiologica Scandinavica*, **91**, 13A–14A.

Fänge, R. and Zapata, A. (1985) Lymphomyeloid system and bodily defenses of hagfish, in *Ontogeny and Phylogeny of the Immune System* (ed. R.A. Good), Vol. 4, Chapter 2. Iwanami Shoten Publisher's Immunology Series (Japanese Edition), pp. 59–76.

Forster, M.E. Davison, W., Satchell, G.H. and Taylor, H.H. (1989) The subcutaneous sinus of the hagfish, *Eptatretus cirrhatus* and its relations to the central circulating blood volume. *Comparative Biochemistry and Physiology (A)*, **93**, 607–12.

Fujii, T., Nakamura, T., Sekizawa, A. and Tomonaga, E. (1922) Isolation and characterization of a protein from hagfish serum that is homologous to the third component of the mammalian complement system. *Journal of Immunology*, **148**, 117–23.

Gilbertson, P., Wotherspoon, J. and Raison, R.L. (1986) Evolutionary development of lymphocyte heterogeneity: leucocyte subpopulations in the Pacific hagfish. *Developmental and Comparative Immunology*, **10**, 1–10.

Good, R.A. and Finstad, J. (1964) Phylogenetic development of transplantation immunity. *Annals of The New York Academy of Science*, **120**, 15–20.

Grodzinski, Z. (1926) Uber das Blutgefäss-system von *Myxine glutinosa* L. *Bulletin international de l'Academie Polonaise des Sciences et des Lettres* (B), 123–55.

Hoffbrand, A.V. and Pettit, J.E. (1980) *Essential Haematology,* Blackwell Science Publications, Oxford.

Holmgren, N. (1950) On the pronephros and the blood in Myxine glutinosa. *Acta Zoologica* (*Stockholm*), **31**, 233–348.

Johansen, K., Fänge, R. and Johannessen, M.W. (1962) Relations between blood, sinus fluid and lymph in *Myxine glutinosa. Comparative Biochemistry and Physiology,* **7**, 23–8.

Johansson, M.-L. (1973) Peroxidase in blood cells of fishes and cyclostomes. *Acta Regiae Societatis Scientarum et litterarum Gothoburgensis, Zoologica,* **8**, 53–6.

Jordan, H.E. (1933) The evolution of blood-forming tissue. *Quarterly Review of Biology,* **8**, 58–76.

Jordan, H.E. (1938) Blood and blood-forming organs of vertebrates, in *Handbook of Hematology,* Vol. 2 (ed. H. Downey), P.B. Hoeber, New York, pp. 715–862.

Jordan, H.E. and Speidel, C.C. (1930) Blood formation in cyclostomes. *American Journal of Anatomy,* **46**, 355–91.

Kohno, S., Nakai, S., Satoh, S., Yoshida, M. and Kobayashi, H. (1986) Chromosome elimination in the Japanese hagfish, *Eptatretus burgeri* (Agnatha, Cyclostomata). *Cytogenet. Cell Genet.,* **41**, 209–14.

Laird, M. (1952) Studies on the Trypanosomes of New Zealand fish. *Proceedings of Zoological Society London,* **121**, 285–309.

Larsson, Å., Johansson-Sjöbeck, M.-L. and Fänge, R. (1976) Comparative study of some haematological and biochemical blood parameters in fishes from the Skagerrak. *Journal of Fish Biology,* **9**, 425–40.

Linna, T.J., Finstad, J. and Good, R.A. (1975) Cell proliferation in epithelial and lymphohematopoietic tissues of cyclostomes. *American Zoologist,* **15**, 29–38.

Linthicum, D.S. (1975) Ultrastructure of hagfish blood leucocytes, in *Immunologic Phylogeny* (eds W.H. Hildeman and A.A. Benedict), *Advances in Experimental Medicine and Biology,* **64**, 241–50. Plenum Press, New York and London.

Malmberg, G. and Fernholm, B. (1991) Locomotion and attachment to the host of *Myxinidocotyle* and *Acanthocotyle* (Monogenea, Acanthocotylidae). *Parasitology Research,* **177**, 415–20.

Mattisson, A. and Fänge, R. (1977) Light- and electronmicroscopic observations on the blood cells of the Atlantic hagfish, *Myxine glutinosa* (L.). *Acta Zoologica* (*Stockholm*), **58**, 205–21.

Morris, J.L. (1981) Structure and function of ciliated peritoneal funnels in the toad kidney (*Bufo marinus*). *Cell Tissue Research,* **217**, 599–610.

Müller, J. (1845) Untersuchungen über die Eingeweide der Fische. Schluss der vergleichenden Anatomie der Myxinoiden. *Abhandlungen der Akademie der Wissenschaften zu Berlin* (1843), 109–70.

Olsson, J.Y. (1973) Activity of delta-aminolevulinic acid dehydrase (ALA-D) in organs of *Myxine glutinosa. Acta Regiae Societatis Scientarum et Litterarum Gothoburgensis, Zoologica,* **8**, 84–5.

Östberg, Y., Fänge, R., Mattisson, A. and Thomas, N.W. (1976) Light and electron microscopical characterization of heterophilic granulocytes in the intestinal wall and islet parenchyma of the hagfish, *Myxine glutinosa* (Cyclostomata). *Acta Zoologica* (*Stockholm*), **57**, 89–102.

Price, G.C. (1910) The structure and function of the adult head kidney of *Bdellostoma stoutii. Journal of Experimental Zoology,* **9**, 849–64.

Riviere, H.B., Cooper, E.L., Reddy, A.L. and Hildemann, W.H. (1975) In search of the hagfish thymus. *American Zoologist,* **15**, 39–49.

Robertson, J.D. (1976) Chemical composition of the body fluids and muscle of the hagfish *Myxine glutinosa* and the rabbit-fish *Chimaera monstrosa. Journal of Zoology,* London, **178**, 261–77.

Sakai, T., Tabata, N. and Suiko, M. (1989) Occurrence of biliverdin IX-alpha in the gall bladder bile of the hagfish, *Eptatretus burgeri. Zoological Science* (*Tokyo*), **6**, 173–6.

Schneider, G. (1897) Ueber die Niere und die Abdominalporen von *Squatina angelus. Anatomische Anzeiger,* **13**, 393.

Schreiner, K.E. (1898) Zur Histologie des Darmkanals. *Bergens Museums Aarbog,* 1–16.

Sekhon, S.S. and Manwell, D.S. (1970) Fine structure of developing hagfish erythrocyte with particular reference to the cytoplasmic organelles. *Journal of Morphology,* **131**, 211–36.

Smith, R.T. (1969) Origin of the spindle cell in Myxine. *The Bulletin Mount Desert Island Biological Laboratory,* **9**, 60.

Strik, J.J.T.W.A. and Fänge, R. (1974) Erythropoietic porphyria in *Myxine glutinosa. Abstract Fed. Verg. Med. Biol. Ver.,* **15**, 345.

Svedberg, T. and Hedenius, A. (1934) The

sedimentation constants of the respiratory proteins. *Biological Bulletin* (*Woods Hole*), **66**, 191–223.

Tanaka, Y., Saito, Y. and Gotho, H. (1981) Vascular architecture and intestinal hematopoietic nests of two cyclostomes, *Eptatretus burgeri* and ammocoetes of *Entosphenus reissneri*: a comparative morphological study. *Journal of Morphology*, **170**, 71–93.

Thoenes, G.H. and Hildemann, W.H. (1970) Immunological responses of Pacific hagfish II. Serum antibody production to soluble antigens. *Development Aspects of Antibody Formation and Structure* (Sterz *et al.*, eds), Vol. 2, pp. 711–26. Czechoslovak Academy of Sciences, Prague; Academic Press, New York and London.

Tomonaga, S., Hirokane, T. and Awaya, K. (1973a) Lymphoid cells in the hagfish. *Zoological Magazine* (*Japan*), **82**, 133–5.

Tomonaga, S., Shinohara, H. and Awaya, K. (1973b) Fine structure of the peripheral blood cells of the hagfish. *Zoological Magazine* (*Japan*), **82**, 211–14.

Tomonaga, S., Hirokane, T., Shinohara, S. and Awaya, K. (1973c) The primitive spleen of the hagfish. *Zoologica Magazine* (*Japan*), **82**, 215–17.

Tomonaga, S., Yamaguchi, K., Ihara, K. and Awaya, K. (1986) Mononuclear phagocytic cells (Kupffer cells) in hagfish liver sinusoids. *Zoological Science*, **3**, 613–20.

Tufts, B.L. and Boutilier, R.G. (1990a) Carbon dioxide transport in agnathan blood: Evidence of erythrocyte chloride ion-bicarbondate ion exchange limitations. *Respiration Physiology*, **80**, 335–48.

Tufts, B.L. and Boutilier, R.G. (1990b) Carbon dioxide transport properties of the blood of a primitive vertebrate, *Myxine glutinosa* (L.). *Experimental Biology* (*Berlin*), **48**, 341–7.

Wells, R.M.G., Forster, M.E. Davison, W., Taylor, H.H., Davie, P.S. and Satchell, G.H. (1986) Blood oxygen transport in the free-swimming hagfish, *Eptatretus cirratus*. *Journal of Experimental Biology*, **123**, 43–53.

Zapata, A., Fänge, R., Mattisson, A. and Villena, A. (1984) Plasma cells in adult hagfish, *Myxine glutinosa*. *Cell Tissue Research*, **235**, 691–3.

19
VOLUME REGULATION IN RED BLOOD CELLS

Niels Dohn and Hans Malte

SUMMARY

Hagfish erythrocytes lack the anion exchanger (band 3 protein) in their membranes. This protein has been suggested to play a key role in the activation of a regulatory volume decrease in teleost and elasmobranch fishes. A total lack of the ability of cellular volume regulation in hagfish erythrocytes supports this notion. The lack of volume regulation is accompanied by an extreme stability of hagfish erythrocytes when exposed to anisotonic media.

19.1 INTRODUCTION

The extracellular fluid of hagfish is isoosmotic or very slightly hyperosmotic to seawater. In fact, the composition of plasma is not very different from that of seawater. Only the concentrations of calcium, magnesium and sulphate are significantly lower in plasma than in seawater (Morris, 1965). This situation resembles that of many marine invertebrates, but is unique among vertebrates where osmotic equilibrium with full strength seawater is always partly obtained by high plasma concentrations of organic substances like urea or TMAO (trimethyl amine oxide). Hagfishes are osmoconformers, i.e. they do not defend their plasma osmotic concentration when transferred to either more dilute or more concentrated seawater. They are normally considered to be stenohaline, and indeed *Myxine* cannot survive acute transfer to less than 70–80% of full strength seawater. This is related to a very slow clearance of the initial osmotic water influx across the gills and body surface (McFarland and Munz, 1965), which is probably due to the limited filtration capacity of the agnathan kidney (Fels *et al.*, this volume; Riegel, this volume). If, however, dilution or concentration is accomplished by small daily changes over several weeks, *Myxine* may be acclimated to salinities ranging from approximately 60% to 150% of full strength seawater (Cholette *et al.*, 1970). During the process of acclimation, the parietal muscle cells regulate their volume by adjusting the intracellular concentration of free amino acids, most notably proline, alanine, leucine and threonine. The ability of the parietal muscle cells to regulate their volume is retained although it is very likely that *Myxine* has been living in an osmotically stable environment for millions of years. Unlike parietal muscle cells, the erythrocytes, however, do not seem to be able to regulate their volume.

19.2 PROPERTIES OF HAGFISH ERYTHROCYTES

The erythrocytes of hagfishes are approximately ellipsoidal in shape and are nucleated. They are larger than teleost red blood cells,

The Biology of Hagfishes. Edited by Jørgen Mørup Jørgensen, Jens Peter Lomholt, Roy E. Weber and Hans Malte. Published in 1998 by Chapman & Hall, London. ISBN 0 412 78530 7.

the two major axes being 25 and 20 μm, and having a volume of 1160 μm³. Even though the blood plasma in which they are suspended is practically pure seawater, the concentrations of the major inorganic ions Na^+, Cl^- and K^+ are not very different from those in teleost fishes (Fincham *et al.*, 1990; Peters and Gros, this volume; Figures 19.2 and 19.4). Unlike in teleosts, however, the concentrations of free α-amino acids are very high (Fincham *et al.*, 1990). In *Eptatretus stoutii* the most abundant amino acids are proline, alanine, leucine and threonine. The erythrocyte membrane of *Eptatretus stoutii* has been shown to be deficient of the anion exchanger (the band 3 protein), spectrin and several other cytoskeletal elements normally found in red blood cells (Ellory *et al.*, 1987). In *Myxine*, Brill *et al.* (1992) found only little binding of the anion exchange inhibitor DIDS (4,4'-diisothiocyanostilbene-2-2'-disulphonate) to the membrane of the erythrocytes, which indicates that the anion exchanger is also absent in this species. The absence of the anion exchanger in the membrane of the red blood cells has consequences for CO_2 and O_2 transport (Fago and Weber, this volume) as well as for cellular volume regulation.

19.3 VOLUME REGULATION

When suspended in hypotonic media, teleost red blood cells (RBCs) initially behave as nearly perfect osmometers undergoing rapid, osmotic cell swelling according to the high membrane permeability to water. Subsequently the cells regulate their volume back towards the initial value by a regulatory volume decrease (RVD) and the final change in cell volume is consistently smaller than expected from van't Hoff's law. The mechanism involved in RVD in teleost and elasmobranch RBCs is the activation of coupled potassium-chloride effluxes and often also of a net efflux of β-amino acids, especially taurine. These fluxes are followed by osmotically obliged water (Hoffmann and Simonsen, 1989; Nikinmaa, 1992). Treating trout RBCs with DIDS inhibits RVD as well as all the ion pathways normally activated by hypotonic swelling, which suggests that the anion exchanger of the erythrocyte membrane plays an essential role in the activation of the response (Garcia-Romeu *et al.*, 1991). It has been shown that expression of rainbow trout band 3 in *Xenopus* oocytes induces chloride and taurine transport, which strengthens the view that band 3 protein plays a key role in the activation of volume regulation (Fievet *et al.*, 1995).

It has recently been demonstrated that, while erythrocytes from the lampreys *Petromyzon marinus* and *Lampetra fluviatilis* are capable of RVD when suspended in hypotonic medium, *Myxine* RBCs appear to be unable to regulate their volume in hypotonic medium (Nikinmaa *et al.*, 1993). In the later investigation, however, the initial osmotic disturbance was obtained by a dilution of the suspension medium of only 20%. Compared to this, skate RBCs did not show any volume regulation when the medium was diluted by 30%, whereas a 50% dilution induced potassium transport activity as well as a volume regulatory response (Dickman and Goldstein, 1990). We therefore examined the osmotic behaviour of *Myxine* RBCs under a larger range of conditions.

When challenging red blood cells from *Myxine* in hypotonic media of differing degrees of hypotonicity it turned out that they were extremely stable, being able to swell to more than 3 times their original volume with very little or no apparent haemolysis. There were no signs of volume recovery when the medium was diluted by 25% (Figure 19.1) in agreement with earlier findings by Nikinmaa and co-workers (1993). Similarly, concentration of the medium by 25% did not elicit a regulatory volume increase. Addition of adrenaline in isotonic medium, which elicits a cell volume increase in trout red blood cells by activating the Na^+/H^+ exchanger (Borgese *et al.*, 1987), was

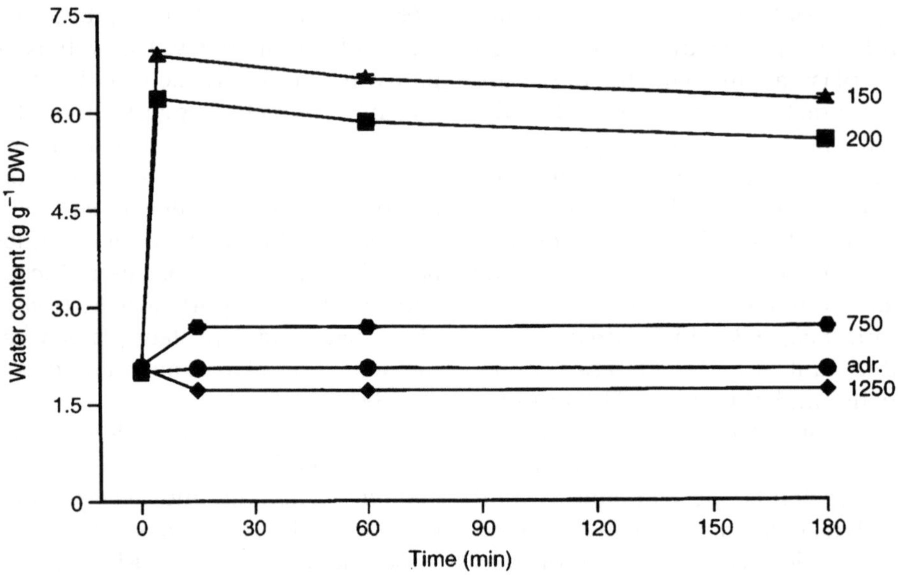

Figure 19.1 Erythrocyte water content as a function of time after an osmotic disturbance or addition of adrenaline. Erythrocytes were initially washed thrice and suspended in an isotonic Ringer solution of 1000 mOsm kg^{-1} in Eschweiler tonometer systems. At time zero either CO_2/HCO_3^--buffered, distilled water or NaCl or adrenaline (10^{-5} M) was added. The approximate, final osmolalities are indicated on each curve. The systems were buffered with CO_2/HCO_3^-, pH = 7.8, T = 10°C. P_{O_2} ≈155 mmHg except in the adrenaline experiment where P_{O_2} = 0 mmHg.

also without any effect. The oxygen tension as well as the pH of the medium did not influence the results of these experiments. Only extreme swelling of the red blood cells elicited a modest recovery in volume. This slight recovery of volume was not reflected in a loss of K^+ and Cl^- from the cells as would be expected if a K^+–Cl^- co-transport pathway had been activated (Figure 19.2). Rather there was a tendency for an increase in Cl^-. Due to the stability of the erythrocytes it was possible to perform long-term *in vitro* experiments with them. In long-term experiments it was still not possible to detect any volume recovery when the osmolality of the medium was halved (not shown). Reducing the osmolality to 1/4 of the original value, however, elicited a significant volume recovery over the following week (Figure 19.3). This was, however, not the result of an activation of a coupled efflux of K^+ and Cl^- as evidenced by Figure 19.4. During the first 24 hours, where the rate of volume recovery was greatest, there was a net influx of Na^+ and Cl^- and a net efflux of K^+ with a net cellular gain of inorganic osmolytes. This could be explained if the extreme degree of swelling led to a general increase in membrane permeability allowing Na^+, K^+ and Cl^- to run down their electrochemical gradients. The same tendency of ion movements is actually present in the short-term experiment (Figure 19.2). Unlike teleost and elasmobranch RBCs, hagfish RBCs contain high intracellular concentrations of α-amino acids, and although no evidence for a volume-activated amino acid pathway has been found (Fincham *et al.*, 1990) it is probable that a temporary, non-specific increase in membrane permeability would result in a loss of amino acids exceeding the net gain of inorganic osmolytes and thus lead to a decrease in cellular water content. The continued

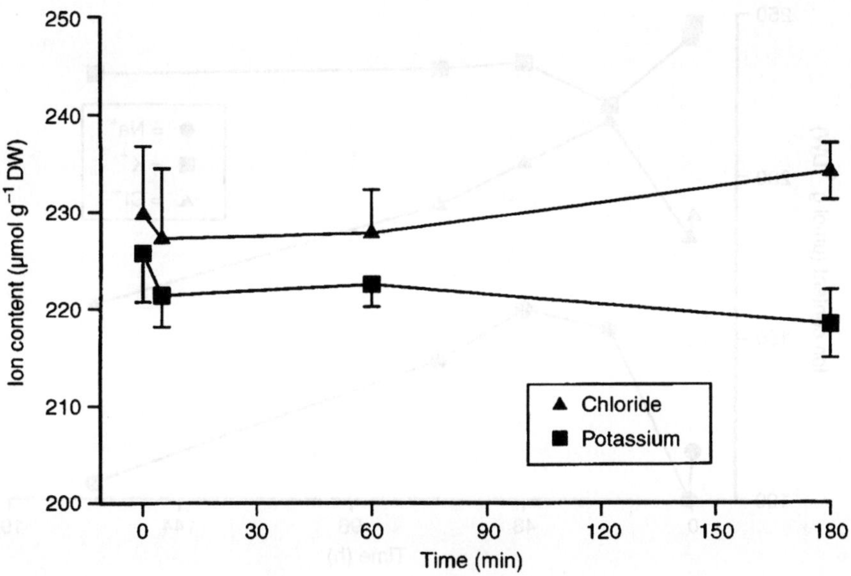

Figure 19.2 Erythrocyte Cl^- and K^+ content as a function of time after the suspending medium had been diluted from 1000 to *c.* 150 mOsm kg^{-1}. Experimental details as in Figure 19.1.

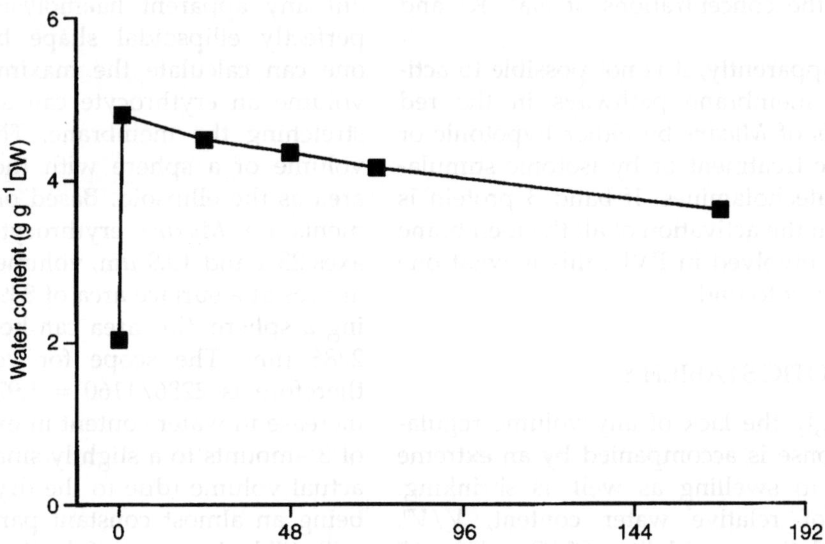

Figure 19.3 Erythrocyte water content as a function of time after dilution of the suspending medium from 1000 to *c.* 250 mOsm kg^{-1}. Note the different time scale in this figure compared to Figure 19.1. Experimental details as in Figure 19.1.

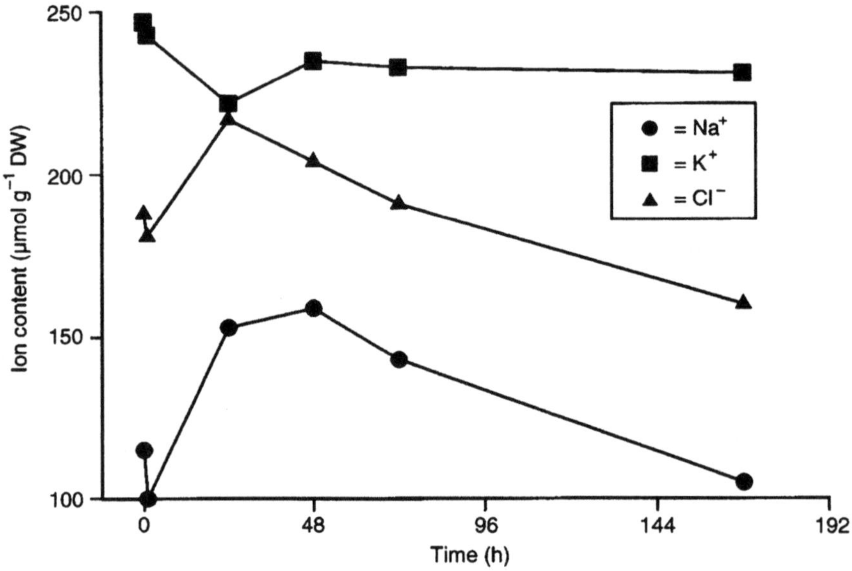

Figure 19.4 Erythrocyte content of K^+, Na^+ and Cl^- as a function of time after dilution of the suspending medium from 1000 to *c.* 250 mOsm kg^{-1}. Note the different time scale compared to Figure 19.2. Experimental details as in Figure 19.1.

recovery in volume from 24 h and onwards is the result of the action of the Na^+/K^+ pump restoring the concentrations of Na^+, K^+ and Cl^-.

Thus, apparently, it is not possible to activate any membrane pathways in the red blood cells of *Myxine* by either hypotonic or hypertonic treatment or by isotonic stimulation by catecholamines. If band 3 protein is essential in the activation of all the membrane pathways involved in RVD, this is what one would expect to find.

19.4 OSMOTIC STABILITY

Interestingly the lack of any volume regulatory response is accompanied by an extreme tolerance to swelling as well as shrinking. Plotting of relative water content, V/V^0, against relative osmolality, O^0/O, where V^0 and O^0 refer to cellular water content and medium osmolality in a reference state (*in vivo* cell volume and medium osmolality), reveals a straight line (Figure 19.5). As evident, it is possible to increase the cellular water content by a factor of more than 3 without any apparent haemolysis. Assuming a perfectly ellipsoidal shape before swelling one can calculate the maximum theoretical volume an erythrocyte can achieve without stretching the membrane. This will be the volume of a sphere with the same surface area as the ellipsoid. Based on our measurements on *Myxine* erythrocytes (two larger axes 25.4 and 19.9 μm, volume 1160 μm^3) one arrives at a surface area of 839 μm^2. If enclosing a sphere, this area can hold a volume of 2286 μm^3. The scope for volume increase therefore is 2286/1160 = 1.97. Although an increase in water content in excess of a factor of 3 amounts to a slightly smaller increase in actual volume (due to the dry cell substance being an almost constant part of the cell) it will still be in excess of the theoretically possible increase with a constant membrane area. The explanation for this apparent discrepancy may be twofold: either the membrane actually is stretched or the surface area is larger

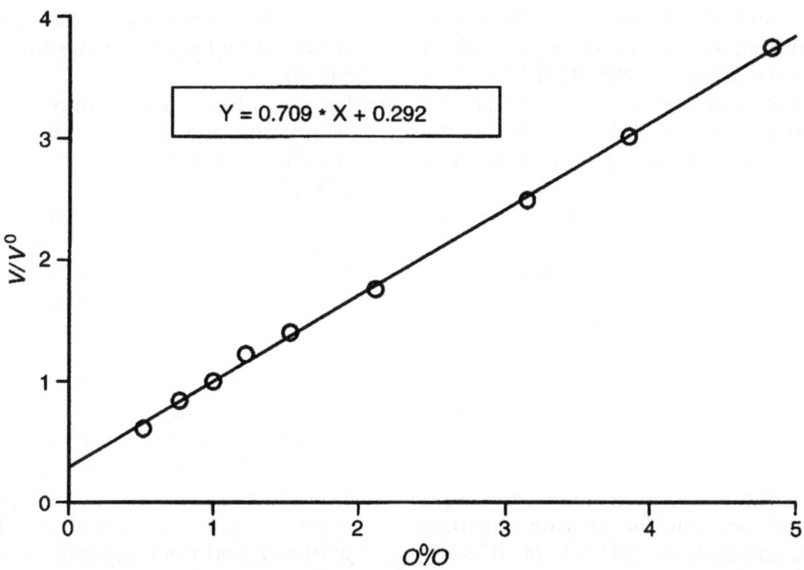

Figure 19.5 A plot showing relative cellular water content, V/V^0, against the relative osmolality of the suspending medium, O^0/O. V^0 and O^0 refer to the cellular water content and medium osmolality of a reference state (*in vivo* isotonic cell water content and plasma osmolality = 1000 mOsm kg^{-1}). At each osmolality the cells were washed three times in ten volumes of the suspending medium before determining the water content.

than for a perfect ellipsoid. The induction of the net fluxes of Na^+ and K^+ (and probably amino acids) down their electrochemical gradients observed by us in extreme swelling may be support for the first explanation. Support for the second explanation also seems at hand since TEM pictures of cross-sections of *Myxine* erythrocytes reveal what appears to be membrane invaginations which would extend its surface area (Welsch, personal communication).

The fact that erythrocytes from hagfishes, being devoid of essential elements of the cytoskeleton, actually show such a high degree of osmotic stability indicates that the cytoskeleton may well stabilize the shape of a cell, but it does not confer osmotic stability to the cell.

Whether the absence of erythrocytic band 3 protein, several cytoskeletal elements and the ability to regulate volume is due to a loss during millions of years of evolution in an osmotically stable environment, or is a truly primitive character is not known. The ability of *Myxine* to regulate acid-base balance via exchanges of Na^+ for H^+ and Cl^- for HCO_3^- in the branchial epithelium (Evans, 1984) indicates the presence of band 3 protein at this site, which favours the first hypothesis.

REFERENCES

Borgese, F., Garcia-Romeu, F. and Motais, R. (1987) Control of cell volume and ion transport by β-adrenergic catecholamines in erythrocytes of rainbow trout. *Salmo Gairdneri. J. Physiol.*, **382**, 123–44.

Brill, S.R., Musch, M.W. and Goldstein, L. (1992) Taurine efflux, band 3 and erythrocyte volume of the hagfish (*Myxine glutinosa*) and lamprey (*Petromyzon marina*). *J. Exp. Zool.*, **264**, 19–25.

Cholette, C., Gagnon, A. and Germain, P. (1970) Isosmotic adaptation in *Myxine glutinosa* L., I: Variations of some parameters and role of the amino acid pool of the muscle cells. *Comp. Biochem. Physiol.*, **33**, 333–46.

Dickman, K.G. and Goldstein, L. (1990) Cell volume regulation by skate erythrocytes: role of potassium. *Amer. J. Physiol.*, **258**, R1217–23.

Evans, D.H. (1984) Gill Na^+/H^+ and Cl^-/HCO_3^- exchange systems evolved before the vertebrates entered fresh water. *J. Exp. Biol.*, **113**, 465–9.

Ellory, J.C., Wolowyk, M.W. and Young, J.D. (1987) Hagfish (*Eptatretus stoutii*) erythrocytes show minimal chloride transport activity. *J. Exp. Biol.*, **129**, 377–83.

Fincham, D.A., Wolowyk, M.W. and Young, J.D. (1990) Characterisation of amino acid transport in red blood cells of a primitive vertebrate, the Pacific hagfish (*Eptatretus stoutii*). *J. Exp. Biol.*, **154**, 355–70.

Fievet, B., Gabillat, N., Borgese, F. and Motais, R. (1995) Expression of band 3 anion exchanger induces chloride current and taurine transport: structure-function analysis. *EMBO* J., **14**, 5158–69.

Garcia-Romeu, F., Cossins, A.R. and Motais, R. (1991) Cell volume regulation by trout erthrocytes: characteristics of the transport systems activated by hypotonic swelling. *J. Physiol.*, **440**, 547–67.

Hoffmann, E.K. and Simonsen, L.O. (1989) Membrane mechanisms in volume and pH regulation in vertebrate cells. *Physiol. Rev.*, **69**, 315–82.

McFarland, W.N. and Munz, F.W. (1965) Regulation of body weight and serum composition by hagfish in various media. *Comp. Biochem. Physiol.*, **14**, 383–98.

Morris, R. (1965) Studies on salt and water balance in *Myxine glutinosa*. *J. Exp. Biol.*, **42**, 359–71.

Nikinmaa, M. (1992) Membrane transport and control of hemoglobin-oxygen affinity in nucleated erythrocytes. *Physiol. Rev.*, **72**, 301–21.

Nikinmaa, M., Tufts, B.L. and Boutilier, R.G. (1993) Volume and pH regulation in agnathan erythrocytes: comparisons between the hagfish, *Myxine glutinosa*, and the lampreys, *Petromyzon marinus and Lampetra fluviatilis*. *J. Comp. Physiol.*, **B163**, 608–13.

20

TRANSPORT OF BICARBONATE, OTHER IONS AND SUBSTRATES ACROSS THE RED BLOOD CELL MEMBRANE OF HAGFISHES

Thomas Peters and Gerolf Gros

SUMMARY

We review here the membrane transport studies that have appeared for the red cells of hagfishes *Myxine glutinosa* and *Eptatretus stoutii*. In the first part the transport of substrates of energy metabolism across the red blood cell membrane is considered. Hagfish red cells possess highly efficient transport systems for glucose, pyruvate and several amino acids. In the case of glucose, this property may be related to the animal's hypoxia tolerance and its capacity for anaerobic glycolysis. Transport systems for small inorganic ions include the Na^+,K^+-ATPase and a K^+–Cl^- symport requiring the presence of Na^+. A striking feature is the absence of a Cl^-–HCO_3^--exchanger, which constitutes in almost all red cells, except those of hagfishes and lampreys, the major pathway for Cl^- and for HCO_3^-. Consequently, Cl^- permeability in hagfish red cells is orders of magnitude lower than in mammalian and most other red cells. We report our mass spectrometric determinations of hagfish red blood cell permeability for HCO_3^- and find its value to be not significantly different from zero. This implies that CO_2 transport in hagfish blood operates quite differently from that in the blood of most other species. Hagfish red cells, due to the presence of intracellular haemoglobin and carbonic anhydrase, can rapidly convert the CO_2 taken up to HCO_3^- and H^+, but they cannot transfer most of the HCO_3^- to the plasma, as red cells of most species do.

20.1 INTRODUCTION

Transport processes across the plasma membrane of red blood cells (RBC) serve a number of functions such as osmotic regulation, maintenance of the membrane potential, pH regulation, uptake of substrates fuelling intracellular metabolism, release of metabolites, and ion transport associated with CO_2 transport by the blood. These transport processes, as far as they have been studied in hagfish RBC, will be reviewed in this chapter.

Erythrocytes of hagfishes and other agnathans are nucleated cells, as are the RBC of fishes, birds, reptiles and amphibians. In contrast, mature mammalian RBC are enucleate. The metabolism of nucleated RBC differs qualitatively and quantitatively from that of RBC that lack a nucleus and other organelles. This is associated with differing transport properties of the cell membranes resulting in

The Biology of Hagfishes. Edited by Jørgen Mørup Jørgensen, Jens Peter Lomholt, Roy E. Weber and Hans Malte. Published in 1998 by Chapman & Hall, London. ISBN 0 412 78530 7.

a match of metabolic and transport characteristics of cells.

At the same temperature, nucleated erythrocytes possess a higher specific oxygen consumption than enucleate mammalian RBC. Fish RBC appear to have an O_2 consumption that is higher than that of RBC of all other vertebrates and may be ten times that of mammalian RBC (Boutilier and Ferguson, 1989). RBC of fishes obtain most of their energy via the Krebs cycle and they use monocarboxylic acids and glucose as well as amino acids as substrates. In contrast, mammalian erythrocytes rely almost exclusively on anaerobic energy production, and, consequently, on glucose utilization. RBC of different fish species vary remarkably with regard to transport properties, and agnathans seem to be special. For agnathan RBC monocarboxylic acids appear to be important fuels, as they are for teleost RBC. Accordingly, RBC of all these species have efficient transport systems for these substrates. However, whereas glucose plays only a minor role in the erythrocytes of several teleosts, it seems to be a main substrate for agnathan RBC. This may be illustrated by two observations of Nikinmaa and Tiihonen (1994) showing that at 20 °C the rate of 3.0-methyl-D-glucose uptake in *Lampetra fluviatilis* RBC is 50 times greater than in carp *Cyprinus carpio* RBC, where the slow glucose uptake is compensated by uptake rates of lactate and pyruvate that are 6 times and 13 times higher, respectively, at identical extracellular substrate concentrations.

This chapter focuses on transport properties of the plasma membrane of hagfish RBC. We will first attempt to review what is known about transport processes involved in the uptake of substrates of energy metabolism. We will then describe the transport mechanisms of small inorganic ions across the red cell membrane and place major emphasis on Cl^-–HCO_3^- permeability of these cells and the consequences for the mode of CO_2 transport in the blood of hagfishes.

20.2 SUBSTRATES OF ENERGY METABOLISM

20.2.1 Glucose

The monosaccharide uptake of hagfish RBC has been investigated in some detail only for the Pacific hagfish, *Eptatretus stoutii*. Ingermann *et al.* (1984, 1985) have shown that glucose uptake by these red cells is very rapid, reaching 50% of equilibrium in 10 s at 10 °C. This uptake occurs primarily by a saturable mechanism which is stereospecific for D-glucose. L-Glucose uptake is very slow, attaining less than 20% of the equilibrium in 4 h, and is attributed to simple diffusion across the membrane. The rapid glucose transport across the hagfish RBC membrane shares several properties with the fast glucose transport system of human RBC. Among these are (1) stereospecificity: hagfish like human RBC accept as substrates, besides α- and β-D-glucose, 2-deoxy-D-glucose, 3-O-methyl-D-glucose, D-mannose, D-galactose and D-xylose, but do not transport α-methyl-D-glucoside, D-sorbitol, D-fructose and L-glucose; (2) sugar transport of hagfish RBC is inhibited, at affinities similar to those for human RBC, by phloretin and cytochalasin B; (3) the hagfish glucose transporter is non-concentrative and establishes identical sugar concentrations inside and outside the red cell; (4) hagfish RBC glucose transport seems to be Na^+-independent because, in contrast to Na^+-dependent glucose transporters, it is not inhibited by harmaline, and shows little if any interaction with α-methyl-D-glucoside unlike the sodium-dependent glucose transporters of mammalian kidney and gut epithelia; (5) glucose uptake is not coupled to H^+ movements (Young *et al.*, 1994).

Young *et al.* (1994), using photoaffinity labelling with ^{3}H-cytochalasin B, identified the RBC glucose transporter of the Pacific hagfish as a protein with a molecular weight of 55 000, which is similar to that of the human RBC glucose transporter (band 4.5, or GLUT 1). Using antibodies against the intact purified human protein and against synthetic peptides corresponding to residues 240–255 or 450–467 of the human RBC glucose transporter, respectively, the 55 kD membrane protein of hagfish red cells was also labelled, again suggesting a remarkable similarity between the glucose transporters of human and hagfish erythrocytes.

What is the physiological significance of the efficient glucose transport system in hagfish RBC, given the absent or much less developed system in teleosts, such as brown trout, rainbow trout and some eels and carps (Nikinmaa and Tiihonen, 1994)? Ingermann *et al.* (1985) hypothesize that this may be related to the hagfish's hypoxia tolerance and its ability to rely on anaerobic glycolysis. They speculate that this property of hagfish red cells may not only serve their own energy requirements under hypoxic conditions but also provide glucose to other tissues. It is interesting in this context to note that in another agnathan, the lamprey, which is less likely to be exposed to hypoxic conditions than the hagfish, glucose transport across the RBC membrane exhibits considerably lower maximal rates than in the Pacific hagfish (Tiihonen and Nikinmaa, 1991; Young *et al.*, 1994).

20.2.2 Monocarboxylic acids

The information on the transport of these substrates in hagfish RBC is scarce. For the Pacific hagfish *Eptatretus stoutii*, Tiihonen *et al.* (1993) have reported a rather rapid uptake kinetics of pyruvate from a 50 μM solution with a half-time of 1.6 min at 10 °C. Within 10 min a 21-fold intracellular enrichment of pyruvate was observed. During this period, some of the pyruvate taken up had been converted intracellularly to lactate, alanine or tricarboxylic acid cycle intermediates. Pyruvate uptake was characterized as inhibitable by the thiol reactive reagent p-chloromercuriphenylsulphonate and by 4,4'-diisothiocyanatostilbene-2,2'-disulphonate (DIDS) but, in contrast to the RBC of several mammalian species, not by the inhibitor of H^+-monocarboxylate symport α-cyano-4-hydroxycinnamate. Pyruvate transport was not coupled to H^+ movements, but depended on the presence of Na^+ and thus presumably constitutes a Na^+-monocarboxylate co-transport system.

These findings suggest that pyruvate – and possibly other monocarboxylates – are important substrates of hagfish RBC. The transport mechanism, however, is quite different from that of several mammalian and teleost RBC, where the Cl^-–HCO_3^--exchanger band 3 and a H^+-monocarboxylate symporter mediate this transport (Deuticke *et al.*, 1978; Tiihonen and Nikinmaa, 1993).

20.2.3 Amino acids

Again, almost all studies concern RBC of the Pacific hagfish *Eptatretus stoutii*. These cells possess a high intracellular concentration of free amino acids, between 100 (Fincham *et al.*, 1990) and 200 mmoles per litre of cell water (Gorkin, 1970). This is 50–100 times the amino acid concentration in the blood plasma. While the red cells of teleosts and elasmobranchs possess high levels of β-amino acids (such as taurine and β-alanine), which are crucially involved in RBC volume regulation, RBC of hagfishes contain essentially α-amino acids (Fincham *et al.*, 1990). In accordance with this, *Myxine glutinosa* RBC, in contrast to cells of teleostean species, exhibit a low capacity for taurine membrane transport (as do *Eptatretus stoutii* RBC; Fincham *et al.*, 1990) and show virtually no regulatory volume decrease upon swelling induced in hypo-osmotic medium (Brill *et al.*, 1992; see also Dohn and Malte, this volume).

Fincham *et al.* (1990) have performed a careful study of the mechanisms of amino acid transport in RBC of *Eptatretus stoutii*. The most abundant intraerythrocytic amino acids contributing to the total concentration of ~100 mmoles per litre of cell water are, in the sequence of decreasing concentrations (numbers in brackets give RBC : plasma ratios):

proline 16.8 mM (173), alanine 8.99 mM (219), leucine 8.76 mM (55), threonine 7.58 mM (67), glutamine 6.93 mM (64), valine 6.27 mM (59). It is apparent that all these amino acids are, like pyruvate, considerably enriched in the intracellular water space compared to plasma. It may be noted that a similar amino acid enrichment has previously been reported for parietal muscle of *Myxine glutinosa* (Cholette and Gagnon, 1973). Fincham *et al.* (1990) found L-alanine, L-leucine and L-histidine uptake to be independent of Na^+. The uptake of L-alanine, which they primarily studied, was also proton-independent, being insensitive to the protonophores CCCP and FCCP. It was shown that L-alanine transfer across the membrane occurs by homoexchange or by exchange against any of a large number of other amino acids. On account of the selectivity of the uptake mechanism for neutral amino acids of intermediate size and its Na^+-independence, Fincham *et al.* (1990) call the carrier a system of the *asc* type. Uptake rates via this system were very fast, 10^4 times higher than those by known *asc* or *ASC* systems of RBC of some mammalian and avian species. L-Alanine transport was not volume-sensitive as incubation of RBC in hypotonic media did not affect L-alanine efflux. Fincham and co-workers assume that a system of the *L*-type is also present in hagfish RBC and responsible for the even faster transport of L-leucine and L-histidine.

Fincham *et al.* (1990) found one amino acid of very high intracellular concentration, whose transport is Na^+-dependent, proline, and a second one of lower concentration, glycine. The uptake of only these amino acids can be driven by the large extra-to-intracellular Na^+ gradient; all others can only move into the cell at the expense of another amino acid leaving the cell. It is speculated that Na^+-driven proline/glycine entry into the cell may be the primary event, which then allows all other α-amino acids to enter the RBC by exchange against proline or glycine and which may constitute the energy source for the others' intracellular accumulation. Definitive evidence for this idea is lacking, however. It should also be noted that the physiological significance of the presence of rapid amino acid transport systems and the high intracellular concentration of these substrates in hagfish RBC is not clear.

20.2.4 Nucleosides

The red cells of *Eptatretus stoutii* possess a slow, saturable and non-concentrative nucleoside uptake system, as demonstrated by Fincham *et al.* (1991) using 2-^{14}C-uridine as a substrate. The uptake of the pyrimidine nucleoside uridine was shown to be inhibitable by the purine nucleosides adenosine and inosine; hence the transport system accepts both types of nucleosides. This system is different from the nucleoside transporter of human and eel RBC in that (1) it is Na^+-independent and (2) dipyridamole is a better inhibitor than nitrobenzylthioinosine (NBMPR). In addition, this transporter of hagfish RBC is inhibited by the thiol reagent PCMBS, which is not the case with human, bovine and rat erythrocytes. Nevertheless, the type of nucleoside transporter found in hagfishes does not appear to be unique; it is very similar in several respects to that observed in rat RBC and in the mammalian CNS. The physiological function of the transporter may be to provide nucleosides for intraerythrocytic energy metabolism.

20.2.5 Amides

The only work dealing with this subject appears to have been published more than 20 years ago. Kaplan *et al.* (1974) reported that red cells of *Myxine glutinosa* like RBC of other fishes and birds seem to take up urea by simple diffusion only, while RBC of amphibia, reptiles and terrestric animals, including man, exhibit a fast urea transport that can be inhibited by phloretin. Whereas this latter mechanism is being studied in great detail (Zhang and Solomon, 1992; Sands *et al.*, 1992; Mannuzu *et al.*, 1993; Olives *et al.*, 1995) no further reports have appeared concerning the former. It appears possible that the lack of an efficient urea transporter in hagfish RBC and the absence of a countercurrent system in the primitive kidney of this animal are functionally related (Macey and Yousef, 1988; Sands *et al.*, 1992).

20.3 SMALL INORGANIC IONS

Hagfish are thought to have evolved entirely in seawater. This is reflected in their extracellular ion composition which has an osmolality close to that of seawater. Nevertheless, their intracellular spaces maintain large ion gradients towards the extracellular space, similar to those seen in other vertebrates. Table 20.1 gives ion concentrations in RBC and plasma of the Pacific hagfish *Eptatretus stoutii* as reported by Fincham *et al.* (1990).

Obviously, a Na^+–K^+-ATPase could be responsible for the gradients of Na^+ and K^+ as it is in other vertebrate cells; and indeed hagfish RBC possess Na^+–K^+-ATPase activity. Ellory and Wolowyk (1991) have reported that part of $^{86}Rb^+$ influx (indicative of K^+ influx) is inhibited by ouabain with an IC_{50} of 10 μM.

In their further studies of K^+ permeation across the *Eptatretus stoutii* RBC membrane, Ellory and Wolowyk (1991) addressed the question whether a Na^+–K^+–$2Cl^-$ co-transport exists in these cells. They found some evidence for the presence of a related but somewhat different co-transporter in the cell membrane of hagfish RBC. Classical inhibitors of this transporter, which is widespread in vertebrates, are furosemide and bumetanide. The authors derive evidence for the presence of a Na^+–K^+–$2Cl^-$ co-transporter from the following observations. (a) A part of the entire $^{86}Rb^+$ influx turned out to be inhibited by both furosemide and bumetanide, classical inhibitors of Na^+–K^+–$2Cl^-$ co-transport, the latter substance having an especially high affinity. Also the IC_{50}s for both drugs are comparable to those observed for inhibition of Na^+–K^+–$2Cl^-$ co-transport in human RBC. (b) The bumetanide-sensitive $^{86}Rb^+$ influx was dependent on the presence of Na^+ in the medium. However, although requiring Na^+, this transport system did not transport Na^+ at all, as shown in stoichiometric measurements using $^{86}Rb^+$, $^{22}Na^+$ and $^{36}Cl^-$, but transported only K^+ (Rb^+) and Cl^- at a ratio of approximately 1. Thus, the hagfish RBC Na^+–K^+–$2Cl^-$ co-transporter is – at least functionally – not identical to that described in many other red cells and tissues.

Cl^- permeability of hagfish red cells is greatly reduced compared to that of other vertebrate RBC. This is due to the virtual absence of the Cl^-–HCO_3^--exchanger or band 3 protein in the membranes of hagfish RBC, as has first been shown by Ellory *et al.* (1987) for *Eptatretus stoutii* RBC using SDS-PAGE gel electrophoresis. This is consistent with the observation of Ellory *et al.* (1987) that $^{36}Cl^-$ uptake of *Eptatretus stoutii* RBC was 6 orders of magnitude lower than in human RBC. The very slow Cl^- influx was further reduced – by ~40% – upon the addition of the band 3 inhibitor H_2-DIDS at a concentration of 20 mmol l^{-1}. The remaining 60% of this flux may represent a Cl^-–K^+ flux mediated by the Na^+–K^+–$2Cl^-$ transporter described above. It may be noted that the absence – or greatly diminished presence – of band 3 protein has been confirmed for *Myxine glutinosa* RBC by Brill *et al.* (1992). They showed that the RBC of the Atlantic hagfish in comparison to RBC from the skate (*Raja erinacea*) exhibit (a) a

Table 20.1 Concentrations of Na^+, K^+, and Cl^- in red blood cells and blood plasma of the Pacific hagfish *Eptatretus stoutii* (from Fincham *et al.*, 1990), and concentration ratios RBC/plasma

	RBC	Plasma	Ratio: RBC/plasma
Na^+	17	400	0.043
K^+	148	7.0	21
Cl^-	151	468	0.32

greatly diminished binding of $^3H_2$–DIDS to their membranes, and (b) a drastically reduced rate of sulphate exchange, which is known also to be mediated by band 3 protein.

20.4 CO_2 TRANSPORT AND BICARBONATE PERMEABILITY IN HAGFISH RED BLOOD CELLS

According to the classical scheme of CO_2 uptake by the blood, CO_2 diffuses from the tissue through the plasma into the RBC, where both haemoglobin and carbonic anhydrase are localized. Together they effect rapid hydration of CO_2 and efficient buffering of one of the reaction products, H^+. This is followed by transfer of a major part of the other reaction product, HCO_3^-, into the plasma in exchange for Cl^-. This latter process is mediated by the Cl^-–HCO_3^- exchanger or band 3 membrane protein and brings HCO_3^- to the plasma space, which constitutes the major site where the CO_2 taken up is stored and transported to the lungs or gills. This anion exchange mechanisms is present in the blood of almost all vertebrates including fish, the only known exceptions being lampreys and hagfishes.

Of the three proteins essential for this classical mode of CO_2 transport by the blood, intraerythrocytic haemoglobin, carbonic anhydrase and band 3 protein, the former two are present (Berglund *et al.*, 1980; Carlsson *et al.*, 1980; Maren *et al.*, 1980), but the latter is missing in the RBC of Cyclostomata. For lampreys, this has been shown by Ohnishi and Asai (1985), Nikinmaa and Railo (1987) and by Tufts and Boutilier (1989). For hagfish, Ellory *et al.* (1987) reported band 3 deficiency in *Eptatretus stoutii* on the basis of SDS-PAGE electrophoresis, and Brill *et al.* (1992) demonstrated a significantly reduced binding of the band 3 inhibitor DIDS to RBC membranes of *Myxine glutinosa* compared to those from skate. Since the band 3 Cl^-–HCO_3^- exchanger is by far the most efficient pathway for Cl^- in most vertebrate RBC, the following observation by Ellory *et al.* (1987) provides further evidence against significant amounts of band 3 protein in hagfish RBC. The rate of influx of $^{36}Cl^-$ into *Eptatretus stoutii* RBC at 11 °C averaged about 2.4 mmole Cl^-/(l cell h), which compares with a maximal rate of 1.5 mole Cl^-/(l cell min) reported for human RBC at 10 °C by Brahm (1977). This would imply a Cl^- permeability in hagfishes that is approximately 40 000 times lower than in human RBC. Despite this very low Cl^- permeation rate Ellory *et al.* (1987) found a further reduction by 40% upon addition of DIDS. Thus, it appears possible that band 3 is not completely absent in hagfishes although its abundance is much lower than in other vertebrates.

Another piece of evidence against the presence of major amounts of band 3 protein in hagfish RBC comes from the inability of suspensions of these cells to significantly buffer added KOH. This contrasts with suspensions of flounder RBC which do so very efficiently (Ellory *et al.*, 1987). We note that the ability of cells to buffer fixed acids or bases added extracellulary depends on the effective permeability of their membranes for H^+ and/or OH^-. In most vertebrate RBC the mechanism that is by far the most efficient for H^+/OH^- transfer involves the Cl^-–HCO_3^- exchanger. This mechanism is called the Jacobs–Stewart cycle (Jacobs and Stewart, 1942) and is depicted in Figure 20.1: if, for example, OH^- have been added extracellularly and intracellular pH, pH_i, is lower in relation to extracellular pH, pH_e, than required for electrochemical equilibrium, H^+ will be transferred from the intra- to the extracellular compartment resulting in a 'buffering' of extracellular OH^-. This is accomplished by a cycle, in which intracellular H^+ reacts with intracellular HCO_3^- to give CO_2 which then diffuses into the extracellular space. There, CO_2 is hydrated producing HCO_3^- and H^+. HCO_3^- is transferred back into the RBC via the Cl^-–HCO_3^- exchanger, leaving H^+ behind. Thus, one cycle of CO_2 exit from and HCO_3^-

entry into the RBC has effectively resulted in a H^+ extrusion. In the 'alkaline challenge' experiment reported by Ellory *et al.* (1987) the Cl^-–HCO_3^- exchanger is responsible for the excellent buffering properties of flounder RBC suspension, and its (near) absence in hagfish RBC results in the poor buffer capacity exhibited by suspensions of these cells. The question whether the buffer responses seen by Ellory *et al.* (1987) were mediated by the Jacobs–Stewart cycle as seen in Figure 20.1 or by a direct Cl^-–OH^- exchange mediated by band 3 protein must remain unanswered. In this kind of experiment, although nominally carried out in the absence of CO_2 and HCO_3^-, both species are usually present in small amounts that may be sufficient to maintain the cycle of Figure 20.1 at an appreciable rate. In any case, these data clearly confirm the very low level of band 3 in the hagfish RBC.

For the lamprey, a nearly complete absence of band 3 protein has been confirmed by a number of CO_2 binding and ion distribution studies in the blood of *Petromyzon marinus* and *Lampetra fluviatilis*:

1. The CO_2 binding curve is much steeper and higher for *Petromyzon* red cells than for its true plasma, suggesting lack of HCO_3^- transfer from RBC to plasma. This situation is even reversed when the Cl^-/OH^- ionophore tri-*n*-propyl tin chloride is added to the blood, simulating the presence of band 3 (Tufts and Boutilier, 1990b).
2. The apparent non-bicarbonate buffer capacity ($\Delta[HCO_3^-]/\Delta$ pH) for *Petromyzon* RBC is about 20 times greater than that for true plasma, which again suggests lack of acid-base exchange between RBC and plasma. As expected, upon addition of the Cl^-/OH^-

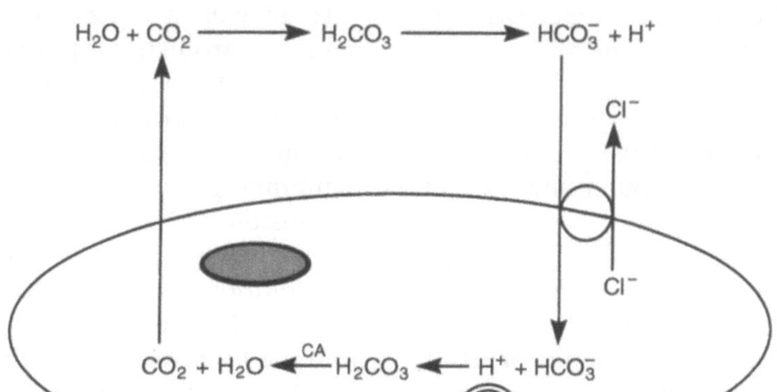

Figure 20.1 Scheme of the efflux of intracellularly produced acid equivalents from a nucleated red blood cell via the Jacobs–Stewart cycle. Bicarbonate (HCO_3^-) enters the cell in exchange with chloride ions (Cl^-). In the cytoplasm HCO_3^- reacts with protons (H^+) to form carbonic acid (H_2CO_3) which is dehydrated to give carbon dioxide (CO_2) and water (H_2O). The latter reaction is catalysed by intracellular carbonic anhydrase (CA). CO_2 diffuses out of the cell, where it is hydrated again and dissociates to protons and bicarbonate. Thus, the net effect of the entire cycle is extrusion of H^+ from the cell. Haemoglobin (Hb) acts as an intracellular buffer.

ionophore the buffer capacities of the two compartments come much closer and differ only by a factor of ~2.

3. The slope of the relationship between pH_i and pH_e is rather shallow when compared to other vertebrate RBC. This holds for the lampreys *Lampetra fluviatilis* (Nikinmaa, 1986) and *Petromyzon marinus* (Tufts and Boutilier, 1990a). It suggests that pH_i is relatively isolated from changing pH_e values due to a deficiency in acid-base exchanges. The slope of pH_i vs pH_e was shown to increase in the case of *Lampetra fluviatilis* by the addition of the protonophore DNP (Nikinmaa, 1986) and in the case of *Petromyzon* by the addition of the Cl^-/OH^- ionophore.
4. The intra- vs extracellular distribution ratios for HCO_3^-, H^+ and Cl^-, which are nearly equivalent for most vertebrate RBC, are vastly discrepant for *Petromyzon* RBC (Tufts and Boutilier, 1990a). Again, this is compatible with the absence of an acid-base exchange system such as the Cl^-–HCO_3^- exchanger. In agreement with this, addition of Cl^-/OH^- ionophore resulted in distribution ratios of the three ions that were similar, and at pH_e 7.6 even identical.

Blood of the hagfish *Myxine glutinosa* studied by Tufts and Boutilier (1990a, b) quite unexpectedly showed a behaviour quite different to that seen in lampreys:

1. The CO_2 binding curve for hagfish RBC is much lower than that for true plasma, suggesting a significant exchange of HCO_3^- between RBC and plasma. The difference became only slightly (and not statistically significantly) more pronounced upon addition of Cl^-/OH^- ionophore.
2. The apparent non-bicarbonate buffer capacities for *Myxine* RBC and true plasma are almost identical.
3. The slope of the relationship pH_i vs pH_e for hagfish blood is 2–3 times greater than it is for *Petromyzon* blood and about equal to that seen in *Petromyzon* blood after addition of Cl^-/OH^- ionophore.
4. Only in the case of intra- vs extracellular ion distribution ratios does *Myxine* blood exhibit a greater ratio for HCO_3^- than for H^+ and Cl^-. However, the HCO_3^- ratio is only 0.8 at pH_e 7.3 while it is ~2 at the same pH_e in *Petromyzon*, and the ratios of H^+ and Cl^- are almost identical at around 0.2 in *Myxine* while they differ drastically in *Petromyzon*
5. A further important finding of Tufts and Boutilier (1990b) for *Myxine* is the lack of an effect of the band 3 inhibitor DIDS on HCO_3^- concentrations in any of the blood compartments.

In conclusion, HCO_3^- seems to be exchanged to a considerable extent between RBC and plasma in *Myxine glutinosa* blood, despite convincing evidence against the presence of significant amounts of the RBC Cl^-–HCO_3^- exchanger and the observation of a very low Cl^- permeability. Is there a HCO_3^- permeation mechanism present in the membranes of these RBC which is different from the Cl^-–HCO_3^- exchanger?

To answer this question, we have used the mass spectrometric approach described by Itada and Forster (1977) that allows one to determine simultaneously the intracellular carbonic anhydrase activity and the membrane permeability for HCO_3^- in cells in suspension. This method uses a thermostatted reaction vessel, which contains a suitable reaction solution including bicarbonate whose O is 2% ^{18}O-labelled. This solution is separated from the high vacuum of a mass spectrometer by a 25 μm thick teflon membrane sitting on a sintered glass disc. As seen in the upper part of Figure 20.2, the labelled HCO_3^- will react to give either labelled water or labelled CO_2. The ^{18}O will be transferred back and forth between HCO_3^- and CO_2, but each time the reaction passes through H_2CO_3 some of the ^{18}O will go into the water, where it is essentially lost from the CO_2-HCO_3^- pool (size ~10^{-2} M) because the pool of water is very much greater (size ~55

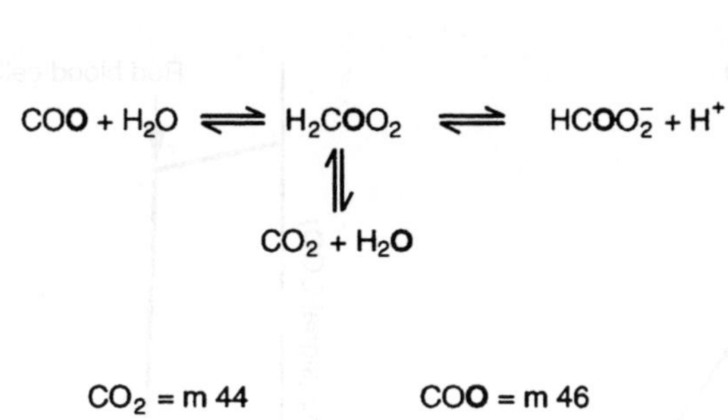

Figure 20.2 Chemical reactions describing the exchange of the oxygen isotope ^{18}O (**O**) between carbon dioxide, carbonic acid and water. This process which is used for measuring the intracellular carbonic anhydrase activity and the bicarbonate permeability of the red cell membrane by mass spectrometry. Singly labelled CO_2 has a molecular mass of 46, unlabelled CO_2 has a molecular mass of 44. Note that the water pool is much greater than the pool of CO_2, H_2CO_3 and HCO_3^-, and therefore nearly all the ^{18}O will end up in the water.

M). All gases dissolved in the reaction medium will diffuse through the teflon membrane to the ion source of the mass spectrometer following a partial pressure gradient that is proportional to the gas concentration in the medium. Thus, we can continuously monitor the concentrations of all the gases in the fluid medium with the mass spectrometer. Figure 20.3a shows schematically a mass spectrometric record, where the signal of labelled CO_2 is seen to decrease continuously because of the loss of ^{18}O into the water pool (first part of Figure 20.3a). Upon addition of carbonic anhydrase with a haemolysate, the decay of ^{18}O-labelled CO_2 occurs much faster but still linearly in a semilogarithmic plot (second part of Figure 20.3a). The upper part of Figure 20.2 suggests the explanation for this: in the presence of carbonic anhydrase the gross reaction rate $CO_2 \leftrightarrow HCO_3^-$ will be accelerated, hence the loss of ^{18}O into the water pool and the decrease in the concentration of labelled CO_2 will be sped up.

Figure 20.3b shows that addition of intact RBC to the reaction medium results in a response of the $C^{18}O^{16}O$ signal that is quite different from its response to carbonic anhydrase in solution: ^{18}O-labelled CO_2 decreases very rapidly in a first phase and then continues to fall more slowly in a second linear phase. Itada and Forster (1977) have shown that this type of response is characteristic for cells and organelles that contain carbonic anhydrase in their interior. The long solid arrow in Figure 20.4 indicates the major event leading to the fast first phase of Figure 20.3b: labelled CO_2 diffuses into the cells added to the medium, where, due to the high intraerythrocytic carbonic anhydrase activity, it is rapidly hydrated to become labelled HCO_3^-. The first phase of $C^{18}O^{16}O$ decay is large, because the process continues until the intracellular HCO_3^-, which is about 20 times the CO_2, is labelled with ^{18}O. At the end of the first phase, the extra- and intracellular CO_2 and the intracellular HCO_3^- possess approximately identical degrees of ^{18}O-labelling. However, in the absence of extracellular carbonic anhydrase, the extracellular rate of dehydration is slow, and therefore extracellular HCO_3^- is still labelled to a much greater extent than intracellular HCO_3^- (and CO_2).

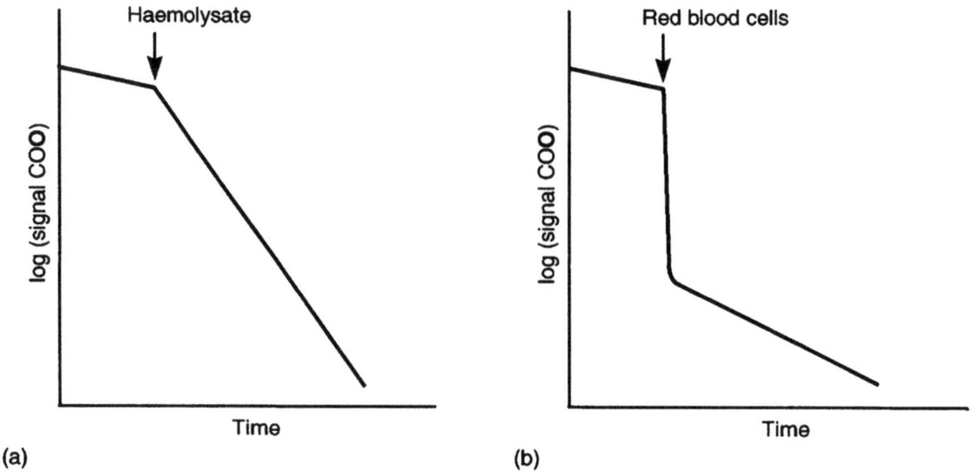

Figure 20.3 Schematic semilogarithmic plots of continuous mass spectrometric recordings of the concentration of singly ^{18}O labelled CO_2 (COO) in the reaction solution. In the initial phase, prior to the addition of the respective sample, the slope of the signal is comparatively small due to the slow uncatalysed loss of ^{18}O into the water pool. Kinetics after addition of carbonic anhydrase in solution or within cells differ depending on whether lysate or a sample with intact cells is investigated. (a) Trace from an experiment carried out with a carbonic anhydrase containing hemolysate which speeds up the disappearance of $CO^{18}O$. Slopes of the trace before and after the addition of the sample can be used to calculate the carbonic anhydrase activity of the sample. (b) Trace of an experiment performed with intact red blood cells. Note the characteristic biphasic kinetics after the addition of the cell suspension. The two phases of this kinetics allow us the calculation of intracellular carbonic anhydrase activity and bicarbonate permeability of the cell membrane.

For this reason, as indicated by the dashed arrows in Figure 20.4, labelled HCO_3^- moves into the cell in exchange for unlabelled HCO_3^- during the second slow phase of the record of Figure 20.3b. While the first fast phase depends critically on intracellular carbonic anhydrase activity, the second slow phase observed after RBC addition is governed by entry of labelled HCO_3^- into the cells. As shown by Itada and Forster (1977), it is possible, therefore, to determine intracellular carbonic anhydrase activity and membrane bicarbonate permeability by evaluating the two phases of $C^{18}O^{16}O$ decay seen in Figure 20.3b after addition of cells. The values obtained for these parameters with this technique agree well with those obtained by other methods. The specific and unique advantage of the mass spectrometric method, however, is that (a) it measures carbonic anhydrase activity within intact cells, and, more importantly, (b) it detects only permeation of the *species* HCO_3^- through the membrane and is not susceptible to the movements of other acids or bases because the entire measurement occurs under conditions of chemical (although not isotopic) equilibrium.

Table 20.2 shows bicarbonate permeabilities, $P_{HCO_3^-}$, of *Myxine glutinosa* RBC determined at 10°C. Shown are 13 single determinations from hagfish RBC that had been washed three times in solution isotonic

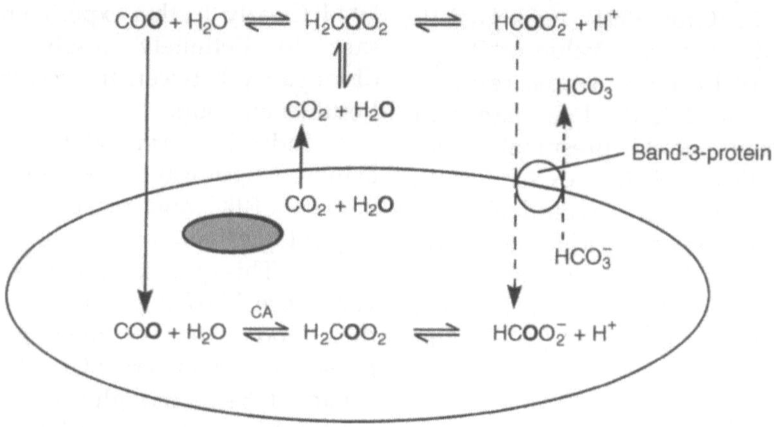

Figure 20.4 Diagram of chemical reactions and diffusion processes involved in the exchange of ^{18}O (**O**) between CO_2, HCO_3^- and H_2O in a suspension of nucleated red cells. Arrows across the cell membrane indicate transport exchange processes between intra- and extracellular space. Dashed arrows indicate transport of HCO_3^- across the membrane of hagfish erythrocytes, which is being measured in this study.

to hagfish extracellular fluid and then suspended in a solution containing 450 mM NaCl and 25 mM (labelled) $NaHCO_3$ (pH 7.4). From the occurrence of positive as well as negative numbers it is clear that $P_{HCO_3^-}$ is indistinguishable from 0. Accordingly, the mean value of $P_{HCO_3^-}$ is -2.5×10^{-6} cm s^{-1} with an S.D. of $\pm$ 6.4×10^{-6} cm s^{-1}. From the range of the figures it appears that the method is able to determine $P_{HCO_3^-}$ values > 10^{-5} cm s^{-1}, an inference that is confirmed by closer analysis. The conclusion then is that the HCO_3^- permeability of *Myxine* RBC is much lower than that of most vertebrates and not distinguishable from zero.

Table 20.3 shows HCO_3^- permeabilities and intraerythrocytic carbonic anhydrase activities for *Myxine*, human and flounder RBC, all determined with the mass spectrometric method at various temperatures. It is apparent that at similar temperatures human as well as flounder RBC exhibit quite high HCO_3^- permeabilities. Similar values can be assumed or have been reported for other fishes and vertebrates (Albers and Goetz, 1985; Heming *et al.*, 1986; Tufts and Randall, 1988; Obaid *et al.*, 1979). Besides the low value of $P_{HCO_3^-}$ for hagfish RBC, Table 20.3 shows that, at the given temperature, *Myxine* RBC exhibit an exceptionally low carbonic anhydrase activity. This raises the question whether CO_2 transport in hagfish RBC may be fundamentally different from that found in most vertebrates and possibly be very similar to that in lampreys.

A possible explanation for the discrepancy between the present result and that of Tufts and Boutilier (1990a, b) may lie in the fact that these authors obtain all their data from RBC that have been equilibrated with a certain pCO_2 for 30 min. If hagfish RBC had just sufficient band 3 protein to allow nearly complete Cl^-–HCO_3^- exchange during this time, one would observe a behaviour similar to that of most vertebrate RBC. If the lampreys had, say, only 10% of the band 3 of hagfishes, this equilibration time might not be sufficient to detect any significant Cl^-–HCO_3^- exchanges. The HCO_3^- permeability of human RBC of ~10^{-3} cm s^{-1} at 37°C is sufficient to achieve a complete Cl^-–HCO_3^- exchange within less than 1 s (Gros, 1991). A rough estimate shows that a maximal value of $P_{HCO_3^-}$ for hagfish RBC of 10^{-5} cm s^{-1} would result in a time required for HCO_3^-–Cl^- exchange of ~100 s. This time would presumably not allow any significant HCO_3^- exchanges in hagfish RBC *in vivo*, but it would allow a complete exchange to occur within an *in vitro* equilibration time of 30 min. To explain the experimental data obtained with lamprey RBC an even lower value of $P_{HCO_3^-}$ in this species than in hagfish would have to be postulated. Clearly, further experimentation is necessary to definitely resolve the apparent discrepancy between the results of Tufts and Boutilier and ours.

Table 20.2 Thirteen single determinations of bicarbonate permeability, $P_{HCO_3^-}$, of washed red blood cells of the hagfish *Myxine glutinosa* at 10°C in a solution with 450 mM NaCl and 25 mM bicarbonate titrated to pH 7.4. The results were obtained by best fits to separate mass spectrometric recordings from 13 samples

Experiment no.	Result: $P_{HCO_3^-}$ (cm s^{-1})
1	-1.248×10^{-5}
2	-1.872×10^{-5}
3	$+4.794 \times 10^{-6}$
4	-1.928×10^{-6}
5	-5.235×10^{-6}
6	$+3.317 \times 10^{-7}$
7	-2.876×10^{-7}
8	-5.677×10^{-7}
9	-1.050×10^{-7}
10	$+1.938 \times 10^{-7}$
11	-9.199×10^{-7}
12	$+2.202 \times 10^{-6}$
13	$+3.913 \times 10^{-7}$
Mean:	-2.495×10^{-6}
SD:	6.363×10^{-6}

If indeed *Myxine glutinosa* RBC do not exhibit significant exchange of HCO_3^- between RBC and plasma, what could the physiological advantage of this be for the animal? This question appears especially relevant since *Myxine glutinosa*, at another site of their body, viz. in their gills, appear to possess a very effective Cl^-–HCO_3^- exchanger (Evans, 1984), although the ability of DIDS to inhibit this system has unfortunately not been tested. The answer may lie in an exceptional sensitivity of O_2 binding of *Myxine* haemoglobin to CO_2 and HCO_3^- (Bauer *et al.*, 1975; Dohn and Malte, this volume; Fago and Weber, this volume). Thus, accumulation of HCO_3^- within the red cell (accompanied, as a consequence, by a somewhat higher postcapillary pCO_2) might be an important mechanism in this animal facilitating the release of O_2 to the tissues.

REFERENCES

Albers, C. and Goetz, K.G. (1985) H^+ and Cl^- ion equilibrium across the red cell membrane in the carp. *Respiration Physiology*, **61**(2), 209–19.

Bauer, C., Engels, U. and Paleus, S. (1975) Oxygen binding to haemoglobins of the primitive vertebrate *Myxine glutinosa* L. *Nature*, **256**(5512), 66–8.

Berglund, L., Carlsson, U. and Kjellstroem, B. (1980) Cyclostome carbonic anhydrase.

Table 20.3 Results of mass spectrometric determinations of bicarbonate permeability ($P_{HCO_3^-}$) and intracellular carbonic anhydrase (CA) activity of red blood cells from hagfishes (*Myxine glutinosa*), flounders (*Platichthys flesus*) and humans at various temperatures. CA activity is defined as the factor by which the uncatalysed rate of CO_2 hydration is accelerated by the enzyme

Temperature	Species	$P_{HCO_3^-}$ (cm s^{-1})	CA activity	*n*
10 °C	Hagfish	approx. 0	6 000	13
10 °C	Human	2.5×10^{-4}	43 000	10
15 °C	Flounder	4.1×10^{-4}	129 000	8
37 °C	Human	9.0×10^{-4}	13 000	8

Purification and some properties of the enzyme from erythrocytes of lamprey. *Acta Chemica Scandinavica, Series B: Organic Chemistry and Biochemistry*, **34**(3), 227–8.

Boutilier, R.G. and Ferguson, R.A. (1989) Nucleated red cell function: metabolism and pH regulation. *Canadian Journal of Zoology*, **67**, 2986–93.

Brahm, J. (1977) Temperature-dependent changes of chloride transport kinetics in human red cells. *Journal of General Physiology*, **70**, 283–306.

Brill, S.R., Musch, M.W. and Goldstein, L. (1992) Taurine efflux, band 3, and erythrocyte volume of the hagfish (*Myxine glutinosa*) and lamprey (*Petromyzon marinus*). *Journal of Experimental Zoology*, **264**, 19–25.

Carlsson, U., Kjellström, B. and Antonsson, B. (1980) Purification and properties of cyclostome carbonic anhydrase from erythrocytes of hagfish. *Biochimica et Biophysica Acta*, **612**, 160–70.

Cholette, C. and Gagnon, A. (1973) Isosmotic adaptation in *Myxine glutinosa* L. – II. Variations of the free amino acids, trimethylamine oxide and potassium of the blood and muscle cells. *Comparative Biochemistry and Physiology A – Comparative Physiology*, **45A**(4), 1009–21.

Deuticke, B., Rickert, I. and Beyer, E. (1978) Stereoselective, SH-dependent transfer of lactate in mammalian erythrocytes. *Biochimica et Biophysica Acta*, **507**, 137–55.

Ellory, J.C. and Wolowyk, M.W. (1991) Evidence for bumetanide-sensitive, Na^+-dependent, partial Na-K-Cl co-transport in red blood cells of a primitive fish. *Canadian Journal of Physiology and Pharmacology*, **69**, 588–91.

Ellory, J.C., Wolowyk, M.W., and Young, J.D. (1987) Hagfish (*Eptatretus stoutii*) erythrocytes show minimal chloride transport activity. *Journal of Experimental Biology*, **129**, 377–83.

Evans, D.H. (1984) Gill Na^+/H^+ and Cl^-/HCO_3^- exchange systems evolved before the vertebrates entered fresh water. *Journal of Experimental Biology*, **113**, 465–9.

Fincham, D.A., Wolowyk, M.W. and Young, J.D. (1990) Characterisation of amino acid transport in red blood cells of a primitive vertebrate, the Pacific hagfish (*Eptatretus stoutii*). *Journal of Experimental Biology*, **154**, 355–70.

Fincham, D.A., Wolowyk, M.W. and Young, J.D. (1991) Nucleoside uptake by red blood cells from a primitive vertebrate, the Pacific hagfish (*Eptatretus stoutii*), is mediated by a nitrobenzylthioinosine-insensitive transport system. *Biochimica et Biophysica Acta*, **1069**(1), 123–6.

Gorkin, A.A. (1970) In-vitro uptake of amino-acids by erthrocytes of the hagfish *Eptatretus stoutii*. *Physiologist*, **13**(3), 210.

Gros, G. (1991) Mechanisms of CO_2 transport in vertebrates. *Verhandlungen der Deutschen Zoologischen Gesellschaft*, **84**, 213–30.

Heming, T.A., Randall, D.J., Boutilier, R.G. *et al.* (1986) Ionic equilibria in red blood cells of rainbow trout (*Salmo gairdneri*): Cl^-, HCO_3^- and H^+. *Respiration Physiology*, **65**, 223–34.

Ingermann, R.L., Hall, R.E., Bissonette *et al.* (1984) Monosaccharide transport into erythrocytes of the Pacific hagfish *Eptatretus stoutii*. *Molecular Physiology*, **6**(5–6), 311–20.

Ingermann, R.L., Bissonnette, J.M. and Hall, R.E. (1985) Sugar uptake by red blood cells, in *Circulation, Respiration, and Metabolism. Current Comparative Approaches* (ed. R. Gilles), Springer-Verlag, Berlin, Heidelberg, New York, Tokyo, pp. 290–300.

Itada, N. and Forster, R.E. (1977) Carbonic anhydrase activity in intact red blood cells measured with ^{18}O exchange. *Journal of Biological Chemistry*, **252**(11), 3881–90.

Jacobs, M.H. and Stewart, D.R. (1942) The role of carbonic anhydrase in certain ion exchanges involving the erythrocyte. *Journal of General Physiology*, **25**, 539–52.

Kaplan, M.A., Hays, L. and Hays, R.M. (1974) Evolution of a facilitated diffusion pathway for amides in the erythrocyte. *American Journal of Physiology*, **226**, 1327–32.

Macey, R.I. and Yousef, L.W. (1988) Osmotic stability of red cells in renal circulation requires rapid urea transport. *American Journal of Physiology*, **254**, C669–74.

Mannuzzu, L.M., Moronne, M.M. and Macey, R.I. (1993) Estimate of the number of urea transport sites in erythrocyte ghosts using a hydrophobic mercurial. *Journal of Membrane Biology*, **133**, 85–97.

Maren, T.H., Friedland, B.R. and Rittmaster, R.S. (1980) Kinetic properties of primitive vertebrate carbonic anhydrases. *Comparative Biochemistry and Physiology B – Comparative Biochemistry*, **67B**, 69–74.

Nikinmaa, M. (1986) Red cell pH of lamprey *Lampetra-fluviatilis* is actively regulated. *Journal of Comparative Physiology B – Biochemical Systemic and Environmental Physiology*, **156**(5), 747–50.

Nikinmaa, M. and Railo, E. (1987) Anion movements across lamprey (*Lampetra fluviatilis*) red cell membrane. *Biochimica et Biophysica Acta*, **899**(1), 134–6.

Nikinmaa, M. and Tiihonen, K. (1994) Substrate transport and utilization in fish erythrocytes. *Acta Physiologica Scandinavica*, **152**, 183–9.

Obaid, A.L., Critz, A.M. and Crandall, E.D. (1979) Kinetics of bicarbonate/chloride exchange in dogfish erythrocytes. *American Journal of Physiology*, **273**(3), R132–8.

Olives, B., Mattei, M.-G., Huet, M. *et al.* (1995) Kidd blood group and urea transport function of human erythrocytes are carried by the same protein. *Journal of Biological Chemistry*, **270**, 15607–10.

Ohnishi, S.T. and Asai, H. (1985) Lamprey erythrocytes lack glycoproteins and anion transport. *Comparative Biochemistry and Physiology B – Comparative Biochemistry*, **81B**(2), 405–8.

Sands, J.M., Gargus, J.J., Fröhlich, O. *et al.* (1992) Urinary concentrating ability in patients with Jk(a-b-) blood type who lack carrier-mediated urea transport. *Journal of the American Society of Nephrology*, **2**, 1689–96.

Tiihonen, K. and Nikinmaa, M. (1991) D-glucose permeability in river lamprey (*Lampetra fluviatilis*) and carp (*Cyprinus carpio*) erythrocytes. *Comparative Biochemistry and Physiology*, **100**(3), 581–4.

Tiihonen, K. and Nikinmaa, M. (1993) Membrane permeability and utilization of L-lactate and pyruvate in carp red blood cells. *Journal of Experimental Biology*, **178**, 161–72.

Tiihonen, K., Yao, S.Y.M., Nikinmaa, M. *et al.* (1993) Erythrocytes from the Pacific hagfish (*Eptatretus stoutii*) transport pyruvate by a concentrative Na^+-dependent mechanism insensitive to inhibition by alpha-cyano-4-hydroxycinnamate. *Biochemistry and Cell Biology*, **71**, Axv.

Tufts, B.L. and Boutilier, R.G. (1989) The absence of rapid chloride-bicarbonate exchange in lamprey erythrocytes: Implications for carbon dioxide transport and ion distribution between plasma and erythrocytes in the blood of *Petromyzon marinus*. *Journal of Experimental Biology*, **144**, 565–76.

Tufts, B.L. and Boutilier, R.G. (1990a) CO_2 transport in agnathan blood: evidence of erythrocyte Cl^-/HCO_3^- exchange limitations. *Respiration Physiology*, **80**, 335–47.

Tufts, B.L. and Boutilier, R.G. (1990b) CO_2 transport properties of the blood of a primitive vertebrate, *Myxine glutinosa* (L.). *Experimental Biology*, **48**(6), 341–7.

Young, J.D., Yao, S.Y.-M., Tse, C.M. *et al.* (1994) Functional and molecular characteristics of a primitive vertebrate glucose transporter: studies of glucose transport by erythrocytes from the Pacific hagfish (*Eptatretus stoutii*). *Journal of Experimental Biology*, **186**, 24–41.

Zhang, Z.H. and Solomon, A.K. (1992) Effect of pCMBS on anion transport in human red cell membranes. *Biochimica et Biophysica Acta*, **1106**, 31–9.

21
HAGFISH HAEMOGLOBINS

Angela Fago and Roy E. Weber

SUMMARY

The properties of the haemoglobins of this phylogenetically ancient group of animals may shed light on evolution of haemoglobin (Hb) function in vertebrates. In contrast to the tetrameric haemoglobin molecules of other vertebrates, whose O_2 affinity is modulated by organic phosphates found in the red blood cells, cyclostome Hbs are monomeric when oxygenated and exhibit no allosteric interaction with phosphates.

Hagfishes possess multiple haemoglobins, involving at least four components. The electrophoretic pattern of the haemoglobins varies significantly among individuals of the same species, reflecting polymorphism. The Hb components are essentially monomeric in the ligated (oxygenated) form. Some components aggregate to dimers and tetramers when deoxygenated. Aggregation is favoured by low pH and high protein concentration.

The Bohr effect (pH modulation of the oxygen affinity) in hagfish haemolysate is small and is due to the formation of (low-affinity) oligomers at low pH and (high-affinity) monomers at high pH. Oxygen binding is virtually non-cooperative, although the isolated tetrameric fraction of *E. burgeri* haemolysate shows haeme–haeme interaction. Bicarbonate ions (formed in the red blood cell by carbonic anhydrase-catalysed hydration of CO_2) act as a potent allosteric effector in *M. glutinosa* haemolysate, causing a significant decrease in the oxygen affinity. This behaviour can be related to the virtual absence of the membrane protein Band III that is implicated in HCO_3^-/Cl^- exchange. This anion-exchanger is similarly lacking in the red cell membranes of *E. stoutii*, suggesting that bicarbonate sensitivity may be a general character of hagfish haemoglobins.

21.1 INTRODUCTION

Haemoglobins that transport O_2 from the respiratory surfaces to the metabolizing tissues are ideal models for studying regulation of protein function. The haemoglobin system of hagfish may represent the early stage in the evolution of cooperative and allosteric interactions in vertebrate haemoglobins. In contrast to the tetrameric haemoglobins from vertebrates where O_2 affinity is regulated by binding of allosteric effectors (like CO_2, chloride, organic phosphates and protons), regulation of oxygen binding in cyclostome haemoglobins is achieved mainly by a mechanism involving aggregation of monomeric haemoglobin components. Although hagfish haemoglobins to some extent resemble those of lampreys, the divergence between hagfishes and lampreys may have already occurred 400–500 million years ago, well before the separation between myoglobin and haemoglobin chains (Goodman, 1981; Liljeqvist *et al.*, 1982), which explains the profound differences between the mechanism of oxygen transport in these two major groups of cyclostomes.

In contrast to cyclostomes (Agnatha), other vertebrates generally have tetrameric

The Biology of Hagfishes. Edited by Jørgen Mørup Jørgensen, Jens Peter Lomholt, Roy E. Weber and Hans Malte. Published in 1998 by Chapman & Hall, London. ISBN 0 412 78530 7.

haemoglobins comprising two α and two β globins, each carrying a haeme group that binds oxygen. Each chain consists of seven or eight α-helical segments (named from A to H; the D helix is missing in the α subunit), which are connected by non-helical ones (AB, BC, etc.). The three-dimensional fold of globins (the so-called 'myoglobin fold') is highly conserved, despite large variations in amino acid sequence in different animal groups: only the proximal His F8 (i.e. the eighth residue of the F helix) and Phe CD1 in the haeme pocket are conserved in all vertebrate haemoglobins (Perutz, 1990). Cooperativity is achieved by a shift of the equilibrium between two alternative quaternary structures, a low-affinity T (*tense*) and a high-affinity R (*relaxed*) state of the deoxygenated and fully oxygenated molecule, respectively. These two conformational states differ mainly by the rotation of two rigid dimeric units ($\alpha_1\beta_1$ and $\alpha_2\beta_2$) relative to each other. The T state is constrained by numerous H-bonds and non-covalent interaction which reduce the affinity for O_2 at the haeme iron. These bonds are broken in the R state. Increases in the concentration of protons (the Bohr effect), CO_2 and organic phosphates (ATP or GTP in fishes, DPG in mammals) result in decreases in oxygen affinity, which enhances oxygen unloading to metabolically active, acid tissues. These allosteric effectors bind preferentially at specific sites in the haemoglobin molecule in the T state, thereby increasing the number of constraints (Perutz, 1970). The decrease in affinity resulting from proton binding (the Bohr effect) facilitates oxygen unloading in acid tissues. Hydrogen ions bind mainly at the C-terminal His HC3 of the β chain and at the N-terminus of the α chain; carbon dioxide forms carbamino compounds with the unprotonated N-termini and organic phosphates bind at the positively charged residues at the interface between the β chains.

The structural mechanisms basic to oxygen binding in hagfish haemoglobins are significantly different. Hagfish haemoglobins are essentially monomeric when oxygenated and tend to aggregate to dimers or tetramers when deoxygenated. Thus, a tetramer⇆dimer⇆monomer equilibrium replaces the T⇆R equilibrium of the tetrameric haemoglobins in controlling cooperativity of oxygen binding and the Bohr effect.

21.2 MULTIPLICITY OF HAEMOGLOBIN COMPONENTS

The number of different haemoglobin components within the same individual ('isohaemoglobins') is generally greater in hagfishes than in other vertebrates: 4–6 different monomeric haemoglobins are present in *Eptatretus* spp. and in *Myxine glutinosa* (Table 21.1). Paleus *et al.* (1971) reported the presence of only three major haemoglobin fractions (I, II, III) in the haemolysate of the Atlantic hagfish *M. glutinosa*, but this may be due to the narrow pH range applied in the isoelectrofocusing purification. Variation in haemoglobin components between individuals of the same species ('allohaemoglobins') indicate genetic polymorphism (Li *et al.*, 1972; Fago and Weber, 1995). Five different phenotypes with four to six haemoglobins encountered in the Pacific hagfish *E. stoutii* have been interpreted to reflect four independent gene loci for monomeric haemoglobins (Ohno and Morrison, 1966). It is not known whether this variability reflects adaptations to environmental conditions or ontogenetic development. However, no relationship between haemoglobin pattern and body weight was observed for *M. glutinosa* (Fago and Weber, 1995).

A similarly high multiplicity of haemoglobins occurs in lampreys, although *Entosphenus japonicus* has only one major component (Dohi *et al.*, 1973), and *Lampetra fluviatilis* has two major components that differ only by the presence of a formyl group at the N-terminal proline (Zelenick *et al.*,

Table 21.1 Multiplicity of hagfish haemoglobins and molecular weight observed under the indicated experimental conditions

Species	No. of components	Mol. wt deoxy	Mol. wt oxy	[Haeme]	pH	Buffer	Method	Reference
M. glutinosa	4–6	20 400	17 400	0.5 mM[1]	8.5	Hepes[2]	Gel filtration	Fago and Weber (1995)
		22 400	20 400	0.5 mM	7.0	Hepes		
		37 100; 20 000	29 500; 15 100	0.5 mM	6.5	Hepes		
		24 500	16 600	0.5 mM	6.5	Hepes, KCl[3]		
		51 300; 19 500	28 200; 15 500	2.0 mM	6.5	Hepes		
E. burgeri	4	2.01 S*; 4.30 S†	2.29 S*	0.37 mM	7.0	Phosphate[4]	Sedimentation equilibrium	Bannai *et al.* (1972)
E. cirrhatus	5	21 000	21 000	0.05 mM[1]	8.5	Tris (?)	Gel filtration	Brittain and Wells (1986)
		32 000; 21 000	21 000	0.05 mM	7.5			
		42 000; 21 000	21 000	0.05 mM	6.0			
E. stoutii	4–6	2.06 S*	1.95 S*	6.0 mM	7.0	Phosphate, NaCl[5]	Sedimentation velocity	Li *et al.* (1972)

* monomer; † tetramer
(1) Applied to the column (2) 0.05 M Hepes (3) 0.05 M Hepes, 0.1 M KCl
(4) 0.1 M phosphate (5) 0.01 M phosphate, 0.2 M NaCl

1979). The isoelectric points of the isohaemoglobins are neutral to alkaline in the hagfish *E. cirrhatus* (Brittain and Wells, 1986) and *M. glutinosa* (Fago and Weber, 1995) and acidic in the sea lamprey *Petromyzon marinus* (Andersen and Gibson, 1971). In this respect sea lamprey haemoglobins resemble invertebrate haemoglobins (Brittain, 1991).

The presence of multiple haemoglobins with different isoelectric points may be favourable by increasing the buffering of protons and ions over a wide range of intracellular pH values and by allowing higher haemoglobin concentration, in accordance with the phase rule that the total haemoglobin concentration in a saturated solution with several components is greater than in a single component solution (Perutz *et al.*, 1959; Weber, 1990). Both these factors may be essential in the haemoglobin system of hagfishes where the regulation of oxygen affinity by binding of protons and other cofactors depends upon protein–protein aggregations.

21.3 AGGREGATIONAL PROPERTIES

The aggregation of hagfish haemoglobins is oxygen-linked. Hagfish haemoglobins are essentially monomeric when oxygenated and may form dimers or tetramers in the deoxygenated state (Table 21.1). Aggregate formation is also pH-dependent: at constant protein concentration, the molecular weights increase with a decrease in pH in both *M. glutinosa* and *E. cirrhatus*. Formation of tetrameric molecules is evident in the deoxygenated haemolysates of *E. burgeri* and *M. glutinosa*, but in the latter species it occurs only at low pH and high haemoglobin concentration, where also the oxygenated form shows slight aggregation. The fact that tetramers have not been observed in the haemolysate of *E. cirrhatus* may be due to the low protein concentration used in gel-filtration studies (Table 21.1). Chloride ions inhibit aggregation in *M. glutinosa* haemoglobin, probably by screening the positive charges involved in the oligomerization process (Fago and Weber, 1995). This may explain why no significant aggregation was observed in *E. stoutii* haemoglobins in the presence of 0.2 M NaCl (Li *et al.*, 1972).

Not all monomers are able to form aggregates, as indicated by the persistence of a low-molecular weight fraction at low pH in the deoxygenated haemolysates from *M. glutinosa* and *E. cirrhatus* (Table 21.1). Of the four haemoglobin components of the Japanese hagfish *E. burgeri* (F1, F2, F3 and F4) only F3 and F4 aggregate to form a heterotetramer, whereas F1 and F2 remain monomeric (Bannai *et al.*, 1972). Self-association of monomers into homodimers (and homotetramers) does not appear to occur in hagfish haemoglobins (Bannai *et al.*, 1972; Bauer *et al.*, 1975; Fago and Weber, 1995), whereas it is common in lampreys (Li and Riggs, 1970; Dohi *et al.*, 1973; Nikinmaa and Weber, 1993). Cooperative oxygen binding also occurs in the homodimeric invertebrate haemoglobin of the clam *Scapharca inaequivalvis* (Chiancone *et al.*, 1981; Weber, 1990; Royer, 1994), whereas in mammalian haemoglobins cooperativity is a characteristic of the tetrameric molecules that is lost upon dissociation into the dimeric half-molecules, $\alpha_1\beta_1$ and $\alpha_2\beta_2$ (Ackers *et al.*, 1992).

21.4 OXYGEN-BINDING PROPERTIES

Reversible association of monomeric haemoglobins with high oxygen affinity into low-affinity dimers and tetramers is basic to the functional properties of hagfish haemoglobins. The equilibrium monomer$\leftrightarrows$dimer$\leftrightarrows$tetramer is shifted to the left by oxygen and to the right by protons, which accounts for both cooperative oxygen binding and the Bohr effect. An analogous allosteric mechanism involving oxygen-linked aggregation among vertebrate haemoglobins is found only in lampreys and in the dogfish *Squalus achantias* (whose haemoglobin is dimeric when oxygenated and

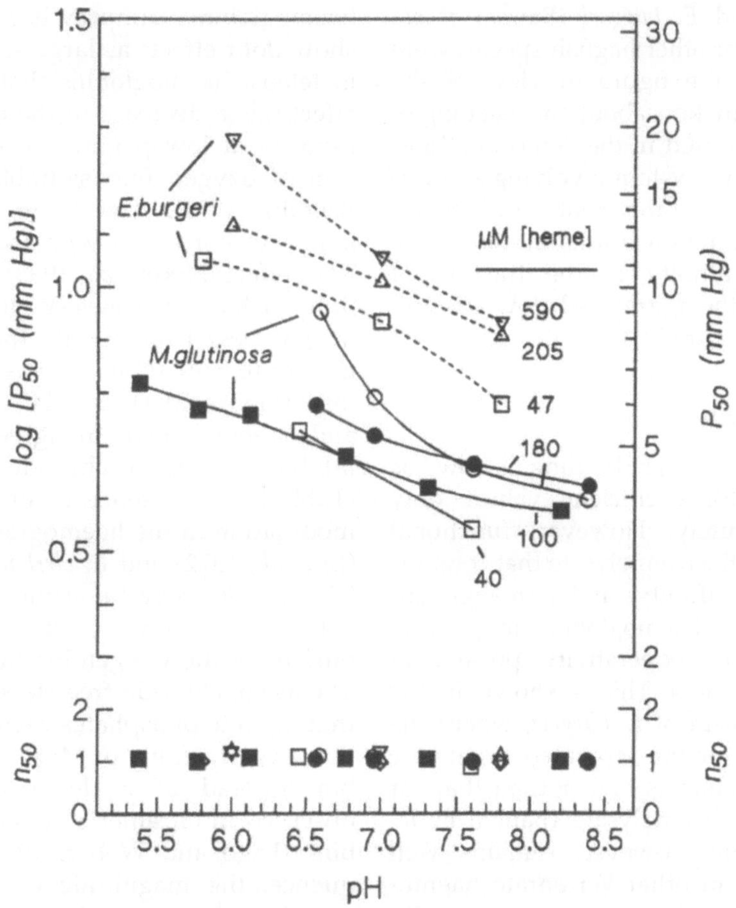

Figure 21.1 Bohr effect and cooperativity of *Myxine glutinosa* and *Eptatretus burgeri* haemolysates, at 20 and 22°C respectively. Oxygen binding experiments for *M. glutinosa* haemoglobin were performed in 100 mM Hepes buffer in the absence (○, ❑) and presence (●) of 100 mM KCl (Fago and Weber, 1995) or in 50 mM bis-Tris buffer, 100 mM NaCl (■) (Bauer *et al.*, 1975). The data for *E. burgeri* haemoglobin are from Bannai *et al.* (1972).

tetrameric when deoxygenated; Fyhn and Sullivan, 1975; Brittain, 1991). Blood oxygen affinity moreover depends on the intrinsic oxygen affinity of the different haemoglobin components. Functional heterogeneity is reported for the multiple haemoglobins of *E. burgeri* (Bannai *et al.*, 1972) and *M. glutinosa* (Bauer *et al.*, 1975; Fago and Weber, 1995) but not for *E. stoutii*, where the major components (A, B and C) have the same oxygen-binding properties (Li *et al.*, 1972). In this latter species, the lower oxygen affinity found in the haemolysate than in the individual components suggests interactions between monomeric haemoglobins as occur in other hagfishes, despite the fact that no aggregation has been observed in sedimentation velocity experiments (Table 21.1).

Figure 21.1 reports the Bohr effects (increase in P_{50} – i.e. the oxygen tension at half saturation – with a pH decrease) and the cooperativity of oxygen binding (expressed as the Hill coefficient, n_{50}) in the haemolysate of *M. glutinosa* (Bauer *et al.*, 1975; Fago and

Weber, 1995) and *E. burgeri* (Bannai *et al.*, 1972). The data for other hagfish species were not included in the figure in view of the absence of information about the haemoglobin concentration used in the determinations. As expected from a system involving association–dissociation of components, an increase in protein concentration results in a decrease in the oxygen affinity as the fraction of haemoglobin in the aggregated (low-affinity) form increases (Figure 21.1).

21.4.1 Cooperativity

Cooperativity of oxygen binding is low, as indicated by Hill coefficient values only slightly above unity. However, functional heterogeneity of the haemolysate that contains aggregating (low-affinity) and non-aggregating (high-affinity) haemoglobins may mask small degrees of cooperativity present in dimers and tetramers. This is shown in the haemoglobin system of *E. burgeri*, where the n_{50} value of 1.3 for the separated tetrameric haemoglobin fraction is higher than that for the unfractioned haemolysate (Bannai *et al.*, 1972). This value, however, remains well below that seen in other vertebrate haemoglobins, where Hill coefficients normally exceed 2. In lamprey haemoglobins n_{50} values as high as 1.7 have been observed (Andersen and Gibson, 1971; Brittain *et al.*, 1989). The low values in hagfishes suggest that the difference in reactivity between the low-affinity oligomers and the high-affinity monomers is not as large as in other haemoglobin systems. The constraints imposed on the molecule upon monomer association are evidently not strong enough to induce a significant decrease in haeme oxygen affinity.

21.4.2 Bohr and anion effect

The Bohr effect observed in hagfish haemoglobins is small ($\phi = \Delta \log P_{50}/\Delta\text{pH} \approx -0.13$), especially when compared to that of lamprey haemoglobins. Lamprey haemoglobins often show Bohr effects as large as those observed in teleost haemoglobins that exhibit a Root effect (i.e. a decrease in the oxygen-carrying capacity at low pH, which facilitates secretion of oxygen into swimbladder and eye; Brittain, 1991). The large Bohr effect in lampreys correlates with the presence of a Root effect (Nikinmaa, 1993) which is absent in hagfish erythrocytes (Wells *et al.*, 1986). In *M. glutinosa* haemolysate the Bohr effect is greater ($\phi = -0.31$) in the absence of chloride ions (Figure 21.1) or ATP and DPG (Fago and Weber, 1995), in agreement with the inhibitory effect of chloride on aggregation (Table 21.1). Absence of organic-phosphate modulation in the haemoglobins of *E. stoutii* (Li *et al.*, 1972) and *E. cirrhatus* (Brittain and Wells, 1986) may be ascribed to the use of chloride-containing Tris or phosphate buffers in the oxygen-binding experiments. The use of chloride-free Hepes buffer reveals that organic phosphates cause an increase in the oxygen affinity of *M. glutinosa* haemoglobin, instead of a decrease, as generally observed in tetrameric vertebrate haemoglobins (Fago and Weber, 1995). As a consequence, the magnitude of the Bohr effect may be larger than that often reported for hagfish haemoglobins. Analysis of hagfish (*E. stoutii*) red blood cells has revealed a high concentration of ADP (up to 3 mM), a variable concentration of ATP and high concentration of an unidentified non-phosphate compound (Bartlett, 1982). It is not known whether ADP or this unknown compound may affect haemoglobin oxygen affinity. ADP only slightly reduces the O_2 affinity of human haemoglobin (Chanutin and Curnish, 1967). Lamprey haemoglobins also appear to lack modulation by chloride and organic phosphates (Brittain, 1991; Nikinmaa and Weber, 1993). Lamprey erythrocytes contain primarily ATP and a substantial pool of DPG, which is found in mammalian but not in hagfish or in teleost red blood cells (Bartlett, 1982).

21.4.3 Temperature effect

An increase in temperature decreases oxygen affinity significantly in hagfish haemoglobins (Fago and Weber, 1995), whereby the different oxygen affinities of *M. glutinosa* and *E. burgeri* haemolysates shown in Figure 21.1 may partly be accounted for by the different temperature of measurements. The apparent heat of oxygenation calculated in *M. glutinosa* haemolysate decreases from -67 to -48 kJ mol^{-1} O_2 as pH decreases from 8.0 to 6.8, indicating that the formation of non-covalent interactions in the pH-dependent aggregation of subunits contributes positively to the overall enthalpy of oxygenation (Fago and Weber, 1995). The temperature effect may be the result of two opposing factors: a negative (exothermic) intrinsic enthalpy of oxygenation at the haeme decreased by a positive (endothermic) heat of aggregate formation at low pH. Similarly, a rise in temperature decreases the oxygen affinity of lamprey haemoglobin (Briehl, 1963). In human and other vertebrate haemoglobins the apparent heat of oxygenation is lowered by the endothermic contribution of the rupture of H-bonds and salt bridges involved in the Bohr effect (Weber and Jensen, 1988).

21.5 EFFECT OF BICARBONATE AND CO_2 TRANSPORT

Aggregation of monomeric haemoglobins in *M. glutinosa* haemolysate was believed to be CO_2-dependent because the Bohr effect and cooperativity increase in the presence of CO_2 (Bauer *et al.*, 1975). It was later established that aggregation is primarily pH-dependent, but nevertheless the pronounced reduction in the oxygen affinity in the presence of CO_2 was confirmed (Fago and Weber, 1995). In Figure 21.2 the effect of CO_2 on the oxygen affinity and cooperativity of *M. glutinosa* haemolysate is reported. The smaller Bohr effect found by Bauer *et al.* (1975) in the presence of CO_2 is probably related to the presence of chloride ions in the buffer. In this species, only one of the haemoglobin components has a free (non-acetylated) proline at the N-terminus (Liljeqvist *et al.*, 1982), but due to its high pKa this would not easily form carbamino compounds, since the intra-erythrocytic pH is about 7.3 (H. Malte, personal comm.). Fago *et al.* (in preparation) have discovered that the modulator of the oxygen affinity in *M. glutinosa* haemolysate is not CO_2 but bicarbonate, formed by the reaction of CO_2 with water and catalysed by the intra-erythrocytic carbonic anhydrase (present in unfractioned haemolysate). The effect of bicarbonate is to stabilize the formation of aggregates, thereby lowering the oxygen affinity. Other anions (chloride and organic phosphates) on the contrary favour dissociation into monomers, which suggests the presence of a specific bicarbonate-binding site. Oxygen-linked binding of bicarbonate has been reported so far only for crocodile haemoglobin (Bauer *et al.*, 1981), but may be a general feature of hagfish haemoglobins, as suggested by the large CO_2-Bohr factor (ϕ = -0.43) in the whole blood of *E. cirrhatus* (Wells *et al.*, 1986). No binding of bicarbonate or CO_2 appears to occur in lamprey haemoglobins (Nikinmaa and Matsoff, 1992; Nikinmaa and Weber, 1993).

The excretion of bicarbonate from red blood cells of *M. glutinosa* may be slower than in other vertebrates, since they almost completely lack the anion-exchanger Band III, which exchanges bicarbonate and chloride ions across the membrane (Tufts and Boutilier, 1990). This may be advantageous for oxygen transport since bicarbonate binds to haemoglobin, acting as an allosteric effector. The temporary increase in the intra-erythrocytic concentration of bicarbonate during passage in the tissues is thus utilized to enhance oxygen release, representing an elegant example of positive feedback-regulation of the oxygen transport by an end-product of metabolism (Bauer *et al.*, 1975).

The aggregation of Hb upon deoxygenation may result in a lower erythrocyte

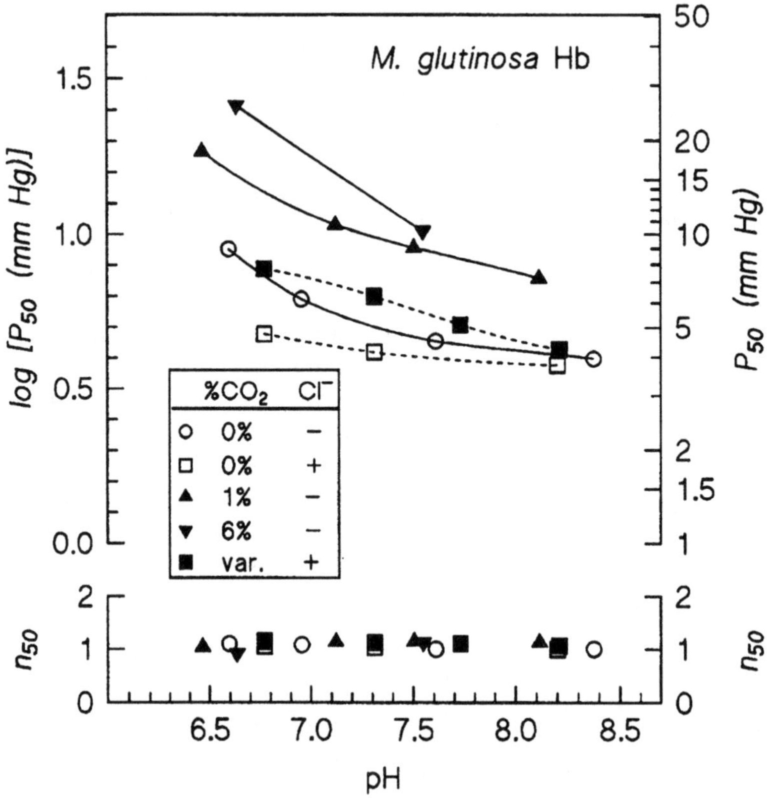

Figure 21.2 Effect of CO_2 on the oxygen binding properties of *Myxine glutinosa* haemolysate in the presence and absence of 100 mM NaCl. Data from Fago and Weber (1995, continuous curves) and Bauer *et al.* (1975; discontinous curves).

osmotic pressure in the tissues than in the gills, which might provide an increased gradient for bicarbonate ions uptake (Andersen, 1971). However, this is unlikely to play a major role since the Hb concentration in the red blood cell (12–15 mM haeme in *E. cirrhatus*; Wells *et al.*, 1986; Wells and Foster, 1989; and 13.7–16.2 mM in *L. fluviatilis*; Nikinmaa, 1993) is low compared to dissolved osmolytes (~1000 mOsmol). *M. glutinosa* erythrocytes can tolerate exceptionally high degrees of swelling without haemolysis and show only very slow recovery to their initial volume (Brill *et al.*, 1992; Nikinmaa *et al.*, 1993; Dohn and Malte, this volume). This unusual property must reflect a radically different erythrocyte membrane skeleton. A novel pattern of membrane proteins has indeed been found in the red blood cells of *E. stoutii* (Ellory *et al.*, 1987), which lack both the anion exchanger (also involved in volume regulation; Brill *et al.*, 1992) and spectrin, two of the major constituents of the membrane skeleton in vertebrate erythrocytes. Hagfish red blood cells are morphologically different from those of other vertebrates, including lampreys, being large in size and having an irregular ellipsoidal shape (Bartlett, 1982; Wells *et al.*, 1986).

```
                          NA              A           AB      B        C      CD      D              E
                                       __________   ___________  _____      _____  _____________
                        1       10         20         30        40        50        60         70        80
E. stoutii B            PIVDQGPLPRLTGGDKXAIRATWXVIYAKF
M. glutinosa III        PITDHGQPPTLSEGDKKAIRESWPQIYKNFEQNSLAVLLEFLKKFPKAQDSFPKFSAKKS  HLEQDPAVKLQAEVIINAVN
P. marinus V            PIVDTGSVAPLSAAEKTKIRSAWAPVYSNYETSGVDILVKFFTSTPAAQEFFPKFKGLTTADQLKKSADVRWHAERIINAVN
L. fluviatilis          PIVDSGSVAPLSAAEKTKIRSAWAPVYSNYETSGVDILVKFFTSTPAAQEFFPKFKGMTSADQLKKSADVRWHAERIINAVN
S. inequivalvis IIA      VADAVAKVCGSEAIKANLRRSWGVLSADIEATGLMLMSNLFTLRPDTKTYFTRLGDVQK   GKANSKLRGHAITLTYALN
Human α-chain                 V LSPADKTNVKAAWGKVGAHAGEYGAEALERMFLSFPTTKTYFPHFDLSH       GSAQVKGHGKKVADALT
Human β-chain                 VHLTPEEKSAVTALWGKV  NVDEVGGEALGRLLVVYPWTQRFFESFGDLSTPDAVMGNPKVKAHGKKVLGAFS
                                  ___________  _____________  _____      _____  _____________
                              NA       A         AB        B          C      CD      D              E


                             EF        F          FG      G                            H          HC
                         ____   __________   __________                       ___________
                              90        100        110       120                   130       140
M. glutinosa III        HTIGLMDKEAAMKKYLKDLSTKHSTEFQVNPDMFKELSAVFVSTM           GGKAAYEKLFSIIATLLRSTYDA
P. marinus V            DAVASMDDTEKMSMKLRDLSGKHAKSFQVDPQYFKVLAAVIADTV          AAGDAGFEKLMSMICILLRSAY
L. fluviatilis          DAVASMDDTEKMSMKLRDLSGKHAKSFQVDPQYFKVLAAVIADTV          AAGDAGFEKLMSMICILLRSAY
S. inequivalvis IIA     NFVDSLDDPSRLKCVVEKFAVNHINRKISGDAFGAIVEPMKETTLKARMGNYYSDDVAGAWAALVGVVQAAL
Human α-chain           NAVAHVDD   MPNALSALSDLHAHKLRVDPVNFKLLSHCLLVTLAAHLPAEFTPAVHASLDKFLASVSTVLTSKYR
Human β-chain           DGLAHLD   NLKGTFATLSELHCDKLHVDPENFRLLGNVLVCVLAHHFGKEFTPPVQAAYQKVVAGVANALAHKYH
                        ____       ________    _____________________        _______________________
                              EF         F      FG          G             GH          H           HC
```

Figure 21.3 Amino acid sequence alignment of haemoglobins from hagfish (*M. glutinosa* Hb III and *E. stoutii* Hb B), lamprey (*P. marinus* Hb V and *L. fluviatilis* Hb), clam (*S. inaequivalvis* Hb II, A chain) and human. Partial sequence of the first 30 amino acid residues of *E. stoutii* Hb B is reported (× = unidentified amino acid residue). The N-terminus of *S. inaequivalvis* is blocked by an acetyl group. The α-helix notations as established for lamprey *P. marinus* Hb V (top; Honzatko *et al.*, 1985) and human haemoglobin (bottom) are reported. Numbering of residues was carried out in accordance to *M. glutinosa* haemoglobin. Amino acid sequences were obtained from the Swiss-Prot protein data bank.

21.6 STRUCTURAL DATA AND POSSIBLE MOLECULAR MECHANISMS

Not much is known about the molecular structure of hagfish haemoglobins. The data available relate to the amino acid sequence of one haemoglobin component from *M. glutinosa* (Liljeqvist *et al.*, 1979, 1982) and the sequence of the first 30 residues of the major component of *E. stoutii* (Li and Riggs, 1972), which are aligned in Figure 21.3 with the sequences of lampreys and clam (*Scapharca inaequivalvis*) haemoglobins and human α and β chains. On the basis of the many differences in primary structure, hagfish and lamprey haemoglobins can be classified neither as α nor as β globins. Cyclostome globins may have diverged before the separation of α- and β-globins (Feng and Doolittle, 1987) or even before the differentiation of globins into myoglobin and haemoglobin chains (Goodman *et al.*, 1975; Goodman, 1981). Nevertheless cyclostome haemoglobins retain the basic overall three-dimensional tertiary structure (the myoglobin fold) adopted by higher vertebrate haemoglobin chains, comprising eight α-helix segments, as shown by x-ray diffraction analysis of the lamprey *P. marinus* haemoglobin V in the monomeric form (Hendrickson *et al.*, 1973; Honzatko *et al.*, 1985). The sequence of *M. glutinosa* haemoglobin more closely resembles those of lampreys so that the same three-dimensional structure can be assumed. The length of the helical and non-helical segments varies considerably in comparison to that of higher vertebrate globins (Figure 21.3). The most important differences found in cyclostome haemoglobins is the presence of an additional segment of nine residues at the N-terminus, an insertion of three residues in the F helix and a large deletion (nine residues) at the GH corner. A D helix is present in cyclostome haemoglobins, as in β-type chains. In *M. glutinosa* haemoglobin at the distal side of the haeme, Gln (71) and Ile (75) replace His E7 and Val E11, which are normally present. This represents a novel haeme-complex among vertebrate haemoglobins (Liljeqvist *et al.*, 1979). The C-terminal residue is Ala, compared to the C-terminal His of the β chain, which is the principal Bohr group of tetrameric haemoglobins.

Although the overall tertiary structure of hagfish and lamprey haemoglobins appears to be maintained as in other haemoglobins, the quaternary arrangement of the interacting monomers may be significantly different. The assembly of α and β subunits in tetrameric vertebrate haemoglobins results in an $\alpha_1\beta_2$ ($\alpha_2\beta_1$) interface involving the FG corners and C helix and an $\alpha_1\beta_1$ ($\alpha_2\beta_2$) contact involving the helices B, G and H (Baldwin and Chothia, 1979). This mode of assembly may explain formation of dimers through an $\alpha_1\beta_2$-like contact between subunits, but not of tetramers, given that large deletions in the G and H helices and the absence of the GH corner in cyclostome haemoglobins would form an unfavourable internal void at the $\alpha_1\beta_1$ contact, which moreover is precluded by the overlapping N-terminal tail (Hendrickson and Love, 1971; Honzatko and Hendrickson, 1986). More plausible appears to be a tetrameric assembly that resembles that of the invertebrate bivalve haemoglobin, where dimer formation would involve mainly the E and F helices, and tetramer formation the A helix and the AB and GH corners of each chain (Royer *et al.*, 1995), although deletion of the GH corner would still imply different subunit arrangements in cyclostome haemoglobins. Inter-subunit communication by direct haeme–haeme contact at the interface between the E and F helices of the two subunits, as occurs in *S. inaequivalvis* haemoglobin, could also be present in cyclostome haemoglobins, supporting the theory that hagfish (and lamprey) haemoglobins may represent the transition between invertebrate and vertebrate haemoglobins (Liljeqvist *et al.*, 1982). In comparison to the α and β chains of higher vertebrate haemoglobins, invertebrate and cyclostome haemoglobins have a N-terminal tail and a longer F helix (Figure 21.3). If the tetrameric subunit arrangement in hagfish (or lamprey) haemoglobins appears to be different from that of higher vertebrates then one can conclude that the evolution of the $\alpha_1\beta_2$ ($\alpha_2\beta_1$) and $\alpha_1\beta_1$ ($\alpha_2\beta_2$) contacts in haemoglobin tetramers does not coincide with the appearance of the early vertebrates but belongs to a subsequent evolutionary stage. This would be in contrast with the assumption that a contact similar to the $\alpha_1\beta_2$ ($\alpha_2\beta_1$) interface of the haemoglobins of higher vertebrates is present in dimers of lamprey haemoglobins (Li and Riggs, 1970; Hendrickson, 1973; Eaton, 1980). X-ray diffraction analysis of crystals of cyclostome haemoglobins in the dimeric and tetrameric form would solve the question.

On the assumption that the Bohr effect is due to a shift in the pK_a of the His residues and that only the proximal and distal His are present in lamprey haemoglobins, Perutz (1990) proposed that in the deoxygenated state, when haemoglobin dimerizes, the distal His may act as a Bohr group by moving out of the haeme pocket and making a salt bridge with a carboxyl group of the other subunit, thereby establishing direct contact between the haeme pockets of two subunits as in clam haemoglobin. This model, however, cannot be operating in the haemoglobin of *M. glutinosa*, where the distal His is replaced by a Gln residue. The identification of the groups which could contribute to the Bohr effect in hagfish haemoglobins is impaired by the lack of detailed information about their three-dimensional structure, which would also give insight into the nature of the bicarbonate-binding site. Oxygen-linked binding of bicarbonate in crocodile haemoglobin occurs at the $\alpha_1\beta_2$ interface (Komiyama *et al.*, 1995); an alternative and unique site might be present in hagfish haemoglobins.

21.7 CONCLUSION

Two major features distinguish cyclostome erythrocytes from those of other vertebrates: (1) the inability to form stable tetramers in the haemoglobins, and (2) the absence of an active anion exchanger across the red cell membrane. Hagfish haemoglobins may not only function as oxygen carriers, but as efficient CO_2 carriers as well. Bicarbonate accumulation in hagfish erythrocytes in the tissue is facilitated by binding to haemoglobin (which is oxygen-linked, thereby enhancing oxygen liberation). Since hagfishes have a low metabolic rate (Malte and Lomholt, this volume) there may be little accumulation of CO_2 and erythrocyte function may be less dependent on anion exchange that is mediated by Band III protein. Oxygen binding without significant Bohr effects and cooperativity may suffice for oxygen transport at very low metabolic rate and low internal oxygen pressures in hagfishes. Considerable venous desaturation after exercise reveals that this oxygen transport system may utilize a large range of the saturation curve (Wells *et al.*, 1986). In lampreys pronounced cooperativities of oxygen binding and Bohr effects are found. Here, in the absence of the anion exchange protein and CO_2- or bicarbonate-binding to the haemoglobin, bicarbonate is accumulated in the red blood cells by the high intracellular pH and is converted back to CO_2 in the gills by liberation of Bohr protons upon haemoglobin oxygenation (Nikinmaa and Weber, 1993). After the evolution of a functional anion-exchanger protein in the erythrocytes of higher vertebrates (making possible CO_2 transport as plasma bicarbonate), oxygen transport has become less dependent on carbon dioxide concentration. This appears to coincide with the use of organic phosphates as major modulators of oxygen binding in vertebrate erythrocytes (Coates, 1975).

REFERENCES

Ackers, G.K., Doyle, M.C., Myers, D. and Daugherty, M.A. (1992) Molecular code for cooperativity in hemoglobin. *Science*, **255**, 54–63.

Andersen, M.E. (1971) Sedimentation equilibrium

experiments on the self-association of hemoglobin from the lamprey *Petromyzon marinus*. *J. Biol. Chem.*, **246**, 4800–6.

Andersen, M.E. and Gibson, Q.H. (1971) A kinetic analysis of the binding of oxygen and carbon monoxide to lamprey hemoglobin (*Petromyzon marinus* and *Petromyzon fluviatilis*). *J. Biol. Biophys.*, **105**, 404–8.

Baldwin, J. and Chothia, C. (1979) Haemoglobin: the structural changes related to ligand binding and its allosteric mechanism. *J. Mol. Biol.*, **129**, 175–220.

Bannai, S., Sugita, Y. and Yoneyama, Y. (1972) Studies on hemoglobin from the hagfish *Eptatretus burgeri*. *J. Biol. Chem.*, **247**, 505–10.

Bartlett, G.R. (1982) Phosphates in red cells of a hagfish and a lamprey. *Comp. Biochem. Physiol.*, **73A**, 141–5.

Bauer, C., Engels, U. and Paleus, S. (1975) Oxygen binding to hemoglobins of the primitive vertebrate *Myxine glutinosa L*. *Nature*, **256**, 66–8.

Bauer, C., Foster, M., Gros, G., Mosca, A., Perrella, M., Rollema, H.R. and Vogel, D. (1981) Analysis of bicarbonate binding to crocodilian haemoglobin. *J. Biol. Chem.*, **256**, 8429–35.

Briehl, R.W. (1963) The relation between the oxygen equilibrium and aggregation of subunits in lamprey haemoglobin. *J. Biol. Chem.*, **238**, 2361–6.

Brill, S.R., Musch, M.W. and Goldstein, L. (1992) Taurine efflux, Band 3 and erythrocyte volume of the hagfish (*Myxine glutinosa*) and lamprey (*Petromyzon marinus*). *J. Exp. Zool.*, **264**, 19–25.

Brittain, T. (1991) Cooperativity and allosteric regulation in non-mammalian vertebrate haemoglobins. *Comp. Biochem. Physiol.*, **99B**, 731–40.

Brittain, T. and Wells, R.M.G. (1986) Characterization of the changes in the state of aggregation induced by ligand binding in the hemoglobin system of a primitive vertebrate, the hagfish *Eptatretus cirrhatus*. *Comp. Biochem. Physiol.*, **85A**, 785–90.

Brittain, T., O'Brien, A.J., Wells, R.M.G. and Baldwin, J. (1989) A study of the role of subunit aggregation in the expression of co-operative ligand binding in the haemoglobin of the lamprey *Mordacia mordax*. *Comp. Biochem. Physiol.*, **93B**, 549–54.

Chanutin, A. and Curnish, R.R. (1967) Effect of organic and inorganic phosphates on the oxygen equilibrium of human erythrocytes. *Arch. Biochem. Biophys.*, **121**, 96.

Chiancone, E., Vecchini, P., Verzili, D., Ascoli, F. and Antonini, E. (1981) Dimeric and tetrameric hemoglobins from the mollusc *Scapharca inaequivalvis*. *J. Mol. Biol.*, **152**, 577–92.

Coates, M.L. (1975) Hemoglobin function in the vertebrates: an evolutionary model. *J. Mol. Evol.*, **6**, 285–307.

Dohi, Y., Sugita, Y. and Yoneyama, Y. (1973) The self-association and oxygen equilibrium of hemoglobin from the lamprey, *Entosphenus japonicus*. *J. Biol. Chem.*, **248**, 2354–63.

Eaton, W.A. (1980) The relationship between coding sequences and function in haemoglobin. *Nature*, **284**, 183–5.

Ellory, J.C., Wolowyk, M.W. and Young, J.D. (1987) Hagfish (*Eptatretus stoutii*) erythrocytes show minimal chloride transport activity. *J. Exp. Biol.*, **129**, 377–83.

Fago, A. and Weber, R.E. (1995) The hemoglobin system of the hagfish *Myxine glutinosa*: aggregation state and functional properties. *Biochim. Biophys. Acta*, **1249**, 109–15.

Feng, D.-F. and Doolittle, R.F. (1987) Progressive sequence alignment as a prerequisite to correct phylogenetic trees. *J. Mol. Evol.*, **25**, 351–60.

Fyhn, U.E.H. and Sullivan, B. (1975) Elasmobranch hemoglobins: dimerization and polymerization in various species. *Comp. Biochem. Physiol.*, **50B**, 119–29.

Goodman, M. (1981) Globin evolution was apparently very rapid in early vertebrates: a reasonable case against the rate-constancy hypothesis. *J. Mol. Evol.*, **17**, 114–20.

Goodman, M., Moore, W. and Matsuda, G. (1975) Darwinian evolution in the genealogy of haemoglobin. *Nature*, **253**, 603–8.

Hendrickson, W.A. (1973) Structural effects accompanying ligand change in crystalline lamprey hemoglobin. *Biochim. Biophys. Acta*, **310**, 32–8.

Hendrickson, W.A. and Love, W.E. (1971) Structure of lamprey haemoglobin. *Nature New Biol.*, **232**, 197–203.

Hendrickson, W.A., Love, W.E. and Karle, J. (1973) Crystal structure analysis of sea lamprey hemoglobin at 2 Å resolution. *J. Mol. Biol.*, **74**, 331–61.

Honzatko, R.B. and Hendrickson, W.A. (1986) Molecular models for the putative dimer of sea lamprey hemoglobin. *Proc. Natl. Acad. Sci. USA*, **83**, 8487–91.

Honzatko, R.B., Hendrickson, W.A. and Love, W.E. (1985) Refinement of a molecular model for lamprey hemoglobin from *Petromyzon marinus*. *J. Mol. Biol.*, **184**, 147–64.

Komiyama, N.H., Miyazaki, G., Tame, J. and

Nagai, K. (1995) Transplanting a unique allosteric effect from crocodile into human haemoglobin. *Nature*, **373**, 244–6.

Li, S.-L. and Riggs, A. (1970) The amino acid sequence of haemoglobin V from the lamprey *Petromyzon marinus*. *J. Biol. Chem.*, **245**, 6149–69.

Li, S.-L. and Riggs, A. (1972) The partial sequence of the first 30 residues from the amino-terminus of hemoglobin B in the hagfish *Eptatretus stoutii*: homology with lamprey hemoglobin. *J. Mol. Evol.*, **1**, 208–10.

Li, S.-L., Tomita, S. and Riggs, A. (1972) The haemoglobins of the Pacific hagfish *Eptatretus stoutii* I. Isolation, characterization and oxygen equilibria. *Biochim. Biophys. Acta*, **278**, 344–54.

Liljeqvist, G., Braunitzer, G. and Paleus, S. (1979) Hämoglobine, XXVII. Die Sequenz des monomeren Hämoglobins III von *Myxine glutinosa* L.: ein neuer Hämkomplex: E7 Glutamin, E11 Isoleucin. *Hoppe-Seyler's Z. Physiol. Chem.*, **360**, 125–35.

Liljeqvist, G., Paleus, S. and Braunitzer, G. (1982) Hemoglobins, XLVIIII. The primary structure of a monomeric hemoglobin from the hagfish, *Myxine glutinosa* L.: evolutionary aspects and comparative studies of the function with special reference to the heme linkage. *J. Mol. Evol.*, **18**, 102–8.

Nikinmaa, M. (1993) Haemoglobin function in intact *Lampetra fluviatilis* erythrocytes. *Respir. Physiol.*, **91**, 283–93.

Nikinmaa, M. and Matsoff, L. (1992) Effects of oxygen saturation on the CO_2 transport properties of *Lampetra* red cells. *Respir. Physiol.*, **87**, 219–30.

Nikinmaa, M. and Weber, R.E. (1993) Gas transport in lamprey erythrocytes, in *The Vertebrate Gas Transport Cascade: Adaptations to Environment and Mode of Life* (ed. J.E.P.W. Bicudo), CRC Press, Boca Raton, pp. 179–87.

Nikinmaa, M., Tufts, B.L. and Boutilier, R.G. (1993) Volume and pH regulation in agnathan erythrocytes: comparisons between the hagfish, *Myxine glutinosa*, and the lampreys, *Petromyzon marinus* and *Lampetra fluviatilis*. *J. Comp. Physiol.* B, **163**, 608–13.

Ohno, S. and Morrison, M. (1966) Multiple gene loci for the monomeric hemoglobin of the hagfish (*Eptatretus stoutii*). *Science*, **154**, 1034–5.

Paleus, S., Vesterberg, O. and Liljeqvist, G. (1971) The hemoglobins of *Myxine glutinosa* L. – I. Preparation and crystallization. Comp. *Biochem. Physiol.*, **39B**, 551–7.

Perutz, M.F. (1970) Stereochemistry of cooperative effects in haemoglobins. *Nature*, **228**, 726–39.

Perutz, M.F. (1990) *Mechanisms of Cooperativity and Allosteric Regulation in Proteins*, Cambridge Univ. Press, Cambridge, pp. 1–101.

Perutz, M.F., Steinrauf, L.K., Stockell, A. and Bangham, A.D. (1959) Chemical and crystallographic study of the two fractions of adult horse haemoglobin. *J. Mol. Biol.*, **1**, 402–4.

Royer, W.E. (1994) High-resolution crystallographic analysis of a cooperative dimeric hemoglobin. *J. Mol. Biol.*, **235**, 657–81.

Royer, W.E, Heard, K.S., Harrington, D.J. and Chiancone, E. (1995) The 2.0 Å crystal structure of *Scapharca* tetrameric haemoglobin: cooperative dimers within an allosteric tetramer. *J. Mol. Biol.*, **253**, 168–86.

Tufts, B.L. and Boutilier, R.G. (1990) CO_2 transport of the blood of a primitive vertebrate, *Myxine glutinosa* (L.). *Exp. Biol.*, **48**, 341–7.

Wells, R.M.G., Foster, M.E., Davison, W., Taylor, H.H., Davie, P.S. and Satchell, G.H. (1986) Blood oxygen transport in the free-swimming hagfish, *Eptatretus cirrhatus*. *J. Exp. Biol.*, **123**, 43–53.

Wells, R.M.G. and Foster, M.E. (1989) Dependence of blood viscosity on haematocrit and shear rate in a primitive vertebrate. *J. Exp. Biol.*, **145**, 483–7.

Weber, R.E. (1990) Functional significance and structural basis of multiple hemoglobins with special reference to ectothermic vertebrates, in *Animal Nutrition and Transport Processes*, Vol. 2: *Transport, Respiration and Excretion: Comparative and Environmental Aspects* (eds J.P. Truchot and B. Lahlou). *Comp. Physiol.* Basel, Karger, pp. 58–75.

Weber, R.E. and Jensen, F.B. (1988) Functional adaptations in hemoglobin from ectothermic vertebrates. *Ann. Rev. Physiol.*, **50**, 161–79.

Zelenick, M., Rudloff, V. and Braunitzer, G. (1979) Hemoglobins, XXX. The amino acid sequence of the monomeric hemoglobin from *Lampetra fluviatilis*. *Hoppe-Seyler's Z. Physiol. Chem.*, **360**, 1879–94.

22
THE HAGFISH IMMUNE SYSTEM

Robert L. Raison and Nicholas J. dos Remedios

SUMMARY

The immune system of hagfishes is characterized by an absence of clearly defined lymphoid tissue such as thymus, bone marrow and spleen, which are key components of the immune systems of other vertebrate species. While hagfishes can respond to allogeneic tissue by mounting a cell-mediated cytotoxic response, there is insufficient experimental evidence to indicate that this reactivity is homologous to the adaptive cell-mediated immune response of higher vertebrates. Furthermore, hagfishes appear incapable of mounting a classical antibody response to foreign antigens, with humoral immunity being mediated by a C3-like complement protein exhibiting opsonic activity. Thus, like their invertebrate ancestors, hagfishes appear to rely on innate as opposed to adaptive immune mechanisms to afford protection from potentially pathogenic organisms.

22.1 INTRODUCTION

Two forms of immune response exist. The innate immune response is non-specific and constitutive in nature and, consequently, response time is short. This type of response is characterized by the involvement of phagocytic cells, natural killer or cytotoxic cells (NK or NCC), proteins of the complement system, broadly acting agglutinins and lytic proteins, and acute phase proteins. In contrast, the adaptive immune response is highly specific and is based on the clonal expansion of lymphocytes. As a consequence, in a primary challenge situation, the response time is delayed until clonal expansion generates a sufficient number of effector cells or molecules. In higher vertebrates the adaptive response is mediated at the cellular level by T and B lymphocytes and at the humoral level by antibodies which can display a range of effector functions. In addition, in most vertebrates, the innate and adaptive systems are not mutually exclusive, and interact with, for example, a critical role being played by phagocytes of the innate response in presenting antigen to the T lymphocytes of the adaptive system.

The presence of an adaptive immune response capability is accompanied by an organizational structure of the lymphoid cells and tissues. Thus all vertebrates above and including bony fishes possess well-defined lymphoid tissue such as thymus, spleen and bone marrow. Although lacking the latter, elasmobranchs possess a thymus and spleen and exhibit both cell-mediated and humoral immune functions. Clear evidence for the divergence of the T and B lymphocyte lineages is available only from the evolutionary level of the teleosts (Lobb and Clem, 1982). While functionally distinct, both cell types are derived from common haematopoietic stem cell progenitors in the bone marrow. The antigen receptors expressed on these cells, immunoglobulin (Ig) on B cells and the T cell receptor (TcR) expressed on T cells, exhibit marked similarities at the protein and gene level (Davis and Bjorkman, 1988). These observations suggest that the T and B cell lineages arose during evolution from a

The Biology of Hagfishes. Edited by Jørgen Mørup Jørgensen, Jens Peter Lomholt, Roy E. Weber and Hans Malte. Published in 1998 by Chapman & Hall, London. ISBN 0 412 78530 7.

common ancestral cell type, and that Ig and TcR diverged from a common ancestral gene encoding a cell surface receptor.

In addition to the specific antigen receptors on T cells and B cells, the adaptive immune response is dependent upon recognition processes involving proteins encoded by the Major Histocompatibility Complex (MHC). These MHC antigens also exhibit structural homologies with Ig, and together with Ig and TcR form the core members of the Immunoglobulin Gene Superfamily. In addition to playing a key role in the functioning of the adaptive immune response the MHC antigens mediate acute allograft rejection by T lymphocytes.

While there is evidence for the existence of histocompatibility systems controlling allorecognition reactions in invertebrates (Raftos *et al.*, 1988) this should not be readily interpreted as indicating the presence of an adaptive immune response system at this level of phylogeny. The nature of these histocompatibility reactions and their interpretation in the context of adaptive immune function has been reviewed by Raftos (1996) and by Smith and Davidson (1992) and will not be further discussed here. The body of evidence indicates that invertebrates possess the ability to mount an innate as opposed to an adaptive immune response.

Hagfishes are modern representatives of the earliest evolved craniates and as such occupy a key position in evolutionary development, linking the vertebrates with their invertebrate ancestors, the protochordates. Thus much of the following discussion of the hagfish immune system is made in the context of this important phylogenetic position as it spans the transition from a reliance on innate immune mechanisms to the development of an integrated immune response incorporating innate and adaptive mechanisms.

22.2 LYMPHOID TISSUE

Hagfishes lack the well-organized primary and secondary lymphoid tissues characteristic of the mammalian immune system. Thus there is an absence of bone marrow, a thymus (Riviere *et al.*, 1975) and clearly defined spleen (Adam, 1963) although the island-like arrangement of lymphoid tissue of the intestinal submucosa has been described as a diffuse or primitive spleen (Mawas, 1922; Tomonaga *et al.*, 1973). Granulocytes are abundant in this tissue which acts as a site of haematopoiesis. Erythropoiesis was not observed in these haematopoietic islands (Tomonaga *et al.*, 1973). Similarly, the pronephros of *Myxine glutinosa* has been described as a primitive lymphoid organ (Fänge, 1966; Zapata *et al.*, 1984) as it contains a large number of basophilic lymphocyte-like cells. In teleosts, the pronephros is an important lymphoid organ, acting as a major site of production of erythroid, lymphoid and myeloid cells, as a filter for entrapment of antigens and as a source of functional T and B lymphocytes and macrophages (Bayne, 1986).

22.3 LYMPHOID CELLS

Unlike higher vertebrates, the blood of hagfishes contains abundant lymphoid haemoblasts which differentiate into several blood cell types (Jordan and Speidel, 1930). Granulocytes are numerous as are various developmental stages of thrombocytes. However, the most abundant cell type after the erythrocyte series is the spindle cell and Jordan and Speidel suggested that the few lymphocyte-like cells observed in the blood represent steps in the differentiation of haemoblasts into spindle cells and thrombocytes.

In an electron microscopy study of hagfish blood leucocytes, Linthicum (1975) demonstrated that granulocytes (50%), lymphocytes (15%) and monocytes make up the major cell types in this population. The monocytes contained various cytoplasmic inclusions and granules characteristic of active phagocytes. The lymphocyte-like cells were similar to mammalian small lymphocytes, with some cells exhibiting a nuclear cleft. Large lymphocytes, proplasmablasts and mature plasma cells were not observed in this study of hagfish blood. However, cells exhibiting the typical morphology of plasma cells were identified in an ultrastructural study of the pronephros of *M. glutinosa* (Zapata *et al.*, 1984). The cells were found in clusters and possessed nuclei containing large masses of chromatin, and cytoplasm filled with rough endoplasmic reticulum and associated mitochondria. Degenerate and precursor cell types were also observed. As plasma cells are the classic antibody secreting cell of higher vertebrates, the morphological identification of a plasma cell type in the hagfish was taken as further evidence that these primitive vertebrates are capable of eliciting an antibody response. However, as indicated below, there is an absence of molecular evidence to support this contention.

Heterogeneity of blood leucocytes from *Eptatretus stoutii* has also been demonstrated by flow cytometry. Analysis of laser scatter properties of buffy coat cells revealed two distinct populations with size and internal organelle characteristics analogous to those of mammalian small lymphocytes and granulocytes/monocytes (Gilbertson *et al.*, 1986). Phenotypic heterogeneity within these cell types was observed using a panel of monoclonal antibodies which revealed subpopulations based on the differential expression of cell surface antigens.

22.4 CELL-MEDIATED IMMUNITY

Despite an apparent lack of primary lymphoid tissue, hagfishes are capable of mounting rudimentary cellular responses to allogeneic stimuli (Hildemann, 1981). Hagfishes have a well-developed capacity to recognize and reject skin allografts (Hildemann and Thoenes, 1969). Rejection is accompanied by several features of cell-mediated immunity: lymphocyte infiltration, haemorrhagic inflammation and the destruction of pigment cells. The rejection of allogeneic skin grafts typically assumes a chronic time course (data summarized in Table 22.1), with a median survival time (MST) for first set grafts of 72 days. An accelerated secondary response with an MST of 28 days was interpreted as evidence for specific immunological memory. However, only limited third-party grafts, which distinguish between specific memory and heightened non-specific responsiveness, were examined. Thus, specific immunological memory would be indicated where third-party allografts were rejected more slowly than second set grafts. Five animals previously sensitized to second-party grafts were given third-party grafts. The graft survival times were 19, 19, 32, 70 and 85 days, respectively. These data were interpreted as indicating the existence of a polymorphic array of transplantation antigens containing many frequently occurring loci. Thus the rapid rejection times (19, 19 and 32 days) were explained by proposed sharing of a limited pool of histocompatibility antigens in the host/donor group. The slower rejection times (70 and 85 days) were consistent with the notion of specificity in the memory response. Unfortunately, the number of third-party grafts undertaken were too few to allow clear interpretation of this phenomenon. While a specific memory response would be consistent with the existence of an adaptive immune response in

Table 22.1 Allograft rejection in the hagfish. Data taken from Hildemann and Thoenes (1969)

Protocol	Rejection time (days)		
	Range	MST	SD
1st set (n = 40)	62–82	72	33
2nd set* (n = 15)	19–40	28	22
2nd set† (n = 9)	nd	14	nd

* Established prior to rejection of 1st set.
† Established 0–56 days post rejection of 1st set.

hagfishes an alternative explanation is that the accelerated second set, and in some cases third set, reactions are due to the up regulation of non-specific effector mechanisms such as those that make up the innate immune response.

The technically demanding grafting experiments undertaken by Hildemann and his colleagues have never been repeated or expanded. However, an *in vitro* correlate of allograft rejection, the mixed lymphocyte reaction (MLR), has been investigated. Proliferation of hagfish leucocytes *in vitro* was observed following stimulation with allogeneic cells (Raison *et al.*, 1987). Dissection of this phenomenon revealed that the responding cells belonged to the small leucocyte population, while the granulocyte/monocyte population acted as the stimulating cells. However, only limited allogeneic pairings were assessed, so the extent and cellular mechanism underlying this allorecognition remains unresolved.

22.5 HUMORAL IMMUNITY

Humoral immunity to a wide range of antigens has been reported in the cyclostomes by a number of investigators. However, molecular characterization of the so-called 'antibodies' yielded a multiplicity of structures. Lamprey 'antibody' to bacteriophage f2 was found to be IgM-like, existing in 14S and 6.6S forms (Marchalonis and Edelman, 1968). The basic subunit, which consisted of non-covalently associated heavy and light chain-like polypeptides, was estimated at 188 kDa (Marchalonis and Cone, 1973). Yet, immunization of lamprey with human erythrocytes yielded an active serum protein of approximately 320 kDa (Pollara *et al.*, 1970). Further characterization revealed a labile molecule consisting of non-covalently associated 75 kD polypeptides (Litman *et al.*, 1970). Litman proposed that this molecule was in fact related to the agglutinins found in invertebrates and this suggestion was later reinforced by studies which distinguished a naturally occurring haemagglutinin from antibody activity in the serum of *Petromyzon marinus* after immunization with human 'O' erythrocytes (Hagen *et al.*, 1985). Induced antigen binding activity was also observed in the sera of hagfishes immunized with either KLH (Thoenes and Hildemann, 1969) or sheep erythrocytes (Linthicum and Hildemann, 1970). However in neither case was the humoral activity clearly distinguished from the naturally occurring agglutinins of the hagfishes.

Subsequent to these studies, hagfishes were shown to respond to immunization with a streptococcal vaccine (Raison *et al.*, 1978a). Prolonged immunization with group A streptococci revealed induced anti-group A carbohydrate binding activity in approximately 25% of the animals. While non-immunized hagfish serum displayed significant binding activity to the group A carbohydrate, absorption and sugar inhibition experiments indicated that the anti-streptococcal activity was distinct from the natural haemagglutinin present in hagfish serum. The inducible serum protein, which was found to exhibit specificity for the rhamnose backbone of the group A carbohydrate, was purified and found to have a polypeptide chain structure similar to that of mammalian IgM; i.e. polypeptides with the molecular characteristics of the heavy and light chains of IgM (Raison *et al.*, 1978b). The subunit dissociated at low concentrations of reducing agent in the presence of denaturing buffer, suggesting that the structure was more labile than that of higher vertebrate immunoglobulin. This finding was consistent with earlier descriptions of the lamprey 'antibody'.

22.6 STRUCTURAL CHARACTERIZATION OF THE HAGFISH 'ANTIBODY'

Using monoclonal and polyclonal antibodies raised against the hagfish anti-streptococcal protein we isolated and characterized the protein from the serum of *E. stoutii*. Analysis revealed the presence of two polypeptides in the heavy chain region migrating with apparent molecular weights of 77 kDa (H1) and 70 kDa (H2) (Hanley *et al.*, 1990) and confirmed a light chain-like component at ~30 kDa. The H chains differed in their amino acid sequences as indicated by the generation of distinct *in situ* peptide maps. Using the isolation and purification techniques described earlier for *E. stoutii*, an antibody-like molecule was also isolated from the serum of *E. burgeri* (Kobayashi *et al.*, 1985). Like the *E. stoutii* molecule, this protein exhibited some immunoglobulin-like features but was more labile than conventional immunoglobulin.

The superficial similarity of the hagfish humoral factor with immunoglobulin remained intriguing, although several features, including the apparent lack of significant disulphide bonding, the appearance of multiple heavy chain-like components and the possible combinations of these chains in the basic subunit distinguished the hagfish molecule. While amino terminal sequence analysis failed to reveal the heterogeneity normally associated with the variable regions of immunoglobulins (Hanley *et al.*, 1990), analysis of peptides derived from hagfish light and heavy chains yielded sequences that could be aligned over short stretches with human and shark immunoglobulin and human T cell receptor polypeptide chains (Varner *et al.*, 1991).

The nature of the relationship of the hagfish serum protein to higher vertebrate humoral factors was resolved through isolation and characterization of the gene that encoded it. In our laboratory, affinity purified serum protein from *E. stoutii* was used to obtain N-terminal and internal amino acid sequence from the larger heavy chain component (Hanley *et al.*, 1992). This information was used to isolate a partial DNA clone from liver RNA. Surprisingly, comparison of the deduced amino acid sequence with known protein sequences failed to reveal any homology with immunoglobulin or immunoglobulin-like molecules. However, striking similarity was seen when the hagfish protein was compared to the mammalian complement components C3, C4 and C5. Kurosawa and his colleagues at Fujita Health University used protein purified by ion exchange and gel permeation chromatography from the serum of *E. burgeri* to obtain partial amino acid sequence data for the heavy and light polypeptide chains (Ishiguro *et al.*, 1992). They subsequently cloned a 2.2 kb cDNA fragment from liver mRNA that encoded the three previously described chains of the hagfish molecule. Again, significant sequence similarities were seen with the three mammalian complement proteins. Comparison of the nucleotide sequences obtained by the two laboratories left no doubt that they were dealing with the same molecule. Furthermore, amino acid sequences obtained for the putative hagfish 'antibody' in three laboratories, and from two species of hagfish (Varner *et al.*, 1991; Hanley *et al.*, 1992; Ishiguro *et al.*, 1992), were clearly encoded in these genes.

The hagfish complement molecule is encoded by a single mRNA species of approximately 5 kb, the three polypeptides apparently arising from post-translational cleavage of a precursor molecule. Sequence comparisons indicate equivalence of the 77 kDa, 70 kDa and 30 kDa hagfish chains with the β, α and γ chains respectively of mammalian C3/C4. Furthermore, post-translational modification of the hagfish protein includes removal of 77 amino acids encoded at the N-terminus of the α chain. This appears analogous to the similarly sized anaphylotoxins cleaved from the α chains of mammalian C3, C4 and C5 (Larsen and Henson, 1983; Fearon and Wong, 1983). Overall sequence comparison shows the greatest similarity of the hagfish protein with mammalian C3, with the hagfish protein also containing a thioester bond (Hanley *et al.*,

1992) which is present in the α chain of the mammalian equivalent. This bond is activated by the cleavage of the C3a and C4a fragments, and plays a key role in the binding of the C3b and C4b fragments to target cells (Fearon and Wong, 1983).

22.7 FUNCTIONAL PROPERTIES OF HAGFISH COMPLEMENT-LIKE PROTEIN (CLP)

The finding that the hagfish CLP, isolated from immunized animals, exhibited binding affinity for carbohydrate groups in the streptococcal cell wall implied a humoral defence role for this molecule. Furthermore, given the relationship of this molecule to mammalian complement components which are involved in enhancement of immune function via opsonization, we examined the potential for CLP to mediate this function. Affinity purified CLP from non-immune animals was shown to bind to the surface of bacteria (streptococci) and yeast and enhance the phagocytosis of these cells by hagfish monocytes (Hanley *et al.*, 1992). The binding and opsonizing properties of CLP were inhibited by co-incubation with rhamnose or mannose, but not by a range of other monosaccharides (Raftos *et al.*, 1992). The specificity for rhamnose was of particular interest as it was consistent with our earlier findings with respect to the protein identified in, and purified from, the serum of hagfishes immunized with streptococci. It is thus evident that CLP exists at significant constitutive levels in the serum of hagfishes and that its production, probably from cells in the liver, can be increased in response to antigen challenge.

The binding and opsonizing properties of hagfish CLP are similar to those displayed by the thioester-containing mammalian complement components C3 and C4 (Fearon and Wong, 1983), as well as the C4-like molecule identified in the lamprey, *Lampetra japonica* (Nonaka *et al.*, 1984). However, while the activated mammalian complement component may covalently attach to carbohydrates via the thioester group (Law and Levine, 1981), this reactivity does not exhibit the restricted sugar specificity seen with CLP. Indeed, the binding of CLP to carbohydrates is lectin-like in that it is divalent cation dependent (Raftos *et al.*, 1992). Thus, a direct link between the presence of the thioester bond on the α-like chain, and the carbohydrate binding specificity of CLP cannot be assumed.

CLP-mediated opsonization occurs via a receptor on the surface of hagfish monocytes (Raison *et al.*, 1994). A monoclonal antibody, raised against monocytes present in the hagfish peripheral blood leucocyte population, inhibited the phagocytosis of CLP-coated target cells. In the reciprocal experiment, CLP blocked the interaction of the monoclonal antibody with monocytes. The monoclonal antibody immunoprecipitated a single polypeptide chain of approximately 100 kDa from the surface of radiolabelled leucocytes. In humans the receptor which mediates the enhanced phagocytosis of C3-coated target cells is CR3, a member of the integrin family of adhesion molecules. Members of this family have been described in invertebrates (Bogaert *et al.*, 1987) and it is tempting to speculate that the opsonic receptor identified on hagfish monocytes may belong to this phylogenetically primitive group of proteins. However, CR3 consists of two subunits and thus, at least superficially, does not appear related to the hagfish receptor. Two other forms of C3 receptor have been identified in mammals (Fearon and Wong, 1983) but the molecular characteristics of these receptors do not immediately suggest a phylogenetic linkage to the hagfish receptor.

22.8 THE COMPLEMENT SYSTEM IN THE HAGFISH

While there is circumstantial, functional evidence for the existence of a rudimentary alternative pathway of complement activity in the invertebrates, no molecules exhibiting structural homology with known complement components have been isolated from invertebrates (Farries and Atkinson, 1991). Thus, the first report of a phylogenetically primitive complement component was that of a C3-like protein from the lamprey (Nonaka *et al.*, 1984). A similar molecule was isolated from the serum of the hagfish and shown to exhibit both primary and secondary structural features in common with human C3 (Fujii *et al.*, 1992). The hagfish C3 consisted of two disulphide-linked polypeptide chains of 115 kDa and 77 kDa. Amino-terminal sequence analysis of the 77 kDa chain of hagfish C3 revealed identity with the 77 kDa chain from the previously described CLP molecule. The apparent structural differences between the C3 molecule isolated from *E. burgeri* by Fujii and the CLP from *E. stoutii* characterized in our laboratory arise from proteolytic cleavage of the 115 kDa α plus γ chain during affinity purification (T. Fujii, personal communication). Thus, in its native form the hagfish complement protein is structurally homologous to mammalian C3. The C3 proteins of lampreys and hagfishes, like the human C3 component, bind to yeast cells and act as a ligand for opsonization by phagocytic cells (Nonaka *et al.*, 1984; Raison *et al.*, 1994; Fujii *et al.*, 1993). Lytic function associated with complement components has not been demonstrated in the cyclostomes, suggesting that they lack the terminal pathway proteins, C5–9 (Farries and Atkinson, 1991).

A protein homologous to the human serum protein α-2-macroglobulin (α2M) has been isolated from the serum of *Eptatretus burgeri* (Osada *et al.*, 1986). α2M is a serum proteinase inhibitor and a member of the complement superfamily of proteins which include C3, C4 and C5. In addition to primary sequence similarities α2M, C3 and C4 are also characterized by the presence of a labile thioester (Chu and Pizzo, 1994). Hagfish α2M consists of a basic subunit of 190 kDa which undergoes conformational change upon complexing with chymotrypsin (Österberet *et al.*, 1991). Functionally, α2M has many diverse roles which include potential binding interactions with proteases, enzymes, cytokines, carbohydrates, cells (via specific receptors) and antigen, as well as playing an important role in inflammation, immuno-modulation and tissue repair (Chu and Pizzo, 1994). This ancient member of the complement superfamily is thought to have branched from the C3/C4 group some 600 million years ago (Dodds and Day, 1993).

22.9 CHEMOTAXIS IN THE HAGFISH

Chemotaxis, the directed migration and recruitment of effector cells, is a fundamental component of the immune and inflammatory responses. A number of chemoattractants have been identified in mammals including IL-8, C5a and the bacterial tripeptide, formyl methionyl leucyl phenylalanine (fMLP). Chemotaxis of phagocytic cells plays a major role in defence against infection by mediating the inflammatory response.

The phylogenetic conservation of chemotactic function has been observed through the ability of mammalian C5a to enhance the migration of leucocytes in the shark (Obenauf and Hyder-Smith, 1985). Recently, we have demonstrated responsiveness of hagfish leucocytes to endogenous and exogenous chemoattractants (Newton *et al.*, 1994). Hagfish monocytes and granular leucocytes migrated in a concentration gradient of both human C5a and LPS-activated hagfish plasma. Thus, activation of hagfish plasma generated a chemotactic product which reacted in a manner analogous to human C5a. In mammals, the 74 amino acid anaphylotoxin C5a, is cleaved from the amino terminus of the α chain of C5. The corresponding chain of hagfish CLP exhibits 32% sequence identity and 43% sequence similarity (i.e. allowing for conservative substitutions) with human C5 (Ishiguro *et al.*, 1992; Hanley *et al.*, 1992). It is therefore likely that LPS activation results in the generation of a C5a-like fragment from the hagfish α chain homologue. This fragment would be the endogenous ligand for a receptor on hagfish leucocytes which also reacts with human C5a.

22.10 CONCLUSION

The organizational and functional properties of the hagfish immune system exhibit several features which distinguish these animals from all other vertebrate classes. These properties include the absence of organized primary lymphoid tissue and an apparent inability to mount adaptive cellular and humoral responses characterized by specificity and memory.

Cell-mediated immune function in the hagfish has been demonstrated both *in vivo* (allograft rejection) and *in vitro* (mixed leucocyte reaction). However, the experimental protocols used have not permitted unequivocal demonstration of specificity and memory, two key features of adaptive cellular responsiveness mediated by T lymphocytes. These cellular immune functions in hagfishes may be mediated by a cell type with properties similar to that of the NK or NCC which exhibit broad specificity for tumour and virally infected cells in higher vertebrates. This is an area that warrants further investigation.

The hagfish serum protein previously described as 'antibody' or 'immunoglobulin' is a complement-related protein bearing strong primary and secondary structural similarities to mammalian C3. The hagfish complement protein may be the evolutionary precursor of the C3, C4, C5 family of complement proteins that are found in higher vertebrates. Thus, humoral immunity in these primitive vertebrates is mediated, at least in part, by a complement protein that functions as broadly reacting opsonin.

Considerable effort has been directed towards the question of the presence of immunoglobulin in the cyclostomes. Despite numerous reports of antibody-like activity in serum after immunization with cellular and soluble antigens, a molecule with definitive structural features of immunoglobulin has not been isolated. A number of techniques have been used in attempts to isolate immunoglobulin genes from cyclostomes. Although the use of mammalian immunoglobulin variable region probes to isolate homologous genes in other species has been highly successful down to, and including, the cartilaginous fishes (Hinds and Litman, 1986), this approach has failed to identify immunoglobulin-like genes in the hagfish. Similarly, the use of phylogenetically conserved blocks of sequence in a PCR strategy has not yielded immunoglobulin-like genes from either lampreys or hagfishes (Ishiguro *et al.*, 1992) despite the success of the approach in isolating MHC genes from carp (Hashimoto *et al.*, 1990) and sharks (Hashimoto *et al.*, 1992). When viewed overall, there is at this time no molecular evidence to support the existence of immunoglobulin-like genes in the most primitive vertebrates. However, given the identification of immunoglobulin-like domain structures in molecules such as amalgam, fasciclin II and haemolin in insects (Sun *et al.*, 1990), and moluscan defence molecule in snails (Hoek *et al.*, 1996), it would appear inevitable that members of the immunoglobulin gene superfamily will ultimately be identified in the cyclostomes.

Taken together, the currently identified immunological functions in the hagfish of cellular cytotoxicity, chemotaxis, and complement enhanced phagocytosis are consistent with the conclusion that hagfishes utilize innate as opposed to adaptive immune mechanisms to repel foreign organisms. This aligns the hagfish more closely with its invertebrate ancestors with respect to immunological properties.

REFERENCES

Adam, H. (1963) Structure and histochemistry of the alimentary canal, in *The Biology of Myxine* (eds A. Brodal and R. Fänge), Scandinavian University Books, Oslo, p. 276.

Bayne, C.J. (1986) Pronepharic leucocytes in *Cyprimus carpio*: isolation, separation and characterization. *Vet. Immunol. Immunopath.*, **12**, 59–67.

Bogaert, T., Brown, N. and Wilcox, M. (1987) The Drosophila PS2 antigen is an invertebrate integrin that, like the fibronectin receptor, becomes localized to muscle attachments. *Cell*, **51**, 929–40.

Chu, C.T. and Pizzo, S.V. (1994) Alpha 2-macroglobulin, complement, and biologicl defense: antigens, growth factors, microbial proteases, and receptor ligation. *Lab. Invest.*, **71**, 792–812.

Davis, M.M. and Bjorkman, P. (1988) T cell antigen receptor genes and T cell recognition. *Nature*, **334**, 395–402.

Dodds, A.W. and Day, A.J. (1993) The phylogeny and evolution of the complement system, in *Complement in Health and Disease* (eds K. Whaley, M. Loos and J.M. Wesner), Kluwer Academic Publishers, Boston, p. 39.

Fänge, R. (1966) Comparative aspects of excretory and lymphoid tissue, in *Phylogeny of Immunity* (eds R.T. Smith, P.A. Miescher and R.A. Good), University of Florida Press, Gainsville, p. 141.

Farries, T.C. and Atkinson, J.P. (1991) Evolution of the complement system. *Immunol. Today*, **12**, 295–300.

Fearon, D.T. and Wong, W.W. (1983) Complement ligand-receptor interactions that mediate biological responses. *Ann. Rev. Immunol.*, **1**, 243–71.

Fujii, T., Nakamura, T. and Tomonaga, S. (1993) Identification and characterization of a variant of the third component of complement (C3) in hagfish serum. *Zool. Sci.*, **10** (Suppl.), 83.

Fujii, T., Nakamura, T., Sekizawa, A. and Tomonaga, S. (1992) Isolation and characterization of a protein from hagfish serum that is homologous to the third component of the mammalian complement system. *J. Immunol.*, **148**, 117–23.

Gilbertson, P., Wotherspoon, J. and Raison, R.L. (1986) Evolutionary development of lymphocyte heterogeneity: leucocyte subpopulations in the Pacific hagfish. *Dev. Comp. Immunol.*, **10**, 1–10.

Hagen, M., Filosa, M.F. and Youson, J.H. (1985) The immune response in adult sea lamprey, *Petromyzon marinus*: the effect of temperature. *Comp. Biochem. Physiol.*, **82A**, 207–10.

Hanley, P.J., Hook, J.W., Raftos, D.A., Gooley, A.A., Trent, R. and Raison, R.L. (1992) Hagfish humoral defense protein exhibits structural and functional homology with mammalian complement proteins. *Proc. Natl. Acad. Sci. USA*, **89**, 7910–14.

Hanley, P.J., Seppelt, I.M., Gooley, A.A., Hook, J.W. and Raison, R.L. (1990) Distinctive Ig H chains in a primitive vertebrate, *Eptatretus stoutii*. *J. Immunol.*, **145**, 3823–8.

Hashimoto, K., Nakanishi, T. and Kurosawa, Y. (1990) Isolation of carp genes encoding major histocompatibility complex antigens. *Proc. Natl. Acad. Sci. USA*, **87**, 6863–7.

Hashimoto, K., Nakanishi, T. and Kurosawa, Y. (1992) Identification of a shark sequence resembling the major histocompatibility complex class I a3 domain. *Proc. Natl. Acad. Sci. USA*, **89**, 2209–12.

Hildemann, W.H. (1981) Immunophylogeny: from sponges to hagfish to mice, in *Frontiers in Immunogenetics* (ed. W.H. Hildemann), Elsevier, North Holland, p. 3.

Hildemann, W.H. and Thoenes, G.H. (1969) Immunological responses of Pacific hagfish. I. Skin transplantation immunity. *Transplantation*, **7**, 506–21.

Hinds, K.R. and Litman, G.W. (1986) Major reorganization of immunoglobulin VH segmental elements during vertebrate evolution. *Nature*, **320**, 546–9.

Hoek, R.M., Smit, A.B., Frings, H., Vink, J.M., de Jong-Brink, M. and Geraerts, W.P.M. (1996) A new Ig-superfamily member, molluscan defence molecule (MDM) from *Lymnaea stagnalis*, is down-regulated during parasitosis. *Eur. J. Immunol.*, **26**, 939–44.

Ishiguro H., Kobayashi, K., Suzuki, M., Titani, K., Tomonaga, S. and Kurosawa, Y. (1992) Isolation of a hagfish gene that encodes a complement component. *EMBO J.*, **11**, 829–37.

Jordan, H.E. and Speidel, C.C. (1930) Blood formation in cyclostomes. *Amer. J. Anat.*, **46**, 355–91.

Kobayashi, K.S., Tomonaga, S. and Hagiwara, K. (1985) Isolation and characterization of immunoglobulin of hagfish, *Eptatretus burgeri*, a primitive vertebrate. *Mol. Immunol.*, **22**, 1091–7.

Larsen, G.L. and Henson, P.M. (1983) Mediators of inflammation. *Ann. Rev. Immunol.*, **1**, 335–60.

Law, S.K. and Levine, R.P. (1981) Binding reaction between the third human complement protein and small molecules. *Biochemistry*, **20**, 7457–63.

Linthicum, D.S. (1975) Ultrastructure of hagfish blood leucocytes, in *Immunological Phylogeny* (eds W.H. Hildemann and A.A. Benedict), Plenum, New York, p. 241.

Linthicum, D.S. and Hildemann, W.H. (1970) Immunologic responses of Pacific hagfish. III. Serum antibodies to cellular antigens. *J. Immunol.*, **105**, 912–18.

Litman, G.W., Frommel, D., Finstad, J., Howell, J., Pollara, B.W. and Good, R.A. (1970) The evolution of the immune response, VIII. Structural studies of the lamprey immunoglobulin. *J. Immunol.*, **105**, 1278–85.

Lobb, C.J. and Clem, L.W. (1982) Fish lymphocytes differ in the expression of surface immunoglobulin. *Dev. Comp. Immunol.*, **6**, 473–9.

Marchalonis, J.J. and Cone, R.E. (1973) The phylogenetic emergence of vertebrate immunity. *Aust. J. Exp. Biol. Med. Sci.*, **51**, 461–88.

Marchalonis, J.J. and Edelman, G.M. (1968) Phylogenetic origins of antibody structure. III. Antibodies in the primary immune response of the sea lamprey, *Petromyzon marinus*. *J. Exp. Med.*, **127**, 891–914.

Mawas, J. (1922) Sur le tissu lymphoide de l'intestine moyenne des *myxinoides* et sur la signification morphologique. *C.R. Acad. Sci., Paris*, **174**, 889–90.

Newton, R.A., Raftos, D.A., Raison, R.L. and Geczy, C.L. (1994) Chemotactic responses of hagfish (Vertebrata, Agnatha) leucocytes. *Dev. Comp. Immunol.*, **18**, 295–304.

Nonaka, M., Fuji, T., Kaidoh, T., Natsuume-Sakai, S., Nonaka, M., Yamaguchi, N. and Takahashi, M. (1984) Purification of a lamprey complement protein homologous to the third component of the mammalian complement system. *J. Immunol.*, **133**, 3242–9.

Obenauf, S.D. and Hyder-Smith, S. (1985) Chemotaxis of nurse shark leucocytes. *Dev. Comp. Immunol.*, **9**, 221–30.

Osada, T., Nishigai, M. and Ikai, A. (1986) Open quaternary structure of the hagfish proteinase inhibitor with similar properties to human α2-macroglobulin. *J. Ultrastruct. Mol. Struct. Res.*, **96**, 136–45.

Österberet, R., Malmensten, B. and Ikai, A. (1991) X-ray scattering of hagfish protease inhibitor, a protein structurally related to complement and α-2-macroglobulin. *Biochem.*, **30**, 7873–8.

Pollara, B.W., Litman, G.W., Finstad, J., Howell, J. and Good, R.A. (1970) The evolution of the immune response. VII. Antibody formation to human 'O' cells and properties of the immunoglobulin in lamprey. *J. Immunol.*, **105**, 738–45.

Raftos, D.A. (1996) Histocompatibility reactions in invertebrates, in *Advances in Comparative and Environmental Physiology*, Vol. 24 (ed. E.L. Cooper), Springer-Verlag, Berlin, p. 78.

Raftos, D.A., Briscoe, D.A. and Tait, N.N. (1988) Mode of recognition of allogeneic tissue in the solitary urochordate, Styela plicata. *Transplantation*, **45**, 1123–6.

Raftos, D.A., Hook, J.W. and Raison, R.L. (1992) Complement-like protein from the phylogenetically primitive vertebrate, *Eptatretus stoutii*, is a humoral opsonin. *Comp. Biochem. Physiol.*, **103B**, 379–84.

Raison, R.L., Coverley, J., Hook, J.W., Towns, P., Weston, K.M. and Raftos, D.A. (1994) A cell-surface opsonic receptor on leucocytes from the phylogenetically primitive vertebrate, *Eptatretus stoutii*. *Immunol. Cell Biol.*, **72**, 326–32.

Raison, R.L., Gilbertson, P. and Wotherspoon, J. (1987) Cellular requirements for mixed leucocyte reactivity in the cyclostome, *Eptatretus stoutii*. *Immunol. Cell Biol.*, **65**, 183–8.

Raison, R.L., Hull, C.J. and Hildemann, W.H. (1978a) Production and specificity of antibodies to streptococci in the Pacific hagfish, *Eptatretus stoutii*. *Dev. Comp. Immunol.*, **2**, 253–62.

Raison, R.L., Hull, C.J. and Hildemann, W.H. (1978b) Characterization of immunoglobulin from the Pacific hagfish, a primitive vertebrate. *Proc. Natl. Acad. Sci. USA*, **75**, 5679–82.

Riviere, H.B., Cooper, E.L., Reddy, A.L. and Hilderman, W.H. (1975) In search of the hagfish thymus. *Amer. Zool.*, **15**, 39–49.

Smith, L.C. and Davidson, E.H. (1992) The echinoid immune system and the phylogenetic occurrence of immune mechanisms in deuterostomes. *Immunol. Today*, **13**, 356–61.

Sun, S.C., Lindstrom, I., Boman, H.G., Faye, I. and Schmidt, O. (1990) Hemolin: an insect immune protein belonging to the immunoglobulin superfamily. *Science*, **250**, 1729–32.

Thoenes, G.H. and Hildemann, W.H. (1969) Immunologic responses of Pacific hagfish. II. Serum antibody production to soluble antigens, in *Developmental Aspects of Antibody Formation and Structure*, Vol. 2 (eds J. Sterzl and I. Riha),

Czechoslova Academy of Science, Prague, pp. 711–21.

Tomonaga, S., Hirokane, T., Shinohara, H. and Awaya, K. (1973) The primitive spleen of the hagfish. *Zool. Mag.*, **82**, 215–17.

Varner, J., Neame, P. and Litman, G.W. (1991) A serum heterodimer from hagfish (*Eptatretus stoutii*) exhibits structural similarity and partial sequence identity with immunoglobulin. *Proc. Natl. Acad. Sci. USA*, **88**, 1746–50.

Zapata, A., Fänge, R., Mattisson, A. and Villena, A. (1984) Plasma cells in adult Atlantic hagfish, *Myxine glutinosa*. *Cell Tissue Research*, **235**, 691–3.

PART NINE

The Uro-genital System

23

THE HAGFISH KIDNEY AS A MODEL TO STUDY RENAL PHYSIOLOGY AND TOXICOLOGY

Lüder M. Fels, Sabine Kastner and Hilmar Stolte

SUMMARY

The kidney of the adult hagfish is a comparatively simple organ consisting of large, segmentally arranged glomeruli and two tubules (archinephric ducts, AND) which drain into the cloaca. The biochemical composition, morphology, and function *in vitro* (permeability of the glomerular barrier to water and proteins) of the glomeruli is similar to that of higher vertebrates. The AND is morphologically characterized by an epithelium with a brush border. The AND reabsorbs molecules such as glucose or amino acids, expresses enzymes, such as cathepsins or acid phosphatase, but seems to be less active than the corresponding segments (proximal tubules) in higher vertebrates.

The kidneys of hagfishes have been validated as an alternative animal model to study renal physiology and toxicology. Animals challenged *in vivo* with cadmium or antibiotic anthracyclines (Adriamycin) revealed an altered glomerular permeability to water and macromolecules, as determined *in vitro* on isolated perfused glomeruli. Metabolic processes such as altered *de novo* synthesis of glomerular protein following treatment with xenobiotics was demonstrated on isolated nephron segments. Such studies that exceed the scope of comparative physiology exemplify how the hagfish kidney can serve as a model to elucidate mechanisms that can be extrapolated to kidney physiology and pathophysiology of higher vertebrates.

23.1 INTRODUCTION

The gills and kidneys play an important role in the osmotic and ionic regulation of the body fluids of aquatic vertebrates. Evolution led to the development of a variety of structurally and functionally different excretory organs. The kidneys of the myxinoids have been described as having a simple structure (Fänge, 1963).

Physiologically myxinoids are different from other vertebrates. For example, they are in osmotic equilibrium with the surrounding sea water, mainly by retaining high extracellular concentrations of inorganic substances in their body fluids (Smith, 1932; Robertson, 1954). This means that these animals have neither a passive water influx like vertebrates such as freshwater teleosts nor an efflux such as marine teleosts. This raises the question whether hagfishes still produce urine and whether their glomeruli are similar in structure and function to those of other vertebrates. The animals have a low blood pressure (Riegel, 1986a). Blood pressure,

The Biology of Hagfishes. Edited by Jørgen Mørup Jørgensen, Jens Peter Lomholt, Roy E. Weber and Hans Malte. Published in 1998 by Chapman & Hall, London. ISBN 0 412 78530 7.

however, is usually considered the driving force for renal filtration. The fact that hagfishes are osmoconformers also raises the question of how important electrolyte regulation is for these animals. The tubular apparatus found in the complex kidneys of other vertebrates is reduced to two ducts. But despite its structural simplicity a variety of functional features known from other species are found in the hagfish tubules.

In the first part of this chapter it will therefore be attempted to give a brief overview from the point of view of comparative physiology on what is known about structure and function of the hagfish kidney. For a more detailed description of glomerular haemodynamics the reader is referred to Chapter 24.

The second part of this chapter will summarize data derived from functional and metabolic *ex vivo/in vitro* studies with isolated hagfish nephron segments. Aquatic vertebrates have developed distinctly specific nephron segments, often to an extreme degree. In hagfishes the glomeruli are about ten times larger than in most mammals. Glomeruli and also parts of the tubular apparatus can easily be microdissected. Therefore these structures lend themselves as models of renal function.

23.2 MORPHOLOGY OF THE HAGFISH KIDNEY

Aquatic vertebrates have developed several diverse kidney types, according to the different requirements of their habitat: marine, brackish water, fresh water. An elaborate organization is required for functions such as ion regulation, osmoregulation and/or retention of urea. The hagfish are exclusively marine and exhibit comparatively simple renal structures that participate in maintaining homeostasis of body fluids (Fänge, 1963; Hentschel and Elger, 1989).

A holonephros is found in young hagfish. In adult animals the cranial part, the pronephros, is reduced (Holmgren, 1950). In the adult animal it is considered a primitive lymphohaemopoietic organ (Zapata *et al.*, 1984). The connecting duct to the more caudal part of the organ degenerates in the adult (Müller, 1875). The caudal region is an opisthonephros, the urine-forming system of adult animals. In the following this review will focus on the latter part of the hagfish kidney.

In the myxinoids the nephron is rudimentary or of a primitive type. It is a kidney without a zonation, consisting of two ducts. The ducts begin 10–20 mm behind the pronephros and are drained into the cloaca. With the exception of short cranial and caudal parts renal corpuscles are segmentally attached to the ducts. The nomenclature for the ducts used in previous publications is not consistent. They were called ureters by Fänge (1963), archinephric ducts by other authors (Hickman and Trump, 1969). They have been compared to the proximal tubule, segment I (Hentschel and Elger, 1989). In the context of this review, the duct system will be referred to as an archinephric duct (AND).

The walls of the ANDs consist of three parts: a columnar epithelium that rests on a basal membrane, connective tissue, and a capillary network (Heath-Eves and McMillan, 1974). The epithelial cells of the AND have a brush border (Figure 23.1). Vesicles on the apical side of the cells indicate reabsorption of macromolecules. The Golgi apparatus and the endoplasmic reticulum are not prominent, and the mitochondria were described as being comparatively small but moderately abundant (Hickman and Trump, 1969). The *zonulae occludentes* between epithelial cells of the AND are composed of five or more strands, occasionally only of one or two (Kühn *et al.*, 1975).

The glomeruli of the hagfish, 30 to 40 along each AND, are exceptionally large. They have a diameter of 500–1500 μm but decrease in size towards the caudal end of the AND. The glomerulus consists of a complex system of capillary loops and it is usually supplied with

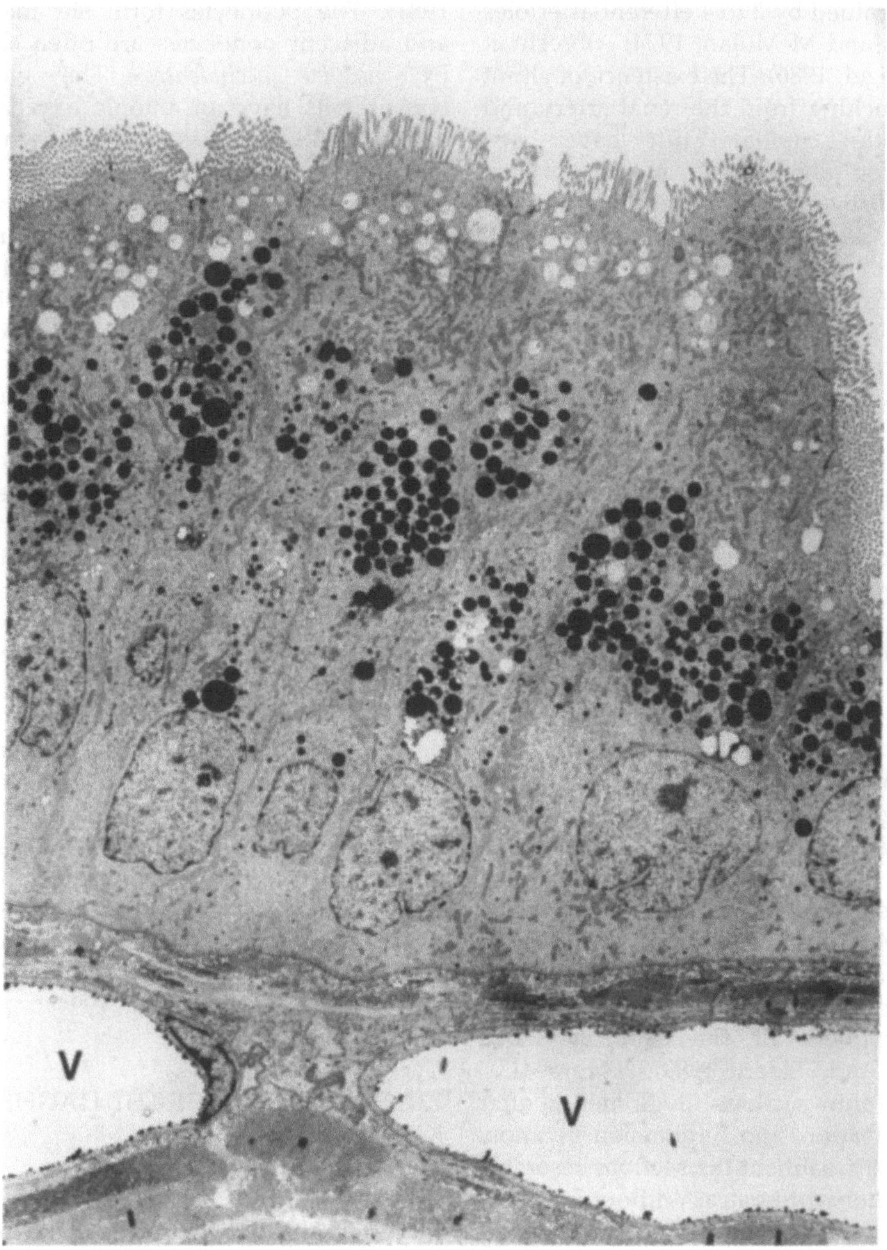

Figure 23.1 Archinephric duct of *Myxine glutinosa*. Brush border on the apical surface of the epithelial cells. An abundant number of granula in the cytoplasm filled with material of different electron density. Basal membrane, connective tissue. V = vessel. ×2400 (Photo courtesy of Dr B. Decker, Hannover, Germany.)

blood from one, sometimes two, afferent arterioles and drained by 2 to 4 efferent arterioles (Heath-Eves and McMillan, 1974; Albrecht *et al.*, 1978; Riegel, 1986). The existence of shunt vessels, branching from the renal artery and bypassing the capillary tuft have been described for about one quarter of the glomeruli (Brown, 1988). Measurements of the filtering surface of the glomerular capillaries taken from serial thin sections of glomeruli fixed under constant perfusion pressure revealed a filtering area of 1.83 ± 0.71 mm^2 per glomerulus for *Myxine glutinosa*. Higher aquatic vertebrates like the flounder, carp or skate have filtering areas – though calculated from standard formulas – ranging from 0.023 to 0.340 mm^2 (Fels *et al.*, 1993). The Bowman's capsule is connected to the archinephric duct by a short neck lined by flattened epithelium. This neck is often connected to a diverticulum of the archinephric duct (Hickman and Trump, 1969).

As in higher vertebrate classes, the glomeruli and the filtration barrier are composed of endothelial cells, mesangial cells and mesangial matrix, podocytes, a basement membrane and parietal cells. The endothelial cells of the glomerular capillary loop have a comparatively scarce fenestration (Heath-Eves and McMillan, 1974; Kühn *et al.*, 1975). There is an abundant mesangium that extends from the central region of the lobules to the periphery of the capillary loops (Hentschel and Elger, 1989). Figure 23.2 shows semi-thin sections of glomeruli after perfusion fixation and immersion fixation. With the first treatment the sections resemble those of higher vertebrates. Without intracapillary pressure mesangial components expand and the glomerular capillaries collapse. The mesangial cells (Figure 23.2c) have numerous processes that extend into the subendothelial space blending with the mesangial matrix (Heath-Eves and McMillan, 1974; Kühn *et al.*, 1975; Reale *et al.*, 1981; Hentschel and Elger, 1989). A very distinct basement membrane is characteristic of the epithelium (Reale *et al.*, 1981). The podocytes form slit membranes and adjacent podocytes are often connected by *maculae occludentes*. The glomerular constituents have an anionic barrier system, as revealed by studies on the brown hagfish, *Paramyxine atami* (Isujii *et al.*, 1984) and the Atlantic hagfish, *Myxine glutinosa* (Decker and Reale, 1991). The anionic groups of proteoglycans are found on both luminal and abluminal surfaces of endothelial cells, within the glomerular basement membrane and on the epithelial cell surface facing the urinary space. Histochemical staining with cuprolinic blue revealed that this cationic dye predominantly binds to the electron dense glomerular basement membrane and along the epithelial surface coat. Digestion experiments with heparitinase I showed that the network of cuprolinic blue glycosaminoglycan precipitates in the glomerular basement membrane was composed of heparan-sulphate (Fels *et al.*, 1994). The finding of proteoglycans in the glomerular basement membrane is in accordance with findings in higher vertebrates. Other structural or extracellular matrix proteins found in the glomerular filtration barrier of higher vertebrates were also identified in the hagfish glomerulus. Fibronectin, laminin and collagen type IV were qualitatively demonstrated in purified glomerular homogenates of *Myxine glutinosa* (Kastner *et al.*, 1994a; Eisenberger *et al.*, 1994).

23.3 PHYSIOLOGY OF THE HAGFISH KIDNEY

23.3.1 Glomerular function

In vertebrates ultrafiltrate is produced in the glomeruli. The tubular apparatus then modifies this fluid by active and passive transport processes (secretion, reabsorption). According to Starling (1899) the driving forces in the formation of ultrafiltrate are hydrostatic and colloid osmotic pressures in the glomerular capillaries and the Bowman's capsule.

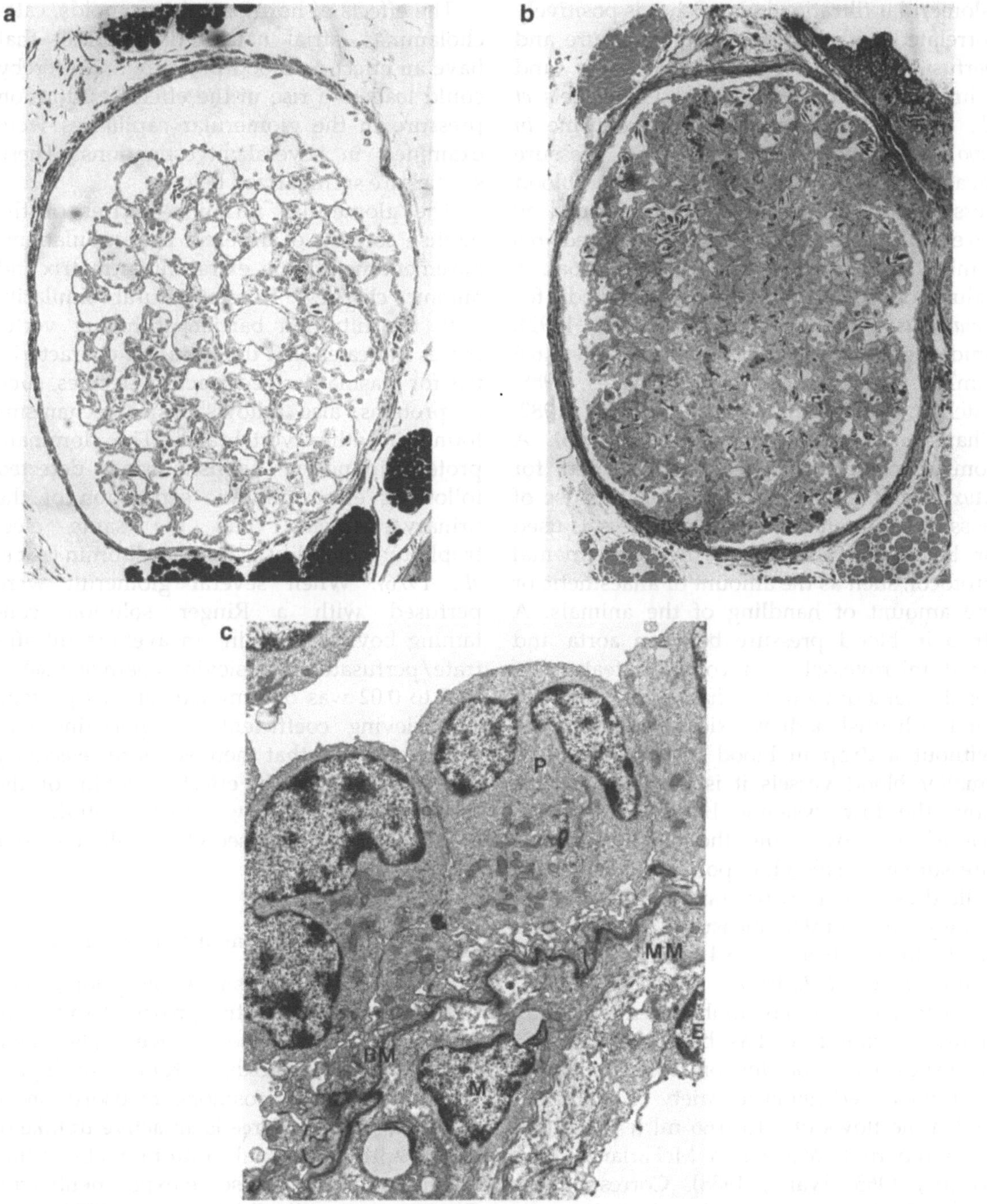

Figure 23.2 Semi-thin sections of a glomerulus of *Myxine glutinosa* after (a) perfusion fixation and (b) immersion fixation ×80. (From Fels *et al.*, 1993, with permission; photo courtesy of Dr B. Decker, Hannover, Germany.) (c) Glomerular filtration barrier of *Myxine glutinosa*. Podocytes (P) rest on a glomerular basement membrane (BM). Mesangial cells (M) surrounded by mesangial matrix (MM). E = Endothelial cell. ×8000. (Photo courtesy of Dr B. Decker, Hannover, Germany.)

Glomerular filtration in hagfishes is positively correlated with both perfusion pressure and perfusion rate (Riegel, 1978; Stolte and Schmidt-Nielsen, 1978; Riegel 1986b; Fels *et al.*, 1993). How hagfishes produce urine *in vivo* still remains unclear. Blood pressure measurement in the aorta and other blood vessels of hagfishes have been performed by several investigators. Riegel measured 6.8 mmHg in *Eptatretus stouti* (Riegel, 1986a). A value of 9.8 mmHg was reported for *Eptatretus cirrhatus* (Forster *et al.*, 1992). Studies on *Myxine glutinosa* reported 3.5 to 6.6 mmHg (Axelsson *et al.*, 1990; Wales, 1988; Satchell, 1986; Carroll and Opdyke, 1982; Chapman *et al.*, 1963; Johansen, 1960). A comparison of the pressures reported for *Myxine glutinosa* is made difficult because of possible differences in the blood vessels used or because of details of the experimental protocol, such as the amount of anaesthetic or the amount of handling of the animals. A drop in blood pressure between aorta and renal microvessels – a common feature of renal haemodynamics – has been described for the hagfish kidney (Riegel, 1986a). Even without a drop in blood pressure towards smaller blood vessels it is hard to imagine how the low systemic blood pressure of hagfish can overcome the colloid osmotic pressure in the blood. Reported values for the colloid osmotic pressure range from approximately 10.5 mmHg measured in *Eptatretus stouti* (Riegel, 1986a) to 5.4 mmHg in *Myxine glutinosa* (Fels *et al.*, 1989b).

Despite these unfavourable pressure conditions, a urine flow has been measured in hagfishes. Urine production in hagfishes has been measured under a variety of conditions and urine flows of 0.16–0.66 ml h^{-1} kg^{-1} have been reported (Munz and McFarland, 1964; Morris, 1965; Wales, 1988). Corresponding values in lampreys or eels adapted to sea water amount to 0.81 ml h^{-1} kg^{-1} (Logan *et al.*, 1980; Schmidt-Nielsen and Renfro, 1975). For the trout in brackish water a urine flow of 0.56 ml h^{-1} kg^{-1} was determined (Elger *et al.*, 1988).

The effects of hormones (eicosanoids, catecholamines, atrial natriuretic peptide) that have an effect on blood pressure and thereby could lead to a rise in the effective filtration pressure in the glomerular capillaries were examined in several investigations. These studies are summarized below.

The glomerular filtration barrier of the hagfish consists of different cell populations, basement membrane, extracellular matrix and anionic charges. This structural similarity with the filtration barrier of higher vertebrates indicates that the filtration characteristics for plasma-derived macromolecules, such as proteins, also follows basic mechanisms found in other vertebrates. The dominant protein found in hagfish urine, detected following electrophoretic separation of the urinary proteins, had the same electrophoretic mobility as human albumin (Alt *et al.*, 1976). When several glomeruli were perfused with a Ringer solution containing bovine albumin, an average ultrafiltrate/perfusate ratio (sieving coefficient,φ) of 0.01 to 0.02 was determined for this protein. The sieving coefficient for globulins was lower, showing that there is a size selectivity (Rost *et al.*, 1983). The effects of toxins on the filtration of albumin was further studied *in vitro* on isolated perfused glomeruli of *Myxine* (see below).

23.3.2 Function of the archinephric duct

The AND of the hagfish has morphologically been compared with the proximal tubule of higher vertebrates (see above). The main function of the proximal tubule of higher vertebrates is the isosmotic reabsorption of fluid. The driving force is an active uptake of NaCl. Although renal sodium reabsorption has been described in some experiments with *Eptatretus stoutii* (McInerney, 1974), the general understanding is that there is no net sodium, chloride and fluid reabsorption in the AND of *Myxine glutinosa* (Rall and Burger, 1967; Munz and McFarland, 1974;

Stolte and Schmidt-Nielsen, 1978). The passive permeability of the AND for sodium, measured *in vitro* as passive sodium influx in isolated segments, is between 5 and 10×10^{-10} mol cm^{-2} s^{-1} (Raguse-Degener, 1988; Fels *et al.*, 1989b). The corresponding values for mammals are much higher. In the rat and rabbit they amount to 315 and 109×10^{-10} mol cm^{-2} s^{-1}, respectively (Ullrich 1964; Kokko, 1971). Measurements in other aquatic vertebrates have not been performed.

The passive efflux and influx of water across the epithelium of the AND shows how permeable this barrier is. The Lp calculated from water efflux was $0.12 \pm 0.01 \times 10^{-10}$ cm s^{-1} Pa^{-1}. For the influx it was slightly higher (Raguse-Degener, 1988; Fels *et al.*, 1989b). The Lp values of tubular segments studied in other species are higher. For the proximal tubules of *Necturus maculosus* it was between 0.15 and 0.46×10^{-10} cm s^{-1} Pa^{-1} (Windhager, 1968; Bentzel *et al.*, 1969). For the rat, which like all mammals has a high volume flux across the epithelium of the proximal tubule, it was 17.8 or 19.6×10^{-10} cm s^{-1} Pa^{-1} (Ullrich *et al.*, 1964; Stolte *et al.*, 1968).

23.4 THE HAGFISH KIDNEY AS A MODEL FOR PHYSIOLOGY, PATHOPHYSIOLOGY AND TOXICOLOGY

The kidney of the hagfish can be used to study physiological and pathophysiological processes *in vivo* and *in vitro*. The specific anatomical features of the hagfish kidney (large, segmentally arranged glomeruli and the existence of two distinct ANDs) make it comparatively easy to dissect nephron segments from these species. Hagfishes are not the only aquatic animals that have been used as models to gain insight into renal function of higher vertebrates. For example, tubules of flounders or the aglomerular kidneys of certain teleosts have been used to study basic mechanisms of renal function (Forster, 1975).

A variety of functional and metabolic studies have been performed with isolated glomeruli or segments of the AND of hagfishes. These studies followed different aims, such as an evaluation of glomerular haemodynamics or permeability for macromolecules following exposure to hormones or heavy metals. Another approach was the *in vitro* characterization of lysosomal enzymes in the AND and the effects of heavy metals. Other experiments addressed the hypothesis that isolated glomeruli of the hagfish can be used as a model to study glomerular metabolism (e.g. protein synthesis and degradation) of vertebrates.

23.4.1 Functional studies

Effects of hormones on glomerular filtration

The effects of various eicosanoids and the prostaglandin synthetase inhibitor indomethacin on blood pressure, urine flow and electrolyte balance was studied *in situ* in *Myxine glutinosa* under light anaesthesia (Wales, 1988). Arachidonic acid, prostaglandin E_1, E_2, A_2, and thromboxane B_2 caused a reduction in urine output. Neither the vasopressor prostaglandin $F_{2\alpha}$ nor the prostaglandin synthetase inhibitor indomethacin had a significant effect on urine production. The hypothesis was put forward that the animals were already filtering at a maximum rate prior to the increase in blood pressure. The eicosanoids did not alter the electrolyte status of *Myxine* (Wales, 1988).

Most studies on glomerular function, such as the one reported above, applied an *in situ* perfusion system. A similar methodological approach was chosen by Riegel (1978) or Stolte and Schmidt-Neilsen (1978) for studies that showed that increased pressures in the arterial or the venous system lead to increases in glomerular filtration. In the experimental set-up of an *in situ* perfusion, possible confounding factors, such as neural stimulation or the action of hormones, cannot always be excluded. Therefore a design to study the characteristics of single isolated glomeruli of

Myxine in vitro was developed. In short, this experimental set-up consists of a cannula and a catheter inserted into the dorsal aorta and one AND of the anaesthetized animal. The kidney can then be perfused with a Ringer solution and the ultrafiltrate can be collected in the catheter. Pressure transducers connected with the cannula and catheter allow the recording of pressure changes in the afferent vessels and the catheter in the AND. Ligatures can be applied to perfuse only one glomerulus. The whole arrangement of a single glomerulus, cannula and catheter can then be transferred into a cooled bathing chamber. In this chamber the isolated glomerulus can be perfused with a defined medium and the ultrafiltrate can be collected in the catheter for biochemical analysis. The effect of substances can be tested by adding them to the perfusate or the bathing medium (Fels *et al.*, 1993).

When glomeruli were perfused in this *in vitro* set-up, a filtration coefficient (K_f) of 0.189 nl s^{-1} $mmHg^{-1}$ was determined for the glomerular capillaries of *Myxine glutinosa*. The perfusate was colloid free, therefore a lower K_f value is to be expected for an *in vivo* situation, when blood containing plasma proteins flows through the glomerular capillaries. The K_f values for lamprey *Lampetra fluviatilis* and the urodele *Necturus* were reported to be 0.028 and 0.10 nl s^{-1} $mmHg^{-1}$, respectively (Renkin and Gilmore, 1973; McVicar and Rankin, 1985). Based on the morphometric measurements of the glomerular filtering surface a hydraulic conductivity (Lp) of 0.618 µl min^{-1} $mmHg^{-1}$ cm^{-2} (7.7×10^{-8} cm s^{-1} Pa^{-1}) was determined for the glomerular filtration barrier of *Myxine* (Fels *et al.*, 1993).

Adrenaline added to the bathing medium led to a higher pressure (P) measured in the afferent vessels of the glomerulus. Consequently, increases in the filtration rate of the single glomerulus (SNGFR) and the filtration fraction (FF) were observed (Table 23.1). K_f remained unchanged (Fels *et al.*, 1993). Noradrenaline had the same effects. A characterization of receptor types revealed that the effect of the catecholamines on kidney function is mediated by α- and β-receptors (Fels *et al.*, 1987). The existence of these receptor types had previously been shown on smooth muscle cells of the ventral hagfish aorta (Reite, 1969).

The atrial natriuretic peptide (ANP) is another hormone that has received considerable attention in renal physiology over recent years. Binding sites for ANP have been demonstrated in kidney and aorta of the hagfish (Kloas *et al.*, 1988; Toop *et al.*, 1995). ANP isolated from hearts and brains of *Myxine* led to a vasodilatation of arterial stripes of the rat preconstricted with angiotensin. However, this ANP did not affect the haemodynamics of isolated perfused glomeruli of *Myxine*. Neither a direct effect on the haemodynamics of isolated glomeruli nor an effect on the

Table 23.1 Changes of pressure in the afferent vessels (P), single nephron glomerular filtration rate (SNGFR), filtration fraction (FF) and filtration coefficient (Kf) following application of adrenaline in the single isolated perfused glomerulus of *Myxine glutinosa*. (Adapted from Fels *et al.*, 1993)

	P (mmHg)	SNGFR (nl min^{-1})	FF (%)	K_f (nl s^{-1} × $mmHg^{-1}$)
Controls[(1)]	4.9 ± 1.2	63 ± 19	13 ± 4	0.236 ± 0.086
Adrenaline[(2)]	7.5 ± 0.5*	110 ± 30*	24 ± 7*	0.305 ± 0.145

Values are the mean ± SD (N = 14), *$p < 0.01$.
[(1)] Ringer solution as a bathing medium.
[(2)] Ringer solution with a concentration of 9×10^{-5} M adrenaline.

haemodynamics of glomerular preparations pretreated with angiotensin could be observed (Fels, unpublished data).

Changes in glomerular function induced by xenobiotics

Cadmium has long been considered to be a toxin damaging the tubular apparatus of kidneys. With growing evidence that the glomerulus is also affected by this heavy metal (Aughey *et al.*, 1984), studies were undertaken to investigate whether there is an accumulation of cadmium in the kidney of the hagfish and whether this affects glomerular permeability for plasma proteins (Fels *et al.*, 1989a).

Hagfishes were treated with 1, 10 or 20 mg $CdCl_2$ kg^{-1} body weight dissolved in saline and injected into the subcutaneous blood sinus. Accumulation of cadmium in different organs was analysed one week later. There was a predominant accumulation in the liver, but concentrations in renal tissue also increased significantly (Table 23.2). Cadmium did not only accumulate in the AND but also in the glomerulus (Figure 23.3). This finding indicates that the glomerulus can indeed be a target of cadmium. At the functional level there was evidence of a cadmium-induced change in glomerular permeability to proteins. The sieving coefficient (φ) for albumin, determined in the *in vitro* preparations of the isolated perfused glomerulus described above, was 0.02 ± 0.01 for control animals. There was a wide variation in φ determined for glomeruli of cadmium-treated animals (Table 23.2). When the normal range of φ was arbitrarily defined as mean ± 2 S.D. of the controls, the percentage of values exceeding the normal range was significantly higher in cadmium-treated animals. A dose–response, that is a correlation of φ with dose of

Table 23.2 Cadmium concentrations of different organs of the hagfish *Myxine glutinosa* and glomerular sieving coefficient (φ) for albumin following a single treatment with 1, 10 or 20 mg $CdCl_2$ kg^{-1} body weight

Dose of cadmium injected (mg kg^{-1} body weight)	Cadmium concentration* (μg g^{-1}) Plasma	AND	Liver	Intestine	Muscle
0	$<0.5 \times 10^{-3}$ (*N* = 9)	0.31 ± 0.33 (*N* = 6)	0.26 ± 0.20 (*N* = 6)	0.33 ± 0.22 (*N* = 4)	<0.03 (*N* = 5)
1	2.60 ± 1.40 (*N* = 5)	2.32 ± 1.40 (*N* = 6)	3.04 ± 0.50 (*N* = 5)	n.d.	0.48 ± 0.19 (*N* = 6)
10	3.94 ± 1.07 (*N* = 5)	15.5 ± 4.26 (*N* = 6)	20.6 ± 4.92 (*N* = 16)	1.81 ± 0.15 (*N* = 2)	1.58 ± 0.64 (*N* = 17)
20	58.3 ± 14.7 (*N* = 12)	42.5 ± 16.6 (*N* = 11)	44.1 ± 17.4 (*N* = 12)	16.6 ± 5.29 (*N* = 6)	10.9 ± 4.80 (*N* = 12)

Dose of cadmium injected (mg kg^{-1} body weight)	Sieving coefficient (φ) median (minimum–maximum)	% exceeding normal reference range of φ†
0	0.02 (0.01–0.09) (*N* = 9)	11
1	0.04 (0.01–0.26) (*N* = 5)	50
10	0.04 (0.01–0.19) (*N* = 4)	50
20	0.04 (0.01–0.16) (*N* = 9)	44

* Concentrations were related to g wet weight of material analysed.
† Reference range defined as $\bar{x} \pm 2$ S.D.
$\bar{x}$ S.D.; N = number of animals; n.d. = not determined

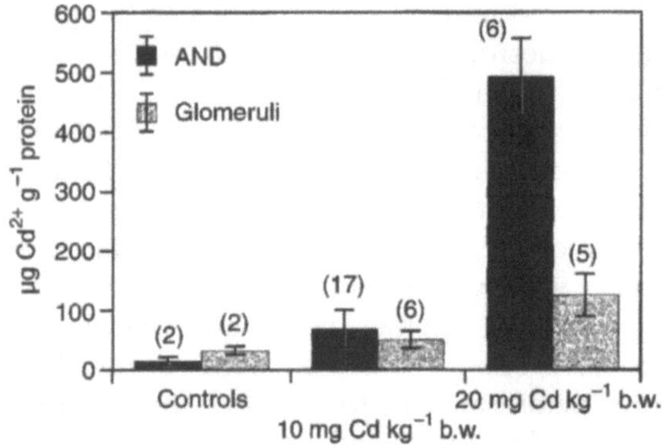

Figure 23.3 Concentration of cadmium (Cd^{2+}) in the archinephric duct (AND) and in glomeruli of *Myxine glutinosa*. Concentrations in untreated control animals or following a single treatment with 10 or 20 mg $CdCl_2$ kg^{-1} body weight (b.w.). Concentrations are expressed in µg g^{-1} protein of dissolved tissue. Numbers in parentheses indicate number of experiments performed, mean ± S.D.

cadmium given, could however not be shown. Using isolated glomeruli of the hagfish, it could be demonstrated at a biochemical and functional level that glomeruli can be a target of cadmium.

The glomeruli of the hagfish have also been used to study the mechanisms of toxicity of drugs. The anthracycline antibiotic Adriamycin (ADR) is a potent cancer chemotherapeutic agent, but its clinical applicability is limited because of toxic side effects. In rodents a nephrotic syndrome can be induced with a single injection of this drug. The generation of free radicals is one mechanism discussed for Adriamycin nephrotoxicity (Bertani *et al.*, 1982). Adriamycin administered to hagfishes induced an increase in the sieving coefficient (φ) for albumin (0.013 in control animals, 0.081 following treatment with Adriamycin) and a decrease in the glomerular hydraulic conductivity (Lp) for water (Barbey *et al.*, 1989). These *in vitro* findings are in accordance with experiments on rats that showed a proteinuria and a decline in glomerular filtration rate induced by Adriamycin (Bertani *et al.*, 1982). When hagfishes were given Adriamycin and *N*-acetyl-cysteine, a scavenger of free radicals, the increase in Lp could be prevented, whereas φ was still found to be elevated (Barbey *et al.*, 1989). These experiments showed that the toxicity of the drug investigated can be ameliorated and that water permeability of the glomerular capillaries is affected by oxidative stress, whereas changes in protein permeability apparently are affected by other or further pathomechanisms.

The AND as a model for tubular toxicity

Properties of the AND that underline its functional similarity mainly with the tubular apparatus of higher vertebrates include the existence of sodium cotransport systems (Flöge *et al.*, 1984), secretion and reabsorption of divalent ions such as calcium and magnesium (Alt *et al.*, 1980; Munz and McFarland, 1964; Morris, 1965) potassium, phosphate and sulphate secretion (Munz and McFarland, 1964; Fänge, 1963; Rall and Burger, 1967), urea diffusion and ammonium secretion (Fels *et al.*, 1989b; Raguse-Degener *et al.*, 1980), glucose and amino acid reabsorption (Alt *et al.*, 1980), and pinocytotic uptake of proteins (Ericson and Seljelid, 1968).

Another similarity is enzymes expressed in the cells of the AND. The lysosomal cysteine proteinases Cathepsin B and L and acid phosphatase (AcPase) have been demonstrated in the AND of hagfishes and the effects of environmental pollutants have been investigated *in vitro*.

Segments of the AND were perfused with Ringer solution via the aorta to remove the blood. AND tissue was dissected and homogenized. Lysosomes were cracked with hypoosmotic treatment and shock freezing. Other phosphatases occurring in the kidney of vertebrates (glucose-6-phosphatase, alkaline phosphatase) and known to use the substrate eventually applied in the kinetic assay were inhibited. AcPase was determined with 4-methylumbelliferyl-phosphate as a substrate

(Olbricht *et al.*, 1984). Enzyme activity was related to the protein content of the AND homogenates. The enzyme assay for AcPase showed a K_m of 0.212 mmol l^{-1} and a V_{max} of 0.66 μM min^{-1} mg^{-1} protein. Lead, copper, cadmium, aluminium and zinc had no effects on the activity of AcPase (concentration range 10^{-3} to 10^{-9} M). $HgCl_2$ inhibited AcPase activity by 60% at a concentration above 10^{-7} M l^{-1} (R. Sievers, Medical School Hannover, pers. communication).

23.4.2 Metabolic studies

Morphometric, biochemical and metabolic characterization of isolated glomeruli of the hagfish

Metabolic changes may precede or even cause functional and morphological alterations of the glomerulus or be a consequence of those (Brendel and Meezan, 1973). Therefore, the metabolism of the glomerulus is of importance if the causes and effects of glomerular disturbances are to be better understood.

The understanding of glomerular metabolism is mainly derived from studies with isolated glomeruli. As a multicellular *in vitro* system isolated glomeruli have the advantage that endothelial, epithelial and mesangial cells are still located in their morphological integrity. Therefore, compared to the culture of single cells, cultured isolated glomeruli are closer to the *in vivo* situation since they represent the balanced metabolic interactions of all three cell types composing the glomerular filtration barrier. In nearly all studies on glomerular metabolism isolated and *in vitro* incubated glomeruli of mammalian kidneys (man, rabbit, rats) have been used (for review see Dousa, 1985; Foidart *et al.*, 1981), whereas glomeruli of lower vertebrates have not been subject to metabolic studies.

Single glomeruli of *Myxine* can easily be obtained by microdissection and cultured *in vitro* for metabolic studies, lending the animal to investigations of glomerular processes.

The basis for the validation of isolated glomeruli of *Myxine glutinosa* as an alternative *in vitro* system is the morphometric and biochemical characterization of the glomeruli and the comparison to glomerular parameters of the established experimental animals. The morphometric characterization of the glomeruli of *Myxine glutinosa* reveals a mean glomerular diameter of 700 μm (Fels *et al.*, 1993), resulting in a geometric volume of 180 nl per glomerulus. The percentage of extracellular volume, estimated as inulin space, of the total glomerular geometric volume amounts to 21%. Compared to mammalian glomeruli, where the percentages range between 61% and 84%, the extracellular space is relatively small in glomeruli of *Myxine glutinosa* (Table 23.3) (Kastner *et al.*, 1994b). Since the glomeruli of *Myxine glutinosa* are described to be very rich in mesangium and mesangial matrix (Figure 23.2) the relatively small glomerular extracellular space could be due to extracellularly accumulated material produced, e.g., by the large number of mesangial cells, causing a reduction of the extracellular volume. Due to the large glomerular size of *Myxine glutinosa*, the protein content is about 70-fold higher and the DNA content about 50-fold higher than in rat glomeruli. The ratio of protein to DNA is slightly higher in glomeruli of *Myxine glutinosa* than in rat glomeruli (8.1 versus 5.9, respectively) (Table 23.4). This high ratio could also be due to the high prevalence of mesangial material and the small extracellular space observed.

In the pathogenesis of various glomerulopathies the accumulation of protein in the glomerulus seems to be one of the major features. Protein accumulation is directly related to the glomerular protein balance, depending either on synthesis or deposition or on the degradation of these proteins (Klahr *et al.*, 1988). These accumulated proteins, either proteins of the extracellular matrix and/or of the glomerular basement membrane, subsequently cause alterations of the glomerular filtration barrier. Therefore, protein synthesis measured as the incorporation of radiolabelled

Table 23.3 Geometrically and isotopically estimated glomerular volumes of *Myxine glutinosa* in comparison to mammals

	Glomerular diameter (μm glomerulus^{-1})	Geometric volume (nl glomerulus^{-1})	Inulin space (nl glomerulus^{-1})	Extracellular space/ glomerular volume (%)
Myxine glutinosa	700 ± 220[1]	180*	38.5 ± 2.62 (N = 31)	21
Rat[2]	135.8	1.31	0.85	65
Rabbit[2]	151.5	1.82	1.11	61
Dog[2]	197.4	4.13	3.46	84
Man[3]	227.0	6.12*	n.d.	n.d.

* Calculated using the mean glomerular diameter (D) and the formula for the volume of a sphere $V = 4/3\,\pi\,(D/2)^3$.
Values are means ± SEM. N = number of measurements. n.d. = not determined.
(1) Fels *et al* (1993); (2) Savin and Terreros (1981); (3) Savin (1983).

Table 23.4 Glomerular protein and DNA contents of *Myxine glutinosa* in comparison to rat glomeruli. (From Kastner *et al*., 1994b)

	Protein content (μg glomerulus^{-1})	DNA content (ng glomerulus^{-1})	Protein/DNA ratio
Myxine glutinosa	3.56 ± 0.09 (N = 228)	437.2 ± 20.0 (N = 76)	8.1
Rat	0.049 ± 0.003 (N = 20)	8.29 ± 0.44 (N = 20)	5.9

Values are means ± SEM. N = number of glomeruli studied.

amino acids into glomerular acid precipitable proteins was chosen as the most important parameter to be determined in isolated glomeruli of *Myxine glutinosa* and to be compared to the conventional *in vitro* system of isolated glomeruli from the rat kidney.

To establish isolated glomeruli as an alternative *in vitro* system, culture medium and culture conditions were adjusted to the requirements of *Myxine glutinosa* to ensure an *in vitro* viability of the isolated glomeruli for at least 12 hours (Kastner *et al*., 1994b). The glomeruli are incubated under gentle shaking at 4–6°C in an incubation medium which is a *Myxine*–Ringer solution (Riegel, 1978) supplemented with components as described by Eagle (1959).

As shown in Table 23.5 the metabolic properties of isolated glomeruli of *Myxine glutinosa* are comparable to those of rat glomeruli. The glomeruli are viable *in vitro* for at least 12 hours. Mammalian glomeruli cultured under similar conditions have been viable for up to 20 hours (Krisko and Walker, 1976). The incorporation rates of radiolabelled amino acids are in the same range as the data obtained from rat glomeruli, with respect to glomerular protein as well as glomerular DNA (Table 23.5). Since the glomeruli of *Myxine glutinosa* reveal comparable morphometric, biochemical and metabolic properties, they represent a model well suited for the study of basic phenomena in glomerular metabolism which may be applicable to other vertebrates as well.

Isolated glomeruli of the hagfish as a tool for pharmaco-toxicological studies

To support a causal relationship between functional changes and metabolic alterations, the effects of ADR on glomerular protein metabolism were determined and related to functional changes described in the previous section.

Table 23.5 Amino acid incorporation into isolated glomeruli of *Myxine glutinosa* in comparison to rat glomeruli. (From Kastner *et al.*, 1994b)

	Amino acid incorporation			
	(DPM/µg protein)		(DPM/µg DNA)	
	4 hours	6 hours	4 hours	6 hours
Myxine glutinosa	140 ± 10 (*N* = 27)	250 ± 22 (*N* = 27)	1091 ± 98 (*N* = 16)	1908 ± 245 (*N* = 16)
Rat	214 ± 13 (*N* = 16)	362 ± 25 (*N* = 14)	1340 ± 84 (*N* = 16)	2282 ± 184 (*N* = 14)

Values are means ± SEM. *N* = number of glomeruli studied.

Compared to untreated animals, treatment of the animal with ADR prior to the dissection of the glomeruli leads to an increase in the incorporation of amino acids into glomerular proteins *in vitro*. After 12 hours' incubation the incorporation rate of radiolabelled amino acids is about 150% of corresponding control values (Figure 23.4). These data are compatible with findings in ADR-treated rats (Kastner *et al.*, 1991). The total uptake of amino acids into glomerular intra- and extracellular spaces, was however significantly reduced in glomeruli of ADR-treated hagfishes.

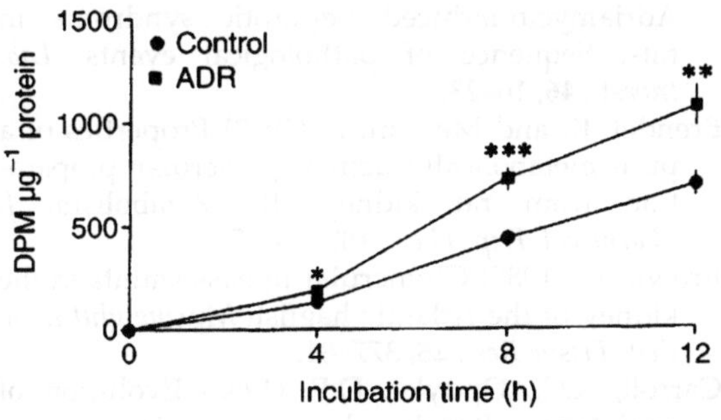

Figure 23.4 Protein synthesis of isolated glomeruli of *Myxine glutinosa*. Glomeruli of controls and animals treated with Adriamycin (ADR) were incubated *in vitro* with ^{3}H amino acids for 12 hours to determine *de novo* protein synthesis. Results are expressed as DMP µg^{-1} protein. Each point is mean±SEM of N = 27 measurements. * p<0.05, ** p<0.01, *** p<0.001 Adriamycin versus controls. (From Kastner *et al.*, 1994b, with permission.)

To elucidate which of the postulated pathomechanisms (free radicals or DNA-interactions) are responsible for the increased protein synthesis and the decreased total glomerular amino acid uptake, animals were treated simultaneously with ADR and the sulphydryl donor *N*-acetylcysteine. The combined treatment of ADR and NAC does not reduce the ADR-induced stimulating effect on protein synthesis. In contrast, the inhibition of amino acid uptake into the glomerulus caused by ADR can be prevented by NAC. It is therefore concluded that (1) ADR has different targets sites within the cell and that (2) different ADR-pathomechanisms are responsible for increased protein synthesis and the decreased amino acid uptake. The increase in protein synthesis is not mediated by free radicals and must therefore be due to other mechanisms. The pathomechanism responsible for the increased protein synthesis is suggested to be rather linked to ADR–DNA interactions or to an impaired balance of protein synthesis and proteolytic degradation. If the balance of glomerular protein synthesis and degradation is disturbed this could lead to an increased net-protein synthesis and finally cause an enrichment of protein in the glomerulus.

This hypothesis is supported by further studies showing decreased proteolytic activities in glomerular homogenates of ADR-treated *Myxine glutinosa* (Kastner *et al.*, 1995).

Relating the metabolic data to the functional changes: ADR-treatment of *Myxine glutinosa* causes a significant increase in protein permeability whereas water permeability, estimated as hydraulic conductivity, is significantly reduced compared to control animals. The decrease in water permeability can be compensated by the sulphydryl donor NAC, whereas NAC has no effect on the elevated protein permeability (Barbey *et al.*, 1989). The decrease in water permeability and amino acid uptake is probably due to the same pathomechanism, presumably free radical mediated damages to epithelial membranes. The postulated membrane damage is further substantiated by elevated levels of malondialdehyde, the final product of lipid peroxidation, detected in the kidney, liver and heart of ADR-treated *Myxine glutinosa* (Barbey *et al.*, 1989). The increase in protein permeability is best explained by a sieving defect of the glomerular basement membrane. As the drug-induced disorders could not be prevented by NAC, the sieving defect is not mediated by free radicals. Instead, it is suggested that ADR directly interferes with the metabolism of the cells composing the filtration barrier for proteins by altering the metabolism of glomerular structural components, which is supported by an increased net protein synthesis due to a decreased proteolytic degradation.

In summary, it could be demonstrated with the *ex vivo/in vitro* studies with isolated glomeruli that ADR has different target sites within the cell such as the cell membranes and the nucleus. It could be further shown that two different ADR pathomechanisms, free radicals and DNA interactions, are responsible for the progression of ADR glomerulopathy. Adriamycin can act *via* these different pathomechanisms thus causing interactions with membranes, enzymes and macromolecules which in the long run might be responsible for functional disturbances of the glomerular filtration barrier.

REFERENCES

Adam, H., Schooner, H. and Walvig, F. (1962) Versuche zur Narkose und Relaxation von *Myxine glutinosa* L. (Cyclostomata, Vertebrata). *Zool. Anz.*, **168**, 217–28.

Albrecht, U., Lametschwandtner, H. and Adam, H. (1978) Die Blutgefäße des Mesonephros von Myxine glutinosa L: Eine rasterelektronenmikroskopische Untersuchung an Korrosionspräparaten. *Zool. Anz. Jena* **200**, 300–8.

Alt, J.M., Raguse-Degener, G., Niermann, U. *et al.* (1976) Renal excretion of protein and nitrogen endproducts in the hagfish, *Myxine glutinosa. Bull. Mt. Desert Isl. Biol. Lab.*, **1**, 1–2.

Alt, J.M., Stolte, H., Eisenbach, G.M. and Walvig, F. (1980) Renal electrolyte and fluid excretion in the Atlantic hagfish, *Myxine glutinosa. J. Exp. Biol.*, **91**, 323–30.

Axelsson, M., Farrell, A.P. and Nilsson, S. (1990) Effects of hypoxia and drugs on the cardiovascular dynamics of the Atlantic hagfish *Myxine glutinosa. J. Exp. Biol.*, **151**, 297–316.

Barbey, M.M., Fels, L.M., Soose, M. *et al.* (1989) Adriamycin effects glomerular renal function: evidence for involvement of oxygen radicals. *Free Rad. Res. Comms.*, **7**, 195–203.

Bentzel, C.J., Parsa, D. and Hare, D.K. (1968) Osmotic flow across proximal tubule of Necturus: correlation of physiologic and anatomic studies. *Amer. J. Physiol.*, **217**, 570–80.

Bertani, T., Poggi, A., Pozzini, R. *et al.* (1982) Adriamycin-induced nephrotic syndrome in rats. Sequence of pathological events. *Lab. Invest.*, **46**, 16–23.

Brendel, K. and Meezan, E. (1973) Properties of a pure metabolically active glomerular preparation from rat kidney. II. Metabolism. *J. Pharmacol. Exp. Ther.*, **187**, 342–51.

Brown, A. (1988) Glomerular bypass shunts in the kidney of the Atlantic hagfish *Myxine glutinosa. Cell. Tissue Res.*, **25**, 377–81.

Carroll, R.C., Opdyke, D.F. (1982) Evolution of angiotensin II-induced catecholamine release. *Amer. J. Physiol.*, **247**, F352–64.

Chapman, C.B., Jensen, D., Wildenthal, K. (1963) On circulatory control mechanisms in the Pacific hagfish. *Circulation Res.*, **12**, 427–40.

Davies, M., Martin, J., Thomas, G.J. *et al.* (1987) Degradation of glomerular extracellular matrices, in *Renal Basement Membranes in Health and Disease* (eds R.G. Price and W.G. Hudson), Academic Press, London, New York, pp. 181–201.

Decker, B. and Reale, E. (1991) The glomerular filtration barrier of the kidney in seven vertebrates classes. Comparative morphological and histochemical observations. *Eur. J. Basic Appl. Histochem.*, **35**, 15–36.

Dousa, T.P. (1985) Glomerular metabolism, in *The Kidney: Physiology and Pathophysiology* (eds S.W. Seldin and G. Giebisch), Raven Press, New York, pp. 645–67.

Eagle, H. (1959) Amino acid metabolism in mammalian cell cultures. *Science*, **130**, 432–7.

Eisenbach, G.M., Weise, R., Hanke, K. *et al.* (1971) Renal handling of protein in the hagfish, *Myxine glutinosa. Bull. Mt Desert Isl. Biol. Lab.*, **6**, 11–15.

Eisenberger, U., Koob-Emund, L. and Stolte, H. (1994) Identification of collagen type IV in the glomerulus of the Atlantic hagfish *Myxine glutinosa. Bull. Mt Desert Isl. Biol. Lab.*, **33**, 24–5.

Elger, B., Rühs, H. and Hentschel, H. (1988) Glomerular permselectivity to serum proteins in rainbow trout (*Salmo gairdneri*). *Amer. J. Physiol.*, **255**, R418–23.

Ericson, J.L.E. and Seljelid, R. (1968) Endocytosis in the ureteric duct epithelium of the hagfish (*Myxine glutinosa* L.). *Z. Zellforsch. Mikrosk. Anat.*, **90**, 263–72.

Fänge, R. (1963) Function and structure of the excretory organs of myxinoids, in *The Biology of Myxine* (eds A. Brodal and R. Fänge), Universitetsforlaget, Oslo, pp. 516–29.

Fels, L.M., Elger, B. and Stolte, H. (1987) Effects of catecholamines on the function of single glomeruli of *Myxine glutinosa* L. (Cyclostomata), in *Proceedings of the German Zoological Society, 80th Meeting in Ulm* (eds F.G. Barth and E.A. Seyfarth), Gustav Fischer Verlag, Stuttgart, p. 162.

Fels, L.M., Barbey, M.M., Elger, B. *et al.* (1989a) The effects of cadmium and Adriamycin on the isolated perfused glomerulus of *Myxine glutinosa (Cyclostomata)*, in *Nephrotoxicity – In vitro to in vivo – Animals to man* (eds P.H. Bach and E.A. Lock), Plenum Press, New York, pp. 75–9.

Fels, L.M., Raguse-Degener, G. and Stolte, H. (1989b) The Archinephron of *Myxine glutinosa* L., in *Comparative Physiology. Structure and Function of the Kidney* (ed. R.K.H. Kinne), Karger, Basel, pp. 73–102.

Fels, L.M., Sanz-Altamira, P.M., Decker, B. *et al.* (1993) Filtration characteristics of the single isolated perfused glomerulus of *Myxine glutinosa. Renal Physiol. Biochem.*, **16**, 276–84.

Fels, L.M., Decker, B., Stolte, H. (1994) Proteoglycans in the filtration barrier of the Atlantic hagfish *Myxine glutinosa. Bull. Mt Desert Isl. Biol. Lab.*, **33**, 22–3.

Flöge, J., Stolte, H. and Kinne, R. (1984) Presence of a sodium-dependent d-glucose transport system in the kidney of the Atlantic hagfish (*Myxine glutinosa*). *J. Comp. Physiol.* B, **15**, 355–64.

Foidar, J.B., Dechenne, C., Dubois, C. *et al.* (1981) Tissue culture of isolated renal glomeruli: present and future, in *Advances in Nephrology* (eds J. Hamburger, J. Crosnier, J.P. Grunfeld and M.H. Maxwell), Yearbook Medical Publishers Inc., Chicago, pp. 267–92.

Forster, R.P. (1975) Structure and function of aglomerular kidneys, in *Excretion, Fortschr. Zool. 23* (ed. A. Wessing), Fischer, Stuttgart, pp. 232–47.

Forster, M.E., Davison, W., Axelsson, M. and Farrel, A.B. (1992) Cardiovascular responses to hypoxia in the hagfish, *Eptatretus cirrhatus. Respir. Physiol.*, **88**, 376–86.

Heath-Eves, M.J. and McMillan, D. (1974) The morphology of the kidney of the Atlantic hagfish, *Myxine glutinosa* (L.). *Amer. J. Anat.*, **139**, 309–34.

Hickman, C.P. and Trump, B. (1969) The kidney, in *Fish Physiology*, Vol. 1 (eds W.S. Hoar and D.J. Randall), Academic Press, London, New York, pp. 93–100.

Holmgren, N. (1950) On the pronephros and blood in *Myxine glutinosa. Acta Zool. Stockholm*, **3**, 233–48.

Isujii, T., Naito, I., Utika, T. *et al.* (1984) The anionic barrier system of the mesonephric renal glomerulus of the brown hagfish, *Paramyxine atami. Anat. Rec.*, **208**, 337–47.

Johansen, K. (1960) Circulation of the hagfish, *Myxine glutinosa. L. Biol. Bull. Mar. Biol. Lab. Lanc.*, **118**, 289–95.

Kastner, S., Fels, L.M., Piippo, S. and Stolte, H. (1994b) Isolated glomeruli of the Atlantic hagfish *Myxine glutinosa* as an alternative *in vitro* model to study glomerular metabolism in pharmaco-toxicology of anticancer drugs. *Comp. Physiol. Biochem.*, **180C**, 349–57.

Kastner, S., Fels, L.M., Koob-Emunds, L., Piippo, S. and Stolte, H. (1995) Effects of Adriamycin on the balance of glomerular protein metabolism: studies on isolated glomeruli of the Atlantic hagfish *Myxine glutinosa. Cell. Physiol. Biochem.*, **5**, 399–407.

Kastner, S., Koob-Emunds, L. and Stolte, H. (1994a)

Extracellular matrix turnover in glomeruli of *Myxine glutinosa*: *In vitro* pharmaco-toxicology studies. *Bull. Mt Desert Isl. Bio. Lab.*, **33**, 19–21.

Kastner, S., Wilks, M.F., Gwinner, W. *et al.* (1991) Metabolic heterogeneity of isolated cortical and juxtamedullary glomeruli in Adriamycin nephrotoxicity. *Renal Physiol. Biochem.*, **14**, 48–54.

Klahr, S., Schreiner, G. and Ichikawa, I. (1988) The progression of renal disease. *N. Engl. J. Med.*, **318**, 1657–66.

Kloas, W., Flügge, G., Fuchs, E. and Stolte, H. (1988) Binding sites for atrial natriuretic peptide in the kidney and aorta of the hagfish (*Myxine glutinosa*). *Comp Biochem. Physiol.*, **91**, 685–8.

Kokko, J.P., Burg, M.B. and Orloff, J. (1971) Characteristics of NaCl and water transport in the renal proximal tubule. *J. Clin. Invest.*, **50**, 69–76.

Kühn, K., Stolte, H. and Reale, E. (1975) The fine structure of the kidney of the hagfish (*Myxine glutinosa* L.): a thin section and freeze-fracture study. *Cell Tissue Res.*, **164**, 201–13.

Logan, A.G., Morris, R. and Rankin, J.C. (1980) A micropunture study of kidney function in the river lamprey *Lempetra fluviatilis* adapted to sea water. *J. Exp. Biol.*, **88**, 239–47.

McInerney, J.E. (1974) Renal sodium reabsorption in the hagfish, *Eptatretus stouti. Comp. Biochem. Physiol.*, **49**, 273–80.

McVicar, A.J. and Rankin, J.C. (1985) Dynamics of glomerular filtration in the river lamprey, *Lampetra fluviatilis L. Amer. J. Physiol.*, **249**, F132–8.

Morris, R. (1965) Studies on salt and water balance in *Myxine glutinosa* (L.). *J. Exp. Biol.*, **42**, 359–71.

Müller, W. (1875) Über das Urogenitalsystem des Amphioxus und der Cyclostomen. *Jena Z. Naturw.*, **9**, 94–129.

Munz, F.W. and McFarland, W.N. (1964) Regulatory function of a primitive vertebrate kidney. *Comp. Biochem. Physiol.*, **13**, 381–400.

Olbricht, C.J., Garg, L.C., Cannon, J.K. and Tisher, C.C. (1984) Acid phosphatase activity in the mammalian nephron. *Amer. J. Physiol.*, **247**, F252–9.

Raguse-Degener, G., Pietschmann, M., Walvig, F. and Stolte, H. (1980) Excretory systems in the hagfish *Myxine glutinosa*, in *Research Animals and Experimental Design in Nephrology. Contributions to Nephrology*, vol. 19 (eds G.M. Berlyne, S. Giovanetti and S. Thomas), Karger, Basel, pp. 1–8.

Raguse-Degener, G. (1988) Untersuchungen zur Funktion des Urnierengangs von *Myxine glutinosa*. Thesis, University of Hannover.

Rall, D.P. and Burger, J.W. (1967) Some aspects of renal secretion in *Myxine. Amer. J. Physiol.*, **212**, 354–6.

Reale, E., Luciano, L., Kühn, K. *et al.* (1981) Glomerular basement membrane and mesangial matrix: a comparative study in different vertebrates. *Renal Physiol.*, **4**, 85–9.

Reite, O.B. (1969) The evolution of vascular smooth muscle responses to histamine and 5-hydroxytryptamine. I. Occurrence of stimulatory actions in fish. *Acta Physiol. Scand.*, **75**, 221–39.

Renkin, E.M. and Gilmore, J.P. (1973) Glomerular filtration, in *Handbook of Physiology. Section 8: Renal Physiology* (eds J. Orloff and R.W. Berliner), American Physiological Society, Washington, pp. 185–248.

Riegel, J.A. (1978) Factors affecting glomerular function in the Pacific hagfish *Eptatretus stouti* (Lockington). *J. Exp. Biol.*, **73**, 261–77.

Riegel, J.A. (1986a) Hydrostatic pressures in glomeruli and renal vasculature of the hagfish *Eptatretus stouti. J. Exp. Biol.*, **123**, 359–71.

Riegel, J.A. (1986b) The absence of arterial pressure effect on filtration by perfused glomeruli of the hagfish, *Eptatretrus stouti. J. Exp. Biol.*, **126**, 361–74.

Robertson, J.D. (1954) The chemical composition of the blood of some aquatic chordates, including members of the tunicates, cyclostomes and osteichtyes. *J. Exp. Biol.*, **31**, 424–42.

Rost, B., Schurek, H.J., Neumann, K.H. and Stolte, J. (1983) Perfusion studies on single glomeruli of the Atlantic hagfish, *Myxine glutinosa* (L.) *Bull. Mt Desert Isl. Biol. Lab.*, **23**, 63–5.

Satchell, G.H. (1986) Cardiac function in the hagfish, *Myxine* (Myxinoidea: Cyclostomata). *Acta Zoologica* (Stockh.), **67**, 115–22.

Savin, V.J. (1983) Ultrafiltration in single isolated human glomeruli. *Kidney Int.*, **24**, 748–53.

Savin, V.J. and Terreros, D.A. (1981) Filtration in single isolated mammalian glomeruli. *Kidney Int.*, **20**, 188–97.

Schmidt-Nielsen, B. and Renfro, J.L. (1975) Kidney function of the American eel *Anguilla rostrata. Amer. J. Physiol.*, **228**, 420–31.

Smith, H.W. (1932) Water regulation and its evolution in the fishes. *Q. Rev. Biol.*, **1**, 1–26.

Starling, E.H. (1899) The glomerular function of the kidney. *J. Physiol. London*, **24**, 317–30.

Stolte, H. and Schmidt-Nielsen, B. (1978) Comparative aspects of fluid and electrolyte regulation by the cyclostome, elasmobranch and lizard kidney, in *Osmotic and Volume Regulation. Alfred Benzon Symposium XI* (eds C.B. Jorgensen and E. Skadhaurer), Munksgaard, Copenhagen, pp. 209–22.

Stolte, H., Brecht, I.P., Wiederholt, M and Hierholzer, K. (1968) Einfluss von Adrenalektomie und Glucocorticoiden auf die Wasserpermeabilität corticaler Nephronabschnitte. *Pflügers Arch.*, **299**, 99–127.

Toop, T., Donald, J.A. and Evans, D.H. (1995) Natriuretic peptide receptors in the kidney and the ventral and dorsal aortae of the Atlantic hagfish *Myxine glutinosa* (Agnatha). *J. Exp. Biol.*, **198**, 1875–82.

Ullrich, K.J., Reimrich, G. and Fuchs, G. (1964) Wasserpermeabilität und transtubulärer Wasserfluß corticaler Nephronabschnitte bei verschiedenen Diuresezuständen. *Pflüger Arch.*, **280**, 99–119.

Windhager, E.E. (1968) Micropuncture techniques and nephron function, in *Butterworths Molecular Biology and Medicine Series* (ed. E.E. Bittar), Butterworth, London.

Wales, N.A.M. (1988) Hormone studies in *Myxine glutinosa*: effects of the eicosanoids arachidonic acid, prostaglandin E_1, E_2, A_2, $F_{2\alpha}$, thromboxane B2 and of indomethacin on plasma cortisol, blood pressure, urine flow and electrolyte balance. *J. Comp. Physiol.*, **B158**, 621–8.

Zapata, A., Fänge, R., Mattison, A. and Villena, A. (1984) Plasma cells in adult Atlantic hagfish, *Myxine glutinosa*. *Cell Tissue Res.*, **235**, 691–3.

24
AN ANALYSIS OF THE FUNCTION OF THE GLOMERULI OF THE HAGFISH MESONEPHRIC KIDNEY

Jay A. Riegel

SUMMARY

Urine produced by glomeruli of the hagfish kidney is colloid-free and approximates a filtrate of the blood plasma. However, conditions necessary for pressure filtration appear to be absent. Average colloid osmotic pressure of the blood plasma is at least double the average hydrostatic pressure in the glomerular capillaries. To clarify this enigma, glomeruli were perfused with colloid-containing Ringer. The principle findings follow: (1) Glomerular capillaries form only part of the perfusion pathway, so that the filtration fraction usually was small (<5%). (2) Glomerular capillaries appeared to be the major source of vascular resistance in the perfused preparation. (3) Perfused glomerular vessels appeared to be of two kinds, which differed in their response to perfusion pressure changes. In *low-pressure-glomerular vessels* (LPGV), pressure did not exceed c. 0.4 kPa when perfused at pressures up to 1.6 kPa. In *high-pressure-glomerular vessels* (HPGV), pressure could exceed 0.4 kPa, especially when the perfusion pressure was elevated to 'unphysiological' levels (i.e. 1.8 to 2 kPa). The HPGV lie at the periphery of the glomeruli and may provide a variable-patency pathway between preglomerular and postglomerular vasculature and between HPGV and the more deeply lying LPGV. (4) Assessments were made of the effects upon single-glomerulus-filtration rate (SGFR) and glomerular-capillary pressure (P_{gc}) of compounds and procedures that stimulate (e.g. theophylline, dilution) or inhibit (e.g. ouabain, 2,4-dinitrophenol) fluid-secreting tissues. Average P_{gc} did not vary significantly in any perfusion, and SGFR was reduced in a proportion of most perfusions probably due to reduced flow through the glomeruli. Adenaline (5×10^{-6} M) appeared to increase flow through the glomerular capillaries. When perfusion fluid contained adrenaline as well as ouabain and 2,4-dinitrophenol, SGFR diminished, and in the case of ouabain, this appeared to be due to a direct effect on the glomerular epithelium.

24.1 INTRODUCTION

The hagfish kidney has been a subject of interest both morphologically and physiologically for well over a century. Because of the apparent simplicity of its structure, it has attracted renal physiologists interested in various aspects of the function of the large glomeruli and the readily accessible kidney ducts. The present chapter will focus on the studies made by the writer of the function of the glomerulus both in the intact animal and

The Biology of Hagfishes. Edited by Jørgen Mørup Jørgensen, Jens Peter Lomholt, Roy E. Weber and Hans Malte. Published in 1998 by Chapman & Hall, London. ISBN 0 412 78530 7.

during perfusion with artificial solutions. As background, the overall structure and function of the hagfish mesonephric kidney will be reviewed briefly. However, for more exhaustive treatments, the reader is referred to the excellent reviews by Fels *et al.* (1989, 1993) and Fels *et al.*, this volume.

Structurally, the mesonephric kidney of the hagfish is composed of a pair of ducts (variously called 'ureters' or 'archinephric ducts') which lie on the dorsal wall of the body cavity on either side of the midline. In the anterior 30 or so segments of the body cavity, each duct is joined, via a short 'neck' segment, by a glomerulus. Figure 24.1 illustrates the nephric structures to be found in one segment of the body cavity of a hagfish and their relationship to the blood supply.

The flow of blood through the kidney follows a fairly complex path which has not been described fully as yet. The major vessels of the arterial supply consist of a pair of arteries in each body segment (segmental arteries) which arise from the dorsal aorta and send branches laterally and dorsally to the body wall. Close to their junction with the dorsal aorta, small renal arteries branch off the segmental arteries and go to the prominent glomeruli which lie nearby. It is clear that the renal arteries (called 'afferent arterioles' by some) serve the capillaries of the glomerulus. The wall of the ureter has an extensive vascular supply and in the living animal it can be seen that blood flows in two series of blood vessels, superficial and deep vessels (Riegel, 1978). These vessels may be interconnected (Albrecht *et al.*, 1978). The blood in the vessels of the ureter wall derives from vessels coming from the glomerulus as well as arteries arising as branches from the segmental arteries lateral to the renal arteries (Riegel, 1986a). Blood drains from the ureter wall through small veins or venules which enter an adjacent postcardinal vein.

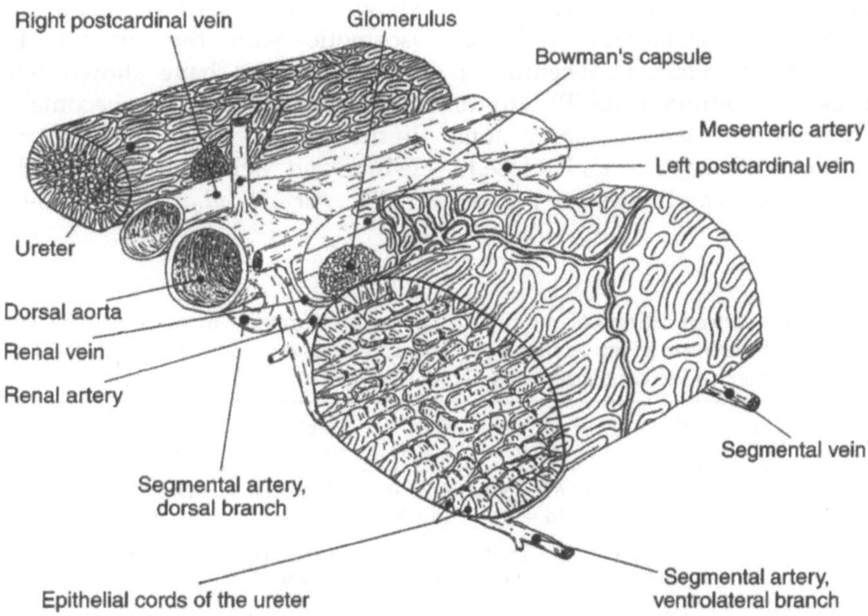

Figure 24.1 Nephric structures and the associated vasculature in one body segment of a hagfish. In the upper part of the diagram the ureter is shown in the collapsed condition. In the lower part of the diagram the ureter is shown fully distended with urine to illustrate its probable function as a urine-storage organ.

The pattern of blood flow through the glomeruli is not clear. However possible patterns of flow based on both morphological and physiological studies will be discussed in the Conclusions section of this chapter.

24.2 FUNCTION OF THE HAGFISH KIDNEY

4.2.1 Studies of kidneys through which blood flows

(a) The nature of the urine of hagfish

Shown in Table 24.1 is a résumé of analyses of the electrolyte composition of ureteral urine and plasma of hagfish compared to analyses of the seawater in which the animals were kept. Although the hagfish is unique amongst marine vertebrates in having a plasma electrolyte concentration very similar to that of seawater, the pattern exhibited in Table 24.1 is characteristic of any active isosmotic marine invertebrates. Whilst the Na and Cl concentrations of the plasma are very close to those of the surrounding seawater, K, and particularly Mg are considerably reduced. The ratios between the plasma concentration and the urine concentration (U/P) are of particular interest. Of the inorganic ions shown, only Na and Cl have an U/P close to 1 and thus are strong candidates for passive distribution between the blood and the ureteral urine. For all of the other inorganic constituents shown (i.e. K, Mg, Ca, P and S), it could be argued that the distribution between the plasma and urine is due to active processes. This would suggest that the wall of the ureter contributes to the nature of the urine by actively absorbing and secreting substances. There is no evidence for active secretion of magnesium in *Myxine glutinosa* (Table 24.1), but in *Eptatretus stoutii*, active secretion is indicated (Riegel, 1978). The ureter possesses mechanisms for the net absorption of glucose and amino acids (Fels *et al.*, 1989). The columnar epithelial cells of the ureter wall bear a resemblance to the cells of the proximal part of the nephrons of other vertebrates. Indeed, as indicated above, many of the functions of the proximal part of the nephron tubule appear to be carried on by the ureter wall. However, two important functions of the proximal tubule are absent in the ureter of hagfishes, namely, the reabsorption of water and the secretion of organic acids. As shown in Table 24.1, the urine of hagfishes is isosmotic with the plasma. Furthermore, several studies have shown that the renal marker, inulin, does not become concentrated in the ureter urine with respect to the plasma (Munz and McFarland, 1964; Rall and Burger, 1967; Alt *et al.*, 1981). The compound

Table 24.1 Electrolyte concentrations of the blood plasma and ureteral urine of *Myxine glutinosa* and of the habitat seawater. Values are given as mean ± standard deviation. *Abbreviation*: U/P ratio = the ratio of the urine and plasma electrolyte concentrations. (Data from Alt *et al.*, 1981)

	Osmolality (mOsm kg^{-1})	Na^+ (mM)	K^+ (mM)	Ca^{2+} (mM)	Mg^{2+} (mM)	Cl^- (mM)	P (mM)	S (mM)
Seawater	898 ± 65 (n = 7)	424 ± 32 (n = 7)	8.9 ± 0.9 (n = 6)	9.7 ± 0.3 (n = 5)	44.6 ± 3.6 (n = 6)	511 ± 27 (n = 6)	1.08	19.8
Plasma	980 ± 104 (n = 11)	439 ± 29 (n = 9)	5.9 ± 0.9 (n = 10)	7.2 ± 1.4 (n = 9)	19.5 ± 6.6 (n = 9)	455 ± 45 (n = 9)	1.0 ± 0.1 (n = 5)	17.6 ± 8.8 (n = 5)
Urine	1053 ± 99 (n = 13)	462 ± 22 (n = 11)	8.6 ± 4.6 (n = 8)	4.2 ± 1.3 (n = 6)	14.7 ± 1.8 (n = 6)	430 ± 82 (n = 7)	2.0 ± 1.9 (n = 5)	17.7 ± 2.2 (n = 5)
U/P ratio	1.07 ± 0.1 (n = 13)	1.06 ± 0.5 (n = 11)	1.5 ± 0.7 (n = 8)	0.5 ± 0.1 (n = 6)	0.77 ± 0.21 (n = 6)	1.0 ± 0.2 (n = 7)	3.1 ± 2.0 (n = 5)	1.15 ± 0.14 (n = 5)

commonly used as an indicator of the presence of the organic-acid secretory pathway, phenol red, is not concentrated in the ureteral urine of hagfishes (Fänge and Krog, 1963; Rall and Burger, 1967, personal observation). Phenol red does enter the urine of hagfishes, presumably by being filtered at the glomeruli. It has been observed by the writer that the red colour (pH > 7) of the urine in the glomeruli-bearing segments of the ureters of phenol-red excreting specimens of *E. stoutii* turns to yellow (pH 6.5 or less) in the posterior ureter. It is likely that acidification occurs there, so that retention of diffusible moieties (such as NH_3) by ionic 'trapping' could occur giving rise passively to U/P greater than 1 for those substances.

(b) Hydrostatic pressures in the kidney and associated vasculature

In Table 24.2 are summarized pressures measured in blood vessels of the kidney and associated vasculature of lightly anaesthetized specimens of *E. stoutii* and *M. glutinosa*. Measurements made in the kidney of *E. stoutii* are more extensive so they will be discussed more fully. Little or no reduction of pressure was found in the segmental artery as compared to the dorsal aorta; pressure in the renal artery was about 80% of the dorsal aorta value. However, the average pressure measured in glomerular capillaries of *E. stoutii* was only about 20% of the dorsal aorta pressure. On the postglomerular side the vessels fell into two distinct groups. In one group, termed by the writer *low-pressure-efferent vessels* (LPEV), the average pressure was intermediate between the average pressure in the glomerular capillaries and the average pressure in the adjacent postcardinal vein. The LPEV were found either entirely within a glomerulus or between a glomerulus and the postcardinal vein. It is likely that these vessels are efferent arterioles or the flow through them is derived from efferent arterioles.

In Table 24.2 are shown data derived from blood vessels which lie on the ventral surface of the ureter and in which pressures comparable to those in the renal artery were seen. In these vessels there was a fully developed arterial pulse. From their position it appeared that these vessels were postglomerular, so it seemed pertinent to investigate them fully. Experiments were made in which pressure in these vessels was monitored, whilst the renal arteries leading to the glomerulus adjacent to which the vessels lay were ligated. In four of five such experiments there was a profound drop in the pressure within the vessels after

Table 24.2 Hydrostatic pressures in the urinary space and various vascular elements associated with the hagfish kidney. Values are expressed as mean ± standard deviation. The number in brackets is the number of measurements. Pressures are given in kilopascals (1 kPa = *c.* 10.2 cm H_2O or *c.* 7.5 mmHg). *Abbreviations*: LPEV = low-pressure-efferent vessel; HPEV = high-pressure-efferent vessel. (*The high pressures measured in the urinary space may have been due to the fact that in most specimens the urinary space was very distended)

	Eptatretus stoutii		*Myxine glutinosa*	
Dorsal aorta	0.90 ± 0.22	(17)	0.52 ± 0.08	(5)
Segmental artery	1.0 ± 0.2	(20)	0.50	(2)
Renal artery	0.78 ± 0.35	(5)	–	
Glomerular capillary	0.21 ± 0.10	(13)	0.33 ± 0.06	(4)
Bowman's space	0.10 ± 0.05	(11)	0.18*± 0.08	(5)
LPEV	0.14 ± 0.10	(2)	0.24	(2)
HPEV	0.77 ± 0.44	(9)	–	
Postcardinal vein	0.04 ± 0.02	(9)	–	

ligation of the renal artery. These vessels were termed by the writer, *high-pressure-efferent vessels* (HPEV, Riegel, 1986a). Blood in the HPEV clearly has a common source (renal artery) with the blood that flows through the glomerular capillaries, yet pressure in the HPEV is very much higher than pressure in the glomerular capillaries. This suggested that the HPEV are themselves vessels that shunt the glomerular capillary network or communicate directly with such vessels. Brown (1988) in her study of the vasculature of the kidney of specimens of *M. glutinosa* described vessels altogether shunting the glomerular capillaries. However, these shunt vessels were found in only 28% of examined glomeruli, so it is doubtful that the vessels described by Brown are the morphological equivalents of HPEV which have been found adjacent to most glomeruli examined by the writer.

Measurements of pressures in the specimens of *M. glutinosa* were made preliminary to the study of pressures in *E. stoutii*. They demonstrate a similar condition to that seen in *E. stoutii*; that is, there is a reduction of pressure between the dorsal aorta/segmental artery and the glomerular capillaries. It is interesting that the pressures in the glomerular capillaries of *M. glutinosa* are comparable to those seen in the glomerular capillaries of *E. stoutii* yet the dorsal aortic pressure in *M. glutinosa* is only about half that of *E. stoutii*.

Just how reliable are blood pressures measured in lightly anaesthetized hagfishes may be judged from the following: pressures measured in the dorsal aorta of specimens of *M. glutinosa* by the writer are comparable to those measured in dorsal aortae of fully conscious specimens of the same species by Axelsson *et al.* (1990). Also, pressures measured in the dorsal aortae of actively swimming specimens of a hagfish congeneric with *E. stoutii*, *E. cirrhatus*, averaged about 1 kPa (Forster *et al.*, 1988), a value very close to that measured in dorsal aortae of lightly anaesthetized specimens of *E. stoutii*.

(c) The overall function of the hagfish kidney

It is probable that the hagfish kidney functions to maintain the electrolytic balance of the animal's body and to recover from the urine substances useful to the animal's metabolism. Whether or not hagfishes utilize their kidneys for volume regulation is problematical. Most studies of hagfishes held in captivity show them to be slightly hypertonic to their habitat water (McInery, 1974). Hagfishes may encounter dilute salinities in their natural habitat, with Japanese and New Zealand species living near river mouths and in estuaries (Hardisty, 1979). Since hagfishes have a high water permeability (Rudy and Wagner, 1970), it would be expected that the kidneys would play a role in fluid-volume adjustments when the animals encounter changes of salinity. A few studies have shown that hagfishes are unable to tolerate direct transfer from full-strength to dilute seawater, but with respect to *E. stoutii*, at least, the animals survive for extended periods if slowly acclimatized (McInery, 1974). Whether or not the kidneys participate in the acclimatization process of *E. stoutii* remains to be investigated.

(d) The formation of the primary urine in hagfishes

The existence of well-developed glomeruli in their kidneys may lead to the conclusion that, as in other glomerular vertebrates, the glomerular fluid (primary urine) is formed by pressure ('ultra') filtration. For the following reasons, this conclusion is unlikely. The hydrostatic pressure in the glomerular capillaries of both *M. glutinosa* and *E. stoutii* is at a maximum of about 0.4 kPa (Table 24.2), whilst the colloid osmotic pressure (COP) of the plasma averages about 1.4 kPa in *E. stoutii* and 0.69 kPa in *M. glutinosa* (Riegel, 1986a, unpublished) or about 0.7 kPa in *M. glutinosa* (Fels *et al.*, 1989). The COP of primary urine taken directly from the Bowman's space of hagfish glomeruli is negligible (Riegel, 1986a), and the hydrostatic pressure in that space is about 0.1 kPa (Table 24.2).

The requirements for pressure filtration in the capillaries of vertebrates were elaborated just over a century ago (1896) by E.H. Starling. That is, the direction of fluid movement across capillary walls is determined by two sets of opposing forces, which are now generally called 'Starling forces'. Forces tending to keep fluid within the capillaries are the water-attracting properties of plasma colloids (COP) and the hydrostatic pressure of the space outside the capillaries. Forces tending to cause fluid to move out of the capillaries are the hydrostatic pressure of the blood within the capillaries and the COP of any colloids in the space outside the capillaries. In most vertebrate glomeruli thus far examined, the balance of the Starling forces most of the time favours the movement of fluid out of the glomerular capillaries so that the primary urine is formed by pressure ('ultra') filtration (e.g. Arendshorst and Gottschalk, 1985). In the hagfish, the Starling forces are never balanced so as to cause fluid to move out of the glomerular capillaries. Nevertheless, hagfish glomeruli do produce primary urine at an appreciable rate under conditions comparable to those under which the measurements in Table 24.2 were made. This has been measured directly in nine specimens of *E. stoutii* in which a glomerulus was cannulated and movement of fluid into the cannula measured: the average single-glomerulus-filtration rate (SGFR) was 58.2 ± 43.1 (s.d.) nl min^{-1} (Riegel, 1986a).

It is clear from this brief summary that urine formation in hagfishes defies the Starling principle; yet, aside from their large size, the glomeruli of hagfishes appear to be identical in their structure to glomeruli of other vertebrates (Kühn *et al.*, 1975). Therefore, further studies have been undertaken in which advantage has been taken of the finding by Stolte and Eisenbach (1973) that hagfish kidneys survive for long periods and produce urine when perfused with Ringer.

24.2.2 Studies of kidneys perfused with artificial solutions

The writer has carried out perfusion studies using hagfish Ringer to which colloid was added sufficient to match the measured average COP of hagfish (*E. stoutii*) plasma. Colloid (Ficoll 70) was added in an effort to match as closely as possible the osmotic conditions normally encountered by hagfish glomerular vasculature. Ficoll 70 is a neutral polysaccharide for which living tissues have a high tolerance (Pharmacia Ltd).

Perfusion studies by Stolte and Eisenbach (1973) on *M. glutinosa* and by the writer (1978) on *E. stoutii* demonstrated that postglomerular backpressure, either through the postcardinal veins or the vessels of the ureter wall, affects SGFR. Consequently, the perfusion arrangement described here was designed to minimize as far as possible the backpressure on the glomerulus (Riegel, 1986b). The success of these precautions may be seen in the following observation: single glomeruli of *E. stoutii* perfused with Ficoll-70-containing Ringer produce urine at rates that fall within the range of values measured by direct cannulation of the Bowman's space of glomeruli through which blood was flowing.

(a) Perfusion of single glomeruli with Ficoll Ringer

Figure 24.2 illustrates a typical perfusion experiment in which an isolated glomerulus of *E. stoutii* was perfused at a known rate with hagfish Ringer that contained Ficoll 70 at a concentration adequate to generate a COP of 1.4 kPa. Simultaneously, pressure in the perfusion line (P_{perf}), pressure in a glomerular capillary (P_{gc}) and SGFR were measured.

The data summarized in Figure 24.2 illustrate several characteristics of perfused glomeruli. Firstly, SGFR was proportional to

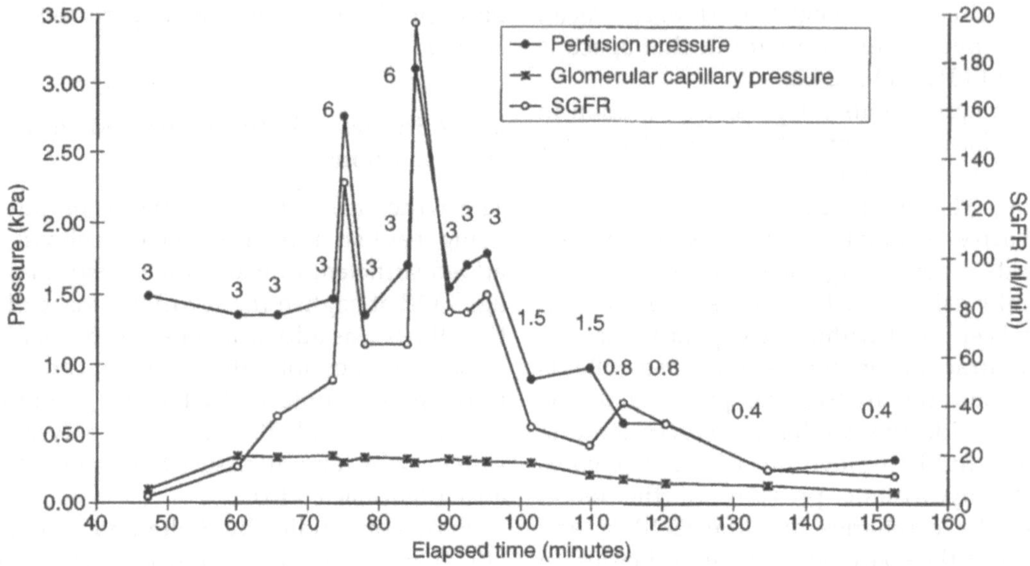

Figure 24.2 Changes in perfusion pressure, pressure in a glomerular capillary, and single-glomerulus-filtration rate (SGFR) measured in a glomerulus isolated and perfused with Ringer containing colloid whose osmotic pressure was 1.4 kPa. Numerals on the diagram show the rate of perfusion in µl min⁻¹.

P_{perf} which, in turn, was proportional to the rate of perfusion (numerals on the figure). Secondly, although the value of P_{gc} varied with long-term variations of perfusion rate, it remained relatively steady during large, short-term alterations of the perfusion rate. For example, between minutes 74 and 75 and again between minutes 84 and 85, the perfusion rate was doubled, giving rise to large elevations of P_{perf}. At the same time SGFR more than doubled, but P_{gc} actually diminished slightly. It must be assumed that the changes of SGFR were underlain by variations of flow through the glomerular capillaries. The fact that P_{gc} changed little suggested that changes of P_{gc} were poor indicators of variations of flow.

(b) Effects of agonists and inhibitors of fluid secretion

It was demonstrated by the writer (1978) that the inhibitors, 2,4-dinitrophenol and ouabain reversibly slowed or halted SGFR of perfused glomeruli without affecting the perfusion pressure. At that time, the anatomical evidence available suggested that the fluid entering the renal artery flowed exclusively through the glomerular capillaries. Therefore, it was concluded that active cellular processes must play a role in the elaboration of the primary urine in hagfishes. Subsequent investigations (Albrecht *et al.*, 1978; Riegel, 1986a; Brown, 1988) have demonstrated that fluid entering the renal artery may shunt the glomerular capillaries. Thus the fluid-secretion argument was weakened.

In more recent studies (unpublished), the writer has examined the effect on P_{perf}, P_{gc} and SGFR of a range of substances that either stimulate or inhibit tissues whose fluid-secreting character is well established (e.g. arthropod Malpighian tubules, amphibian urinary bladders or mammalian gall bladders).

Shown in Table 24.3 are averaged values of the SGFR, P_{perf} and P_{gc} during times when single glomeruli were perfused through dual

Table 24.3 Perfusion of isolated glomeruli of hagfish alternately with control Ringer or an experimental Ringer. Experimental Ringer were identical to control Ringer, except as indicated by their description. Data are given as mean ± standard deviation followed by the number of estimations in parentheses (*n*). Paired values (in **bold** face) are significantly different from each other (P = 0.01 or less). The ratio in brackets beneath the Ringer description is the number of perfusions in which SGFR was reduced relative to the total number of perfusions using that Ringer

		SGFR (nl min^{-1})	P_{perf} (kPa)	P_{gc} (kPa)
Control	Control	72.7 ± 69.5 (75)	0.94 ± 0.33 (85)	0.36 ± 0.34 (64)
(0/10)	Experimental	85.0 ± 82.6 (78)	1.01 ± 0.43 (79)	0.34 ± 0.34 (70)
Na-free	Control	54.8 ± 21.4 (13)	1.03 ± 0.15 (15)	not measured
(0/3)	Experimental	71.1 ± 41.7 (14)	1.03 ± 0.13 (14)	not measured
K-free	Control	64.9 ± 34.4 (11)	1.16 ± 0.16 (13)	not measured
(1/3)	Experimental	66.7 ± 25.3 (15)	1.17 ± 0.17 (15)	not measured
80%	Control	25.2 ± 11.3 (12)	0.82 ± 0.33 (15)	0.33 ± 0.11 (15)
(3/6)	Experimental	19.4 ± 12.1 (9)	0.96 ± 0.28 (9)	0.33 ± 0.39 (26)
Papaverine	Control	**72.1 ± 52.2 (34)**	1.30 ± 0.35 (40)	0.25 ± 0.21 (22)
(5/5)	Experimental	**34.8 ± 21.6 (18)**	1.17 ± 0.30 (18)	0.20 ± 0.26 (6)
Theophylline	Control	41.1 ± 21.2 (14)	1.02 ± 0.27 (13)	0.07 ± 0.06 (24)
(3/3)	Experimental	26.7 ± 18.1 (9)	1.02 ± 0.13 (9)	0.09 ± 0.07 (8)
Ouabain	Control	**27.6 ± 17.3 (23)**	0.85 ± 0.34 (24)	0.22 ± 0.20 (48)
(8/12)	Experimental	**10.7 ± 9.49 (15)**	0.89 ± 0.30 (15)	0.23 ± 0.16 (62)
2,4-Dinitrophenol	Control	**40 ± 33.8 (42)**	0.92 ± 0.21 (45)	0.36 ± 0.28 (34)
(14/14)	Experimental	**22.3 ± 20.3 (30)**	0.86 ± 0.29 (30)	0.31 ± 0.15 (37)
Choline	Control	**58.7 ± 49.9 (15)**	0.86 ± 0.16 (19)	0.17 ± 0.11 (17)
(4/5)	Experimental	**11.6 ± 11.1 (9)**	0.88 ± 0.25 (9)	0.26 ± 0.11 (7)

microcannulae alternately with either control perfusion fluid (Ringer containing Ficoll 70 to make a COP of *c.* 1.4 kPa) or experimental perfusion fluids. The experimental perfusion fluids, in addition to Ficoll 70, either (a) contained substances (high K, high Na, theophylline) or were subject to procedures (dilution) that normally stimulate fluid-secreting tissues or (b) contained substances (ouabain, 2,4-dinitrophenol, choline) that normally inhibit fluid secretion. Also tested was the vasodilator, papaverine. The various procedures had no effect on average P_{perf} or average P_{gc}. Furthermore, substances or procedures that stimulate fluid secretion in other tissues had no discernible effect on the average value of any measured parameter. However, the inhibitors of fluid secretion, ouabain, 2,4-dinitrophenol and choline, as well as the vasodilator, papaverine, all significantly reduced average SGFR. Averaged values may be misleading, in that they obscure the fact that in a proportion of all perfusions involving other than control and Na-free Ringer there was a reduction of SGFR. Furthermore, in not all perfusions involving ouabain and choline was the SGFR reduced. The figures in brackets under the descriptions of the various perfusion media represent the ratios of the numbers of perfusions in which SGFR was reduced relative to the total number of perfusions using that medium.

The fact that in a proportion of most perfusions SGFR was reduced led to the following conclusion: although fluid secretion could not be excluded, it is likely that the main factor contributing to the reduction of SGFR was a relative diminution of flow through the glomerular capillaries. Somewhere in the perfused preparation there must exist a mechanism that detects changes in the composition of the fluid perfusing the preparation, and reacts by shunting flow away from the glomerular capillaries.

(c) Flow in blood vessels of the glomerular capillary tuft

Hydrostatic pressures in blood vessels of perfused glomeruli were measured (Figure 24.3). These measurements demonstrated that even though changes of composition of the fluid perfusing the glomerular blood vessels brought about changes of SGFR this was not reflected consistently by pressure changes. However, these quite extensive measurements revealed the presence within the glomerular vasculature of what appeared to be two kinds of vessels with respect to their response to pressure. The first kind of vessel responded to increments of P_{perf} like the vessel illustrated in Figure 24.2. That is, as P_{perf} was raised, P_{gc} rose to a maximum of about 0.4 kPa. Such vessels were termed *low-pressure-glomerular vesels* (LPGV). In vessels of the second kind P_{gc} rose to values which sometimes matched the P_{perf} as the perfusion pressure was increased to 'unphysiological' levels of 1.8 to 2.0 kPa. These vessels were termed *high-pressure-glomerular vessels* (HPGV).

In Figure 24.3a (solid circles) are summarized measurements of P_{lpgv} in relation to P_{perf} in glomerular blood vessels identified as LPGV by the fact that pressure in them would not rise higher than *c.* 0.4 kPa when P_{perf} was varied between 0.2 and 2.0 kPa. There was no significant correlation between P_{perf} and P_{lpgv}. The open circles of Figure 24.3a summarize measurements of P_{hpgv} in relation to P_{perf} in glomerular blood vessels identified as HPGV. Pressure in HPGV would rise to levels in excess of 0.4 kPa when P_{perf} was varied between 0.2 and 2 kPa. There was a significant correlation between pressure in the HPGV and P_{perf}.

The existence of the HPGV may provide additional evidence for a high-pressure vascular pathway through the hagfish glomerulus found in previous studies (Riegel, 1986a). The HPGV seemed to be confined to the periphery of the glomerular vascular tuft, whilst the LPGV were found deeper within the vascular tuft. It is possible that the HPGV are the glomerular vessels in which Albrecht *et al.* (1978) observed constriction ('Einschnürungen') at the junctions they made with blood vessels lying more deeply in the glomerular vascular tuft of hagfishes.

With the exception of the HPGV described above, pressures measured in the glomerular capillaries did not vary very much despite substantial changes in perfusion pressure. Even in the HPGV pressures greater than 0.4 to 0.5 kPa generally were seen only when the perfusion pressure was elevated above 1.2 to 1.4 kPa, the 'physiological' level. Whilst pressure in the glomerular capillaries did not vary very much when P_{perf} was varied between 0.2 and 1.4 kPa, the average SGFR rose steeply (see Figure 24.2) over that range of perfusion pressures. From the foregoing and the data shown in Figure 24.3a, it seems clear that under physiological conditions pressure in the glomerular capillaries does not determine the rate of SGFR. However, the rate of urine flow is strongly correlated with the rate of perfusion. Therefore, the writer has concluded that it is the *volume* of fluid flowing through the glomerular capillaries that is the major determinant of SGFR. Very likely the pressures measured in the glomerular capillaries are a manifestation of adjustments to the volume of the glomerular vascular tuft.

At present, reliable direct measurement of flow in the glomerular capillaries does not seem to be a practical possibility (Steinhausen *et al.*, 1990). Consequently estimates of that flow must be got by indirect means. In the studies summarized in Figure 24.3 it was found that the proportion of fluid perfused through glomeruli that became urine (the single-glomerulus-filtration fraction or SGFF) bore a positive correlation with the vascular resistance (P_{perf}/perfusion rate) of the perfused preparation. This is illustrated in

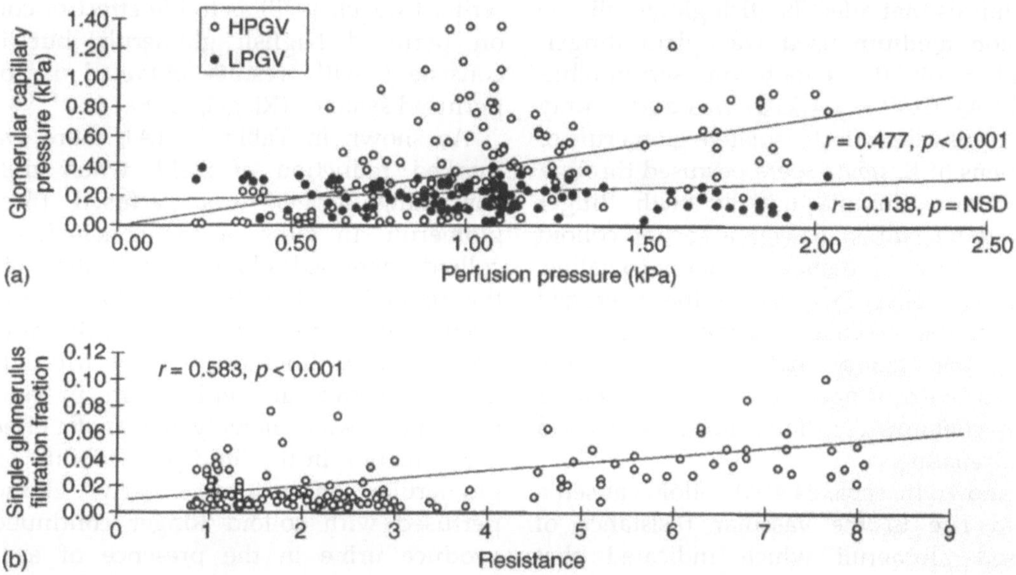

Figure 24.3 Glomeruli isolated and perfused with Ringer containing colloid whose osmotic pressure was 1.4 kPa. (a) Perfusion pressure in relation to glomerular capillary pressure measured in capillaries designated as high-pressure-glomerular vessels (HPGV) and low-pressure-glomerular vessels (LPGV). (b) Single-glomerulus-filtration fraction (SGFF = single-glomerulus-filtration rate/the perfusion rate) in relation to the vascular resistance (= perfusion pressure/perfusion rate) of isolated, perfused glomeruli.

Figure 24.3b. This finding suggested that the glomerular capillaries represent the greatest resistance to flow in perfused preparations.

(d) The effect of catchecholamines on perfusion pressure and SGFR

Fels *et al.* (1987) observed that the catecholamines, adrenaline and noradrenaline affected the function of perfused glomeruli. The P_{perf} was raised and SGFR was increased when pharmacological doses (M^{-5}) of adrenaline and noradrenaline were placed in the medium of isolated, perfused glomeruli of *Myxine glutinosa*. Recently, the writer has studied in detail the effects of pharmacological doses of adrenaline and noradrenaline on glomerular function. These studies (unpublished) confirm those of Fels *et al.* (1987) in that adrenaline caused an increase of SGFR and P_{perf} when glomeruli of *Eptatretus stoutii* were isolated and perfused with Ringer containing it. However, noradrenaline consistently only elevated P_{perf}.

When glomeruli were perfused at a constant rate with Ringer containing 5 μM adrenaline there was a consistent effect in elevating both P_{perf} and SGFR. It was reasoned that this result was unlikely to be due to changes of pressure, therefore adrenaline must have caused increased flow through the glomerular capillaries, thus increasing SGFR. This result raised the possibility that adrenaline could be used as a research tool to increase flow through the glomerular capillaries whilst simultaneously studying the effects of fluid-secretion perturbants and other compounds on factors important to glomerular function. Therefore studies were made in which glomeruli were perfused with Ringer that contained colloid and the perturbants of fluid secretion, ouabain and 2,4-dinitrophenol as well as adrenaline.

In order to more clearly discern the effects

of chemicals that affect hagfish glomeruli, the perfusion medium used was plain Ringer. Colloid (Ficoll 70 + 0.5% bovine serum albumin (BSA); COP = 1.4 kPa) was added only when testing its effect. Isolated glomeruli of specimens of *E. stoutii* were perfused through a dual microcannula initially with Ringer followed by colloid Ringer alone or colloid Ringer which contained 5 μM adrenaline. Shown in Table 24.4(A) are the averaged values of the vascular resistance and SGFF during the times that glomeruli were perfused with Ringer (control) and either Ringer containing colloid alone or colloid plus adrenaline.

As shown in Table 24.4(A) colloid caused a marked rise in the vascular resistance of perfused glomeruli which indicated that colloid had the effect of increasing flow through the glomerular capillaries. This result disagrees with a previous conclusion of the writer (Riegel, 1978) as to the effect of colloid on perfused hagfish glomeruli, but it is consistent with results obtained in other perfused systems (Riegel, 1978).

As shown in Table 24.4(A), there was a marked reduction of SGFF when colloid-containing Ringer was perfused through glomeruli. In view of the conclusion that colloid increased glomerular-capillary flow, the reduction of SGFF must have resulted from a direct effect of colloid on the filtering epithelium. Colloid could have affected the glomerular-capillary epithelium in one or both of two ways, namely an osmotic effect or the reduction in the fluid permeability of the glomerular epithelium. Clearly, glomeruli perfused with colloid Ringer continued to produce urine in the presence of a large disparity between P_{gc} and perfusion-fluid COP. Therefore, reduction of the glomerula-capillary-fluid permeability must have been

Table 24.4 Changes of vascular resistance and SGFF of single glomeruli perfused initially with Ringer (control), followed by various experimental Ringer: (A) colloid Ringer or colloid Ringer plus adrenaline; (B) Ringer plus adrenaline or Ringer plus adrenaline plus either 2,4-dinitrophenol or ouabain. Values for resistance and SGFF represent averages of percentage change that occurred after switching from control to the experimental perfusion channels. Values are expressed as mean ± standard deviation. The number of glomeruli perfused is indicated by the figures in brackets. Mean values in bold are significantly different ($p < 0.001$) from control mean values

	Control	*Experimental*
A (1) 0.5% BSA + Ficoll 70 ($n = 7$)		
Resistance	104 ± 9	**128 ± 18**
SGFF	99 ± 15	**27 ± 30**
(2) 0.5% BSA + Ficoll 70 + 5 μM adrenaline ($n = 5$)		
Resistance	112 ± 12	**164 ± 43**
SGFF	109 ± 14	**48 ± 16**
B (1) Ringer + 5 μM adrenaline ($n = 6$)		
Resistance	102 ± 7	**123 ± 25**
SGFF	94 ± 24	**123 ± 26**
(2) 100 μM 2,4-dinitrophenol + 5 μM adrenaline ($n = 7$)		
Resistance	102 ± 14	100 ± 18
SGFF	110 ± 21	**65 ± 1**
(3) 10 μM ouabain + 5 μM adrenaline ($n = 6$)		
Resistance	101 ± 9	**117 ± 24**
SGFF	105 ± 11	**83 ± 11**

the most important factor in the reduction of SGFR. When glomeruli were perfused with Ringer that contained adrenaline as well as colloid, there were relatively large (and very significant) increases in both the vascular resistance and SGFF relative to the increases when colloid Ringer was perfused (Table 24.4(A)). This result probably was due solely to an effect on glomerular flow by adrenaline.

The results shown in Table 24.4(B) derive from experiments in which glomeruli were perfused through dual microcannulae initially with Ringer, followed by Ringer that contained adrenaline alone or in combination with 2,4-dinitrophenol or ouabain. Adrenaline had the effect of increasing both the vascular resistance and SGFF of perfused glomeruli, which probably indicates that glomerular capillary flow increased. When the perfusion fluid contained 2,4-dinitriphenol, as well as adrenaline, the increase in vascular resistance did not occur, but there was a reduction of SGFF. It is difficult to interpret this result because it would be expected that if the reduction of SGFF was due to a reduction of flow through glomerular capillaries the vascular resistance would have fallen. However, it is possible that dinitrophenol has a generalized effect on the glomerular capillaries blocking the action both of adrenaline and whatever it is that causes urine to be formed.

When glomeruli were perfused with Ringer that contained ouabain as well as adrenaline, there was a rise of the vascular resistance and a fall of SGFF. This result suggested that despite increased flow through the glomerular capillaries there was a reduction of urine formation, probably due to an effect of ouabain on the urine-forming mechanism. This is the clearest evidence obtained thus far in perfusion studies that fluid secretion plays a role in primary urine formation by hagfish glomeruli.

Hagfish glomeruli perfused with Ringer that contained colloid had an average SGFR of 83.0 ± 89.1 nl min^{-1}. However, when colloid was deleted from the Ringer, the average SGFR was 420 ± 188 nl min^{-1}, and when 5 μM adrenaline was added to the Ringer, the average SGFR was 572 ± 216 nl min^{-1}. These data make it clear that when the constraints imposed by colloid on urine formation by perfused glomeruli were removed, SGFR increased dramatically. Therefore, when the balance of Starling forces favour it, glomerular capillaries are capable of forming urine at a very rapid rate. This would suggest that hagfish glomerular capillaries do not represent a curious departure from the vertebrate norm, but in fact, will function like any other capillary when the pressure conditions are suitable.

24.3 CONCLUSIONS

It was concluded in a previous section that the main determinant of glomerular filtration in the hagfish is the volume of fluid that flows through the glomerular vasculature. The vascular tuft of the hagfish glomerulus is formed from a rete or network of interbranching vessels (Albrecht *et al.*, 1978; Brown, 1988). This fact was observed repeatedly and use was made of it to ascertain the correct placement of pressure-sensing electrodes in the studies here discussed. An electrode filled with saline containing the dye, lissamine green, was placed in a blood vessel and its location ascertained by discharging a bolus of the saline from the tip by applying pressure to the electrode. If the tip of the electrode was located in a vessel, the green colour of the saline usually would spread rapidly through a web of vessels. This observation, taken with the observation that the hagfish glomerular vascular tuft expands and contracts readily, suggests the following. Perhaps flow through the glomerulus is adjusted so as to maintain the volume of the vascular bed, rather than a particular pressure. If this is true, then it suggests that flow through the glomerulus is far more complex than presently envisaged and explains why simple models do not adequately describe glomerular filtration in vertebrates (Steinhausen *et al.*, 1990). As far as the hagfish glomerulus is concerned, it seems clear that some mechanism other than pressure filtration accounts for the formation of the primary urine. The fact that the inhibitor of Na/K-activated ATPase, ouabain, inhibits SGFR, even after taking into account possible alterations of flow through the glomerulus, suggests that active fluid secretion must play a role in the hagfish.

Although the fundamental question of how the hagfish kidney forms the primary urine still awaits a definitive answer, its overall function seems fairly clear. The hagfish kidney functions to keep the animal in electrolytic balance, and, at the same time, there are mechanisms in the ureter wall which perform many of the usual functions of kidneys. That is, metabolites such as glucose and amino acids are recovered and substances such as ammonia and phosphorus- and sulphur-containing compounds are excreted. That the ureter apparently lacks the ability to rid the plasma of organic acids is of little consequence since the large liver of the animal has a well-developed ability to do so (Fänge and Krog, 1963).

REFERENCES

Albrecht, U., Lametschwandtner, A. and Adam, H. (1978). Die Blutgefässe des Mesonephros von *Myxine glutinosa* L. *Zool. Anz. Jena*, **200**, 300–8.

Alt, J.M., Stolte, H., Eisenbach, G.M. and Walvig, F. (1981). Renal electrolyte and fluid excretion in the Atlantic hagfish *Myxine glutinosa*. *J. Exp. Biol.*, **91**, 323–30.

Arendshorst, W.J. and Gottschalk, C.W. (1985). Glomerular ultrafiltration dynamics: historical perspective. *Amer. J. Physiol.*, **248**, F163–74.

Axelsson, M., Farrell, A.P. and Nilsson, S. (1990). Effects of hypoxia and drugs on the cardiovascular dynamics of the Atlantic hagfish *Myxine glutinosa*. *J. Exp. Biol.*, **151**, 297–316.

Brown, J.A. (1988) Glomerular bypass shunts in the kidney of the Atlantic hagfish, *Myxine glutinosa*. *Cell Tissue Res.*, **253**, 377–81.

Fänge, R. and Krog, J. (1963) Inability of the kidney of the hagfish to secrete phenol red. *Nature, Lond.*, **17**, 713.

Fels, L.M., Elger, B. and Stolte, H. (1987) Effects of catecholamines on the function of single glomeruli of *Myxine glutinosa* L. (Cyclostomata), in *Proceedings of the German Zoological Society, 80th Meeting in Ulm* (eds F.G. Barth and E.A. Seyfarth), Gustav Fischer Verlag, Stuttgart.

Fels, L.M., Raguse-Degener, G. and Stolte, H. (1989) The archinephron of *Myxine glutinosa* L. (Cyclostomata), in *Structure and Function of the Kidney*, Vol. 1, *Comparative Physiology* (ed. R.K.H. Kinne), Karger, Basel, pp. 73–102.

Fels, L.K., Sanz-Altamira, P.M., Decker, B., Elger, B. and Stolte, H. (1993) Filtration characteristics of the single isolated perfused glomerulus of *Myxine glutinosa*, in *Renal Physiology and Biochemistry*, Vol. 16, (eds G.M. Berlyne and F. Lang), Karger, Basel, pp. 276–84.

Forster, M.E., Davie, P.S., Davison, W., Satchell, G.H. and Wells, R.M.G. (1988) Blood pressures and heart rates in swimming hagfish. *Comp. Biochem. Physiol.*, **89A**, 247–50.

Hardisty, M.W. (1979) *Biology of the Cyclostomes*, Chapman & Hall, London.

Health-Eves, M.J. and McMillan, D.B. (1974) The morphology of the kidney of the Atlantic hagfish, *Myxine glutinosa* (L.). *Amer. J. Anat.*, **139**, 309–34.

Kühn, K., Stolte, H. and Reale, E. (1975) The fine structure of the kidney of the hagfish *Myxine glutinosa*. *Cell Tissue Res.*, **164**, 201–13.

McInery, J.W. (1974) Renal sodium reabsorption in the hagfish, *Eptatretus stouti*. *Comp. Biochem. Physiol.*, **49A**, 273–80.

Munz, F.W. and McFarland, W.N. (1964) Regulatory function of a primitive vertebrate kidney. *Comp. Biochem. Physiol.*, **13**, 381–400.

Rall, D.P. and Burger, J.W. (1967) Some aspects of renal secretion in *Myxine*. *Amer. J. Physiol.*, **212**, 354–6.

Riegel, J.A. (1978) Factors affecting glomerular function in the Pacific hagfish *Eptatretus stouti* (Lockington). *J. Exp. Biol.*, **73**, 261–77.

Riegel, J.A. (1986a) Hydrostatic pressures in glomeruli and renal vasculature of the hagfish, *Eptatretus stouti*. *J. Exp. Biol.*, **123**, 359–71.

Riegel, J.A. (1986b) The absence of an arterial pressure effect on filtration by perfused glomeruli of the hagfish, *Eptatretus stouti* (Lockington). *J. Exp. Biol.*, **126**, 361–74.

Rudy, P.P. and Wagner, R.C. (1970) Water permeability in the Pacific hagfish, *Polistotrema stoutii* and the staghorn sculpin, *Leptococcus armatus*. *Comp. Biochem. Physiol.*, **34**, 399–403.

Starling, E.H. (1896) Physiological factors involved in the causation of dropsy. Lecture II: The absorption of fluids from the connective tissue spaces. *Lancet*, **1**, 1331–4.

Steinhausen, M., Endlich, K.H. and Wiegman, D.L. (1990) Glomerular blood flow. *Kidney Int.*, **38**, 769–84.

Stolte, H. and Eisenbach, G.M. (1973) Single nephron filtration rate in the hagfish *Myxine glutinosa*. *Bull. Mt Desert Isl. Biol. Lab.*, **13**, 120–1.

25
GONADS AND REPRODUCTION IN HAGFISHES

Robert A. Patzner

SUMMARY

Reproductive organs and germ cells of hagfishes were mainly studied in three species of hagfish: *Myxine glutinosa, Eptatretus stoutii* and *Eptatretus burgeri*. Especially noteworthy is the scanty number of eggs and the small amount of sperm in all of them. A yearly synchronous reproductive cycle was only found in the Japanese hagfish *E. burgeri*. A certain percentage of hagfish is sexually sterile and a functional hermaphroditism in 0.1% (or even less) in *E. burgeri* and in *E. stoutii* is presumed. There is no copulatory organ in any species of hagfish and the mode of fertilization is still not clear. Investigation on the adenohypophysial region in *M. glutinosa* has revealed the presence of agranular cells with unclear function. Only in *E. burgeri* some influence of the adenohypophysis on the gonads was revealed. The GnRH-like system of myxinoids is considered as a possible homologue of the caudal neuromodulatory GnRH system of higher vertebrates. In *M. glutinosa, E. stoutii* and *E. burgeri* indications for the production of steroids by the gonads have been brought forth.

25.1 INTRODUCTION

As hagfishes are the most primitive vertebrates, studies on their reproduction are essential for the understanding of phylogenetic aspects of the reproduction in all vertebrates. However, knowledge of their reproductive physiology, endocrinology and spawning behaviour is very sparse or even completely missing.

For more than 100 years authors have been occupied with studies on the reproduction of myxinoids. Walvig (1963) gave a review of those papers published until the sixties. Thereafter several scientists from all over the world studied details of reproductive organs and reproductive processes in several species of hagfish. Many questions remain unsolved however and the award of the Royal Danish Academy of Sciences, which was announced in 1860 to solve the riddles of reproduction in *Myxine*, has not yet been given to anybody.

Studies of hagfish gonads and reproduction are more or less limited to three species: *Myxine glutinosa* living off the northern Atlantic coast, *Eptatretus stoutii* on the east Pacific coast and *Eptatretus burgeri* on the west Pacific coast.

25.2 REPRODUCTIVE ORGANS AND GERM CELLS

The gonads of hagfishes are situated in the peritoneal cavity. The anlage for the ovary is found in the anterior part of the gonads, the one for the testis in the posterior part. In the process of sex differentiation either the cranial part develops (the animal becomes female) or the caudal part develops (the animal becomes male). In some cases none of

The Biology of Hagfishes. Edited by Jørgen Mørup Jørgensen, Jens Peter Lomholt, Roy E. Weber and Hans Malte. Published in 1998 by Chapman & Hall, London. ISBN 0 412 78530 7.

them develops (the animal is considered sterile) and in a few cases both parts develop (the animal becomes a (functional?) hermaphrodite (Patzner and Adam, 1976).

25.2.1 Sex differentiation

The problem of sex differentiation in hagfishes was discussed in detail by Gorbman (1990). In studying the gonads of *M. glutinosa*, several older authors considered these animals protandric hermaphrodites (Cunningham, 1886; Nansen, 1887). Later some described *Myxine* not as protandric but as dioecious (Conel, 1931; Schreiner, 1955; Walvig, 1963). Especially by investigating small *E. stoutii* with a body length under 20 cm, Gorbman (1990) came to the conclusion that gonadal differentiation in *E. stoutii* is juvenile progynous and does not follow the sequence described by the earlier authors.

25.2.2 Ovary and oogenesis

The ovary of hagfishes is situated within the peritoneal cavity (Figure 25.1). The ripe eggs break up their follicles as well as the peritoneal cover of the ovary and are released directly into the coelomic cavity. From there they are deposited through the cloaca. There is no oviduct. The anterior end of the ovary is at the level of the liver. Its length equals about three quarters of the distance between the liver and the cloaca (Figure 25.1). Initially the gonads of hagfishes are paired organs, but the left part of the ovary atrophies during early development (Felix and Bühler, 1906). Schreiner (1955) describes and shows pictures of some adult *M. glutinosa* with paired gonads. As this development is very rare one has to consider the unpaired right part of the ovary as regular. The ovary of the hagfishes is attached to the intestine by a mesovarium throughout its total length (Figure 25.1). The mesovarium is a lamellar structure consisting of a connective tissue matrix covered by serosal epithelial cells (Tsuneki and Gorbman, 1977a). The oogenesis of hagfish was described on the basis of electron microscopical studies in *M. glutinosa* (Patzner, 1973, 1974, 1975) and *E. stoutii* (Tsuneki and Gorbman, 1977a). The process of oogenesis can be divided into two steps: In the course of the first the oogonia coming from the germinal epithelium develop into oocytes with a diameter of 1 to 2 mm. At the beginning of the oogenesis the oocytes migrate from the outer margin of the mesentery into the mesenchyme (Figure 25.1). In the early stages of this development the big bubble-shaped nucleus is situated in the middle of the cell. Later it moves towards the animal pole. The envelope of the growing egg is composed of five layers (Lyngnes, 1930, 1936; Schreiner, 1955): (1) the zona pellucida, (2) the follicular epithelium = granulosa, (3) the *membrana propria* = *zona intima*, (4) a capsule of connective tissue, consisting of a *theca interna* and a *theca externa*, and (5) the peritoneum. When the eggs have reached a diameter of 1 to 2 mm they enter into a resting period. This stage has been defined as *ovum expectans* by Patzner (1974). Only after a preceding generation of eggs has been ovulated do these resting eggs resume growth and a limited number of them reach a length of about 20 mm. In this second step of development a great part of the egg content is composed of yolk elements, consisting of a central body and an outer layer, surrounded by a membrane. In the almost mature eggs vacuoles with a diameter of about 8 μm are found. The *zona pellucida* reaches its maximum width of about 6 μm when the eggs are 5 to 8 mm long. Then it gets thinner again. At the end of the development of the eggs the follicular epithelium has become very thick, especially at the poles. After reaching the final length the egg shell is formed between the follicular epithelium and the egg cell itself. It starts its growth on both poles of the egg. Later these caps grow towards each other. Only after the egg cell is completely surrounded by the shell do the anchor filaments start to develop at both poles of the

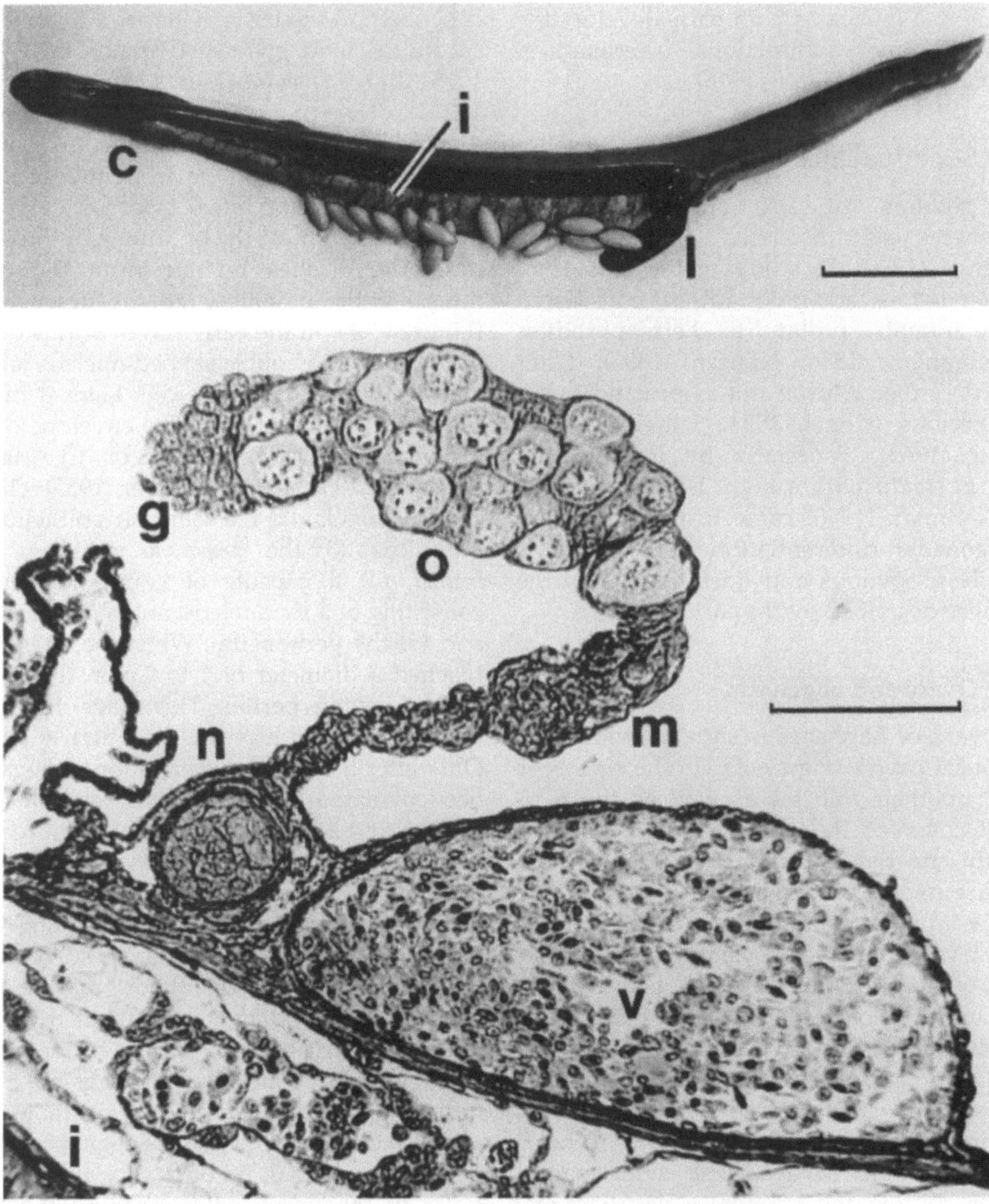

Figure 25.1 Ovary of the hagfish. *Top*: Female *Myxine glutinosa* with nearly mature eggs in the peritoneal cavity (scale shows 5 cm). The ovary of the hagfish is an unpaired organ on the right side of the intestine (i). l = liver, c = cloaca (after Patzner, 1982). *Bottom*: Cross-section through the gonad of a 15 cm long *M. glutinosa* (scale shows 0.1 mm). The ovary is attached to the intestine (i) by a mesovarium (m) throughout its total length. One can distinguish between the germinal epithelium (g) and the group of oocytes (o). A nerve (n) and the *vena supraintestinalis* (v) run along the intestine (after Patzner, 1974).

egg. The material of the filaments is secreted by the follicular epithelium (Lyngnes, 1930).

The ripening of the eggs and the production of large amounts of yolk, respectively, are dependent on the liver in *E. burgeri* as well as in *M. glutinosa* (Patzner, 1980a; Patzner and Adam, 1981). In *M. glutinosa* the hepatosomatic indices (HSI) were studied in different groups of animals (Patzner and Adam, 1981). Females with eggs longer than 10 mm show a statistically significant difference in their HSI from all other groups (females with small eggs, females with ovulated follicles, males, juveniles and sterile animals). In *E. burgeri,* which has an annual reproductive cycle, the HSI shows a large increase from December until migration to deeper water in July. From the beginning of April to the end of August the difference between male and female HSI is statistically significant (Patzner, 1980a).

Yu *et al.* (1980, 1981) studied the protein metabolism in liver and plasma during vitellogenesis and its control by steroid hormones in *E. stoutii.* They found that vitellogenic females have considerably higher levels of radioactive newly-labelled protein in their plasma than non-vitellogenic ones and males. This indicates that the amount of circulating vitellogenic protein concentrations is small compared to non-vitellogenic plasma proteins (Yu *et al.,* 1980). This protein represents a vitellogenin, since it (1) contains phosphorus, (2) is incorporated into egg yolk, and (3) can be induced to appear in the plasma by implanting estrogens.

25.2.3 The ripe egg

Within the course of one reproductive cycle only few eggs get mature. In 79 *M. glutinosa* that had large eggs, the number of eggs varied between 6 and 32, with a mean of 18 (Patzner, 1982). No correlation was observed between the body length and the number of eggs. Female *E. burgeri* (n = 73) had a mean number of 32 eggs varying from 13 to 67. In contrast to *Myxine* a distinct correlation between the number of large eggs in the ovary and the length of the females in *E. burgeri* could be found (Patzner, 1982).

In *E. stouti* between 12 and 45 large oocytes were observed in one single gonad (Koch *et al.,* 1993). There is no indication that the number of eggs is correlated to body length in this species.

Completely ripe eggs could never be obtained from freshly caught hagfishes, neither in *M. glutinosa* nor in *E. burgeri* (Patzner, 1982). Only when animals were kept in aquaria for two months or more were mature eggs found in the ovary, free in the peritoneal cavity or, rarely, at the bottom of the aquarium. Similar facts are reported from *E. stoutii* by Koch *et al.* (1993) but the authors do not mention how long they kept the animals in the aquaria. Fernholm (1975) obtained deposited eggs by keeping mature *E. burgeri* in cages in the sea for about two months.

When the eggs are released from the cloaca into the water, they are surrounded by a mucous layer and the anchor filaments are soft. After some hours the mucus dissolves and the filaments become harder.

The length of the egg of most hagfish species, excluding the anchor filaments, is around 20 mm, varying from 14 to 25 mm. The width measures around 8 mm, varying from 7.5 mm to 10 mm. In the middle region of the egg the shell is about 50 to 90 μm thick. Dean (1900) gives detailed measurements on the eggs of *M. glutinosa* and *E. stoutii.* The length of the eggs of *E. carlhubbsi* measure 77 mm (Martini, personal communication). In mature eggs a thin suture, the opercular ring, can be observed at the shell near the animal pole (Figure 25.2). Along this line the shell breaks when the young hagfish hatches (Walvig, 1963).

A remarkable structure on the eggs of hagfishes are the anchor filaments (Figure 25.2). These tuft-like appendages are formed on both poles of the egg of most hagfish species. The only exception known up to now are the eggs of *E. carlhubbsi* which lack anchor filaments (Martini, personal communication). The single filaments are between 3.2 and 4.9 mm long and have an anchor-shaped apical end. The number of filaments varies around 50 but is always smaller on the vegetative pole. Jensen (1901) describes six different types of filaments. On both ends of laid eggs the filaments are enveloped by a gel-like matrix intertwined with large cytoskeletal biopolymers (keratin-like intermediate filaments) of possible holocrine origin (Koch *et al.*, 1991).

To guarantee fertilization of eggs with a thick outer membrane, they have a pore called the micropyle, a structure found in all fishes. Fernholm (1975) was the first to study the eggs of *E. burgeri* by scanning electron microscopy. More detailed studies were later made by Kosmath *et al.* (1981a) in *M. glutinosa* and *E. burgeri*, and by Koch and Spitzer (1992) and Koch *et al.* (1993) in *E. stoutii*. The hagfish egg contains a single micropylar opening (Figure 25.2), located at the animal pole. It is situated at the bottom of a micropylar funnel or micropylar cup which lies between the anchor filaments. The diameter of the funnel differs in the three species: in *M. glutinosa* it measures 375 μm, in *E. stoutii* 325 μm and in *E. burgeri* 219 μm. The base of the funnel shows a hexagonal, honeycomb-like structure. The edges of these combs are frayed in *M. glutinosa* (Figure 25.2) but smooth in both *Eptatretus* species. The diameter of the micropylar canal measures 4.2 μm in *M. glutinosa*, 4.7 μm in *E. burgeri* and around 4 μm in *E. stoutii*. The area around the opening is structureless or only slightly structured, at least in *M. glutinosa* and *E. burgeri* (Kosmath *et al.*, 1981a).

In Japan there exist two patents describing the potential of hagfish eggs as a commercial food (Koch, personal communication). Up to now only very few fertilized eggs containing embryos of hagfishes have been discovered (Fernholm, 1969; Patzner, 1980b, 1982; Wicht and Tusch, this volume).

25.2.4 Regressive processes in the ovary

Lyngnes (1936) made a detailed study on degenerative processes in the ovary of *M. glutinosa*. Tsuneki and Gorbman (1977a) and Kosmath *et al.* (1983a, b, c, 1984a, b) completed these data with electron microscopical and histochemical investigations. Within the hagfish ovary one has to distinguish between the degeneration of the post-ovulatory follicles and several atretic processes.

During the regression of the post-ovulatory follicles (*corpora lutea*) a phase of reorganization and a phase of regression can be discerned (Lyngnes, 1936). In the cells of the *theca* secretory granules are formed during the degeneration. The regression of the granulosa cells is characterized by the appearance of lysosomal residual bodies and a yellow pigment. At the end of the regression phase, residues of cells are removed by blood. The net of blood vessels widens itself, originating from the *theca externa* towards the direction of the *theca interna* and the follicular epithelium. The endothelium of the capillaries is not fenestrated. In *M. glutinosa* no morphological indications for hormone-producing cells were found in the post-ovulatory follicles (Kosmath *et al.*, 1984a). Tsuneki and Gorbman (1977a) attained the same results in *E. stoutii*. Histochemical investigations of the post-ovulatory follicles revealed that the granulosa cells and the theca cells reorganize immediately after the ovulation and develop a high amount of lipid droplets (Tsuneki and Gorbman, 1977a; Kosmath *et al.*, 1984b). The required conditions for steroid synthesis – cholesterol as precursor, and a positive proof of 3β-hydroxysteroid dehydrogenase – were not found in the cells of the ovulated follicles by histochemical methods in *M. glutinosa* (Fernholm, 1972b; Kosmath *et al.*, 1984b). Schützinger *et al.* (1987) also found very low concentrations of estrogens in the plasma of *M. glutinosa* with ovulated follicles larger than 4 mm and ovulated follicles between 2 and 4 mm length.

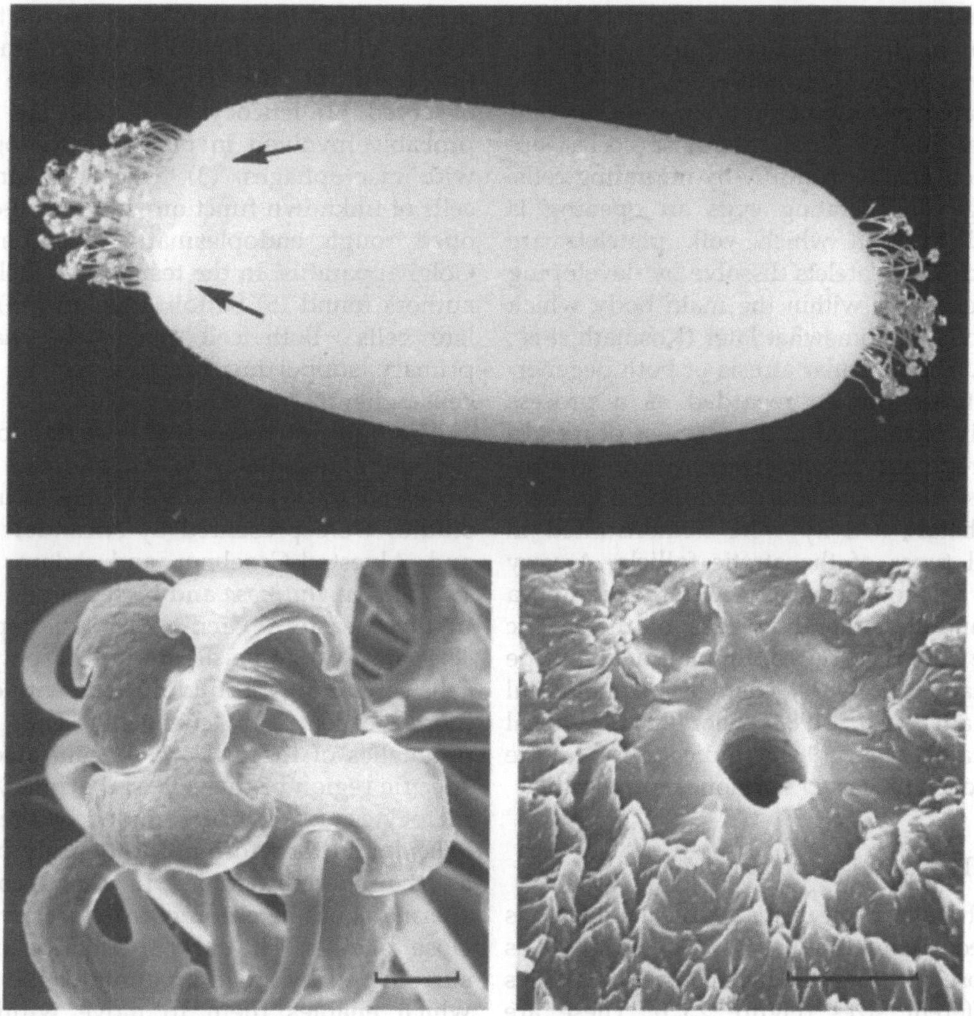

Figure 25.2 Mature egg of the hagfish. *Top*: The egg of *Eptatretus burgeri* has a length of 18.5 mm without the anchor filaments. Near the animal pole (left side) lies the opercular ring (arrows) along which the shell breaks when the young hagfish hatches (after Patzner, 1980b). *Bottom left*: Apical end of anchor filaments of a *Myxine* egg (scale shows 50 µm). Six different types of filaments are described (after Patzner, 1975). *Bottom right*: Outer opening of the micropylar canal of a *Myxine* egg surrounded by the hexagonal structures of the micropylar funnel (scale shows 10 µm). The hagfish egg contains one micropylar opening, located at the animal pole. It is situated at the bottom of the micropylar funnel which lies between the anchor filaments (after Kosmath *et al.*, 1981a).

In the atretic follicles (*corpora atretica*) one distinguishes between atretic follicles of the open involutionary type and of the closed involutionary type (Lyngnes, 1936). In the first ones which occur in larger follicles, an opening in the wall of the follicle is formed through which granulosa cells and yolk platelets are emitted. Migrating cells of the theca layer invade the follicular lumen and absorb phagocytotically residues of granulosa cells. On the other hand, atretic follicles of the closed involutionary type – occurring in smaller follicles – show yolk platelets which remain in the follicular lumen and are dissolved there (Kosmath *et al.*, 1983c). The granulated residue of the yolk platelets and the residue of the granulosa cells are absorbed phagocytotically by migrating cells. In large degenerating eggs an opening is formed through which yolk platelets are emitted. The platelets dissolve by developing small channels within the main body which may collapse somewhat later (Kosmath *et al.*, 1983a). The follicular atresia of both degenerating types can be regarded as a process exclusively devoted to the purpose of resorbing atretic oocytes. By histochemical investigations aryl sulphatase activity and acid phosphatase activity were demonstrated in all cell types of the atretic follicle. A very strong reaction was observed in the granulosa cells and the migrating cells of the atretic follicle of the closed involutionary type (Kosmath *et al.*, 1983c). Neither morphological (Kosmath *et al.*, 1983b) nor histochemical (Kosmath *et al.*, 1983c) indications for the production of steroid hormones were found.

25.2.5 Testis and spermatogenesis

Like the ovary the testis of the hagfishes is situated within the peritoneal cavity. It is composed of lobules which consist of follicles of different sizes (Figure 25.3). These are encapsulated by a thin layer of connective tissue. In one follicle only one developmental stage of spermatogenesis is found. In most species of hagfish, some follicles contain spermatogonia, others spermatocytes, spermatids and rarely spermatozoa. Only *E. burgeri* has a synchronous development within all the testicular follices (Patzner, 1977a).

The fine structure of the interstitial tissue of the hagfish testis was studied by Tsuneki and Gorbman (1977b). They found (1) interstitial cells which were morphologically equivalent to Leydig cells of higher vertebrates as they have a smooth endoplasmatic reticulum and tubular mitochondria. It is probable that they are the source of testosterone which was found in the plasma and the testis of hagfish. Furthermore they observed (2) leucocytes of vascular origin probably involved in phagocytosis together with macrophages, (3) fibroblasts, and (4) cells of unknown function with a well-developed rough endoplasmatic reticulum and Golgi apparatus. In the testicular follicles the authors found (5) Sertoli cells, and (6) 'stellate cells'. Both cell types may have a primary supporting function for the germ cells rather than a secretory role.

The light microscopical investigations on the spermatogenesis in *M. glutinosa* were reviewed by Walvig (1963). Fine structural studies were presented by Nicander (1968) and Alvestad-Graebner and Adam (1976, 1977) in *M. glutinosa* and by Jespersen (1975) in three eastern Pacific *Eptatretus* species. The spermatogonia are round or polyedric cells containing a round nucleus with one or more excentrical nucleoli. The other organelles of the cell are concentrated in a specific region of the cell. The primary spermatocytes are each surrounded by the *liquor folliculi* and therefore not as densely packed as the spermatogonia. In the secondary spermatozoa the chromosomes are arranged close together. The early spermatids are still surrounded by small amounts of the liquor which enables them to move within the follicle. The cells are spherical and measure 13 to 14 µm in diameter. First the nucleus is situated in the centre of the cell, later it moves towards the anterior pole. At this stage the acrosome is formed between the nucleus and the plasma membrane. The secondary spermatids are longitudinal with a longitudinal nucleus. They are characterized by numerous microtubules parallel to the axis of the cell. The acrosome is disc-like and covers the anterior end of the nucleus and the centriole becomes situated on one side of the nucleus. The outgrowing flagellum is surrounded by a few elongated mitochondria.

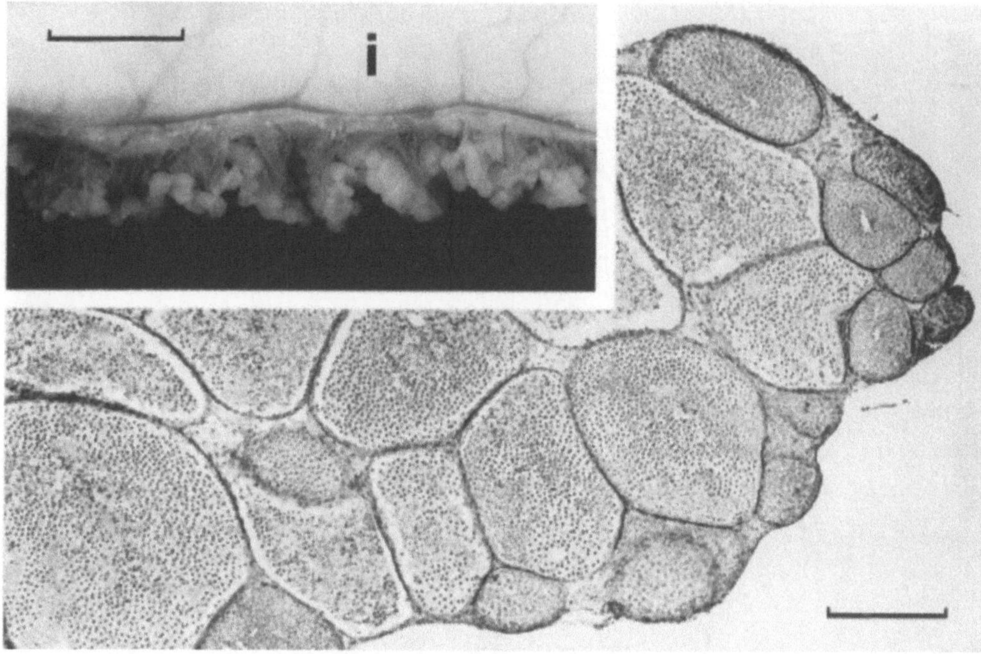

Figure 25.3 Testis of the hagfish. Cross-section through a testis of *Eptatretus burgeri* (scale shows 250 μm). The testis which is situated within the peritoneal cavity is composed of follicular lobules of different sizes. In one lobule only one developmental stage of spermatogenesis is found (after Patzner, 1982). *Insert*: Mature testis of *Paramyxine atami* (scale shows 10 mm). The testicular lobules are fixed to the intestine (i) by a mesenterium (after Patzner, 1982).

25.2.6 The ripe spermatozoon

Jespersen (1975) found among 1000 *M. glutinosa* only one animal containing motile sperm. A higher frequency of mature males was found by her in *E. stoutii* and *E. deani*. In more than 2000 males of *M. glutinosa* and in *E. burgeri* none had a mature testis (Patzner, 1982). Only in a single specimen of another species, the Japanese *Paramyxine atami*, ripe testis follicles containing some spermatozoa were observed (Patzner, 1982) (Figure 25.3). Morisawa (1995) kept *E. burgeri* in a seawater tank during the spawning season and in this way he obtained testes with mature spermatozoa.

Up to now the fine structure of mature hagfish sperm has only been investigated in *E. stouti* by Jespersen (1975) and in *E. burgeri* by Morisawa (1995) (Figure 25.4). Some information on the *Myxine* spermatozoon was given by Nicander (1968, 1970). Measurements of spermatozoa of the three different species are given in Table 25.1. In total it is about 40 to 65 μm long and consists of a head with an acrosome, a midpiece and a 9 + 2 flagellum lacking accessory fibres. The nucleus is elongated with a tapering end on each side. The chromatin is highly condensed. The centriole and the base of the flagellum are situated near the posterior end of the nucleus. The cap-shaped acrosome is only about 1 μm long and covers a subacrosomal substance and the anterior tip of the nucleus (Figure 25.4). The shape of the acrosome seems to show variations in the different species of hagfish (Nicander, 1970; Jespersen, 1975).

Spermatozoa of hagfishes have similar features to those of chondrichthyes (acrosome, very elongated nucleus, well-developed

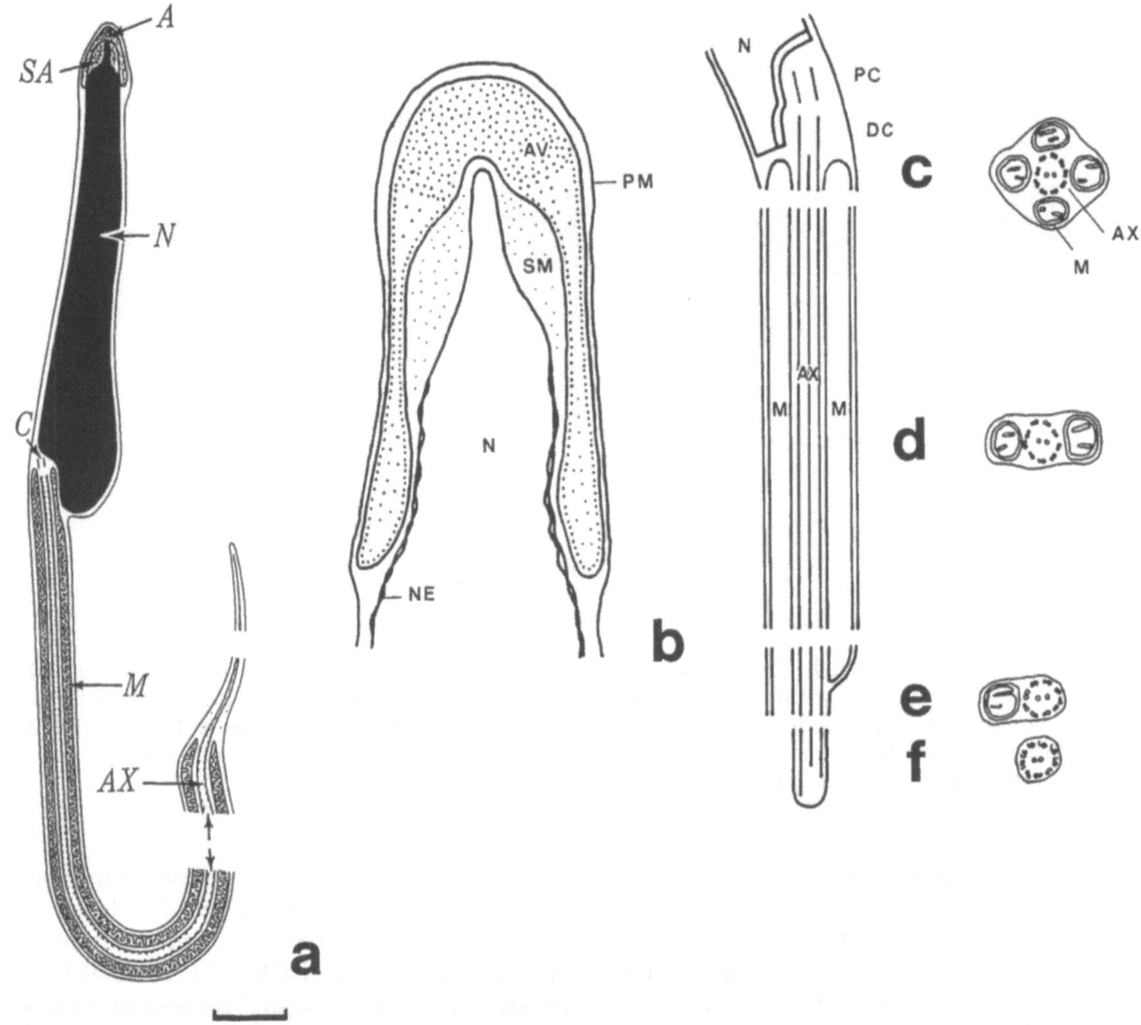

Figure 25.4 Mature spermatozoa of the hagfish. Schematic drawings of *Eptatretus* spermatozoa. (a) *E. stoutii* (scale shows 2 µm) (from Jespersen, 1975) and (b) to (f) *E. burgeri* (from Morisawa, 1995). (b) acrosomal region, (c) to (f) longitudinal and cross-section of the flagellum. A = AV = acrosome, AX = axoneme, C = centriole, DC = distal centriole, M = mitochondria, N = nucleus, NE = nuclear envelope, PC = proximal centriole, PM = plasma membrane, SA = SM = subacrosomal material.

midpiece). Differences exist especially in the number of mitochondria: with almost similar dimensions, the midpiece of chondrostean spermatozoa contains numerous small mitochondria (Jamieson, 1991), while that of *Eptatretus* has only four. In *E. burgeri* two of the mitochondria, and in the other *Eptatretus* species all mitochondria, are extremely elongated. Spermatozoa of *Lampetra* are different as they have an endonuclear canal and a smaller midpiece region. The spermatozoa of teleosts differ from those of hagfishes: they have no acrosomes, their nuclei are less elongated, mostly only spherical or ovoid, and the midpiece is small (only 2 to 4 µm long).

The formation of an acrosome in hagfishes

Table 25.1 Measurements of mature spermatozoa in three species of hagfish

	M. glutinosa (Walvig, 1963)	*E. stoutii* (Jespersen, 1975)	*E. burgeri* (Morisawa, 1995)
Head length	10–12 μm	12 μm	8–10 μm
Head width	2.5–3 μm	1.5 μm	0.5–1.2 μm
Midpiece length	25–30 + 14–18 μm*	20 μm	35 μm†
Endpiece length	4–6 μm	?	5 μm†
Acrosome length	?	1 μm	1.2 μm†
Total length	60–65 μm	40–60 μm	45–50 μm

* After Schreiner and Schreiner (1908), the midpiece consists of a main piece and a middle piece.
† Measured on photographs of Morisawa (1995).

is a particularity as usually there is a high positive correlation between the presence of egg micropyles and the absence of an acrosome (Jamieson, 1991). Holostei and Teleostei have eggs with a micropyle but their spermatozoa lack an acrosome whereas lamprey eggs have no micropyle but their spermatozoa have an acrosome. Only the sturgeon, *Acipenser*, has eggs with (several) micropyles and spermatozoa with an acrosome, like hagfishes (Cherr and Clark, 1984). The acrosome, the very elongated nucleus and the well-developed and long midpiece region of hagfish spermatozoa are features characteristic of internally fertilizing fishes. Even in the more simply constructed teleost sperm cells, those with internal fertilization show tendencies for elongation of the head and for an increase of the midpiece region (Stanley, 1969; Grier, 1973).

In teleosts the sperm motility is of short duration due to the low efficiency of the tricarboxylic acid cycle and the oxidative phosphorylation rate (Jamieson, 1991). This may be a consequence of the small midpiece region with only few mitochondria. Species with a well-developed midpiece, as internal fertilizers, have the ability to move for an extended period (Gardiner, 1978). Therefore the well-developed midpiece region of the hagfish spermatozoon indicates a long period of motility.

25.2.7 Sterile animals and hermaphrodites

In sexually sterile hagfishes two groups may be distinguished: The first is completely sterile; neither a testis nor an ovary can be found; in the second, gonads consist of a testicular and ovarial part developed abnormally and exhibiting degenerative traits. Lyngnes (1931) and Schreiner (1955) give detailed descriptions of sterile animals in *Myxine* which are summarized by Walvig (1963).

Between 1971 and 1980 up to 1300 *M. glutinosa* were caught in Norway and their gonads were studied by the author. In 215 adult individuals the gonads were studied thoroughly with regard to sterility. After sex determination 14.4% of the animals could be considered sterile. Of these, 77.4% had no gonads, and 22.6% had an ovary as well as a testis each with strong signs of degeneration. In more than 1000 gonads of *M. glutinosa* no hint of functional hermaphroditism could be found (Patzner, 1982).

E. burgeri living in Japanese waters is the only known hagfish species having a clear reproductive cycle with a definitive spawning season (Patzner, 1977a, 1978). This means that at a certain time of the year all animals attain the same stage of gonadal development (see below). Out of 916 animals caught in 1975 and 1976, 52.2% turned out to be female (Patzner, 1982). In 160 of these the ovary was studied more thoroughly. In 1.2% of the animals all eggs were degenerated and 3.7% had essentially smaller eggs than the other animals. In the male specimens around 10% of the animals had a less developed testis with a low gonadosomatic index. However, in histological examinations no unusual signs of degeneration were observed.

Only one single individual out of the 916 *Eptatretus* caught in February 1976 had developed a testis as well as an ovary. The eggs had a length of 10.5 mm, which accords with the average size of the eggs of normally developed females. Also the stages of the spermatogenesis in the testicular part agrees with those of other males from the same season. So a functional hermaphroditism in 0.1% (or even less) of *E. burgeri* was presumed (Patzner, 1977b). Gorbman (1983, 1990) depicts the gonad of an adult hermaphroditic *E. stoutii*. He found two overt hermaphrodites out of approximately 3000 adult animals. He could show clearly that there is no overlap between the structure of the testis and of the ovary. The large eggs in this animal measure between 14 and 18 mm, though their number is small and the number of atretic egg follicles in the mesovarium is high.

25.2.8 The cloaca

There is no copulatory organ in hagfishes. No detectable differences in the anatomy of the cloaca between male and female were found in *M. glutinosa* (Kosmath *et al.*, 1981b). On the other hand, Conel (1931) and Tsuneki *et al.* (1985) found a sex difference in size of the cloacal glands in *E. stoutii* and *E. burgeri* before the breeding season. But it appears that there are some differences in the anatomy of the cloacal gland between *Eptatretus* and *Myxine* (Burne, 1898; Kosmath *et al.*, 1981b; Tsuneki *et al.*, 1985).

An approximately 10 mm long slit at the ventral side of the body leads into the cloacal chamber. Four openings meet and empty into it. Two of them, the *rectum* below and the *ductus coelomaticus*, serving as genital pore, above are placed vertically above each other. A pair of smaller openings (*ducti segmentalii*) is situated on a papilla (Kosmath *et al.*, 1981b). The cloacal gland is a long tube filled with mucus, surrounded by a thin layer of connective tissue and a thin layer of striated muscle (Tsuneki *et al.*, 1985). Like the lateral mucous glands in the epidermis it consists of large mucous gland cells, thread cells and undifferentiated cells (Kosmath *et al.*, 1981b). In male *Eptatretus* the gland gets larger towards the spawning season, suggesting that their function is related to reproduction (Tsuneki *et al.*, 1985). It is probable that sperm is released as a sperm mass mixed with mucus. Such a slimy sperm mass may adhere to the surface of deposited eggs or even be incorporated into the female cloaca. It can be maintained that a sperm groove, as exists in the cloaca of female Necturus, or a ciliary channel, as is found in the caudal region of male *Necturus* (Dawson, 1922; Wahlert, 1953/54), is totally absent in *Myxine*. The cloacal epithelium neither develops a spatial separation by folds nor is there any ciliation in the caudal and dorsal part of the cloacal chamber of female myxinoids. Therefore female myxinoids do not show any structures which would allow transportation of sperm into the abdominal cavity. Thus, internal fertilization in hagfishes does not seem to be possible (Kosmath *et al.*, 1981b).

25.3 REPRODUCTIVE CYCLE

Walvig (1963) did not mention any seasonal reproductive cycle in *M. glutinosa*. To verify this, a total of 582 ovaries of *Myxine* were investigated in spring (March, May) over a period of three years (Patzner, 1982). Thereby the stages of the ovary were divided into 4 groups: (1) females after ovulation containing ovulated follicles larger than 4 mm, (2) ovulated follicles with a length of 3 to 4 mm, (3) eggs with a length of 4 to 10 mm, and (4) eggs larger than 10 mm. It turned out that in the spring of three different years none of the stages was found to dominate (Patzner, 1982). Therefore it can be stated that there is no synchronized reproductive cycle in *M. glutinosa*. The same was found in the eastern Pacific hagfish, *E. stoutii* (Matty *et al.*, 1976).

In contrast to *M. glutinosa* and *E. stoutii*, a yearly reproductive cycle was found in the Japanese hagfish *E. burgeri* (Kobayashi *et al.*, 1972; Patzner, 1977a, 1978; Tsuneki *et al.*, 1983). After spawning in mid October the animals migrate back from about 100 m deep water to shallow areas near the coast. In November male and female gonads begin a period of growth. The gonadosomatic index of male *Eptatretus* increases sevenfold until spawning in the beginning of October. In the female *Eptatretus* the 1 to 2 mm large eggs start their regular growth in November. In September of the next year when the final egg length of 22 to 25 mm is reached, the shell develops within the follicle. Only when it is completely formed are anchor filaments built on both poles of the eggs. After ovulation the eggs remain a certain time within the peritoneal cavity, and they are then all deposited together. As far as we know, *E. burgeri* is the only hagfish having a seasonal reproductive cycle.

The absence of synchronized spawning is not necessarily a particularly primitive trait. There are other lower vertebrates that show no definite spawning season (Wourms, 1977).

25.4 SPAWNING

At present, little if anything is known about mating or deposition and fertilization of the eggs in any species of hagfish. As myxinoids lack copulatory organs, a fertilization outside the body can be assumed (Kosmath *et al.*, 1981b). Also the fine structure of sperm supports this opinion (see section 25.2.6). Especially noteworthy is the scanty number of eggs and their large volume. This is analogous to some elasmobranchs which, however, all have an internal fertilization (Hoar, 1969). Equally astonishing is the small size of the testis and the scanty amount of sperm. The low number seems to be inadequate for a successful fertilization in the open water. Hagfishes hide in burrows in the soft muddy bottom (for details see Martini, this volume). After Fernholm's (1974) and the author's observations on *E. burgeri* in the shallow waters of Japan the animals prefer ready-made holes but can also make a new burrow by swimming into the mud. Sometimes two or three hagfishes are present within one opening. However, no observations were made in the spawning areas of *E. burgeri* which are at more than 100 m depth. Because of the very small number of eggs and sperm it is obvious that a male and a female have to get in close contact for fertilization to occur. However, it is not clear if the males search for the females or vice versa. The localization could only be possible by the olfactory organ, as it is completely dark where they live and they have almost no eyes. The olfactory organs are well developed but there seems to be no sexual differences in size (Ross, 1963) which would favour one sex finding the other. It is a question whether the small amounts of steroid hormones produced in the ovary (see section 25.6) are sufficient to act as a pheromone (Stacey *et al.*, 1986).

Sex ratio is strongly unequal in *M. glutinosa* (Nansen, 1887; Conel, 1931) and in *E. stoutii* (Gorbman and Dickhoff, 1978; Gorbman, 1983). In these species the female sex is more frequent than the male. However, in *E. burgeri* the sex ratio is nearly equal (female to male = 52 to 48) (Patzner, 1982).

It does not seem possible to catch female hagfishes with completely ripe eggs by baited traps. The few freshly caught males containing ripe sperm had such a small amount of spermatozoa in their testis that it is assumed that these are leftovers from the previous spawning (Jespersen, 1975; Patzner, 1982). The reason may be that shortly before spawning animals stop eating and therefore cannot be caught by baited traps. This is known from several teleost fishes (Smigielski, 1975). Walvig (1967/68) does not exclude a spawning migration into other areas in *M. glutinosa* as an explanation of the fact that ripe animals are not trapped. On the other hand it is known that *E. burgeri* migrates into deeper water prior to spawning (Kobayashi *et al.*, 1972) but can be trapped there just before they become completely mature (Patzner, personal observation).

Until now the Japanese *E. burgeri* is the only species of hagfish in which a synchronous gonadal cycle could be demonstrated. *Eptatretus cirrhatus* from New Zealand is the only other species which also lives in shallow coastal waters (Fernholm and Holmberg, 1975). However, it is not known whether *E. cirrhatus* has a synchronous gonadal cycle and whether it migrates to deeper water to spawn.

25.5 HORMONAL CONTROL OF THE GONADS

Not only in elasmobranch and teleost fishes (Pickford and Atz, 1957) but also in lampreys (Falkmer *et al.*, 1974) the mechanisms of regulation of the function of the gonads by the adenohypophysis are rather well understood. In these animals it was clearly demonstrated that the presence of the adenohypophysis is necessary for the completion of reproductive processes. Whilst in lampreys clear indications for the secretion of gonadotropins already exist (Larsen, 1973), the function of the adenohypophysis in hagfishes is largely unclear. This organ obviously presents the most primitive form of organization of the vertebrate hypophysis (Adam, 1963; Kobayashi and Uemura, 1972).

The gonadotropin-producing cells in the adenohypophysis in vertebrates are agranular, PAS-positive cells (Pickford and Atz, 1957). Fine structural investigations of the adenohypophysial region in *M. glutinosa* have revealed the presence of agranular cells (Fernholm, 1972c, Schultz *et al.*, 1979; Patzner *et al.*, 1982). Their function has remained unclear, however. In *E. burgeri* histological studies exhibited very few PAS-positive cells (Patzner, 1982).

Matty *et al.* (1976) came to the conclusion that the gonads of *E. stoutii* are not regulated by the adenohypophysis. However, they raised the question whether this might be the fact in *E. burgeri*, having a distinct cycle of reproduction in contrast to *E. stoutii*. Patzner and Ichikawa (1977) performed a hypophysectomy in *E. burgeri* and could show that a certain gonadotrophic factor exists within the adenohypophysis. Shortly after the onset of the reproductive cycle, hypophyses of male animals were ectomized. After three months in the aquarium it was shown that in sham operated and control animals the gonadosomatic index was normally increased. In the ectomized *Eptatretus* no increase could be found, however. In these animals the progress of the spermatogenesis was not stopped but showed a strong delay in comparison to the sham operated animals.

A certain influence of the adenohypophysis on the gonads in *E. burgeri* was also revealed by a radioreceptor assay in the course of which a testis of a pigeon for an assay of FSH and a testis of a rat of LH were used as receptors (Patzner, Ishii and Adachi, not published). They were incubated with ^{125}I-FSH and ^{125}I-LH, respectively, and an extract of glycoproteins of the *Eptatretus* hypophysis. It was shown that small amounts of FSH and LH-like substances were found in the adenohypophysis. However, it could not be clarified how far the adenohypophysis is necessary for the completion of reproduction.

Tsuneki (1976) examined the effects of estradiol and testosterone on the pituitary and the gonads of *E. burgeri*. None of the steroids caused a conspicuous change in the cells of the adenohypophysis. Testosterone had an effect neither on the gonads nor on the testis or the ovary. Estradiol caused a degeneration of the ovary but this might be due to a pharmacological effect. These results suggested the absence of an adenohypophysio-gonadal feedback system in *E. burgeri*.

Earlier attempts to identify gonadotropin-releasing hormone (GnRH) in the brain of hagfishes showed that either no reactivity was present in *E. stoutii* (Crim *et al.*, 1979; Sherwood and Sower, 1985) and in *E. burgeri* (Nozaki and Kobayashi, 1979), or that only small amounts of GnRH-like peptides were found in *E. stoutii* (Jackson, 1980), in *E. hexatrema* (King and Millar, 1980) and in *M. glutinosa* (Blähser *et al.*, 1989; Knox and Sower, 1991). In a recent study Braun *et al.* (1995) described the distribution of GnRH-like immunoreactivity in the brain of *E. stoutii*. Their findings indicate that the brain-pituitary axis of hagfishes is more similar to that of other vertebrates than had been previously suggested. They considered the GnRH-like system of myxinoids as a possible homologue of the caudal neuromodulatory GnRH system of higher vertebrates.

25.6 STEROID PRODUCTION IN THE GONADS

Fernholm (1972a, b) could not demonstrate a production of steroid hormones in the ovary of *M. glutinosa*. However, Callard and Fedele (not published, in Lance and Callard, 1978), Hansson *et al.* (1979) and Schützinger *et al.* (1987) could find sexual steroids in *M. glutinosa* by using more accurate methods. Also in *E. stoutii* (Matty *et al.*, 1976; Gorbman and Dickhoff, 1978; Weisbart *et al.*, 1980) as well as in *E. burgeri* (Hirose *et al.*, 1975; Inano *et al.*, 1976; Tamaoki, 1979) indications for the production of steroids by the gonads were brought forth. Kime and Hews (1980) and Kime *et al.* (1980) identified the following metabolites of testosterone in the testis of *M. glutinosa*: androstenedione, 6α-hydroxytestosterone, 5α-androstane-3β, 7α, 17-triol, 5α-androstane-3β,6β,17β-triol, and a conjugate of testosterone. Testosterone was identified as a metabolite of progesterone. However, the amounts of these sexual steroids are very low. They only have one thousand of the concentration of most other vertebrates (Gorbman, 1979).

ACKNOWLEDGEMENTS

This study was made possible by several grants of the Austrian 'Fonds zur Förderung der wissenschaftlichen Forschung' and by scholarships of the Norwegian 'Kgl. Norske Utenriksdepartement' and of the Japan Society for the Promotion of Science. Especially I want to thank Hans Adam (University of Salzburg), Hideshi Kobayashi (Tokyo University) and Finn Walvig (Oslo University) for their help and comments in different phases of the work. Furthermore, I acknowledge the help of Ingrid Alvestad-Graebner, Aubrey Gorbman, Gabriele Erhart, Bo Fernholm, Tomoyuki Ichikawa, Susumo Ishii, Ilse Kosmath, Franz Lahnsteiner, Masumi Nozaki, Anne-Marie Patzner, Siegfried Schützinger, Peter Simonsberger, Kazuhiko Tsuneki, Johanna Üblagger, and many others.

REFERENCES

Adam, H. (1963) The pituitary gland, in *The Biology of Myxine* (eds A. Brodal and R. Fänge), Universitetsforlaget, Oslo, pp. 459–76.

Alvestad-Graebner, I. and Adam, H. (1976) Relationship between the chromatoid body and the acrosomal system in early spermatids in *Myxine glutinosa* L. *Cell and Tissue Research*, **174**, 427–30.

Alvestad-Graebner, I. and Adam, H. (1977) Zur Feinstruktur der spermatogenetischen Stadien von *Myxine glutinosa* L. (Cyclostomata). *Zoologica Scripta*, **6**, 113–26.

Blähser, S., King, J.A. and Kuenzel, W.J. (1989) Testing of arg-8-gonadotropin-releasing hormone-directed antisera by immunological and immunocytochemical methods for use in comparative studies. *Histochemistry*, **93**, 39–48.

Braun, C.B., Wicht, H. and Northcutt, R.G. (1995) Distribution of gonadotropin-releasing hormone immunoreactivity in the brain of the Pacific hagfish *Eptatretus stoutii* (Craniata: Myxinoidea). *Journal of Comparative Neurology*, **353**, 464–76.

Burne, R.H. (1898) The 'porus genitalis' in the Myxinidae. *Journal of the Linnean Society of London*, **26**, 487–95.

Cherr, G.M. and Clark, W.H. (1984) An acrosome reaction in sperm from the white sturgeon, Acipenser transmontanus. *Journal of Experimental Zoology*, **232**, 129–39.

Conel, J.L. (1931) The genital system of the Myxinoidea: a study based on notes and drawings of these organs in *Bdellosotoma* made by Bashford Dean, in *The Bashford Dean Memorial Volume: Archaic Fishes* (ed. E.W. Gudger), American Museum of Natural History, New York, pp. 67–102.

Crim, J., Urano, A. and Gorbman, A. (1979) Immunocytochemical studies of luteinizing hormone-releasing hormones in brain of agnathan fishes. I. Comparisons of adult Pacific lamprey (*Entosphenus tridentata*) and the Pacific hagfish (*Eptatretus stouti*). *General and Comparative Endocrinology*, **37**, 294–305.

Cunningham, J.T. (1886) On the structure and development of the reproductive elements in *Myxine glutinosa* L. *Quarterly Journal of Microscopical Science*, **27**, 49–76.

Dawson, A.B. (1922) The cloaca and the cloacal glands of the male Necturus. *Journal of Morphology*, **36**, 447–63.

Dean, B. (1900) The egg of the hagfish *Myxine glutinosa*, Linnaeus. *New York Academy of Sciences*, **2**, 33–45.

Falkmer, S., Thomas, N.W. and Boquist, L. (1974) Endocrinology of Cyclostomata, in *Chemical Zoology*, Vol. 8 (eds Florkin and Scheer), pp. 195–257.

Felix, W. and Bühler, A. (1906) Die Entwicklung der Keimdrüsen und ihrer Ausfuhrgänge, in *Handbuch der vergleichenden und experimentellen Entwicklungslehre der Wirbeltiere* (ed. O. Hertwig), Vol. III, Part I.

Fernholm, B. (1969) A third embryo of *Myxine*. Considerations on hypophysial ontogeny and phylogeny. *Acta Zoologica (Stockholm)*, **50**, 169–77.

Fernholm, B. (1972a) *Pituitary and ovary of the Atlantic hagfish. An endocrinological investigation.* Thesis, University of Stockholm.

Fernholm, B. (1972b) Is there any steroid hormone formation in the ovary of the hagfish, *Myxine glutinosa*? *Acta Zoologica (Stockholm)*, **53**, 235–42.

Fernholm, B. (1972c) The ultrastructure of the adenohypophysis of *Myxine glutinosa*. *Zeitschrift für Zellforschung*, **132**, 451–72.

Fernholm, B. (1974) Diurnal variations in the behaviour of the hagfish *Eptatretus burgeri*. *Marine Biology*, **27**, 351–6.

Fernholm, B. (1975) Ovulation and eggs of the hagfish *Eptatretus burgeri*. *Acta Zoologica (Stockholm)*, 56, 199–204.

Fernholm, B. and Holmberg, K. (1975) The eyes in three genera of hagfish (*Eptatretus, Paramyxine and Myxine*) – a case of degenerative evolution. Vision Research, **15**, 253–9.

Gardinier, D.M. (1978) Utilization of extracellular glucose by spermatozoa of two viviparous fishes. *Comparative Biochemistry and Physiology*, **59**, 165–8.

Gorbman, A. (1979) Endocrine regulatory patterns in Agnatha: primitive or degenerate? *International Symposium of Hormones and Evolution in Japan*, p. 10.

Gorbman, A. (1983) Reproduction in cyclostome fishes and its regulation, in *Fish Physiology*, Vol. IX (eds W.S. Hoar, D.J. Randall and E.M. Donaldson), Academic Press, New York, pp. 1–30.

Gorbman, A. (1990) Sex differentiation in the hagfish *Eptatretus stoutii*. *General and Comparative Endocrinology*, **77**, 309–23.

Gorbman, A. and Dickhoff, W.W. (1978) Endocrine control of reproduction in hagfish, in *Comparative Endocrinology* (eds J. Gaillard and H.H. Boer), Elsevier, Amsterdam, pp. 49–54.

Grier, H.J. (1973) Aspects of germinal cyst and sperm development in *Poecilia latipinna* (Teleostei: Poeciliidae). *Journal of Morphology*, **146**, 229–50.

Hansson, T., Rafter, J. and Gustafsson, J.A. (1979) A comparative study on the hepatic in vitro metabolism of 4-androstene-3,17-dione in the hagfish, *Myxine glutinosa*, the dogfish, *Squalus acanthias*, and the rainbow trout, *Salmo gairnerii*. *General and Comparative Endocrinology*, **37**, 240–5.

Hirose, K., Tamaoki, B., Fernholm, B. and Kobayashi, H. (1975) *In vitro* bioconversions of steroids in the mature ovary of the hagfish, *Eptatretus burgeri*. *Comparative Biochemistry and Physiology*, **51B**, 403–8.

Hoar, W.S. (1969) Reproduction, in *Fish Physiology*, Vol. III (eds W.S. Hoar and D.J. Randall), Academic Press, New York, London, pp. 1–72.

Inano, H., Mori, K. and Tamaoki, B. (1976) *In vitro* metabolism of testosterone in hepatic tissue of a hagfish, *Eptatretus burgeri*. *General and Comparative Endocrinology*, **30**, 358–66.

Jackson, I.M.D. (1980) Distribution and evolutionary significance of the hypophysiotropic hormones of the hypothalamus. *Frontiers of Hormone Research*, **6**, 35–69.

Jamieson, B.G.M. (1991) *Fish Evolution and Systematics: Evidence from Spermatozoa*, Cambridge University Press, Cambridge.

Jensen, A.S. (1901) Om Slimaalens Æg. *Videnskabelige Meddeleser fra naturhistorisk Forening i Kobenhaven for Aaret 1900*, pp. 1–14.

Jespersen, A. (1975) Fine structure of spermiogenesis in Eastern Pacific species of hagfish (Myxinidae). *Acta Zoologica (Stockholm)*, **56**, 189–98.

Kime, D.E. and Hews, E.A. (1980) Steroid biosynthesis by the ovary of the hagfish *Myxine glutinosa*. *General and Comparative Endocrinology*, **42**, 71–75.

Kime, D.E., Hews, E.A. and Rafter, J. (1980) Steroid biosynthesis of the hagfish *Myxine glutinosa*. *General and Comparative Endocrinology*, **41**, 8–13.

King, J.A. and Millar, R.P. (1980) Comparative aspects of luteinizing hormone-releasing hormone structure and function in vertebrate phylogeny. *Endocrinology*, **106**, 707–17.

Knox, C.J. and Sower, S.A. (1991) Multiple forms of gonadotropin-releasing hormone (GnRH) in four species of teleosts. *American Zoologist*, **31**, 20A.

Kobayashi, H. and Uemura, H. (1972) The neurohypophysis of the hagfish, *Eptatretus burgeri* (Girard). *General and Comparative Endocrinology, Supplement*, **3**, 114–24.

Kobayashi, H., Ichikawa, T., Suzuki, H. and Sekimoto, M. (1972) Seasonal migration of the hagfish *Eptatretus burgeri*. *Japanese Journal of Ichthyology*, **19**, 191–4.

Koch, E.A., Spitzer, R.H., Pithawalla, R.B., Wilson, L.J. and Castillos, F.A. III (1991) Metabolic-morphologic events at a late stage of oogenesis in the hagfish. *Journal of Cell Biology*, **115**, 47A.

Koch, E.A. and Spitzer, R.H. (1992) Three-dimensional view of the micropylar end of the hagfish egg at late developmental stages. *American Zoologist*, **32**, 82A.

Koch, E.A., Spitzer, R.H., Pithawalla, R.B., Castillos, F.A. III and Wilson, L.J. (1993) The hagfish oocyte at late stages of oogenesis: structural and metabolic events at the micropylar region. *Tissue and Cell*, **25**, 259–73.

Kosmath, I., Patzner, R.A. and Adam, H. (1981a) The structure of the micropyles of the eggs of *Myxine glutinosa* and *Eptatretus burgeri* (Cyclostomata). *Zoologischer Anzeiger* (Jena), **206**, 273–8.

Kosmath, I., Patzner, R.A. and Adam, H. (1981b) The cloaca of *Myxine glutinosa* (Cyclostomata): a scanning electron microscopical and histochemical investigation. *Zeitschrift für mikroskopisch-anatomische Forschung*, **95**, 936–42.

Kosmath, I., Patzner, R.A. and Adam, H. (1983a) Regression im Ovar von *Myxine glutinosa* L. (Cyclostomata). I. Abbau der Dotterplättchen in den atretischen Follikeln. *Zoologischer Anzeiger* (Jena), **211**, 273–6.

Kosmath, I., Patzner, R.A. and Adam, H. (1983b) Regression im Ovar von *Myxine glutinosa* L. (Cyclostomata). II. Elektronenmikroskopische Untersuchungen an den atretischen Follikeln. *Zeitschrift für mikroskopisch-anatomische Forschung*, **97**, 667–74.

Kosmath, I., Patzner, R.A. and Adam, H. (1983c) Regression im Ovar von *Myxine glutinosa* L. (Cyclostomata). IV. Histochemische Untersuchungen an den atretischen Follikeln. *Zeitschrift für mikroskopisch-anatomische Forschung*, **97**, 941–7.

Kosmath, I., Patzner, R.A. and Adam, H. (1984a) Regression im Ovar von *Myxine glutinosa* L. (Cyclostomata). III. Untersuchungen der Feinstruktur post-ovulatorischer Follikel. *Zoologischer Anzeiger (Jena)*, **212**, 129–38.

Kosmath, I., Patzner, R.A. and Adam, H. (1984b) Regression im Ovar von *Myxine glutinosa* L. (Cyclostomata). V. Histochemische Untersuchungen an den post-ovulatorischen Follikeln. *Zoologischer Anzeiger (Jena)*, **212**, 309–16.

Lance, V. and Callard, I.P. (1978) Hormonal control of ovarian steroidogenesis in nonmammalian vertebrates, in *The Vertebrate Ovary* (ed. R.E. Jones), Plenum, New York, pp. 361–407.

Larsen, L.O. (1973) *Development in adult freshwater river lamprey and its hormonal control. Starvation, sexual maturation and natural death.* Thesis, University of Copenhagen.

Lyngnes, R. (1930) Beiträge zur Kenntnis von *Myxine glutinosa* L. I. Über die Entwicklung der Eihülle bei *Myxine glutinosa*. *Zeitschrift für Morphologie und Ökologie der Tiere*, **19**, 591–608.

Lyngnes, R. (1931) Über atretische und hypertrophische Gebilde im Ovarium der *Myxine glutinosa* L. *Biologisches Zentralblatt*, **51**, 437–41.

Lyngnes, R. (1936) Rückbildung der ovulierten und nicht ovulierten Follikel im Ovarium der *Myxine glutinosa* L. *Skrifter norske Videnskaps-Academi i Oslo l. matematisk-naturvidenskapelige Klasse*, **4**, 1–116.

Matty, A.J., Tsuneki, K., Diskhoff, W.W. and

Gorbman, A. (1976) Thyroid and gonadal function in a hypophysectomized hagfish, *Eptatretus stoutii. General and Comparative Endocrinology*, **30**, 500–16.

Morisawa, S. (1995) Fine structure of spermatozoa of the hagfish *Eptatretus burgeri* (Agnatha). *Biological Bulletin*, **189**, 6–12.

Nansen, F. (1887) A proterandric hermaphrodite (*Myxine glutinosa* L.) amongst the vertebrates. *Bergens Museums Aarsberetning*, **7**, 3–39.

Nicander, L. (1968) Gametogenesis and the ultrastructure of germ cells in vertebrates. *Proceedings of the 6th International Congress of Artificial Reproduction of Animals, Paris*, **1**, 87–107.

Nicander, L. (1970) Comparative studies on the fine structure in vertebrate spermatozoa, in *Comparative Spermatology* (ed. B. Bacchetti), Academic Press, New York, pp. 47–55.

Nozaki, M. and Kobayashi, H. (1979) Distribution of LHRH-like substance in the vertebrate brain as revealed by immunohistochemistry. *Archivum Histologicum Japonicum*, **42**, 201–19.

Patzner, R.A. (1973) *Progression und Regression während der Oogenese von Myxine glutinosa* L. (*Myxinoidea, Cyclostomata*). Thesis, University of Salzburg.

Patzner, R.A. (1974) Die frühen Stadien der Oogenese bei *Myxine glutinosa* L. (Cyclostomata). Licht- und elektronenmikroskopische Untersuchungen. *Norwegian Journal of Zoology*, **22**, 81–93.

Patzner, R.A. (1975) Die fortschreitende Entwicklung und Reifung der Eier von *Myxine glutinosa* L. (Cyclostomata). Licht- und elektronenmikroskopische Untersuchungen. *Norwegian Journal of Zoology*, **23**, 111–20.

Patzner, R.A. (1977a) Cyclical changes in the testis of the hagfish *Eptatretus burgeri* (Cyclostomata). *Acta Zoological* (*Stockholm*), **58**, 223–6.

Patzner, R.A. (1977b) Ein Fall von funktionellem Hermaphroditismus bei *Eptatretus burgeri* (Myxinoidea, Cyclostomata)? *Sitzungsberichte der Österreichischen Akademie der Wissenschaften, mathematisch naturwissenschaftliche Klasse*, **186**, 425–7.

Patzner, R.A. (1978) Cyclical changes in the ovary of the hagfish *Eptatretus burgeri* (Cyclostomata). *Acta Zoologica* (*Stockholm*), **59**, 57–61.

Patzner, R.A. (1980a) Cyclical changes in weight and fat content of the liver and their relationship to reproduction in the hagfish *Eptatretus burgeri* (Cyclostomata). *Acta Zoologica* (*Stockholm*), **61**, 157–60.

Patzner, R.A. (1980b) *Die Reproduktionsbiologie der Schleimaale (Myxinoidea, Cyclostomata) unter besonderer Berücksichtigung von Bau and Funktion der Adenohypophyse.* Habilitationsschrift, University of Salzburg.

Patzner, R.A. (1982) Die Reproduktion der Myxinoiden. Ein Vergleich von Myxine glutinosa und *Eptatretus burgeri. Zoologischer Anzeiger* (*Jena*), **208**, 132–44.

Patzner, R.A. and Adam, H. (1976) The gonadal development of the female Atlantic hagfish, *Myxine glutinosa* L. *Meeting of the Japanese Society of Comparative Endocrinology in Gifu, Japan.*

Patzner, R.A. and Ichikawa, T. (1977) Effects of hypophysectomy on the testis of the hagfish, *Eptatretus burgeri* Girard (Cyclostomata). *Zoologischer Anzeiger* (*Jena*), **199**, 371–80.

Patzner, R.A. and Adam, H. (1981) Changes in weight of the liver and the relationship to reproduction in the hagfish *Myxine glutinosa* (Cyclostomata). *Journal of the Marine Biological Association* (*United Kingdom*), **61**, 461–4.

Patzner, R.A., Erhart, G. and Adam, H. (1982) Cell types in the adenohypophysis of the hagfish *Myxine glutinosa* (Cyclostomata). *Cell and Tissue Research*, **223**, 583–92.

Pickford, G.E. and Atz, J.W. (1957) *The Physiology of the Pituitary Gland of Fishes*, Zoological Society, New York.

Ross, D.M. (1963) The sense organs of *Myxine glutinosa*, in *The Biology of Myxine* (eds A. Brodal and R. Fänge), Universitetsforlaget, Oslo, pp. 150–60.

Schreiner, K.E. (1955) Studies on the gonad of *Myxine glutinosa* L. *Universitet i Bergen, Arbok 1955, Naturvidenskapelige Rekke*, **8**, 1–40.

Schreiner, A. and Schreiner, K.E. (1908) Zur Spermienbildung der Myxinoiden. Über die Entwicklung der männlichen Geschlechtszellen von *Myxine glutinosa* L., III. Archiv für Zellforschung, **1**, 152–231.

Schultz, H.J., Patzner, R.A. and Adam, H. (1979) Fine structure of the agranular adenohypophysial cells in the hagfish, *Myxine glutinosa* (Cyclostomata). *Cell and Tissue Research*, **204**, 67–75.

Schützinger, S., Choi, H.S., Patzner, R.A. and Adam, H. (1987) Estrogens in plasma of the hagfish, *Myxine glutinosa* (Cyclostomata). *Acta Zoologica* (*Stockholm*), **68**, 263–6.

Sherwood, N.M. and Sower, S.A. (1985) A new family member for gonadotropin-releasing hormone. *Neuropeptides*, **6**, 205–14.

Smigielski, A.S. (1975) Hormonal-induced ovulation in the winterflounder, *Pseudopleuronectes americanus*. *Fishery Bulletin*, **73**, 431–8.

Stacey, N.E., Kyle, A.L. and Liley, N.R. (1986) Fish reproductive pheromones, in *Chemical Signals in Vertebrates*, Vol. 4 (eds D. Duvall, D. Müller-Schwarze and R.M. Silverstein), Plenum, New York, pp. 117–33.

Stanley, H.P. (1969) An electron microscope study in the teleost fish *Oligocottus maculosus*. *Journal of Ultrastructural Research*, **27**, 230–43.

Tamaoki, B.I. (1979) Comparative endocrinology of steroidogenesis. *International Symposium of Hormones and Evolution in Japan*, p. 30.

Tsuneki, K. (1976) Effects of estradiol and testosterone in the hagfish *Eptatretus burgeri*. *Acta Zoologica (Stockholm)*, **57**, 137–46.

Tsuneki, K. and Gorbman, A. (1977a) Ultrastructure of the ovary of the hagfish, *Eptatretus stouti*. *Acta Zoologica (Stockholm)*, **58**, 27–40.

Tsuneki, K. and Gorbman, A. (1977b) Ultrastructure of the testicular interstitial tissue of the hagfish, *Eptatretus stoutii*. *Acta Zoologica (Stockholm)*, **58**, 17–25.

Tsuneki, K., Ouji, M. and Saito, H. (1983) Seasonal migration and gonadal changes in the hagfish *Eptatretus burgeri*. *Japanese Journal of Ichthyology*, **29**, 429–40.

Tsuneki, K., Suzuki, A. and Ouji, M. (1985) Sex difference in the cloacal gland in the hagfish, *Eptatretus burgeri*, and its possible significance in reproduction. *Acta Zoologica (Stockholm)*, **66**, 151–8.

Wahlert, G. (1953/54) Eileiter, Laich und Kloake der Salamandriden. *Zoologische Jahrbücher, Abteilung für Anatomie und Ontogenie der Tiere*, **73**, 276–324.

Walvig, F. (1963) The gonads and the formation of the sexual cells, in *The Biology of Myxine* (eds A. Brodal and R. Fänge), Universitetsforlaget, Oslo, pp. 530–80.

Walvig, F. (1967/68) Experimental marking of hagfish (*Myxine glutinosa*). *Nytt Magasin for Zoologi* (Oslo), **15**, 35–9.

Weisbart, M., Dickhoff, W.W., Gorbman, A. and Idler, D.R. (1980) The presence of steroids in the sera of the Pacific hagfish, *Eptatretus stoutii* and the sea lamprey, *Petromyzon marinus*. *General and Comparative Endocrinology*, **41**, 506–19.

Wourms, J.P. (1977) Reproduction and development in chondrichthyan fishes. *American Zoologist*, **17**, 397–410.

Yu, J.Y.-L., Dickhoff, W.W., Inui, Y. and Gorbman, A. (1980) Sexual patterns of protein metabolism in liver and plasma of hagfish, *Eptatretus stoutii* with special reference to vitellogenesis. *Comparative Biochemistry and Physiology*, **65B**, 111–17.

Yu, J.Y.-L., Dickhoff, W.W., Swanson, P. and Gorbman, A. (1981) Vitellogenesis and its hormonal regulation in the Pacific hagfish, *Eptatretus stouti* L. *General and Comparative Endocrinology*, **43**, 492–502.

PART TEN

The Endocrine System

26

THE ENDOCRINE SYSTEM OF HAGFISHES

Michael C. Thorndyke and Sture Falkmer

SUMMARY

There have been substantial advances in hagfish endocrinology in recent years although restricted to rather specialized areas. Much of the new information has been supported by the availability of new molecular techniques and, as usual for virtually all advances in hagfish biology, has been driven by a continuing appreciation of their unique phylogenetic position and the valuable insights they may contribute to our understanding of evolution. In the pituitary, emphasis has been placed on neurosecretory activity, especially for hypothalamic gonadotrophin-releasing hormone (GnRH) and arginine vasotocin. The most dramatic advances have come in the area of pancreatic hormones where both insulin and insulin-like growth factors (IGFs) have been identified and the full sequence of *Myxine* insulin confirmed. Distribution is widespread and includes neurones in the brain as well as pancreatic islets. Other endocrine advances are less well documented particularly with respect to biological functions. Of particular note is the identification of substance P and a native natriuretic peptide both as potent vasoregulatory factors.

26.1 INTRODUCTION

In the predecessor to this volume (Brodal and Fänge, 1963) the endocrine system featured simply as an appraisal of the structural organization of the pituitary which relied simply on a histochemical description of cell types, reference to the thyroid was a matter of recording radioactive iodine accumulation and binding to a thyroidal protein while the pancreatic system was similarly an account based on histology and anatomy. There was no available information on the chemical characterization and specific cellular localization of hormones or neurohormones in the previous work, least of all any indication of function.

In the intervening years since that original review, enormous advances have been made in the field of endocrinology and neuroendocrinology throughout the animal kingdom. Much of this has centred on the application of modern molecular and cellular techniques and has resulted in the identification of many hitherto unknown regulatory molecules (Baulieu and Kelly, 1990).

Another feature of this progress has been the realization that the division between the endocrine system and nervous system is largely artificial. Thus: many neurones secrete their products directly into the blood system; it is common to find the same molecules synthesized by both endocrine cells and neurones; some gut endocrine cells produce regulatory molecules which have a purely local (paracrine) role.

Such overlap and blurring of boundaries is

The Biology of Hagfishes. Edited by Jørgen Mørup Jørgensen, Jens Peter Lomholt, Roy E. Weber and Hans Malte. Published in 1998 by Chapman & Hall, London. ISBN 0 412 78530 7.

reflected in the organization of the chapters and topics in this volume. Thus details of the role of the pituitary, adrenocorticotropic hormone (ACTH) and other hormonal control systems in the regulation of catecholamine secretion is discussed in Chapter 27 by Bernier and Perry (this volume). In Chapter 30 Nilsson and Holmgren (this volume) present a thorough review of chromaffin tissues and an illustration of the way in which the autonomic nervous system frequently produces neuropeptides which have counterparts in endocrine cells with many indeed serving a paracrine (or neurocrine) function. Chapters 28 and 29 review ontogeny and the central nervous system in this respect also.

In spite of its highly specialized nature and allegedly primitive features, the hagfish has been at the forefront of research in the gastro-entero-pancreatic (GEP) system. Thus the structural characterization of hagfish insulin was amongst the first of the non-mammalian insulins to be published whilst the organization and putative function of the gastro-entero-pancreatic axis in *Myxine* has made a substantial contribution to our understanding of the evolution and phylogeny of gut endocrine systems in vertebrates (Thorndyke and Falkmer, 1985).

Finally, since *Myxine* occupies such a unique and dominant position at the base of the vertebrate evolutionary tree it is not surprising that there have been a number of reports on the occurrence (but not often structures) of other peptide hormones, known to have a widespread distribution in more advanced vertebrates.

In this chapter we shall concentrate on some more recent advances in our understanding of the neuroendocrine role of the pituitary with particular emphasis on the control of reproductive activity (see also Patzner, this volume) and the evolutionary/phylogenetic implications of the neurohypophyseal products of the posterior pituitary. In addition we shall review the present state of knowledge with regard to the GEP system together with an appraisal of some new molecules and mechanisms in hagfish endocrine control.

26.2 THE PITUITARY

26.2.1 General morphology

As in all vertebrates, the pituitary of hagfish comprises a neural lobe (neurohypophysis) and an epithelial lobe (adenohypophysis). See also Part IX (this volume). There is not the close anatomical relationship between neurohypophysis and adenohypophysis which is typical of more advanced vertebrates and which is responsible for much of its functional sophistication. Lampreys show a similar pattern although here a somewhat closer association between the lobes is evident together with the additional presence of an intermediate lobe (Figure 26.1). Nevertheless, the extensive vascular connections between the neurohypophysis and adenohypophysis in hagfishes suggest a close functional complementarity.

26.2.2 Gonadotrophin-releasing hormone (GnRH)

The GnRHs are a rapidly expanding family of regulatory molecules found primarily in vertebrates and which play a central role in the control of reproductive activity (for recent reviews see King and Millar, 1995; Zandberger *et al.*, 1995). Initially, the presence of GnRH in hagfish brain was the subject of conjecture. Immunoreactive GnRH was originally detected on the basis of radioimmunoassay and chromatography (King and Millar, 1980; Jackson, 1980). However later ICC studies failed to substantiate these early claims in spite of extensive reports in GnRH in lamprey brain (see Sower *et al.*, 1995 for review). Fortunately this problem has recently been revisited in an in-depth study using ICC, HPLC and RIA in combination with a large panel of specific antibodies (Sower *et al.*, 1995). Localization of GnRH-like

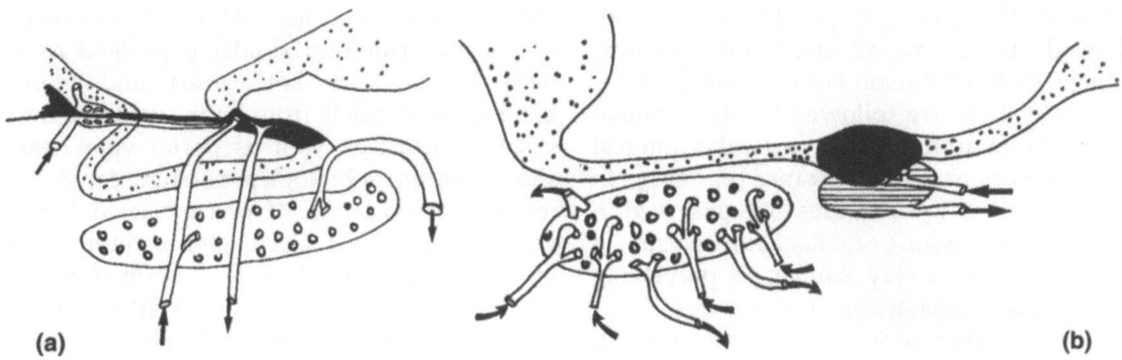

Figure 26.1 Diagrams to illustrate the anatomical organization of the pituitary in *Myxine* (a) and lamprey (b) with reference to their vascularization. Stipple: floor of third ventricle; solid: neural lobe; circles: adenohypophysis; cross hatching: neurointermediate lobe. Arrows indicate direction of blood flow. Drawings by W.S. Thorndyke and E.K. Thorndyke.

material was confirmed for the neurohypophysis exclusively. Other brain areas proved negative with all antibodies tested. This finding may well be a peculiarity of the Atlantic hagfish since an equally recent ICC study of the Pacific hagfish (Braun *et al.*, 1995) has suggested a rather more extensive distribution of GnRH-positive cell bodies and fibres in the brain, including putative neurovascular and CSF contacts of a type typical in many other vertebrates. HPLC and RIA data from these studies indicated two forms of GnRH in hagfish brain, the major peak showing complementarity with cGnRH II, the minor peak with lamprey GnRH III. In the absence of conclusive evidence, it appears most likely that the hagfish neurohypophysis elaborates a molecule more like lamprey GnRH III than any other form and the weakness of the ICC immunoreactivity may indicate that hagfish GnRH is even so rather different from that of the lamprey. This should not be totally unexpected since such a difference has also been shown between hagfish and lamprey insulins (see below).

26.2.3 Neurohypophyseal peptides

The vertebrate neurohypophysis synthesizes members of a family of nonapeptides characterized by their remarkable structural conservation from agnathans to mammals. In *Myxine*, only the founder member arginine vasotocin (AVT) has been described and extensive immunoreactive perikarya are detectable ventrocaudal to the hippocampus primordium and around the post-optic commissure (Nozaki and Gorbman, 1983). Most perikarya show a bilateral distribution above paired prehypophyseal arteries which must represent putative release sites. Perikarya are found also close to capillary walls in the prehypophyseal vascular plexus. Early reports on the vascularization of the hagfish hypophysis proved to be puzzling since an extensive arterial network was located in a prehypophyseal position (Gorbman *et al.*, 1983). However, the description of AVT-positive cells in this region gives credence to the idea that this may also operate as a release site.

Recent molecular studies have confirmed earlier suggestions that a single neurohypophyseal peptide is present in *Myxine* (Suzuki *et al.*, 1995). They also indicate only 46% similarity between the AVT cDNAs of *Lampetra japonica* and *Eptatretus burgeri*, suggesting some phylogenetic distance between the two groups. However, both showed similar overall structural organization of their precursor with: signal peptide, AVT, Gly–Lys–Arg followed by the terminal neurophysin domain. Of particular interest and evolutionary significance is that in *Lampetra*, all 15 Lys residues are conserved in the central region of its neurophysin, a feature which is very similar to previously characterized gnathostome sequences. In *E. burgeri*, however, at least two insertions and one deletion are seen in this conserved central region, a pattern very similar to another hagfish, *E. stoutii*. This would suggest that the hagfish AVT precursor is less closely related to gnathostome AVT than its lamprey counterpart.

In vertebrates, neurohypophyseal peptides perform a variety of functions, most of which relate to the regulation of smooth muscle activity either in the vasculature or that associated with reproductive and allied events, or they are implicated in osmoregulation.

The function of AVT in hagfishes is far from clear. Osmoregulation is not a prominent feature of hagfish life and it seems likely that vascular control will be the most promising area for investigation. In this respect, a more topical discovery has stolen the hagfish endocrine headlines in relation to control of vascular tone. This is considered next.

26.2.4 Atrial natriuretic peptide

Recent years have seen a great expansion in our knowledge of a previously unsuspected source of regulatory peptides, namely the myoendocrine cells found in the atria of many animal species. The biologically active product of these cells has been variously named cardiodilatin, atriopeptin, atrial natriuretic peptide, or atrial natriuretic factor according to laboratory and species of description. Furthermore, as with many other peptides, similar, if not identical molecules have been located in the CNS. For convenience the term atrial natriuretic peptide (ANP) will be used here. *Myxine* is no exception to this rule and similar peptide(s) have been localized to both heart and brain. Bioassay of extracts from brain, ventricle and systemic heart as well as portal vein heart show significant biological potency, causing a marked vasodilation of precontracted blood vessels (see below). Curiously, secretory cells in the hagfish heart were described many years before the discovery of ANP and their significance passed unnoticed until the recent interest in ANP phylogeny. Reinecke *et al.* (1987) reported granule-rich cells in the interstitium as well as myoendocrine cells amongst the myocardiocytes of *Myxine* heart. Their localization and morphology agrees well with those similarly described in lampreys (Shibata and Yamamoto, 1976). In addition, for *Myxine*, an extensive distribution of immunoreactive neurones was found in the brain and spinal cord (Reinecke *et al.*, 1987). Localizations included large perikarya in the primordium while smaller cells and fibres were detectable in hippocampus, thalamus and hypothalamus. When extracts from both heart(s) and brain were tested by bioassay there was clear evidence of potent activity. Initial experiments indicated a dose-dependent vasorelaxant effect on rabbit aorta which had been pre-contracted with acetylcholine. When similar extracts were applied to native *Myxine* ventral aorta *in vivo*, a smaller, but clearly dose-dependent, vasorelaxation could be demonstrated but with greater potency (Figure 26.2a).

The heart of *Myxine* apparently lacks innervation, and in view of the demonstrable effect of *Myxine* ANP extracts on *Myxine* tissues it seems highly probable that ANP released from myoendocrine cells could play a part in the local regulation of cardiac activity. Unfortunately, initial *in vitro* studies on *Myxine* heart have proved inconclusive (Reinecke, 1989). Equally, it is possible that ANP released into the circulation from myoendocrine cells might have a more distant, typically humoral, role. This distinct

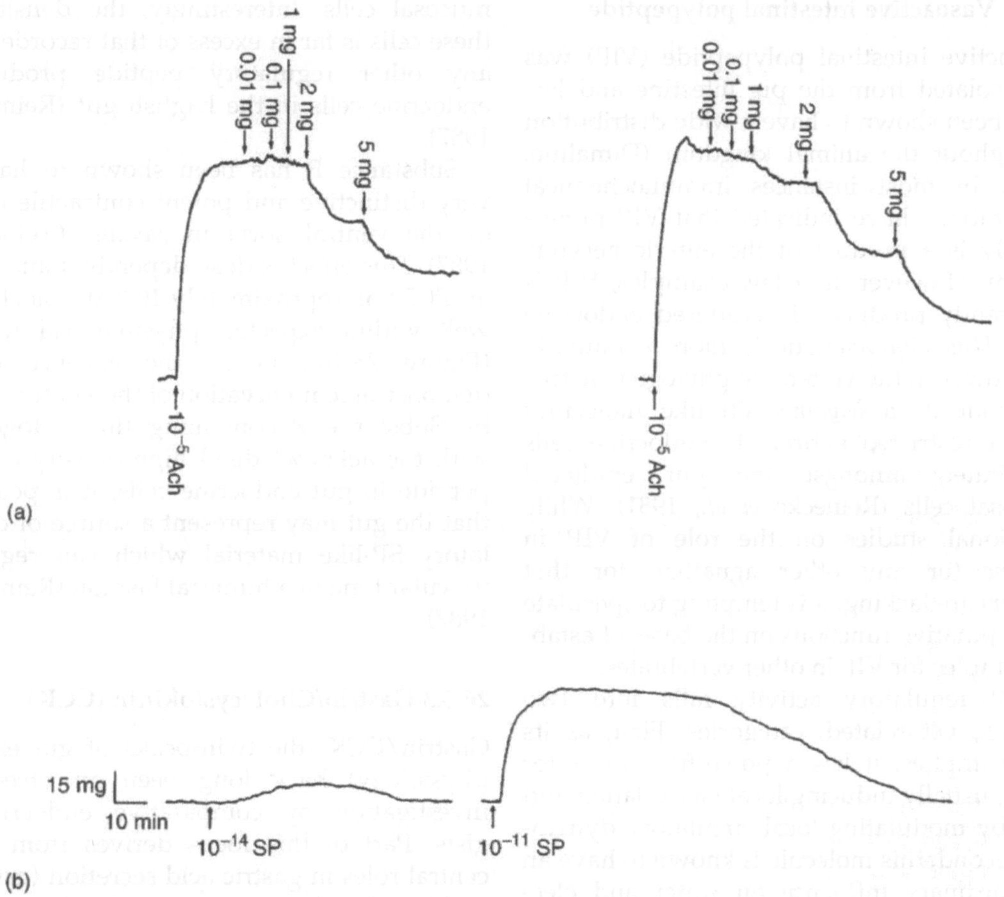

Figure 26.2 (a) Dose-dependent vasorelaxant effects of purified ventricular (left) and brain (right) extracts (mg/wet weight) on precontracted *Myxine* ventral aorta (after Reinecke, 1987). (b) Response *in vitro* of rings from *Myxine* ventral aorta to substance P (after Reinecke *et al*., 1993).

possibility is highlighted by the recent description of specific, saturable binding sites for radiolabelled ANP on the respiratory lamellar epithelium of *Myxine* gills (Toop *et al*., 1995). Moreover, evidence from these authors has also suggested that this peptide may operate through a guanylate cyclase-linked receptor pathway similar to that found in mammals. A role for *Myxine* ANP in the regulation of gill blood flow remains, then, a clear possibility. The potential role of CNS-produced ANP is more problematic and as yet there are no indications of a putative role for this material.

26.3 GASTRO-ENTERO-PANCREATIC PEPTIDES

The neurohormonal profile of the vertebrate gut is now extensive and includes peptides produced by both classical endocrine cells which line the gut mucosa as well as an extensive array of molecules elaborated by the autonomic innervation of the gastrointestinal tract. Putative regulatory neuropeptides from the gut autonomic system are reviewed elsewhere (Nilsson and Holmgren, this volume). Here only the products of endocrine-like cells will be considered.

26.3.1 Vasoactive intestinal polypeptide

Vasoactive intestinal polypeptide (VIP) was first isolated from the pig intestine and has since been shown to have a wide distribution throughout the animal kingdom (Dimaline, 1989). In most instances immunochemical localizations have indicated that VIP preferentially is a product of the enteric nervous system. However, in a few examples, VIP is apparently produced in scattered endocrine cells. This characteristic is more commonly seen lower in the vertebrate phylogenetic tree and indeed in *Myxine*, VIP-like molecules appear restricted entirely to endocrine cells distributed amongst the gut epithelial mucosal cells (Reinecke *et al.*, 1981). While functional studies on the role of VIP in *Myxine* (or any other agnathan for that matter) are lacking, it is tempting to speculate upon putative functions on the basis of established roles for VIP in other vertebrates.

VIP regulatory activity falls into two distinct, yet related, categories. First, as its name implies, it has a powerful vasomotor effect, usually inducing local vasodilation and thereby modulating local circulatory dynamics. Second, this molecule is known to have an extraordinary influence on water and electrolyte secretion, especially where it is associated with the gastrointestinal tract.

In view of the well-known secretory activity of the hagfish gut, there would appear to be a distinct possibility that the native VIP molecule produced by the gut endocrine cells discussed above is involved in the regulation of fluid and electrolyte flow in the gastrointestinal tract of *Myxine*.

26.3.2 Substance P

Substance P was one of the earliest regulatory peptides shown to have a dual distribution in both neurones and endocrine cells (Pernow, 1983). However, as with VIP, the distribution of substance P in *Myxine* seems to be restricted entirely to typical endocrine cells which lie scattered amongst the epithelial mucosal cells. Interestingly, the density of these cells is far in excess of that recorded for any other regulatory peptide producing endocrine cells in the hagfish gut (Reinecke, 1987).

Substance P has been shown to have a very distinctive and potent contractile effect on the ventral aorta in *Myxine* (Reinecke, 1987). This effect is dose-dependent and with an EC50 of approximately 10^{-13} M it is clearly well within expected physiological ranges (Figure 26.2b). Given the absence of a demonstrable innervation of the ventral aorta by Substance P containing fibres, together with the acknowledged high density of this peptide in gut endocrine cells, it is possible that the gut may represent a source of circulatory SP-like material which can regulate vascular tone in a humoral fashion (Reinecke, 1987).

26.3.3 Gastrin/Cholecystokinin (CCK)

Gastrin/CCK (the twin-peaks of gut extract physiology) have long been an object of investigation by comparative endocrinologists. Part of this focus derives from their central roles in gastric acid secretion (gastrin) and gall bladder contraction/pancreatic enzyme secretion (CCK). Another reason centres on an interest in their biosynthetic pathways and genetic relationships. They are remarkably similar molecules with a common c-terminal pentapeptide sequence (Figure 26.3) and it is this which has led to so much speculation concerning their evolutionary and phylogenetic interrelationships. Molecular and genetic evidence is strongly suggestive their origin from a common ancestral gene (Liddle, 1994) and recent evidence from the protochordate *Ciona intestinalis* points to the existence of such a gene in primitive chordates. An octapeptide named cionin was identified in *Ciona* following extensive reports of gastrin/CCK-like material in this and related sea squirts by ICC (Thorndyke and Dockray, 1986; Johnsen and Rehfeld, 1990).

Cionin is of great interest because it has a sulphated tyrosine in a position typical of both gastrin and CCK (Figure 26.3). Analysis of full length cDNA clones suggests that cionin resembles more closely CCK than gastrin and it is clear that the appearance of separate gastrin and CCK molecules phylogenetically must post-date the protochordates. Unfortunately, as we note below, rather little is known about these molecules in agnathans, a group which on the basis of the *Ciona* data is likely to provide rather pertinent information on the evolution of these two gut peptides. Bioassay evidence using hagfish gut extracts points to the presence of a molecule which is more CCK-like, certainly in terms of its ability to stimulate guinea-pig gall bladder muscle strips in a laminar-flow superfusion system (Vigna, 1979).

In spite of uncertainties regarding the chemical structure of the native hagfish molecule(s), a most important consideration is the potential normal physiological role this molecule might have in *Myxine* itself. To date mainly *in vitro* approaches have been adopted and by default these have relied upon the use of synthetic mammalian peptides since the structure of the native hagfish peptide remains unknown. Using a combination of *in vitro* organ bath techniques, Vigna and Gorbman (1979) were able to investigate the response of gall bladder and intestinal muscle to a variety of CCK- and gastrin-related molecules. In addition these authors measured intestinal enzyme secretion (lipase and alkaline phosphatase) in response to CCK and several other gastrointestinal hormones.

Surprisingly, although the muscle preparations responded well to acetylcholine, confirming their viability, all tests with CCK, gastrin or analogues thereof, proved negative. *In vivo* experiments designed to monitor gall bladder contraction following injection of CCK were also negative. In contrast, porcine CCK_{33} potently stimulated intestinal lipase secretion while secretin, VIP and glucagon were ineffective. No elevation of alkaline phosphatase was noted. These are surprising findings because it confirms *Myxine* as the only vertebrate with a gall bladder (so far examined) which does not respond to CCK. It is possible, of course, that the native hagfish CCK-like molecule is sufficiently different from other vertebrate forms to make ligand–receptor interaction either weak or impossible. The stimulation of intestinal lipase secretion intriguingly foreshadows the situation in higher vertebrates where a discrete pancreas is present. In hagfishes, the enzyme secreting cells are scattered in the intestinal mucosa and represent the phylogenetic base line. If the recognized evolutionary sequence is followed, such cells migrate and become concentrated as a discrete pancreatic exocrine organ, connected to the gut by way of the pancreatic duct (Thorndyke and Falkmer, 1985). Thus in hagfishes, the regulation of intestinal enzyme-secreting cell activity anticipates their concentration in the pancreas. Presumably in this relatively simple/primitive state, control is local and paracrine, an accepted primitive mechanism. It is clear that isolation and characterization of the native molecule(s) is now urgent.

CCK	Ser-	Asp-	Arg-	Asp-	Tyr-	Met-	Gly-	Trp-	Met-	Asp-	Phe-	NH_2
Gastrin	Glu-	Glu-	Glu-	Glu-	Ala-	Tyr-	Gly-	Trp-	Met-	Asp-	Phe-	NH_2
Cionin				Asn-	Tyr-	Tyr-	Gly-	Trp-	Met-	Asp-	Phe-	NH_2

Figure 26.3 Amino acid sequences (aligned for maximum correspondence) of cholecystokinin (CCK), gastrin and cionin. Only the c-terminal biologically active regions of CCK and gastrin are shown.

26.3.4 The islet organ

The alimentary tract of *Myxine* consists of a simple, non-convoluted tube (Falkmer, 1993) and the opening of the common bile duct is marked by a small papilla, corresponding to the papilla of Vater in higher vertebrates, just a few millimetres from the oesophageal-gut junction. The gall bladder is located in a position close to the region where the two liver lobes (cranial and caudal) meet and is served by two main hepatic bile ducts. The common bile duct is just a few millimetres long and at its junction with the gut at the papilla, a small whitish swelling can be seen. This is the islet organ. Its size is related to the size of the hagfish; in a large specimen (about 35 cm body length), it can have a diameter of 1–2 mm and a wet weight of about 2–3 mg. Exceptionally, it can be enlarged and discoloured when it has been transformed into a hamartoma or a genuine islet-cell adenoma or carcinoma. Such a tumour arising in the hagfish islet organ is – next to primary neoplasms of the liver – the second most common tumour lesion in this jawless fish (see Falkmer, this volume).

In the phylogenetic evolution of the islets of Langerhans, the hagfish islet organ occupies a unique position (Falkmer, 1995). In protochordates and in invertebrates the endocrine cells producing islet hormones appear as normal endocrine cells in the gut mucosa (Falkmer *et al.*, 1975; Thorndyke and Falkmer, 1985). The hagfish islet represents the first appearance in phylogeny of a discrete aggregation of gut hormone-producing cells in isolation from enzyme-secreting exocrine cells. A further unique feature is that it represents a two-hormone islet organ; its parenchymal cells are an almost pure population of insulin cells (Figure 26.4a); accompanied by a few somatostatin-producing cells (Falkmer, 1995). Most somatostatin cells occur

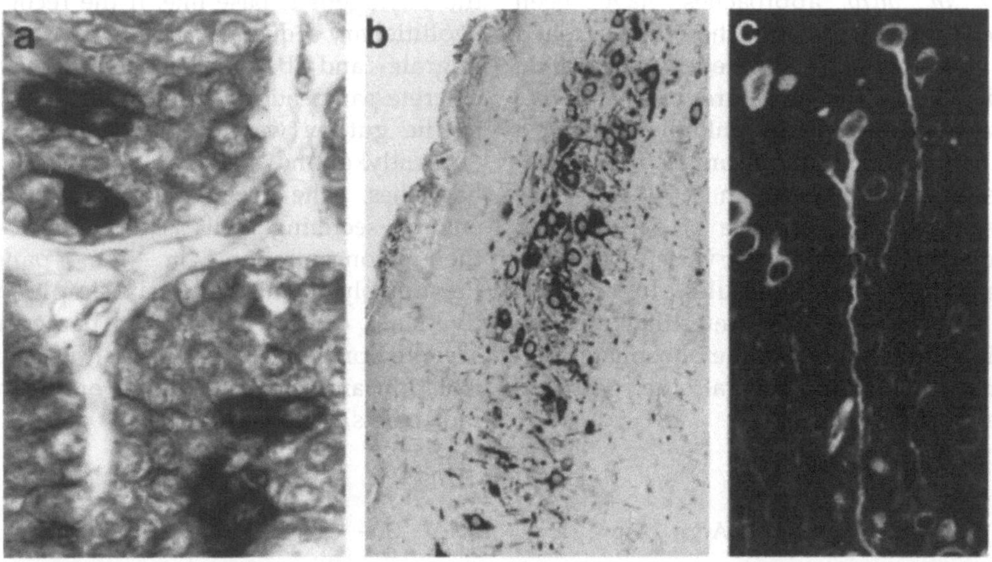

Figure 26.4 Insulin and insulin-like immunoreactivity in the islet (a) and brain (b) of *Myxine* stained with anti-salmon insulin sera (3797, Plisetskaya). (a) Group of islet cells revealed using the PAP procedure, ×1000. (b) A collection of insulin-immunoreactive neurones in the cutaneous column of the brain (PAP, ×250). (c) Some cells have spectacularly long processes (FITC, ×200). Micrographs modified and added to from Thorndyke *et al.*, 1989.

in the bile duct and gut mucosa, while the insulin cells are restricted to the islet organ and the bile duct epithelium. The hagfish islet organ is not densely innervated but is well vascularized with an extensive capillary bed and blood flow through the islet, as measured by a microsphere technique, is high (Jansson and Falkmer, 1997).

The parenchymal cells producing molecules belonging to the remaining two islet hormone families, glucagon and pancreatic polypeptide, remain as open endocrine cells in the gut mucosa (Falkmer, 1995). It is noteworthy that great variation can occur between individual hagfish; some exhibit practically no endocrine cells while, in others, such cells can be numerous.

26.3.5 Insulin

Hagfish insulin is one of the most extensively studied of all vertebrate insulins, perhaps second only to porcine insulin. Conservation of structure is high (Figure 26.5) although there are important differences; for example, hagfish insulin lacks histidine at position 10 in the B-chain. This affects the ability of the molecule to form hexamers. In fact it forms only dimers because the absence of this imidazole group prevents the binding of zinc which is essential for hexamer formation. Surprisingly, hagfish and lamprey insulins also show several differences (Figure 26.5) and indeed they are no more similar to each other than they are to teleost or even mammalian insulins (Plisetskaya *et al.*, 1988). This finding presumably reflects the ancient divergence of these two extant agnathan groups.

The hagfish insulin gene is organized in much the same way as other vertebrate insulin genes. Thus the sequence of hagfish pre-pro-insulin shows structural features remarkably similar to those of other vertebrates, an indication that the biosynthetic pathway and intracellular processing must be a highly conserved feature (Emdin *et al.*, 1985). In spite of this there are (as noted above) a number of substitutions. For example, 19 amino acid residues of 51 are different from pig insulin. Of these, 16, mostly located in the B chain, have not been seen in other vertebrates. For a more detailed assessment of insulin evolution in general see Emdin *et al.* (1985).

The noted differences are also reflected in

A-chain

					5					10					15					20	
Myxine	G	I	V	E	Q	C	C	H	K	R	C	S	I	Y	N	L	Q	N	Y	C	N
Lamprey	-	-	-	-	-	-	-	-	R	K	-	-	-	-	D	M	E	-	-	-	-
Salmon	-	-	-	-	-	-	-	-	-	P	-	N	-	F	D	-	-	-	-	-	-
Pig	-	-	-	-	-	-	-	T	S	I	-	-	L	-	Q	-	-	-	-	-	-

B-chain

					5					10					15					20					25					30	
Myxine	R	T	T	G	H	L	C	G	K	D	L	V	N	A	L	Y	I	A	C	G	V	R	G	F	F	Y	D	P	T	K	M
Lamprey	A	G	G	T	-	-	-	-	S	H	-	-	E	-	-	-	V	V	-	-	D	-	-	-	-	-	T	-	S	-	T
Salmon	-	A	A	Q	-	-	-	-	S	H	-	-	D	-	-	-	L	V	-	-	E	K	-	-	-	-	N	-	K		
Pig	F	V	N	Q	-	-	-	-	S	H	-	-	E	-	-	-	L	V	-	-	E	-	-	-	-	-	T	-	K	A	

Figure 26.5 Amino acid sequences of both A and B chains of *Myxine*, lamprey, salmon and pig insulin. Hyphens indicate common residues.

the bioactivity of hagfish insulin in mammalian systems. For example, although it binds with 25% efficiency to rat fat cells, it has only 5% activity (Emdin and Falkmer, 1977) indicating that it is acting as a partial agonist on the receptor.

It is now clear that insulin is not the only member of the insulin family produced by the cells of the islet parenchyma. High-molecular weight forms of an insulin-like growth factor (IGF) have been identified following the use of degenerate primers to mammalian IGFs (Chan *et al.*, 1993). These studies indicate that insulin and the IGFs must have diverged prior to the emergence of the agnathans. Interestingly, the single *Myxine* IGF molecule has similarities to both IGF-I and IGF-II in gnathostomes suggesting that these diverged after the jawed vertebrates appeared (Chan *et al.*, 1993; Nagamatsu *et al.*, 1991). Recent evidence from the elasmobranch, *Squalus acanthias*, demonstrates the presence of two molecules, one with IGF-I- and the other with IGF-II-like properties (Dugay *et al.*, 1995). Clearly, the prototypical IGF molecule, perhaps as illustrated by the *Myxine* IGF, must have duplicated and diverged after the radiation of the modern agnathans but before the emergence of the extant gnathostomes.

Concurrent with the studies cited above on *Myxine* islet IGF, a large IGF-like molecule (2–4 kDa) was localized to the intestine, islet, serum and brain of *Myxine* using a combination of chromatography, RIA and ICC (Reinecke *et al.*, 1991). This confirmed earlier immunocytochemical and chromatographic identification of large insulin-like molecules (6 kDa and 66 kDa) in the brain of *Myxine* (Thorndyke *et al.*, 1989). These molecules are expressed by large numbers of cell foci distributed widely throughout the brain of both *Myxine* and *Eptatretus* (Figure 26.4b, c) (Thorndyke *et al.*, 1989).

The biological significance of all these findings is not yet clear but the recent identification of highly specific binding sites for IGFs in the hagfish brain together with somewhat less specific sites in liver and gut support the idea of a widespread role in growth and metabolism similar to that established in other vertebrate systems.

The second of the islet hormones, somatostatin, is well known, both in the islet organ and in the gut. Yields recovered following radioimmunoassay of tissue extracts showed that the islet organ contained 7900 ng g^{-1} wet weight and the gut 5 ng g^{-1} wet. The form of somatostatin produced in *Myxine* shows some interesting features. An unusual somatostatin peptide comprising 34 amino-acid residues, instead of the normal 14 (Conlon *et al.*, 1988) suggests that an anomalous processing pathway exists and almost 75% of somatostatin immunoreactivity in the islet, bile duct and gut endocrine cells consists of this prosomatostatin peptide. The remaining 25% is the normal somatostatin-14; with amino acid sequence identical to that of the mammalian hormone (Conlon *et al.*, 1988).

26.4 INSULIN AND METABOLISM IN HAGFISH

The comparatively early characterization of hagfish insulin (Petersen *et al.*, 1975) did not lead immediately to a massive expansion of interest in its regulatory properties in *Myxine*. This was due partly to the lack of sufficient quantities of the native peptide together with the absence of a suitable, homologous, immunoassay.

Early work employed synthetic mammalian insulins which despite showing only 62% homology with the hagfish molecule, were sufficiently identical in key positions to provide a research tool of some value. Initial studies indicated that intramuscular injection of mammalian insulin into hagfish produced long-lasting hypoglycaemia and depressed plasma amino nitrogen levels (Inui and Gorbman, 1977).

Since this time there have been a number of investigations on the endocrine regulation of carbohydrate metabolism in hagfish (as well as in lamprey) and a complex picture has emerged. The following represent key points and details of the primary experiments may be found in two substantial reviews (Plisetskaya, 1985; Emdin *et al.*, 1985).

1. Skeletal muscle rather than the liver is the main storage (and metabolism) site for both glycogen and fat in hagfishes (and lampreys). For example, glycogen concentration in hagfish muscle is about twice as high as that found in the liver (Emdin, 1982a, b).
2. Hagfish (and lamprey) brains also have appreciable stores of glycogen and with the recognized presence also of insulin-like molecules in the brain (see above) it seems likely that carbohydrate metabolism in the hagfish CNS may be a largely independent phenomenon.
3. Glucose administration will stimulate islet insulin release *in vitro* in a dose-dependent manner (Emdin, 1982a).
4. In starved hagfishes, blood glucose and plasma insulin levels fell. When these animals were subsequently loaded with glucose, insulin level rose dramatically and thus provided *in vivo* corroboration of the *in vitro* response (Emdin, 1982b).
5. Native insulin enhanced the uptake of labelled metabolites into muscle rather than liver (Emdin *et al.*, 1985).

Inconsistencies in these patterns are also reported: for example administration of anti-insulin serum to the Atlantic hagfish *M. glutinosa* has no effect whereas the same experiment on the Pacific hagfish *E. stoutii* caused an elevation of plasma glucose (Plisetskaya, 1985).

In higher vertebrates insulin is not the sole regulator of carbohydrate metabolism; thyroid hormones, pancreatic glucagon, catecholamines as well as other regulatory molecules all have a part to play. The same is almost certainly true for hagfishes although here our understanding of hormonal interplay is far from clear.

26.5 OTHER REGULATORY PEPTIDES

Other neurohormonal peptides have been detected in the *Myxine* gut but their concentrations usually are rather low and to date there has not been a systematic investigation, utilizing a broad spectrum of peptide hormone antisera (Falkmer, 1993). Most of the data that are available are of an immunocytochemical nature and only rarely have they been corroborated with radioimmunochemical or biochemical observations. An exception is an early study of the enteroinsular axis. Here it was shown by combined radioimmunochemical and immunocytochemical investigation that GIP (glucose-dependent insulinotropic peptide or gastric inhibitory peptide) is present in the hagfish gut, but not in its islet organ (Falkmer *et al.*, 1980).

There are some immunochemical data available about the nature of a few other peptide hormones present in *Myxine* gut. Members of the glucagon and tachykinin families have been studied (Conlon and Falkmer, 1989). A feature common to all was that the concentrations found were 10 to 100 times less than the corresponding concentrations in the rat intestine. Another difference was that when the glucagon and tachykinin immunoreactive peptides were resolved by means of HPLC, their retention times were appreciably different from those of the corresponding mammalian peptides, suggesting the presence of molecules with rather different characteristics compared to other vertebrates.

Finally, reports exist on the potential hormonal regulation of calcium in the New Zealand hagfish *Eptatretus cirrhatus*. Data indicate that this hagfish regulates its plasma calcium concentration to a level which is approximately 50% of that of seawater (Forster and Fenwick, 1994). Preliminary experiments using isolated hagfish gill pouches challenged with an extract prepared from the corpuscles of Stannius of the eel *Anguilla dieffenbachii*, indicated that, in contrast to teleosts, regulation is achieved by stimulating calcium efflux rather than inhibiting influx (Forster and Fenwick, 1994).

26.6 EPILOGUE

It should be abundantly clear from this brief review that endocrine systems in hagfishes remain largely unexplored. In spite of getting off to such a splendid start with the isolation and full characterization of insulin in *Myxine*, there have been rather few major advances in recent years. Much of this must stem from the pressure to study organisms of more direct economic importance and this approach has witnessed great leaps forwards in our understanding of the neuroendocrine systems of commercially valuable fish such as salmon, trout and a range of marine teleosts. We must draw some comfort from the current political interest in biodiversity and an emerging whole ecosystem approach to conservation and trust that this will provide stimulus for more work in this challenging area.

ACKNOWLEDGEMENTS

The authors would like to thank Chris Edwards for her help with the manuscript and unique ability to interpret MCT's hieroglyphics. Thanks also to Zyg Podhorodecki for the preparation of figures. MCT is especially grateful to FC without whom the manuscript would not have been written.

REFERENCES

Baulieu, E.E. and Kelly, P.A. (1990) *Hormones: From Molecules to Disease*, Chapman & Hall, New York, London, 708 pp.

Braun, C.B., Wicht, H. and Northcutt, R.G. (1995) Distribution of gonadotrophin-releasing hormone immunoreactivity in the brain of the Pacific hagfish *Eptatretus stoutii* (Craniata, Myxinoidea). *J. Comp. Neurol.*, **353**, 464–76.

Brodal, A. and Fänge, R. (1963) *The Biology of Myxine*, Universitetsforlaget, Oslo, 588 pp.

Chan, S.J., Cao, Q.-P., Nagamatsu, S. and Steiner, D.F. (1993) Insulin and insulin-like growth factor genes in fishes and other primitive chordates, in *Biochemistry and Molecular Biology of Fishes*, Vol. 2 (eds P. Hochachka and T. Mommsen), Elsevier, pp. 407–17.

Conlon, J.M. and Falkmer, S. (1989) Neurohormonal peptides in the gut of the Atlantic hagfish (*Myxine glutinosa*) detected using antisera raised against mammalian regulatory peptides. *Gen. Comp. Endocrinol.*, **76**, 292–300.

Conlon, J.M., Askensten, U., Falkmer, S. and Thim, L. (1988) Primary structures of somatostatins from the islet organ of the hagfish suggest an anomalous pathway of posttranslational processing of prosomatostatin-1. *Endocrinology*, **122**, 1855–9.

Dimaline, R. (1989) Vasoactive intestinal peptide, in *The Comparative Physiology of Regulatory Peptides* (ed. S. Holmgren), Chapman & Hall, London, pp. 150–73.

Dugay, S.J., Chan, S.J., Mommsen, T.P. and Steiner, D.F. (1995) Divergence of insulin-like growth factors I and II in the elasmobranch *Squalus acanthias*. *Febs Letters*, **371**, 69–72.

Emdin, S.O. and Falkmer, S. (1977) Phylogeny of insulin. *Acta Paediat-Scand. Suppl.*, **270**, 15–25.

Emdin, S.O. (1982a) Effects of hagfish insulin in the Atlantic hagfish *Myxine glutinosa*: in vivo metabolism of 14C-glucose and 14C-leucine and studies on starvation and glucose loading. *Gen. Comp. Endocrinol.*, **47**, 414–25.

Emdin, S.O. (1982b) Insulin release in the Atlantic hagfish *Myxine glutinosa, in vitro*. *Gen. Comp. Endocrinol.*, **48**, 333–41.

Emdin, S.O., Steiner, D.F., Chan, S.J. and Falkmer, S. (1985) Hagfish insulin: evolution of insulin, in *Evolutionary Biology of Primitive Fishes* (eds R.E. Foreman, A. Gorbman, J.M. Dodd and R. Olsson), New York, Plenum Press, 363–78.

Falkmer, S. (1993) Phylogeny and ontogeny of the neuroendocrine cells of the gastrointestinal tract. *Endocr. Metabol. Clin. N. Amer.*, **22**, 731–52.

Falkmer, S. (1995) Origin of parenchymal cells of the endocrine pancreas: some phylogenetic and ontogenetic aspects. *Front. Gastrointest. Res.*, **23**, 2–29.

Falkmer, S., Cutfield, J.F., Cutfield, S.M., Dodson, G.G., Gliemann, J., Gammeltoft, S., Marques, M., Peterson, J.D., Steiner, D.F., Dundby, F., Emdin, S.O., Havu, N., Östberg, Y. and Windbladh, L. (1975) Comparative endocrinology of insulin and glucagon production. *Amer. Zool.*, **15**, *Suppl* **1**, 255–270.

Falkmer, S., Ebert, R., Arnold, R. and Creutzfeldt, W. (1980) Some phylogenetic aspects on the enteroinsular axis with particular reference to the appearance of the gastric inhibitory polypeptide. *Front. Horm. Res.*, **7**, 1–6.

Forster, M.E. and Fenwick, J.C. (1994) Stimulation of calcium efflux from the hagfish *Eptatretus cirrhatus* gill pouch by an extract of corpuscles of Stannius from an eel (*Anguilla dieffenbachii*)-teleostei. *Gen. Comp. Endocrinol.*, **94**, 92–103.

Gorbman, A., Dickhoff, W., Vigna, S.R., Clark, N.B. and Ralph, C.L. (1983) *Comparative Endocrinology*, J. Wiley & Sons, New York, 572 pp.

Inui, Y. and Gorbman, A. (1977) Sensitivity of Pacific hagfish, *Eptatretus stoutii*, to mammalian insulin. *Gen. Comp. Endocrinol.*, **33**, 423–7.

Jackson, I.M.D. (1980) Distribution and evolutionary significance of the hypophysiotropic hormones of the hypothalamus. *Front. Horm. Res.*, **7**, 36–69.

Jansson, L. and Falkmer, S. (1997) Islet organ blood flow in a Cyclostome, *Myxine glutinosa*. *Amer. J. Physiol.*, **00**, 000–000. In press.

Johnsen, A.H. and Rehfeld, J.F. (1990) Cionin: a disulfotyrosyl hybrid of cholecystokinin and gastrin from the neural ganglion of the protochordate *Ciona intestinalis*. *J. Biol. Chem.*, **265**, 3054–8.

King, J.A. and Millar, R.P. (1980) Comparative aspects of luteinizing hormone-releasing hormone structure and function in vertebrate phylogeny. *Endocrinology*, **106**, 707–17.

King, J.A. and Millar, R.P. (1995) Evolutionary aspects of gonadotrophin-releasing hormone and its receptors. *Cellular and Molecular Neurobiology*, **15**, 5–23.

Liddle, R.A. (1994) Cholecystokinin, in *Gut Peptides: Biochemistry and Physiology* (eds J.H. Walsh and G.J. Dockray), Raven Press, New York, pp. 175–216.

Nagamatsu, S., Chan, S.J., Falkmer, S. and Steiner, D.F. (1991) Evolution of the insulin gene superfamily. Sequence of a preproinsulin-like growth factor cDNA from the Atlantic hagfish. *J. Biol. Chem.*, **266**, 2397–402.

Nozaki, M. and Gorbman, A. (1983) Immunocytochemical localization of somatostatin and vasotocin in the brain of the Pacific hagfish *Eptatretus stouti*. *Cell Tissue Res.*, **229**, 541–50.

Pernow, B. (1983) Substance P. *Pharmacol. Rev.*, **25**, 85–141.

Petersen, J.D., Steiner, D.F., Emdin, S.O. and Falkmer, S. (1975) The amino acid sequence of the insulin from a primitive vertebrate, the Atlantic hagfish (*Myxine glutinosa*). *J. Biol. Chem.*, **250**, 5183–91.

Plisetskaya, E.M. (1985) Some aspects of hormonal regulation of metabolism in agnathans, in *Evolutionary Biology of Primitive Fishes* (eds R.E. Foreman, A. Gorbman, J.M. Dodd and R. Olsson), Plenum Press, New York, pp. 339–61.

Plisetskaya, E.M. Pollock, H.G., Elliott, W.M., Youson, J.N. and Andrews, P.C. (1988) Isolation and structure of lamprey (*Petromyzon marinus*) insulin. *Gen. Comp. Endocrinol.*, **69**, 46–55.

Reinecke, M. (1987) Substance P is a vasoactive hormone in the Atlantic hagfish *Myxine glutinosa* (Cyclostomata). *Gen. Comp. Endocrinol.*, **66**, 291–6.

Reinecke, M. (1989) Atrial natriuretic peptides – localization, structure and phylogeny, in S. Holmgren (ed.), *The Comparative Physiology of Regulatory Peptides*, Chapman & Hall, London and New York, pp. 5–33.

Reinecke, M., Betzler, D. and Forsmann, W.-G. (1987) Immunocytochemistry of cardiac polypeptide hormones (Cardiodilatin/atrial natriuretic polypeptide) in brain and hearts of *Myxine glutinosa* (Cyclostomata). *Histochemistry*, **86**, 233–9.

Reinecke, M., Betzler, M., Aoki, A. and Forsmann, W.-G. (1993) Atrial natriuretic peptides (ANP) in fish heart, in *Fish Ecotoxicology and Ecophysiology* (eds T. Braunbeck, W. Hanke and H. Segner), VCH Verlagsgesellschaft mbh, D-6940 Weinheurn, Germany, pp. 285–404.

Reinecke, M., Drakenberg, K., Falkmer, S. and Sara, V.R. (1991) Presence of IGF-1-like peptides in the neuroendocrine system of the Atlantic hagfish, *Myxine glutinosa* (Cyclostomata): evidence derived by chromatography radioimmunoassay and immunocytochemistry. *Histochemistry*, **96**, 191–6.

Reinecke, M., Schluter, P., Yanaihara, N. and Forsmann, W.-G. (1981) VIP immunoreactivity in enteric nerves and endocrine cells of the vertebrate gut. *Peptides*, **2**, Suppl. 2, 149–56.

Shibata, Y. and Yamamoto, T. (1976) Fine structure and cytochemistry of specific granules in the lamprey atrium. *Cell Tissue Res.*, **172**, 487–501.

Sower, S.A., Nozaki, M., Knox, C.J. and Gorbman, A. (1995) The occurrence and distribution of GnRH in the brain of Atlantic hagfish, an Agnathan, determined by chromatography and immunocytochemistry. *Gen. Comp. Endocrinol.*, **97**, 300–7.

Suzuki, M., Kubokawa, M., Nagasawa, K. and Vrano, A. (1995) Sequence analysis of vasotocin cDNAs of the lamprey, *Lampetra japonica* and the hagfish *Eptatretus burgeri* – Evolution of cyclostome vasotocin precursors. *J. Mol. Endocrinol.*, **14**, 67–77.

Thorndyke, M.C. and Dockray, G.J. (1986) Identification and localization of material with gastrin-like immunoreactivity in the neural ganglion of a protochordate *Ciona intestinalis*. *Regulatory Peptides*, **16**, 269–79.

Thorndyke, M.C. and Falkmer, S. (1985) Evolution of gastro-entero-pancreatic endocrine systems in lower vertebrates, in *Evolutionary Biology of Primitive Fishes* (eds R.E. Foreman, A. Gorbman, J.M. Dodd and R. Olsson), Plenum Press, New York, pp. 379–400.

Thorndyke, M.C., Purvis, D. and Plisetskaya, E.M. (1989) Insulin-like immunoreactivity in the brain of two hagfishes, *Eptatretus stouti* and *Myxine glutinosa*. *Gen. Comp. Endocrinol.*, **76**, 371–81.

Toop, D., Donald, J.A. and Evans, D.H. (1995) Localization and characteristics of natriuretic peptide receptors in the gills of the Atlantic hagfish *Myxine glutinosa* (Agnatha). *J. Exp. Biol.*, **195**, 117–26.

Vigna, S.R. (1979) Distinction between cholecystokinin-like and gastrin-like biological activities extracted from gastrointestinal tissues of some lower vertebrates. *Gen. Comp. Endocrinol.*, **39**, 512–20.

Vigna, S.R. and Gorbman, A. (1979) Stimulation of intestinal lipase secretion by porcine cholecystokinin in the hagfish, *Eptatretus stoutii*. *Gen. Comp. Endocrinol.*, **38**, 356–9.

Zandeberger, M.A., Kah, O., Bogerd, J., Pente, J. and Goos, H.J.T. (1995) Expression and distribution of two gonadotrophin-releasing hormones in the catfish brain. *Neuroendocrinology*, **62**, 571–7.

27

THE CONTROL OF CATECHOLAMINE SECRETION IN HAGFISHES

Nicholas J. Bernier and Steve F. Perry

SUMMARY

The systemic heart, portal heart, and posterior cardinal vein of hagfishes store large quantities of catecholamine in chromaffin cells. However, unlike other vertebrates, the chromaffin tissue of hagfishes appear to lack extrinsic innervation. Although carbachol, a cholinergic agonist, elicits a dose-dependent release of catecholamines *in situ*, *in vivo* there is no evidence that the control of catecholamine release may be achieved through cholinergic mechanisms. Evidence presented in this chapter suggests that this may be achieved through hormonal and/or paracrine means by specific non-cholinergic secretagogues. While both serotonin and ACTH stimulate catecholamine secretion *in situ*, angiotensin II and histamine, potent secretagogues in other vertebrates, do not appear to elicit catecholamine release. *In vivo* and *in situ* evidence also suggests that adenosine may be an important modulator of catecholamine secretion. Although the specific mechanisms of catecholamine secretion during stress *in vivo* have yet to be characterized, serotonin, ACTH and adenosine may all be involved in the overall control of catecholamine release in hagfishes.

27.1 INTRODUCTION

Hagfishes store large quantities of the catecholamines, noradrenaline (NA) and adrenaline (AD), in their systemic and portal hearts, and posterior cardinal vein (PCV; Östlund, 1954; Augustinsson *et al.*, 1956; Östlund *et al.*, 1960; Euler and Fänge, 1961; Perry *et al.*, 1993). The results of early morphological and pharmacological studies of these catecholamine-storing cells revealed a striking resemblance between them and the granular chromaffin cells of the mammalian adrenal medulla (Östlund *et al.*, 1960; Bloom *et al.*, 1961). Jönsson (1983) demonstrated that the chromaffin tissues from the systemic and portal hearts of the Atlantic hagfish, *Myxine glutinosa*, contain the enzymes dopamine-β-hydroxylase (DβH) and phenylethanolamine-*N*-methyl transferase (PNMT). Reid *et al.* (1995) also observed tyrosine hydroxylase (TH)-like immunoreactivity within a distinct population of cells from *Myxine glutinosa* hearts. TH, DβH and PNMT are three key enzymes in the biosynthesis of catecholamines (see Randall and Perry, 1992).

In general the control of catecholamine release in vertebrates may be achieved through both cholinergic and non-cholinergic mechanisms (Livett and Marley, 1993). Although in *Eptatretus stoutii*, ganglion cells have been described in the vicinity of the systemic heart (Hirsch *et al.*, 1964), overall the systemic heart of *Eptatretus stoutii* (Greene, 1902; Jensen, 1965) and *Myxine glutinosa* (Östlund, 1954; Augustinsson *et al.*, 1956; Johnsson and Axelsson, 1996) appear to be

The Biology of Hagfishes. Edited by Jørgen Mørup Jørgensen, Jens Peter Lomholt, Roy E. Weber and Hans Malte. Published in 1998 by Chapman & Hall, London. ISBN 0 412 78530 7.

insensitive to cholinergic agonists and lack an extrinsic innervation. Furthermore, the autonomic nervous system is poorly developed in hagfishes (Campbell, 1970; Nilsson, 1983), and there is no evidence that the PCV is innervated. Therefore, contrary to that found in other vertebrates, cholinergic control of the chromaffin tissues in hagfishes is unlikely, and the nature of the mechanisms involved in the control of catecholamine secretion is still unclear (Perry *et al.*, 1993). This chapter summarizes the recent progress made in elucidating potential secretagogues and suggests some possible mechanisms of catecholamine release during stress *in vivo*.

27.2 CATECHOLAMINE STORAGE LEVELS

The levels of stored catecholamines and the ratio of NA to AD varies considerably among the different sites of storage (Figure 27.1). The principle catecholamine storage site in hagfishes is the systemic heart, followed by the portal heart, and lastly by the PCV (Östlund, 1954; Augustinsson *et al.*, 1956; Östlund *et al.*, 1960; Euler and Fänge, 1961; Perry *et al.*, 1993). In *Myxine glutinosa*, the catecholamine content of the systemic heart is approximately 3 times greater than it is in the PCV (Perry *et al.*, 1993). Overall, the concentrations of NA in these tissues are comparable to the storage concentrations observed in several teleosts (Perry *et al.*, 1993). Whereas the overall NA/AD storage ratio in the systemic heart is close to 1, the auricle contains mainly NA and the ventricle mainly AD (Perry *et al.*, 1993; Euler and Fänge, 1961; Östlund *et al.*, 1960). The portal heart and the PCV contain primarily NA (Euler and Fänge, 1961; Östlund *et al.*, 1960; Perry *et al.*, 1993). The high PCV NA/AD ratio of *Myxine glutinosa* (approximately 26; Perry *et al.*, 1993) contrasts greatly with that of the rainbow trout (approximately 0.4; Reid and Perry, 1994). In general, species differences in the development and distribution of NA and AD cells is thought to be determined by the spatial relationship between the chromaffin cells and the interrenal cells (Livett and Marley, 1993). In mammals, this relationship is usually explained by the control that the interrenal glucocorticoids exert over the expression of PNMT, the enzyme which catalyses the methylation of NA to AD (Livett and Marley, 1993). There is evidence, however, that the expression of PNMT in lower vertebrates is insensitive to glucocorticoids (Wurtman *et al.*, 1968; Jönsson *et al.*, 1983; Reid *et al.*, 1996). Regardless, numerous paracrine interactions are possible between the chromaffin cells and the interrenal cells in vertebrates (Bornstein *et al.*, 1994; Reid *et al.*, 1996), and these may, in part, explain the heterogeneous distribution of catecholamine storage sites in hagfishes.

Experiments designed to elucidate potential secretagogues of catecholamine release in hagfishes have used an *in situ* saline-perfused systemic heart/PCV preparation (Figure 27.1; Perry *et al.*, 1993; Bernier and Perry, 1996). Although this preparation is a reliable tool for evaluating the involvement of potential secretagogues, it does not include the possible catecholamine contribution from the chromaffin tissue of the portal heart, and it combines the contribution from the systemic heart's auricle and ventricle to that of the PCV. Given the heterogeneity of storage levels and NA/AD ratio between the different sites of catecholamine storage described above, one should be careful in generalizing the potential implication of the secretagogues described in this chapter to all the catecholamine storage sites in hagfishes.

27.3 NON-CHOLINERGIC CONTROL OF CATECHOLAMINE SECRETION

27.3.1 Potassium

In contrast to dogfishes (*Squalus acanthias*), in which small exercise-induced increases in plasma [K^+] appear to supplement the neuronal mechanism of catecholamine

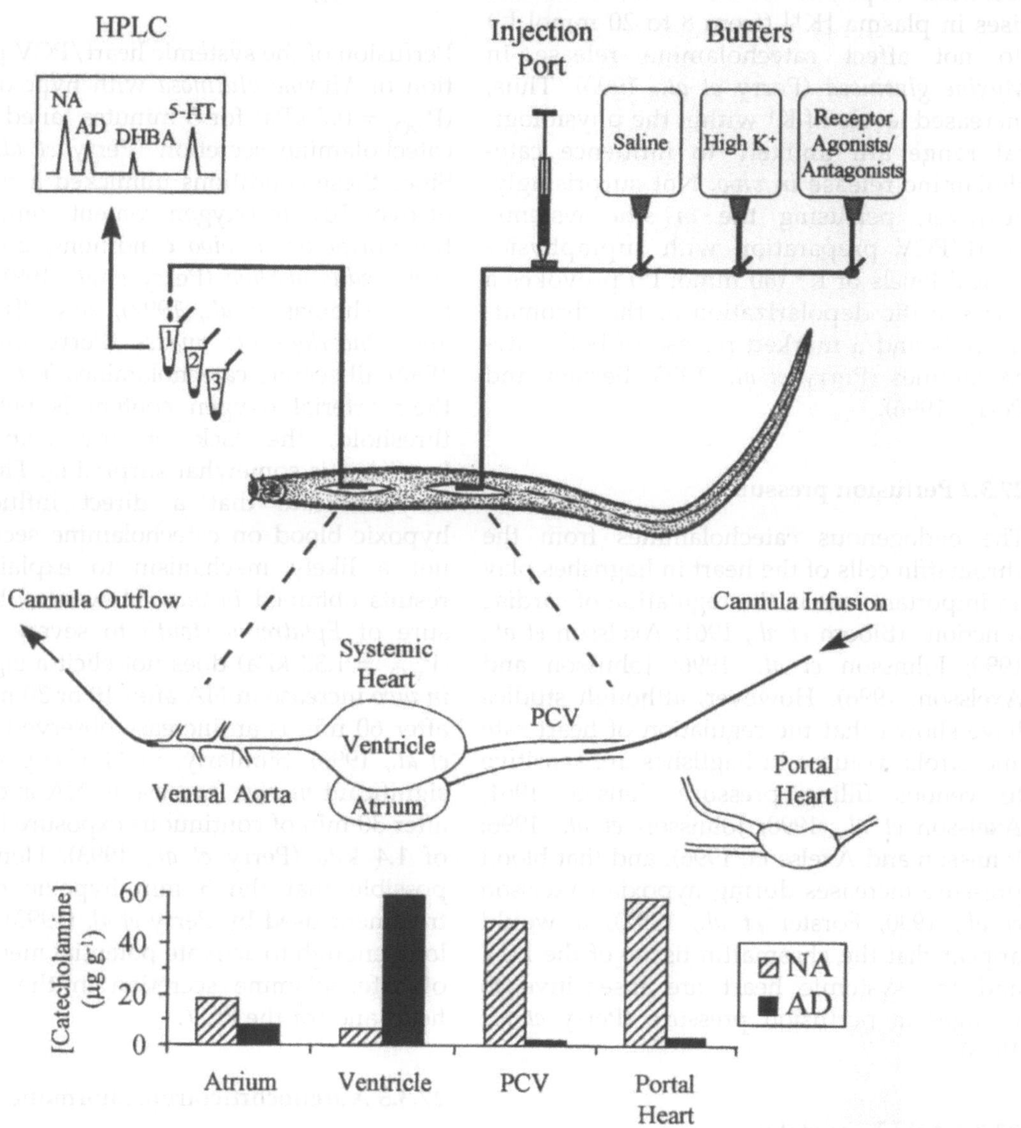

Figure 27.1 Schematic diagram of the *in situ* saline-perfused hagfish systemic heart/posterior cardinal vein (PCV) preparation, the distribution of chromaffin tissues, and the catecholamine concentration of these tissues. The preparation can be perfused with either hagfish saline, saline with added K^+, or saline modified with receptor blockers. Potential secretagogues are added to the preparation via the injection port. The inflowing cannula is positioned in the PCV, and the outflowing cannula is in the ventral aorta. The effluent is collected over a given time sequence and subsequently assessed for catecholamine (NA, noradrenaline; AD, adrenaline; DHBA, 3,4-dihydroxybenzylamine 'internal HPLC standard'; 5-HT, serotonin) content using HPLC. The catecholamine concentrations (NA + AD) of the chromaffin tissues perfused by the *in situ* preparation (the atrium and ventricle of the systemic heart and the PCV), as well as those not perfused by the preparation (the portal heart), are taken from Euler and Fänge (1961) and Perry *et al.* (1993).

secretion (Opdyke *et al.*, 1982), moderate rises in plasma [K^+] (from 8 to 20 mmol l^{-1}) do not affect catecholamine release in *Myxine glutinosa* (Perry *et al.*, 1993). Thus, increased levels of K^+ within the physiological range are unlikely to influence catecholamine release *in vivo*. Not surprisingly, however, perfusing the *in situ* systemic heart/PCV preparation with supraphysiological levels of K^+ (60 mmol l^{-1}) provokes a non-specific depolarization of the chromaffin cells and a marked release of both catecholamines (Perry *et al.*, 1993; Bernier and Perry, 1996).

27.3.2 Perfusion pressure

The endogenous catecholamines from the chromaffin cells of the heart in hagfishes play an important role in the regulation of cardiac functions (Bloom *et al.*, 1961; Axelsson *et al.*, 1990; Johnsson *et al.*, 1996; Johnsson and Axelsson, 1996). However, although studies have shown that the regulation of heart rate and stroke volume in hagfishes are sensitive to venous filling pressure (Jensen, 1961; Axelsson *et al.*, 1990; Johnsson *et al.*, 1996; Johnsson and Axelsson, 1996), and that blood pressure increases during hypoxia (Axelsson *et al.*, 1990; Forster *et al.*, 1992), it would appear that the chromaffin tissue of the PCV and the systemic heart are insensitive to changes in perfusion pressure (Perry *et al.*, 1993).

27.3.3 Acid-base states

The chromaffin cells of hagfishes are also unresponsive to regional acidosis. Lowering the pH of the saline perfusing the systemic heart and PCV of *Myxine glutinosa* from 8.1 to 7.0 failed to elicit catecholamine secretion (Perry *et al.*, 1993). This is similar to observations in teleosts, where acidosis *in vivo* is not a direct stimulus for catecholamine secretion (Perry *et al.*, 1989; Randall and Perry, 1992).

27.3.4 Oxygen status

Perfusion of the systemic heart/PCV preparation of *Myxine glutinosa* with hypoxic saline (P_{wO_2} = 0.7 kPa) for 5 minutes failed to elicit catecholamine secretion (Perry *et al.*, 1993). Since these conditions mimicked a reduction of over 50% in oxygen content compared to the normoxic *in vivo* conditions, and given that *Gadus morhua* (Perry *et al.*, 1991), *Salmo fario* (Thomas *et al.*, 1992), *Anguilla rostrata* and *Oncorhynchus mykiss* (Perry and Reid, 1992) all secrete catecholamines *in vivo* when their arterial oxygen content is below this threshold, the lack of any response in hagfishes is somewhat surprising. This result may indicate that a direct influence of hypoxic blood on catecholamine secretion is not a likely mechanism to explain other results obtained *in vivo*. However, the exposure of *Eptatretus stoutii* to severe hypoxia (P_{wO_2} = 1.33 kPa) does not elicit a significant *in vivo* increase in NA after 10 or 30 min; only after 60 min is an increase observed (Bernier *et al.*, 1996). Similarly, in *Myxine glutinosa*, a significant *in vivo* increase in NA is observed after 30 min of continuous exposure to a P_{wO_2} of 1.4 kPa (Perry *et al.*, 1993). Hence, it is possible that the 5 min hypoxic exposure treatment used by Perry *et al.* (1993) was not long enough to activate potential mechanisms of catecholamine secretion in the systemic heart and/or the PCV.

27.3.5 Adrenocorticotropic hormone

Using an *in situ* systemic heart/PCV preparation, Perry *et al.* (1993) showed that a pituitary extract from *Gadus morhua* caused a marked release of catecholamines from the chromaffin tissue of *Myxine glutinosa*, and hypothesized that the most likely pituitary hormone in the extract responsible for catecholamine release was adrenocorticotropic hormone (ACTH). This hypothesis was supported by Bernier and Perry (1996), who observed a significant increase in the NA and

AD secretion rate of *Myxine glutinosa* after the addition of 7.5 IU kg^{-1} of porcine ACTH to the inflowing saline of a similar *in situ* preparation (Figure 27.2). Although ACTH favoured the secretion of NA over AD, relative to the control conditions, the NA/AD ratio significantly decreased following ACTH injections (Bernier and Perry, 1996). This result suggests that in hagfishes, as in other vertebrates, AD and NA are stored in different chromaffin cell types (Accordi, 1991; Chritton *et al.*, 1991; Reid and Perry, 1994), and the secretion elicited by ACTH, as in trout, may arise primarily from adrenaline-storing cells (Reid *et al.*, 1996).

Rainbow trout is the only other vertebrate where ACTH is known to stimulate catecholamine secretion (Reid *et al.*, 1996). In this species, while porcine ACTH elicits the release of both catecholamines from an *in situ* PCV preparation, only the release of adrenaline occurs in a dose-dependent manner (Reid *et al.*, 1996). Also, extracts of trout pituitary given *in situ*, and intra-arterial injections of porcine ACTH given *in vivo*, both cause an elevation of plasma AD, but fail to increase NA levels in the trout (Reid *et al.*, 1996). So, while ACTH favours the secretion of NA over AD in hagfishes, the opposite is true in rainbow trout. A likely explanation

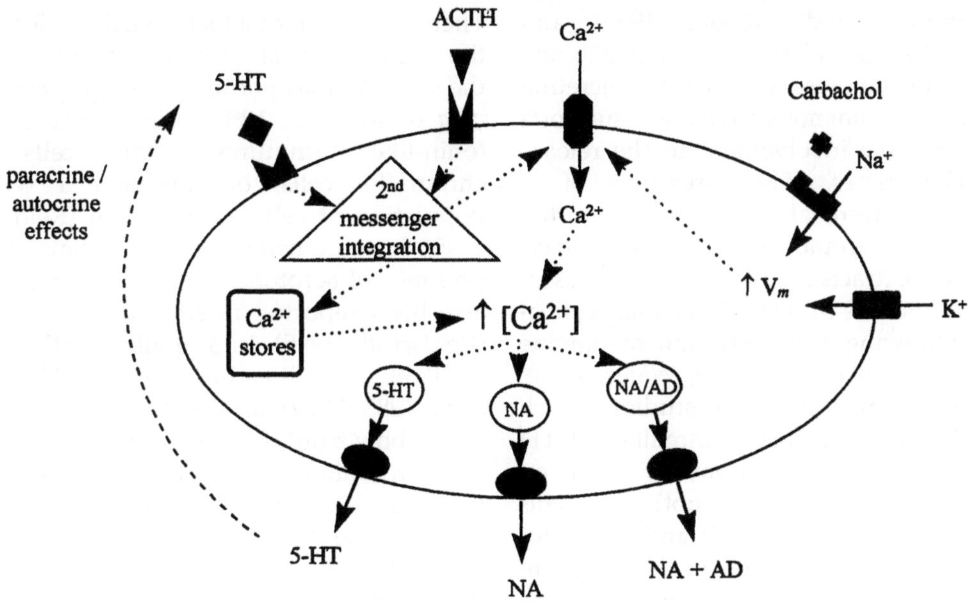

Figure 27.2 Schematic diagram of the presumptive mechanisms of catecholamine secretion in a generic chromaffin cell of a hagfish. Although the intracellular signalling pathways responsible for exocytosis have yet to be identified in hagfish chromaffin cells, this diagram is based on existing vertebrate models (Burgoyne, 1991; Furimsky *et al.*, 1996). Stimulation of serotonergic (5-HT) or adrenocorticotropic hormone (ACTH) receptors on the surface of chromaffin cells elicit the activation of secondary messenger signals which lead to a rise in $[Ca^{2+}]_i$. Similarly, stimulation of the nicotinic receptors by the cholinergic agonist carbachol is associated with an increase in permeability of the cell membrane to both Na^+ and K^+, an increase in membrane potential (V_m), and a rise in $[Ca^{2+}]_i$. The sources of calcium include entry from the extracellular fluid and release from internal stores. The rise in $[Ca^{2+}]_i$ is the physiological stimulus for exocytosis of the secretory granules containing noradrenaline (NA), adrenaline (AD) and serotonin. The secretion of serotonin may also lead to modulation of catecholamine release via paracrine or autocrine effects.

for this difference may lie in the known differences in catecholamine storage between the two species (see section on catecholamine storage levels above).

The *in situ* stimulatory effect of ACTH on the secretory rate of both catecholamines in hagfishes was unaffected by pretreatment with the ganglionic blocker, hexamethonium, or by the serotonergic receptor antagonist, methysergide (Bernier and Perry, 1996). Similar results were observed in trout (Reid *et al.*, 1996), and they suggest that ACTH is not exerting its effects by interacting with either of these receptor types.

The ACTH-like activity of *Myxine glutinosa*'s pituitary gland (Buckingham *et al.*, 1985), and the morphological, pharmacological (Fernholm and Olsson, 1969) and immunohistochemical evidence (pers. comm. S.G. Reid) for the presence of ACTH secreting cells in the adenohypophysis supports ACTH's possible involvement in the release of catecholamines. The presence of a hypophysial–interrenal axis is also supported by the corticotropic activity of *Myxine* hypophysial extracts on mammals (Jørgensen, 1976), and the rise in hagfish plasma corticosteroids following administration of porcine ACTH (Idler and Burton, 1976; Wales, 1988). Other immunohistochemical studies which also used antisera against mammalian ACTH in the hagfish adenohypophysis, however, obtained negative results in both *Eptatretus burgeri* (Jirikowski *et al.*, 1984) and *Eptatretus stoutii* (Nozaki, 1985). Moreover, Buckingham *et al.* (1985) observed substantial differences between the dose-response curves of mammalian ACTH and the pituitary extracts of hagfish in a cytochemical bioassay. Hence, these conflicting observations suggest that hagfish ACTH may be structurally distinct from mammalian ACTH. Also, while ACTH stimulates catecholamine release *in situ* (Bernier and Perry, 1996), the addition of ACTH to isolated heart preparations of *Myxine glutinosa* has no chronotropic effect, and by inference may not affect their endogenous catecholamine stores (Johnsson and Axelsson, 1996).

Although the interrenal tissue of *Myxine glutinosa* is known to be dispersed throughout the pronephric and pericardial regions (Idler and Burton, 1976), the extent to which these interrenal cells are associated with chromaffin cells is not known (Hardisty, 1979). Hence, it is possible that ACTH elicits catecholamine secretion indirectly through ACTH-induced corticosteroid secretion. In trout, however, the secretory effects of ACTH on the chromaffin tissue occur independently of cortisol (Reid *et al.*, 1996).

27.3.6 Serotonin

There is immunohistochemical evidence for the presence of serotonin-containing cells in the systemic and portal hearts of *Myxine glutinosa* (Reid *et al.*, 1995). Whether these serotonin-like immunoreactive cells are chromaffin cells, or represent a separate population of cells, is unknown (Reid *et al.*, 1995). Immunohistochemical evidence for the presence of serotonin in chromaffin cells has also been reported in several teleost species (Reid *et al.*, 1995), in amphibians (Delarue *et al.*, 1988), as well as in mammals (Brownfield *et al.*, 1985; Holzwarth and Brownfield, 1985). In rainbow trout, Fritsche *et al.* (1993) showed that serotonin is stored in high concentrations (44.61 ± 5.96 μg g^{-1} tissue) in the anterior region of the PCV within the head kidney.

In *Myxine glutinosa*, serotonin stimulates the secretion rate of both NA and AD when injected in the *in situ* perfused systemic heart/PCV preparation (Figure 27.2; Bernier and Perry, 1996). The stimulatory effects of serotonin in hagfishes primarily elicit the secretion of NA. Methysergide, a serotonergic receptor antagonist, abolishes the serotonin-induced release of AD and NA (Bernier and Perry, 1996). This result suggest that serotonergic methysergide-sensitive receptors exist on both AD- and NA-chromaffin cells in the hagfish.

Perfusing the *in situ* heart/PCV preparation with a saline solution containing a depolarizing [K^+] (60 mmol l^{-1}), as observed previously with catecholamine secretion, provokes a marked release of serotonin. Injection of a 100 nmol kg^{-1} dose of teleost angiotensin II ([Asn^1, Val^5]Ang II) into the *in situ* preparation also elicits a significant release of serotonin. Since angiotensin II does not elicit catecholamine secretion (see angiotensin section below), this response appears to be specific to the control of serotonin release. Meanwhile, known secretatogues of catecholamine release in hagfishes, ACTH and carbachol, have no effect on the [serotonin] of the perfusate (Bernier and Perry, 1996). Hence, although non-specific depolarization of chromaffin cells elicits the release of both catecholamines and serotonin, the mechanisms involved in the control of catecholamine and serotonin secretion may be different. These results also suggest that the serotonin-containing cells of the systemic heart may represent a different population of cells than the catecholamine-containing chromaffin cells.

The serotonin stored in the systemic heart of hagfishes, in addition to a paracrine/autocrine role in the stimulation of catecholamine secretion, may also have other physiological functions. This is supported by the observations that serotonin has inotropic and chronotropic effects on the systemic heart (Augustinsson *et al.*, 1956), and potential vasoactive actions on the gill vasculature of *Myxine glutinosa* (Sundin *et al.*, 1994).

Serotonin also stimulates catecholamine secretion in the rainbow trout, both *in situ* and *in vivo* (Fritsche *et al.*, 1993). In trout, however, like ACTH, serotonin triggers the preferential release of AD over NA, and *in situ*, methysergide only blocks the serotonin-induced release of AD (Fritsche *et al.*, 1993). Here again (see ACTH section above) a likely explanation for the differences may lie in the known differences in catecholamine storage between the two species.

27.3.7 Angiotensin

Although angiotensins elicit a pressor response in *Myxine glutinosa* (Carroll and Opdyke, 1982), and angiotensin-converting enzyme-like activity has been measured in the liver and plasma of *Eptatretus stoutii* (Lipke and Olson, 1988), a complete renin-angiotensin system has not yet been identified in hagfish (Taylor, 1977; Nishimura, 1985). Since the pressor activity of mammalian angiotensin II (Ang II) in hagfishes can be completely abolished by adrenergic receptor blockade, Carroll and Opdyke (1982) concluded that this response was mediated entirely by catecholamines. However, whereas angiotensins have been shown to stimulate the secretion of catecholamins from elasmobranchs to mammals (Carroll and Opdyke, 1982), they have no effect on the basal secretion rate of NA and AD in the *in situ* systemic heart/PCV preparation of *Myxine glutinosa* (Bernier and Perry, 1996). Differences between the results of the two studies cannot be explained by differences in Ang II dosage. While Carroll and Opdyke (1982) used a dose of 1.91 nmol kg^{-1} of mammalian Ang II, Bernier and Perry (1996) used an even greater dose of 100 nmol kg^{-1} of teleost Ang II ([Asn^1, Val^5]Ang II), the dose which causes maximal catecholamine secretion from the chromaffin tissue of rainbow trout (Bernier and Perry, 1997). Differences in the amino acid sequence of the Ang II used in the two studies also are not likely to explain the contradictory results, since among vertebrates diverse species respond to homologous and non-homologous angiotensins (Silldorff and Stephens, 1992; Bernier and Perry, 1997). One possible explanation is that the Ang II-induced pressor response observed in *Myxine glutinosa* by Carroll and Opdyke (1982) was mediated through actions of Ang II on the sympathetic nerve system of vertebrates (Wilson, 1984; Reid, 1992). However, the adrenergic autonomic nerve fibers of the hagfish vasculature appear to be poorly developed (Reite, 1969;

Nilsson, 1983). Thus, although Ang II exerts a pressure response, there is no direct evidence for its role in catecholamine release in hagfishes.

27.3.8 Histamine

There is chromatographic evidence, but no immunohistochemical evidence (Reid *et al.*, 1995), that the systemic and portal hearts of *Myxine glutinosa* contain histamine (Augustinsson *et al.*, 1956). However, intravascular injections of histamine yield only weak and inconsistent effects on the systemic and branchial vasculature, and no effect on the activity of the heart in *Myxine glutinosa* and *Polistotrema* (= *Eptatretus*) *stoutii* (Reite, 1969). Moreover, the effects of histamine on the branchial blood vessels of *Myxine glutinosa* are similar to those obtained with AD and NA in the same preparation, and Reite (1969) suggested that these may be elicited either by direct stimulation of adrenergic receptors or indirectly by release of endogenously stored catecholamines. In mammals, histamine does elicit a substantial secretory response from the adrenal medulla, and in bovine adrenal chromaffin cells, histamine is the most potent non-cholinergic secretagogue (Burgoyne, 1991). However, although histamine elicits the release of catecholamines from *in vitro* perfused rat adrenals with an EC_{50} of 3 μM (Borges, 1994), the administration of a 300 μM dose of histamine, in the *in situ* systemic heart/PCV preparation of *Myxine glutinosa*, failed to stimulate catecholamine secretion (Bernier and Perry, 1996). The control secretion rate of either NA or AD did not change in response to 10 min perfusion periods of saline solutions containing histamine concentration of 0.3, 3, 30, or 300 μM (Bernier and Perry, 1996). Although these results do not shed any light on the possible physiological significance of the histamine found in the hearts of hagfishes, it seems unlikely that histamine is involved in catecholamine secretion *in vivo*. Alternatively, the histamine found in the heart of hagfishes may modulate, as in mammals, the gene expression of the regulatory enzymes involved in the biosynthesis of catecholamins (Wan *et al.*, 1989; Bunn *et al.*, 1995).

27.4 CHOLINERGIC CONTROL OF CATECHOLAMINE SECRETION

Bolus injections of the cholinergic receptor agonist carbachol (10^{-5} or 10^{-4} mol kg^{-1}) in the *in situ* systemic heart/PCV preparation of *Myxine glutinosa* elicits dose-dependent increases in the release of both catecholamins (Figure 27.2; Perry *et al.*, 1993; Bernier and Perry, 1996). NA, as with the other secretagogues discussed in this chapter, is the predominant catecholamine secreted in response to carbachol. The ganglionic blocking agent selective for nicotinic receptors, hexamethonium, abolishes the release of both AD and NA in response to injections of carbachol in *Myxine glutinosa* (Bernier and Perry, 1996). Since carbachol is a non-specific cholinergic receptor agonist which stimulates both nicotinic *and* muscarinic receptors, these results suggest that the secretion from adrenaline- and noradrenaline-containing chromaffin cells of *Myxine glutinosa* is controlled exclusively by nicotinic receptors. This situation is similar to the one observed with the chromaffin tissue of eels (*Anguilla rostrata*; Reid and Perry, 1994) and cod (*Gadus morhua*; Nilsson, 1983), but differs from the situation in the rainbow trout where adrenaline-storing chromaffin cells also appear to have muscarinic receptors (Reid and Perry, 1994). Since there is no evidence that the systemic heart, portal heart, or PCV of hagfishes receive an extrinsic innervation (Greene, 1902; Östlund, 1954; Augustinsson *et al.*, 1956; Jensen, 1965; Johnsson and Axelsson, 1996), the presence of nicotinic receptors on the chromaffin cells from the systemic heart or the PCV of *Myxine glutinosa* may only reflect their common embryological origin as sympathetic neurons (Burgoyne, 1991).

27.5 MODULATION OF CATECHOLAMINE SECRETION

It has been known since 1965 that adenosine is formed from the breakdown of ATP which is released in parallel with catecholamines from the mammalian chromaffin cells (Douglas *et al.*, 1965). More recently investigators have proposed that adenosine has a physiological role in catecholamine secretion by acting as a negative feedback modulator of release from the adrenal medulla (Kumakura, 1984; Chern *et al.*, 1987, 1992). At the cellular level, adenosine inhibits catecholamine secretion from bovine adrenal medullary cells by reducing agonist-evoked calcium fluxes across the plasma membrane (Chern *et al.*, 1987). In mammals, support for these modulatory attributes of adenosine comes from an *in vivo* study where the adenosine receptor antagonist DPSPX (1,3-dipropyl-8-(p-sulphophenyl)xanthine) stimulates adrenaline release from the adrenal medulla in hypotensive rats (Tseng *et al.*, 1994). Similarly, hypoxic rainbow trout treated with the adenosine receptor antagonist theophylline, showed a 16-fold higher plasma AD concentration than their sham counterparts (Bernier *et al.*, 1996). Adenosine may also play a role in the control of catecholamine secretion from

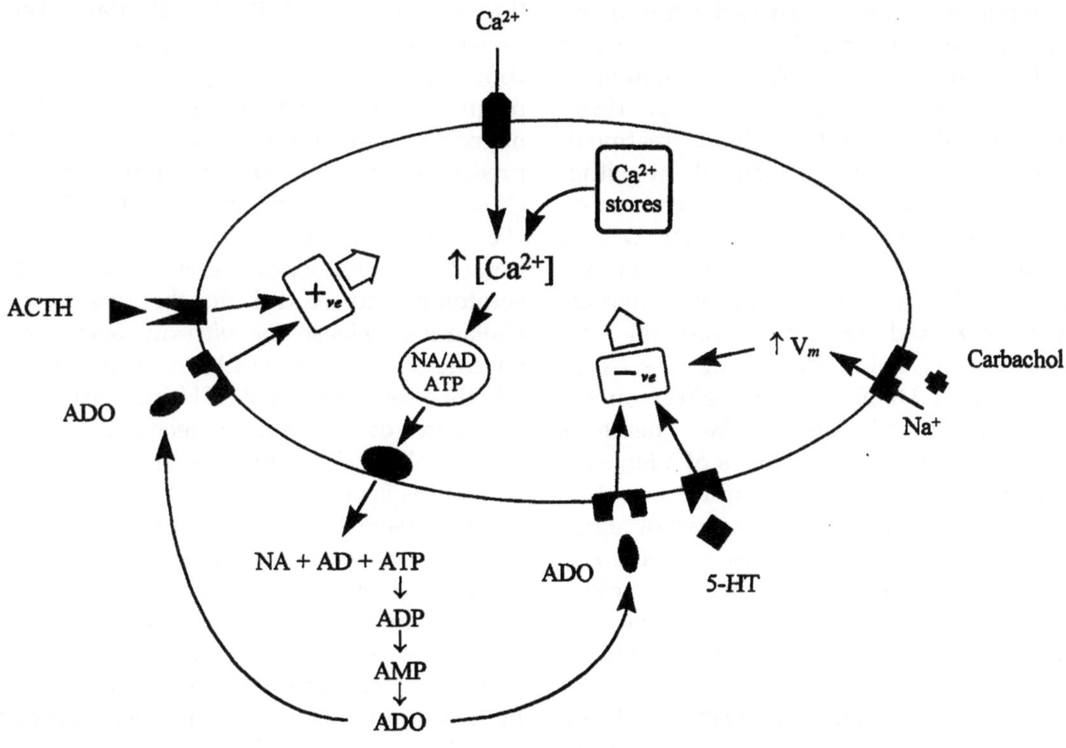

Figure 27.3 Schematic diagram of the presumptive modulatory effects of adenosine (ADO) on catecholamine secretion in a generic chromaffin cell of a hagfish. The exocytosis of secretory granules containing noradrenaline (NA), adrenaline (AD), and ATP, leads to the formation of ADO in the intercellular space. During secretion elicited by either serotonin (5-HT) or carbachol, adenosine has negative ($-_{ve}$) modulatory effects on catecholamine secretion. During secretion elicited by adrenocorticotropic hormone (ACTH), adenosine has positive ($+_{ve}$) modulatory effects on catecholamine secretion. See Figure 27.2 for a more complete description of the secretory mechanisms involved in catecholamine release.

the chromaffin tissue in hagfishes (Figure 27.3; Bernier *et al.*, 1996; Bernier and Perry, 1996). In *Eptatretus stoutii*, whereas exposure to a P_{wO_2} of 10 torr for 60 min had no effect on the plasma AD concentration, it increased the latter by 3.8-fold in hypoxic hagfishes pretreated with the adenosine receptor antagonist theophylline (Bernier *et al.*, 1996).

Using isolated bovine adrenal chromaffin cells, Chern *et al.* (1992) showed that the adenosine agonist PIA (N^6-L-phenylisopropyladenosine) inhibits catecholamine secretion induced by the nicotinic agonist DMPP (1,1-dimethyl-4-phenylpiperazinium). Similarly, in the *in situ* perfused systemic heart/PCV preparation of *Myxine glutinosa*, the rates of catecholamine secretion induced by the cholinergic receptor agonist carbachol, or by serotonin, were inhibited by the adenosine receptor agonist NECA (5-*N*′-ethylcarboxyadenosine; Bernier and Perry, 1996). Moreover, relative to the NECA treatment, the AD secretion rates induced by either carbachol or serotonin were both increased by the adenosine receptor antagonist DPSPX (Bernier and Perry, 1996). In contrast, however, the adenosine receptor agonist NECA increased the AD secretion rate induced by ACTH injections, while the receptor antagonist DPSPX had no effect on the latter (Bernier and Perry, 1996). Therefore, although the modulatory effects of adenosine on the pattern of catecholamine secretion induced by serotonin and carbachol agree with the inhibitory properties attributed to adenosine by Chern and his co-workers (Chern *et al.*, 1987, 1992), the interactions between adenosine and ACTH in *Myxine glutinosa* are more difficult to explain.

In the *in situ* systemic heart/PCV preparation of *Myxine glutinosa*, irrespective of the secretagogue used to elicit catecholamine secretion from the chromaffin tissue, adenosine agonist and antagonists had greater modulatory effects on the AD secretion rate than on the NA secretion rate (Bernier and Perry, 1996). These observations are consistent with the *in vivo* results obtained in *Eptatretus stoutii*, rainbow trout (Bernier *et al.*, 1996) and in rats (Tseng *et al.*, 1994) and, together, they suggest that the modulatory effects of adenosine may be mostly aimed at the adrenaline-storing cells.

27.6 CONSIDERATION OF PHYSIOLOGICAL MECHANISMS *IN VIVO*

Among the non-cholinergic secretagogues of catecholamine release investigated so far in hagfishes, only serotonin and ACTH have the potential of controlling catecholamine secretion *in vivo*. Another lower vertebrate where serotonin or ACTH are known to stimulate catecholamine secretion is rainbow trout (Fritsche *et al.*, 1993; Reid *et al.*, 1996). While catecholamine secretion appears to be controlled entirely through non-cholinergic means in hagfishes (Bernier and Perry, 1996), neuronal stimulation is believed to be the predominant mechanism initiating catecholamine secretion in teleosts (Nilsson, 1983). In both species, the relative contribution of non-cholinergic secretagogues, such as serotonin and ACTH, to the overall catecholamine release *in vivo* in response to specific stresses has not been explored. Also, whether serotonin or ACTH indiscriminately stimulate catecholamine secretion or only play a role under some specific physiological conditions is not known.

In hagfishes and rainbow trout, ACTH and serotonin directly stimulate catecholamine secretion (Bernier and Perry, 1996; Reid *et al.*, 1995). While serotonin may stimulate catecholamine secretion directly in mammals (Feniuk *et al.*, 1980; Chaouloff *et al.*, 1992), both serotonin and ACTH also indirectly stimulate catecholamine secretion in higher vertebrates. In dogs, the release of catecholamines from the adrenal medulla in response to carotid body hypoxia, is partly mediated by the corticosteroids secreted from the adrenal cortex under the control of ACTH (Critchley *et al.*, 1982). Meanwhile, serotonin stimulates catecholamine secretion from the adrenal medulla of the rat by modulating the sympathetic nervous activity to the chromaffin tissue (see review by Chaouloff, 1993). Both ACTH and serotonin are stored in the chromaffin cells of the adrenal medulla of mammals (Mazzocchi *et al.*, 1994; Holzwarth and Brownfield, 1985), however, their role in adrenal catecholamine release, if any, is not known. On the other hand, the ACTH and serotonin of the mammalian adrenal medulla have several different stimulatory effects on the corticosteroid production of the adrenal cortex (Mazzocchi *et al.*, 1994; Lefebvre *et al.*, 1992).

It is possible that the direct stimulation of catecholamine secretion by ACTH and serotonin in hagfishes represents more ancestral features of the non-cholinergic mechanisms involved in the regulation of the chromaffin tissue in vertebrates. Consistent with this hypothesis is the observation that ACTH directly stimulates catecholamine secretion in the haemocytes of the freshwater snail *Planorbarius corneus* (Ottaviani *et al.*, 1992). The ancestral stress response of invertebrates appears to be mediated by a CRF-ACTH-biogenic amine axis (Ottaviani *et al.*, 1992). This basic axis, although involving different structures, may also be an important mechanism mediating the stress response in hagfishes.

The serotonin-containing cells found in the systemic heart of *Myxine glutinosa* (see serotonin section above) may be homologous to the neuroepithelial cells (NECs) found throughout vertebrates. NECs found in respiratory epithelia, from fish gills to human lungs, are characterized by their ability to synthesize and store serotonin in chromaffin granular vesicles, and are sensitive to oxygen (Bailly *et al.*, 1992). While these often synapse with surrounding nerves, non-innervated NECs have also been reported (Bailly *et al.*, 1992). Although the basis for this homology remains to be established, the possible chemoreceptor role of NECs in vertebrates may also be shared by the serotonin-containing cells of hagfishes.

In both fishes and mammals, exposure to hypoxic conditions causes a decrease in the number of NECs and an increased exocytosis of their indolamine-storing vesicles (Dunel-Erb *et al.*, 1982; Lauweryns and Van Lommel, 1982). Although no such morphological investigation has been carried out in hagfishes, it is possible that serotonin may be involved in a paracrine stimulation of catecholamine secretion during severe hypoxia. The results obtained from severe hypoxic exposure of *Eptatretus stoutii* and *Myxine glutinosa in vivo* suggest that NA is, by far, the principal catecholamine secreted, and that secretion of AD may be inhibited by adenosine (Bernier *et al.*, 1996; Perry *et al.*, 1993). Similarly, the NA/AD ratio elicited by serotonin *in situ* (12.0 ± 5.1) is significantly greater than that elicited by ACTH (2.0 ± 0.7). Furthermore, the adenosine agonist NECA inhibits the adrenaline secretion elicited by serotonin (Bernier and Perry, 1996). Although the evidence is indirect, we suggest that at least during severe hypoxic exposure, serotonin may be released from the serotonin-storing cells of the systemic heart of hagfishes and mediate catecholamine secretion.

Given the role ascribed to catecholamines in the control of cardiac activity (Axelsson *et al.*, 1990; Johnsson *et al.*, 1996; Johnsson and Axelsson, 1996), the inhibitory effects of adenosine on the pattern of adrenaline secretion induced by serotonin (Bernier and Perry, 1996) may be important in protecting the heart from adrenergic over-stimulation. Although adenosine injections do not have any direct effects on the normoxic heart (Axelsson *et al.*, 1990), adenosine receptor agonists and antagonists, both *in vivo* and *in situ*, do not have any effects on the basal release of catecholamines from the chromaffin tissues (Bernier *et al.*, 1996; Bernier and Perry, 1996).

In general, hagfishes, like other vertebrates, probably respond to stressful conditions, or conditions of increased physiological demand, by recruiting several integrated mechanisms which, together, help maintain homeostasis (Axelrod and Reisine, 1984). Serotonin, ACTH and adenosine may all be involved in mediating catecholamine secretion in hagfishes during periods of acute stress. How are these signals, and undoubtedly other signals, specifically integrated for each physiological condition? This area of research has not yet been addressed and, therefore, remains a rich avenue of investigation.

27.7 CONCLUSION

The evidence presented in this chapter indicates that the control of catecholamine secretion in hagfishes may be achieved through non-cholinergic means. Both serotonin and ACTH stimulate catecholamine secretion in an *in situ* saline perfused systemic heart/PCV preparation. *In vivo* and *in situ* evidence also suggest that adenosine may be an important modulator of catecholamine secretion. As a result, non-cholinergic control of adrenal catecholamine secretion and multiple mechanisms of catecholamine synthesis and secretion have now been observed in all of the major vertebrate groups. However, although more than 70 years have elapsed since the discovery of non-cholinergic secretagogues in mammals, the cellular mechanisms involved and the physiological conditions under which these may play a role are still poorly characterized (Livett and Marley, 1993).

Potent secretagogues in other vertebrates, Ang II and histamine, do not appear to elicit catecholamine secretion in hagfishes. Although a number of other potential secretagogues or modulators, such as arginine vasotocin, bradykinin, or natriuretic peptides remain to be investigated, it appears that the control of catecholamine secretion in hagfishes may be relatively simple in comparison to that in higher vertebrates (Burgoyne, 1991). Whether these are primitive or regressive features of hagfishes is uncertain. However, given this relative simplicity, hagfish chromaffin cells may be one of the best models by which to study the control of catecholamine secretion.

REFERENCES

Accordi, F. (1991) The chromaffin cells of urodele amphibians. *J. Anat.*, **179**, 1–8.

Augustinsson, K.-B., Fänge, R., Johnels, A. *et al.* (1956) Histological, physiological and biochemical studies on the heart of two cyclostomes, hagfish (*Myxine*) and lamprey (*Lampetra*). *J. Physiol.*, **131**, 257–76.

Axelrod, J. and Reisine, T.D. (1984) Stress hormones: their interaction and regulation. *Science*, **224**, 452–9.

Axelsson, M., Farrell, A.P. and Nilsson, S. (1990) Effects of hypoxia and drugs on the cardiovascular dynamics of the Atlantic hagfish, *Myxine glutinosa*. *J. Exp. Biol.*, **151**, 297–316.

Bailly, Y., Dunel-Erb, S. and Laurent, P. (1992) The neuroepithelial cells of the fish gill filament: indolamine-immunocytochemistry and innervation. *Anat. Rec.*, **233**, 143–61.

Bernier, N.J. and Perry, S.F. (1996) Control of catecholamine and serotonin release from the chromaffin tissue of the Atlantic hagfish. *J. Exp. Biol.*, **199**, 2485–97.

Bernier, N.J. and Perry, S.F. (1997) Angiotensins stimulate catecholamine release from the chromaffin tissue of the rainbow trout. *Amer. J. Physiol.* (In press).

Bernier, N.J., Fuentes, J. and Randall, D.J. (1996) Adenosine receptor blockade and hypoxia tolerance in rainbow trout and Pacific hagfish. II. Effects on plasma catecholamins and erythrocytes. *J. Exp. Biol.*, **199**, 497–507.

Bloom, G., Östlund, E., Euler, U.S.v. *et al.* (1961) Studies on catecholamine-containing granules of specific cells in cyclostomes hearts. *Acta Physiol. Scand.*, **53**, Suppl. 185, 1–34.

Borges, R. (1994) Histamine H1 receptor activation mediates the preferential release of adrenaline in the rat adrenal gland. *Life Sciences*, **54**, 631–40.

Bornstein, S.R., Gonzalez-Hernandez, J.A., Ehrhart-Bornstein, M. *et al.* (1994) Intimate contact of chromaffin and cortical cells within the human adrenal gland forms the cellular basis for important intraadrenal interactions. *J. Clin. Endocrinol. Metab.*, **78**, 225–33.

Brownfield, M.S., Poff, B.C. and Holzwarth, M.A. (1985) Ultrastructural immunocytochemical co-localization of serotonin and PNMT in adrenal medullary vesicles. *Histochemistry*, **83**, 41–6.

Buckingham, J.C., Leach, J.H., Plisetskaya, E. *et al.* (1985) Corticotrophin-like bioactivity in the pituitary gland and brain of the Pacific hagfish, *Eptatretus stoutii*. *Gen. Comp. Endocrinol.*, **57**, 434–7.

Bunn, S.J., Sim, A.T.R., Herd, L.M. *et al.* (1995) Tyrosine hydroxylase phosphorylation in bovine adrenal chromaffin cells: the role of intracellular Ca^{2+} in the histamine H1 receptor-stimulated phosphorylation of Ser^{8}, Ser^{19}, Ser^{31}, and Ser^{40}. *J. Neurochem.*, **64**, 1370–8.

Burgoyne, R.D. (1991) Control of exocytosis in adrenal chromaffin cells. *Biochim. Biophys. Acta.*, **1071**, 174–202.

Campbell, G. (1970) Autonomic nervous systems, in *Fish Physiology, Vol. IV* (eds W.S. Hoar and D.J. Randall), Academic Press, London, pp. 109–32.

Carroll, R.G. and Opdyke, D.F. (1982) Evolution of angiotensin II-induced catecholamine release. *Amer. J. Physiol.*, **243**, R54–69.

Chaouloff, F. (1993) Physiopharmacological interactions between stress hormones and central serotonergic systems. *Brain Res. Rev.*, **18**, 1–32.

Chaouloff, F., Gunn, S.H. and Young, J.B. (1992) Central 5-hydroxytryptamine$_2$ receptors are involved in the adrenal catecholamine-releasing and hyperglycemic effects of the 5-hydroxytryptamine indirect agonist d-fenfluramine in the conscious rat. *J. Pharmacol. Exp. Ther.*, **260**, 1008–16.

Chern, Y.-J., Bott, M., Chu, P.-J. *et al.* (1992) The adenosine analogue N^6-L-phenylisopropyladenosine inhibits catecholamine secretion from bovine adrenal medulla cells by inhibiting calcium influx. *J. Neurochem.*, **59**, 1399–1404.

Chern, Y.-J., Herrera, M., Kao, L.S. *et al.* (1987) Inhibition of catecholamine secretion from bovine chromaffin cells by adenine nucleotides and adenosine. *J. Neurochem.*, **48**, 1573–6.

Chritton, S.L., Dousa, M.K., Yaksh, T.. *et al.* (1991) Nicotinic- and muscarinic-evoked release of canine adrenal catecholamines and peptides. *Amer. J. Physiol.*, **260**, R589–99.

Critchley, J.A.J.H., Ellis, P., Henderson, C.G. *et al.* (1982) The role of the pituitary-adrenocortical axis in reflex responses of the adrenal medulla of the dog. *J. Physiol.*, **323**, 533–41.

Delarue, C., Leboulenger, F., Morra, M. *et al.* (1988) Immunohistochemical and biochemical evidence for the presence of serotonin in amphibian adrenal chromaffin cells. *Brain Res.*, **459**, 17–26.

Douglas, W.W., Poisner, A.M. and Rubin, R.P. (1965) Efflux of adenine nucleotides from perfused adrenal glands exposed to nicotine and other chromaffin cell stimulants. *J. Physiol.*, **183**, 130–7.

Dunel-Erb, S., Bailly, Y. and Laurent, P. (1982) Neuroepithelial cells in fish gill primary lamellae. *J. Appl. Physiol.*, **53**(6), 1342–53.

Euler, U.S.v. and Fänge, R. (1961) Catecholamines in nerves and organs of *Myxine glutinosa, Squalus acanthius, and Gadus callarias. Gen. Comp. Endocrinol.*, **1**, 191–4.

Feniuk, W., Hare, J. and Humphrey, P.P.A. (1980) An analysis of the mechanism of 5-hydroxytryptamine-induced vasopressor responses in gangion-blocked anaesthetized dogs. *J. Pharm. Pharmacol.*, **33**, 155–60.

Fernholm, B. and Olsson, R. (1969) A cytopharmacological study of the *Myxine* adenohypophysis. *Gen. Comp. Endocrinol.*, **13**, 336–56.

Forster, M.E., Davison, W., Axelsson, M. *et al.* (1992) Cardiovascular responses to hypoxia in the hagfish, *Eptatretus cirrhatus*. *Respir. Physiol.*, **88**, 373–86.

Fritsche, R., Reid, S.G., Thomas, S. *et al.* (1993) Serotonin-mediated release of catecholamines in the rainbow trout *Oncorhynchus mykiss*. J. *Exp. Biol.*, **178**, 191–204.

Furimsky, M., Moon, T.W. and Perry, S.F. (1996) Calcium signaling in isolated single chromaffin cells of the rainbow trout (*Oncorhynchus mykiss*). *J. Comp. Physiol.* B., **166**, 396–404.

Greene, C.W. (1902) Contribution to the physiology of the California hagfish *Polistrotema stoutii* II. The absence of regulative nerves for the systemic heart. *Amer. J. Physiol.*, **6**, 318–24.

Hardisty, M.W. (1979) *Biology of the Cyclostomes*, Chapman & Hall, London.

Hirsch, E.F., Jellinek, M. and Cooper, T. (1964) Innervation of the systemic heart of the California hagfish. *Circul. Res.*, **14**, 212–17.

Holzwarth, M.A. and Brownfield, M.S. (1985) Serotonin coexists with epinephrine in rat adrenal medulary cells. *Neuroendocrinology*, **41**, 230–6.

Idler, D.R. and Burton, M.P.M. (1976) The pronephroi as the site of presumptive interrenal cells in the hagfish *Myxine glutinosa L. Comp. Biochem. Physiol.*, **53A**, 73–7.

Jensen, D. (1961) Cardioregulation in an aneural heart. *Comp. Biochem. Physiol.*, **2**, 181–201.

Jensen, D. (1965) The aneural heart of the hagfish. *Ann. N.Y. Acad. Sci.*, **127**, 443–58.

Jirikowski, G., Erhart, G., Grimmelikhuijzen, C.J.P. *et al.* (1984) FMRF-amide-like immunoreactivity in brain and pituitary of the hagfish *Eptatretus burgeri* (Cyclostomata). *Cell Tissue Res.*, **237**, 363–6.

Johnsson, M. and Axelsson, M. (1996) Control of the systemic heart and the portal heart of *Myxine glutinosa. J. Exp. Biol.*, **199**, 1429–34.

Johnsson, M., Axelsson, M., Davidson, W. *et al.* (1996) Effects of preload and afterload on the performance of the *in situ* perfused portal heart of the New Zealand hagfish *Eptatretus cirrhatus. J. Exp. Biol.*, **199**, 401–5.

Jönsson, A.-C. (1983) Catecholamine formation *in vitro* in the systemic and portal hearts of the Atlantic hagfish, *Myxine glutinosa. Mol. Physiol.*, **3**, 297–304.

Jönsson, A.-C., Wahlqvist, I. and Hansson, T. (1983) Effects of hypophysectomy and cortisol on the catecholamine biosynthesis and catecholamine content in chromaffin tissue from rainbow trout, *Salmo gairdneri. Gen. Comp. Endocrinol.*, **51**, 278–85.

Jørgensen, C.B. (1976) Sub-mammalian vertebrate hypothalamic-pituitary-adrenal interrelationships, in *General, Comparative and Clincal Endocrinology of the Adrenal Cortex* (eds I. Chester Jones and I.W. Henderson), Academic Press, London, pp. 143–206.

Kumakura, K. (1984) Possible involvement of ATP and its metabolites in the function of adrenal chromaffin cells, in *Dynamics of Neurotransmitter Function* (ed. I. Hanin), Raven Press, New York, pp. 271–80.

Lauweryns, j.M. and Van Lommel, A. (1982) Morphometric analysis of hypoxia-induced synaptic activity in intrapulmonary neuroepithelial bodies. *Cell Tissue Res.*, **226**, 210–14.

Lefebvre, H., Contesse, V., Delarue, C. *et al.* (1992) Serotonin-induced stimulation of cortisol secretion from human adrenocortical tissue is mediated through activation of a serotonin4 receptor subtype. *Neuroscience*, **47**, 999–1007.

Lipke, D.W. and Olson, K.R. (1988) Distribution of angiotensin-converting enzyme-like activity in vertebrate tissues. *Physiol. Zool.*, **61**, 420–8.

Livett, B.G. and Marley, P.D. (1993) Noncholinergic control of adrenal catecholamine secretion. *J. Anat.*, **183**, 277–89.

Mazzocchi, G., Malendowicz, .K., Markowska, A. *et al.* (1994) Effect of hypophysectomy on corticotropin-releasing hormone and adrenocorticotropin immunoreactivities in the rat adrenal gland. *Mol. Cell. Neurosci.*, **5**, 345–9.

Nilsson, S. (1983) *Autonomic Nerve Function in the Vertebrates*, Springer-Verlag, Berlin.

Nishimura, H. (1985) Endocrine control of renal handling of solutes and water in vertebrates. *Renal Physiol.*, **8**, 279–300.

Nozaki, M. (1985) Tissue distribution of hormonal peptides in primitive fishes, in *Evolutionary Biology of Primitive Fishes* (eds R.E. Foreman, A. Gorbman, J.M. Dodd and R. Olsson), Plenum Press, New York, pp. 433–54.

Opdyke, D.F., Carroll, R.G. and Keller, N.E. (1982) Catecholamine release and blood pressure changes induced by exercise in dogfish. *Amer. J. Physiol.*, **242**, R306–10.

Östlund, E. (1954) The distribution of catecholamines in lower animals and their effect on the heart. *Acta Physiol., Scand.*, **31**, Suppl 12, 1–67.

Östlund, E., Bloom, G., Adams-Ray, J. *et al.* (1960) Storage and release of catecholamines, and the occurrence of a specific submicroscopic granulation in hearts of cyclostomes. *Nature*, **188**, 324–5.

Ottaviani, E., Caselgrandi, E., Petraglia, F. *et al.* (1992) Stress response in the freshwater snail *Planorbarius corneus* (L.) (Gastropoda, Pulmonata): interaction between CRF, ACTH, and biogenic amines. *Gen. Comp. Endocrinol.*, **87**, 354–60.

Perry, S.F. and Reid, S.D. (1992) Relationship between blood O_2 content and catecholamine levels during hypoxia in rainbow trout and American eel. *Amer. J. Physiol.*, **32**, R240–9.

Perry, S.F., Fritsche, R., Kinkead, R. *et al.* (1991) Control of catecholamine release *in vivo* and *in situ* in the Atlantic cod (*Gadus morhua*) during hypoxia. *J. Exp. Biol.*, **155**, 549–66.

Perry, S.F., Fritsche, R. and Thomas, S. (1993) Storage and release of catecholamine from the chromaffin tissue of the Atlantic hagfish *Myxine glutinosa. J. Exp. Biol.*, **183**, 165–84.

Perry, S.F., Kinkead, R., Gallaugher, P. *et al.* (1989) Evidence that hypoxemia promotes catecholamine release during hypercapnic acidosis in rainbow trout (*Salmo gairdneri*). *Respir. Physiol.*, **77**, 351–64.

Randall, D.J. and Perry, S.F. (1992) Catecholamines, in *Fish Physiology*, Vol. XIIB (eds D.J. Randall, W.S. Hoar and A. Farrell), Academic Press, New York, pp. 255–300.

Reid, I.A. (1992) Interactions between ANG II, sympathetic nervous system, and baroreceptor reflexes in regulation of blood pressure. *Amer. J. Physiol.*, **262**, E763–78.

Reid, S.G. and Perry, S.F. (1994) Storage and differential release of catecholamines in rainbow trout (*Oncorhynchus mykiss*) and American eel (*Anguilla rostrata*). *Physiol. Zool.*, **67**, 216–37.

Reid, S.G., Fritsche, R. and Jönsson, A.-C. (1995) Immunohistochemical localization of bioactive peptides and amines associated with the chromaffin tissue of five species of fish. *Cell Tissue Res.*, **280**, 499–512.

Reid, S.G., Vijayan, M.M. and Perry, S.F. (1996) Modulation of catecholamine storage and release by the pituitary-interrenal axis in the rainbow trout (*Oncorhynchus mykiss*). *J. Comp. Physiol. B.*, **165**, 665–76.

Reite, O.B. (1969) The evolution of vascular smooth muscle responses to histamine and 5-hydroxytryptamine. I. Occurrence of stimulatory actions in fish. *Acta Physiol. Scand.*, **75**, 221–39.

Silldorff, E.P. and Stephens, G.A. (1992) The pressor response to exogenous angiotensin I and its blockade by angiotensin II analogues in the American alligator. *Gen. Comp. Endocrinol.*, **87**, 141–8.

Sundin, L., Axelsson, M., Nilsson, S. *et al.* (1994) Evidence of regulatory mechanisms for the distribution of blood between the arterial and the venous compartments in the hagfish gill pouch. *J. Exp. Biol.*, **190**, 281–6.

Taylor, A.A. (1977) Comparative physiology of the renin-angiotensin system. *Fedn. Proc.*, **36**, 1776–80.

Thomas, S., Perry, S.F., Pennec, Y. *et al.* (1992) Metabolic alkalosis and the response of the trout, *Salmo fario*, to acute severe hypoxia. *Respir. Physiol.*, **87**, 91–104.

Tseng, C.-J., Ho, W.-Y., Lin, H.-C. *et al.* (1994) Modulatory effects of endogenous adenosine on epinephrine secretion from the adrenal medulla of the rat. *Hypertension*, **24**, 714–18.

Waies, N.A.M. (1988) Hormone studies in *Myxine glutinosa*: effects of the eicosanoids arachidonic acid, prostaglandin E_1, E_2, A_2, F_2a, thromboxane B_2 and indomethacin on plasma cortisol, blood pressure, urine flow and electrolyte balance. *J. Comp. Physiol. B.*, **158**, 621–6.

Wan, D.C.-C., Marley, P.D. and Livett, B.G. (1989) Histamine activates proenkephalin A mRNA but not phenylethanolamine *N*-methyltransferase mRNA expression in cultured bovine adrenal chromaffin cells. *European J. Pharmacol.*, **172**, 117–29.

Wilson, J.X. (1984) Coevolution of the renin-angiotensin system and the nervous control of blood circulation. *Can. J. Zool.*, **62**, 137–47.

Wurtman, R.J., Axelrod, J., Veseli, E.S. *et al.* (1968) Species differences in inducibility of phenylethanolamine-N-methyltransferase. *Endocrinology*, **82**, 584–90.

PART ELEVEN

The Nervous System

28

ONTOGENY OF THE HEAD AND NERVOUS SYSTEM OF MYXINOIDS

Helmut Wicht and Udo Tusch

SUMMARY

The ontogeny of myxinoids is not well documented and there are no experimental embryological data. The present contribution summarizes the available descriptive data with emphasis on the development of head and nervous system. Myxinoids lay large and polylecithalic eggs, cleavage is partial and discoidal. The endodermal pharynx initially develops a large number of pharyngeal pouches. The anterior pouches degenerate and the posterior pouches are transformed into the definite gill slits. The buccal and nasal cavities develop from an unusually shaped stomodeum that displays large lateral recesses reminiscent of typical (endodermal) pharyngeal pouches. The neuraxis develops by invagination of a neural plate and initially contains an extensive ventricular system that is reduced to a system of narrow canals and vestigial structures in adults. A well-developed dorsal neural crest and ectodermal placodes contribute to the formation of the ganglia of the cranial nerves; an additional ventral neural crest that gives rise to the optic vesicle has been observed in the rostral neural tube. A lens placode is present in the ectoderm overlying the optic cup, but disappears later without giving rise to a lens. The dorsal and ventral roots of the spinal nerves fuse in an unusual way to form mixed spinal nerves. Chorioid plexuses never develop, and the presence of a pineal anlage is doubtful. The neural crest probably contributes to the formation of a very dense pre- and postotic head mesenchyme which gives rise to cartilages and muscles, there are no anlagen of external eye muscles.

28.1 INTRODUCTION

This review has two objectives: to summarize what is known about the ontogeny of the head and nervous system of myxinoids and, as important, to point out what is not known. We will briefly describe the very early development from cleavage to neurulation, then summarize the specific ontogeny of the head, branchial region, and nervous system. First, however, we provide a few paragraphs on the history of embryological research on hagfishes.

In many respects, myxinoids are ideal organisms for ontogenetic studies. The eggs (Figure 28.1A) and the embryos are quite large – average egg length in *Eptatretus stoutii* is about 23 mm (Dean, 1899) – and embryonic development is relatively fast, at least compared to lampreys which take several years to develop and metamorphose. Hagfishes are direct developers. Dean (1899) estimated that the minimal time between (external) fertilization and hatching is about 8 weeks in *Eptatretus stoutii*, but there are great variations in the pace of development among individual embryos (Worthington, 1905a).

The Biology of Hagfishes. Edited by Jørgen Mørup Jørgensen, Jens Peter Lomholt, Roy E. Weber and Hans Malte. Published in 1998 by Chapman & Hall, London. ISBN 0 412 78530 7.

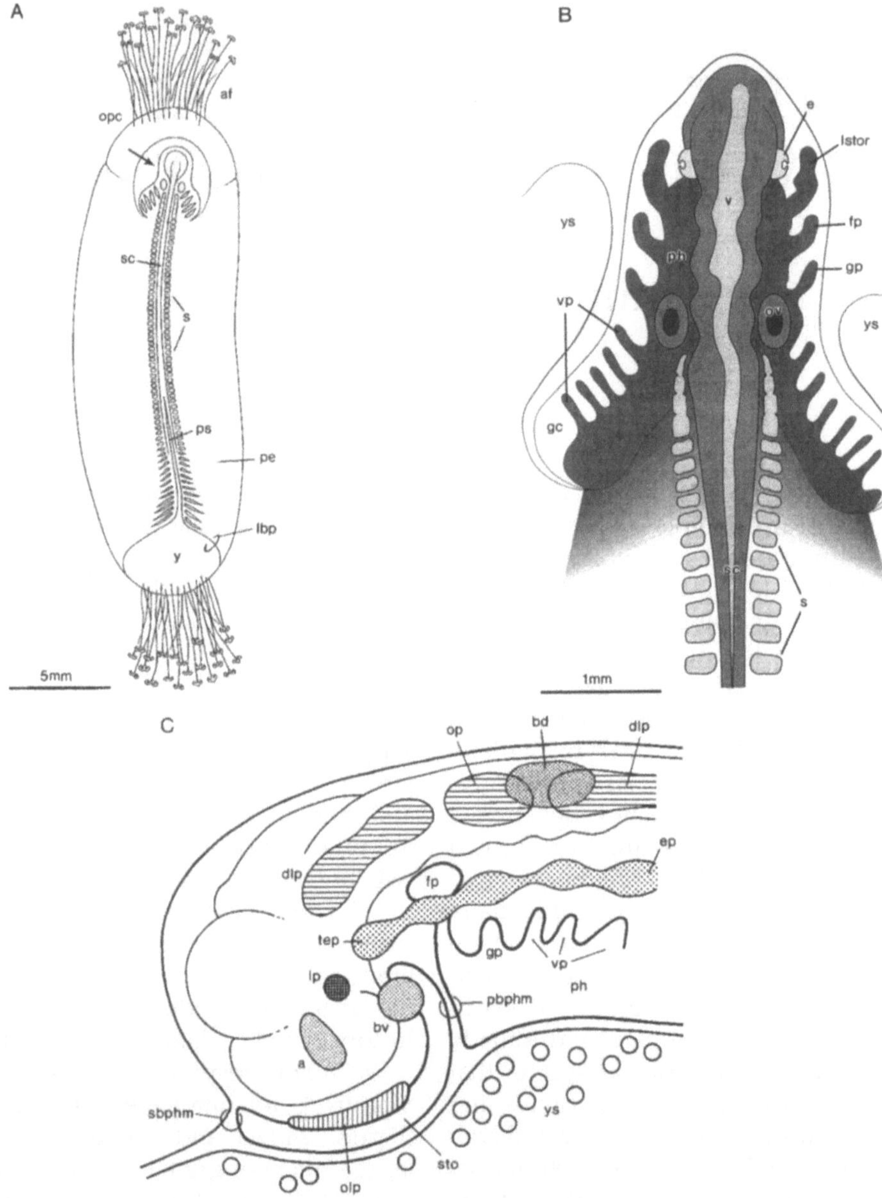

Figure 28.1 Drawings of dorsal views of an entire egg (A) and a head (B), and of a lateral view of a head (C) of early embryos of *Eptatretus stoutii*, to illustrate the overall appearance and topographical relationships of embryonic structures. (A) An egg and an early embryo of *Eptatretus stoutii*, modified from Dean (1899). The shell is somewhat translucent and the developing embryo is visible as a whitish streak in the fresh egg, but many of the details shown in this drawing are visible only after the egg shell has been removed. At the animal pole (top), a circular sulcus delimits the opercular cap (opc) which will open at the time of hatching. The anchor filaments (af) at the vegetative and the animal poles connect the eggs of an individual female to form clusters of 20 or more eggs. The arrow points at a depression in the yolk underneath the developing head. (B) Dorsal view of the head and anterior body of an early embryo, redrawn from Dean (1899). The pharyngeal and stomodeal cavities (dark shading) show a large number

Body length at hatching is between 45 and 65 mm (Dean, 1899; Worthington, 1905a); sexually mature specimens of *Eptatretus stoutii* measure between 250 and 650 mm (Reid, 1990). Juvenile and adult hagfishes are abundant in waters close to many scientific institutions, and they can easily be obtained in large numbers by trapping.

Fertilized eggs, on the other hand, are difficult to obtain. Hagfishes can easily be held in captivity for extended periods, but they do not reproduce, even though large numbers (several dozen) of ripe eggs are found in many captured females. In the wild, the eggs are deposited in clusters of about 20, held together by the so-called anchor filaments at the poles of the 'bean' or 'sausage-shaped' (Dean, 1899, Figure 28.1A) eggs, and harvesting eggs from the ocean floor is a laborious and unreliable process. As far as we can determine, E.C. Price was the first to obtain embryos (three) of eptatretid hagfishes (Price, 1896), followed by Bashford Dean (Dean *et al.*, 1897), Franz Doflein, a collaborator of Carl von Kupffer (Doflein, 1898), and Le Roy Conel (Conel, 1931a, 1942). In each case, eggs of *Eptatretus stoutii* were collected by fishermen in Monterey Bay (California), but their techniques (if specified) varied.

> The mature myxinoid, having taken a hook of the trawl line, secretes an enormous quantity of thick and viscid slime, and if by chance during this process its writhing body comes in contact with a string of eggs these may become enslimed, and thus, together with the fish, be brought to the surface. Such an accident, it may readily be understood, is decidedly uncommon . . . (Dean *et al.*, 1897)

Another, even more 'unlikely' method is reported by Worthington (1905a):

> In gathering the eggs . . . advantage is taken of the fact that they are eaten by the males. The fishermen set traps or lines for the fish, and when they are caught, hold them firmly by the head with one hand and 'strip' the body with the other, thus forcing out the eggs and any newly hatched young, for they are also eaten. Under these circumstances, only a very small portion of the eggs obtained are good for histological study, as most of them have been more or less digested.

The material obtained by these researchers is, to our knowledge, the only source of information on the embryology of eptatretid hagfishes, as no one has found embryos since Conel in 1930. Many of the embryos from the early Monterey Bay expeditions were processed histologically and are now located in the 'Dean–Conel' collection at the Museum of Comparative Zoology (Harvard University, Cambridge, Massachusetts, see Wicht and Northcutt (1995) for additional details). Notably, the embryos collected for Franz Doflein, which were transferred to Germany and formed the basis of Carl von

of laterally directed diverticula. The posterior diverticula are incorporated into the gill-collars (gc) that flank the head laterally. These diverticula represent the vagal pharyngeal pouches (vp), the anterior diverticula are the facial (fp) and the glossopharyngeal pouch (gp, also see Figure 28.2E). The anteriormost diverticulum (lstor) is the lateral recess of the stomodeal cavity (also see Figure 28.2A, B). A continuous row of somites (s) and an undivided but segmented string of mesoderm flank the neural tube caudal to the otic vesicle (ov), notably, there are no preotic somites or somitomeres. The ventricular system (v) can be seen shining through the neural tube which, at this stage, is already closed. (C) Highly schematic lateral view of an embryonic head resting on the yolk sac (ys), to clarify general topographical relationships, such as relative positions of placodal systems and buccopharyngeal membranes. Many structures that do not occur simultaneously in actual embryos appear together in this drawing. See text for detailed description.

Kupffer's (1900, 1906) description of myxinoid development, have disappeared (Conel, 1931b).

The development of myxinid hagfishes (family Myxinidae) is even less well documented. In 1862, 'the Academy of Sciences in Copenhagen offered a prize for the discovery of the manner of reproduction and development of *Myxine glutinosa*' (cited from Conel, 1931a), and this prize has never been claimed. Müller (1875, cited from Holmgren, 1946) may have been the first to obtain and inspect a fertilized egg of *Myxine glutinosa* (from the stomach of a cod), but Holmgren (1946) obtained the first material suitable for histological processing. Over a period of more than 20 years, he managed to collect about 130 eggs of *Myxine glutinosa* which were occasionally caught in trawling nets; three of them were fertilized and contained advanced embryos (also see Fernholm, 1969).

We will review the results of these early researchers on hagfish embryology and provide our own observations of some of the material in the collection in the context of current comparative embryological questions. Unless specified otherwise, the observations are based on embryos of *Eptatretus stoutii*, as data for *Myxine glutinosa* are far fewer.

28.1.1 List of abbreviations used in Figures 28.1–28.4

I	olfactory nerve
II	optic nerve
a	placode of nervus lateralis 'a'
af	anchor filaments
alar	alar plate
aol	octavolateral area
au	auricula
basal	basal plate
bd	dorsal placode of nervus lateralis 'b'
bo	olfactory bulb
bv	ventral placode of nervus lateralis 'b'
chd	chorda dorsalis
di	diencephalon
dlp	dorsolateral placode
dnc	dorsal neural crest
e	eye
ep	epibranchial placode
form.ret.	reticular formation
fp	facial (first, hyomandibular) pharyngeal pouch
gc	gill-collar
gp	glossopharyngeal (second, thyroidean) pharyngeal pouch
inf	infundibulum
lbp	lip of blastopore
lob.vag.	viscerosensory area
lp	lens placode
lstor	lateral stomodeal recess
mes	mesencephalon
nc	neural crest
oc	optic cup
olp	olfactory placode
op	otic placode
opc	opercular cap
ov	otic vesicle
p	pallium
pe	periblast
pbphm	primary buccopharyngeal membrane
ph	pharynx
ps	primitive streak
rdvmes	dorsal recess of mesencephalic ventricle
rhomb	rhombencephalon
rlvhy	lateral recess of hypothalamic ventricle
rvvmes	ventral recess of mesencephalic ventricle
s	somites
s.lim.	sulcus limitans of His
sbphm	secondary buccopharyngeal membrane
sc	spinal cord
sto	stomodeum
tel	telencephalon
tep	trigeminal epibranchial placode
v	ventricle
V-Xmot.	branchiomotor column
ve	velar apparatus
vhy	hypothalamic ventricle
vinf	infundibular ventricle
vl(d)	diencephalic lateral ventricle
vl(t)	telencephalic lateral ventricle
vmes	mesencephalic ventricle (Sylvian aqueduct)
vnc	ventral neural crest
vp	vagal pharyngeal pouches
vpo	preoptic ventricle
vq	fourth (rhombencephalic) ventricle
Vsens.	sensory nucleus of the trigeminal nerve
vsh	subhabenular ventricle
y	yolk
ys	yolk sac

28.2 STAGING

The development of *Eptatretus stoutii* requires about 2 months from fertilization to hatching (see above). No staging system is available for hagfishes: therefore, we will simply distinguish between very early embryos (embryoblastic stages, i.e. germ layer formation, neurulation), early embryos (organogenetic stages) and late embryos (histogenetic stages).

28.3 CLEAVAGE, GASTRULATION

The earliest developmental stages were described by Dean (1899) and Gudger and Smith (1931), based on inspection of entire eggs and on microscopical data. The eggs are polylecithalic, thus, cleavage is partial and discoidal. The germinal disc forms underneath the opercular cap of the egg and rapidly expands towards the vegetative pole of the egg, covering the yolk with a continuous sheet of flat epithelial cells that Dean called the periblast (Figure 28.1A). The circular rim of the periblast defines the blastopore which is closed as the growing periblast reaches the vegetative pole of the egg. While the periblast is still growing towards the vegetative pole, the embryoblast (Dean's 'definite embryo') begins to form in the region below the opercular cap. At the time the blastopore closes, the embryoblast is already quite differentiated; head and neural axis are easily visible, and gill slits have appeared (Figure 28.1A).

The relation of the periblast to the tissues of the embryoblast is unknown and problematic. Dean's observations imply that the embryoblast differentiates from the initially uniform periblast, but this differentiation is not documented histologically. Furthermore, the relation of the periblast to the germ layers of the embryo is unclear. In sections through the early embryoblast, the periblast can be seen as a flat epithelial layer of cells separating the embryoblast from the yolk beneath (Dean, 1899, arrowheads in Figure 28.2A). Von Kupffer (1900) interpreted this layer of periblast underneath the embryo as embryonic (yolk sac-) endoderm, whereas Dean (1899) assumed that the endoderm arises in a different fashion: he observed a double-layered embryoblast above the periblast and interpreted the dorsal, cuboidal epithelial layer of the embryoblast as the first anlage of the (neuro-) ectoderm and the more ventral, mesenchymal layer between the periblast and the ectoderm as the 'mes-endoderm'. We still do not know whether the endoderm forms by delamination (as implied by Dean) or by epiboly (as implied by von Kupffer).

28.4 FORMATION OF MESODERM

As noted above, Dean (1899) assumed that both endo- and mesoderm originated by separation of a mesenchymal 'mes-endoderm' from the epithelial ectoderm above. In later embryonic stages, he observed a primitive streak extending from the caudal tip of the neural plate to the dorsal lip of the blastopore. He did not examine this streak microscopically, but his observations seem to imply that the paraxial mesoderm derives from an invagination of material through this primitive streak, as somites become visible on both sides of the streak in slightly older embryos (Figure 28.1A). Von Kupffer (1900) did not comment on the origin of the paraxial mesoderm, but he observed that the chorda dorsalis is incorporated into the dorsal roof of the archenteron in the presumptive head region of very young embryos. It is not clear, however, whether the chorda develops from the endodermal roof of the archenteron or whether this incorporation is a secondary and transient phenomenon, as is the case in other craniates.

28.5 NEURULATION

Dean (1899), von Kupffer (1900) and Conel (1929) agree on the mode of formation of the neural plate and tube. The first structures to become visible are the medullary folds, which

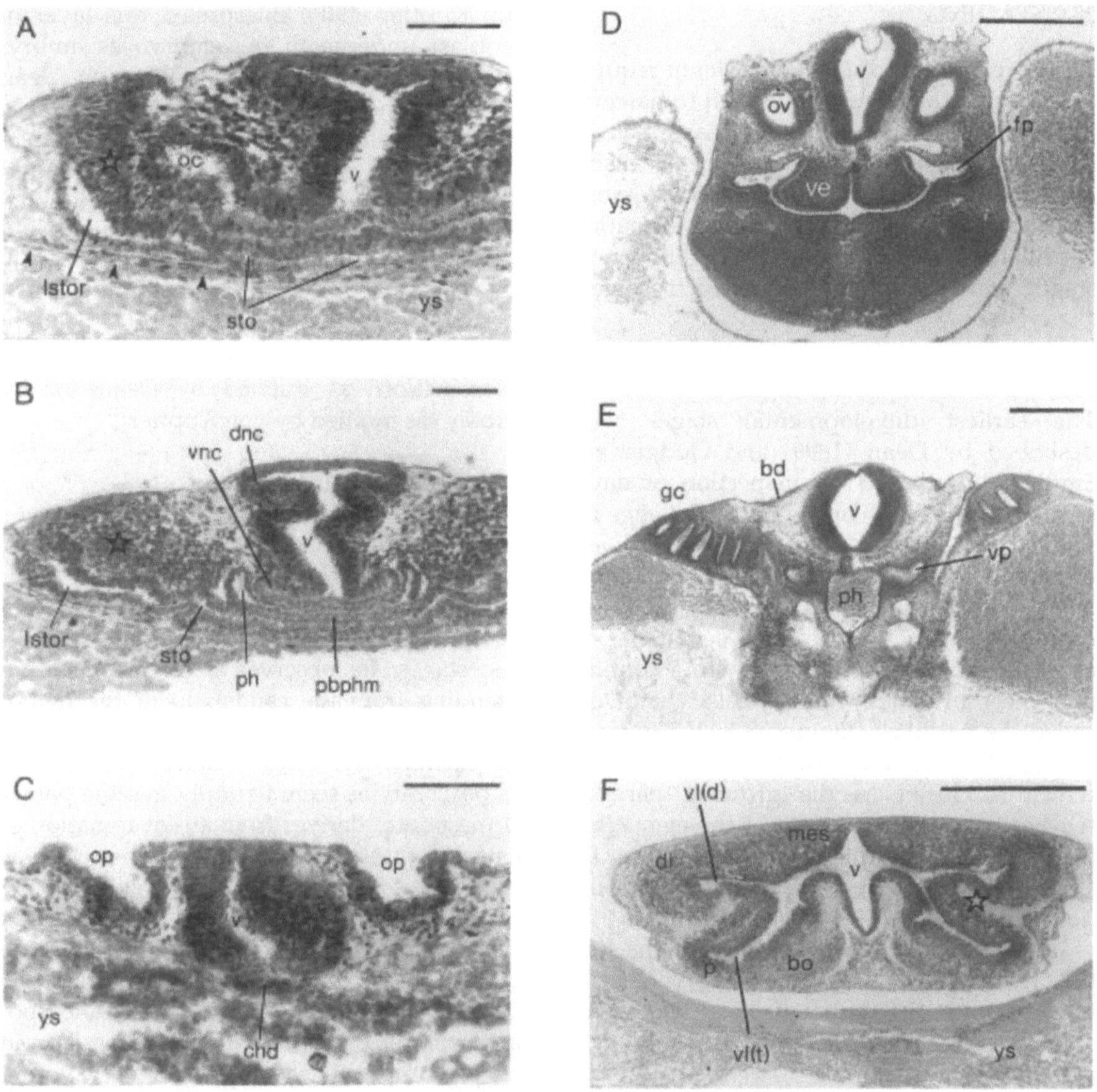

Figure 28.2 Photomicrographs of transverse sections of embryos of *Eptatretus stoutii* showing sections from the 'Dean–Conel' collection also illustrate the general state of preservation of the material. Frames A–C are fairly typical of much of the material and show poor fixation and fading of the staining. Frames D, E and F are exceptionally well-preserved and well-stained embryos. Scale bars in A–C equal 100 μm; scale bars in D–F equal 500 μm. (A) Transverse section through the head of an early embryo at the level of the optic cup (oc). Note the slit-like stomodeal cavity (sto) and the large lateral recesses with a distinct lumen (lstor) that are continuous with that cavity. Lateral to the optic cup and dorsal to the lateral recesses of the stomodeum, the ectoderm displays a conspicuous thickening (asterisk) which may represent the lens placode and/or the trigeminal epibranchial placode of von Kupffer (1900). In later developmental stages, the lens placode is clearly present as a separate entity (see Wicht and Northcutt, 1995). The arrowheads point at a sheet of flat epithelial cell which line the yolk sac (ys) and which represent Dean's (1899) periblast and von Kupffer's (1900) yolk-sac endoderm. (B) Same embryo as in A, slightly more

enclose a rather narrow medullary plate. The neural tube forms by invagination of that plate; the medullary folds meet in the dorsal midline, thus closing the tube and giving rise to the ventricular cavity. It is not known whether the posterior neural tube in the tail region develops by the same process, or whether it develops by 'secondary neurulation', as is the case in many other craniates (Griffith *et al.*, 1992). Interestingly, both von Kupffer (1900) and Conel (1929) reported that, in contrast to the order in other craniates, the anterior neural tube (presumptive brain) closes earlier than the neural tube in the presumptive spinal region.

From a comparative point of view, this information on the mode of neurulation is highly interesting, since the neural tube of lampreys and teleosts forms in a fundamentally different manner. In these groups, the medullary plate gives rise to a solid neural 'keel', and the ventricular cavities develop secondarily. The mode of neurulation in hagfishes might appear to indicate that invagination is the primitive mode of neural tube formation among craniates (Dean, 1899), except that the next relevant out-group of craniates, the cephalochordates, display yet another mode of neural tube formation. In cephalochordates, the neural plate is overgrown by general ectoderm and displaced ventrally as a plate. The folding of that plate and the formation of the neural tube occur secondarily, after the ectoderm has closed above the neural plate (Hertwig, 1890). Thus, the primitive mode of neurulation among craniates cannot be firmly established from an out-group comparison with cephalochordates.

28.6 NEURAL CREST

The mode of formation of the neural crest is unusual. In the anterior neural tube of early embryos, von Kupffer (1900) observed 'dorsale *und* ventrale Neuralleisten' (dorsal *and* ventral neural crests, Figure 28.2B), and he noted that the ventral neural crest was a rostral and caudal continuation of the optic vesicle. The dorsal neural crest, which corresponds to the

posterior section at the level of the primary buccopharyngeal membrane (pbphm). Again, note the slit-like stomodeal cavity (sto) and its lateral recess (lstor). The pharyngeal cavity (ph) is visible between the base of the neural tube and the stomodeal cavity; note that it is not continuous with the stomodeal cavity. The roof of the stomodeal cavity and the floor of the pharyngeal cavity form a two-layered membrane (pbphm, the primary buccopharyngeal membrane of von Kupffer, 1900) which separates the two lumina. The neural tube appears dorsoventrally compressed (possibly an artifact of fixation) and displays distinct dorsal and ventral ridges (dnc, vnc) that were interpreted by von Kupffer (1900) as 'dorsale und ventrale Neuralleisten'. The 'ventrale Neuralleiste' (vnc) is rostrally continuous with the optic stalk and vesicle as shown in A. The asterisk marks a dense condensation of mesenchyme medially adjacent to the lateral stomodeal recess. (C) An early embryo slightly younger than the one depicted in A and B showing the otic placodes (op) that have begun to sink in to form the otic vesicles. (D) An early embryo, but older than the one shown in A and B. Transverse section at the level of the otic vesicle (ov) and the facial pharyngeal pouches (fp). The developing velar apparatus (ve) can be seen in the dorsal wall of the pharyngeal cavity. (E) Same embryo as in D; transverse section through the region of the anterior vagal pharyngeal pouch (vp) and the gill-collar (gc). Note the ectodermal thickening (bd) laterally adjacent to the neural tube which represents one of the lateral line placodes of Wicht and Northcutt (1995). (F) Late embryo; section through the anterior forehead and the forebrain. Note that there are two pairs of lateral evaginations (vl[t] and vl[d]) which arise from the unpaired prosencephalic vesicle (v). The embryo shown in this photomicrograph is approximately at the same stage as the embryo from which the drawing of the brain in Figure 28.4A was prepared; at this stage, the prosencephalic vesicle is still bent sharply downwards. After the reversal of the bending of the prosencephalon (see text), the evaginations which are dorsal at this stage (vl[d]) will come to lie posterior to the ventral evaginations (vl[t], compare to Figure 28.4B). The asterisk identifies the developing central prosencephalic nuclear complex.

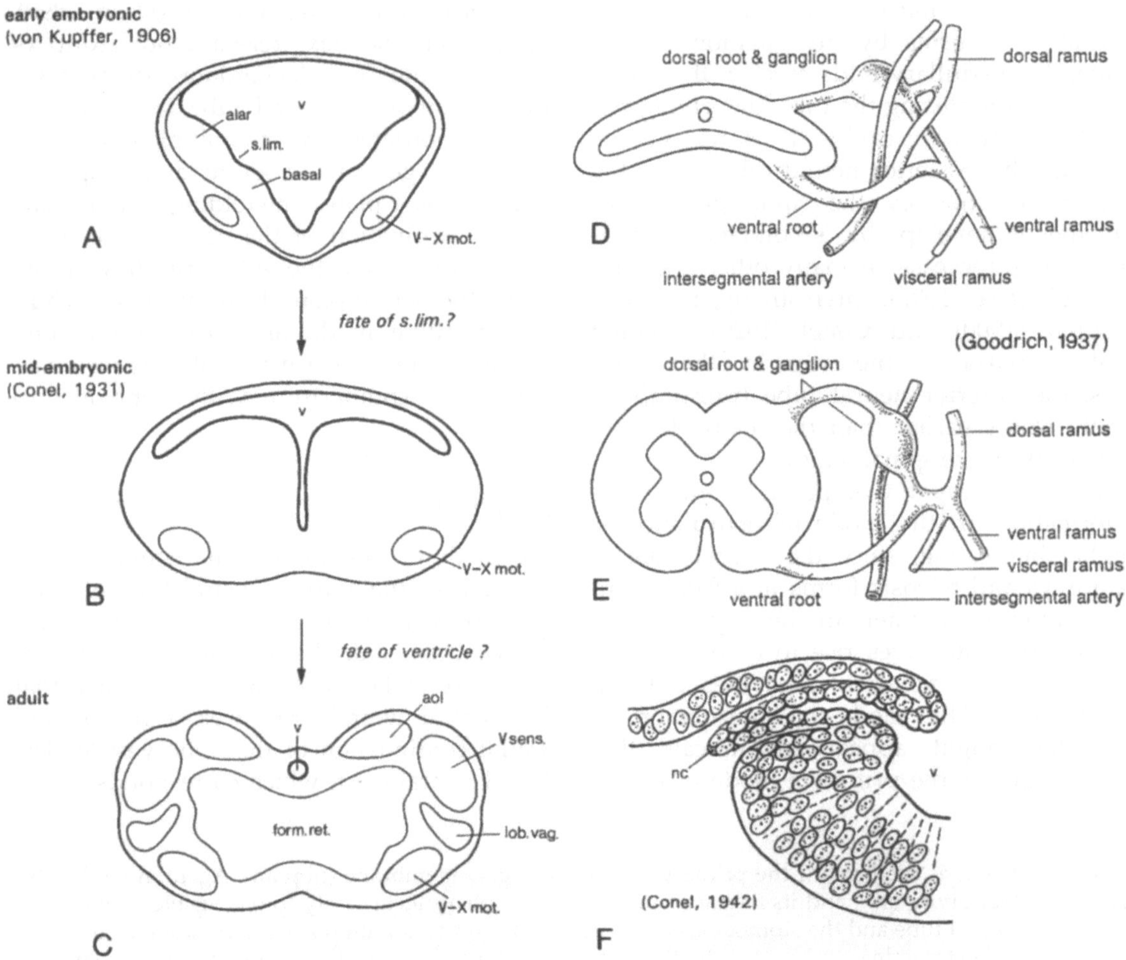

Figure 28.3 (A–C) Diagrammatical transverse sections through the rhombencephalon of an early embryo (A), a late embryo (B), and of an adult (C) depicting the gross-morphological transformations. From von Kupffer (1906) and Conel (1929), redrawn. See text for details. (D and E) A diagrammatic comparison of the mixed spinal nerves in hagfishes (D) and gnathostomes (E). Note the different position of the (embryonic) intersegmental artery with respect to the dorsal and ventral roots, indicating that dorsal and ventral roots from different segmental levels fuse in hagfishes and gnathostomes to form mixed spinal nerves. Based on Goodrich (1937). (F) The neural plate of in the trunk region of a very early embryo, showing an epithelial 'pocket' of neural crest. From Conel (1942).

neural crest in the modern sense, initially is an epithelial sheet of cells surrounding a central lumen which communicates with the lumen of the ventricular system. These pouches or ridges are supposed to arise from the alar plate (Conel, 1942, Figure 28.3F). Conel called the neural crest the 'ganglionic vesicles' and made a number of highly interesting claims regarding their formation and fate. First, he claimed that *all* ganglia of *all* cranial nerves (with the possible exception of the ganglion of the eighth nerve) derive from those 'ganglionic vesicles'; secondly, he claimed that the optic cup (and von Kupffer's ventral neural crest) were simply the rostralmost of these 'ganglionic vesicles', displaced ventrally by the cephalic flexure. Essentially, he claimed that the eye was a serial homologue of the crest-derived cranial nerve ganglia.

It is clear from von Kupffer's (1900) work that the first claim is not substantiated. Placodes are present in various locations in the ectoderm of the head (see below), and there is good histological and circumstantial evidence (von Kupffer, 1900; Wicht and Northcutt, 1995, see below) that those placodes contribute to the formation of cranial ganglia. Conel's second claim, however, although it may seem far-fetched, merits consideration, because it offers an explanation for the simultaneous occurrence of crest-derived tissues (i.e. an elaborate head skeleton) and paired lateral eyes during the evolution of craniates. Furthermore, it could explain the apparent absence of neural crest from the anterior end of the neural tube (Noden, 1991): the crest may have been 'consumed' in the formation of the optic vesicles. Conel's claim, therefore, should be critically re-examined.

The initial epithelial pockets of the dorsal neural crest transform into a mesenchyme and contribute to the formation of the ganglia of the cranial nerves (von Kupffer, 1900, see below, section on placodes). The (probable) further contribution of migrating crest cells to structures of the head (skeletal elements, other supportive tissues) is unknown.

28.7 PLACODES

Placodes are clearly present in various locations in the head ectoderm (von Kupffer, 1900; Stockard, 1907; Wicht and Northcutt, 1995; see Figure 28.1C for the following description of placodal systems). An unpaired nasal placode is found in the epithelium lining the ventral aspect of the head (von Kupffer, 1900), but there are numerous problems connected with the interpretation of this placode and its derivatives (see below, section on stomodeal region).

Dean (1899) and Conel (1929) claimed that there was no lens placode. However, a lens placode in the ectoderm adjacent to the developing optic cup is clearly present in *Eptatretus stoutii*, although it disappears soon after its formation (Price, 1896; von Kupffer, 1900, Stockard, 1907; Wicht and Northcutt, 1995; Figure 28.1C). Adult *Eptatretus stoutii* do not possess lenses, and the functional and evolutionary significance of the lens placode is enigmatic. It would be highly interesting to know whether a lens placode is present in developing *Myxine glutinosa*, as these animals in the adult stage have a visual system much more reduced than that of *Eptatretus stoutii* (Fernholm and Holmberg, 1975). The available embryos of *Myxine glutinosa* (Holmgren, 1946) are too advanced, however, to resolve this question.

Epibranchial placodes have been observed in the surface ectoderm of the branchial region (von Kupffer, 1900; Stockard, 1906a; Wicht and Northcutt, 1995), and, as is the case in other craniates, these placodes give rise to ganglia associated with the branchial cranial nerves. Interestingly, von Kupffer (1900) claimed that parts of the trigeminal ganglionic complex derive from a trigeminal epibranchial placode, located above the large lateral recesses of the stomodeal cavity (Figure 28.1C and 28.2A), in *both* hagfishes and lampreys and that no fewer than four small ganglia arise from a trigeminal epibranchial placode and become associated with the ganglion of the trigeminal nerve. This epibranchial placodal component of the trigeminus was supposed by von Kupffer to give rise to fibres accompanying the medial (palatine or pharyngeal) and mandibular rami of the fifth nerve.

In gnathostomes, trigeminal epibranchial placodes do not occur, and the trigeminal nerve does not have a chemosensory component, but von Kupffer's observations of lampreys and hagfishes would imply that an epibranchial placode overlying the mandibular arch is a plesiomorphic craniate feature. We do not know how epibranchial placodes and the chemosensory system interact during ontogeny, but there is circumstantial evidence in gnathostomes (Landacre, 1908, and Stone, 1922, both cited from Starck, 1975) that the epibranchial placodes may be the source of ganglion cells that innervate non-olfactory chemoreceptors, i.e. gustatory organs. The presence of trigeminal epibranchial placodes may therefore explain the fact that both internal and external chemoreceptors of hagfishes are supplied by rami of the fifth nerve (Nishizawa *et al.*, 1988; Braun and Northcutt, this volume). However, the situation is further complicated by the fact that the chemoreceptors of hagfishes may not be homologous to the taste-buds of other craniates (Braun and Northcutt, this volume), nor is it clear whether they are innervated by ganglion cells that derive from the epibranchial placode of the trigeminal nerve.

The presence of lateral line and octaval placodes is evident from the work of von Kupffer (1900) and Wicht and Northcutt (1995), even though Dean (1899) initially claimed that lateral line placodes were absent. Three such placodes are present (Figure 28.1C) and have a unique developmental fate. One placode is found in the preocular region ('a' in Figure 28.1C), another is located in the postocular region ('bv' in Figure 28.1C), and the third is found laterally and caudally adjacent to the otic vesicle ('bd' in Figures 28.1C, 28.2E). Circumstantial evidence indicates that the preocular placode ('a') gives rise to the ganglion and nervus lateralis 'a' of Worthington (1905b, anterior lateral line nerve or nervus buccalis of other authors) and that the posterior placodes ('bd' and 'bv') give rise to Worthington's nervus lateralis 'b' (posterior lateral line nerve of other authors), which does have two ganglia (Wicht and Northcutt, 1995; Braun and Northcutt, 1997). However, the placodes also give rise to neuromast-precursors (von Kupffer, 1900; Wicht and Northcutt, 1995) which later disintegrate and apparently transform into the system of pre- and postocular skin grooves found in the head of eptatretid hagfishes. These skin grooves are innervated by nervus lateralis 'a' and 'b' (Kishida *et al.*, 1987; Wicht and Northcutt, 1995; Braun and Northcutt, 1997; Braun and Northcutt, this volume; also see Fernholm, 1985).

There is a single octaval placode (von Kupffer, 1900; Figure 28.2C) which gives rise to the inner ear and the anterior ('utricular') and posterior ('saccular') ganglia of the eighth nerve.

It should be noted that von Kupffer (1900) also reported an earlier set of placodes, prior to the appearance of the above lateral line placodes. These placodes are described as longitudinal thickenings in the ectoderm, located anterior and posterior to the octaval placode (Figure 28.1C). The posterior placode extends back to the level of the vagal pharyngeal pouches; the anterior placode extends forward to the trigeminal region. Von Kupffer called these the 'dorsolaterale Plakoden', and they are not identical to the lateral line placodes described above. The relationship of these placodes is better understood if one considers von Kupffer's concept of a (branchial) cranial nerve and its components: every branchial nerve has an epibranchial placodal component which contributes the viscerosensory ganglion cells. The ganglion cells of the somatosensory component are derived from neural crest ('Ganglienleiste') and form the 'Hauptganglion' of the respective cranial nerve. The 'dorsolaterale Plakoden' contribute the special somatosensory (lateral line) ganglion cells. In contrast to the modern understanding of lateral line development, where both receptors and ganglion cells of the lateral line are derived from lateral line placodes (Northcutt, 1992a, b; Northcutt *et al.*, 1994), this concept allows von Kupffer to identify separate embryonic sources for the ganglion cells ('dorsolaterale

Plakoden') and the receptors of the lateral line (his buccal, supra-, and infraorbital placodes, 'a', 'bv' and 'bd'-placodes of Wicht and Northcutt, 1995). Whether this is actually the case in hagfishes is unknown. Although the scant embryonic material has been interpreted according to the modern concept (Wicht and Northcutt, 1995), von Kupffer's theory could be valid for hagfishes, though certainly not for gnathostomes (Northcutt *et al.*, 1994). If so, the entire lateralis system of hagfishes may be non-homologous to the system in other craniates.

Finally, a profundal trigeminal placode may be present. Von Kupffer (1900) claimed that the trigeminal ganglion receives a placodal contribution from the anteriormost part of his preotic 'dorsolaterale Plakode', but it is not known whether this corresponds to the profundal placode of gnathostomes or to a lateral line placode.

28.8 PARAXIAL MESODERM IN THE HEAD

Numerous well-defined somites occur in the trunk but do not continue into the head. The most rostral individualized somites are found adjacent to the caudal portion of the medulla oblongata; more rostrally, there is an undivided string of paraxial mesoderm that terminates at the posterior pole of the otic vesicle. This mesoderm does, however, show traces of serially repeated constrictions (Dean, 1899) and they may therefore represent somitomeres (Figure 28.1B). Von Kupffer noted that the preotic paraxial mesoderm never assumes an epithelial organization (i.e. there are no head cavities). In an early embryo, where the neural crest had not yet migrated, he observed several distinct 'condensations of a very dense mesenchyme' adjacent to the optic cup, the lateral stomodeal recess (Figure 28.2B), and the pharyngeal pouches. Their fate is unknown, and it is not clear whether they represent preotic somitomeres. In later stages, after the migration of the neural crest, the entire head is filled with an extremely dense mesenchyme (Wicht and Northcutt, 1995; Figure 28.2D and E), and it is impossible to determine the contribution of crest or 'somitomeres' to this mesenchyme and its derivatives with non-experimental methods. It is noteworthy, however, that all students of myxinoid development agree that no anlagen of external eye muscles are ever formed.

28.9 PHARYNGEAL AND BRANCHIAL APPARATUS

Adult myxinoids possess up to 15 pairs of gills (Worthington, 1905a), and their development is remarkable. Initially, the developing branchial apparatus, i.e. the endodermal pharyngeal pouches and corresponding ectodermal pockets, is located on the sides of the head in the ventral vicinity of the otic region. Secondarily, the posterior set of pouches is shifted backwards. For some time, the backshifting pouches assume the form of a 'gill-collar' flanking the head laterally (Figures 28.1A, B, 28.2E); later this collar is incorporated into the ventrolateral region of the anterior third of the body (Dean, 1899). As the posterior part of the branchial apparatus backshifts, the anterior pharyngeal pouches undergo a unique fate.

The first facial, or hyomandibular, pair of pouches is initially very large, projecting upwards from the pharynx and ending in a distinct vesicle underneath the rostral part of the otic capsule (Figures 28.1C, 28.2D). The next two pouches (i.e. the second, or glossopharyngeal, pouch and the foremost vagal pouch) are relatively small tube-like diverticula arising from the dorsal and lateral walls of the pharynx. As the next, or fourth, pair of pouches gives rise to the first pair of gills, the first three pairs of pouches degenerate, even before the formation of branchial apertures (Allis, 1903; Stockard, 1906a).

In spite of the early disappearance of pharyngeal pouches 1–3, numerous skeletal elements develop where they were located, and these skeletal elements resemble typical branchial arches (Neumayer, 1938; Holmgren, 1946; data from *Myxine glutinosa*), connecting the base of the neurocranium dorsally with an unpaired ventral basal cartilage.

These observations clearly indicate a fundamental difference between the pharyngeal apparatus of myxinoids and that of other craniates. In the latter, the first five pharyngeal pouches and arches play an important and fairly constant role in the generation of structures of the adult head, but the fate and even the number of the posterior pouches is variable. Ultimately, this constancy has led to many efforts to relate those arches and pouches and other serially repeated elements of the craniate head, resulting in segmental theories of the vertebrate head (e.g. Goodrich, 1930). In myxinoids, by contrast, the anterior pouches disappear quite early, and their role in the 'patterning' of head structures is poorly understood, because it is very difficult – if not impossible – to trace the adult structures back to their original arches.

The development of the thyroid gland was described by Stockard (1906b). The adult gland consists of follicular tissue located around the ventral aorta in the branchial region. Stockard observed that the gland develops from the floor of the pharynx, and he noted that the gland does not resemble a typical endostyle in any stage of development.

28.10 STOMODEAL REGION, BUCCAL CAVITY, NOSE AND ADENOHYPOPHYSIS

Adult myxinoids possess an unpaired external nasal duct (the prenasal sinus) which opens to the exterior at the tip of the head. Internally, the duct continues into the nasal cavity immediately rostral to the brain. The so-called nasopharyngeal duct then continues under the base of the forebrain and ultimately opens into the roof the velar chamber, slightly rostral to the velar apparatus (see Braun and Northcutt; Malte and Lomholt, both in this volume). The adenohypophyseal tissue is formed by small clusters of cells embedded in the connective tissue of the roof of the nasopharyngeal duct underlying the base of the diencephalon and the neurohypophysis. Fernholm (1969) has shown that the adenohypophysis actually develops from the roof of the nasopharyngeal duct in *Myxine glutinosa*.

There is no agreement with regard to the early embryonic origin of these structures, and the main controversy revolves around the question of whether a 'true' (i.e. ectodermal) stomodeum is present underneath the developing head (see Figure 28.1C). Von Kupffer (1900) claimed that such a stomodeum, formed by an ectodermal fold reaching under the head process, is present early in development. The lumen inside this fold is initially continuous with the extra-embryonic space: i.e. the structure labeled 'sbphm' in Figure 28.1C is initially absent according to von Kupffer (see below). He interpreted a thickening in the ectodermal roof of the stomodeum at the base of the prosencephalic vesicle as the olfactory placode and believed that this placode and the adjacent ectoderm gave rise to the nasal ducts, the olfactory epithelium, and the adenohypophysis, while the remainder of the stomodeal cavity was transformed into the buccal cavity. Caudally, this stomodeal cavity is separated from the tip of the (endodermal) pharynx by the 'primäre Rachenhaut' (primary bucco-pharyngeal membrane, Figures 28.1C, 28.2B). However, von Kupffer's stomodeal cavity and its derivatives lose their connection with the exterior later in ontogeny: the exterior openings of mouth and nasopharyngeal duct are closed by a membrane which ruptures only shortly before hatching. Von Kupffer interpreted this membrane as a 'secundäre Rachenhaut' (secondary bucco-pharyngeal membrane, Figure 28.1C, sbphm).

Gorbman (1983) came to a radically different conclusion. He also observed a stomodeal cavity ventral to the developing head, but, in contrast to von Kupffer, he claimed that this cavity was never in open contact with the extra-embryonic space. In other words, he claimed that the ectodermal lining of the head does not extend caudally beyond the 'secundäre Rachenhaut' of von Kupffer, and he concluded that the entire stomodeum was actually a structure derived from the rostral-most part of the pharynx, i.e. from endoderm. Thus, Gorbman assumed that the olfactory placode and its derivatives (i.e. the prenasal sinus, the olfactory epithelia, the adenohypophysis and the nasopharyngeal duct) and the entire buccal cavity are of endodermal origin. He refuted von Kupffer's concept of a '*secundäre* Rachenhaut', confirming the presence of such a membrane (which had been observed by Stockard [1906a] as well) but interpreting it as the *primary* and sole buccopharyngeal membrane.

The stomodeum of hagfishes possesses large lateral recesses (Stockard, 1906a, Figure 28.1B, 28.2A, B) that are directed laterally and dorsally towards the dorsal surface of the developing head. These lateral recesses are indeed reminiscent of pharyngeal pouches, and Stockard (1906a) repeatedly stressed that 'these conditions suggest forcibly . . . the idea of the origin of the vertebrate mouth from a pair of gill slits'. However, it is not clear how these (paired) lateral recesses relate to the (unpaired) definite mouth opening of adult myxinoids. As noted above, the mouth opens only immediately prior to hatching by the rupture of von Kupffer's 'secundäre Rachenhaut'.

Gorbman's interpretation of stomodeal ontogeny in hagfishes may seem daring, because it implies a dual embryonic origin of the adenohypophysis and the olfactory epithelium among craniates and violates the theory of germ layer specificity. However, the assumption that the (neuro-) ectodermal tissues are the sole source of receptors has been falsified recently: Barlow and Northcutt (1995) have shown that amphibian chemoreceptors (taste buds) derive from endoderm. The same may therefore be true for the olfactory chemoreceptors of hagfishes. Gorbman's view also complements Stockard's (1906a, see above) on the origin of the mouth cavity. Furthermore, it offers an explanation for the presence of trigeminal epibranchial placodes in hagfishes (see above, section on placodes): if such placodes were induced by the endodermal pharyngeal pouches, then an endodermal stomodeal cavity with large lateral recesses might be responsible for the formation of trigeminal epibranchial placodes. It should be stressed that these suggestions are highly speculative. Both von Kupffer (1900) and Gorbman (1983) examined histological material that is less than ideal by modern standards. Furthermore, the identification of individual germ layers in very early embryos is by no mean easy or exact, as noted above (section on cleavage and gastrulation). The whole issue clearly deserves a close re-examination.

28.11 BRAIN

The most comprehensive studies of myxinoid brain development are Conel's (1929, 1931b) papers on *Eptatretus stoutii* and von Kupffer's (1900, 1906) on the same species. Early in development, even before the closure of the neural folds, the brain displays three indistinct vesicles identified by Conel (1929) as the rhomb-, mes-, and prosencephalon. As noted above, the neural folds close in a rostral to caudal sequence, and a true anterior neuropore cannot be recognized. At this stage, the longitudinal axis of the brain simply continues the axis of the spinal cord, and the lamina terminalis is located at the anterior end of the neural tube. The rhombencephalon displays a number of rhombomeres. Conel (1929) depicts three, while von Kupffer (1900) depicts five in an older embryo (Figure 28.4A). Their fate and their relation to the nuclei of the cranial nerves are unknown.

In older stages the longitudinal axis of the prosencephalic vesicle bends progressively ventrally, resulting in the formation of the ventral cephalic flexure. The prosencephalic

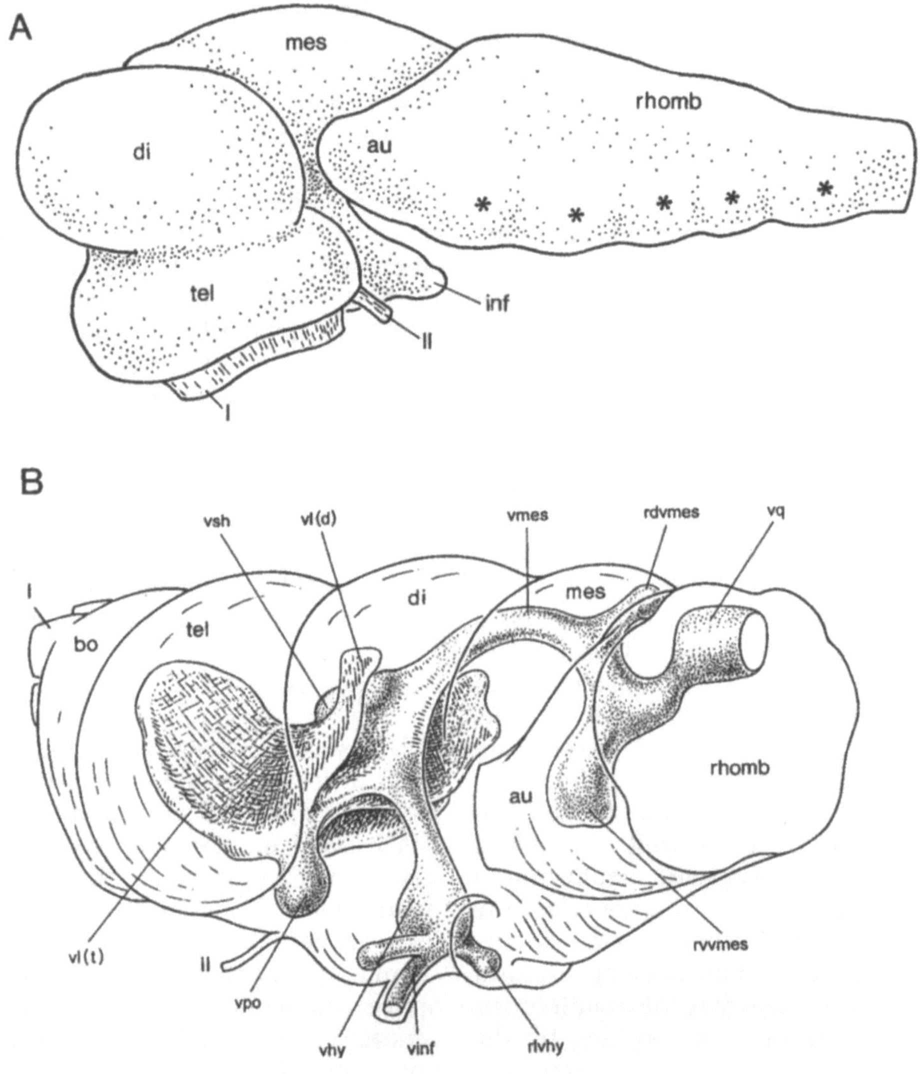

Figure 28.4 Lateral views of the brain of an early embryo (A) and a juvenile (B) hagfish. (A) Drawing of a lateral view of the brain of an early embryo, after von Kupffer (1906) and Conel (1929). Note the sharp ventrally directed flexure of the prosencephalic vesicle and the position of the olfactory nerve (I) which is located ventral to the telencephalic hemispheres at this stage, but rostral after the straightening of the neural axis (compare to figure B). The prosencephalic vesicle shows dorsal (diencephalic, di) and ventral (telencephalic, tel) evaginations. The asterisks identify rhombomeres. Combined from von Kupffer (1906) and Conel (1929). (B) A reconstruction of the ventricular system (solid body) and the brain (outlines) of a juvenile (105 mm) Pacific hagfish (*Eptatretus stoutii*), based on serial horizontal sections and viewed from a position lateral, posterior and slightly ventral to the brain. Anterior is to the left. Those parts of the ventricular system that retain a lumen and a typical ependyma in the juvenile brain are represented by stippling in the reconstruction. The parts that lose their lumen during development and are characterized by a transformed ependyma (see text) are represented by the cross-hatched areas. Note the presence of the anterior, telencephalic (vl[t]) and posterior, diencephalic (vl[d]) lateral ventricles. From Wicht (1996), slightly modified.

vesicle undergoes a ventral deflection of almost 180 degrees (Figure 28.4A), so that the former tip of the neuraxis is ultimately located ventral to the rhombencephalon. Still later in development, this bending is reversed, and the adult brain shows a more or less longitudinal arrangement of the derivatives of the individual vesicles (Figure 28.4B).

During the formation of the cephalic flexure, the prosencephalic vesicle expands rapidly and gives rise to paired dorsal and ventral evaginations (Figures 28.2F, 28.4A). The dorsal evaginations form the thalamic part of the diencephalon; the ventral evaginations form the telencephalic hemispheres. Thus, one may say that hagfishes possess four lateral ventricles and hemispheres: two posterior (diencephalic) ventricles and two anterior (telencephalic) ventricles (Tusch *et al.*, 1995). In the process of the reversal of the prosencephalic deflection (see above), another set of events complicates the morphology of the prosencephalon. First, the posterior walls of the telencephalic vesicles and the anterior walls of the diencephalic vesicles fuse, and the combined matrix zones give rise to a large nuclear complex in the centre of the forebrain, termed the central prosencephalic nucleus by Wicht and Northcutt (1992a, medial pallium of Jansen, 1930 and Conel, 1931b, Figure 28.2F). Secondly, the walls of the prosencephalic vesicle increase massively in thickness, and the majority of the neuroblasts migrate from their periventricular site of origin towards the meningeal surface of the brain. In the telencephalon, this migration ultimately leads to the formation of a layered cortex that may contain as many as five (*Eptatretus stoutii*, Wicht and Northcutt, 1992a) or eight (some cortical areas of *Myxine glutinosa*, Wicht, pers. obs.) alternating layers of grey and white matter. The growth of the brain walls also leads to the obliteration of most ventricular cavities in the forebrain. However, their former site can still be identified, since strings and rows of transformed ependymal cells remain in their original position after the closure of the ventricles (Edinger, 1906; Wicht and Northcutt, 1992a). A reconstruction of the ventricular system of a juvenile hagfish (body length 105 mm), as shown in Figure 28.4B, clearly shows the remnants of the diencephalic and the telencephalic lateral ventricles. This mode of development of the prosencephalic vesicle differs radically from that in any other craniate, including lampreys. Lampreys show a relatively restricted, paired lateral evagination from the prosencephalic vesicle, resulting in the formation of small paired telencephalic hemispheres. The thalamic grisea, and even parts of the telencephalic pallium, do not participate in that evagination and remain in the unevaginated part of the neural tube, thus flanking the third ventricle (Johnston, 1912; Herrick, 1921; also see Wicht and Northcutt, 1992a). The formation of the hemispheres in gnathostomes follows still other and diverse patterns, since inversions and eversions may be superimposed on the process of evagination. As with the formation of the neural tube, the primitive mode of forebrain development cannot be determined from an out-group comparison. It is clear, however, that Herrick's (1921) claim regarding the primitive nature of the pattern in lampreys is not supported.

The ontogeny of the eyes and the hypophysis was described above, and our description of the prosencephalic development will conclude with some remarks on the pineal organ, the terminal nerve, and prosomeres. In 1888 Beard (cited in Retzius, 1893) observed an epiphysis in a single adult specimen of *Myxine glutinosa*, but all other authors agree on the absence of a pineal organ in adult hagfishes. Von Kupffer (1900, 1906) and Conel (1929, 1931b) described a small recess in the roof of the embryonic diencephalic ventricle as an 'epiphysis', but we believe this is doubtful. A terminal nerve and ganglion are lacking in both adults (Wicht and Northcutt, 1992b; Braun *et al.*, 1995) and embryos. Holmgren (1946) observed three 'bulges' in the walls of the developing diencephalon of

Myxine glutinosa (partes frontalis, medialis, and caudalis thalami) that might correspond to prosomeres one to three of Puelles and Rubenstein (1993), but his identification is rather tentative.

The mesencephalon undergoes relatively few gross morphological changes during ontogeny. It more or less retains the character of a tube, even though the lumen of the ventricular cavity is reduced to a narrow canal with a dorsal recess (Figure 28.4B). As in other parts of the brain, there is extensive neuronal migration towards the meningeal surface.

The rhombencephalon, however, undergoes drastic morphological changes (Figure 28.3A, B, C). After the closure of the neural folds, the rhombencephalon has the shape of a tube, and von Kupffer (1906) observed a typical sulcus limitans and alar and basal plates (Figure 28.3A). However, in further development, these boundaries are almost entirely obscured. The topological transformations that lead to the appearance of the adult rhombencephalon (see Ronan and Northcutt, this volume) are documented in the work of Conel (1929, 1931b, Figure 28.3A, B, C), but their implications for the interpretation of this structure are not clear. Again, massive neuronal migration, which also occurs in the rhombencephalon, further confounds the issue. All of the efferent and afferent nuclear centres of the medulla are found in a more or less submeningeal position, including those nuclei that typically retain a periventricular position in other craniates (Fig. 28.3C). Thus, even the viscerosensory nuclei of the solitary/communis system migrate towards the meningeal surface in hagfishes (see Matsuda *et al.*, 1991, Figure 28.3C). Furthermore, the rhombencephalon develops paired lateral outgrowths (the auricula) which initially contain a ventricular cavity. These auricula flank the mesencephalon laterally and contain the anterior parts of the sensory and motor nuclei of the trigeminal nerve. Conel (1931b) observed an anterior rhombic lip at the rhomb-mesencephalic boundary. He understandably concluded that this anlage might represent a cerebellum, but the corresponding region in adults does not show any characteristics of a cerebellum. That is, it does not receive any commissural octaval afferents (Amemiya *et al.* 1985); instead, it receives commissural viscerosensory fibres (Matsuda *et al.*, 1991). A cerebellum is therefore entirely absent in hagfishes. In accordance with the lack of development of the external eye muscles, rhomb- and mesencephalic somatomotor nuclei and their corresponding nerves never develop (Conel, 1931b; Holmgren, 1946; data for *Myxine glutinosa*).

It is noteworthy, that the chorioid plexuses, which are lacking in adults, are also lacking during development. In some areas, particularly in the prosencephalon (Holmgren, 1946, data for *Myxine glutinosa*) and in the rhombencephalon (Conel, 1931b, Figure 28.3B), the roof plates develop into thin epithelial sheets, but these sheets remain in intimate contract with the overlying ectoderm. The typical pattern of (mesodermal) vascularization and folding of those sheets never occurs, and chorioid plexuses thus never form.

28.12 SPINAL CORD AND NERVES

The spinal cord is initially tube-like, as is the entire neuraxis (von Kupffer, 1900). As shown by von Kupffer (1906), this tubular cord gradually transforms into the ribbon-like cord of the adult. As in the brain, the walls of the initially wide and slit-like ventricle fuse in a dorsal to ventral sequence, leaving in the adult only a small central canal which represents a remnant of the ventral part of the spinal ventricle. Notably, this small central canal is again subdivided into a more dorsal canal and a more ventral canal. Reissner's fibre occurs in the ventral canal, and the epithelium lining the dorsal canal shows signs of a secretory activity (Wicht, unpubl. obs.).

Dorsal and ventral roots of spinal nerves are clearly present; the dorsal root ganglia develop from neural crest (von Kupffer, 1906). As in the head (see above), the neural crest remains an epithelial tissue, even after the neural tube has closed, and the 'pockets' of neural crest cells have a lumen that communicates with the lumen of the spinal ventricle (Figure 28.3F). As in gnathostomes, but not in lampreys, dorsal and ventral roots fuse to form mixed spinal nerves, except in the tail region (Allen, 1917). The embryological study of Goodrich (1937) has shown, however, that the mode of fusion is different in gnathostomes and hagfishes (see Figure 28.3D, E), and it must be concluded that the two groups have acquired mixed spinal nerves independently. It is noteworthy that the spinal nerves of hagfishes carry a chemosensory (Braun and Northcutt, this volume) and even a photosensory component (Ross, 1963). In this context, it is also interesting to note that the ganglia of the spinal nerves are not homogeneous. The dorsal rami of the spinal nerves in the tail region of *Myxine glutinosa* carry small ganglia that are separate from the main ganglia of the dorsal roots (Allen, 1917; Peters, 1963; data for *Myxine glutinosa*). Similarly, two distinct ganglia have been observed in the most anterior (occipito-spinal) spinal nerve of *Eptatretus stoutii* (Worthington, 1905b), and single ganglion cells are dispersed in the peripheral rami of the anteriormost spinal nerves (Allen, 1917). It is, of course, unclear, whether this division of the cranial nerve ganglia has anything to do with the presence of 'additional' modalities, but it suggests a multiple embryonic origin of the ganglia of the spinal nerves. Intraspinal and intracerebral sensory cells (cells of Rohon-Beard, 'Hinterzellen') have not been reported in any developmental stage of hagfishes.

28.13 CRANIAL NERVES, SUBCUTANEOUS NERVE PLEXUS

We have outlined above the general mode of origin of the ganglia of the cranial nerves from placodal and crest-derived tissues and have stressed that the somatomotor nerves (i.e. the oculomotor, trochlear and abducent), as well as the terminal nerve, are entirely lacking in all developmental stages.

We will not attempt a complete description of the distribution and development of the cranial nerves in myxinoids; for this, the reader is referred to Worthington (1905b), Lindström (1949) and Peters (1963). The development of the lateralis system has recently been re-described by Wicht and Northcutt (1995). We will therefore review some of the major differences in the organization of the cranial nerves between myxinoids and other craniates in the light of ontogenetical data.

The trigeminal nerve of myxinoids is unique in having a chemoreceptive component (see above and see Braun and Northcutt, this volume). Furthermore, it has a large ramus palatinus directed to the roof of the buccal cavity, the nasal cavity, the nasopharyngeal duct and the prenasal sinus. Such a ramus palatinus is absent from the trigeminal complex of other craniates, including lampreys (Lindström, 1949). These characters may be connected with the presence of epibranchial trigeminal placodes and the possible endodermal origin of the 'pharyngeal territory' of the fifth nerve (see above). Unlike in other craniates, in hagfishes the trigeminal nerve may thus be considered a 'complete' branchial nerve in the sense of Sewertzoff (1911), Allis (1920) and Norris (1924), since it carries a chemosensory component and a 'ramus pharyngeus'. However, as tempting as this interpretation may seem (because it could serve as a basis for further speculation on the evolutionary origin of the mouth), there are numerous problems connected with it. It is by no means certain that there is any inductive relationship between endoderm, epibranchial placodes, and chemoreceptors (see above); thus, it is unclear whether the trigeminal epibranchial placode reported by von Kupffer (1900) is a serial homologue of the posterior (facial, glossopharyngeal and vagal) epibranchial placodes. It is also not certain whether the palatine ramus of the trigeminus actually derives from epibranchial placodes. Furthermore, it is possible that the entire gustatory chemosensory system of hagfishes is non-homologous to that of other craniates (see Braun and Northcutt, this volume), and the fact that it receives parts of its innervation from the trigeminal nerve may be used to support that claim.

The facial nerve of adult hagfishes is rather small, possibly due to the early suppression of the second branchial pouch. It does not participate in the sensory innervation of the skin but innervates several muscles associated with the basal cartilage. It carries one to two (Lindström, 1949; Peters, 1963; data for *Myxine glutinosa*) small distal ganglia, which probably arise from a facial epibranchial placode (Wicht and Northcutt, 1995). Thus, it may have a viscerosensory component, but neither a typical ramus pharyngeus nor a central sensory projection of the facial nerve has been observed so far. It is interesting that von Kupffer (1900) observed an additional, very large proximal ganglion of the seventh nerve in embryos; this ganglion is supposed to derive from neural crest. There is no trace of that ganglion in the adults, but it is possible that material from that ganglion becomes incorporated into the adjacent octaval ganglia (Lindström, 1949).

Similarly, the glossopharyngeal nerve is hard to identify, and this may also be due to the early reduction of its branchial pouch. In adults, the glossopharyngeus may be represented by a nerve closely associated with the vagus (ramus pharyngeus) which innervates the pharyngeal region of the first true gill (Worthington, 1905b; Peters, 1963). In embryos, however, a glossopharyngeal epibranchial placode and ganglion are clearly present (Wicht and Northcutt, 1995), and it may therefore be assumed that the association with the vagus is a secondary phenomenon.

There is almost no embryological information on the development of the vagal nerve. Vagal epibranchial placodes are present (von Kupffer, 1900; Wicht and Northcutt, 1995), but their number and fate are undetermined. In adults, the vagal/glossopharyngeal complex is large and innervates branchial regions of the pharynx. Interestingly, neither the vagal nor the glossopharyngeal nerve seem to have skin-branches. The innervation of the skin caudal to the territory of the trigeminus is carried out by rami of the anterior spinal nerves (Braun and Northcutt, this volume).

The skin of adult hagfishes displays an extensive subcutaneous nerve plexus which also contains large ganglion cells. This plexus is particularly dense in the head and around the slime glands. The embryonic source of this neuronal tissue is unknown.

28.14 CONCLUDING REMARKS

At best, the available embryological data on myxinoids form a coarse meshwork. The knots in that meshwork are loosely tied by non-experimental data. The holes in the meshwork can be filled only by speculation, and the available information allows practically any hypothesis, ranging from an independent origin of the entire mandibular/premandibular head in myxinoids to rigid models of germ-layer specificity or branchiomery. Rather than support a particular view, we conclude this review by echoing previous authors. Myxinoids and their unusual ontogeny 'deserve further study by modern procedures and from modern viewpoints' (Gorbman and Tamarin, 1985).

ACKNOWLEDGEMENTS

We thank Chris Braun for many discussions and useful hints, special thanks to Glenn Northcutt for hosting us in his lab at the Scripps Institution of Oceanography in La Jolla, CA. Barbara Rupp's help with the drawings is greatly appreciated, special thanks to Mary Sue Northcutt for help with the manuscript. Our own original research reported in this review was supported by the Dr Senckenbergische Stiftung (Frankfurt).

REFERENCES

Allen, W.F. (1917) Distribution of the spinal nerves in *Polistotrema* and some special studies on the development of spinal nerves. *Journal of Comparative Neurology*, **28**, 137–213.

Allis, E.P. (1903) On certain features of the cranial anatomy of *Bdellostoma dombeyi*. *Anatomischer Anzeiger*, **23**, 259–81 and 321–39.

Allis, E.P. (1920) The branches of the branchial nerves of fishes, with special reference to Polyodon spathula. *Journal of Comparative Neurology*, **32**, 137–53.

Amemiya, F., Kishida, R., Goris, R.C., Onishi, H. and Kusunoki, T. (1985) Primary vestibular projections in the hagfish, *Eptatretus burgeri*. *Brain Research*, **337**, 73–9.

Barlow, L. and Northcutt, R.G. (1995) Embryonic origin of amphibian taste buds. *Developmental Biology*, **169**, 273–85.

Braun, C.B., Wicht, H. and Northcutt, R.G. (1995) Distribution of gonadotropin-releasing hormone immunoreactivity in the brain of the Pacific hagfish, *Eptatretus stoutii* (Craniata: Myxinoidea). *Journal of Comparative Neurology*, **353**, 131–43.

Braun, C.B. and Northcutt, R.G. (1997) The lateral line system of hagfishes (Craniata: Myxinoidea). *Acta Zoologica* (in press).

Conel, J.L. (1929) The development of the brain of *Bdellostoma stoutii*. I. External growth changes. *Journal of Comparative Neurology*, **47**, 343–403.

Conel, J.L. (1931a) The genital system of the Myxinoidea: a study based on notes and drawings of these organs in *Bdellostoma* made by Bashford Dean, in *The Bashford Dean Memorial Volume Archaic Fishes* (ed. E.W. Gudger), American Museum of Natural History New York, New York, pp. 63–102.

Conel, J.L. (1931b) The development of the brain of *Bdellostoma stoutii*. II. Internal growth changes. *Journal of Comparative Neurology*, **52**, 365–499.

Conel, J.L. (1942) The origin of the neural crest. *Journal of Comparative Neurology*, **76**, 191–215.

Dean, B. (1899) On the embryology of *Bdellostoma stoutii*. A general account of myxinoid development from the egg and segmentation to hatching, in *Festschrift zum siebenzigsten Geburtstag von Carl von Kupffer*, Gustav Fischer Verlag, Jena, pp. 221–76.

Dean, B., Harrington, N.R., Calkins, G.N. and Griffin, B.B. (1897) The Columbia University zoölogical expedition of 1896. With a brief account of the work of collecting in Puget Sound and on the Pacific coast. *Transactions of the New York Academy of Sciences*, **16**, 33–42.

Doflein, F. (1898) Bericht über eine wissenschaftliche Reise nach Californien. (Mittheilungen über die Erlangung von Eiern und Embryonen von *Bdellostoma*). *Sitzungsberichte der Gesellschaft für Morphologie und Physiologie in München*, **14**, 105–18.

Edinger, L. (1906) Über das Gehirn von *Myxine glutinosa*. Abhandlungen der königlich preussischen Akademie der Wissenschaften aus dem Jahre 1906. Verlag der königlichen Akademie der Wissenschaften, Berlin, pp. 1–36.

Fernholm, B. (1969) A third embryo of *Myxine*: considerations on hypophyseal ontogeny and phylogeny. *Acta Zoologica*, **50**, 169–77.

Fernholm, B. (1985) The lateral line system of cyclostomes, in *Evolutionary Biology of Primitive Fishes* (eds R.E. Foreman, A. Gorbman, J.M. Dodd and R. Olsson), Plenum Press, New York, pp. 113–22.

Fernholm, B. and Holmberg, K. (1975) The eyes in three genera of hagfish (*Eptatretus, Paramyxine* and *Myxine*): a case of degenerative evolution. Vision Research, **15**, 253–9.

Goodrich, E.S. (1930) *Studies on the structure and development of vertebrates*. Macmillan, London (Reprint 1986: The University of Chicago Press, Chicago).

Goodrich, E.S. (1937) On the spinal nerves of the Myxinoidea. *Quarterly Journal of Microscopical Sciences*, **80**, 153–8.

Gorbman, A. (1983) Early development of the hagfish pituitary gland: evidence for an endodermal origin of the adenohypophysis. *American Zoologist*, **23**, 639–54.

Gorbman, A. and Tamarin, A. (1985) Early development of oral, olfactory, and adenohypophyseal structures of agnathans and its evolutionary implications, in *Evolutionary Biology of Primitive Fishes* (eds R.E. Foreman, A. Gorbman, J.M. Dodd and R. Olsson), Plenum Press, New York, pp. 165–85.

Griffith, C.M., Wiley, M.J. and Sanders, E.J. (1992) The vertebrate tail bud: three germ layers from one tissue. *Anatomy and Embryology*, 101–13.

Gudger, E.W. and Smith, B.G. (1931) The segmentation of the egg of the myxinoid, *Bdellostoma stoutii*, based on the drawings of the late Bashford Dean, in *The Bashford Dean Memorial Volume Archaic Fishes* (ed. E.W. Gudger), American Museum of Natural History New York, New York, pp. 47–57.

Herrick, C.J. (1921) A sketch of the origin of the cerebral hemispheres. *Journal of Comparative Neurology*, **32**, 429–54.

Hertwig, O. (1890) *Lehrbuch der Entwicklungsgeschichte des Menschen und der Wirbelthiere*, Gustav Fischer Verlag, Jena.

Holmgen, N. (1946) On two embryos of *Myxine glutinosa*. *Acta Zoologica*, **27**, 1–90.

Jansen, J. (1930) The brain of *Myxine glutinosa*. *Journal of Comparative Neurology*, **49**, 359–507.

Johnston, J.B. (1912) The telencephalon in cyclostomes. *Journal of Comparative Neurology*, **22**, 341–404.

Kishida, R., Goris, R.C., Nishizawa, H., Koyama, H., Kodota, T. and Amemiya, F. (1987) Primary neurons of the lateral line nerves and their central projections in hagfishes. *Journal of Comparative Neurology*, **264**, 303–10.

Kupffer, C. von (1900) *Studien zur vergleichenden Entwicklungsgeschichte des Kopfes der Kranioten.* Heft 4: *Zur Kopfentwicklung von Bdellostoma.* Verlag von J.F. Lehmann, München.

Kupffer, C. von (1906) Ontogenie des Zentralnervensystems, in *Handbuch der vergleichenden und experimentellen Entwickelungslehre der Wirbeltiere*, Vol. 11/3 (ed. O. Hertwig), Gustav Fischer Verlag, Jena.

Lindström, T. (1949) On the cranial nerves of the cyclostomes with special reference to *n. trigeminus*. *Acta Zoologica*, **30**, 315–458.

Matsuda, H., Goris, R.C. and Kishida, R. (1991) Afferent and efferent projections of the glossopharyngeal-vagal nerve in the hagfish. *Journal of Comparative Neurology*, **311**, 520–30.

Neumayer, L. (1938) Die Entwicklung des Kopfskelettes von *Bdellostoma st.* L. Archivio *Italiano di Anatomia e di Embriologia*, **40** (Suppl.), 1–222.

Nishizawa, H., Kishida, R., Kodota, T. and Goris, R.C. (1988) Somatotopic organization of the primary sensory trigeminal neurons in the hagfish, *Eptatretus burgeri*. *Journal of Comparative Neurology*, **267**, 281–95.

Noden, D.M. (1991) Vertebrate craniofacial development: the relation between ontogenetic process and morphological outcome. *Brain, Behavior and Evolution*, **38**, 190–225.

Norris, H.W. (1924) Branchial nerve homologies. *Zeitschrift für Morphologie unde Anthropologie*, **24**, 211–26.

Northcutt, R.G. (1992a) The phylogeny of octavolateralis ontogenies: a reaffirmation of Garstang's phylogenetic hypothesis, in *The Evolutionary Biology of Hearing* (eds A. Popper, D. Webster and R. Fay), Springer Verlag, New York, pp. 21–47.

Northcutt, R.G. (1992b) Distribution and innervation of lateral line organs in the axolotl. *Journal of Comparative Neurology*, **325**, 95–123.

Northcutt, R.G., Catania, K.C. and Criley, K.C. (1994) Development of the lateral line organs in the axolotl. *Journal of Comparative Neurology*, **340**, 480–514.

Peters, A. (1963) The peripheral nervous system, in *The Biology of Myxine* (eds A. Brodal and R. Fänge), Universitetsforlaget, Oslo, pp. 92–123.

Price G.C. (1896) Some points in the development of a myxinoid (*Bdellostoma stoutii* Lockington). *Anatomischer Anzeiger*, **12** (Suppl.), 81–6.

Puelles, L. and Rubenstein, J.L.R. (1993) Expression patterns of homeobox and other putative regulatory genes in the embryonic mouse forebrain suggest a neuromeric organization. *Trends in Neurosciences*, **16**, 472–9.

Reid, R. (1990) *Research on the fishery and the biology of the hagfish.* Research report prepared for the California Environmental Protection Agency, Contract Number A800–185.

Retzius, A. (1893) Das Gehirn und das Auge von *Myxine. Biologische Untersuchungen, Neue Folge*, **5**, 55–68.

Ross, D.M. (1963) The sense organs of *Myxine glutinosa*, in *The Biology of Myxine* (eds A. Brodal and R. Fänge), Universitetsforlaget, Oslo, pp. 151–60.

Sewertzoff, A.N. (1911) Die Kiemenbogennerven der Fische. *Anatomischer Anzeiger*, **38**, 487–94.

Starck, D. (1975) Embryologie, Thieme Verlag, Stuttgart.

Stockard, C. (1906a) The development of the mouth and gills in *Bdellostoma stoutii*. *American Journal of Anatomy*, **5**, 481–517.

Stockard, C. (1906b) The development of the thyroid gland in *Bdellostoma stoutii*. *Anatomischer Anzeiger*, **29**, 91–9.

Stockard, C. (1907) The embryonic history of the lens in *Bdellostoma stoutii* in relation to recent experiments. *American Journal of Anatomy*, **6**, 511–15.

Tusch, U., Wicht, H. and Korf, H.-W. (1995)

Observations on the topology of the forebrain of the Pacific hagfish, *Eptatretus stoutii*. *Journal of Anatomy*, **187**, 227–8.

Wicht, H. (1996) The brains of lampreys and hagfishes – characteristics, characters, and comparisons. *Brain, Behavior and Evolution*, **48**, 248–61.

Wicht, H. and Northcutt, R.G. (1992a) The forebrain of the Pacific hagfish: A cladistic reconstruction of the ancestral craniate forebrain. *Brain, Behavior and Evolution*, **40**, 25–64.

Wicht, H. and Northcutt, R.G. (1992b) FMRFamide-like immunoreactivity in the brain of the Pacific hagfish, *Eptatretus stoutii* (Myxinoidea). *Cell and Tissue Research*, **270**, 443–9.

Wicht, H. and Northcutt, R.G. (1995) Ontogeny of the head of the Pacific hagfish (*Eptatretus stoutii*, Myxinoidea): development of the lateral line system. *Philosophical Transactions of the Royal Society London, Series B*, **349**, 119–34.

Worthington, J. (1905a) Contribution to our knowledge of the myxinoids. *American Naturalist*, **39**, 625–63.

Worthington, J. (1905b) The descriptive anatomy of the brain and cranial nerves of *Bdellostoma dombeyi*. *Quarterly Journal of Microscopical Sciences*, **49**, 137–81.

29
THE CENTRAL NERVOUS SYSTEM OF HAGFISHES

Mark Ronan and R. Glenn Northcutt

SUMMARY

A brain and spinal cord constitute the central nervous system of hagfishes, the extant sister group of lampreys and gnathostomes among the craniates. While the spinal cord of adult hagfishes appears similar to the rather flattened, ribbon-like cord of lampreys, the brains of hagfishes differ markedly from the brains of the other craniates. A massive forebrain includes olfactory bulbs, twin telencephalic hemispheres and a diencephalon. The forebrain tapers caudally into a wedge-shaped mesencephalon which fills a part of the space between trigeminal prominences of the rostrolateral medulla. The spinal cord continues caudally from the end of the medulla throughout most of the body length. Adult myxinoid brains reveal a greatly reduced ventricular system. Another distinguishing characteristic is the complexity of cytoarchitecture within the thick walls of the brain, particularly in the forebrain. Especially notable in the forebrain are the organization of the telencephalic pallium into multiple cellular and fibre laminae and the presence of a curious, multipartite central prosencephalic nucleus extending from the telencephalon into the diencephalon. The midbrain and medulla of hagfishes, more than the forebrain, resemble comparable regions in the brains of lampreys and gnathostomes, but even here unusual features prevail. The roof of the medulla is thick; all motor nuclei are located at the ventrolateral margin of the medulla, and the region is dominated by a large trigeminal system. As in lampreys, there is little sign of a cerebellum. The unique organization of myxinoid brains, long known from descriptive analyses, has more recently become the subject of experimental studies which show the pattern of retinal, lateral line, inner ear, trigeminal, and vagal projections while also defining many of the afferent and efferent connections of the olfactory bulbs, the pallial laminae, the mesencephalic tectum and part of the central prosencephalic nucleus. Such studies not only reveal the organization of the brain and spinal cord in hagfishes, but also further our understanding of the evolution of the central nervous system in craniates.

29.1 INTRODUCTION

The central nervous system of hagfishes consists of a brain and spinal cord, and, in this most general of contexts, it is therefore identical to the central nervous systems of lampreys and jawed vertebrates. Such an assessment supports the basic, but not surprising, conclusion that a central nervous system consisting of an integrated union of brain and spinal cord is a shared character inherited from the common ancestor of all extant craniates. Closer inspection of

The Biology of Hagfishes. Edited by Jørgen Mørup Jørgensen, Jens Peter Lomholt, Roy E. Weber and Hans Malte. Published in 1998 by Chapman & Hall, London. ISBN 0 412 78530 7.

hagfishes reveals their brains and spinal cords as combinations of features – some with obvious homologies in other craniates and some different enough in anatomy or chemistry to pose real problems of interpretation. Much the same could be said of any lamprey or gnathostome brain; however, what particularly distinguishes the neurology of hagfishes from that of their fellow craniates is a collection of features that are notable for their seeming strangeness. Whether apparent at first glance or encountered only after careful examination, these features of myxinoid brains and spinal cords lend a real sense of the curious to the least familiar and least understood central nervous system among all the craniates.

In hagfishes, as in lampreys and some gnathostomes, it is the spinal cord that comprises the largest major division of the central nervous system. The spinal cord of hagfishes resembles the cord of lampreys far more than it resembles the cord of gnathostomes, but its organization is recognizable as that of all extant craniates, indicating that the spinal cord retains more characteristics of the embryonic neural tube than does the brain. In all craniates, including hagfishes, the rostral end of the neural tube undergoes the greatest ontogenetic and phylogenetic transformation to produce a brain that enables an organism to cope successfully with the demands of its environment, but one that develops within the restraints imposed by an inherited pattern of neural development and therefore reflects the history of brains in a species.

Hagfish brains have long been a subject of study (see, for example, Worthington, 1905; Holmgren, 1919; Jansen, 1930). Early on, it was clear that while the brain of each craniate species is unique, the brains of hagfishes have elevated uniqueness to a remarkable degree and closely resemble no other craniate brains at all. For a time, there were even basic disagreements about the identity of the major brain regions. In time, however, a consensus of overall brain organization emerged, and subsequent analysis focused on finer points of structure, connections and histochemistry. While the reader is referred to the older literature, the more recent reports constitute the primary material for this review. They provide insights into the central nervous system of hagfishes and are of interest not only for the information they provide on these intriguing animals but also for furthering our understanding of the central nervous system in the only living sister group of lampreys and jawed vertebrates.

The first part of this chapter describes major divisions of the central nervous system and considers each major brain division in a rostral-to-caudal order from the olfactory bulbs to the medulla before concluding with the spinal cord.

29.1.1 LIST OF ABBREVIATIONS USED IN FIGURES 30.1–30.7

ANa	anterior (or 'utricular') ramus of the acustic nerve
ANp	posterior (or 'saccular') ramus of the acustic nerve
bfb	basal forebrain bundle
CG	central gray of mesencephalon
D	diencephalon
df	dorsal funiculus
DFN	dorsal funicular nucleus
dl	dorsolateral reticular nucleus
DLR	dorsolateral reticular nucleus
fr	fasciculus retroflexus
HA	habenula
HAl	lateral nucleus of the habenular complex
HAle	left habenula of habenular complex
HAr	right habenula of habenular complex
HY	hypothalamus
HYinf	infundibular nucleus of hypothalamus
Inp	interpeduncular nucleus
LLNa	lateral line nerve A
LLNb	lateral line nerve B
lot	lateral olfactory tract
lotd	lateral olfactory tract (deep portion)
lots	lateral olfactory tract (superficial portion)
mc	medullary commissure
mlf	medial longitudinal fasciculus
ME	medulla
MT	mesencephalic tectum

NC	central prosencephalic nucleus
NCd	dorsal subnucleus of the central prosencephalic nucleus
NCm	medial subnucleus of the central prosencephalic nucleus
NCvl	ventrolateral subnucleus of the central prosencephalic nucleus
Nmlf	nucleus of the medial longitudinal fasciculus
Npc	nucleus of the posterior commissure
OB	olfactory bulb
OBci	olfactory bulb internal cellular layer
OBg	olfactory bulb glomerular layer
OBm	olfactory bulb mitral cell layer
OBp	olfactory bulb periglomerular layer
OL	octavolateralis area
ON	optic nerve
P1	pallial layer 1
P2	pallial layer 2
P2co	pallial layer 2 (compact part)
P2l	pallial layer 2 (lateral part)
P2mc	pallial layer 2 (magnocellular part)
P3	pallial layer 3
P4	pallial layer 4
P4l	pallial layer 4 (lateral part)
P4m	pallial layer 4 (medial part)
P5	pallial layer 5
P5co	pallial layer 5 (compact part)
P5cp	pallial layer 5 (parvocellular portion)
Pallium	pallial layers of telencephalic hemisphere
pc	posterior commissure
PO	preoptic area
poc	postoptic commissure
POd	preoptic area (dorsal nucleus)
POe	preoptic area (external nucleus)
POim	preoptic area (intermediate nucleus)
POp	preoptic area (periventricular nucleus)
PT	pretectal nucleus
PV	perivagal nucleus
RET	reticular nucleus
SC	spinal cord
Se	septum
SG	spinal gray matter
sl	spinal lemniscus
SON	spino-occipital nerve
SN	spinal nerve
St	striatum
TE	telencephalon
TEG	mesencephalic tegmentum
THa	anterior nucleus of thalamus
THdi	diffuse nucleus of thalamus
THe	external nucleus of thalamus
THico	intracommissural nucleus of thalamus
THi	interior nucleus of thalamus
THpco	paracommissural nuclei of thalamus
THsh	subhabenular nucleus of thalamus
THt	triangular nucleus of thalamus
TP	posterior tuberculum
UG	ganglion of anterior ('utricular') ramus of acustic nerve
VMR	ventromedial reticular nucleus
vca	cerebral aqueduct
vinf	infundibular ventricle
vinfl	lateral infundibular ventricle
vsh	subhabenular ventricle
VT	ventral thalamus
II	optic nerve
V	trigeminal nerve
Vm	trigeminal motor nucleus
Vmm	magnocellular trigeminal motor nucleus
Vmp	parvocellular trigeminal motor nucleus
Vnm	motor ramus of the trigeminal nerve
Vs	trigeminal sensory nucleus
VII	facial nerve
VIIm	facial motor nucleus
Xs	vagal sensory nucleus
Xm	vagal motor nucleus
IX–X	glossopharyngeal and vagal nerves
IX–Xm	glossopharyngeal and vagal motor nucleus

29.2 THE CENTRAL NERVOUS SYSTEM

29.2.1 Olfactory bulb

The two olfactory bulbs are prominent structures that make a significant contribution to the mass of the telencephalon amounting to nearly 40% of the total telencephalic volume and weight in *Myxine* (Platel and Delfini, 1981). Each olfactory bulb abuts the rostral end of one telencephalic hemisphere but is separated from the hemisphere externally by a distinct circumferential sulcus (Figure 29.1). During development, the bulbs appear to partially invaginate into the hemisphere (Conel, 1931). However, the cytoarchitecture of the olfactory bulb distinguishes its internal, posterior border with the cerebral hemispheres (Figure 29.2A). A series of successive, rostrally-curved layers comprise the olfactory bulbs (Jansen, 1930; Wicht and Northcutt, 1992a). Located most rostrally is a layer of primary olfactory afferents. These olfactory

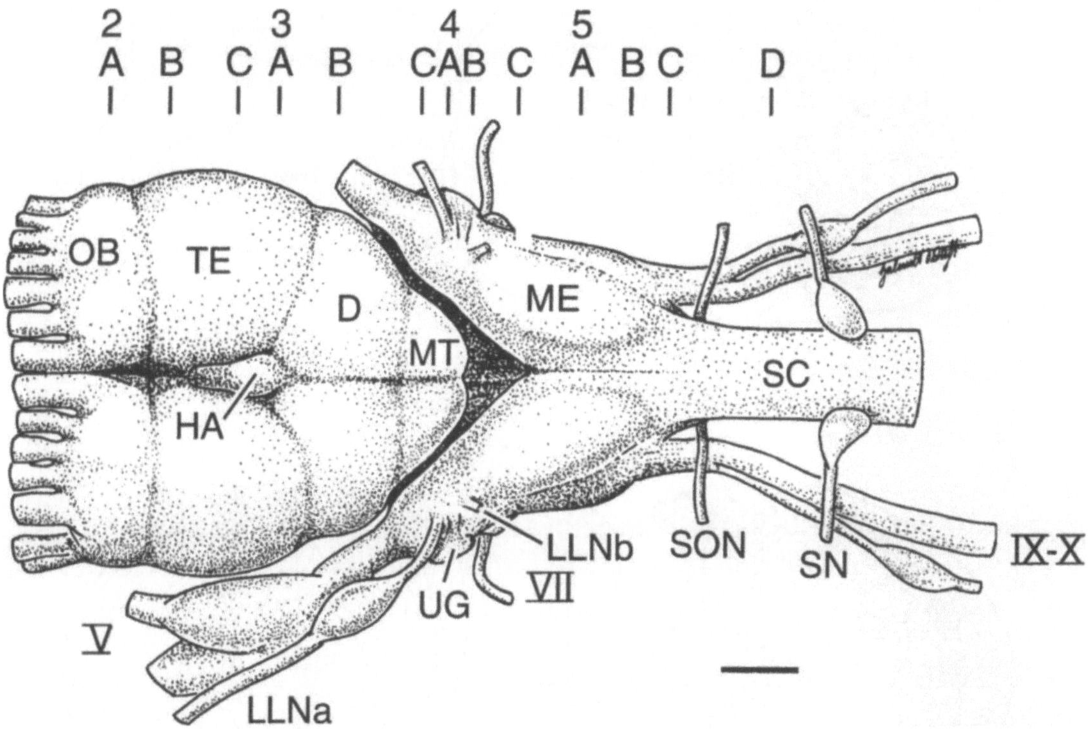

Figure 29.1 Dorsal view of the brain of an adult hagfish (*Eptatretus stoutii*) shows the major divisions of the brain and cranial nerves. Also shown are the levels of the transverse sections in Figures 2–5. Bar scale equals 1 mm.

inputs arise in rod-shaped sensory neurons of the olfactory mucosa, and the olfactory nerves consist of several thin and thick bundles of fine-diameter axons. Olfactory nerve fibres enter the dorsal and lateral margins of the olfactory bulbs and terminate widely in glomeruli surrounded by mitral cells and bipolar cells. Together, these neurons and glomeruli comprise a glomerular region that caps the rostral end of the olfactory bulb and extends caudally along an irregular front before giving way to a thin, cell-rich periglomerular layer recognized by Wicht and Northcutt (1992a). Jansen (1930) also described a cell-poor molecular layer in this region. A region of large mitral cells interspersed with smaller stellate cells occupies much of the bulb caudal to the periglomerular layer. Deep within the olfactory bulb at its caudal limit is the internal cellular layer (Figure 29.2B). This layer was earlier identified as the anterior olfactory nucleus by Jansen (1930). While such a nucleus does lie between the olfactory bulbs and the telencephalon proper in mammals, there is insufficient information in hagfishes regarding projections from the olfactory bulbs and the internal cellular layer to support such a homology (Wicht and Northcutt, 1992a).

Dendrites of the many olfactory bulb neurons, particularly those of the mitral cells, contact the olfactory nerve fibres in the glomeruli. Bipolar cells and mitral cells are among the cells seen in the olfactory bulb layers. The general structure of the olfactory system in hagfishes thus resembles the pattern seen in lampreys and gnathostomes. Primary olfactory fibres terminate in

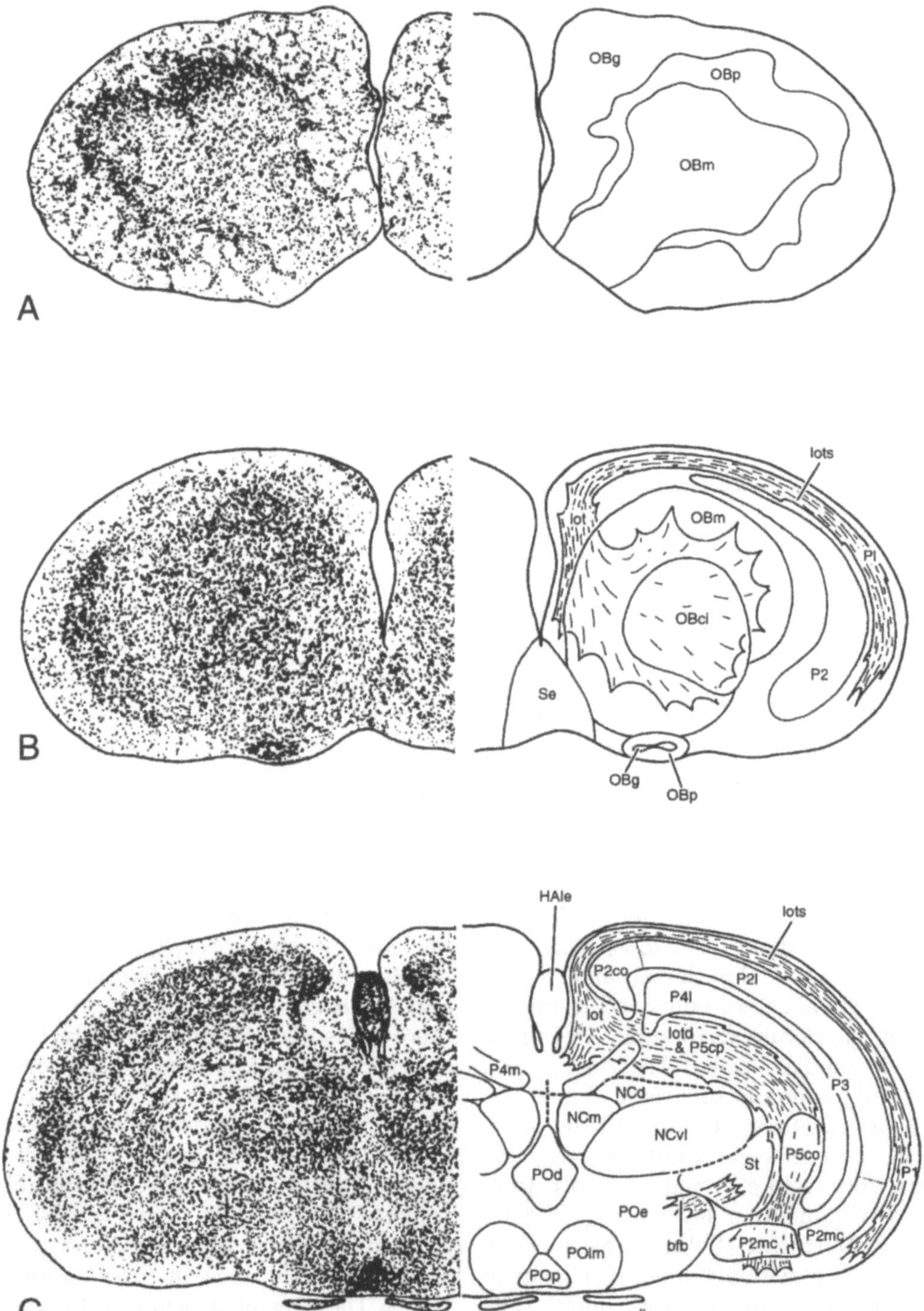

Figure 29.2 Transverse sections through the rostral forebrain of an adult hagfish (*Eptatretus stoutii*). Bar scale equals 0.5 mm. Figure modified from Wicht and Northcutt (1992a).

glomeruli where they synapse on mitral cells and other neurons of the olfactory bulbs. Presumably, these cells in turn convey olfactory input to targets in the telencephalic hemispheres.

29.2.2 Telencephalic hemispheres

The massive hemispheres of the telencephalon and the complexity of telencephalic organization are significant hallmarks of myxinoid brains. A wide diversity of neuronal groups form the telencephalic hemispheres, but it is difficult to relate telencephalic organization in myxinoids to that seen in lampreys and gnathostomes. An additional defining characteristic of the brains of adult myxinoids is the marked reduction of the ventricles. Vestiges of the ventricular system may, however, provide valuable insights into the identity of some neuronal groups (Wicht and Northcutt, 1992a).

A preoptic recess in the rostroventral telencephalon (Figure 29.2C) is the most obvious vestige of the ventricular system in the hemispheres (Jansen, 1930; Adam, 1963). The opposed borders of an obliterated ventricular lumen extend dorsally from the preoptic recess and are continuous with vestigial lateral ventricles in the telencephalic hemispheres. The disposition of the vestigial ventricles is indicated by thin sheets of cells outlining tiny residual spaces and the borders of cytoarchitectural discontinuities (Edinger, 1906; Wicht and Northcutt, 1992a). A small ventricular space continues caudally from the union of the vestigial lateral ventricles, beneath the habenular complex, into the diencephalon where a second set of vestigial lateral ventricles mirror the shape of the more rostral telencephalic ventricles (Wicht and Tusch, this volume).

The nuclear groups of the telencephalic hemispheres can be apportioned into prospective pallial and subpallial populations. The pallium is notable for the large region it occupies in the dorsolateral hemisphere and for the fact that it comprises a layered cortex. The following description is based on Jansen (1930) and Wicht and Northcutt (1992a; 1994). Structures of the telencephalic hemispheres are seen in Figures 29.2C and 29.3A. Lamina 1, the most superficial layer of the pallium, consists primarily of fibres, many of which originate in the olfactory bulbs. The thick and complex lamina 2 is a neuronal layer, composed of four divisions – a rostrodorsal parvocellular region, a large lateral portion with neurons oriented perpendicular to the brain surface, a dorsomedial region of densely packed cells, and a ventrolateral magnocellular division. Lamina 3 is dominated by efferent pallial fibres. Lamina 4 contains a larger lateral and a smaller dorsomedial component. The deep lamina 5 is a mixed layer of cells and fibres. Cells of lamina 5 are organized into a large parvocellular component and a smaller, compact component, located ventrolaterally. Many of the fibres in lamina 5 also originate in the olfactory bulb. The subpallium includes the septum and the striatum. Although it lacks several of the histochemical signatures of septal regions in gnathostome brains, a nucleus identified as the septum lies along the ventral midline of the hemispheres, immediately caudal to the olfactory bulbs and rostral to the preoptic recess. An area identified as a possible striatal region is associated with the lateral forebrain bundle and lies deep to the pallial laminae where it surrounds the rostral, lateral and ventral edges of the vestigial lateral ventricles in the telencephalon.

A prominent central prosencephalic nucleus occupies a position at the heart of the telencephalic hemisphere, but it cannot be readily designated as pallial or subpallial and in fact, may be more diencephalic than telencephalic in origin. As noted by Kusunoki *et al.* (1981) and later described by Wicht and Northcutt (1992a), the central prosencephalic nucleus can be divided into several major subnuclei on cytoarchitectural,

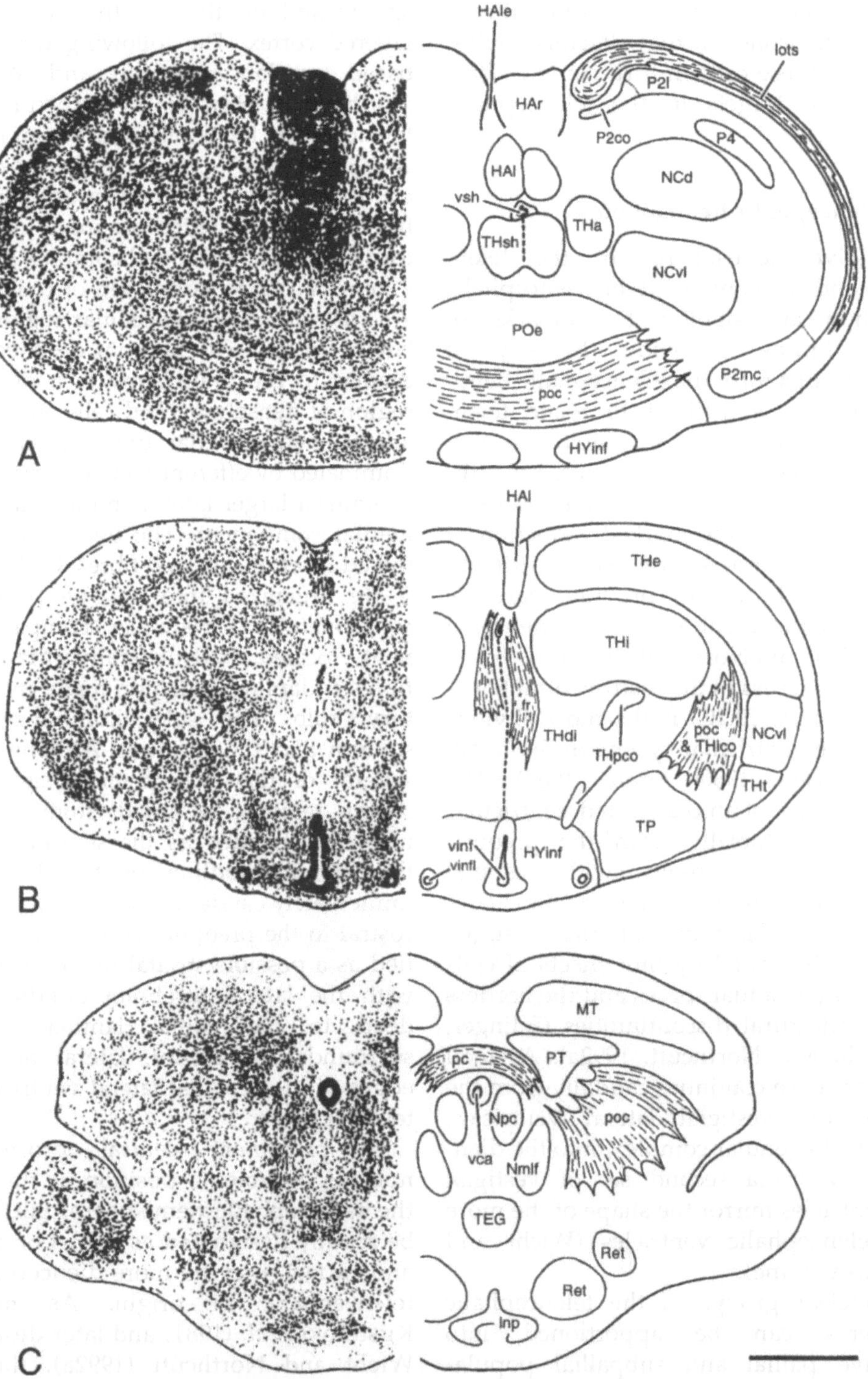

Figure 29.3 Transverse sections through the caudal forebrain and midbrain of an adult hagfish (*Eptatretus stoutii*). Bar scale equals 0.5 mm. Figure modified from Wicht and Northcutt (1992a).

immunohistochemical and connectional grounds. The medial subnucleus occupies the rostral prosencephalic nucleus and comprises densely packed, intensely staining neurons. The ventrolateral subnucleus, the largest division of the central prosencephalic nucleus, extends caudally from the rostral border of the central prosencephalic nucleus in a ventrolateral direction, reaching well into the diencephalon. The dorsolateral subnucleus begins rostrally near the two other subnuclei but continues dorsolaterally where its expanded caudal terminus lies close to lamina 5 and the medial part of pallial lamina 4.

The preoptic area consists of several distinct regions which together occupy a locus in close proximity to the major structures of the basal forebrain (Wicht and Northcutt, 1992a). The preoptic nuclei lie caudal to the septum and rostral to the hypothalamus. The ventrolateral subnucleus of the central prosencephalic nucleus, the striatum, and the magnocellular part of pallial lamina 2 border the preoptic area along an arc that extends from a dorsomedial to a lateral position. The posterior third of the preoptic area is filled by the large bundle of fibres that constitute the postoptic commissure.

29.2.3 Diencephalon

Prominent sulci demarcate the twin lobes of the diencephalon from the more rostral telencephalic hemispheres and the more caudal, tapered lobe of the mesencephalon (Figure 29.1). No pineal or parapineal organs exist in hagfishes, but the dorsal diencephalon is the location of a well-developed habenular complex. Ventrally, an infundibular stalk and a flared neurohypophysis is continuous with a modest hypothalamic region. The internal structure of the diencephalon, like that of the telencephalon, is massive, with much-reduced ventricular spaces surrounded by thick walls comprising a large number of neuronal groups (Figure 29.3A, B). This account is based on Wicht and Northcutt (1992a).

In the epithalamus, the habenular complex consists of asymmetrical right and left habenular bodies and, at more ventral and caudal levels, a lateral habenular nucleus. Two notable commissural pathways lie within the habenular complex or in close proximity to it. A commissure that interconnects the olfactory bulbs passes beneath the rostral habenular complex, and a massive habenular commissure crosses through the caudal portion of the habenular complex. A pair of vestigial ventricles, directed laterally like the telencephalic remnants of lateral ventricles but distinct from them, are present in the diencephalon caudal to the central prosencephalic nucleus and anterior thalamic nucleus (Wicht, 1996). A small, cylindrical third ventricle continues caudally beneath the interbulbar commissure into the mesencephalon while a narrow ventricular recess extends ventrally along the midline of the infundibulum before it splays out at several points at the level of the ventral infundibulum and neurohypophysis.

The postoptic commissure and the optic tract contact the ventral midline in the rostral diencephalon and then curve dorsally and caudally towards the border of the diencephalon and mesencephalon. A series of thalamic nuclei lie dorsal to the optic tract and the postoptic commissure. In the rostral diencephalon, a subhabenular nucleus lies medial to the anterior thalamic nucleus. More caudally, an internal thalamic nucleus replaces the subhabenular nucleus, and it lies within the curve of a large external thalamic nucleus near the surface of the diencephalon. In the caudal diencephalon, the posterior commissure arches across the ventricle. Nearby lies a nucleus of the posterior commissure. A densely celled pretectal area is located above the posterior commissure and medial to the external thalamic nucleus.

A second group of nuclei form a more ventral tier in the thalamus. These nuclei include a medially located diffuse thalamic nucleus and a number of only slightly more distinct nuclei located medial to the postoptic commissure and within the commissure itself. The region of the posterior tuberculum is found in the ventral diencephalon amid several large fibre tracts. Most ventrally and medially is a relatively undifferentiated infundibular hypothalamic nucleus.

29.2.4 Mesencephalon

In sagittal and transverse sections, the mesencephalon is approximately wedge-shaped with a wide, rounded dorsal region and a narrow ventral region (Jansen, 1930). The left and right lobes of the mesencephalic roof lie behind the diencephalon in the angle formed by the diverging horns of the medulla (Figures 29.1, 29.3C, 29.4A–B). The mesencephalon is smaller than the diencephalon, but it is present in both *Eptatretus* and *Myxine*. This is true despite marked differences in the development of the eyes in the two genera (Fernholm and Holmberg, 1975). The ventricular spaces of the mesencephalon are reduced but clearly visible. A cerebral aqueduct, located in the dorsal third of the midbrain, extends caudally from the diencephalon and merges in the caudal mesencephalon with vertically aligned ventricular recesses, one of which protrudes well into the dorsal mesencephalon while its counterpart reaches into the ventral mesencephalon (Jansen, 1930; Wicht and Tusch, this volume). The ependyma lining the cerebral aqueduct forms a large subcommissural organ from which a well-developed Reissner fibre extends through the ventricular spaces of the hindbrain and the central canal of the spinal cord.

In the dorsal mesencephalon of eptatretid hagfishes, a layer of ependymal cells and neuroblasts lies adjacent to the ventricles (Iwahori *et al.*, 1996). More superficially is an extensive region designated as the central grey of the mesencephalon. A cap-like optic tectum lies above the central grey. A recent Golgi study of the dorsal midbrain in *Eptatretus* distinguishes a periventricular layer and a thick layer of cells and fibres within the mesencephalic central grey which consists of fibre bundles intermingled with a dense meshwork of small, medium and large neurons (Iwahori *et al.*, 1996). The dorsal part of the cell and fibre layer corresponds to the mesencephalic region designated here as the optic tectum. No portion of the cell and fibre layer, including its dorsal part, evidences the marked lamination characteristic of the optic tectum and the superior colliculi in many vertebrates. A marginal layer is present on the superficial border of the dorsal mesencephalon. Jansen (1930) observed a similar organization in *Myxine*. This is noteworthy, in that the eyes of myxinid hagfishes are both more rudimentary than those of eptatretid hagfishes and are buried beneath muscle as well as skin (Holmberg, 1970; Fernholm and Holmberg, 1975). Clearly, eye structure and its impact on visual function are not associated with marked changes in the dorsal mesencephalon of hagfishes.

A mesencephalic tegmental area is present in the ventral mesencephalon below the level of the cerebral aqueduct. Particularly notable in the ventral midbrain are bilateral clusters of very large cells in the rostral tegmentum which comprise the nuclei of the medial longitudinal fasciculus. Near the rostroventral margin of the midbrain, Jansen (1930) described a mesencephalic interpeduncular nucleus. More recently, connectional data have led to the identification of this nucleus as a segment of the posterior tuberculum (Wicht and Northcutt, 1992a; Wicht and Nieuwenhuys, 1997).

29.2.5 Medulla

The hindbrain of early-stage hagfish embryos bears a strong resemblance to the rhomboid-shaped hindbrain of lampreys and jawed vertebrates (Conel, 1931). However, later developmental processes reduce the similarity, and the morphology and cytoarchitecture

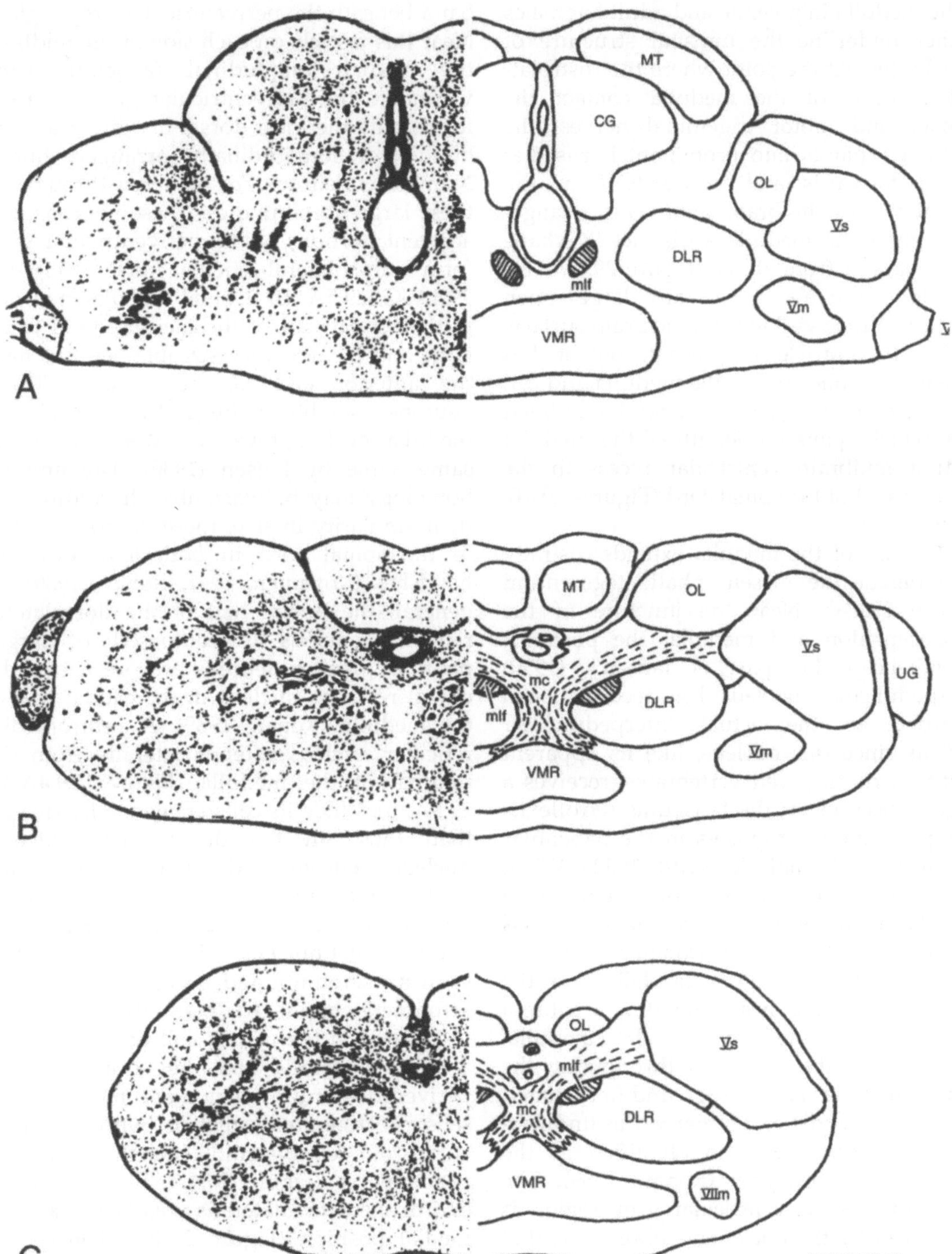

Figure 29.4 Transverse sections through the midbrain and medulla of an adult hagfish (*Eptatretus stoutii*). Bar scale equals 0.5 mm.

of the medulla in juvenile and adult hagfishes further underline the unusual structure of their brains. At the point where the rostrolateral margins of the medulla contact the sensory and motor trigeminal nerves, the medulla expands into prominent horns that extend forwards and outwards from the central axis of the brain stem. This arrangement gives the medulla a distinct 'Y'-shape when viewed from above (Figure 29.1). Also notable is the absence of any large, open fourth ventricle visible from the brain surface. A fourth ventricle is present, but it lies beneath the solid roof of the medulla and can be seen there as a narrow, ependymal-lined tube which spans the length of the medulla from a midbrain ventricular recess to the central canal of the spinal cord (Figures 29.4B, C; 29.5A–D).

The base of the medulla extends rostrally to undercut the mesencephalic tegmentum (Figure 29.4A). Near the juncture of the mesencephalon and medulla, the posterior (rhombencephalic) part of Jansen's (1930) interpeduncular nucleus has recently been regarded as the actual interpeduncular nucleus since this nucleus, like its apparent counterparts in jawed vertebrates, receives a large projection via the fasciculus retroflexus that presumably originates in the habenular complex (Wicht and Northcutt, 1992a; Wicht and Nieuwenhuys, 1997). Concentrations of reticular neurons are situated laterally and dorsally to the interpeduncular nucleus (Figure 29.3C). Continuing caudally from the tegmentum of the mesencephalon, a column of nuclei fill the core of the medulla. Surrounding the ventricle throughout the entire medulla is a large, undifferentiated region of grey matter. Formerly, this undifferentiated region was also identified as the medullary central grey (Ronan, 1989). The shape of this periventricular grey region is influenced by the nuclei and fibre tracts that abut it. The medullary commissure, a fascicle of decussating fibres which is most prominent in the rostral medulla, forms a 'V'-shaped band beneath the periventricular grey region. Near this fascicle on each side of the midline, the medial longitudinal fasciculus runs ventral to the periventricular grey where it forms only the most dorsal margin of a large bundle of longitudinally arranged fibres. Numerous neurons are scattered throughout these large fibre bundles. These neurons and adjacent portions of the periventricular grey form a dorsolateral reticular nucleus (Figures 29.4A–C; 29.5A, B). Dorsal reticular neurons in the rostral half of the medulla tend to be large and include very sizeable neurons near the anterior end of the nucleus. These neurons resemble the huge Müller cells in the medulla of lampreys and were given the same name by Jansen (1930). The implied homology may be warranted. In addition to their similarity in size, these neurons project to the spinal cord in both lampreys and hagfishes (Rovainen, 1967; Ronan, 1989). In general, the dorsal half of the dorsolateral reticular nucleus is composed of larger neurons; smaller neurons are present in the dorsal reticular nucleus in the caudal half of the medulla (Ronan, 1989). A ventromedial reticular nucleus extends laterally from the midline of the medulla (Figures 29.4A–C; 29.5A, B). While its cells are generally smaller than those in the dorsomedial reticular nucleus, neurons in the anterior half of the ventromedial nucleus are somewhat larger than those in the caudal half. Within the ventromedial nucleus, cells along the midline stain for serotonin and may correspond to a serotonergic raphe (Kadota, 1991).

Located in the rostral half of the medulla, an octavolateralis area lies dorsal to the periventricular grey region and medial to the sensory trigeminal nucleus with which it curves out rostrolaterally into the trigeminal prominences (Figure 29.4A, B). The caudal end of the octavolateralis area lies in a more medial position (Figure 29.4C). The central part of the octavolateralis area contains loosely packed neurons, while denser concentrations of cells fill its medial, ventral and

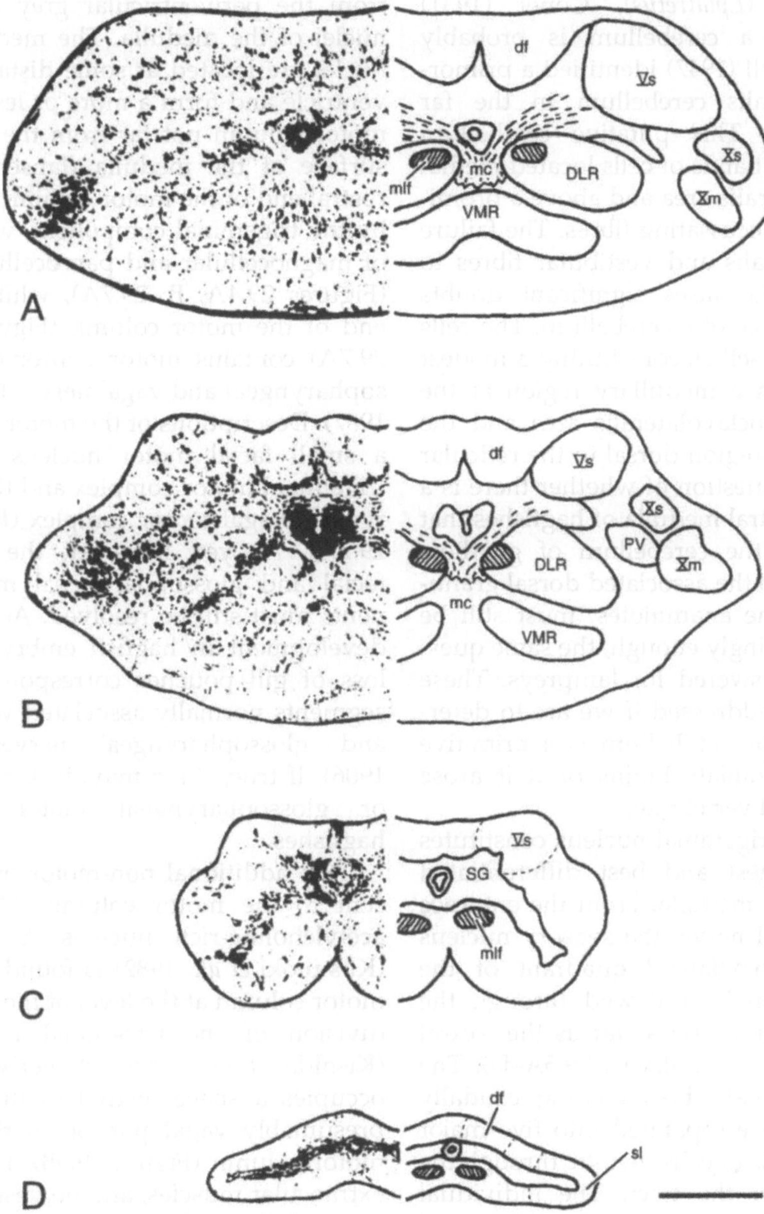

Figure 29.5 Transverse sections through the medulla and rostral spinal cord of an adult hagfish (*Eptatretus stoutii*). Bar scale equals 0.5 mm.

lateral margins. Large neurons are clustered at one site in the rostral octavolateralis area and at a second site more caudally. Fibres cross a molecular region along the dorsal margin of the octavolateralis area.

There is no obvious cerebellum in the brains of hagfishes. Whether a small cerebellar component is present or not has been the subject of disagreement. Jansen (1930) and Bone (1963) found no cerebellum in *Myxine*.

In *Bdellostoma* (*Eptatretus*), Conel (1931) concluded that a cerebellum is probably absent, but Larsell (1947) identified a primordial octavolateralis cerebellum in the far rostral medulla. This putative cerebellum consisted of thin bands of cells located medial to the octavolateralis area and above a prominent fascicle of decussating fibres. The failure of primary lateralis and vestibular fibres to reach these cells raises significant doubts about the existence of a cerebellum. The cells described by Larsell as constituting a modest cerebellum lie in a medullary region at the juncture of the octavolateralis area and the undifferentiated region dorsal to the reticular formation. The question of whether there is a region in the rostral medulla of hagfishes that corresponds to the cerebellum of gnathostomes, or even to the associated dorsal granular ridge of some anamniotes, must still be resolved. Intriguingly enough, the same question must be answered for lampreys. These points must be addressed if we are to determine whether the cerebellum is a primitive component of craniate brains or if it arose only in the jawed vertebrates.

The sensory trigeminal nucleus constitutes one of the largest and best differentiated structures in the medulla. From the entrance of the trigeminal nerve, the sensory nucleus occupies the dorsolateral quadrant of the medulla and can be followed through the medulla in this position as far as the rostral spinal cord (Figures 29.4A–C; 29.5A–D). The primary trigeminal fibres passing caudally along the tract are separated into five major fascicles by bands of cells that arc through the interior space of the tract. The individual fascicles shift somewhat in position and diminish in size in the caudal medulla before becoming indiscernible at spinal levels. In the caudal half of the medulla, a vagal sensory nucleus, far smaller than its trigeminal counterpart, lies ventral to the lowermost fascicle of the sensory trigeminal nucleus (Figure 29.5A, B).

A band of cells extends ventrolaterally from the periventricular grey to the motor nuclei of the medulla. The medullary motor nuclei are located at some distance from the ventricle and form a more or less continuous motor column not far from the ventrolateral surface of the medulla (Jansen, 1930). The rostral end of the motor column is dominated by the trigeminal component, which consists of magnocellular and parvocellular divisions (Figures 29.4A, B; 29.7A), while the caudal end of the motor column (Figures 29.5A, B; 29.7A) contains motor neurons of the glossopharyngeal and vagal nerves (Kishida *et al.*, 1987). Descriptions of the motor column place a small facial motor nucleus between the trigeminal motor complex and the glossopharyngeal-vagal motor complex (Figures 29.4C; 29.7A). As yet, however, the existence of facial and glossopharyngeal motor components must still be resolved. Accounts of gill development in hagfish embryos indicate a loss of gill pouches corresponding to head segments normally associated with the facial and glossopharyngeal nerves (Stockard, 1906). If true, there may, in fact, be no facial or glossopharyngeal motor nucleus in hagfishes.

Two additional non-motor nuclei lie adjacent to the motor column. The enigmatic, acetylcholine-rich nucleus A of Kusunoki (Kusunoki *et al.*, 1982) is found lateral to the motor column at the level of the parvocellular division of the trigeminal motor nucleus (Kishida *et al.*, 1986). A perivagal nucleus occupies a space medial to the caudal and presumably vagal portion of the medullary motor column (Figure 29.5B). Hagfishes lack extraocular muscles, and no vestigial counterparts of the oculomotor, trochlear and abducent motor nuclei have been described.

In addition to the trigeminal nerves, several other cranial nerves communicate with the medulla (Jansen, 1930; Lindström, 1949; Peters 1963). At least two lateral line nerves are present (Figure 29.1). Identified early on as 'nervus acusticus a and b' (Worthington, 1905) and subsequently labelled as the anterior and posterior lateral line nerves (Kishida *et al.*, 1987), more recent terminology refers to these nerves as lateral line nerves A and B (Braun and Northcutt, this volume). Between the two lateral line nerves, the anterior and posterior rami of the acoustic nerve convey sensory input from the toroidal inner ear, although the nature of the sensory input remains to be determined. The anterior acoustic ramus is also known as the utricular nerve while the more caudal saccular nerve corresponds to the posterior acoustic ramus. A cranial nerve complex, traditionally identified as the combined glossopharyngeal and vagal nerves, attaches to the caudal medulla. The majority of these nerve fibres appear to constitute the vagal nerve; only the most rostral fibres have been regarded as a glossopharyngeal component. A small and hard-to-find facial nerve has been described exiting the brain near the 'utricular' and 'saccular' ganglia of the acoustic nerve rami. Again, it is worth noting that according to one scenario of head development in hagfishes, the loss of particular gill pouches may mean that there is no facial nerve in myxinoids and the glossopharyngeal-vagal nerve may be entirely vagal.

29.2.6 Spinal cord

While the spinal cord is a far less complicated structure than the brain, it nonetheless comprises a very substantial portion of the central nervous system, extending for many centimetres through the branchial region, trunk and tail of a large adult hagfish. In jawed vertebrates and in lampreys, the caudal end of the fourth ventricle at the obex provides a conspicuous landmark for the transition of the medulla into the spinal cord. No comparable landmark is afforded by the small, internal fourth ventricle of adult hagfish brains. Instead, there is a gradual reduction in the size of the caudal medulla until the constant width of the spinal cord is attained. During this transition, the neuraxis diminishes in depth even more than in width, as both the grey matter, now concentrated in the centre, and the surrounding fibre tracts, including the prominent dorsal funiculi and trigeminal sensory tracts, decrease in cross-sectional area. Eventually, the spinal cord assumes the approximate form of a flattened, inverted 'V' throughout most of its length. In the core of the spinal cord, narrow bands of neuronal and glial grey matter extend laterally from the small central canal while a matrix consisting of neuronal dendrites, non-myelinated axons and occasional neurons makes up the thick outer margin of the spinal cord (Bone, 1963). The 'white matter' of the spinal cord can be described according to location as consisting of dorsal, lateral and ventral funiculi. The longitudinal column of the dorsal funiculus continues rostrally from spinal levels into the caudal part of the medulla where it lies adjacent to the dorsal midline (Figure 29.5A–D). Cells interspersed with fibres in the dorsal funiculus were termed by Amemiya (1983) as the dorsal funicular nucleus. Along most of the cord's length, dorsal and ventral spinal nerves respectively enter the dorsolateral and ventrolateral surfaces of the cord. Similarly flattened spinal cords are seen in adult and larval lampreys. However, the spinal cord of hagfish embryos is much rounder than the cord of adults (Conel, 1931; Holmgren, 1946). The flattening of the cord in hagfishes occurs during development.

Bone (1963) described several types of bipolar and multipolar neurons. Most appear to be interneurons networked into intersegmental or ascending circuits, but conspicuously large cells in the ventral and lateral grey matter possess axons that enter the roots of the ventral spinal nerves and so appear to be motor neurons. Rami of the ventral roots distribute exclusively to myotomal muscles and the segmental slime glands (Peters, 1963). No intramedullary sensory neurons such as the dorsal cells of lampreys and the Rohon-Beard cells of some anamniotes have been found in hagfishes. In hagfishes, as in many other anamniotes, dendrites of many spinal neurons radiate widely into the spinal fasciculi. Longitudinally running axons of various diameters fill the ventral, lateral and dorsal funiculi of the spinal cord. A cluster of large-diameter axons, identified as the medial longitudinal fasciculus, is found adjacent to the midline in the ventral spinal funiculi.

Except for the most rostral portion of the spinal cord, two spinal nerves connect the spinal cord with each body segment (Allen, 1917; Peters, 1963). As noted, a series of rootlets emerging from the ventrolateral surface of the spinal cord combine to form a single nerve whose rami are regarded as motor nerves. Ganglia of the dorsal spinal nerves contain neuronal somata whose peripheral processes distribute between the myotomes to sensory endings in skin and perhaps in muscles. The first and perhaps the second most rostral spinal nerves have been called spino-occipital nerves, in view of their structural differences with caudal nerves, their innervation of head structures and their location near the transition of the brain and spinal cord (Worthington, 1905; Lindstrom 1949). The dorsal and ventral spinal nerves fuse to form spinal nerves resembling the mixed sensory and motor nerves of jawed vertebrates. However, the anatomical details of the fusion suggest that spinal nerves in hagfishes may actually resemble the separate dorsal and ventral spinal nerves of lampreys more than the mixed nerves of gnathostomes, the superficial resemblance of conjoined sensory and motor nerves in hagfishes and jawed vertebrates being likely a case of parallelism (Goodrich, 1937).

29.3 CONNECTIONS AND COMMENTS

In the following section, known connections of the central nervous system are reviewed and pertinent aspects of hagfish neurobiology, including possible evolutionary relationships of brain regions, are assessed. The discussion of particular sites in major brain regions proceeds through the brain in a rostral to caudal direction before concluding with the spinal cord.

29.3.1 Olfactory bulb (Figures 29.6A; 29.7A, B)

Several factors – the deep benthic ethological niche occupied by hagfishes, the reduced visual and lateral line systems, and the large size of the olfactory bulbs – implicate the olfactory system as one of prime importance in the sensory world of hagfishes. Consistent with this outlook is the extent of secondary olfactory projections to widespread regions of the forebrain by way of several olfactory pathways (Wicht and Northcutt, 1993). A small, medial olfactory tract projects to the septal nucleus, and limited commissural projections also pass between the olfactory bulbs along this tract. A limited ventral olfactory tract joins the lateral forebrain bundle and continues to a more caudal location where a few fibres terminate in the preoptic area, the hypothalamus and the triangular nucleus of the thalamus. The lateral olfactory tract constitutes the largest projection of olfactory bulb efferents to the telencephalic hemispheres. The superficial subdivision of this tract primarily forms pallial lamina 1, while providing olfactory input to pallial lamina 2. Lateral olfactory tracts located superficially in the telencephalon are commonly present in craniates; however, the lateral olfactory tract of hagfishes also includes a unique deep division. The deep division of the lateral olfactory tract passes through the striatal region and the deepest layer of the pallium and gives off a few fibres that reach pallial lamina 3 and the lateral part of lamina 4. Heavy projections terminate in the magnocellular part of pallial lamina 2 and the medial part of pallial lamina 4. Secondary olfactory projections also reach the dorsal division of the central prosencephalic nucleus. Although heaviest ipsilaterally, olfactory projections are bilateral and

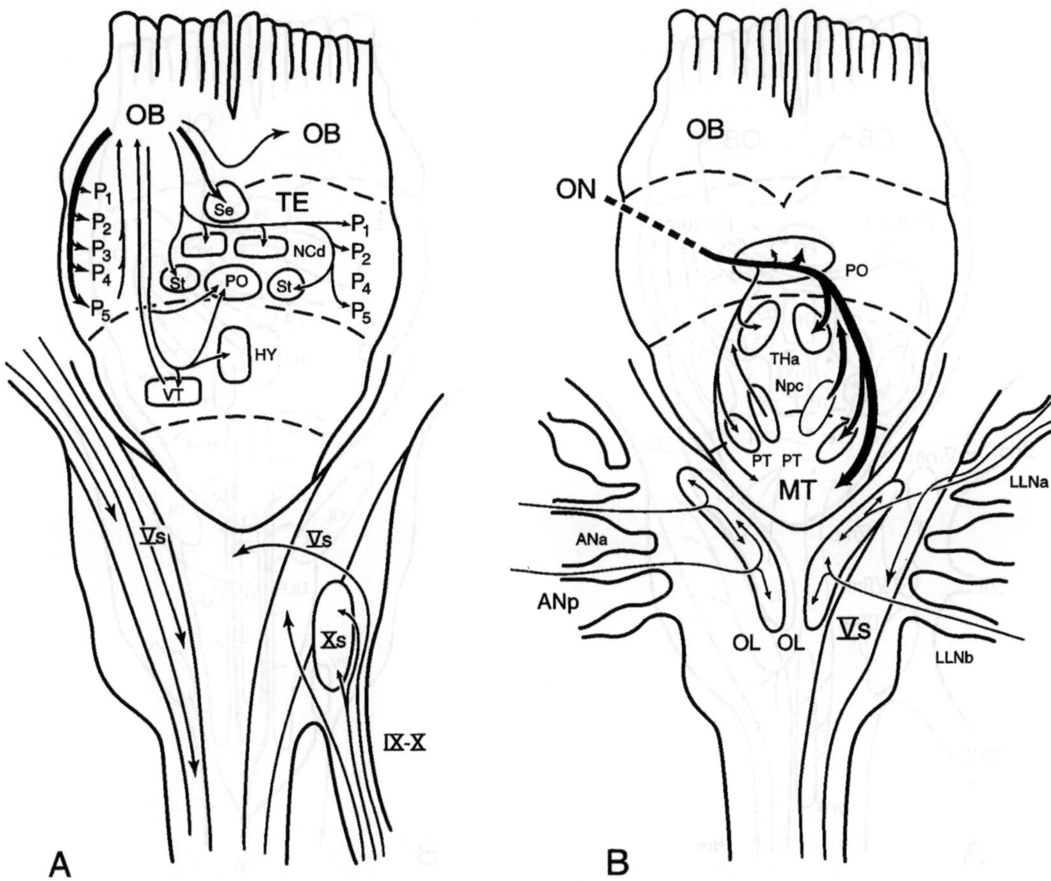

Figure 29.6 Dorsal view of the brain of an adult hagfish (*Eptatretus stoutii*) shows the outline of the brain and the rostral spinal cord and indicates the general location of several nuclei within the central nervous system. (A) Projections of the olfactory bulb to regions of the forebrain are shown. Also shown are the primary projections of the trigeminal nerve to the left side of the medulla and spinal cord and the primary projections of the glossopharyngeal-vagal nerve to the right side of the medulla. Larger arrows indicate heavier projections. (B) Projections of the retina to the forebrain and midbrain and the source of projections to the retina are shown. Also shown are the primary projections of the acoustic nerve rami to the left side of the medulla and the primary projections of lateral line nerves A and B to the right side of the medulla. Larger arrows indicate heavier projections.

reach the opposite hemisphere through the interbulbar commissure and habenular commissures. Neurons in the ipsilateral lateral pallium and in the ventral thalamus near the ventral olfactory tract project back to the olfactory bulbs.

Several points can be addressed with the available information (Wicht and Northcutt, 1993). First, there is no sign of a terminal nerve in hagfishes. This separate cranial nerve is associated with the olfactory nerve in some gnathostomes and also includes a projection to the retina. Its presence is indicated by positive reaction for FMRFamide and gonadotropin-releasing hormone. Despite immunopositive tests for FMRFamide and gonadotropin-releasing hormone in part of the hypothalamus, and

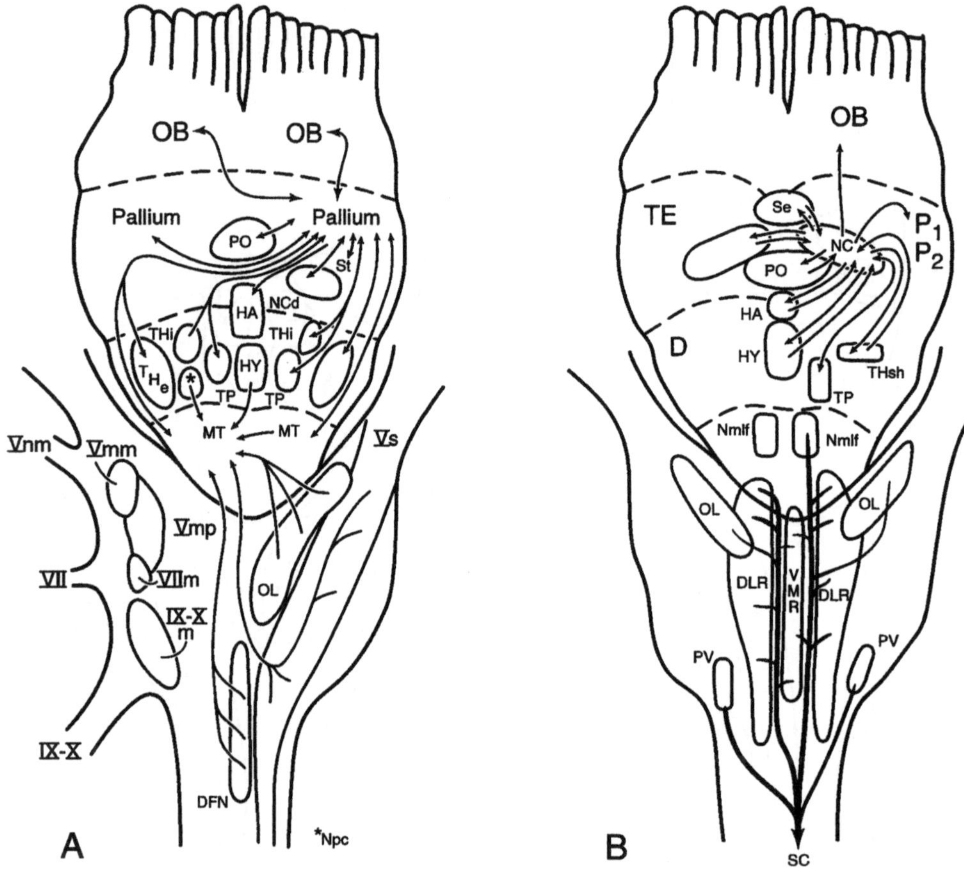

Figure 29.7 Dorsal view of the brain of an adult hagfish (*Eptatretus stoutii*) shows the outline of the brain and the rostral spinal cord and indicates the general location of several nuclei within the central nervous system. (A) Connections of the telencephalic pallial laminae and the mesencephalic tectum are shown. Also shown is the location of motor nuclei on the left side of the medulla. (B) Connections of the central prosencephalic nucleus are shown. Also shown in the midbrain and medulla are the nuclei of origin of the descending spinal projections.

FMRFamide-positive fibres in the olfactory bulb, no hypothalamic projection to the olfactory bulb has been demonstrated (Wicht and Northcutt, 1992b; Braun *et al.*, 1995). Moreover, no retinopetal fibres are associated with the ventral olfactory tract (Wicht and Northcutt, 1990). Secondly, the widespread projections from the olfactory bulb to the telencephalic hemispheres exceed the secondary olfactory projections present in gnathostomes and even the extensive secondary olfactory projections present in lampreys. Given the phylogenetic distribution of secondary and higher-order olfactory projections in hagfishes, lampreys and gnathostomes, the olfactory bulb (and possibly the anterior olfactory nucleus) probably projected to widespread areas of the telencephalic hemispheres in their common ancestor (Wicht and Northcutt, 1992a).

29.3.2 Telencephalic pallial laminae (Figures 29.6A; 29.7A, B)

The laminar pallium is one of the most remarkable forebrain features in hagfishes. The only known sensory input to this elaborate structure thus far disclosed by experimental anatomical studies consists of a secondary olfactory projection from both olfactory bulbs (Wicht and Northcutt, 1993). Multiple pallial injections of HRP involving large portions of all five laminae, disclose several other pallial afferents (H. Wicht and R.G. Northcutt, unpublished observations). Cells in all five pallial layers give rise to projections within the ipsilateral telencephalic hemisphere. Commissural fibres originate in laminae 2 and 4 of the contralateral hemisphere and pass through the interbulbar and habenular commissures. Additional pallial afferents originate in the dorsal division of the central prosencephalic nucleus, the preoptic area, the posterior tuberal area and the anterior nucleus of the dorsal thalamus. The ipsilateral internal and external thalamic nuclei of the dorsal thalamus are a major source of pallial afferents, although the sorts of information that ascend from any part of the diencephalon to the pallium are not known.

Efferents of the pallium resemble, in some respects, the secondary olfactory projections, especially the superficial and deep parts of the lateral olfactory tracts. Still, projections revealed by HRP injections of the pallium are not identical to the olfactory projections and cannot be interpreted simply as labelled collaterals of olfactory bulb neurons (H. Wicht and R.G. Northcutt, unpublished observations). There are unique projections revealed by pallial injections: (i) a small projection ends in the habenula; (ii) projections to the preoptic area and postoptic commissure caudally pass through the striatum and, in fact, terminate in the ventral striatum; (iii) fibres reach the terminal field external to the magnocellular division of pallial layer 2 by travelling around the circumference of the hemisphere; (iv) bundles of fibres converge on the dorsal division of the central prosencephalic nucleus from wide expanses of the pallium; (v) pallial efferents terminate in the lateral segment of the internal thalamic nucleus and the margin of the external thalamic nucleus with a conspicuously dense terminal field on the ventrolateral surface of the latter nucleus; (vi) a limited number of efferents attain the margin of the dorsal midbrain, including the ipsilateral optic tectum.

No matter what other inputs reach the pallium, the olfactory system alone provides a massive source of afferents. Second-order olfactory fibres reach all five pallial laminae. Unlike the pallia of many gnathostomes, the pallium of myxinoids may be dominated by olfaction, and this remarkably elaborate pallium may have evolved to extract pertinent information from what for hagfishes must be a crucial stimulus channel. At least a portion of the pallium must be homologous to the olfactory lateral pallium of jawed vertebrates (Wicht and Northcutt, 1993). It remains to be determined whether the entire pallium in myxinoids is a homologue of the lateral pallia in other craniates or whether it also contains homologues of the dorsal and medial pallia (Wicht and Northcutt, 1992a). Available data appear to be compatible with at least two answers to this question. The finding that afferents ascend to the pallial laminae from the dorsal thalamus in *Eptatretus* suggests that dorsal and medial pallial elements may be present in hagfishes (H. Wicht and Northcutt, unpublished observations). Pallial efferents also reach the hypothalamus, a connection comparable to the fornix, and this also suggests the presence of a medial pallial homologue (Amemiya and Northcutt, 1995). Much of the myxinoid pallium may be a lateral pallium, but the medial part of lamina 4 (and the nearby dorsal subnucleus of the central prosencephalic nucleus) may be a medial pallial counterpart. On the other hand, note that the projections ascending from the dorsal thalamus terminate diffusely in the pallial laminae with no single focus, and the secondary olfactory fibres have widespread endings in the pallium (Wicht and Northcutt, 1993 and unpublished observations). The pallium of hagfishes includes an extensive element that corresponds to the olfacto-recipient lateral pallium of gnathostomes; however, homologues of the medial and dorsal pallia remain to be identified. If present, medial and dorsal pallial regions in myxinoids may not have fully differentiated from the lateral pallium (Wicht and Nieuwenhuys, 1997). Whatever the evolutionary history of olfactory input to the telencephalon in myxinoids, the extent of the telencephalic centres receiving olfactory input implies a paramount importance of olfaction for hagfishes. Moreover, the extent of secondary olfactory projections to the lateral pallium, and possibly to the medial and dorsal pallia, in hagfishes and to the lateral pallium and portions of the medial and dorsal pallia in lampreys, argue for widespread olfacto-pallial projections in the common ancestor or agnathans and gnathostomes (Northcutt and Puzdrowski, 1988; Wicht and Northcutt, 1993).

29.3.3 Central prosencephalic nucleus (Figures 29.6A, 29.7B)

The central prosencephalic nucleus comprises a major part of the forebrain while also constituting one of its foremost enigmas. Such basic considerations as the make-up, embryonic origins, connections and functions of this nucleus remain areas of inquiry. In recent years, a number of new insights have become available for consideration (Wicht and Northcutt, 1992a, 1994; Amemiya and Northcutt, 1996).

Contiguous medial, dorsolateral and ventrolateral subnuclei make up the central prosencephalic nucleus. Previously, the central prosencephalic nucleus has been compared to a number of forebrain components. Jansen (1930) and Conel (1931), among others, regarded it as a medial pallial homologue. Alternative interpretations have focused on dorsal and ventral thalamic homologies (Crosby and Schnitzlein, 1974; Wicht and Northcutt, 1992a). High concentrations of acetylcholinesterase have also led to its identification as a possible striatum (Wächtler, 1975; Northcutt, 1981). The recent insights into the connections of the central prosencephalic nucleus have not provided a clear answer as to what, if any, structure in other craniates might be homologous to it, but, together with the available topological, embryological and immunohistochemical data, these insights exclude several possibilities while pointing towards others.

All three subnuclei of the central prosencephalic nucleus lie caudal to the remnants of the telencephalic lateral ventricles (Wicht and Northcutt, 1992a). In addition, tracer injections into the central prosencephalic nucleus involving the medial, ventrolateral and dorsolateral subnuclei reveal a diverse pattern of connections (Amemiya and Northcutt, 1996). Ascending efferents project to the mitral layer of the olfactory bulb, septum and preoptic area. Descending efferents turn ventrally to terminate in the hypothalamus and dorsally to terminate in the habenula, the subhabenular nucleus of the thalamus and the dorsomedial margins of pallial laminae 1 and 2. Other descending projections reach the nucleus of the medial longitudinal fasciculus and the posterior tubercle. Commissural efferents terminate in all three subnuclei of the contralateral central prosencephalic nucleus. Following the injections, a small number of retrogradely filled cells were observed within the contralateral ventrolateral division of the central prosencephalic nucleus and also within the septum, preoptic area, hypothalamus and subhabenular nucleus of the thalamus, indicating the existence of reciprocal connections between the latter nuclei and the central prosencephalic nucleus. The dorsal division receives olfactory bulb efferents via the lateral olfactory tract (Wicht and Northcutt, 1993). Such connections are more extensive than typical of a medial pallial, striatal or thalamic homologue alone (Amemiya and Northcutt, 1996). Moreover, while a portion of the central prosencephalic nucleus is telencephalic in origin, the medial and ventrolateral subnuclei are derived from a diencephalic embryonic precursor (Conel, 1931; Holmgren, 1946). Chemoarchitectonics also suggest that the three contiguous subnuclei of the central prosencephalic nucleus may not constitute a single unified structure. All three stain for substance P, serotonin, and L-enkephalin, but the medial and ventrolateral nuclei exhibit high levels of acetylcholinesterase (Wicht and Northcutt, 1994). In contrast, the dorsal subnucleus is comparatively poor in acetylcholinesterase and serotonin; however, it is particularly rich in L-enkephalin-containing cells.

This pattern does not support a close comparison with the medial pallium (Wicht and Northcutt, 1994). Moreover, the central prosencephalic nucleus is not a likely homologue of the striatum, despite intense acetylcholinesterase staining (Amemiya and Northcutt, 1996). The central prosencephalic nucleus exhibits limited catecholaminergic inputs, lacks extensive thalamic afferents and shares few connections with the lateral forebrain bundle – all common striatal characteristics (Wicht and Northcutt, 1994). Although not without its own difficulties, a better case can be made for a more lateral subpallial group as a counterpart of the striatum on histochemical and connectional grounds (Wächtler, 1975; Wicht and Northcutt, 1994).

The central prosencephalic nucleus, as a single structure, may have no single homologue in other craniates, and homologies with a number of pallial, subpallial and diencephalic structures can be considered (Wicht and Northcutt, 1992a). The most recent connectional data suggests that the medial and ventrolateral subnuclei may correspond to the rostral ventral thalamus or the thalamic eminence of other craniates, and the dorsal subnucleus bears some histochemical and cytoarchitectural similarities to the medial pallia of other craniates (Amemiya and Northcutt, 1996).

29.3.4 Dorsal mesencephalon (Figures 29.6B, 29.7A)

The simplified structure of the eyes, their location beneath the skin and the modest diameter of the optic nerve imply that the visual system of hagfishes does not constitute a robust sensory modality. Nevertheless, tracings of the retinal projections in *Eptatretus burgeri* (Kusunoki and Amemiya, 1983) and *Eptatretus stoutii* (Wicht and Northcutt, 1990) demonstrated that retinal efferents are indeed present and terminate in several forebrain and midbrain cell groups. Retrograde staining in *Eptatretus stoutii* also revealed the existence of retinopetal projections (Wicht and Northcutt, 1990).

Labelled retinal efferents travel from the retina to the base of the forebrain in long, thin optic nerves to enter the preoptic area where a partial decussation of retinal efferents occurs. Approximately 90% of the fibres decussate, but optic projections terminate symmetrically in diencephalic and mesencephalic nuclei. Some optic fibres terminate in the preoptic area; other fibres pass dorsolaterally and caudally to enter the optic tracts located within the postoptic commissures. Optic fibres may provide input to cells distributed along the course of the postoptic commissure. In the rostral diencephalon, optic fibres exit the tract and terminate in the anterior thalamic nucleus. Fibres continuing caudally reach the pretectal area, the heaviest and most distinct terminal field of optic fibres, and the roof of the mesencephalon, where they extend dorsally along the lateral aspect of the central grey and terminate among the cells of the optic tectum (cellular and fibre layer) and in the marginal layer.

Retinal projections tend to terminate superficially in the dorsal mesencephalon by largely avoiding the central grey region. Although the available evidence is quite limited, there may be a tendency for mesencephalic afferents from the trigeminal sensory nucleus and spinal cord to follow a more medial course in the dorsal mesencephalon (Amemiya, 1983; Ronan, 1988). There is no indication as yet that retinal and non-retinal projections to the dorsal mesencephalon segregate into superficial and central terminal fields as is the case in a number of vertebrates. Nor is there any information regarding a retinotopic pattern of termination of optic fibres in the dorsal mesencephalon, a near universal trait of gnathostomes.

Following Di-I injections of the eye, retrogradely labelled cells are observed in the nucleus of the posterior commissure and in an indistinct area lateral to the nucleus of the posterior commissure. The filling of such cells cannot be readily attributed to transneuronal labelling; instead the cells must be regarded as true retinal afferent neurons (Wicht and Northcutt, 1990). As most retinal efferents terminate contralaterally, most retinal afferents project to the contralateral eye. Functional studies of the dorsal mesencephalon are needed to disclose the presence of any sign of sensory segregation within the tectum and to reveal, for that matter, the modalities that actually convey sensory input to the region.

Fernholm and Holmberg (1975) found that complexity of eye structure in three hagfish species corresponds with the level of ambient light in their habitat. Species living in the shallowest, best-illuminated waters exhibited the greatest degree of eye development; species living in the deepest and darkest environment possessed more rudimentary eyes. The eyes of *Myxine* are not only less elaborate than those of *Eptatretus*, they lie beneath layers of skin, blood and muscle. Myxinid hagfishes might be expected to operate a much reduced visual system; however, experimental studies of *Myxine* reveal retinal projections very similar to those seen in *Eptatretus* (H. Wicht, personal communication). The multiple targets of the retinal efferents and the existence of retinal afferents imply that this is not a fully degenerate visual system, though, in view of visual system organization in lampreys and gnathostomes, it may formerly have been more elaborate. Only functional studies can disclose how it works, but the location of the eyes indicates that the visual system of these animals is not concerned with processing formed images but functions primarily to detect levels of illumination. It is possible that retinopetal fibres help to set the sensitivity of retinal photoreceptors.

Connections of the dorsal mesencephalon in eptatretid hagfishes are similar to a number of connections seen in other craniates. Mesencephalic afferents arise in the sensory trigeminal nucleus, the octavolateralis area, the dorsal funicular nucleus at the caudal end of the medulla and the rostral spinal cord (Amemiya 1983; Ronan, 1988). Second-order trigeminal inputs to the optic tectum and second-order lateral line inputs to the tectum and torus semicircularis are common in jawed vertebrates and lampreys. Further studies will be necessary to disclose whether the dorsal mesencephalon of hagfishes contains a homologue of the torus semicircularis as a recipient zone of second-order lateral line fibres. Forebrain afferents to the dorsal mesencephalon originate in the caudolateral pallium, which projects bilaterally to the tectum, the hypothalamus, and a small thalamic nucleus located medial to the postoptic commissure (Amemiya, 1983; Wicht and Northcutt, personal communication). To date, the limited information regarding tectal efferents in hagfishes shows a projection reaching the magnocellular portion of lamina 2 in the caudal ventrolateral pallium. Reciprocal projections thus exist between the tectum and a limited portion of the pallium. Such a connection is not unprecedented; cells in the medial pallium of bullfrogs do project to the rostromedial tectum (Northcutt and Ronan, 1992). It is unlikely that these pallial cells in hagfishes belong to a medial pallial homologue (Wicht and Northcutt, 1992a). A true appreciation of such a projection awaits a more complete understanding of the pallium in myxinoids. There is no sign of a tectospinal projection, even to the most rostral spinal levels (Ronan, 1989). Although present in some amphibians and reptiles, this pathway has not generally been found in lampreys or jawed fishes, animals lacking a neck and independent movements of the head.

29.3.5 Trigeminal nerve projections (Figure 29.6A, B)

Several major rami combine to form the large sensory trigeminal nerve (Lindström, 1949). The ophthalmic nerve innervates structures on the dorsal rostral head and is comparable to the ophthalmic ramus of other craniates. The external and dental/velobuccal nerves innervate more rostral and ventral head structures and correspond to the maxillary and mandibular rami of the trigeminal in other craniates. The central processes of primary trigeminal fibres in *Eptatretus burgeri* (Nishizawa *et al.*, 1988) and in *Eptatretus stoutii* (Ronan, 1989) course into the sensory trigeminal nucleus of the medulla and reach as far as the rostral spinal cord. Afferent fibres in the trigeminal nucleus do not bifurcate into ascending and descending branches as they do in other craniates, presumably due the entrance of the afferents at the rostral margin of the medulla. Primary trigeminal afferents are somatotopically organized into five large fascicles. Ophthalmic afferents are located dorsomedially, dentate/velobucal afferents most ventrolaterally, and external afferents in between. Trigeminal afferents of the ophthalmic and external nerves terminate both as fine-diameter fibres in an external fascicle and as thick-diameter fibres in a deeper fascicle. Thin-diameter fibres appear to be trigeminal cutaneous afferents whereas thick-diameter fibres seem to be trigeminal chemosensory afferents that innervate 'taste' receptors on the barbels and head, a condition not seen in other craniates (Nishizawa *et al.*, 1988). Neurons in the cellular bands that separate the fascicles of primary trigeminal afferents may constitute a distributed nucleus of the sensory tract, since these cells are the source of an ascending projection to the dorsal midbrain (Ronan, 1988). Restriction of afferents in particular rami to particular fascicles is not absolute; moreover, at least some of the neurons in the bands of cells between the fascicles send dendrites into adjacent fascicles. Both of these factors would seem to compromise the striking somatotopic organization of the primary trigeminal afferents. Like lampreys, hagfishes possess no mesencephalic root of the trigeminal nerve. Hagfishes possess toothed plates that function effectively as jaws, but they are not homologous to the jaws of gnathostomes. The sensory mesencephalic root of the trigeminal system assists in gnathostome jaw function and may have arisen with gnathostome jaws.

The extent of the sensory trigeminal nucleus from the rostral medulla to at least rostral spinal levels is similar in eptatretid hagfishes, lampreys and gnathostomes. The trigeminal system in hagfishes is, however, notable for its size. The fact that it occupies such a large fraction of the medulla's cross-sectional area implies that these animals rely heavily on this system. The presence of a chemosensory component that supplements a presumed tactile component is like nothing in other craniates, though arguably it is very adaptive. Processing of chemical and tactile inputs would appear crucial for an organism with limited visual capabilities to locate sexual partners and food items in its benthic environment. An adaptive expansion of the fascicles of trigeminal afferents in myxinoid evolution may have dispersed neurons in what may once have been a more localized sensory trigeminal nucleus into the bands of cells that now border the fascicles. Whatever the origin of these bands, they underline the somatotopy of the trigeminal afferents. The somatotopy of the sensory trigeminal nucleus in eptatretids is the reverse of that seen in jawed vertebrates, where the ophthalmic afferents are most ventrolaterally while maxillary and mandibular afferents (counterparts of the external and dentate/velobuccal afferents, respectively) are located dorsomedially in the tract (Nishizawa *et al.*, 1988). The tract itself is located more laterally and ventrally in gnathostomes. Nishizawa *et al.* (1988) argue that these differences are due to developmental processes in gnathostomes, in which a thinning of the medulla's roof and an outward eversion of its alar plate at the sides of the fourth ventricle carry the sensory trigeminal nucleus to a more ventrolateral location and also invert the relative order of the trigeminal fibres. No comparable change occurs in hagfishes, leaving the sensory trigeminal nucleus in its original dorsomedial location, with the original orientation of trigeminal fascicles, and also preserving the smaller fourth ventricle with its thicker roof.

29.3.6 Octavolateralis area (Figures 29.6B; 29.7A, B)

Primary afferents in the anterior and posterior rami of the acoustic nerve project centrally from the inner ear (Amemiya *et al.*, 1985). The axons of small-celled neurons in the ganglia of the two rami cross medially over the dorsolateral surface of the medulla external to the sensory trigeminal nucleus. In the brain, the fibres pass along the margin of the sensory trigeminal nucleus and terminate in the ventral part of the octavolateralis area where they bifurcate and send branches rostrally and caudally. Acoustic fibres entering the ventral octavolateralis area from the ganglia of the anterior and posterior rami overlap extensively and exhibit no obvious somatotopy. Efferents to the ear have not been reported (Amemiya *et al.*, 1985); however, the presence of a few cholinergic fibres in the sensory structures of the inner ear holds out the possibility that cholinergic efferents, like those found in gnathostomes, also exist in hagfishes (J. Jørgensen, personal communication).

A long-standing uncertainty exists concerning the identification of the lateral line nerves. Tracing studies in *Eptatretus burgeri* demonstrated primary afferent fibres in the lateral line nerves (Kishida *et al.*, 1987). Thin-diameter afferents enter the medulla through the large root of lateral line nerve A and the smaller root of lateral line nerve B. Their compact somata are located in the multiple ganglia of the lateral lines nerves and also in the ganglion of the anterior acoustic ('utricular') ramus. Lateral line fibres reach the medial part of the octavolateralis nucleus, where they bifurcate into ascending and descending branches prior to terminating with considerable overlap between the afferents of lateral line nerves A and B. Lateral line fibres innervating skin in front of the eye (nerve A) lie somewhat dorsal to those that innervate skin behind the eye (nerve B). In eptatretid hagfishes, a series of short, shallow grooves are present in the skin in front of and behind the eyes. Sensory receptors in these skin structures are innervated by lateral line nerve fibres (Kishida *et al.*, 1987). The skin of *Myxine* exhibits no such skin structures and must still be examined for lateral line projections to the brain. Additional primary afferents in lateral line nerve A project to the sensory trigeminal nucleus where they terminate mostly in the ventrolateral fascicle. These afferents may convey input from chemo- and somatosensory receptors in the skin of the head (Braun and Northcutt, this volume). Thus, in hagfishes, at least one lateral line nerve, as well as some trigeminal nerves, appears to carry chemosensory input to the medulla.

Currently, the octavolateralis area has several sources of confirmed or suspected input: lateral line fibres terminate medially;

acoustic afferents terminate ventrally; ascending spinal projections pass through the overlying neuropil and may also provide input. No known source of afferents provides input to the cells and neuropil located in the remaining central part of the octavolateralis area. Additional types of primary sensory afferents to this region cannot be expected. It is not clear whether octavolateralis efferents to the mesencephalic tectum originate in some of the cells located in the medial part of the central region (Amemiya, 1983). Although no prominent counterpart of the cerebellum is anywhere in evidence in hagfishes, the central zone of the octavolateralis area remains one of the most likely of the possible cerebellar components in hagfishes. If any cerebellar component does exist in hagfishes, it may be only a small part of the cerebellum associated with lateral line structures rather than the cerebellum proper.

29.3.7 Glossopharyngeal and vagal projections (Figure 29.6A)

Mixed sensory and motor fibres of the glossopharyngeal-vagal nerves enter the posterior medulla of *Eptatretus burgeri* (Matsuda *et al.*, 1991). Thin-diameter afferents terminate in a vagal sensory nucleus found between the motor nucleus and the sensory trigeminal nucleus. Additional sensory fibres continue rostrally in a fascicle adjacent to the sensory trigeminal nucleus, and a number of the sensory afferents end in the ventrolateral bundle of the trigeminal nucleus. Far rostral in the medullary horns the remaining fascicle of glossopharyngeal fibres turns dorsomedially beneath the trigeminal sensory nucleus and then slightly caudally as a few fibres decussate in a small commissure above the medullary ventricle. Somata of the primary afferent neurons are present in the glossopharyngeal ganglion and are also sparsely scattered in the brain near the entrance of the nerve roots.

Thick-diameter axons originate in the combined glossopharyngeal-vagal motor nucleus lying close to the surface of the caudal medulla. Most motor axons depart the brain directly by entering the nearby nerve, but other axons depart the nucleus and detour towards the central medulla where they make a hairpin turn and only then proceed ventrolaterally to exit the brain. The somata of motor neurons are confined to the motor nucleus. Dendrites of the large, multipolar neurons in the motor nucleus extend dorsally into the overlying vagal sensory nucleus and medially into the nearby reticular formation and the ascending spinal efferents which run through the reticular formation (Matsuda *et al.*, 1991). The glossopharyngeal-vagal motor neurons may thus receive primary visceral sensory input via the entering IX–X afferents and second-order cutaneous sensory input by way of spinal projections in the reticular formation and thus are in position to initiate branchiomeric and autonomic responses in a fairly direct reflex arc.

29.3.8 Descending spinal projections (Figure 29.7B)

Several cell groups in the brain stem project to at least the rostral levels of the spinal cord (Ronan, 1989). The disposition of the reticular nuclei is unusual in hagfishes, perhaps reflecting the organization of the medulla, but the origins of descending spinal projections generally resemble the pattern seen in lampreys and gnathostomes. The very large neurons of the nucleus of the medial longitudinal fasciculus in the midbrain tegmentum are the source of the thick diameter axons that extend along the ventral midline of the medulla and the rostral spinal cord. Most descending spinal projections originate in the dorsolateral and ventromedial nuclei of the medullary reticular formation. Ipsilateral and contralateral reticulospinal fibres descend from large and small reticular neurons, including cells in the putative raphe along the midline of the ventromedial reticular nucleus.

Fibres descending from the octavolateralis areas may constitute a type of vestibulospinal system. These axons come from fairly large neurons in the posterior part of the contralateral octavolateralis area and in the anterior part of the ipsilateral octavolateralis area. The spinal afferents from the magnocellular components of the octavolateralis areas in hagfishes are similar to vestibulospinal projections in some amniotes and the spinal efferents that arise in the intermediate and posterior octavomotor nuclei of lampreys. All of these projections may mediate turning movements of the head in response to vestibular stimulation.

The perivagal nucleus is one unique feature of descending spinal projections evident in eptatretid hagfishes. It contributes descending efferents to the contralateral spinal cord (Ronan, 1989). The nucleus is not a motor nucleus, nor is it a target of vagal sensory afferents (Matsuda *et al.*, 1991). By reason of its position, it may link vagal motor function with spinal cord activity.

29.3.9 Spinal cord

The number of sensory and motor spinal nerves that innervate a single body segment is not known. There can be little doubt, however, that the dorsolateral and ventrolateral spinal nerves of hagfishes can be assigned predominant, if not exclusive, sensory and motor functions, respectively. Following injections of tracers into trunk muscles, labelled axons appear to extend out from large spinal neurons through the ventral spinal nerves (unpublished observations). Labelled sensory afferents enter the cord via the dorsal nerves and bifurcate into ascending and descending branches in the dorsal funiculus (Nansen, 1886; Ronan and Northcutt, 1990). The afferents entering more rostral spinal roots join the dorsal funiculus along its lateral edge. Sensory afferent projections to the spinal cord carry information arising in photoreceptors (Newth and Ross, 1955; Steven, 1955) and probably in cutaneous and proprioceptive endorgans located in the skin. Moreover, primary spinal afferents, like trigeminal afferents, terminate in two distinct tracts within the spinal cord, one of which appears to convey input from taste receptors widely distributed in the skin.

Spinal projections ascend to the brain from the rostral spinal cord in two pathways: one in the dorsal funiculus and a second, called the spinal lemniscus, in the lateral and ventral funiculi of the spinal cord (Ronan and Northcutt, 1990). The dorsal funicular projection extends along the midline of the rostral spinal cord. As in lampreys and a number of fishes, amphibians and reptiles, this projection continues well into the medulla. In eptatretid hagfishes, spinal efferents in the dorsal funicular pathway continue along the midline of the caudal medulla before arching laterally over the octavolateralis area and the sensory trigeminal nucleus in the rostral medulla. Some fibres ascend into the horns at the forward end of the medulla. Many ascending fibres in the spinal lemniscus arise in spinal neurons, mostly contralateral to their termination in the brain. The course of the spinal lemniscus proceeds through the ventral extent of the dorsolateral reticular formation in the medulla and continues into the central grey and optic tectum of the midbrain.

Clearly, ascending spinal efferents, including at least one contingent of primary afferents in the dorsal funiculus and perhaps a second in the sensory trigeminal nucleus, are in position to synapse with brain neurons in the reticular formation, motor column of the medulla and the midbrain. To date, studies of central pattern generators in the spinal cords of jawless craniates have focused on the neural networks responsible for swimming in lampreys. Swimming movements induced by illumination of the skin in hagfishes, which occur even after animals have been beheaded (Newth and Ross, 1955), indicate that hagfishes possess networks of spinal neurons capable of initiating and maintaining coordinated locomotory movements.

29.4 CONCLUSIONS

Numerous characters distinguish the central nervous systems of hagfishes. Among the notable characters are the much reduced ventricular system, the relatively large size of the forebrain, the lamination within the pallium, the complexity of the central prosencephalic nucleus, the large habenular complex, the absence of pineal organs and the questionable presence of a cerebellum, the striking size of secondary olfactory and primary trigeminal projections, the existence of retinofugal and retinopetal fibres in visually limited animals, and a long, flattened spinal cord. Some morphological, connectional and histochemical characters are shared with other craniates; some characters appear to be unique to hagfishes; however, a great many features remain to be explained before they can be used to test to elucidate the organization and workings of myxinoid nervous systems. Continued study promises not only a better understanding of the intriguing brains of hagfishes, but also offers the prospect of important insights into the evolution of all craniate brains.

ACKNOWLEDGEMENTS

The authors express their appreciation to the organizers for the opportunity to summarize information concerning the central nervous system of hagfishes and thank C.B. Braun and H. Wicht for their comments on the manuscript. We are also grateful to M.S. Northcutt for her helpful editorial assistance. This work was supported in part by grants NS 24669 and NS 24869 to RGN.

REFERENCES

Adam, H. (1963) Brain, ventricles, ependyma and related structures, in *The Biology of Myxine* (eds A. Brodal and R. Fänge) Universitetsforlaget, Oslo, pp. 137–49.

Allen, B.M. (1917) The eye of *Bdellostoma stouti*. *Anatomisher Anzeiger*, **26**, 208–11.

Amemiya, F. (1983) Afferent connections of the tectum mesencephali in the hagfish, *Eptatretus burgeri*: an HRP study. *Journal für Hirnforschung*, **24**, 255–63.

Amemiya, F., Kishida, R., Goris, R.C., Onishi, H. and Kusunoki, T. (1985) Primary vestibular projections in the hagfish, *Eptatretus burgeri*. *Brain Research*, **337**, 73–9.

Amemiya, F. and Northcutt, R.G. (1996) Afferent and efferent connections of the central prosencephalic nucleus in the Pacific hagfish. *Brain, Behavior and Evolution*, **47**, 149–55.

Bone, Q. (1963) The central nervous system, in *The Biology of Myxine* (eds A. Brodal and R. Fänge) Universitetsforlaget, Oslo, p. 50–91.

Braun, C.B., Wicht, H. and Northcutt, R.G. (1995) Gonadotropin-releasing hormone systems in the brain of the Pacific hagfish, *Eptatretus stoutii*. *Journal of Comparative Neurology*, **353**, 464–76.

Conel, J.L. (1931) The development of the brain of *Bdellostoma stoutii*. II. Internal growth changes. *Journal of Comparative Neurology*, **52**, 365–501.

Crosby, E.C. and Schnitzlein, H.N. (1974) The comparative anatomy of the telencephalon of the hagfish, *Myxine glutinosa*. *Journal für Hirnforschung*, **15**, 211–36.

Edinger, L. (1906) Über das Gehirn von *Myxine glutinosa*. Abhandlungen der königlichen preussischen Akademie der Wissenschaften aus dem Jahre 1906. Verlag der königlichen Akademie der Wissenschaften, Berlin, pp. 1–36.

Fernholm, B. and Holmberg, K. (1975) The eyes in three genera of hagfish (*Eptatretus, Paramyxine* and *Myxine*): a case of degenerative evolution. *Vision Research*, **15**, 253–9.

Goodrich, E.S. (1937) On the spinal nerves of the Myxinoidea. *Quarterly Journal of Microscopical Science*, **80**, 153–8.

Holmberg, K. (1970) The hagfish retina: fine structure of retinal cells in *Myxine glutinosa*, L., with special reference to receptor and epithelial cells. *Zeitschrift für Zellforschung*, **111**, 519–38.

Holmgren, N. (1919) Zur Anatomie des Gehirns von *Myxine*. *Kungliga Svenska Vetenskapsakademiens Handlingar*, **60**, 1–96.

Holmgren, N. (1946) On two embryos of *Myxine glutinosa*. *Acta Zoologica*, **27**, 1–90.

Iwahori, N., Nakamura, K. and Tsuda, A. (1996) Neuronal organization of the optic tectum in the hagfish, *Eptatretus burgeri*. *Anatomy and Embryology*, **193**, 271–9.

Jansen, J. (1930) The brain of *Myxine glutinosa*. *Journal of Comparative Neurology*, **49**, 359–507.

Kadota, T. (1991) Distribution of 5-HT (serotonin)

immunoreactivity in the central nervous system of the inshore hagfish, *Eptatretus burgeri* (Cyclostomata). *Cell and Tissue Research*, **266**, 107–16.

Kishida, R., Goris, R.C., Nishizawa, H., Koyama, H., Kadota, T. and Amemiya, F. (1987) Primary neurons of the lateral line nerves and their central projections in hagfishes. *Journal of Comparative Neurology*, **264**, 303–10.

Kusunoki, T., Kadota, T. and Kishida, R. (1981) Chemoarchitectonics of the forebrain of the hagfish, *Eptatretus burgeri*. *Journal für Hirnforschung*, **22**, 285–98.

Kusunoki, T., Kadota, T. and Kishida, R. (1982) Chemoarchitectonics of the brain stem of the hagfish, *Eptatretus burgeri*, with special reference to the primordial cerebellum. *Journal für Hirnforschung*, **23**, 109–19.

Kusunoki, T. and Amemiya, F. (1983) Retinal projections in the hagfish, *Eptatretus burgeri*. *Brain Research*, **262**, 295–8.

Larsell, O. (1947) The cerebellum of myxinoids and petromyzontids including developmental stages in the lampreys. *Journal of Comparative Neurology*, **86**, 395–445.

Lindström, T. (1949) On the cranial nerves of the cyclostomes with special reference to *n. trigeminus*. *Acta Zoologica*, **30**, 315–458.

Matsuda, H., Goris, R.C. and Kishida, R. (1991) Afferent and efferent projections of the glossopharyngeal-vagal nerve in the hagfish. *Journal of Comparative Neurology*, **311**, 520–30.

Nansen, F. (1886) The structure and combination of the histological elements of the central nervous system. *Bergens Museums Aarsberetning*, pp. 29–215.

Newth, D.R. and Ross, D.M. (1955) On the reaction to light of *Myxine glutinosa* L. *Journal of Experimental Biology*, **32**, 4–21.

Nishizawa, H., Kishida, R., Kadota, T. and Goris, R.C. (1988) Somatotopic organization of the primary sensory trigeminal neurons in the hagfish, *Eptatretus burgeri*. *Journal of Comparative Neurology*, **267**, 281–95.

Northcutt, R.G. (1981) Evolution of the telencephalon in nonmammals. *Annual Review of Neuroscience*, **4**, 301–50.

Northcutt, R.G. and Puzdrowski, R.L. (1988) Projections of the olfactory bulb and nervus terminalis in the silver lamprey. *Brain, Behavior and Evolution*, **32**, 96–107.

Northcutt, R.G. and Ronan, M. (1992) Afferent and efferent connections of the bullfrog medial pallium. *Brain, Behavior and Evolution*, **40**, 1–16.

Peters, A. (1963) The peripheral nervous system. In *The Biology of Myxine* (eds A. Brodal and R. Fänge) Universitetsforlaget, Oslo, pp. 92–123.

Platel, R. and Delfini, C. (1981) L'encephéphalization chez la *Myxine* (*Myxine glutinosa* L.). Analyse quantifiée des principales subdivisiones encéphaliques. *Cahiers de Biologie Marine*, **22**, 407–30.

Ronan, M. (1988) The sensory trigeminal tract of Pacific hagfish. Primary afferent projections and neurons of the tract nucleus. *Brain, Behavior and Evolution*, **32**, 169–80.

Ronan, M. (1989) Origins of the descending spinal projections in petromyzontid and myxinoid agnathans. *Journal of Comparative Neurology*, **281**, 54–68.

Ronan, M. and Northcutt, R.G. (1990) Projections ascending from the spinal cord to the brain in petromyzontid and myxinoid agnathans. *Journal of Comparative Neurology*, **291**, 491–508.

Rovainen, C. (1967) Physiological and anatomical studies on large neurons of the central nervous system of the sea lamprey (*Petromyzon marinus*). I. Muller and Mauthner cells. *Journal of Neurophysiology*, **30**, 1000–23.

Steven, D.M. (1955) Experiments on the light sense of the hag, *Myxine glutinosa* L. *Journal of Experimental Biology*, **32**, 22–38.

Stockard, C.R. (1906) The development of the mouth and gills in *Bdellostoma stoutii*. *American Journal of Anatomy*, **5**, 481–517.

Wächtler, K. (1975) The distribution of acetylcholinesterase in the cyclostome brain. II. *Myxine glutinosa*. *Cell and Tissue Research*, **159**, 109–20.

Wicht, H. and Nieuwenhuys, R. (1997) Hagfishes, in *The Central Nervous System of Vertebrates* (eds R. Nieuwenhuys, H.J. ten Donkelaar and C. Nicolson), Springer-Verlag, Heidelberg.

Wicht, H. and Northcutt, R.G. (1990) Retinofugal and retinopetal projections in the Pacific hagfish, *Eptatretus stouti*. *Brain, Behavior and Evolution*, **36**, 315–28.

Wicht, H. and Northcutt, R.G. (1992a) The forebrain of the Pacific hagfish: a cladistic reconstruction of the ancestral craniate forebrain. *Brain, Behavior and Evolution*, **40**, 25–64.

Wicht, H. and Northcutt, R.G. (1992b) FMRFamide-like immunoactivity in the brain of the Pacific hagfish, *Eptatretus stoutii* (Myxinoidea). *Cell and Tissue Research*, **270**, 443–9.

Wicht, H. and Northcutt, R.G. (1993) Secondary

olfactory projections and pallial topography in the Pacific hagfish, *Eptatretus stoutii*. *Journal of Comparative Neurology*, **337**, 529–42.

Wicht, H. and Northcutt, R.G. (1994) An immunohistochemical study of the telencephalon and the diencephalon in a myxinoid jawless fish, the Pacific hagfish, *Eptatretus stoutii*. *Brain, Behavior and Evolution*, **43**, 140–67.

Worthington, J. (1905) The descriptive anatomy of the brain and cranial nerves of *Bdellostoma dombeyi*. *Quarterly Journal of Microscopical Science*, **49**, 137–81.

30
THE AUTONOMIC NERVOUS SYSTEM AND CHROMAFFIN TISSUE IN HAGFISHES

Stefan Nilsson and Susanne Holmgren

SUMMARY

The autonomic nervous system of cyclostomes appears rudimentary compared to that of the gnathostomes, and autonomic neurones can sometimes be hard to distinguish from sensory neurones. Vagal pathways are present in both lampreys and hagfishes; the left and right vagus (X) unite to form a single ganglionated nerve along the dorsal side of the intestine. Fibres run to the gut of hagfishes, but the function of this innervation is unclear. In fact, the only organ in *Myxine* for which a distinct vagal innervation has been shown is the gallbladder.

Sympathetic chains are absent in the cyclostomes, although scattered neurone clusters occur along the cardinal veins in the abdominal cavity of lampreys. A unique feature is the subcutaneous ganglionated nerve plexus where spinal, possibly autonomic, fibres may be involved in the control of effectors in the skin. Another remarkable feature of the systemic and portal hearts of cyclostomes is the presence of specialized endocardial cells that store catecholamines. Contrary to the hearts of other fishes, including that of lampreys, the hagfish heart receives no extrinsic innervation, although intracardiac neurones have been described.

The hagfish autonomic nervous system is not an 'early template' of a control system, but possibly a rudiment of a more elaborate system in the earliest vertebrates. The structure of the autonomic nervous system of hagfishes is not well known, and there is a flagrant lack of knowledge about the autonomic nerve functions. Thus, further physiological experiments are essential to our understanding of the autonomic nerve function in the hagfishes.

30.1 INTRODUCTION

Hagfishes may in many respects be regarded as the most ancient of the extant vertebrates, and the development of their autonomic nervous system confirms this view. With the exception of a relatively well-developed autonomic innervation of the gallbladder, the autonomic nervous system, and its functions, are difficult to identify. The autonomic nervous system of the hagfish has been extensively reviewed by Fänge *et al.* (1963a), and further information about the cyclostome autonomic nervous system can be found in reviews by Nicol (1952), Campbell (1970), Pick (1970), Nilsson (1983) and Nilsson and Holmgren (1994).

30.2 ANATOMICAL CONSIDERATIONS

The autonomic nervous system of cyclostomes is poorly developed compared to

The Biology of Hagfishes. Edited by Jørgen Mørup Jørgensen, Jens Peter Lomholt, Roy E. Weber and Hans Malte. Published in 1998 by Chapman & Hall, London. ISBN 0 412 78530 7.

that of the gnathostomes, and knowledge of its function is fragmentary. Neurones that are part of an autonomic nervous system from a functional point of view, sometimes cannot be distinguished from sensory neurones.

30.2.1 Cranial autonomic (parasympathetic) pathways

In mammals, nerve fibres belonging to the autonomic nervous system run in cranial nerves III (oculomotor), VII (facial), IX (glossopharyngeal) and X (vagus), but in teleosts the outflow is restricted to nerve III and X (see Nilsson, 1983; Gibbins, 1994). In lampreys, autonomic fibres that are probably involved in the control of the vasculature of the gills and brain have been postulated in both the facial (VII) and the glossopharyngeal (IX), although there are no distinct ganglia in these nerves (Johnston, 1908; Hirt, 1934; Lindström, 1949; Johnels, 1956; Pick, 1970; Iijima and Wasano, 1980; Nakao, 1981). In hagfishes, the presence of a separate glossopharyngeal (IX) nerve has been debated (see Peters, 1963; Matsuda *et al.*, 1991).

The eyes of lampreys are small, and a distinct ciliary ganglion is absent although scattered autonomic neurones have been described in the oculomotor nerve (Tretjakoff, 1927). In Myxinoids, the eye is degenerate and there are neither autonomic oculomotor (III) pathways, nor a ciliary ganglion.

In both lampreys and hagfishes, the left and right vagus (X) unite to form a plexus near the *constrictor cardiae* muscle and continue as a single ganglionated nerve, the *nervus intestinalis impar*, which runs along the dorsal side of the intestine. Autonomic fibres to the viscera emerge from the plexus and along the length of the *nervus intestinalis impar*. The targets and function of the vagal fibres are still somewhat unclear. In lampreys, vagal fibres control the heart (Zwaardemaker, 1924; Augustinsson *et al.*, 1956; Nakao *et al.*, 1981). An extrinsic cardiac innervation is absent in myxinoids (Greene, 1902; Carlson, 1906; Augustinsson *et al.*, 1956), although ganglion cells have been described in the vicinity of the heart of *Eptatretus stoutii* (Hirsch *et al.*, 1964; see later).

Vagal fibres also run to the gut, although the function of this innervation is not well understood (Nicol, 1952; Campbell, 1970).

30.2.2 Spinal automatic (sympathetic) pathways

Segmented sympathetic chains are absent in both lampreys and hagfishes; however, scattered neurone clusters that may be regarded as autonomic ganglia occur along the cardinal veins in the abdominal cavity of lampreys (Johnels, 1956). The spinal roots are incompletely united in myxinoids and not united at all in lampetroids, and continue as separate nerves. Spinal autonomic pathways in lampetroids may run in both dorsal and ventral spinal nerves, although the anatomical distinction from sensory nerves is, again, uncertain (Nicol, 1952; Johnels, 1956; Romer, 1962; Campbell, 1970). A unique feature of the cyclostomes is a subcutaneous ganglionated nerve plexus, where spinal fibres that may be regarded as autonomic could be involved in the control of mucus secretion, cutaneous blood flow or colour change (Bone, 1963; Campbell, 1970).

Leont'eva (1966) demonstrated catecholamine-containing fibres in blood vessels of the lamprey, suggesting a functional control of the vasculature reminiscent of the sympathetic (adrenergic) control found in other vertebrates.

30.2.3 Enteric nervous system

The enteric nervous system of vertebrates is by definition the part of the autonomic nervous system which is confined to the gut, with nerve cell bodies situated within or in close vicinity to the gut wall. The nerves project to the different parts of the gut wall, controlling motility, secretion and blood flow.

The neurones may also interconnect through networks of fibres forming plexuses such as the myenteric plexus and the submucous plexus (see Furness and Costa, 1987). A well-developed enteric nervous system is present in all investigated fishes, amphibians, reptiles, birds and mammals, but there are no descriptions of proper enteric nerve nets in cyclostomes. In *Myxine*, several nerve cells are present along the course of the vagus nerve (Brandt, 1922), and in *Eptatretus* we have similarly found occasional serotonergic ganglion cells in the bundles of serotonergic fibres along and within the intestinal wall (see below 30.3.4.c; Holmgren and Nilsson, unpublished). Whether these cells may be considered part of an enteric nervous system is, however, uncertain. It has been suggested that *Myxine* has remained at a stage of early migration of neuronal cells in the phylogenetic development of the autonomic nervous system (Brandt, 1922).

30.2.4 Chromaffin tissue

The definition of chromaffin cells as outlined by Coupland (1965, 1972) states that these cells (1) develop from neuroectoderm, (2) are innervated by preganglionic 'sympathetic' fibres, (3) synthesize and release catecholamines and (4) store enough catecholamines to produce a chromaffin reaction. In cyclostomes, chromaffin cells have been described in the walls of arteries and veins of both lampreys and hagfishes (Giacomini, 1902a, b; Gaskell, 1912). A remarkable feature of the cyclostome systemic heart is the presence of specialized cells that store catecholamines in the endocardium (Figure 30.1A and Table 30.1). However, there is some confusion as to the use of the term 'chromaffin' to describe catecholamine-storing cells in hagfishes. In fact the true chromaffin reaction, i.e. the formation of adrenochromes as a result of oxidation with dichromate solutions (Kohn, 1902) may be weak, especially when applied to formalin-treated material (Augustinsson *et al.*, 1956; Giacomini, 1902a, b). Furthermore, the hagfish heart receives no extrinsic innervation (see later), so an innervation of the cardiac cells by preganglionic fibres is inconceivable. Using the Falck–Hillarp technique to visualize monoamines, it has, on the other hand, been possible to demonstrate amine stores, thus confirming the 'chromaffin-like' properties of the specialized intracardiac cells (Figure 30.1A–B; Dahl *et al.*, 1971; Shibata and Yamamoto, 1976). The term 'chromaffin' is used here simply to describe non-neuronal, catecholamine-storing cells. The mechanisms of catecholamine release from chromaffin cells in hagfishes are discussed by Bernier and Perry (this volume).

The specialized endocardial chromaffin cells are also present in the myxinoid portal heart (Figure 30.1B and Table 30.1). A similar feature occurs in all genera of lungfish (*Protopterus*, Abrahamsson *et al.*, 1979; Scheuermann, 1970; *Lepidosiren*, Axelsson *et al.*, 1989; *Neoceratodus*, Fritsche *et al.*, 1993), but is less obvious in other vertebrates. In *Petromyzon*, a substantial increase in the plasma levels of catecholamines occurs after 'stress': adrenaline increases from 0.19 to 1.08 μg ml^{-1} (1.0 to 5.9 μmol $litre^{-1}$) plasma and noradrenaline from 0.29 to 1.36 μg ml^{-1} (1.7 to 8.0 μmol $litre^{-1}$) plasma (Mazeaud, 1971). In experiments with *Myxine* exposed to 30 min severe hypoxia, Perry *et al.* (1993) demonstrated an increase in the plasma concentration of noradrenaline from 3.2 to 10.8 nmol $litre^{-1}$, while the adrenaline concentration remained virtually unchanged. However, little is known about the mechanism of control of the catecholamine release from chromaffin stores in the cyclostomes.

30.3 FUNCTIONS OF THE HAGFISH AUTONOMIC NERVOUS SYSTEM

30.3.1 The systemic heart

The systemic heart of hagfishes, sometimes referred to as the branchial heart (*cor venosum*

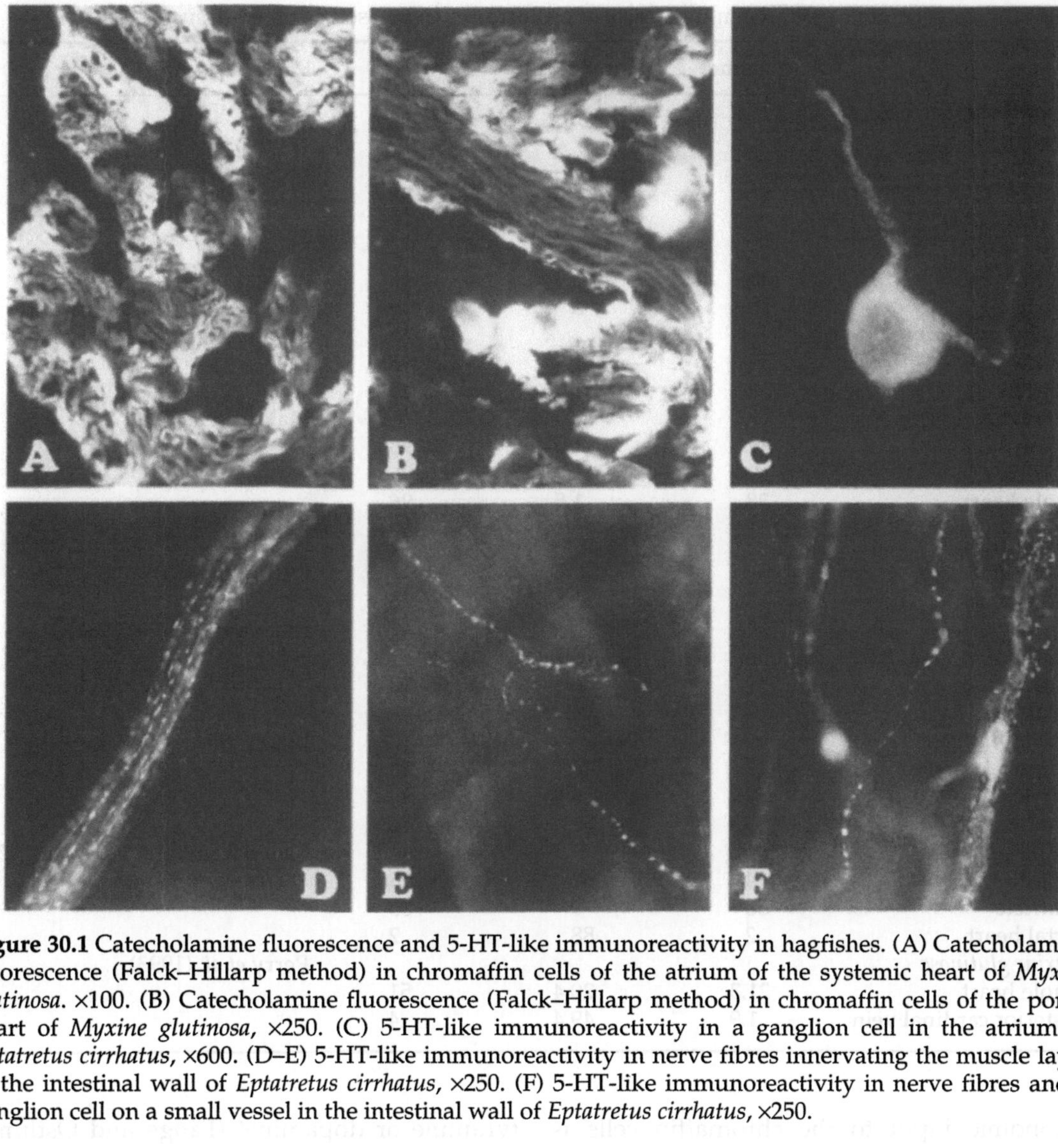

Figure 30.1 Catecholamine fluorescence and 5-HT-like immunoreactivity in hagfishes. (A) Catecholamine fluorescence (Falck–Hillarp method) in chromaffin cells of the atrium of the systemic heart of *Myxine glutinosa*. ×100. (B) Catecholamine fluorescence (Falck–Hillarp method) in chromaffin cells of the portal heart of *Myxine glutinosa*, ×250. (C) 5-HT-like immunoreactivity in a ganglion cell in the atrium of *Eptatretus cirrhatus*, ×600. (D–E) 5-HT-like immunoreactivity in nerve fibres innervating the muscle layer of the intestinal wall of *Eptatretus cirrhatus*, ×250. (F) 5-HT-like immunoreactivity in nerve fibres and a ganglion cell on a small vessel in the intestinal wall of *Eptatretus cirrhatus*, ×250.

branchiale; Bloom *et al.*, 1963), consists of a sinus venosus, atrium and ventricle resembling the heart of gnathostome fish. Contrary to the hearts of other fishes and, indeed to that of lampreys, the hagfish heart receives no extrinsic innervation, although intracardiac nerve cells have been described (Figure 30.1C; Greene, 1902; Carlson, 1904; Augustinsson *et al.*, 1956; Hirsch *et al.*, 1964; Caravita and Coscia, 1966; Beringer and Hadek, 1973). Thus, the control of heart rate and stroke volume depends on mechanisms other than the autonomic nervous system.

The hagfish heart stores large quantities of catecholamines (Table 30.1), but the mechanisms that control the release of the stored amines are poorly understood. The cholinergic agonist carbachol produces release of catecholamines, but since the heart lacks an innervation, the 'normal' pattern of preganglionic

Table 30.1 Catecholamine storage in the hearts of cyclostomes expressed as μg g^{-1} tissue

Species	Adrenaline	Noradrenaline	%Adr	References
Lampetroids				
Lampetra fluviatilis				Stabrovskii (1967)
Atrium	127.1	16	89	
Ventricle	81	11.6	87	
Blood vessels	1.2	5	19	
Lampetra fluviatilis				Bloom *et al.* (1962)
Atrium	130	6.3	95	
Ventricle	28	0	100	
Petromyzon marinus				Bloom *et al.* (1962)
Atrium	51	1.95	96	
Ventricle	9.8	1.33	88	
Petromyzon marinus				Mazeaud (1972)
Atrium	89.4	1.5	98	
Ventricle	16.7	0.7	96	
Whole heart	38	1.5	96	
Myxinoids				
Myxine glutinosa				Östlund (1954)
Whole heart	5	0.8	86	
Myxine glutinosa				Euler and Fänge (1961)
Atrium	8.1	18	31	
Ventricle	59	6.5	90	
Portal heart	3.1	58	5	
Myxine glutinosa				Bloom *et al.* (1962)
Atrium	13	47	22	
Ventricle	49	6.2	89	
Portal heart	3.4	53	6	
Myxine glutinosa				Lagerstrand and Nilsson (1973)
Atrium	14	62	18	
Ventricle	50	9	85	
Portal heart	2	88	2	
Myxine glutinosa				Perry *et al.* (1993)
Whole heart	21.3	20.4	51	
Posterior cardinal vein	1.9	49.4	4	

autonomic input to the chromaffin cells is unfeasible. Catecholamine release can be generated by injection of pituitary extract (Perry *et al.*, 1993), and the possible control mechanisms also include an influence by substances such as adenosine which has been shown to affect the release (Bernier *et al.*, 1996; Bernier and Perry, this volume), and also affects cardiac function (Axelsson *et al.*, 1990).

In experiments with the isolated heart of *Myxine glutinosa*, '. . . no distinct responses are obtained . . . with adrenaline, noradrenaline, tyramine or dopamine' (Fänge and Östlund, 1954). Indeed, the isolated heart of *Myxine* seems remarkably insensitive to transmitters or hormones that produce marked effects on the hearts of other vertebrates. Similarly, acetylcholine produced no inhibitory action on the heart of *Myxine* (Fänge and Östlund, 1954). However, following pretreatment of the heart with dihydroergotomine, an alkaloid derivative from the fungus, *Claviceps purpurea*, and better known for its α-adrenoceptor antagonistic properties, the isolated

Myxine heart becomes sensitive to both adrenaline and noradrenaline (Fänge and Östlund, 1954). A similar response can be seen after pretreatment with reserpine, a drug that acts by preventing the granular uptake mechanism in adrenergic cells and thus causes depletion of catecholamine stores. Reserpine caused pronounced bradycardia in the isolated perfused heart of *Myxine*, an effect that could be reversed by addition of noradrenaline or adrenaline (Bloom *et al.*, 1961).

The function of an adrenergic cardiac tonus, conceivably dependent on the intracardiac stores of catecholamines, is further illustrated by the very severe bradycardia caused by β-adrenoceptor antagonists such as sotalol (Axelsson *et al.*, 1990; Johnsson and Axelsson, 1996).

Contrary to the equivocal effects of drugs *in vitro*, *in vivo* experiments show pronounced cardiovascular effects of both adrenaline (Figure 30.2) and acetylcholine. Adrenaline injections produce tachycardia and increased stroke volume in both *Eptatretus* and *Myxine*. The resulting increase in cardiac output generates an increase of both the ventral and dorsal aortic blood pressure, despite a reduced vascular resistance in both systemic and branchial vasculature (Axelsson *et al.*, 1990; Forster *et al.*, 1992). A dose of 10 nmol kg^{-1} of acetylcholine generated a slight increase in heart rate and cardiac output in *Myxine*, while significant cardiac effects of a higher dose (100 nmol kg^{-1}) were lacking (Axelsson *et al.*, 1990).

Although conclusive evidence regarding the cardiac control in hagfishes is still lacking, it seems clear that the intracardiac chromaffin cells provide an adrenergic tonus that is necessary for the normal function of the heart. The chromaffin cells are not innervated, and a control of the release via cholinergic nerves is therefore out of the question. The control may instead occur either via circulating factors, via intrinsic neurones, or locally due to direct chemical and/or mechanical effects on the amine-storing cells. One potent stimulus of catecholamine release is hypoxia, even to the extent that the levels of circulating amines is affected. A challenging idea would be to regard the chromaffin cells as a kind of original chemoreceptors (oxygen receptors), directly affecting the surrounding cardiac muscle cells via paracrine mechanisms.

30.3.2 The portal heart

Similar to the systemic heart, the portal heart of *Myxine* stores catecholamines in specialized cells (Figure 30.1B). Direct electrical stimulation of the portal heart produces acceleration, but when the strength of the stimulation is increased there is an increase of the tonus of the preparation (Carlson, 1904). Mechanical stimulation increases the rate of the *Myxine* portal heart (Johansen, 1960; Fänge *et al.*, 1963b).

In a recent study of *Eptatretus cirrhatus*, Johnsson *et al.* (1996) demonstrated a positive chronotropic effect of input pressure in the *in situ* perfused portal heart. Bolus injections of adrenaline (0.1 ml at 10^{-3} M) produced a small increase in heart rate (from 42 to 45.5 beats min^{-1}). The β-adrenoceptor antagonist sotalol, which could be expected to affect a possible adrenergic tonic stimulation of heart, produced a significant decrease of the heart rate.

Similar observations were made in *Myxine glutinosa*, where the β-adrenoceptor antagonist sotalol reduced portal heart rate (from 19.7 to 15.5 beats min^{-1}). Adrenaline by itself produced no change in the heart rate, but after sotalol treatment higher concentrations of adrenaline (10^{-6}–10^{-5} M) produced a small acceleration of both the systemic and the portal heart (Johnsson and Axelsson, 1996). The results demonstrate an endogenous adrenergic tonus affecting the portal heart, not much different from that concluded for the systemic heart.

30.3.3 The vasculature

(a) The branchial vasculature

Acetylcholine produces a rapid and marked increase in the branchial vascular resistance

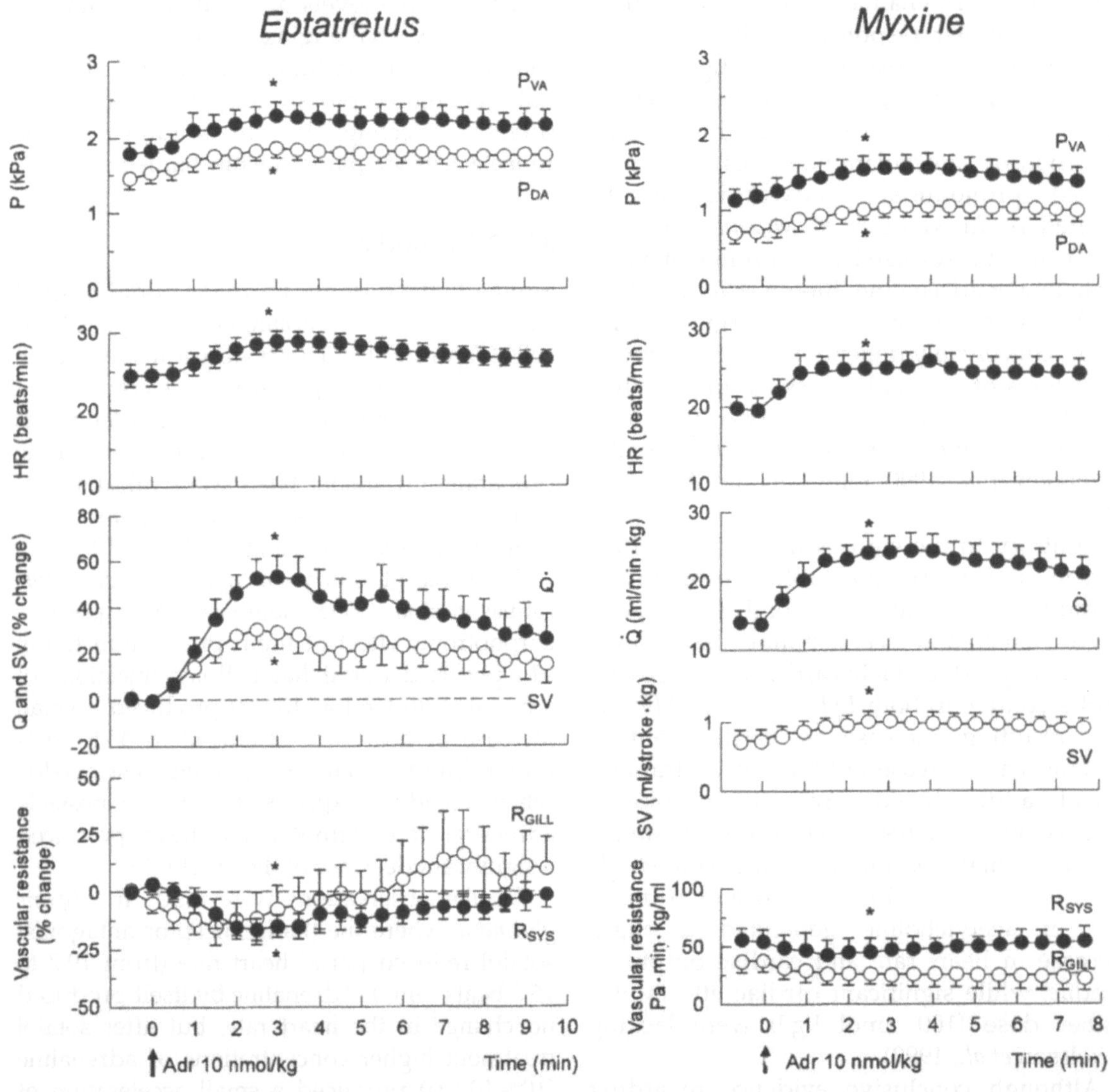

Figure 30.2 Effects of adrenaline injection (10 nmol kg^{-1}) *in vivo* on cardiovascular parameters in *Eptatretus cirrhatus* (left) and *Myxine glutinosa* (right). The panels show simultaneous recordings of arterial blood pressures (P_{VA}, ventral aortic blood presure; P_{DA}, dorsal aortic blood pressure), heart rate (HR, beats min^{-1}), cardiac output (*Q*, percent change (*Eptatretus*) or ml min^{-1} kg^{-1} (*Myxine*)), stroke volume (SV, percent change (*Eptatretus*) or ml stroke^{-1} kg^{-1} (*Myxine*)), and calculated vascular resistances in the branchial and systemic vasculature, respectively (R_{GILL} and R_{SYS}, percent change (*Eptatretus*) or Pa min kg ml^{-1} (*Myxine*)). Note marked stimulation of the heart (both HR and SV) in both species, which is strong enough to produce an increase in both P_{VA} and P_{DA} despite the reduced vasculare resistances. Redrawn from Forster *et al.* (1992) (*Eptatretus*) and Axelsson *et al.* (1990) (*Myxine*) and reproduced with permission from Elsevier Science-NL, Sara Burgerhartstraat 25, 1055 KV Amsterdam, The Netherlands.

of *Myxine glutinosa*, *Eptatretus cirrhatus* and *Polistotrema* (= *Eptatretus*) *stoutii*, and both α- and β-adrenoceptor-mediated responses to catecholamines have been demonstrated in *Myxine* (Reite, 1969; Axelsson *et al.*, 1990). In perfusion experiments with *Eptatretus*, Sundin *et al.* (1994) demonstrated a potential for a differential distribution of blood between an arterio-arterial and an arterio-venous pathway. The role of an adrenergic control of the branchial vasculature *in vivo* is not known (see earlier, section 2.1).

(b) The systemic vasculature

In lampreys, nerve fibres from the spinal autonomic system appear to innervate blood vessels (Tretjakoff, 1927; Johnels, 1956), and some of these fibres are known to be adrenergic (Leont'eva, 1966; Govyrin, 1977). Little is known about the possible vascular innervation in hagfishes, but both catecholamines and acetylcholine induce elevated blood pressure *in vivo* (Figure 30.2). In perfused systemic vascular beds of *Myxine*, both adrenaline and noradrenaline produced increased vascular resistance (Reite, 1969), but *in vivo* the catecholamine effect is due to an increase in cardiac output rather than an increase of the systemic vascular resistance (Axelsson *et al.*, 1990; Forster *et al.*, 1992).

30.3.4 The alimentary canal

Hagfishes have no stomach. The intestine forms a straight tube along the body. There is a thin outer smooth muscle layer of the intestine, the intestinal *muscularis serosa*, which presumably is involved in the propulsion of food, although this is considerably supported by movements of the body (Patterson and Fair, 1933). The liver and gallbladder are large, and the muscle layer of the gallbladder is comparatively well developed. As described above (section 2.1), the intestine is innervated by the thick *ramus intestinalis impar*, formed from the fused intestinal branches of the right and the left vagus (Brandt, 1922). The gallbladder receives an extensive vagal innervation via the two hepatic nerves and the *plexus hepaticus* (Fänge and Johnels, 1958).

The intestine of most vertebrates is extensively innervated by extrinsic and intrinsic (enteric) pathways comprising, e.g., cholinergic, adrenergic, serotonergic, nitrergic and several types of peptidergic nerves. The same types of nerve fibres have been looked for in hagfishes and, in summary, only cholinergic nerves (as indicated by the presence of the cholinergic enzymes acetylcholine esterase and choline acetyltransferase) and serotonergic nerves (Figure 30.1D–F) have so far been demonstrated. There are few studies made on the effects of putative neutrotransmitters on the motility of the gut, and no studies so far on the influence on secretory mechanisms, which may be the most important function of hagfish gut.

(a) Cholinergic mechanisms

Adam (1965), by using the cholinesterase method of Gomori, stained fine nerve endings forming an intestinal plexus in *Myxine*. These results were confirmed by Hallbäck (1973) using the staining method of Koelle as modified by Naik (1963); the acetylcholinesterase activity was localized to the intestinal branch of the vagus nerve, to thin nerve fibres appearing to branch from the intestinal branch to the smooth muscles of the intestine, and to the muscularis layer itself. However, the specificity of cholinesterase as a marker of cholinergic neurones has been questioned, and the acetylcholine synthesizing enzyme choline acetyltransferase may be a more reliable marker. Low levels of this enzyme are present in extracts of the vagus nerve and the intestine, suggesting a sparse cholinergic innervation of the gut (Holmgren and Fänge, 1981).

Cholinergic agonists contract smooth muscle of the *Myxine* intestine, by acting on receptors which appear to be of the muscarinic type, since the effect is blocked by atropine but not by the nicotinic receptor antagonist mecamylamine (Figure 30.4; Fänge, 1948; Holmgren and Fänge, 1981). This would be in agreement with the cholinergic effect on the gastro-intestinal canal of most other vertebrates. However, the pharmacological properties of the blockade of the receptor are somewhat different from the classical theoretical case, indicating slightly different properties of the cyclostome cholinoceptor.

(b) Adrenergic mechanisms

Small amounts of catecholamines, predominantly noradrenaline, are present in the prebranchial part of the vagus (Euler and Fänge, 1961), but it has not been conclusively determined whether these catecholamines belong to neurones innervating the gills, the intestine or the gallbladder, and no further evidence of an adrenergic innervation of the intestine has been reported so far. Adrenergic effects *in vivo* may as well be due to circulating catecholamines (see above) as to release from adrenergic neurones.

Fänge (1948) obtained an inhibitory effect on the tonus of intestinal smooth muscle in response to adrenaline, while Holmgren and Fänge (1981) report inconsistent effects ranging from a pronounced relaxation, over mixed excitatory and inhibitory effects, to a pronounced contraction. It was concluded that two different populations of adrenergic receptors may be present, one mediating the contraction, the other mediating relaxation. This would be in contrast to most other vertebrates, where adrenergic stimulation causes a relaxation of intestinal smooth muscle (Burnstock, 1969; Nilsson, 1983).

(c) Serotonergic nerves

The presence of serotonergic (5-HT-containing) neurones in the intestine of *Myxine* was established by Goodrich *et al.* (1980), using autoradiography and measurements of tissue concentrations of 5-HT. These results have later been confirmed by the use of immunohistochemistry (Figure 30.1D–F; Holmgren, unpublished). There is a dense innervation by 5-HT-containing nerve fibres of the whole intestine, most conspicuous in the submucosa and the mucosa, but also prominent in the muscle layer. Occasional nerve cells are present in the nerve fibre bundles. The same dense serotenergic innervation is found in all other cyclostomes studied: *Eptatretus cirrhatus* (Holmgren and Nilsson, unpublished), *Lampetra* and *Petromyzon* (Holmgren, unpublished). So far, there are no reports on the physiological significance of these nerves.

(d) Nitric oxide-releasing nerves

Numerous recent studies have shown that nitric oxide (NO) may be formed by neurones, and may diffuse into target cells and react with cytoplasmatic receptors, thereby causing an effect on the effector cells. This ability can be attributed to a large population of enteric nerves in fishes, amphibians, reptiles, birds and mammals (see Olsson and Holmgren, 1997, and Olsson, unpublished (reptiles)), but in a study of the intestine of *Myxine* no such neurones could be identifed (Olsson and Karila, 1995).

(e) Peptidergic innervation

Several antisera which have demonstrated the presence of peptides in gut neurones in other vertebrates have been tested in *Myxine glutinosa* (Table 30.2), and in other cyclostomes, and there are remarkably few positive findings of peptidergic neurones in the cyclostome gut. In *Myxine*, antisera raised against bombesin, calcitonin gene-related peptide (GCRP), cholecystokinin (CCK8, CCK39), enkephalin, FMRF, galanin, gastrin releasing peptide (GRP), helospectin, peptide YY (PYY), neuropeptide Y (NPY),

Table 30.2 Results of immunohistochemical surveys of gut, gallbladder and heart of the two cyclostomes *Myxine glutinosa* and *Eptatretus cirrhatus*. The antisera used were raised against bombesin- and gastrin-releasing peptide (BM/GRP), calcitonin gene-related peptide (CGRP), endothelin (ENDO), enkephalin (ENK), peptide phe-met-arg-phe (FMRF), galanin (GAL), gastrin and CCK (G/CCK), neuropeptide Y and peptide YY (NPY/PYY), neurotensin (NT), somatostatin (SOM), substance P (SP), neurokinin A (NKA), vasoactive intestinal polypeptide (VIP), helospectin (HELO), pituitary adenylate-cyclase activating peptide (PACAP), dopamine β-hydroxylase (DBH), tyrosine hydroxylase (TH) or serotonin (SER), o = endocrine cells, + = nerve fibres, nc = nerve cells. The density of fibres and cells were graded on a scale from zero to +++ or ooo, respectively. - denotes absence of immunoreactivity. × = immunoreactive cells reported by Reid *et al.* (1995).

Myxine glutinosa	BM/GRP	CGRP	ENDO	ENK	FMRF	GAL	G/CCK39*	NPY/PYY	NT	SOM	SP	NKA	VIP	HELO	PACAP	DBH	TH	SER
gut: Ant. intestine	ooo	-		-	-	-	-	-	-	-	-	o	o	o	o	-		++
mid. intestine	o			-	o	-	-	-	-	o	-	o	o	-	o	-		+++
rectum 1	ooo	-		-	-	-	-	-	-	o	-	-	o	-	o	-		+++
gallbladder	-			-		-		-		-	-	-	-	-	-	-		+++
heart				-	-	-		-		-	-		-		-	-	×	×

Eptatretus cirrhatus	BM/GRP	CGRP	ENDO	ENK	FMRF	GAL	G/CCK39	NPY/PYY	NT	SOM	SP	NKA	VIP	GELO	PACAP	DBH	SER
gut: Ant. intestine	-	-	-	-	o	-	-	-	-	-	-	o	o	o	oo	-	++(+)
mid. intestine	-	-	-	-	o	-	-	-	-	-	-	o	o	o	oo	-	++
rectum 1	-	-	-	-	-	-	-	-	-	-	-	-	o	(!)	o	-	++(+) nc
gallbladder	-	-	-	-	-	-	-	-	-	-	-	-	-	-	-	-	++
heart			-														
atrium	-	-		-	-	-	-	-	-	-	-	-	-	-	-	-	+++ nc
portal heart	-	-	-	-	-	-	-	-	-	-	-	-	-	-	-	-	-
vent. aorta	-	-		-	-	-	-	-	-	-	-	-	-	-	-	-	-
ventricle	-	-	-	-	-	-	-	-	-	-	-	-	-	-	-	-	-+?

neurotensin, pituitary adenylate cyclase-activating peptide (PACAP), somatostatin, substance P, and vasoactive intestinal polypeptide (VIP) all failed to demonstrate immunoreactive material in neurones (Bjenning and Holmgren, 1988; Jensen and Holmgren, 1991; Holmgren, unpublished). Conlon and Falkmer (1989) report low immunoreactivity levels of somatostatin, CCK/gastrin, substance P and neurokinin A (NKA) in extracts of the *Myxine* gut. The somatostatin immunoreactivity may be attributed to a low number of endocrine cells present in the intestinal mucosa (Holmgren, unpublished), while further studies are needed to establish the origin of the gastrin/CCK- and the SP/NKA-like material. The immunohistochemical study also revealed endocrine cells in the intestinal mucosa of *Myxine*, which showed VIP-, helospectin-, and PACAP-like immunoreactivity (Holmgren, unpublished).

Amongst the neuropeptides, only tachykinins have been tested for an effect on the *Myxine* gut. Substance P, eledoisin, kassinin and physalaemin were added to isolated preparations of the intestinal wall, but were without a recordable effect on tension or motility (Jensen and Holmgren, 1991).

The same immunohistochemical study has been performed on *Eptatretus cirrhatus*, with identical results, except that a few FMFR-immunoreactive endocrine cells were observed (Table 30.2; Holmgren and Nilsson, unpublished). No somatostatin-immunoreactive endocrine cells were found, but an early radioimmunoassay study demonstrated the presence of somatostatin-like material in intestinal extracts, with a proposed function in the control of insulin release (Stewart *et al.*, 1978).

Similar immunohistochemical studies on *Lampetra* and *Petromyzon* have revealed CGRP-immunoreactive nerves and bombesin/GRP-immunoreactive nerves, in addition to the dense innervation by 5-HT-immunoreactive nerves, in the intestine of both species (Holmgren, unpublished). Tachykinins, which commonly occur in neurones in other vertebrates, appear to be present in endocrine cells but had no recordable effect on isolated strip preparations of *Lampetra* gut smooth muscle (Jensen and Holmgren, 1991).

30.3.5 The gallbladder

The gallbladder of *Myxine* is well developed, with a comparatively strong muscle layer. The gallbladder is innervated by postganglionic fibres from the aggregation of vagal ganglion cells in the region of the gallbladder duct (Figure 30.3). When stimulated electrically, the vagal branches to the intestine cause a contraction of the gallbladder. The same response is obtained when smaller branches closer to the gallbladder are stimulated, and the effects could be mimicked by cholinergic agonists (Figure 30.4) (Fänge and Johnels, 1958; Holmgren and Fänge, 1981). Histochemically, the same innervation pattern as of the intestine is observed; 5-HT-containing fibres occur frequently, but no adrenergic, peptidergic or NO-synthesizing neurones are found (Holmgren, unpublished). Pharmacologically, the gallbladder smooth muscle reacts in much the same way as the intestinal muscle, with a clear cholinergic/muscarinic control mechanism, in this case with a high cholinesterase activity (Figure 30.4). The responses to adrenaline and 5-HT are occasional and mainly excitatory (Holmgren and Fänge, 1981).

30.4 CONCLUDING REMARKS

In conclusion, the hagfish autonomic nervous system appears poorly developed compared to other fishes and even to the lampreys. The autonomic nervous system of hagfishes cannot be regarded as an 'early template' of this control system, but rather as a vestigial version of what was perhaps once functioning in the earliest vertebrates.

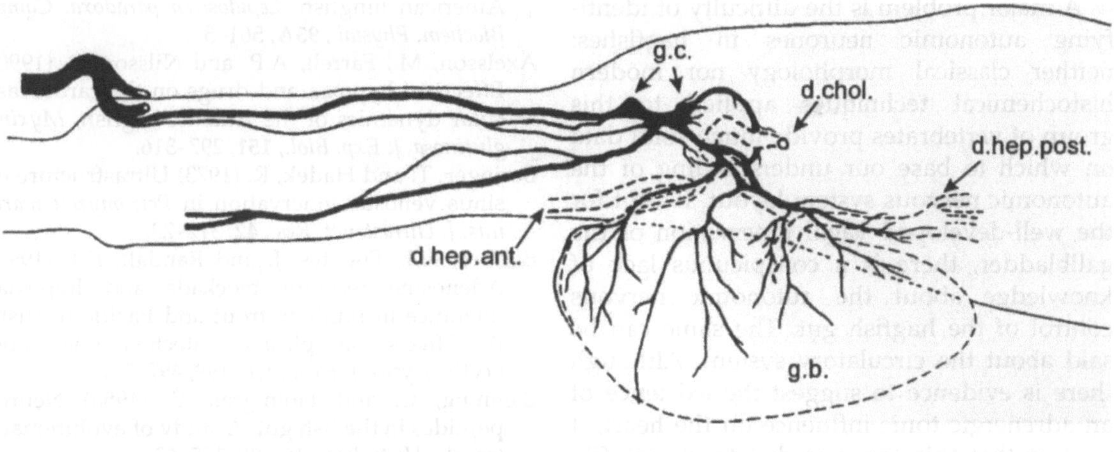

Figure 30.3 Innervation of the gallbladder and hepatic ducts of *Myxine glutinosa*. Ventral view showing the left intestinal branch of the vagus nerve, with the gut in the background. *Abbreviations*: d.chol, *ductus choledochus*; d.hep.ant, *ductus hepaticus anterior*; d.hep.post, *ductus hepaticus posterior*, g.b., gallbladder; g.c., gall bladder ganglia. (From Fänge and Johnels, 1958; reproduced with permission of the Royal Swedish Academy of Sciences.)

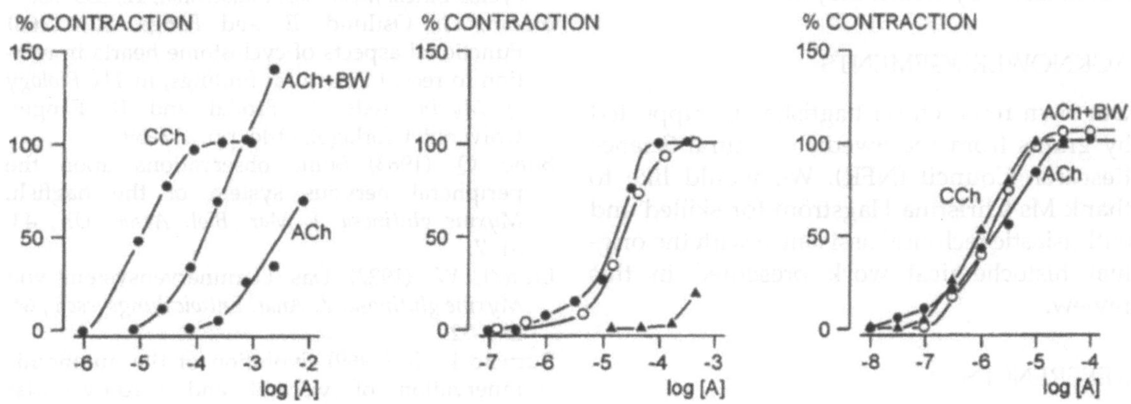

Figure 30.4 Recordings of the effects of cholinergic drugs on the gallbladder and intestine of *Myxine glutinosa*. *Left panel*: Concentration-response curves for carbachol (Ch), acetylcholine (ACh) and acetylcholine after pretreatment with the specific cholinesterase inhibitor BW284 C51 (10^{-6} M; ACh+BW) from one gallbladder preparation. Responses are shown as percent of the maximal contraction force produced by carbachol, and the agonist concentrations (log [A]) refer to the logarithm of the concentration in mol litre^{-1}. Note especially the enhancement of the acetylcholine effect after inhibition of the cholinesterase with BW284 C51. *Middle panel*: Concentration–response curves for carbachol (CCh) on a gallbladder preparation before and after addition of the muscarinic cholinoceptor antagonist atropine • = control; o = atropine 10^{-8} M; Δ = atropine 3×10^{-8} M. *Right panel*: Concentration–response curves for carbachol (CCh), acetylcholine (ACh) and acetylcholine after pretreatment with the specific cholinesterase inhibitor BW284 C51 (10^{-6} M; ACh+BW) from one intestinal muscularis-serosa preparation. Note that the effect of BW284 C51 is small or absent, suggesting that the cholinesterase activity of the preparation is negligible. (From Holmgren and Fänge, 1981; reproduced with permission from Elsevier Science-NL, Sara Burgerhartstraat 25, 1055 KV Amsterdam, The Netherlands.)

A major problem is the difficulty of identifying autonomic neurones in hagfishes: neither classical morphology nor modern histochemical techniques applied to this group of vertebrates provide much solid data on which to base our understanding of the autonomic nervous system layout. Except for the well-developed vagal innervation of the gallbladder, there is a conspicuous lack of knowledge about the autonomic nervous control of the hagfish gut. The same can be said about the circulatory system. Although there is evidence to suggest the existence of an adrenergic tonic influence on the heart, it is clear that this tonus is due to chromaffin cells rather than adrenergic nerves and little is known about the modulation of cardiac activity. In particular, physiological experiments are sorely needed to supply information about the nature of the neurones and the mechanisms by which they act.

ACKNOWLEDGEMENTS

Our own research on hagfishes is supported by grants from the Swedish Natural Science Research Council (NFR). We would like to thank Ms Christina Hagström for skilled and enthusiastic technical assistance with the original histochemical work presented in this review.

REFERENCES

Abrahamsson, T., Holmgren, S., Nilsson, S. *et al.* (1979) On the chromaffin system of the African lungfish, *Protopterus aethiopicus*. *Acta Physiol. Scand.*, **107**, 135–9.

Adam, H. (1965) Eine Beiträge zur Kenntnis des Darmes von *Myxine glutinosa* L. (Cyclostomata). *Zool. Anz.*, Suppl. Bd. Verhandl. D. Zoologen in Kiel, 23.

Augustinsson, K.-B., Fänge, R., Johnels, A. *et al.* (1956) Histological, physiological and biochemical studies on the heart of two cyclostomes, hagfish (*Myxine*) and lamprey (*Lampetra*). *J. Physiol.*, **131**, 257–76.

Axelsson, M., Abe, A.S., Bicudo, J.E.P.W. *et al.* (1989) On the cardiac control in the South American lungfish, *Lepidosiren paradoxa*. *Comp. Biochem. Physiol.*, **93A**, 561–5.

Axelsson, M., Farrell, A.P. and Nilsson, S. (1990) Effects of hypoxia and drugs on the cardiovascular dynamics of the Atlantic hagfish, *Myxine glutinosa*. *J. Exp. Biol.*, **151**, 297–316.

Beringer, T. and Hadek, R. (1973) Ultrastructure of sinus venosus innervation in *Petromyzon marinus*. *J. Ultrastruct. Res.*, **42**, 312–23.

Bernier, N.J., Fuentes, J. and Randall, D.J. (1996) Adenosine receptor blockade and hypoxia-tolerance in rainbow trout and Pacific hagfish. II. Effects on plasma catecholamines and erythrocytes. *J. Exp. Biol.*, **199**, 497–507.

Bjenning, C. and Holmgren, S. (1988) Neuropeptides in the fish gut. A study of evolutionary trends. *Histochemistry*, **88**, 155–63.

Bloom, G., Östlund, E., Euler, U.S.v. *et al.* (1961) Studies on catecholamine-containing granules of specific cells in cyclostome hearts. *Acta Physiol. Scand.*, **53**, Suppl 185, 1–34.

Bloom, G., Östlund, E. and Euler, U.S.v. (1962) A specific granular secretory cell in the heart of cyclostomes. *Mem. Soc. Endocrinol.*, **12**, 255–63.

Bloom, G., Östlund, E. and Fänge, R. (1963) Functional aspects of cyclostome hearts in relation to recent structural findings, in *The Biology of Myxine* (eds A. Brodal and R. Fänge), Universitetsforlaget, Oslo, pp. 317–39.

Bone, Q. (1963) Some observations upon the peripheral nervous system of the hagfish, *Myxine glutinosa*. *J. Mar. Biol. Assoc.* UK, **43**, 31–7.

Brandt, W. (1922) Das Darmnervensystem von *Myxine glutinosa*. *Z. Anat. Entwicklungsgesch.*, **64**, 284–92.

Burnstock, G. (1969) Evolution of the autonomic innervation of visceral and cardiovascular systems in vertebrates. *Pharmacol. Rev.*, **21**, 247–324.

Campbell, G. (1970) Autonomic nervous system, in *Fish physiology*, Vol. IV (eds W.S. Hoar and D.J. Randall), Academic Press, New York, London, pp. 109–32.

Caravita, S. and Coscia, L. (1966) Les cellules chromaffines du coeur de la lamproie (*Lampetra zanandreai*). Étude au microscope éléctronique avant et après un traitement à la réserpine. *Arch. Biol.*, **77**, 723–53.

Carlson, A.J. (1904) Contributions to the physiology of the heart of the California hagfish (*Bdellostoma dombeyi*). *Z. Allg. Physiol.*, **4**, 259–88.

Carlson, A.J. (1906) The presence of cardioregulative nerves in the lampreys. *Amer. J. Physiol.*, **16**, 230–2.

Conlon, J.M. and Falkmer, S. (1989) Neurohormonal peptides in the gut of the Atlantic hagfish (*Myxine glutinosa*) detected using antisera raised against mammalian regulatory peptides. *Gen. Comp. Endocrinol.*, **76**, 292–300.

Coupland, R.E. (1965) *The Natural History of the Chromaffin cell*, Longman, London.

Coupland, R.E. (1972) The chromaffin system, in *Handbook of Experimental Pharmacology*, Vol. 33. Catecholamines (eds M. Blaschko and E. Muscholl), Springer, Berlin, Heidelberg, New York.

Dahl, E., Ehinger, B., Falck, B. *et al.* (1971) On the monoamine-storing cells in the heart of *Lampetra fluviatilis* and *L. planeri* (Cyclostomata). *Gen. Comp. Endocrinol.*, **17**, 241–6.

Euler, U.S.v. and Fänge, R. (1961) Catecholamines in nerves and organs of *Myxine glutinosa*, *Squalus acanthias*, and *Gadus callarias*. *Gen. Comp. Endocrinol.*, **1**, 191–4.

Fänge, R. (1948) Effect of drugs on the intestine of a vertebrate without sympathetic nervous system. *Arkiv Zool.*, **40A**, 1–9.

Fänge, R. and Johnels, A.G. (1958) An autonomic nerve plexus control of the gall bladder in *Myxine*. *Acta Zool.*, **39**, 1–8.

Fänge, R., Johnels, A.G. and Enger, P.S. (1963a) The autonomic nervous system, in *The Biology of Myxine* (eds A. Brodal and R. Fänge), Universitetsforlaget, Oslo, pp. 124–36.

Fänge, R., Bloom, G. and Östlund, E. (1963b) The portal vein heart of Myxinoids, in *The Biology of Myxine* (eds A. Brodal and R. Fänge), Universitetsforlaget, Oslo, pp. 340–51.

Fänge, R. and Östlund, E. (1954) The effects of adrenaline, noradrenaline, tyramine and other drugs on the isolated heart from marine vertebrates and a *cephalopod* (*Eledone cirrosa*). *Acta Zool. (Stockholm)*, **35**, 289–305.

Forster, M.E., Davison, W., Axelsson, M. *et al.* (1992) Cardiovascular responses to hypoxia in the hagfish, *Eptatretus cirrhatus*. *Respir. Physiol.*, **88**, 373–86.

Fritsche, R., Axelsson, M., Franklin, C.E. *et al.* (1993) Respiratory and cardiovascular responses to hypoxia in the Australian lungfish. *Respir. Physiol.*, **94**, 173–87.

Furness, J.B. and Costa, M. (1987) *The Enteric Nervous System*, Churchill Livingstone, Edinburgh.

Gaskell, J.F. (1912) The distribution and physiological action of the suprarenal medullary tissue in Petromyzon fluviatilis. *J. Physiol.* (London), **44**, 59–67.

Giacomini, E. (1902a) Contributo alla conoscenza delle capsule surrenali dei ciclostomi. Sulle capsule surrenali dei missinoidi. *R. Accad Sci. Bologna*, **7**, 135–40.

Giacomini, E. (1902b) Contributo alla conoscenza delle capsule surrenali dei ciclostomi. Sulle capsule surrenali dei petromizonti. *Monit. Zool. Ital.*, **13**, 143–62.

Gibbins, I.L. (1994) Comparative anatomy and evolution of the autonomic nervous system, in *Comparative Physiology and Evolution of the Autonomic Nervous System* (series ed., G. Burnstock), (eds S. Nilsson and S. Holmgren), Harwood Academic Publishers, Chur, Switzerland, pp. 1–67.

Goodrich, J.T., Bernd, P., Sherman, D.L. and Gershon, M.D. (1980) Phylogeny of enteric serotonergic neurons. *J. Comp. Neurol.*, **190**, 15–28.

Govyrin, V.A. (1977) Development of vasomotor adrenergic innervation in onto- and phylogenesis. *J. Evol. Biochem. Physiol.*, **13**, 614–20.

Greene, C.W. (1902) Contributions to the physiology of the California hagfish, *Polistotrema stoutii*. II. The absence of regulative nerves for the systemic heart. *Amer. J. Physiol.*, **6**, 318–24.

Hallbäck, D.-A. (1973) Acetylcholinesterase-containing structures in the intestine of *Myxine glutinosa* L., in *Myxine glutinosa: Biochemistry, Physiology and Structure. Acta Regiae Societatis Scientiarum et Litterarum Gothoburgensis, Zoologica* 8 (ed. R. Fänge), Kungl. Vetenskaps- och Vitterhetsamhället, Göteborg, pp. 24–5.

Hirsch, E.F., Jellinek, M. and Cooper, T. (1964) Innervation of the hagfish heart. *Circ.* Res., **14**, 212–17.

Hirt, A. (1934) Die vergleichende Anatomie des sympathischen Nervensystems, in *Handbuch der vergleichende Anatomie der Wirbeltiere*, Band 2, Teil 1 (ed. L. Bolk), Urban & Schwartzenberg, Berlin, pp. 685–776.

Holmgren, S. and Fänge, R. (1981) Effects of cholinergic drugs on the intestine and gallbladder of the hagfish, *Myxine glutinosa* L., with a report on the inconsistent effects of catecholamines. *Mar. Biol. Lett.*, **2**, 265–77.

Iijima, T. and Wasano, T. (1980) A histochemical and ultrastructural study of serotonin-containing nerves in cerebral blood vessels of the lamprey. *Anat. Rec.*, **198**, 671–80.

Jensen, J. and Holmgren, S. (1991) Tachykinins and intestinal motility in different fish groups. *Gen. Comp. Endocrinol.*, **83**, 388–96.

Johansen, K. (1960) Circulation in the hagfish *Myxine glutinosa* L. *Biol. Bull.*, **118**, 289–95.

Johnels, A.G. (1956) On the peripheral autonomic nervous system of the trunk region of *Lampetra planeri*. *Acta Zool. (Stockholm)*, **37**, 251–86.

Johnsson, M. and Axelsson, M. (1996) Control of the systemic heart and the portal heart of *Myxine glutinosa*. *J. Exp. Biol.*, **199**, 1429–34.

Johnsson, M., Axelsson, M., Davison, W. *et al.* (1996) Effects of preload and afterload on the performance of the *in situ* perfused portal heart of the New Zealand hagfish, *Eptatretus cirrhatus*. *J. Exp. Biol.*, **199**, 401–5.

Johnston, J.B. (1908) Additional notes on the cranial nerves of petromyzonts. *J. Comp. Neurol.*, **18**, 569–608.

Kohn, A. (1902) Das chromaffin Gewebe. *Ergerb. Anat. Entwicklungsgesch.*, **12**, 253–348.

Lagerstrand, G. and Nilsson, S. (1973) Effects of 6-hydroxydopamine on the catecholamine levels in the systemic and portal hearts of *Myxine glutinosa*, in *Myxine glutinosa: Biochemistry, Physiology and Structure. Acta Regiae Societatis Scientiarum et Litterarum Gothoburgensis*, Zoologica 8 (ed R. Fänge), Kungl. Vetenskaps- och Vitterhetsammhället.

Leont'eva, G.R. (1966) Distribution of catecholamines in blood vessel walls of cyclostomes, fishes, amphibians and reptiles. *J. Evol. Biochem. Physiol.*, **2**, 31–6.

Lindström, T. (1949) On the cranial nerves of the cyclostomes with special reference to n. trigeminus. *Acta Zool. (Stockholm)*, **30**, 315–458.

Matsuda, H., Goris, R.C. and Kishida, R. (1991) Afferent and efferent projections of the glossopharyngeal-vagal nerve in the hagfish. *J. Comp. Neurol.*, **311**, 520–30.

Mazeaud, M.M. (1971) Recherches sur la biosynthése, la sécrétion et le catabolisme de l'adrénaline et de la noradrénaline chez quelques espèces de cyclostomes et de poissons, Ph.D. Thesis, Université de Paris, Paris.

Mazeaud, M. (1972) Epinephrine biosynthesis in *Petromyzon marinus* (Cyclostoma) and *Salmo gairdneri* (teleost). *Comp. Gen. Pharmacol.*, **32**, 457–68.

Naik, N.T. (1963) Technical variations in Koelle's histochemical method for localizing cholinesterase activity. *Proc. Soc. Exp. Biol.*, **104**, 89–100.

Nakao, T. (1981) An electron microscopic study on the innervation of the gill filaments of a lamprey, *Lampetra japonica*. *J. Morphol.*, **169**, 325–36.

Nakao, T., Susuki, S. and Saito, M. (1981) An electron microscopic study of the cardiac innervation in larval lamprey. *Anat. Rec.*, **199**, 555–63.

Nicol, J.A.C. (1952) Autonomic nervous systems in lower chordates. *Biol. Rev.*, **27**, 1–49.

Nilsson, S. (1983) *Autonomic Nerve Function in the Vertebrates*, Springer-Verlag, Berlin, Heidelberg, New York.

Nilsson, S. and Holmgren, S. (1994) *Comparative Physiology and Evolution of the Autonomic Nervous System*, Harwood Academic Publishers, Chur, Switzerland, 376pp.

Olsson, C. and Holmgren, S. (1997) Nitric oxide in the fish gut. *Comp. Biochem. Physiol.* In press.

Olsson, C. and Karila, P. (1995) Coexistence of NADPH-diaphorase and vasoactive intestinal polypeptide in the enteric nervous system of the Atlantic cod (*Gadus morhua*) and the spiny dogfish (*Squalus acanthias*). Cell and Tissue Res., **280**, 297–305.

Östlund, E. (1954) The distribution of catecholamines in lower animals and their effect on the heart. *Acta Physiol. Scand.*, **31**, Suppl. 112, 1–67.

Patterson, T.L. and Fair, E. (1933) The action of the vagus on the stomach-intestine in the hagfish. Comparative studies VIII. *J. Cell. Comp.* Physiol., **3**, 113–19.

Perry, S.F., Fritsche, R. and Thomas, S. (1993) Storage and release of catecholamines from the chromaffin tissue of the Atlantic hagfish, *Myxine glutinosa*. *J. Exp. Biol.*, **183**, 165–84.

Peters, A. (1963) The peripheral nervous system, in *The Biology of Myxine* (eds A. Brodal and R. Fänge), Universitetsforlaget, Oslo, pp. 92–123.

Pick, J. (1970) *The Autonomic Nervous System. Morphological, Comparative, Clinical and Surgical Aspects*, Lippincott, Philadelphia, Toronto.

Reite, O.B. (1969) The evolution of vascular smooth muscle responses to histamine and 5-hydroxytryptamine. I. Occurrence of stimulatory actions in fish. *Acta Physiol. Scand.*, **75**, 221–39.

Romer, A.S. (1962) *The Vertebrate Body* (3rd edn), Saunders Company, Philadelphia, London.

Scheuermann, D.W. (1979) Untersuchungen hinsichtlich der Innervation des Sinus venosus und des Aurikels von *Protopterus annectens*. *Acta Morphol. Neerl. Scand.*, **17**, 231–2.

Shibata, Y. and Yamamoto, T. (1976) Fine structure and cytochemistry of specific granules in the lamprey atrium. *Cell Tissue Res.*, **172**, 487–502.

Stabrovskii, E.M. (1967) The distribution of adrenaline and noradrenaline in the organs of the baltic lamprey *Lampetra fluviatilis* at rest and during various functional stresses. *J. Evol. Biochem. Physiol.*, **3**, 216–21.

Stewart, J.K., Goodner, C.J., Koerker, D.J., Gorbman, A., Ensinck, J. and Kaufman, M. (1978) Evidence for a biological role of somatostatin in the Pacific hagfish, *Eptatretus stoutii*. *Gen. Comp. Endocrinol.*, **36**, 408–14.

Sundin, L., Axelsson, M., Davison, W. *et al.* (1994) Evidence of regulatory mechanisms for the distribution of blood between the arterial and venous compartments in the hagfish gill pouch. *J. Exp. Biol.*, **190**, 281–6.

Tretjakoff, D. (1927) Das peripherische Nervensystem des Flußneunauges. *Z. Wiss. Bull.*, **129**, 359–452.

Zwaardemaker, H.H.L. (1924) Action du nerf vague et radio activite. *Arch. Neerl. Physiol.*, **9**, 213–28.

PART TWELVE

The Sensory Organs

31

SKIN SENSORY ORGANS IN THE ATLANTIC HAGFISH *MYXINE GLUTINOSA*

Monika von Düring and Karl H. Andres

SUMMARY

Slice series of total bodies of *Myxine glutinosa* were documented with the reflecting light method. With the help of semithin serial sections and electron micrographs the integument of the Atlantic hagfish is described with respect to the topography and the fine structural organization of the epidermal, dermal and hypodermal nerve fibre plexus. The nerve fibre plexus of the dermal and hypodermal layer contain small ganglion cells with axodendritic and axosomatic synapses. These ganglion cells seem to represent a part of the peripheral autonomic nervous system.

With respect to the topography and the structural organization of the microenvironment of the various sensory nerve terminals in the skin of the head, the nasal and oral barbels and the tail fin, different types of mechano- and chemoreceptors are described. Different axon diameters indicate the fast- and slow-conducting systems which are to be differentiated.

One of the types of intraepidermal nerve terminals is synaptically linked to a distinct epidermal cell. These epidermal cells bear distinct microvilli on their apical membrane and show accumulations of small clear vesicles in their basal cytoplasm. The other type of intraepidermal nerve endings terminate within the intercellular clefts of the epidermal cell layer.

Various types of mechano- and chemoreceptors of the trigeminal nerve which compose the highly differentiated innervation pattern of the tentacles indicate that the tentacles function as a complex receptor organ with stereognostic and chemoperception. The tail fin represents an afferent innervation supplied by spinal nerves and intraepidermal and dermal receptors comparable to the skin of the head. The further observation of two different types of afferents located in the mucous connective tissue envelope of each of the radii pterygiales makes the tail fin likely to be further involved in the propriosensory system to supply the complex motor behaviour.

31.1 INTRODUCTION

In general, peripheral axons of sensory ganglion cells are synaptically linked to a secondary sensory cell or exhibit free nerve terminals. These axon terminals exhibit in the receptive or transductive areas a receptor matrix associated to the axon membrane combined with the lack of a Schwann cell covering or with an encapsulation (Andres and v. Düring, 1972). In *Myxine* both types of tissue receptors occur (Table 31.1), the free nerve terminals predominate.

The Biology of Hagfishes. Edited by Jørgen Mørup Jørgensen, Jens Peter Lomholt, Roy E. Weber and Hans Malte. Published in 1998 by Chapman & Hall, London. ISBN 0 412 78530 7.

Table 31.1 Skin receptors in *Myxine glutinosa*

Type of nerve termina	Type of tissue (localization)
Free nerve terminal	Epidermis
	Dermis
	Fascia
	Mucous connective tissue
	Perichondrial connective tissue layer
Secondary sensory cell/afferent axon	Epidermis

There is scant information in the literature about the skin sensory innervation of *Myxine glutinosa*. Worthington (1905) observed an extraordinary mechanosensitivity of the nasal and oral tentacles in *Myxine* in behavioural experiments. Ayers and Worthington (1907) gave the first histological evidence of epidermal free nerve terminals and sensory cells with surrounding nerve fibres in the tentacles of *Bdellostoma dombeyi*. Epithelial sensory cells were later described in the tentacles of *Myxine glutinosa* by Schreiner (1918) and recently confirmed in the tentacles of *Myxine glutinosa* and *Eptatretus burgeri* by scanning and transmission electron microscopy (Patzner *et al.*, 1977; Georgieva *et al.*, 1979). The authors supposed that the assembly of these cells named 'sensory bud' represent taste organs, based on their observation of cells bearing apical microvilli. Synapses have so far never been observed. Taste cells in all classes of vertebrates are associated to afferent nerve fibres.

A number of short lateral lines occur on the head in the Pacific hagfish *Eptatretus stoutii*, *Eptatretus burgeri* and *Bdellostoma* (Ayers and Worthington, 1907; Fernholm, 1983; Kishida *et al.*, 1987). In *Myxine glutinosa*, however, ampullary and neuromast organs of the lateral line system, functional eyes and electroreceptive organs are lacking (Ross, 1963; Bullock *et al.*, 19823, Fernholm, 1983; Andres and v. Düring, 1993a, b).

The complex locomotion behaviour of *Myxine glutinosa* as gliding, figure-of-eight, vermiform and swimming movements requires an adequate afferent proprioceptive system (Adam, 1960; Strahan, 1963; Ross, 1963). The observation of a large trigeminal ganglion with thousands of afferent nerve fibres in *Myxine* and the evidence of different tissue receptors including proprioreceptors indicates a highly organized afferent system (Jansen, 1930; Lindström, 1949; Andres and v. Düring, 1993a, b; v. Düring and Andres, 1994). Due to physiological recordings from skin nerves McVean (1989) distinguished velocity-sensitive receptors and displacement-sensitive receptors. He observed a different distribution of both receptors in the skin of the head and body of *Myxine glutinosa*.

31.2 METHODS

Myxine glutinosa were fixed by the perfusion technique via the left ventricle of the heart with the fixative of Karnovsky (1965) and adapted to the osmolarity of the seawater by the addition of sodium chloride. Slice series of the total bodies were photographically documented after osmification with the reflecting light method (Andres and v. Düring, 1981, 1993a) to guarantee the topographical orientation during the embedding procedure. Series of alternative semithin and ultrathin sections were cut with an Ultracut E. The semithin sections were stained with 1% toluidine blue (pH 9.3). Photodocumentation was performed with a Zeiss Axiophot. Ultrathin sections were stained with uranyl acetate followed by lead citrate and examined with a Philips 300 electron microscope (Reynolds, 1963).

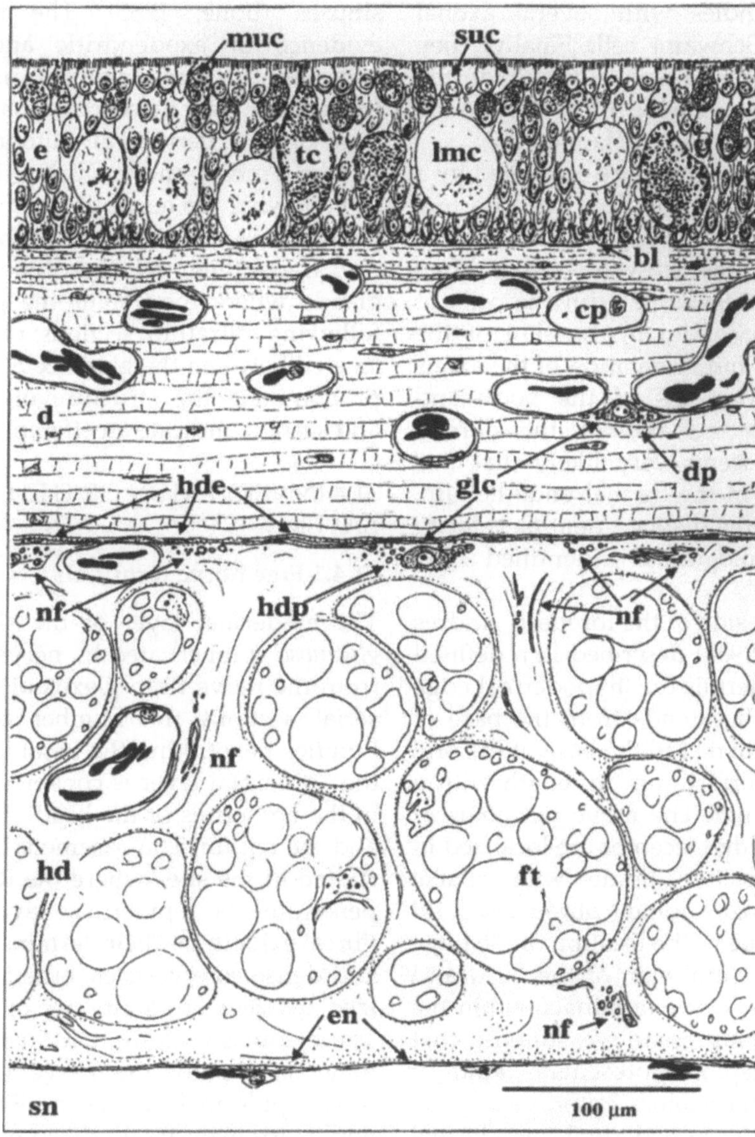

Figure 31.1 Schematic representation of the skin from the trunk. The dermis (d) exhibits a dense capillary network (cp) and is separated from the hypodermis (hd) by the dermal endothelium (hde). e, Epidermis; dp, dermal nerve fibre plexus; hdp, hypodermal nerve fibre plexus; glc, ganglion cells, ft, fat cells; en, sinus endothelium lining the subcutaneous sinus (sn); nf, nerve fibres; suc, superficial epithelial cell; muc, mucous cell; tc, thread cell; lmc, large mucous cell; bl, basement lamina. (From Andres and v. Düring, 1993, *Cell and Tissue Research*, 274, 353–66. With permission from Springer-Verlag GmbH and Co. KG).

31.3 NERVE FIBRE PLEXUS OF THE SKIN OF *MYXINE GLUTINOSA*

Skin nerves in *Myxine glutinosa* are unmyelinated, lack a perineural sheath and are very numerous at least in the skin of head, tentacles and tail fin. Large and small axons, ranging from 1 to 8 μm in diameter, form distinct hypodermal and dermal nerve fibre plexus. Nerve fibres leave the plexus, arborize and form Remak bundles with several axonal profiles in one Schwann cell. Finally, they terminate in the collagenous tissue of the different skin layers (Figure 31.1), close to the vessels of the dermis or to melanocytes.

Dermis and hypodermis are separated by a single layer of cells called the dermal endothelium (Whitear *et al.*, 1980; Welsch, this volume). Nerve fibres, Schwann cells and free axon terminals intermingle with the cells of the dermal endothelium in the skin of the head. The course and orientation of the axons are parallel to the axis of the body and perpendicular to the axis of the dermal endothelium. It seems very likely that this receptor, located in the dermal endothelium, represents either the displacement or velocity receptor physiologically identified by McVean (1989).

On the ventral side of the forehead Andres and v. Düring (1993a) described in a defined area of the hypodermis the 'hypodermal edge receptor organ'. It extends from the base of the ventral nasal barbel to the base of the oral barbel and measures 8 mm in length with a diameter of 0.2 mm. The nerve terminals of the hypodermal edge receptor are localized in a mucous connective tissue with sparse collagenous fibres, some fibrocytes and Schwann cells. Due to the location of the edge receptor in the ventral skin of the head and the location of the nerve terminals within the mucous collagenous tissue we suppose that this type of receptor represents a kind of slowly adapting mechanoreceptor.

Additionally the hypodermal and dermal nerve fibre plexus contain a great number of small neuronal cell bodies primarily observed on light microscopical level (Retzius, 1890; Bone, 1963) (Figures 31.1, 31.2B). These neuronal cell bodies occur in all skin regions of the body except the tips of the tentacles. The function of these neurons is still unclear, but it was suggested that they may be involved in the innervation of epidermal gland cells, hypodermal slime glands and vessels including the hypodermal venous sinuses (Bone, 1963). The morphological evidence of axodendritic and axosomatic synapses on these neurons underlined the hypothesis that these cells may represent a part of the peripheral autonomic nervous system (Andres and v. Düring, 1993b).

Following the nerve fibres of the peripheral ganglia cells in our serial semithin sections we can trace them to the hypodermal slime glands where they enter the glands. In the following ultrathin sections these axons are located between the gland cells. Clear vesicles predominate the structure of the axoplasm and seem to indicate an efferent function.

31.4 INNERVATION OF THE EPIDERMAL LAYER

31.4.1 Free nerve terminals

The epidermal layer of the skin of *Myxine glutinosa* is innervated by nerve fibres arising from the nerve fibre plexus of the dermis. In serial sections the number of nerve fibre bundles penetrating the basal lamina to enter the epidermal layer is conspicuous especially in the epidermis of the head, the six tentacles and the tail fin. The basement lamina is interrupted at the sites where the Remak bundles penetrate the epidermal layer. The nerve fibres exhibit in their terminal course clear and dense core vesicles, others are characterized by their large amount of mitochondria. The cytoplasm of the terminal Schwann cell may accompany the nerve fibres into the basal epidermal cell layer. Finally, the free nerve terminals are localized in the intercellular clefts of the epidermal cells or deeply invaginated into their cell cytoplasm (Figure 31.3C, D, E). Up to now all immuno-cytochemical studies failed to demonstrate various neurotransmitters and neuromodulators. Their superficial location in the epidermal layer makes them suitable candidates for chemoreceptors. In this context it is interesting to mention that the tips of the paired anterior fin rays of the pectoral fins of the searobin *Trigla lucerna* exhibit a comparable

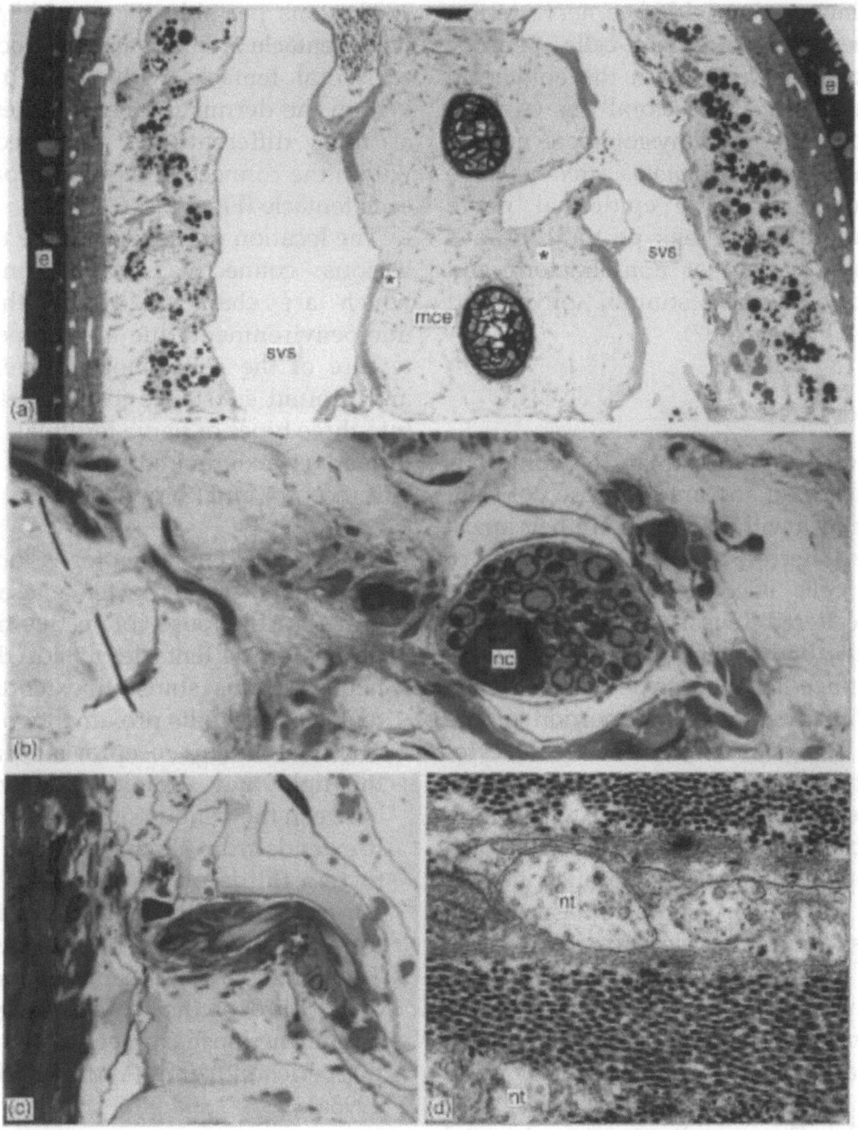

Figure 31.2 (A) Tangential section through the tail fin with two radii pterygoidei (rp) surrounded by the mucous connective tissue envelope (mce). Epidermal layer (e); dermal layer with distinct capillaries (c). The hypodermal layer exhibits fat cells (fc) and a loose connective tissue. Subcutaneous venous sinus (svs). Note the dense outer rim of the mucous connective tissue to the subcutaneous venous vessel. The location of the nerve fibre terminals of the mucous connective envelope is marked by a large asterisk, and the location of the nerve fibre terminals of the dense collagenous rim of the mucous connective envelope are marked by a small asterisk. Semithin section, bar = 100 µm. (B) Nerve fibre bundle with a nerve cell body (nc) in the hypodermal layer of the tail fin. Semithin section, bar = 10 µm. (C) The location of the nerve fibre terminal (large asterisk) within the mucous connective tissue envelope of a radii pterygei of the tail fin. Note the large sensory axon. The small asterisk marks the location of the nerve terminal in figure D. Semithin section, bar = 10 µm. (D) Nerve fibre terminal (nt) within the dense collagenous tissue surrounding the circumference of the mucous connective tissue compartment. Compare with figures A and C for the location of the nerve terminals. Electron micrograph, bar = 1 µm.

epidermal innervation with free nerve terminals between the epidermal cells. Sensory cells are not observed within the epidermal layer of the tips of the pectoral fins (Andres and v. Düring, 1985). Physiological experiments of the searobin *Prionotus carolinus* have already shown that the epidermal nerve terminals in the tips of their pectoral fin rays respond to a variety of non-olfactory and non-gustatory chemical stimulii (Silver and Finger, 1984; Finger, 1987).

31.4.2 Sensory cells

Sensory buds are numerous in the epidermis of the tentacle skin and are composed of slender sensory cells with microvilli on their apical cell surface (Figure 31.3A) and two types of supporting cells (Georgieva *et al*, 1979). In addition, scattered sensory buds occur elsewhere on the body surface (Blackstad, 1963). Bundles of fine filaments can be followed from the microvilli into the apical portion of the cytoplasm. The outer rim of the apical cytoplasm of the sensory cell contains numerous microtubules, which are oriented to the long axis of the cell. In its basal part the sensory cell forms synapses to nerve fibre profiles invaginating into the sensory cell cytoplasm. Small clear vesicles accumulate close to a presynaptic membrane (Figures 31.3B and 31.4). The arrangement of the sensory cell neurite association is reminiscent of the synapse between taste cells and sensory nerve fibres of other vertebrates (Andres, 1974; v. Düring and Andres, 1976). These results give experimental evidence that the sensory cells described by Schreiner and Georgieva represent *Myxine*'s taste or gustatory cells.

31.5 INNERVATION OF THE DERMAL LAYER

31.5.1 Sensory dermal receptors of the tentacles

Branches of the ophthalmic nerve supply the four nasal and two oral tentacles with about 2800 axons per side (Figure 31.5A, B). Each nasal tentacle is innervated by about 800 and each oral tentacle with about 1200 axons. Within the dermis the nerve fibres terminate at three different and restricted locations within the connective tissue compartments of each tentacle (Figure 31.6).

The location of the nerve fibre terminals in various connective tissue compartments, which are characterized by their special microenvironment due to composition and texture of the collagenous tissue, fibrocytes and ground substance justify a classification into three types of nerve terminals. They were termed: (1) external cuff receptor, (2) internal cuff receptor and (3) perichondrial receptor.

1. The external cuff receptor is located in the circumference of the tentacle and is integrated in the loose connective tissue of the dermis of the tentacle outside the subcutaneous venous sinus. It extends from the tentacle base to its proximal two thirds.
2. The internal cuff receptor is located within the tight and straight collagenous tissue between the venous sinus and the cartilage. It extends from the base of the tentacle up to two thirds of the length of the barbel. Large axons with up to 8 μm in diameter form the internal cuff receptor terminals.
3. The perichondrial receptor is located in the distal third of the tentacle length and lies within the loose connective tissue fibres of the perichondrium (Andres and v. Düring, 1993a).

The nerve terminals of all three types of receptors lack a Schwann cell envelope in their most distal part (Figure 31.5C and D). The function of these terminals is not clear. The close connection of the collagenous fibres to the nerve terminals is comparable to receptive areas of defined mechanoreceptors of other vertebrates such as the Golgi tendon organs, the Ruffini corpuscles or the sinus hair receptors (Andres, 1966; Chambers *et al.*, 1972; Schoultz and Swett, 1972; Andres and

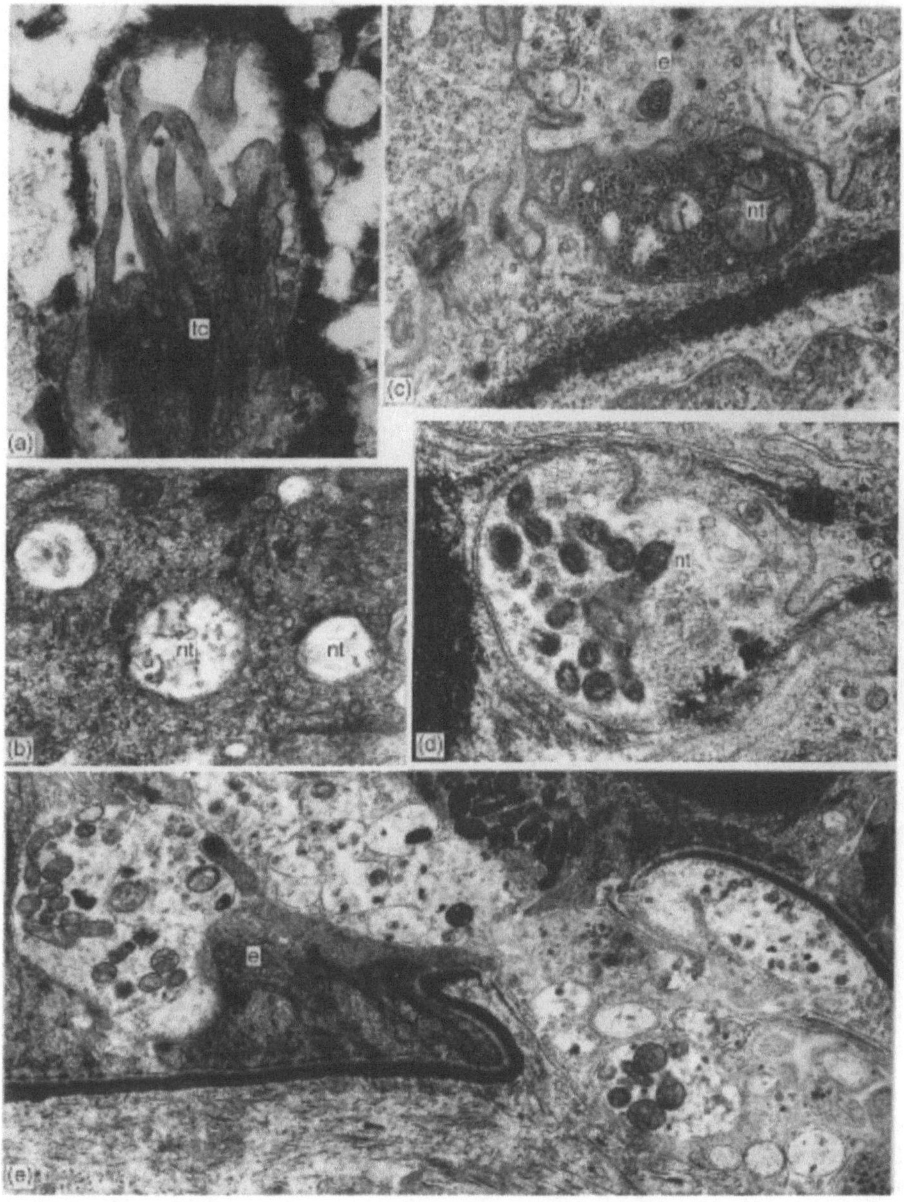

Figure 31.3 (A) Apical portion of a taste cell (tc) with microvilli surrounded by secreted material of the epidermal secretory cells. Note the filaments inside the taste cell. Electron micrograph, bar = 1 μm. (B) Basal part of a taste cell with the synaptical contact to the afferent nerve fibre (nt). Note the accumulation of clear synaptic vesicles close to the presynaptic membrane. Electron micrograph, bar = 1 μm. (C) Small nerve terminal (nt) with accumulations of glycogen from the epidermal cells (e) of the tentacle tip. Electron micrograph, bar = 1 μm. (D) Large nerve terminal (nt) in the epidermal layer of the tentacle tip with accumulations of mitochondria and clear vesicles, surrounded by the cytoplasm of the epidermal cell (e). Electron micrograph, bar = 1 μm. (E) Remak bundle penetrating the basement lamina (bl) of the epidermal layer (e). Note the numerous nerve fibres and nerve terminals containing various clear and dense core vesicles and mitochondria; collagen fibrils (cf) of dermal layer. Electron micrograph, bar = 1 μm.

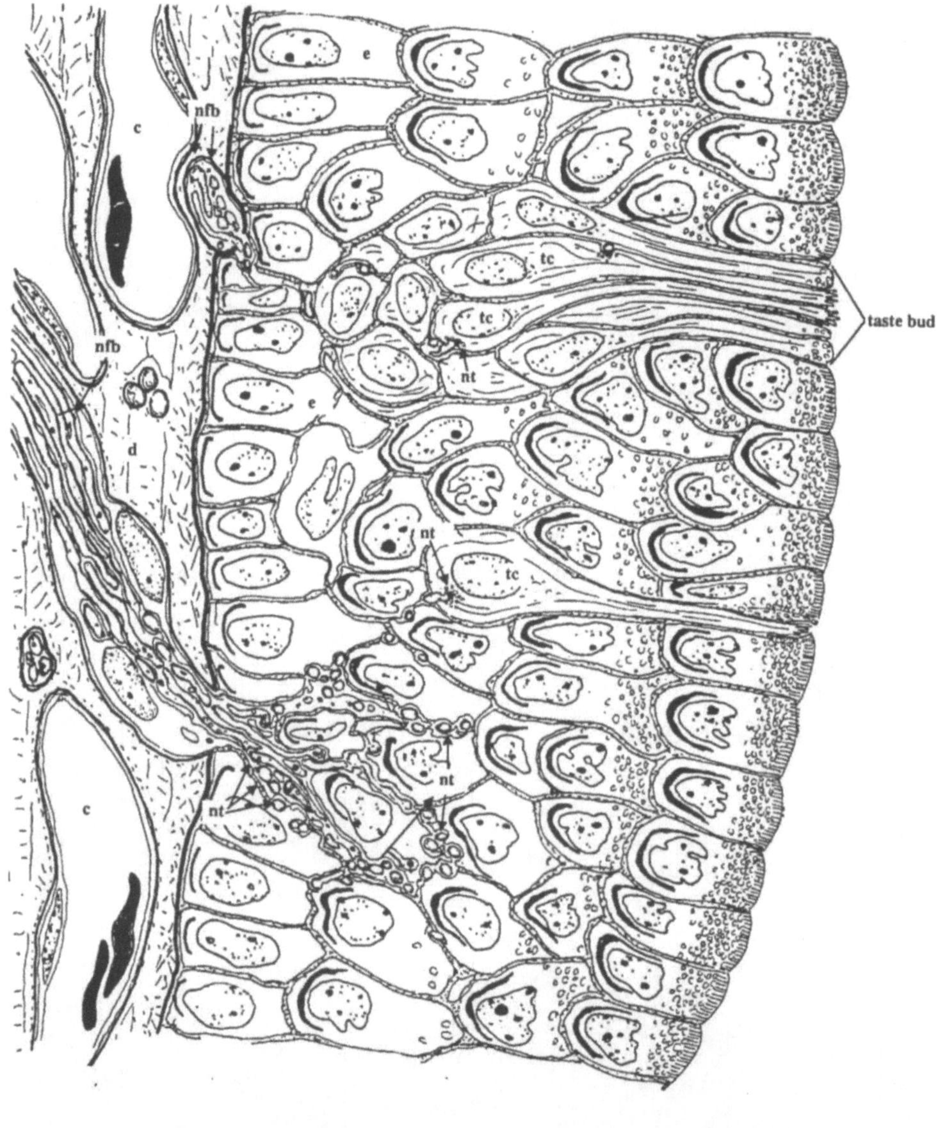

Figure 31.4 Schematic representation of the innervation of a tentacle epidermis with a taste bud and free intra-epidermal nerve terminals. Supporting cells of the taste bud are not indicated. Nerve terminals of the taste cells (nt) and of the free epidermal nerve fibres (nt); taste cell (tc). Epidermal cell (e); dermis (d); nerve fibre bundle (nfb); dermal capillary (c). Location see inset in Figure 31.6.

v. Düring, 1985). Any polarized arrangement of receptive structures may indicate that, depending on its configuration, either stretch or pressure may be transmitted to the transductive areas. Therefore the three receptor types in the tentacles of *Myxine glutinosa* may represent mechanoreceptors. Physiological experiments are necessary to define the functional properties of the described receptor complexes using an experimental device which is specially adapted to the local organization of the microenvironment.

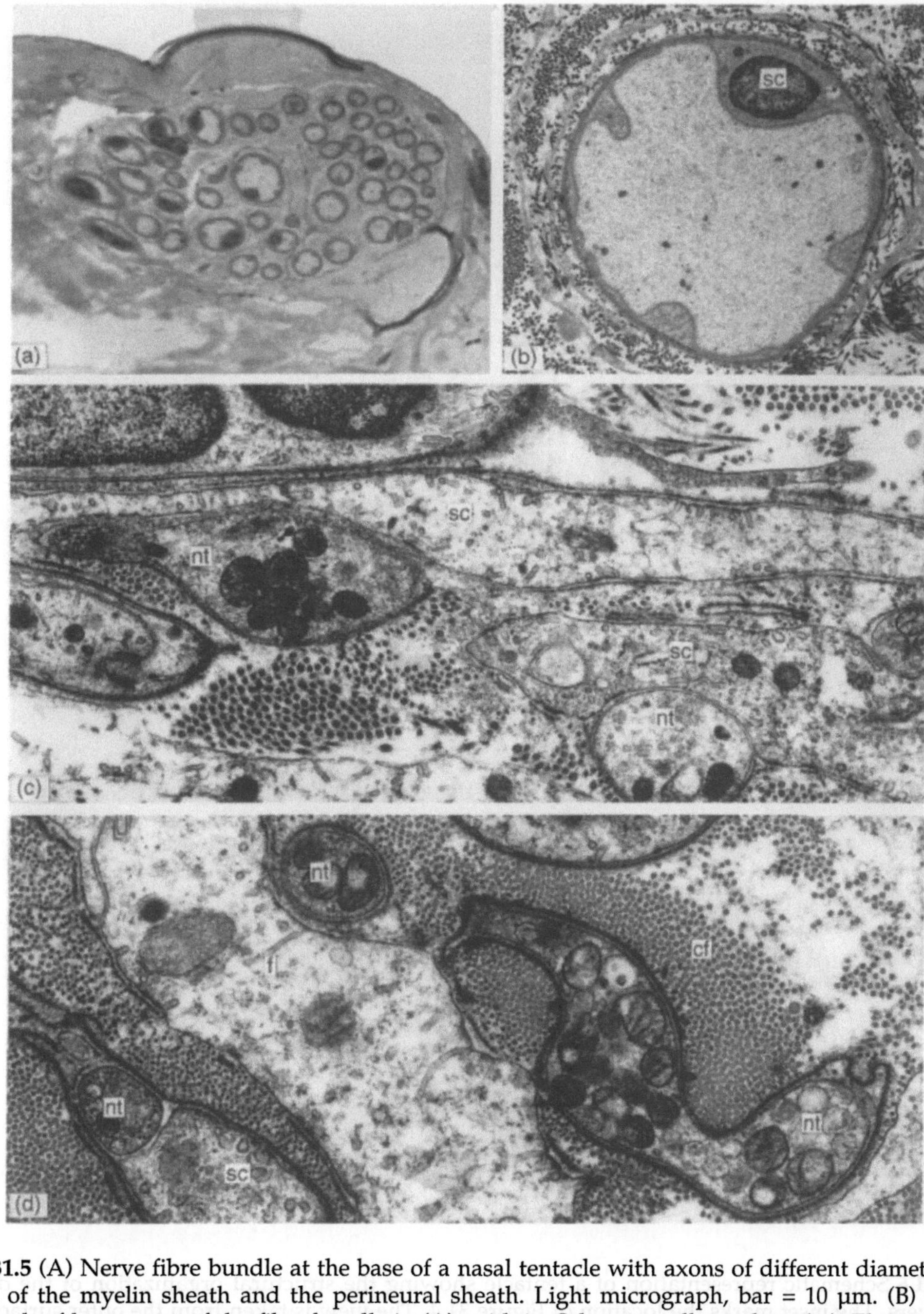

Figure 31.5 (A) Nerve fibre bundle at the base of a nasal tentacle with axons of different diameter. Note the lack of the myelin sheath and the perineural sheath. Light micrograph, bar = 10 μm. (B) Electron micrograph of large axon of the fibre bundle in (A), with its Schwann cell envelope (sc). Electron micrograph, bar = 1 μm. (C) Segment of the external cuff receptor of a nasal tentacle of *Myxine*. Note the basal lamina of the nerve terminals (nt) directly facing the loose connective tissue; Schwann cell (sc), collagen fibrils (cf). Electron micrograph, bar = 1 μm. (D) Segment of the internal cuff receptor of a nasal tentacle. The nerve terminals (nt) are only covered by their basement lamina, collagen fibrils of the microenvironment (cf); Schwann cell (sc); fibrocyte (f). Electron micrograph, bar = 1 μm.

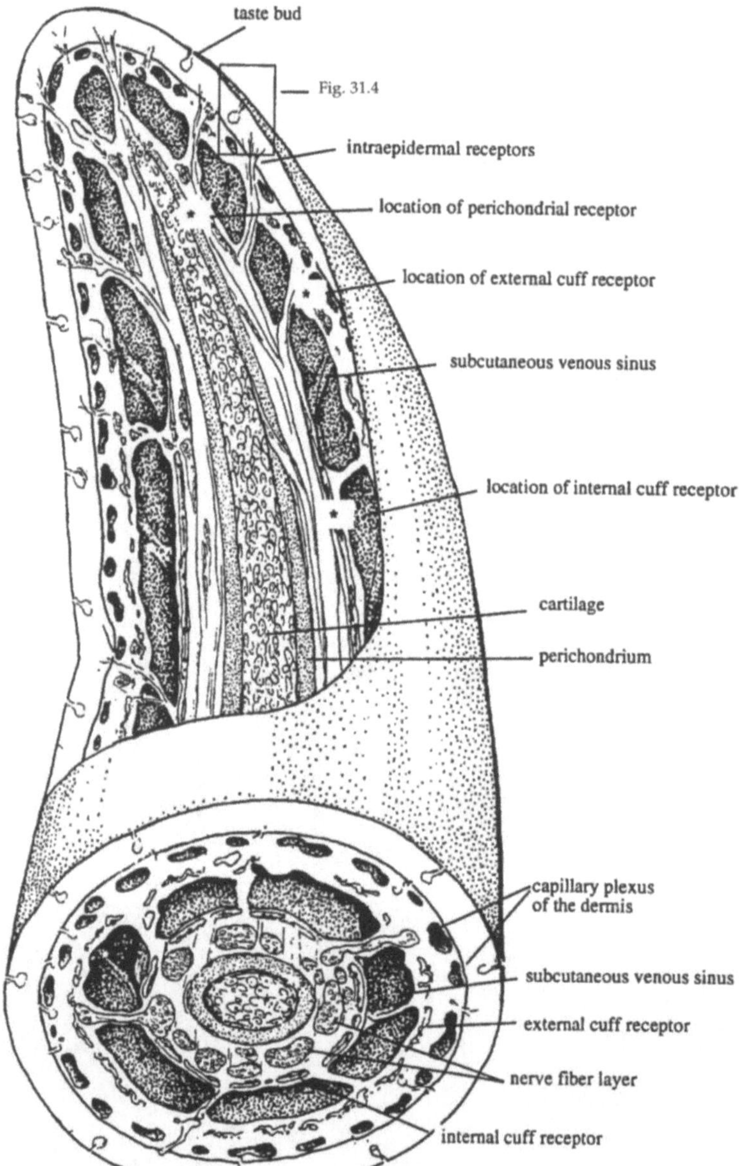

Figure 31.6 Schematic representation of a tentacle showing the structural organization of the different tissue layers. The inset marks the location of Figure 31.4. The view is taken from the outer surface of the tip of the tentacle to the inner cartilaginous ray at the base. Note the extent of the subcutaneous venous sinus up to the tentacle's tip. The location of the different types of sensory receptors is indicated with asterisks.

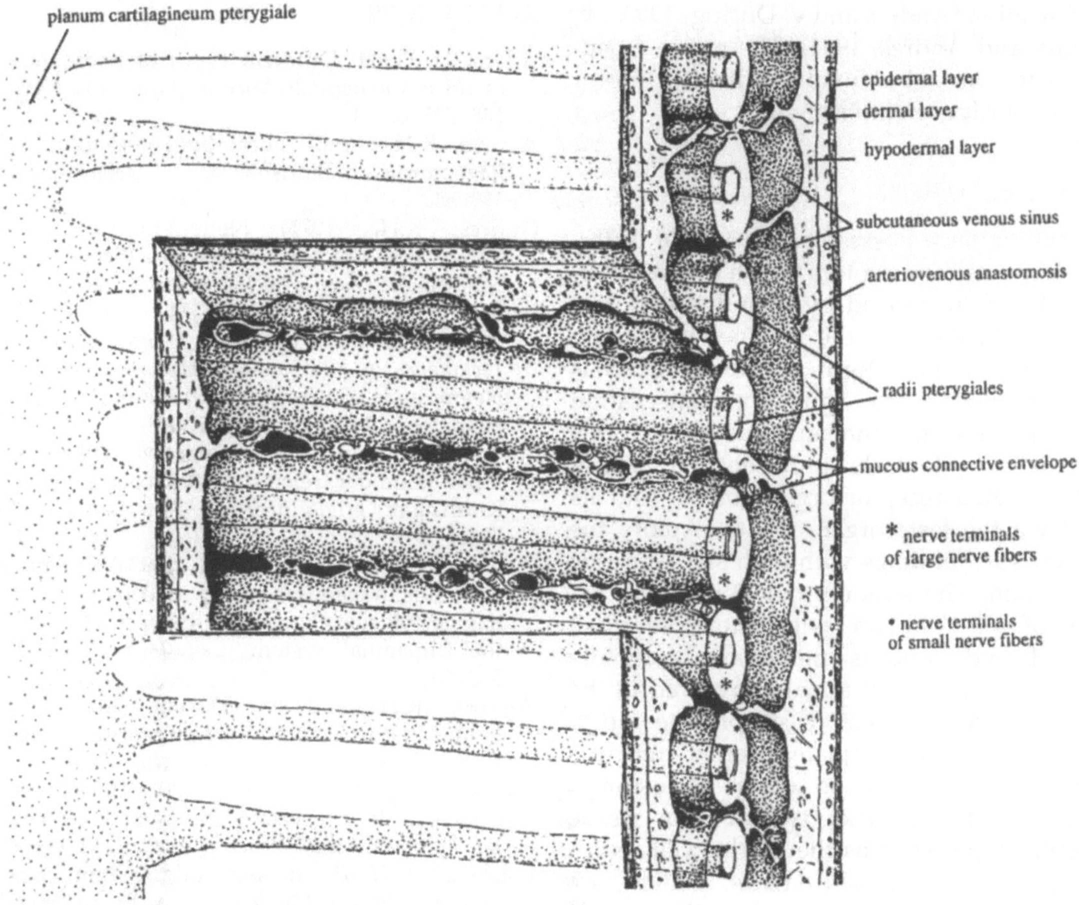

Figure 31.7 Schematic representation of the radii pterygiales of the tail fin with the structural organization of their mucous connective envelope including the location of their afferent nerve terminals (asterisks according to Figure 31.2A). Arteriovenous anastomoses to the subcutaneous venous sinus are numerous.

31.5.2 Sensory dermal receptors in the tail fin

The tail region of *Myxine glutinosa* consists of a median planum cartilagineum pterygiale from which the radii pterygiales extent radially to the dorsal and ventral skin surface. A broad sheath of mucous connective tissue surrounds each of these radii up to their end (Figures 31.2A, 31.7). The organization of the skin with the subcutaneous venous sinus resembles the other regions of the body.

The tail skin is supplied by sensory axons of the spinal nerves. Apart from the innervation of the skin the mucous connective tissue envelope of each of the radii pterygiales is also innervated (Figures 31.2A, 31.7). We observed free nerve terminals at two different locations. Smaller axons terminate in the dense collagenous fibres close to the venous sinus (Figure 31.2C, D), larger axons arborize and terminate within the mucous connective tissue resembling the lamellated receptors of the body fascia in their fine structural organization (Andres and v. Düring, 1993a; v. Düring and Andres, 1994). These two receptor types may be involved in the motor control of *Myxine glutinosa*.

31.6 CONCLUSION

Myxine glutinosa possesses a variety of different skin sensory receptors which are morphologically characterized by their location and the organization of the microenvironment. The innervation pattern of the tentacles is comparable to the complex innervation pattern of the mammalian sinus hair. In this context, the tentacle represents a highly differentiated receptor organ able to find its food via the taste organ and to explore the nearest surroundings with high stereognostic perception. The striking and dense innervation of the tail fin with intraepidermal, dermal and mucous membrane receptors indicates that the tail may be involved in the propriosensory system to supply the motor system with the adequate afferent information. In comparison to other cyclostomes *Myxine glutinosa* lacks typical skin tissue receptor types such as the Merkel cell receptor complex and neuromasts, which are common in the lamprey's skin (Fox *et al.*, 1980; Whitear and Lane, 1981a, b; Reuter, 1984). *Myxine* lacks these skin receptors; on the other hand *Myxine* has elaborated epidermal free nerve endings, dermal receptors and compound sensory organs indicating a specialization of the cutaneous afferent nervous system reflecting the special habitat of *Myxine glutinosa*.

ACKNOWLEDGEMENTS

We are grateful to Professor Strömberg of the Marine Biological Institute, Kristineberg, Sweden, for procurement of living hagfishes and laboratory space. We thank Mrs Luzie Augustinowski, Mrs Katja Knippschild and Mrs Petra Parakenings for skilled technical and photographical assistance.

REFERENCES

Adam, H. (1969) Different types of body movements in the hagfish, *Myxine glutinosa* L. *Nature*, **188**, 595–6.

Andres, K.H. (1966) Über die Feinstruktur der Rezeptoren an Sinushaaren. *Z. Zellforsch.*, **75**, 339–65.

Andres, K.H. (1975) Neue morphologische Grundlagen zur Physiologie des Riechens und Schmeckens, *Arch. Oto-RhinoLaryng.*, **210**, 1–14.

Andres, K.H. and v. Düring, M. (1981) General methods for characterization of brain regions, in *Techniques in Neuroanatomical Research* (eds C. Heym and W. Forssmann), Springer, Berlin, Heidelberg, New York, pp. 100–8.

Andres, K.H. and v. Düring, M. (1985) Zur Innervation der freien Brustflossenstrahlen vom Knurrhahn (Trigla lucerna). *Verh. Anat. Ges.*, **79**, 503.

Andres, K.H. and v. During, M. (1993a) Cutaneous and subcutaneous sensory receptors of the hagfish *Myxine glutinosa* with special respect to the trigeminal system. *Cell Tissue Res.*, **274**, 353–66.

Andres, K.H. and v. During, M. (1993b) Lamellated receptors in the skin of the hagfish, *Myxine glutinosa. Neurosc. Lett.*, **151**, 74–6.

Ayers, H. and Worthington, J. (1907) The skin end-organs of the trigeminus and lateralis nerves of *Bdellostoma dombeyi*, *Amer. J. Anat.*, **7**, 327–36.

Blackstad, T. (1963) The skin and the slime glands, in *The Biology of Myxine* (eds A. Brodal and R. Fänge), Universitetsforlaget, Oslo, pp.195–230.

Bone, Q. (1963). Some observations upon the peripheral nervous system of the hagfish, *Myxine glutinosa*, *J. Mar. Biol. Assoc.*, UK, **43**, 31–47.

Bullock, T.H., Northcutt, R.G. and Bodznick, D.A. (1982) Evolution of electroreception. *Trends Neurosci.*, **5**, 50–3.

Chambers, M.R., Andres, K.H., v. Düring, M. and Iggo, A. (1972) The structure and function of slowly adapting type II mechanoreceptor in hairy skin. *Q.J. Exp. Physiol*, **57**, 417–45.

v. Düring, M. and Andres, K.H. (1976) The ultrastructure of taste and touch receptors of the frog's taste organ. *Cell Tissue Res.*, **165**, 185–98.

v. Düring, M. and Andres, K.H. (1994) Topography and fine structure of proprioceptors in the hagfish, *Myxine glutinosa*. *Europ. J. Morphol.*, **32**, 248–56.

Georgieva, V., Patzner, R.A. and Adam, H. (1979) Transmissions- und elektronenmikroskopische

Untersuchung an den Sinnesknospen der Tentakeln von *Myxine glutinosa*. L. (Cyclostomata). *Zool. Scripta*, **8**, 61–7.

Fernholm, B. (1983) The lateral line of cyclosomes, *NATO ASI Ser. A*, **103**, 113–22.

Finer, T.E. (1987) Organization of chemosensory systems within the brains of bony fishes, in *Sensory Biology of Aquatic Animals* (eds J. Atema, R.R. Ray, A.N. Popper, W.N. Tavolga), Springer-Verlag, pp. 339–63.

Fox, H., Lane, B.E. and Whitear, M. (1980) Sensory nerve endings and receptors in fish and amphibians, in *The Skin of Vertebrates* (eds R.I.C. Spearman and P.A. Riley), Academic Press, pp. 271–81.

Jansen, J. (1930) *The brain of Myxine glutinosa*, J. Comp. Neurol., **49**, 359–507.

Karnovsky, M.J. (1965) A formaldehyde-glutaraldehyde fixative of high osmolarity for use in electron microscopy, *J. Cell. Biol.*, **27**, 137A.

Lindström, T. (1949) On the cranial nerves of the cyclosomes, with special reference to n. trigeminus. *Acta Zool.* (Stockh), **30**, 315–45.

McVean, A. (1989) Velocity and displacement receptors in the skin of the hagfish *Myxine glutinosa*. *J. Zool. Lond.*, **219**, 251–67.

Kishida, R., Goris, R.C., Nishizawa, H., Koyama, H., Kadota, T. and Amemiya, F. (1987) Primary neurons of the lateral line nerves and their central projections in hagfishes, *J. Comp. Neurol.*, **264**, 303–10.

Patzner, R.A., Georgieva, V. and Adam, H. (1997) Sinnerzellen an den Tentakeln der Schleimaale *Myxine glutinosa* und *Eptatretus burgeri* (Cyclostomata). Sitzungsber Österr. Akad. Wiss., **5**, 77–79.

Retzius, G. (1890) Über die Ganglienzellen der Cerebrospinalgalglien und über subcutane Ganglienzellen bei *Myxine glutinosa*. *Biol. Untersuch.*, **3**, 41–52.

Reuter, K. (1984) Chemoreceptors, in *Biology of the Integument*, Vol. 2, Vertebrates (eds J. Bereiter-Hahn, A.G. Matoltsy and K.S. Richards), Springer-Verlag, Berlin, pp. 586–99.

Reynolds, E.S. (1963) The use of lead citrate of high pH as an electron-opaque stain in electron microscopy, *J. Cell. Biol.*, **17**, 208–12.

Ross, D.M. (1963) The sense organs of *Myxine glutinosa* L., in *The Biology of Myxine* (eds A. Brodal, R. Fänge), Universitetsforlaget, Oslo, pp. 150–60.

Schoultz, T.W. and Swett, J.E. (1972). The fine structure of the Golgi-tendon organ. *J. Neurocytol.*, **1**, 1–26.

Silver, W.L. and Finger, T.E. (1984) Electrophysiological examination of a non-olfactory, non-gustatory chemosense in the searobin, *Prionotus carolinus*, J. *Comp. Physiol.* A, **154**, 167–74.

Schreiner, K.E. (1918) Zur Kenntnis der Zellgranula. Untersuchungen über den feineren Bau der Haut von *Myxine glutinosa*. I. Teil. Zweite Hälfte. *Arch. Mikr. Anat.*, **92**, 1–63.

Strahan, R. (1963) The behavior of *Myxine* and other myxinoids, in *The Biology of Myxine* (eds A. Brodal, R. Fänge), Universitetsforlaget, Oslo; pp 22–32.

Whitear, M., Mittal, A.K. and Lane, E.B. (1980) Endothelial layers in fish skin, *J. Fish Biol.*, **17**, 43–65.

Whitear, M. and Lane, E.B (1981a) Bar synapses in the end buds of lamprey skin. *Cell Tissue Res.*, **216**, 445–8.

Whitear, M. and Lane, E.B. (1981b) Fine structure of Merkel cells in lampreys. *Cell Tissue Res.*, **220**, 139–51.

Worthington, J. (1905) Contributions to our knowledge of the myxinoids. *Amer. Nat.*, **39**, 625–63.

32
CUTANEOUS EXTERORECEPTORS AND THEIR INNERVATION IN HAGFISHES

Christopher B. Braun and R. Glenn Northcutt

SUMMARY

Earlier portrayals of hagfish biology and behaviour are inconsistent with emerging evidence of hagfish sensory capabilities and ecology. Hagfishes may be characterized as chemosensory specialists, perhaps actively preying on invertebrates in the benthic and endobenthic habitat. Two sensory systems are examined in detail: the lateral line system and a chemoreceptive system which is unique to hagfishes. The lateral line system is composed of a small number of sensory patches with a simpler organization than the neuromasts of vertebrate lateral line systems. The simplicity of eptatretid lateral line systems is argued to be derived via regressive evolution in relation to burrowing habits. Potentially primitive features of the lateral line system of hagfish may be limited to the morphology of the receptor cells themselves. This simplicity and apparently paltry sensory capability is contrasted with an elaborate cutaneous chemosensory system: the Schreiner organ system. Schreiner organs are compound organs of sensory and support cells whose cytology resembles similar cell types in vertebrate taste buds. Schreiner organs are distributed quite densely on the tentacles, snout, prenasal sinus and nasopharyngeal duct, and at more modest densities in the pharynx and the epidermis of the head, trunk and tail. Schreiner organs are innervated by branches of the trigeminal, vagal and spinal nerves. This advanced sensory system rivals the most elaborate gustatory systems of vertebrates (e.g. siluriform teleosts) and is interpreted as a convergent adaptation to a dark and turbid habitat. Analysis of both the lateral line and Schreiner organ systems indicates that the sensory world of hagfishes is far richer than previously assumed.

32.1 INTRODUCTION

> The Myxinoid senses and the brain to which they pass their information tell us that *Myxine*'s life is as empty as it is obscure.
> Thus, from the poor development of some senses, distance receptors in particular, we know that *Myxine* is a creature with few responses and limited activities.
>
> (Ross, 1963)

The life of myxinoids, as pictured by most biologists, is that of sluggish, perhaps degenerate vermin, lying in zen emptiness between infrequent bouts of gorging on carcasses which happen to fall to the sea floor. Ross's (1963) chapter on the sensory system of *Myxine* in *The Biology of Myxine*, edited by Brodal and Fänge (1963), accurately characterizes the then current and still popular image of hagfishes. This image of hagfishes must be partly due to their sluggish behaviour in

The Biology of Hagfishes. Edited by Jørgen Mørup Jørgensen, Jens Peter Lomholt, Roy E. Weber and Hans Malte. Published in 1998 by Chapman & Hall, London. ISBN 0 412 78530 7.

aquaria and rapacious attacks on moribund fishes, but also perhaps to our visual bias as primates. Despite the poor development of the visual system of hagfishes, two or three other sensory systems collect information on their environment, including the distant receptive systems of olfaction, chemosensation and lateral line, and much information on the morphology of these systems has been collected since 1963 (see Braun, 1996, for a general overview of hagfish sensory biology).

Biologists, particularly morphologists, have studied myxinoids so intensively for the past two centuries that Stockard (1906, p. 483) could correctly complain that the literature was 'cumbersome'. Despite this attention, all investigations of morphology, organismal function and evolutionary significance have been performed in nearly complete ignorance of the development, life history and ecology of hagfishes.

Fortunately, recent investigations have shed light on these broader aspects of hagfish biology. The commercial importance of the hagfish fishery, exploited by the Japanese and Korean industries, has prompted several surveys (Gorbman *et al.*, 1990; Barss, 1993), providing new information on basic features of hagfish ecology, previously guessed at by only anecdotal evidence (Worthington, 1905; Foss, 1962, 1968; Strahan, 1963; Tambs-Lyche, 1969; McInerney and Evans, 1970; Howard, 1982). Still, ecological data on hagfishes (as with so many other species) are scattered and information on population size, gut contents, or potential predators is difficult to locate. For references and a comprehensive review of the subject, see Martini (this volume). In short, it is clear that hagfish populations are quite large, despite an obviously low fecundity, and a habitat generally low in primary productivity. Although hinted at by a few earlier studies (Bigelow and Schroeder, 1953; Strahan, 1963; Howard, 1982), recent dietary analysis show that hagfishes are unselective feeders, but the prevalence of invertebrate remains in hagfish guts suggests a predatory habit. It is also questionable if enough biomass would pass, as carcasses, to the benthos to permit the existence of large numbers of hagfish scavengers dependent on such food. The popular image of hagfishes as scavengers is in need of revision, and must incorporate the possibility that these animals are active predators, on both endobenthic and benthic invertebrates (mainly polychaete worms).

Recent studies of the development and morphology of the nervous system have also sparked a need to reinvent the popular notion of hagfishes. While some simplicities of the nervous system of hagfishes (e.g. the absence of a cerebellum and extrinsic or intrinsic ocular muscles) may be primitive features (Braun, 1996; Ronan and Northcutt, this volume), others appear to be secondarily derived, such as the unusual lateral line system (Braun and Northcutt, 1997), or the poorly laminated neural retina (Fernholm and Holmberg, 1975; Locket and Jørgensen, this volume). While the sensory apparatus of hagfishes is more limited than that of lampreys or gnathostomes (Braun, 1996), the anatomy of the brain itself clearly exhibits a higher level of complexity than lampreys or the probable vertebrate morphotype (Wicht and Northcutt, 1992). Ronan and Northcutt (this volume) summarize the unique features of the hagfish brain, but it is worth repeating that many of the unusual and complex neural assemblages (such as the laminated olfactory cortex) of the hagfish brain are of sensory function, and indeed relate to 'distance receptors'.

Thus, despite their few responses and limited activities in aquaria, hagfishes are not easily dismissed as inactive, insensitive creatures. Ross's four 'chief activities' of arousal, food-finding, withdrawal, and mating (1963; p. 159) are in fact, the chief activities of all species, including our own. Hagfish manage to excel at these tasks in a stygian environment with limited or no primary productivity. They are attracted to carcasses in a short time, and from some distance, and may actively locate and prey upon invertebrates in the endobenthic fauna. They are supplied with a limited number of sperm and eggs (Walvig, 1963; Hardisty, 1979), suggesting complex mating behaviours or some form of parental investment to ensure fertilization. It is difficult to believe, as Ross did, that olfaction and touch are the only modalities subserving all these functions. Only in the ignorance of biologists is the life of hagfish empty or obscure.

This chapter will focus on two previously neglected hagfish sensory modalities. The lateral line system, a primitive craniate system which has been greatly reduced and modified in hagfishes, perhaps in relation to their endobenthic habits; and a cutaneous and nasopharyngeal chemoreceptive system which rivals the elaborate gustatory systems of cyprinid or siluriform teleost fishes. Both of these unique features of hagfishes appear well suited to hagfish life as we would now know it. Local hydrodynamic pressure gradients—spatial and temporal patterns of water flow, the stimuli for the lateral line system—are not expected to be faithfully maintained beneath the benthic substrate, but diffusion of chemical stimuli (amino acids or ionic molecules) would provide excellent cues in a dark and turbid habitat. For each of these systems, the available data on the morphology of the receptors, innervation and central projections will be summarized separately. A final section will integrate these data into a broader biological context by exploring functional and evolutionary scenarios of both systems and the ecology of hagfishes.

32.2 THE LATERAL LINE SYSTEM

The lateral line system is a primitive craniate sensory system present in some form in hagfishes, lampreys and nearly all primitively aquatic gnathostomes (Northcutt, 1989). Most of these taxa have two lateral line modalities, the detection of very small electric fields and local hydrodynamic pressure differences. Electric fields are detected by small multicellular organs and electrosensory information is processed in parallel to the mechanosensory (or hydrodynamic) lateral line information (Bodznick and Northcutt, 1980; McCormick and Braford, 1988; Bodznick, 1989; Montgomery *et al.*, 1995). Electroreceptors themselves have a diversity of morphology, and appear to have evolved repeatedly in independent groups, although electroreception is a primitive vertebrate feature (Bodznick and Northcutt, 1981; Northcutt, 1986; see also Bullock and Heiligenberg, 1986). Hagfishes do not appear to be electroreceptive. There are no sensory receptors of the skin similar to either lamprey or gnathostome electroreceptors and no physiological responses to small electric fields have been detected (Bullock *et al.*, 1983; Ronan, 1986). There is no obvious reason to believe that hagfishes lost all trace of electroreceptors, so it is parsimonious to assume that electroreception evolved after the divergence of hagfishes. The mechanosensory lateral line system (hereafter the lateral line system), however, is variably developed in hagfishes and is best known from *Eptatretus stoutii* (Braun and Northcutt, 1997). Although it is unlike the lateral line systems of lampreys and gnathostomes, it does exhibit some shared primitive characteristics. Other aspects of the hagfish lateral line system are best explained as secondary simplifications unique to hagfishes.

The literature on the lateral line system in hagfishes is, quite expectedly, confused. Although several research papers have focused on the lateral line system of hagfishes (Ayers and Worthington, 1907; Fernholm, 1985; Kishida *et al.*, 1987; Braun and Northcutt, 1997), most reviews and textbooks discount the existence of a lateral line system in hagfishes (Ross, 1963; Løvtrup, 1977; Hardisty, 1979; Forey and Janvier, 1993; Forey, 1996). Some of this discrepancy may be due to systematic variation between the myxinids commonly studied by Europeans, and the eptatretids studied by Americans and Japanese (Fernholm, 1985), but some confusion has also been created by inconsistencies in the data reported in the papers cited above.

Worthington (1906), and her adviser Ayers (Ayers and Worthington, 1907), first described the peripheral sensory organs and cranial nerves they believed homologous to the lateral line system of vertebrates. According to Ayers and Worthington (1907), the peripheral lateral line system of *Eptatretus* is composed of a small number of narrow and shallow epidermal grooves, each containing a long ridge-like spindle cell neuromast. They called these whole structures canals, and described how certain fixatives caused them to actually split in vertical cross-section, forming intra-epidermal canals which they called tubes. It is reasonable to assume from their writings that Ayers and Worthington recognized that these tubes were artefactual, and represented 'definitive line[s] of weakness' (1907, p. 335). These incipient divisions may have contributed to Ayers and Worthington's belief that the lateral line system they described is 'embryonic' (p. 336), with the lines of weakness being precursors to the cavitation they may have believed formed lateral line canals. Ayers and Worthington (1907) also observed numerous epidermal neuromasts, not associated with these 'canals', distributed on the head. We now believe these epidermal neuromasts are identical to the 'endbuds' described by Schreiner (1919) and they will be discussed as chemoreceptors in the next section.

It is unfortunate that Ayers and Worthington (1907) chose to call these grooves canals, and that they dedicated half of their description to artefactual cavities which they believed demonstrated some points of fine structure of these organs. While the grooves do exist in many eptatretid species (McMillan and Wisner, 1984; Wisner and McMillan, 1990), no hagfishes possess true canals, and all subsequent publications (Plate, 1914; Fernholm, 1985) have stated that the canals described by Ayers and Worthington (1907) are artefacts. In fact, the intra-epidermal structures were called tubes by Ayers and Worthington, and their use of the term canal, to describe the lateral line grooves, is a common problem in much of the descriptions of lateral line systems, particularly those older descriptions by systematists and palaeontologists who are unable to examine any remains of the epidermis and are therefore unable or fail to recognize the distinction between superficial lines of neuromasts and canals within the dermis. This 'controversy' over artefact, and the systematic differences between *Eptatretus* and *Myxine*, may have contributed to the obscurity of the hagfish lateral line system in the nine decades following Ayers and Worthington's description (1907).

Only three recent papers have contributed new data on the lateral line system of eptatretids. Fernholm (1985) used electron microscopy to examine the grooves and Kishida *et al.* (1987) used horseradish peroxidase injections to document the central targets of the lateral line nerves. Our own work (Braun and Northcutt, 1997) used scanning electron microscopy and immunohistochemistry to document the morphology and distribution of the lateral line receptors and the nerves that innervate them.

Fernholm (1985) was unable to find any specialized cells within the grooves, although he did note that the apical microvilli in the grooves were longer than those of the general skin surface. He found no neurites in the epidermis and saw no evidence that the grooves were innervated. He reached the obvious conclusion that the grooves (which were quite real and found in all eptatretids he examined) must be a vestigial remnant of a degenerate lateral line system.

Kishida *et al.* (1987) found that the lateral line nerves (described by Worthington in 1906) project to a medial portion of the ipsilateral area acousticolateralis (hereafter the area octavolateralis).[1] This area is in a reasonably similar topological position to homologous areas in all craniates (see Ronan and Northcutt, this volume). Kishida *et al.* (1987) were cautious with their evolutionary speculations, but on the basis of the simple ipsilateral projections, and the lack of efferent innervation in hagfishes and lampreys (Yamada, 1973), they concluded that the hagfish lateral line system is in a primitive state.

32.2.1 Eptatretid lateral line organ distribution

All eptatretids thus far examined have two or three sets of epidermal grooves on each side of the head, one rostral to the unpigmented eyespot, and one or two caudal to the eyespot (McMillan and Wisner, 1984; Wisner and McMillan, 1990). Similar grooves have not been seen in any species of the genus *Myxine*, and it is presumed that this genus does not have any lateral line system. The precise distribution of grooves has only been described in *E. stoutii* (Braun and Northcutt, 1997), and the following description pertains to that species.

Although the exact number, position and orientation of the grooves is individually variable, a typical pattern of lateral line organ distribution is shown by the black patches in Figure 32.1B. Four roughly horizontal grooves are present rostral to the unpigmented epidermis overlaying the eye. This preoptic series of horizontal grooves, arranged in a diagonal row, slanting from rostrodorsal to caudoventral, has been observed in all individuals we have ever examined of *E. stoutii*. In a sample of 38 individuals caught off the coast of La Jolla, CA, the mean number of grooves in the preoptic series was 5 per side (SE ± 1). Caudal to the eyespot, there are also a number of grooves in the epidermis. In Figure 32.1B, the three vertical grooves compose the dorsal postoptic series. Vertical grooves in this position have been found in nearly all individuals of *E. stoutii* we have examined. In our sample of 38 individuals, the mean number of grooves in the dorsal postoptic series was 3 (SE ± 1). The third series of grooves is directly ventral to the dorsal postoptic group, and is thus called the ventral postoptic series. This cluster of parallel horizontal grooves was present in all individuals of *E. stoutii* we have examined and is perhaps the most consistent set of grooves, in their position, orientation and distribution amongst eptatretid species (Wisner and McMillan, 1990). In our sample of 38 animals, the mean number of ventral postoptic grooves was 3 (SE ± 1).

32.2.2 Groove morphology

Figure 32.2 shows the apical and cellular morphology of hagfish lateral line organs. Putative receptor cells within the grooves contain a substance immunoreactive to an antibody generated against acetylated alpha-tubulin (Braun and Northcutt, 1997), as shown in Figures 32.2A and B. The organization of these organs, or grooves is clearly different from the polarized strip of individually polarized hair cells found in neuromasts.

The groove overlies a shallow depression in the dermis, as described by Ayers and Worthington (1907), and has a cryptic organization (Figure 32.2B). In conventional histological material it is impossible to recognize any specialized cells. Those identified by immunohistochemical staining are indistinguishable in size and general appearance of the cell bodies in section, from the small mucous cells of the epithelium. In the intact grooves (Figure 32.2A), the apices of the

[1] These authors do not cite Fernholm (1985). They were apparently unaware that Fernholm failed to find receptive cells or groove innervation. While it is unfortunate that they may not have had access to the NATO publication of Fernholm's paper, we are fortunate that they were never tempted by it to preclude publication of their findings.

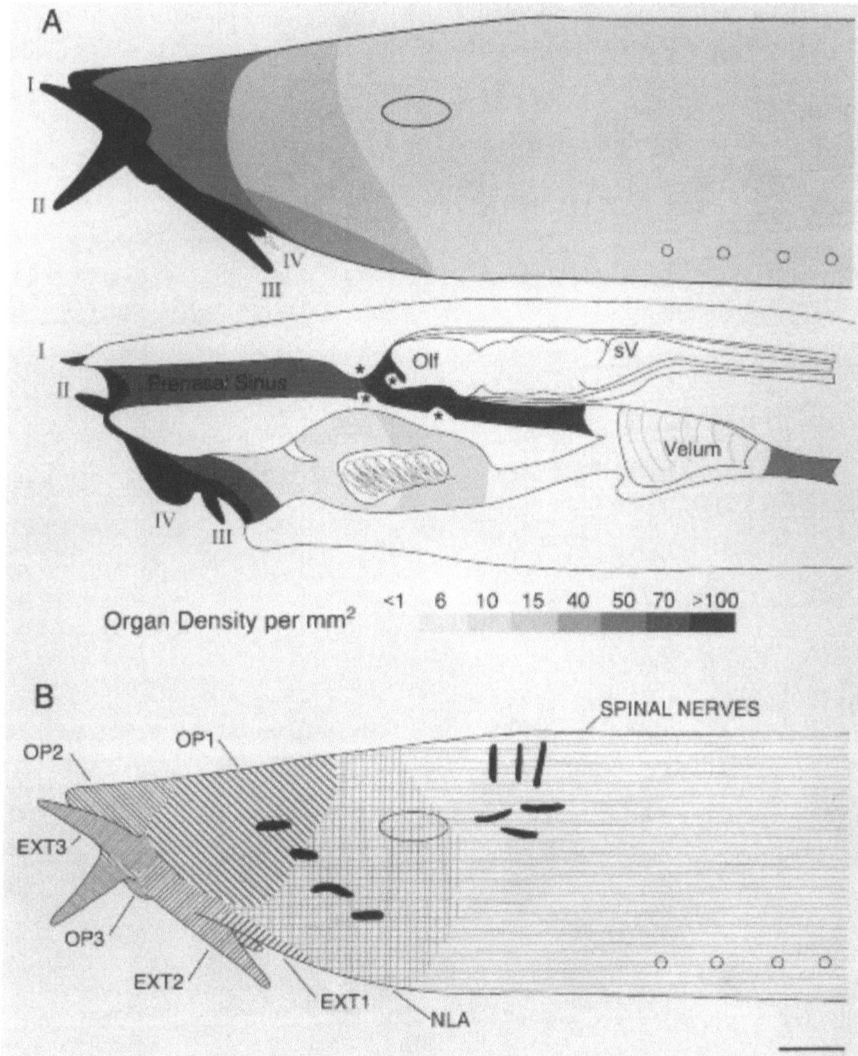

Figure 32.1(A) Semi-schematic representation of the density of Schreiner organs on the outer (upper) and inner (lower) epithelial surfaces of *Eptatretus stoutii*. Darker shades of grey indicate greater densities of Schreiner organs, as indicated by the key. The unpigmented eye spot is indicated by the large oval and the openings of the segmental slime glands are denoted with smaller circles. The flaps lining the olfactory chamber are indicated with asterisks (see Figure 32.3) and are regions with a high density of solitary sensory cells as well as Schreiner organs. I–IV: Tentacles I–IV. Olf: Olfactory organ. sV: Sensory trigeminal region of the medulla oblongata. (B) Semi-schematic representation of the innervation territories of the nerves which innervate the epidermis on the head of *Eptatretus stoutii*. The position of the lateral line grooves of that species are shown by the oblong black shapes. Trigeminally innervated territories are shown by diagonal hatching. Regions innervated by ophthalmic branches of the trigeminal are shaded with upper left to lower right hatching. Regions innervated by external branches are shaded with upper right to lower left hatching. The region of skin that is innervated by the cutaneous components of the nervus lateralis A (NLA = nervus buccalis in *Myxine glutinosa*) is indicated by a crossed hatch pattern. The remainder of the epidermal surface is innervated by cutaneous rami of all spinal nerves (horizontal hatching). OP1–3: Ophthalmic nerves 1–3. EXT1–3: External nerves 1–3. Scale for both A and B equals 0.5 cm.

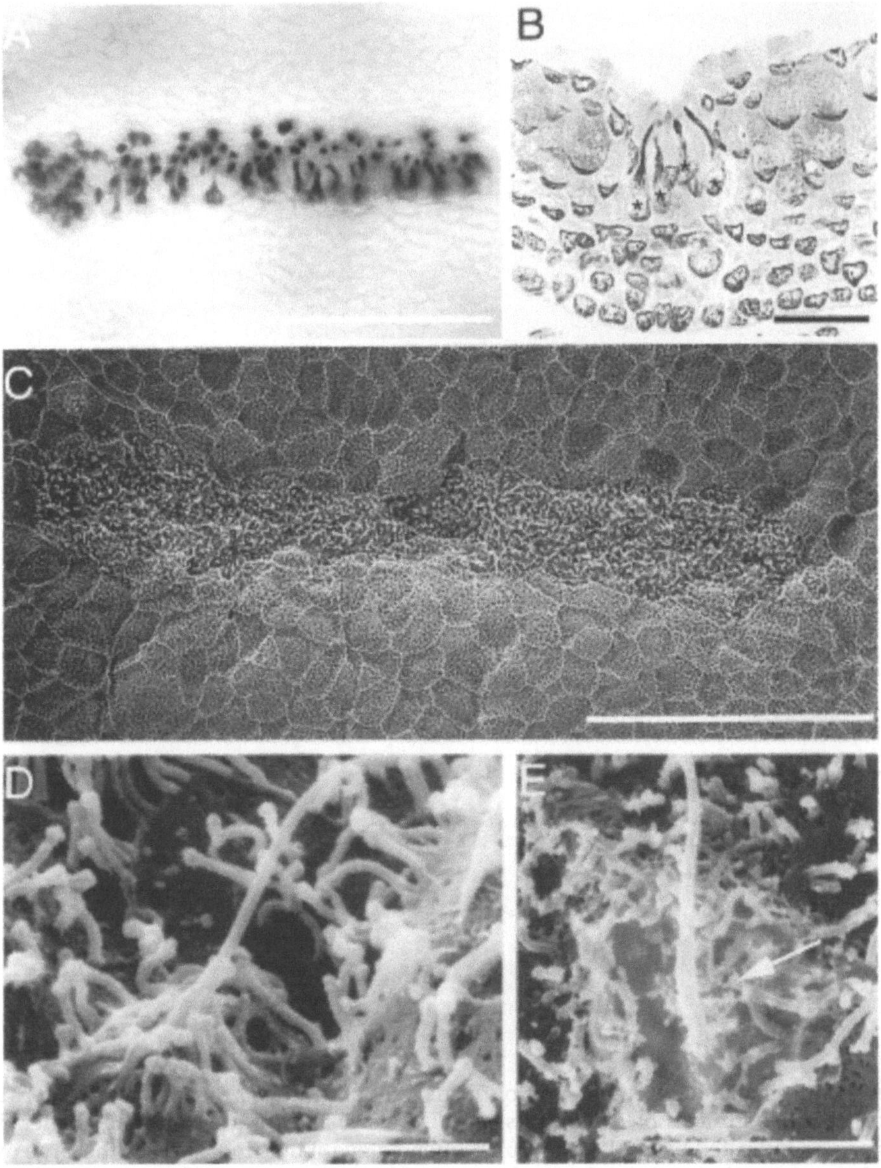

Figure 32.2 Lateral line receptors in *Eptatretus stoutii*. (A) External view of flat-mounted epidermis stained by immunohistochemical localization of acetylated α-tubulin. The flask-shaped receptor cells are seen throughout the groove. Scale bar equals 50 µm. (B) Transverse section (perpendicular to both the skin surface and the long axis of the groove) of a similarly prepared lateral line groove. The apices of four individual receptor cells are clearly stained black. Their cell bodies are marked with asterisks. Scale bar equals 25 µm. (C) Scanning electron micrograph of a lateral line groove. The groove is clearly distinct from the surrounding epidermal cell surfaces and has a somewhat irregular outline. Scale bar equals 50 µm. (D) Scanning electron micrograph of one type of specialized apex found within the lateral line grooves of *Eptatretus stoutii*. A long kinocillium is surrounded by a corolla of microvilli. Scale bar equals 2 µm. (E) Scanning electron micrograph of a second type of specialized apex found within the lateral line grooves of *Eptatretus stoutii*. A long kinocillium is flanked on one side (arrow) by microvilli. Scale bar equals 2 µm.

immunostained cells do not completely line the groove, and it is likely that the other cells lining the groove are distinct from the small mucous cells, possibly as modified support cells. The morphology of the tubulin-containing cells was very consistent, however, and does not indicate multiple receptive cell types. The cells appear to arise from the germinal layer of the epithelium (Spitzer *et al.*, 1979) and exhibit tubulin immunoreactivity shortly after leaving the most basal layer of cells. Immunoreactive cells are located within all layers of the epidermis and the most superficial cells in a groove often stain darkly in conventional histological material and appear pycnotic, suggesting that cell-turnover may be occurring in the groove. No evidence of a cupula has been found. Within the groove, a small number of cells clearly possess a long kinociliary process. Due perhaps both to preparation artefacts (breakage) or to their actual scarcity, few of these cells have been observed. Most of the cells within the groove are immunopositive for acetylated tubulin, but the relationship between immunoreactive cells and the hair cells has not been definitively established. Both features are indicative of sensory hair cells, and transcellular labelling following carbocyanine dye application indicates that the immunoreactive cells all receive innervation from the lateral line nerves (Braun and Northcutt, 1997).

Of those kinociliated cells observed, two types of apices can be differentiated. On some of these cells, a long kinocilium is ringed by a dense corolla of microvilli (Figure 32.2D), while others have a kinocilium with few surrounding microvilli and a small number of eccentrically located microvilli (Figure 32.2E). It is worth noting that hair cells of the myxinoid *macula communis* also fall into two types, one with the typical eccentric array of stereovilli of decreasing height, and one with a corolla of microvilli or stereovilli (Jørgensen, this volume). A single cilium surrounded by a corolla of microvilli is the primitive arrangement for all chordate sensory cells (Jørgensen, 1989; Braun and Northcutt, 1997), thus the presence of these cells in hagfish octavolateralis organs may be a retention of a primitive trait. It must be cautioned that very few of either cells have been reliably identified in the lateral line grooves of *E. stoutii*, and it is not clear that they reflect multiple cell types within the grooves. There are no other data, either immunohistochemical or cytological, to support such a division.

Unfortunately there have been no transmission electron microscopic examinations of the fine structure of these grooves since the work of Fernholm (1985). It is likely that the long projections seen on many cells within the grooves (Figure 32.2D and E) are cilia, with a central arrangement of microtubules similar to those of neuromast kinocilia (Jørgensen, 1989). Although the cells become transcellularly labelled after lipophilic dye application to the innervating cranial nerves (Braun and Northcutt, 1997), no details of synaptic organization or structure are known. It is likely that ultrastructural analysis will also shed light on the possible existence of multiple receptive cell types.

The lateral line grooves range in size from small patches (35 μm diameter) to long trenches nearly 5 mm long. The density of the receptor cells is fairly consistent, with an average of 215 immunoreactive cells per linear mm of groove ($N = 15$, SE $\pm$ 18). Groove length is positively correlated with animal size and it appears that the grooves continue to grow throughout the life of the animal. Irregularly shaped grooves appear to represent fusions of previously separate grooves, probably post-embryonically.

32.2.3 Groove innervation

The same technique used to visualize the receptive cells in histological preparation also stains all nerve axons and all ganglion cells in hagfishes. Intact preparations of isolated

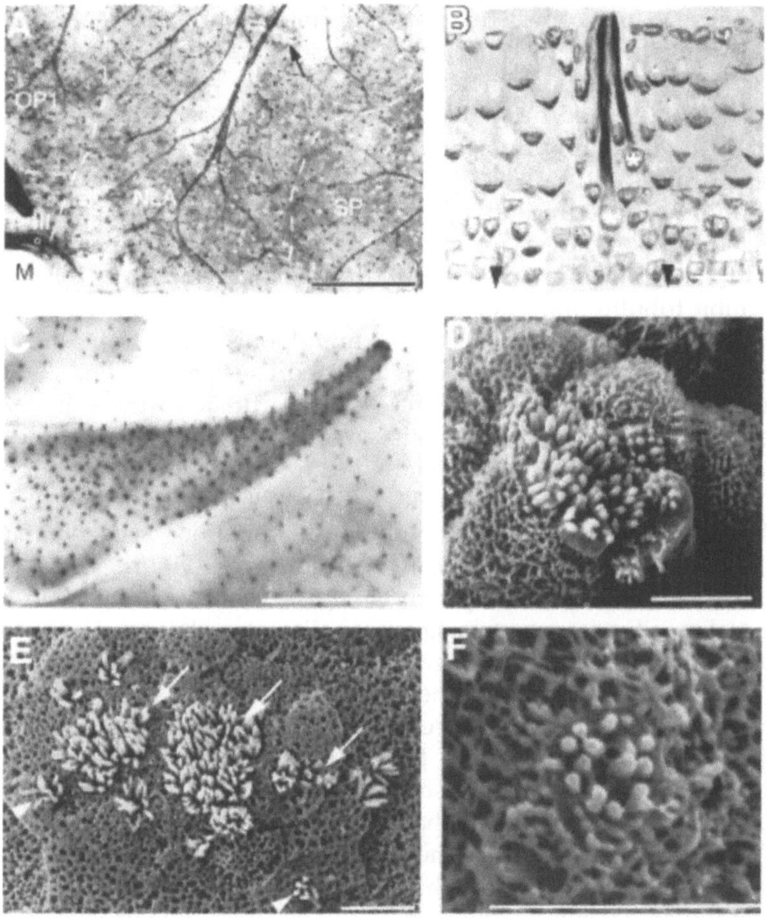

Figure 32.3 Schreiner organ morphology, distribution and innervation in *Eptatretus stoutii.* (A) Photomicrograph of the internal surface of the skin of *Eptatretus stoutii,* lateral to the opening of the mouth (M). The tip of tentacle III is also visible (III). All nerve fibres are stained black by immunohistochemical localization of acetylated α-tubulin. Fibres of the first ramus of nervus ophthalmicus (OP1), nervus lateralis A (NLA) and spinal rami (SP) can be seen and their respective innervation territories are demarcated by dashed lines. A lateral line groove innervated by NLA is also visible (arrow). Scale bar equals 2 mm. (B) Transverse section (perpendicular to the skin surface) through an epidermal Schreiner organ after acetylated α-tubulin immunohistochemical staining. Putative receptor cells are stained black. Note that the nuclei (asterisks) of the stained cells lie well above the basement lamina of the epidermis (arrowheads). Scale bar equals 25 µm. (C) A low magnification photomicrograph of tentacle III and the nearby epidermis after acetylated α-tubulin immunostaining. Note the high density of Schreiner organs (small black dots) on the tentacle and somewhat lower densities on the nearby head skin. Scale bar equals 2 mm. (D) Scanning electron micrograph of a Schreiner organ apex in the pharynx of *Eptatretus stoutii.* In the pharynx, most Schreiner organs form a mound 5–20 µm above the surrounding epithelial surface. Scale bar equals 5 µm. (E) Scanning electron micrograph of 'untidy' Schreiner organ apices from the rim of the prenasal sinus. The apices of most of the cells form three clusters (arrows), perhaps indicative of three organs. Other cells project apices to the regions nearby (arrowheads). It is unclear if these dispersed apices should be considered individual sensory cells or merely scattered apices of a single organ. Scale bar equals 5 µm. (F) Scanning electron micrograph of a solitary apex which resembles those found within the sensory surface of Schreiner organs. The cell pictured here is from the pharyngeal epithelium. Scale bar equals 5 µm.

epithelia have been used to map the innervation of all cutaneous surfaces (Figures 32.1B, 32.3A).

The grooves are innervated by ramules of two ganglionated cranial nerves. Ramules of the lateral line nerves penetrate the hypodermis and ramify in the dermis beneath the groove. The ramifications result in a dermal plexus of small (1–10) fibre bundles which is somewhat larger than the groove itself. Underneath the groove and in the surrounding regions, there are small (5–12 μm) disruptions in the basement membrane of the epidermis, where the nerve fibres likely enter the epidermis. There is currently no information available on the nature of the neurite receptor cell association or the course of individual nerve fibres within the epidermis.

Those grooves located rostral to the eye are innervated by fibres of the *nervus lateralis A*, a cranial nerve closely associated with the trigeminal complex. In addition to supplying the preoptic set of grooves, this nerve also carries fibres that terminate throughout a large patch of the epidermis of the rostrolateral snout (Figure 32.1B). This nerve has three central targets, the area octavolateralis and two separate regions of the trigeminal complex (Kishida *et al.*, 1987; Nishizawa *et al.*, 1988). These three regions subserve three separate modalities: lateral line, chemoreceptive, and general cutaneous sensation. In *Myxine*, a similar nerve, *nervus buccalis*, has been described and innervates a similar region of the snout of this genus. However, no lateral line receptors have been found, so it is likely that the *nervus buccalis* only carries two of the three modalities carried by *nervus lateralis A*.

The grooves caudal to the eye are innervated by a small nerve, *nervus lateralis B*, which is independent of all other cranial nerves. Unlike *nervus lateralis A*, this nerve carries a single component, innervating only the lateral line grooves, with no other ramifications in the epidermis. Centrally, *nervus lateralis B* only projects to the area octavolateralis (Kishida *et al.*, 1987). This nerve has not been found in *Myxine*.

32.3 CUTANEOUS CHEMORECEPTION

Retzius (1892) described clusters of sensory cells in the skin of *Myxine glutinosa*, which he likened to the taste buds of fishes. In 1919, Schreiner also observed these organs in *Myxine*, noting that they were present in the skin of the tentacles, head and body. Like Retzius, he was struck by the resemblance of these organs to vertebrate taste buds. Schreiner did describe one significant difference between these 'Sinnesknospen' or sensory buds, and taste buds in vertebrates. While taste buds generally span the distance from the outer surface of the epidermis to the basement membrane, the sensory buds on the tentacles of *M. glutinosa* did not, residing instead in the upper half of the epidermis only (Figure 32.3B). Nonetheless, Schreiner (1919) believed that these organs represented the primitive form of taste buds.

Lindström (1949) realized that these organs, if present on the tentacles and head, must be innervated by branches of the trigeminal (V) nerve. The taste buds of vertebrates are always innervated by branches of the facial (VII), glossopharyngeal (IX) and vagal (X) nerves. In light of this distinction, he argued that these organs were very unlikely to be homologous to vertebrate taste buds. Lindström (1949) believed that these organs could best be interpreted as part of the somatic sensory system and it was probable that hagfishes did not possess any endbuds (his terminology for external taste buds).

More recently, Patzner *et al.* (1977) and Georgieva *et al.* (1979) examined the fine structure of these sensory buds, particularly in comparison to vertebrate taste buds. They found enough similarities to claim that the sensory buds are chemoreceptive, but the differences they found restrained the authors from inferring a direct evolutionary relationship between taste buds and sensory buds.

Braun (1994) has detailed numerous differences between this system of sensory organs and all gustatory systems of vertebrates. Aside from gross similarities in shape and size, there are no substantial reasons to believe that these organs are homologous to taste buds. We therefore propose that the use of terms like endbuds or sensory buds is misleading and these organs should be designated Schreiner organs, in honour of the early microscopist who documented so many details of their cytology and distribution (Schreiner, 1919).

32.3.1 Organ morphology

The morphology of Schreiner organs is illustrated in Figure 32.3. They are roughly pear-shaped aggregates of cells whose nuclei reside in the middle portions of the epidermis (Figure 32.3B). The apex of each cell ascends to the surface, generally with a fairly direct course, congregating at the apex of the organ to form a receptive surface approximately 5–10 μm in diameter (Figure 32.3D). The apices of the receptive cells bear tufts of microvilli (Figure 32.3D–F) which Georgieva *et al.* (1979) called stereocilia. Schreiner organs are somewhat morphologically variable, depending on their location. In the prenasal sinus, velar cavity and pharynx, they form prominent (5–15 μm) mounds, while in the mouth and on the tentacles and epidermis, their morphology ranges from small mounds to slightly sunken pits. These differences may only be related to the differences in thickness of the surrounding epithelia, as the cell size and shape within Schreiner organs is generally consistent. The apical surface itself is also variable, as can be seen in Figure 32.3D and E. In most cases, the organs form a concise cluster of sensory apices (Figure 32.3D). In other cases, the apex is more dispersed, with occasional stray apices, up to 1 μm away from the main cluster of apices. Solitary cells, which resemble the cells within Schreiner organs, both apically (Figure 32.3F) and immunohistochemically (both are immunoreactive to an antibody generated against acetylated α-tubulin, Figure 32.3B), are found in various locations, particularly the epithelia surrounding the olfactory chamber and in the nasopharyngeal duct. It is likely that these cells are solitary chemosensory cells (see below), like those described in most classes of vertebrates (Lane and Whitear, 1982; Kotrschal, 1995). In hagfishes, however, ultrastructural studies are needed to determine if all Schreiner organs are compound organs with specific relations between multiple cell types, or if haphazard conglomerations of otherwise solitary cells also occur. Georgieva *et al.* (1979) demonstrated that the apices of all the cells in the organs are bound by desmosomes, so it is possible that the receptor cells are ensheathed by support cells, as in the taste buds of vertebrates (Jakubowski and Whitear, 1990). However this question should be re-examined, particularly in regions where the solitary cells are plentiful, and it is often impossible to distinguish between clusters of solitary cells and compound organs.

More ultrastructural data on these organs and cells are clearly needed, but some information was detailed by Georgieva *et al.* (1979) for the Schreiner organs on the tentacles of *Myxine glutinosa*. These authors described three classes of cells within the Schreiner organ, types I–III. The sensory cells (type I), which possess a bundle of apical microvilli (stereocilia of Georgieva *et al.*, 1979), have electron lucent cytoplasm and apical concentrations of fibrils, fibril bundles (*fibrillenstrukturen*) and small light vesicles. Microtubules and mitochondria are abundant, as is the endoplasmic reticulum. Cell type II, or the supporting cell is not apparent in our immunohistochemically stained material. It bears surface microvilli which do not appear very different from those of the surrounding epidermis, although the microvilli may be slightly longer. The cytoplasm is packed with lucent granules and contains few organelles. The third cell type (III, or the secretory cell) described by these authors was similar to the small mucous cells of the epidermis. It contains large amounts of endoplasmic reticulum and Golgi apparatus and a high concentration of mitochondria.

32.3.2 Organ distribution

Figure 32.1A illustrates the distribution of Schreiner organs on the head of *Eptatretus stoutii*.[2] They are distributed throughout the epidermis, from the tips of the barbels to the trailing edge of the caudal fin. The greatest densities are found on the barbels (over 250 mm^{-2}), the nasopharyngeal duct (over 400 mm^{-2}) and in the prenasal sinus (a more modest 75–100 mm^{-2}). The regions surrounding the mouth and snout also have high densities of Schreiner organs. The density is reduced caudally, particularly dorsally, but from the middle of the head rearward, they are found at a fairly uniform (6–11 buds per mm^2) density. A conservative estimate of the total number of external Schreiner organs in an adult hagfish is over 180 000 organs. Adult *Ictaluris natalis*, the catfish exemplar of gustatory fishes, possess approximately 175 000 external taste buds (Atema, 1971).

In addition to this extensive external sampling surface, Schreiner organs are also present in huge numbers internally. The elongate prenasal sinus of *E. stoutii* contains numerous Schreiner organs. The olfactory chamber and the flaps which surround it rostrally and ventrocaudally are lined with an epithelium densely (> 500 mm^{-2}) packed with Schreiner organs. Only the lateral, medial and midventral faces of the olfactory lamellae themselves are free of Schreiner organs. This high density of organs continues caudally into the nasopharyngeal duct, sharply tapering off dorsally and caudally before the velar chamber. The epithelia of the velum and velar chamber do not appear to contain Schreiner organs, but solitary acetylated-tubulin reactive cells are present. In the pharynx itself, Schreiner organs are again found at moderate densities (50–75 mm^{-2}) through the branchial region. Although the mouth opening and oral tentacles contain high densities of organs, the epithelia of the oral plates and the dorsal wall of the buccal cavity are characterized by few small Schreiner organs.

The high density of organs in the prenasal sinus indicates that this structure is an important sampling organ (Figure 32.4). Interestingly, *Myxine glutinosa* in contrast to *E. stoutii* has thousands of solitary cells in the epithelium of the prenasal sinus, with relatively few discretely organized organs. In both species the flaps surrounding the olfactory organ also contain large numbers of solitary sensory cells. The density of cells in such regions can be so high as to give the epithelium the appearance of a receptive sheet. In some individuals, Schreiner organs are also clearly present in these epithelia, but in many cases it is impossible without ultrastructural data to distinguish between clusters of individual cells and compound organs.

32.3.3 Innervation and central processing of Schreiner organs

The most obvious vertebrate parallel to the Schreiner organ system is the system of external taste buds found in many bony fishes, particularly the cyprinid and siluriform fishes (e.g. Atema, 1971; Gomahr *et al.*, 1992). In bony fishes, external taste buds are innervated by facial rami which fuse with other cutaneous (primarily trigeminal) rami or form

[2] All of the specific data cited below on Schreiner organ densities are for *E. stoutii*. Although Schreiner (1919) described these organs from the skin and tentacles of *Myxine glutinosa*, von Düring and Andres (this volume) caution that, while present, Schreiner organs are far fewer in this species. Our (SEM and immunohistochemical) observations on more limited numbers of specimens do indicate that this system is similarly extensive in both species, although there are some potentially interesting differences in the morphology and distribution of Schreiner organs.

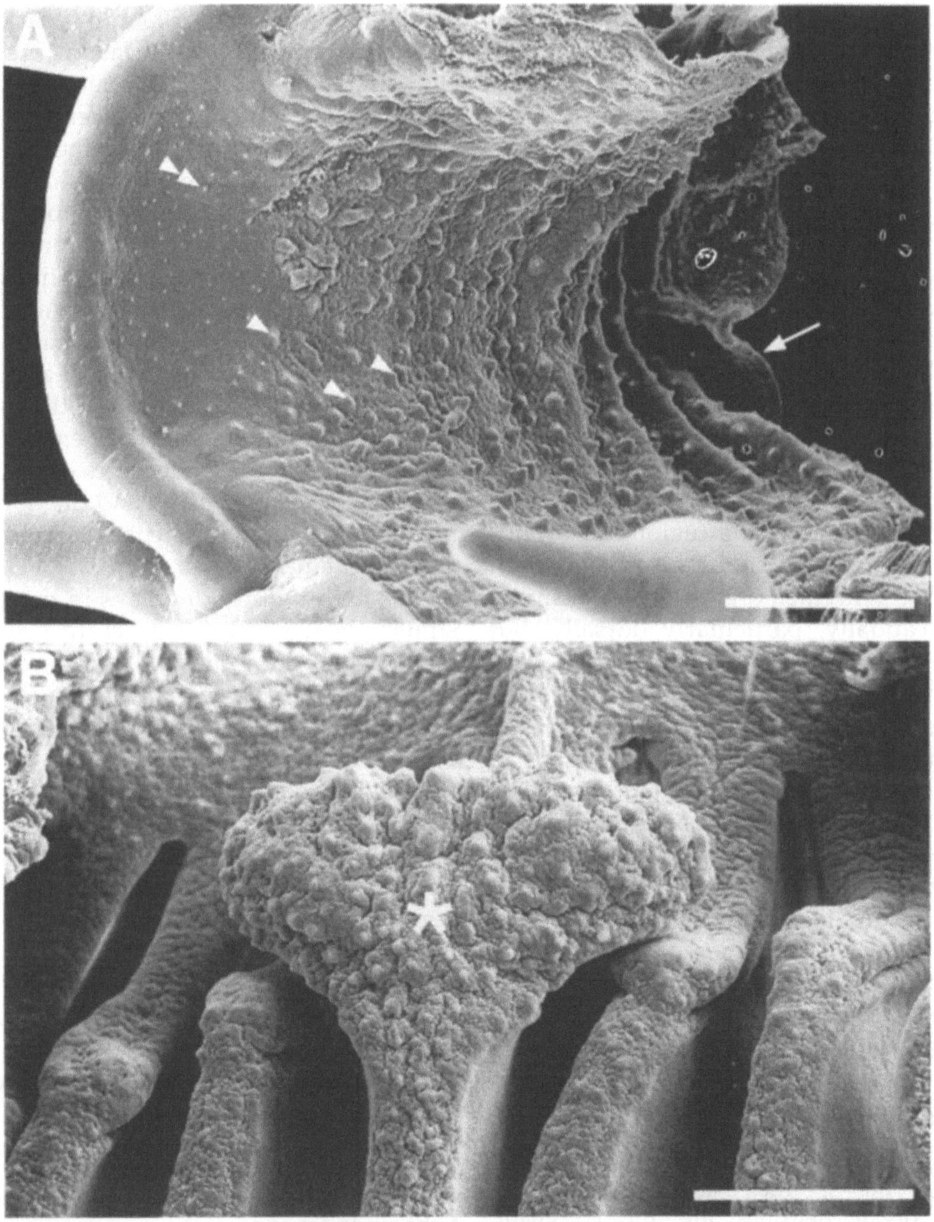

Figure 32.4 Scanning electron micrographs of the prenasal sinus and nasal chamber in *Eptatretus stoutii*. (A) Dorsal wall of the prenasal sinus. The numerous Schreiner organs (arrowheads) are readily identifiable because of the mounds they form in the caudal four-fifths of this epithelium. Schreiner organs with flush apices (in the rostral portion of the prenasal sinus and the epidermis of the head and trunk) can be seen as smaller white spots (double arrowheads). The valve rostral to the olfactory organ can also be seen (arrow) although it has been distorted by the preparation of the specimen. Scale bar equals 1 mm. (B) Ventral view of the olfactory organ. Olfactory and respiratory water enters the olfactory chamber rostrally from the prenasal sinus (up in this figure), flowing over the vertical lamellae of the olfactory organ. The central lamella bears a rostral prominence (asterisk) which contains a high density of Schreiner organs and solitary chemosensory cells. Scale bar equals 400 µm.

a recurrent ramus (often fused with a ramus of the posterior lateral line nerve) which runs the length of the body innervating taste buds on the trunk and tail. Within the rostral pharynx, taste buds are also innervated by the facial nerve. Taste buds in the caudal pharynx are typically innervated by branches of the glossopharyngeal and vagal nerves. Centrally, all gustatory afferents combine to form the solitary tract with its associated nuclei in the medulla (Finger, 1987). Schreiner organs clearly differ from taste buds in their morphology and from the primitive oropharyngeal distribution of taste buds in vertebrates (Braun, 1996). The innervation and central projections of Schreiner organs, however, are even more unusual and indicate that Schreiner organs are not homologous to vertebrate taste buds.

Schreiner organs are ubiquitous epithelial sensory receptors in hagfishes. That is, they are present in nearly all epithelia exposed to the aquatic milieu. Their innervation is correspondingly diverse. Description of the innervation of this system is complicated by the differences between hagfishes and vertebrates. There have been disagreements about the very existence of facial (Worthington, 1906; Lindström, 1949; Nishizawa *et al.*, 1988) and glossopharyngeal nerves (Stockard, 1906; Johnston, 1908) in hagfishes, and it is now clear that their corresponding pharyngeal pouches degenerate in early ontogeny (Stockard, 1906; Wicht and Tusch, this volume). The trigeminal and vagal nerves may be more easily compared to those of vertebrates. The condition of the sensory facial and glossopharyngeal neurons, and the homologies of any particular nerve, branch or ramus await a full reconstruction of the course and central and peripheral targets of all ganglion cells in hagfishes. However, some preliminary hypotheses may be inferred from the available evidence (Worthington, 1906; Lindström, 1949; Kishida *et al.*, 1987; Nishizawa *et al.*, 1988; Ronan, 1988; Matsuda *et al.*, 1991), and these will be described below.

Historically, interpretation of the facial complex of hagfishes has been difficult. Worthington (1906) described a small facial nerve which provides motor and proprioceptive innervation to several muscles of the jaw and velum. Like Lindström (1949), we have been unable to confirm Worthington's observation of a cutaneous ramus in the vicinity of the fourth tentacle, a region innervated by ramule 1 of the external ramus of the trigeminal nerve. The facial nerve of Worthington (1906) does not innervate Schreiner organs.

Although the existence of a distinct glossopharyngeal nerve has been questioned (Johnston, 1908), the nerve that Worthington described as the combined glossopharyngeal and vagal nerves runs caudally from its exit from the vagal lobe to the pharynx. At the caudal end of the velar region, segmental branches leave the nerve trunk and innervate the prebranchial pharynx. A single branch is given off to each gill pouch and provides sensory innervation to the gill epithelium, afferent and efferent branchial ducts, and presumably the pharynx proper. The muscular sheath surrounding the pharynx, branchial ducts and gills also receive motor innervation from this nerve. Schreiner organs within the pharynx are presumably innervated by fibres within IX/X. The details of pharyngeal Schreiner organ innervation await experimental demonstration, and the organization of the glossopharyngeal and vagal neurons deserves further study.

All sensory nerve branches of the hagfish trigeminal (ophthalmic, palatine, external, dentary, velobuccal and buccal, in the terminology of Lindström, 1949) innervate epithelia which contain Schreiner organs (Figure 32.1B). This includes the epidermis of the snout, the barbels, the prenasal sinus, and the mucosa of the mouth, velum and velar chamber (with perhaps only solitary chemosensory cells) and rostral pharynx. The regions of highest Schreiner organ density are innervated by the ophthalmic and external rami (barbels and snout), and the palatine ramus (prenasal sinus and nasopharyngeal duct). The remainder of the Schreiner organs are innervated by cutaneous rami of spinal nerves (*rr. cutanous superior, medius*, and *inferior* of Allen, 1917).

Thus, unlike the gustatory system of vertebrates, where taste buds are innervated by a few nerves (VII, IX, X), the Schreiner organ system of hagfishes receives a diffuse projection from all cutaneous rami (trigeminal and spinal) and perhaps from the glossopharyngeal/vagal nerve. The central projections of these nerves are similarly distinct from the gustatory nuclei (the solitary complex: Finger, 1987) of vertebrates. The central projections of most trigeminal rami in hagfishes have been determined experimentally (Nishizawa *et al.*, 1988; Ronan, 1988), and these data provide the basis for the following interpretation of the central processing of Schreiner organs.

All cutaneous rami of trigeminal (Nishizawa *et al.*, 1988; Ronan 1988) and spinal nerves (C.B.B., pers. obs.) contain at least two classes of afferent fibres (coarse and fine fibres). Upon entering the central nervous system, these fibre classes segregate into two fasciculi running from the hindbrain (in the sensory trigeminal nucleus of Jansen, 1930) through the spinal cord (in the dorsolateral funiculus), although discreet fasciculation of the different fibre types is not evident caudal to the second or third spinal segment. In the hindbrain, each of these tracts comprises a complete somatotopic map of the snout and tentacles (Nishizawa *et al.*, 1988; Ronan, 1988). Nishizawa *et al.* (1988) therefore concluded that each map represents a different cutaneous sensory modality. All trigeminal rami carry both Schreiner organ information and general cutaneous sensitivity (pain, touch, etc.), so it has not been possible to determine which map contains Schreiner organ information and which contains general somatosensory information. Based on the relative intensity of projections of rami which contain predominantly Schreiner organ afferents (such as the palatine), Nishizawa *et al.* (1988) concluded that the coarser fibres form a somatotopic map of chemosensory ('endbuds', after Georgieva *et al.*, 1979) information. The entire trigeminal complex (both maps plus a representation of the oropharyngeal mucosa) was compared to the nucleus of the descending trigeminal tract of vertebrates, in light of its relationship with the spinal cord (Nishizawa *et al.*, 1988) and its efferent projections to the optic tectum (Ronan, 1988).

Spinal afferents merge with the trigeminal afferents rostrally and both fasciculi of the descending trigeminal tract are continued into and through the length of the spinal cord, in the same manner that Lissauer's tract and the dorsal horn of the spinal cord continue to descending trigeminal tract and nucleus in vertebrates (Ariëns Kappers *et al.*, 1936). Thus external (and nasopharyngeal) Schreiner organs are represented in a continuous column which spans the length of the neuraxis in parallel to other somatosensory information.

In summary, the details of the innervation and central projections of Schreiner organs are quite different from the innervation of taste buds in vertebrates. In all vertebrates, taste buds are innervated by branches of the facial, glossopharyngeal and vagal nerves. Those vertebrates which possess extra-oral taste buds also possess derived external and recurrent rami of the facial nerve that innervate the taste buds of the lips, head, and trunk. In contrast, Schreiner organs are primarily innervated by trigeminal and spinal nerves. The primary representation of Schreiner organs is not a medullary branchiomeric nucleus of the facial, glossopharyngeal and vagal nerves, but a continuous nucleus of the hindbrain and spinal cord. We feel that this is the strongest evidence that the Schreiner organs have no more than a superficial resemblance to taste buds and they should not be considered homologous.

32.3.4 Solitary chemosensory cells

Solitary cells similar to the receptor cells of Schreiner organs have been seen in both acetylated tubulin immunohistochemical material and with scanning electron microscopy (Figure 32.3F). These solitary cells have been seen primarily internally, although they are also present at the tips of the barbels. They are most common, in *Eptatretus*, on the fleshy folds surrounding the olfactory chamber and on a heart-shaped rostral projection of the medial olfactory lamella (marked with an asterisk in Figure 32.4B), in the nasopharyngeal duct, and on the velum. These areas, especially the folds, also contain Schreiner organs and are often so packed with sensory cells and organs as to form a receptive sheet. In all of these locations, it is difficult to tell the difference between Schreiner organs and haphazard conglomerations of solitary cells. Transmission electron microscopy is needed to determine the relationships of the various cell types in these regions. The function of these solitary cells is entirely unknown, as are their potential stimulants. Solitary chemosensory cells in lampreys and some teleosts are specifically tuned to particular components of fish mucous, such as sialic acid (Baatrup and Døving, 1985; Peters *et al.*, 1987; Kotrschal, 1995). By analogy to these cells, and those in Schreiner organs, we hypothesize that the solitary cells of hagfishes are also chemosensory. The apical surface of these cells is similar to oligovillous cells of lampreys (Whitear and Lane, 1983) and to those of elasmobranchs (Whitear and Moate, 1994), and it is possible that solitary chemosensory cells of all craniates are homologous (Braun, 1996).

32.4 ECOLOGICAL AND EVOLUTIONARY CONSIDERATIONS

Our new assessment of hagfish sensory biology must include two previously unrecognized modalities. The lateral line and the cutaneous chemoreceptive systems of hagfishes display an interesting contrast: while the lateral line system of eptatretids seems meager and degenerate, the system of Schreiner organs is elaborate and unique. The lateral line system exemplifies what has interested biologists in hagfishes, their unusual features only interpretable as primitive or degenerate, or perhaps more accurately as a mixture of both. The Schreiner organ system, on the other hand, is surprising because we know so little about hagfishes and their behaviour. Its elaboration and potential sophistication invalidate our image of hagfishes as either simply primitive or degenerate craniates.

The lateral line system of craniates is a primitive sense which is important in the lives of many species. In the few groups extensively studied, the lateral line system is used for prey detection, predator avoidance and orientation via detection of slight spatial and temporal disturbances in hydrodynamic pressure (Dijkgraaf, 1962; Coombs and Janssen, 1990; Coombs and Montgomery, 1996). While we know nothing of the relevant stimuli or biological role of the lateral line system in hagfishes, it is clear from the number of receptors, their organization and the central processing apparatus that the system in hagfishes is far simpler. On the other hand, it is not clear that this simplicity reflects the primitive characters of the earliest lateral line systems or if it is a specialization of hagfishes. If the latter possibility is correct, the lateral line systems of the earliest craniate may have been more similar to those of vertebrates than those of hagfishes.

Arguments for the primitive status of the lateral system of hagfishes can be based on the relative simplicity of the receptors. Lateral line grooves are loosely organized trenches filled with solitary receptor cells without the apical polarization typical of vertebrate mechanosensitive hair cells. Neuromasts, on the other hand, are complex multicellular organs, with multiple cell types, and with all sensory cells organized in maculae with a discrete axis of best sensitivity (Flock, 1971). The receptive apex of the lateral line sensory cells in *E. stoutii* are similar to many ciliated sensory cells of protochordates (Bone and

Ryan, 1978; Bone and Best, 1978; Stokes and Holland, 1995) in that they all bear a central cilium, surrounded by a corolla of microvilli. Each chordate hair cell type, however, differs in the exact position and packing of the surrounding microvilli, but it is likely that this arrangement of a central cilium surrounded by microvilli is the primitive condition of sensory-ciliated cells in chordates (Jørgensen, 1989). Contemporary objective techniques for the reconstruction of evolutionary sequences (Eldredge and Cracraft, 1980; Wiley, 1981) recognize that primitive characters contain no information about the relatedness of organisms or structures which retain the primitive state. Thus the similarity between hagfish lateral line receptor cells and the ciliated receptor cells of protochordates does not indicate any particular relationship between these cells, aside from the fact that they are all chordate sensory hair cells. It may indicate however that this primitive cellular morphology was also shared by some of the earliest craniates.

Like the myxinoid visual system, there are difficulties with equating simplicity with primitive features in the lateral line system. Myxinid hagfishes have no lateral line systems and the distribution and extent of lateral line organs varies in eptatretids. Any primitive features of the lateral line system in hagfishes must have been inherited from the ancestors of craniates but lost altogether in some hagfish lineages. However, if some hagfishes have reasons to lose the lateral line system, might there not be similar pressures on other hagfishes? Regressive evolution may often be said to follow trends, forming morphoclines of primitive (in this case well developed) to derived (in this case degenerate) structures (Lankester, 1880; Eigenmann, 1909; Fernholm and Holmberg, 1975). The eyes of hagfishes probably follow such a trend (Locket and Jørgensen, this volume). In this case, the well-developed and sophisticated eyes of eptatretids more closely reflects the primitive condition, while the simpler eyes of myxinids are derived. This difference has been correlated with the difference in depths the two genera typically inhabit (Fernholm and Holmberg, 1975), the greater darkness of the myxinid habitat giving free reign to regressive evolution by any of the mechanisms most frequently proposed (drift, metabolic constraints, selection: Kane and Richardson, 1986). This proposition has not been tested by examination of the range of variation within each genus, but it does appear to integrate all of the available data.

The lateral line system of hagfishes might follow a similar trend. Hagfish in general are burrowers, spending much of their time buried up to their snouts in soft clay or silt. The appropriate stimulus for a neuromast is, strictly speaking, displacement of the cupula. It is reasonable to assume that skin pressed to clay permits little water flow past superficial lateral line organs. As darkness negates the use of vision, burrowing may preclude the use of hydrodynamic receptors. While both genera of hagfish burrow, there are anecdotal data that suggest eptatretids spend less time in burrows and may prefer rocky substrata (Worthington, 1905; Martini, this volume). This difference may explain the difference in distribution of lateral line systems in the two genera. Regressive evolution has been in progress since the ancestors of hagfishes burrowed in the substrates, and has culminated in the complete loss of the system in myxinids, while the eptatretids have retained some vestige of a system that might still function while above the substrate.

Like the eyes of eptatretids however, there are still indications that the eptatretid lateral line organs have also been affected by regressive trends. The sensory cells may have the primitive morphology of octavolateralis hair cells, but they are scarcely and haphazardly distributed. The organs are also pseudo-randomly distributed and show variation in size, shape, number and position (Braun and Northcutt, 1997). There is little obvious stability in the development of the lateral line in several species of hagfishes. It is difficult to believe that this randomness has been maintained in the roughly 500 million years since the system arose. Thus it is possible that simple arrays of sensory cells and the unstereotyped distribution of organs are regressive features of hagfishes and not reflective of the primitive nature of the lateral line system. Discrete lines of well organized and polarized neuromasts are equally likely to have been features of the earliest lateral line systems. These features of hagfish lateral line systems, disorganized grooves haphazardly distributed on the head, could either be primitive for craniates, or derived from a system more reminiscent of the vertebrate condition (lines of neuromasts on the head and trunk). The functional argument based on hagfish habitats and distribution support a degenerate trend, as does the variation in organ distribution (Wilkens, 1993). It is therefore difficult to assume that any aspects of the eptatretid lateral line system reflect a retention of primitive features. The lateral line system may be just what Ross (1963) predicted, 'the poor remnant that might be expected if the lateral line organs of hagfish were undergoing the same kind of reduction seen in the eye . . .' (p. 155).

The Schreiner organ system, however, could not have been predicted by Ross, and certainly questions his and many other biologists' image of hagfish life. The vertebrate gustatory system may be defined as a system of chemosensory organs primitively confined to the oropharynx innervated by branchiomeric nerves VII, IX, and X, and represented in the region of the hindbrain associated with other sensations from the oropharyngeal mucosa (Herrick, 1906; Ariëns Kappers *et al.*, 1936). It is not at all clear that hagfishes have such a system, due to the difficulties in interpreting the condition of nerves VII, IX and X. If hagfishes do not possess a gustatory system, it is possible that such a system was not present in the earliest craniates and arose later, perhaps with the origin of the vertebrates (Braun, 1996).

The Schreiner organ system, however, may be a gustatory analogue, evolved under the same pressures that shaped the most elaborate gustatory systems of vertebrates: darkness, turbidity and detection of prey buried within the substrate. Unlike gustatory systems, the Schreiner organ system is innervated by trigeminal and spinal nerves. The Schreiner organ system is not represented in the viscerosensory zone of the hindbrain, but rather a 'somatosensory' zone of the hindbrain and spinal cord. Even the organs themselves are dissimilar to taste buds. All of these differences, in the absence of all but gross similarities between receptors, make it difficult to accept a hypothesis of homology between taste buds and Schreiner organs. The similarities, chemosensory cytology (Georgieva *et al.*, 1979), and putative functional and ecological roles, must therefore be interpreted as convergent.

A well-developed long-range chemosensory system could complement the olfactory system in a dark, turbid habitat where hagfishes hunt their prey, be it distantly scattered carcasses or cryptic endobenthic invertebrates. Teleost fishes which inhabit similar niches often possess the most highly developed gustatory systems (Gomahr *et al.*, 1992). Although experimental data are desperately lacking, we may infer that hagfishes use their Schreiner organs for similar purposes, orientation to food (Bardach *et al.*, 1967), and evaluation of foodstuff and respiratory water (Kapoor *et al.*, 1975). In contrast to the lateral line system, the Schreiner organ system forces us to question a view of hagfishes as sluggish, degenerate worms with little need for sensory input. The complexity of the Schreiner organ system and its obvious elaboration surprise because they indicate that the life of hagfishes is obscure, but not empty. It reminds us of how little we know about the things that have enabled the hagfish lineage to survive repeated mass extinctions and thrive today, hundreds of millions of years after their origin.

ACKNOWLEDGEMENTS

We are most grateful to the organizers of this conference for the opportunity to participate in this symposium and contribute this chapter. We also thank H. Wicht, C. Wong, L. Barlow, G. Schlosser, and S. Pinca for editorial comments and suggestions on earlier versions of this manuscript. C.B.B. thanks the Department of Neurosciences (U.C.S.D.) and the meeting organizers for financial support. Portions of this research were supported by N.I.H. Grants NS 24669 and NS 24869 to R.G.N. and training grant GM 08107 to C.B.B.

REFERENCES

Allen, W.F. (1917) Distribution of the spinal nerves in *Polistotrema* and some special studies on the development of spinal nerves. *J. Comp. Neurol.*, **28**: 137–213.

Ariëns Kappers, C.U., Huber, G.C. and Crosby, E.C. (1936) *The Comparative Anatomy of the Nervous System of Vertebrates, including Man.* Reprinted in 1967 by Hafner Publishing Company, New York.

Atema, J. (1971) Structures and functions of the sense of taste in the catfish (*Ictalurus natalis*). *Brain Behav. Evol.*, **4**, 273–94.

Ayers, H., and Worthington, J. (1907) The skin end-organs of the trigeminus and lateralis nerves of *Bdellostoma dombeyi. Amer. J. Anat.*, **7**, 327–36.

Baatrup, E. and Døving, K.B. (1985) Physiological studies on soliary receptors of the oral disc papillae in the adult brook lamprey, *Lampetra planeri* (Bloch). *Chem. Senses*, **10**(4), 559–66.

Bardach, J.E., Todd, J.H. and Crickmer, R. (1967) Orientation by taste in fish of the genus *Ictalurus*. Science, **155**, 1276–8.

Barss, W.H. (1993) Pacific hagfish, *Eptatretus stoutii*, and black hagfish, *E. Deani*: the Oregon fishery and port sampling observations 1988–92. *Mar. Fish. Rev.*, **55**, 19–30.

Bigelow, H.B. and Schroeder, W.C. (1953) Fishes of the gulf of Maine. *U.S. Fish Wildl. Serv. Fish. Bull.*, **53**(74).

Bodznick, D. (1989) Comparisons between electrosensory and mechanosensory lateral line systems, in *The Mechanosensory Lateral Line: Neurobiology and Evolution* (eds. S. Coombs, P. Görner and H. Munz), Springer-Verlag, New York, pp. 653–85.

Bodznick, D. and Northcutt, R.G. (1980) Segregation of electro- and mechanoreceptive inputs to the elasmobranch medulla. *Brain Res.*, **195**, 313–21.

Bodznick, D. and Northcutt, R.G. (1981) Electroreception in lampreys: evidence that the earliest vertebrates were electroreceptive. *Science, 212*, 465–7.

Bone, Q. and Best, A.C.G. (1978) Ciliated sensory cells in amphioxus (Branchiostoma). *J. Mar. Biol. Assoc. U.K.*, **58**, 479–86.

Bone, Q., and Ryan, K.P. (1978) Cupular sense organs in *Ciona* (Tunicata: Ascidiacea). *J. Zool. (Lond.)*, **186**, 417–29.

Braun, C.B. (1994) A novel cutaneous sensory system in hagfish. Soc. *Neurosci. Abs.*, **20**, 1418.

Braun, C.B. (1996) The sensory biology of the living jawless fishes: a phylogenetic assessment. *Brain Behav. Evol.*, **48**, 262–76.

Braun, C.B. and Northcutt, R.G. (1997) The lateral line system of hagfishes (Craniata: Myxinoidea). *Acta Zool.* (Stockh.), in press.

Brodal, A. and Fänge, R. (eds) (1963) *The Biology of Myxine*, Scandinavian University Books, Oslo.

Bullock, T.H., Bodznick, D.A. and Northcutt, R.G. (1983) The phylogenetic distribution of electroreception: evidence for convergent evolution of a primitive vertebrate sense modality. *Brain Res. Rev.*, **6**, 25–46.

Bullock, T.H. and Heiligenberg, W. (eds) (1986) *Electroreception*, Wiley, New York.

Coombs, S. and Janssen, J. (1990) Water flow detection by the mechanosensory lateral line (eds W.C. Stebbins and M.A. Berkley) in *Comparative perception: Complex signals, Vol. II*, John Wiley & Sons, New York, pp. 89–124.

Coombs, S. and Montgomery J. (1996) The enigmatic lateral line system, in *Comparative Hearing: Fish and Amphibians* (eds A.N. Popper and R.R. Fay). Springer-Verlag (in press).

Dijkgraaf, S. (1962) The functioning and significance of the lateral-line organs. *Biol. Rev.*, **38**, 51–105.

Eigenmann, C.H. (1909) *Cave vertebrates of America: A study in degenerative evolution*, The Carnegie Institution of Washington, Washington, D.C.

Eldredge, N. and Cracraft, J. (1980) *Phylogenetic Patterns and the Evolutionary Process: Method and Theory in Comparative Biology*, Columbia University Press, New York.

Fernholm, B. (1985) The lateral line system of cyclostomes. In *Evolutionary Biology of Primitive Fishes* (eds R. Foreman, A. Gorbman, and J. Dodd), Plenum Press, New York, pp. 113–22.

Fernholm, B. and Holmberg, K. (1975) The eyes in three genera of hagfish (*Eptatretus, Paramyxine, and Myxine*) – a case of degenerate evolution. Vis. Res., **15**, 253–9.

Finger, T.E. (1987) Gustatory nuclei and pathways in the central nervous system, in *Neurobiology of Taste and Smell* (eds T.E. Finger and W.L. Silver). John Wiley & Sons, New York, pp. 331–53.

Flock, Å. (1971) Sensory transduction in hair cells, in *Handbook of Sensory Physiology, Vol 1. Principles of Receptor Physiology* (eds W.S. Hoar and D.J. Randall). Springer-Verlag, Berlin, pp. 241–264.

Forey, P.L. (1996) Agnathans recent and fossil, and the origin of jawed vertebrates. *Rev. Fish Biol. Fish.*, **5**(3), 267–303.

Forey, P. and Janvier, P. (1993) Agnathans and the origins of jawed vertebrates. *Nature*, **361**, 129–34.

Foss, G. (1962) Some observations on the ecology of *Myxine glutinosa* L. *Sarsia*, **7**, 17–22.

Foss, G. (1968) Behaviour of *Myxine glutinosa* L. in natural habitat: investigation of the mud biotope by a suction technique. *Sarsia*, **31**, 1–14.

Georgieva, V., Patzner, R.A. and Adam, H. (1979) Transmissions- und rasterelektronenmikroskopische Untersuchung an den Sinnesknospen der Tentakeln von *Myxine glutinosa* L. (Cyclostomata). *Zool. Scripta*, **8**, 61–7.

Gomahr, A., Palzenberger, M. and Kotrschal, K. (1992) Density and distribution of external taste buds in cyprinids. *Env. Biol. Fishes*, **33**, 125–34.

Gorbman, A., Kobayashi, H., Honma, Y. and Matsuyama, M. (1990) The hagfishery of Japan. *Fisheries*, 15(4), 12–18.

Hardisty, M.W. (1979) *Biology of the Cyclostomes*, Chapman & Hall, London.

Herrick, C.J. (1906) On the centers for taste and touch in the medulla oblongata of fishes. *J. Comp. Neurol.*, **16**, 403–40.

Howard, F.G. (1982) . . . Of shrimps and sea anemones; Of prawns and other things. . . . *Scottish Fish. Bull.*, **47**, 39–40.

Jakubowski, M. and Whitear, M. (1990) Comparative morphology and cytology of taste buds in Teleosts. *Z. Mikrosk.-anat. Forsch.*, **104**, 529–60.

Jansen, J. (1930) The brain of *Myxine glutinosa*. *J. Comp. Neurol.*, **49**(3), 359–507.

Johnston, J.B. (1908) A note on the presence or absence of the glosso-pharyngeal nerve in myxinoids. *Anat. Rec.*, **2**(6), 233–9.

Jørgensen, J.M. (1989) Evolution of octavolateralis sensory cells, in *The Mechanosensory Lateral Line: Neurobiology and Evolution* (eds S. Coombs, P. Görner and H. Münz). Springer-Verlag, New York, pp. 115–45.

Kane, T. and Richardson, R. (1986) Regressive evolution: an historical perspective. *NSS Bulletin*, **47**(2), 71–7.

Kapoor, B.G., Evans, H.E. and Pevzner, R.A. (1975) The gustatory system in fish. *Adv. Mar. Biol.*, **13**, 53–108.

Kishida, R., Goris, R.C., Nishizawa, H., Koyama, H., Kadota, T. and Amemyia, F. (1987) Primary neurons of the lateral line nerves and their central projections in hagfishes. *J. Comp. Neurol.*, **264**: 303–10.

Kotrschal, K. (1995) Ecomorphology of solitary chemosensory cell systems in fish: a review. *Env. Biol. Fishes*, **44**, 143–55.

Lane, E.B. and Whitear, M. (1982) Sensory structures at the surface of fish skin I. Putative chemoreceptors. *Zool. J. Linn. Soc.*, **75**, 141–51.

Lankester, E.R. (1880) *Degeneration. A Chapter in Darwinism*. Reprinted in *The Sources of Science*, No. 15 (1967): *The Interpretation of Animal Form*, Johnson Reprint Corporation, New York.

Lindström, T. (1949) On the cranial nerves of the cyclostomes with special reference to *N. Trigeminus*. *Acta Zool. (Stockh.)*, **30**, 316–458.

Løvtrup, S. (1977) *The Phylogeny of the Vertebrata*, Wiley, London.

Matsuda, H., Goris, R.C. and Kishida, R. (1991) Afferent and efferent projections of the glossopharyngeal-vagal nerve in the hagfish. *J. Comp. Neurol.*, **311**, 520–30.

McMillan, C.B. and Wisner, R.L. (1984) Three new species of seven-gilled hagfishes (Myxinidae, *Eptatretus*) from the Pacific Ocean. *Proc. California Academy of Sciences*, **43**(16), 249–67.

McCormick, C.A. and Braford Jr, M.R. (1988) Central connections of the octavolateralis system: evolutionary considerations, in *Sensory Biology of Aquatic Animals* (eds J. Atema, R.R. Fay, A.N. Popper and W.N. Tavolga), Springer-Verlag, New York, pp. 733–56.

McInerney, J.E. and Evans, D.O. (1970) Habitat characteristics of the Pacific hagfish, *Polistotrema stoutii*. *J. Fish. Res. Bd Canada*, **27**(5), 966–8.

Montgomery, J.C., Coombs, S., Conley, R.A. and Bodznick, D. (1995) Hindbrain sensory processing in lateral line, electrosensory and auditory systems: a comparative overview of anatomical and functional similarities. *Aud. Neurosci.*, **1**, 207–31.

Nishizawa, H., Kishida, R., Kadota, T. and Goris, R.C. (1988) Somatotopic organization of the primary sensory trigeminal neurons in the Hagfish, *Eptatretus burgeri. J. Comp. Neurol.*, **267**, 281–95.

Northcutt, R.G. (1986) Electroreception in non-teleost bony fishes, in *Electroreception* (eds T.H. Bullock and W. Heiligenberg), John Wiley & Sons, New York, pp. 257–86.

Northcutt, R.G. (1989) The phylogenetic distribution and innervation of craniate mechanoreceptive lateral lines, in *The Mechanosensory Lateral Line: Neurobiology and Evolution* (eds S. Coombs, P. Görner and H. Münz), Springer-Verlag, New York, pp. 17–78.

Patzner, R.A., Georgieva, V. and Adam, H. (1977) Sinneszellen und den Tentakeln der Schleimaale *Myxine glutinosa* und *Eptatretus burgeri* (Cyclostomata). Eine rasterelektronenoptische Untersuchung. *Anz. Öster. Akad. Wissen.*, **5**, 77–9.

Plate, L. (1924) *Allgemeine Zoologie und Abstammungslehre: Die Sinnesorgane der Tiere*, Verlag von Gustav Fisher, Jena.

Peters, R., Van Steenderen, G. and Kotrschal, K. (1987) A chemoreceptive function for the anterior dorsal fin in rocklings (*Gaidropsrus* and *Ciliata*: Teleostei: Gadidae): Electrophysiological evidence. *J. Mar. Biol. Assoc. U.K.*, **67**, 819–23.

Retzius, G. (1892) Ueber die Sensiblen Nervenendigungen in den Epithelien bei den Wirbelthieren. *Biol. Untersuch.*, Neue Folge, **4**, 37–44.

Ronan, M. (1986) Electroreception in cyclostomes, in *Electroreception* (eds T.H. Bullock and W. Heiligenberg), John Wiley & Sons, New York, pp. 209–24.

Ronan, M. (1988) The sensory trigeminal tract of the Pacific hagfish: primary afferent projections and neurons of the tract nucleus. *Brain Behav. Evol.*, **32**, 169–80.

Ross, D.M. (1963) The Sense Organs of *Myxine glutinosa* L., in *The Biology of Myxine* (eds A. Brodal and R. Fänge), Scandinavian University Books, Oslo, pp. 150–60.

Schreiner, K.R. (1919) Zur Kenntis der Zellgranula. Untersuchungen über den feineren Bau der Haut von *Myxine glutinosa. Archiv. Mikr. Anat.*, **92**(1), 1–63.

Spitzer, R.H., Downing, S.W. and Koch, E.A. (1979) Metabolic-morphologic events in the integument of the Pacific hagfish (*Eptatretus stoutii*). *Cell Tissue Res.*, **197**, 235–55.

Stockard, C.R. (1906) The development of the mouth and gills in *Bdellostoma stoutii. Amer. J. Anat.*, **5**, 481–517.

Stokes, M.D. and Holland, N.D. (1995) Embryos and larvae of a lancelet, *Branchiostoma floridae*, from hatching through metamorphosis: Growth in the laboratory and external morphology. *Acta Zool. (Stockh.)*, **76**, 105–20.

Strahan, R. (1963) The behaviour of myxinoids. *Acta Zool. (Stockh.)*, **44**, 73–102.

Tambs–Lyche, H. (1969) Notes on the distribution and ecology of *Myxine glutinosa* L. *Fisk. skrifter serie havundersøkelser*, **15**, 279–84.

Walvig, F. (1963) The gonads and the formation of the sexual cells, in *The Biology of Myxine* (eds A. Brodal and R. Fänge), Scandinavian University Books, Oslo, pp. 530–80.

Whitear, M. and Lane, E.B. (1983) Oligovillous cells of the epidermis: sensory elements of lamprey skin. *J. Zool. (Lond.)*, **199**, 359–84.

Whitear, M. and Moate, R.M. (1994) Chemosensory cells in the oral epithelium of *Raja clavata* (Chondrichthyes). *J. Zool. (Lond.)*, **232**, 295–312.

Wicht, H. and Northcutt, R.G. (1992) The forebrain of the Pacific hagfish: a cladistic reconstruction of the ancestral craniate forebrain. *Brain Behav. Evol.*, **40**, 25–64.

Wiley, E.O. (1981) *Phylogenetics: The Theory and Practice of Phylogenetic Systematics*, Wiley, New York.

Wilkens, H. (1993) Neutrale mutationen und evolutionäre fortenwicklung. *Z. Zool. Syst. Evolut.-Forsch.*, **31**, 98–109.

Wisner, R.L. and McMillan, C.B. (1990) Three new species of hagfishes, genus *Eptatretus* (Cyclostoma, *Myxinidea*), from the Pacific coast of north America, with new data on *E. deani* and *E. stoutii. Fish. Bull., U.S.*, **88**, 787–804.

Worthington, J. (1905) Contributions to our knowledge of the myxinoids. *Amer. Nat.*, **39**, 625–63.

Worthington, J. (1906) The descriptive anatomy of the brain and cranial nerves of *Bdellostoma dombeyi. Q.J. Microsc. Sci.*, **49**, 137–81.

Yamada, Y. (1973) Fine structure of the ordinary lateral line organ, I. The neuromast of lamprey *Entosphenus japonicus. J. Ultrastruc. Res.*, **43**, 1–17.

33
THE OLFACTORY SYSTEM OF HAGFISHES

Kjell B. Døving

SUMMARY

A single anterior nostril situated above the mouth leads to the olfactory organ of hagfishes. The olfactory sensory epithelium is found just anterior to the olfactory bulb. The sensory cells are distributed on both sides of seven olfactory lamellae that are oriented in parallel to the median plane and attached to the dorsal roof of the olfactory cavity. The receptor neurones are bipolar sensory cells with axons terminating in the olfactory bulb. The receptor neurones are of two types, one type is equipped with microvilli on their distal terminal swelling, the other type has cilia. Isolated receptor neurones have diameters around 2 µm and the total length including cell body and dendrite varies between 40 and 135 µm. Recordings of the EOG (electro-olfactogram) indicate that the receptor neurones are sensitive to amino acids.

Tagging experiments indicate that hagfishes can return to the original capture site after displacements. The role of the olfactory organ of hagfishes to perform these migrations and aggregations is discussed.

33.1 INTRODUCTION

The olfactory organ of *Myxine* is well developed and seems to play an important part in finding food and in return to home site. The morphology of the olfactory organ has been described by several authors (Müller, 1834, 1840; Parker, 1883; Cole, 1913; Peters, 1963; Roos, 1963), but the more recent studies on *Myxine glutinosa* by Thiesen (1973) and on *Eptatretus stoutii, Eptatretus deani* and *Myxine circifrons* by Theisen (1976) are to be consulted for details on the fine morphology. Histological studies have been presented by Retzius (1880).

33.2 GROSS MORPHOLOGY

One single nostril leads to the olfactory organ in the hagfish. The nostril is situated on the terminal of the head above the mouth. The exterior part of the nostril is surrounded by a dorsomedial lip and two nasal tentacles, one on each side. From the nostril a prenasal sinus leads to the olfactory organ which is situated just anterior to the brain. The duct continues behind the olfactory organ as a nasopharyngeal duct (Figure 33.1A).

Just anterior to the olfactory organ is found a valve of soft tissue. It is attached to the wall of the nasal duct and consists of a circular fold. The free edge of the valve is bent posteriorly, and in connection with the valve a longitudinal medial fold extends forward. This valve may function as a device to divert the waterflow into the spaces between the olfactory lamellae.

The nasal cavity is almost completely filled

The Biology of Hagfishes. Edited by Jørgen Mørup Jørgensen, Jens Peter Lomholt, Roy E. Weber and Hans Malte. Published in 1998 by Chapman & Hall, London. ISBN 0 412 78530 7.

by the seven olfactory lamellae (Figure 33.1B). Each lamella is triangular in shape. The olfactory lamellae are oriented in parallel to the body axis and fastened to the dorsal roof of the olfactory cavity. The olfactory epithelium covers both sides of the seven lamellae and the lateral walls of the olfactory cavity. The ventral ridge of the lamellae are covered with indifferent epithelium.

33.3 THE MORPHOLOGY OF THE OLFACTORY RECEPTOR NEURONES

The olfactory receptor neurones are bipolar primary sensory cells with the cell somata situated in the sensory epithelium. The dendrite extends to the epithelial surface where it forms a slight terminal swelling, the olfactory knob. The axon of the receptor neurone parts from the cell somata. Theisen (1973) noted that the olfactory cells are separated by sustentacular cells that spaced out the receptor neurones in a regular manner (see Figure 33.5). It is only in the deeper part of the epithelium that sensory cell dendrites may occasionally be in contact with each other.

Isolated receptor neurones are easily obtained by exposing small pieces of olfactory epithelia to a papain in hagfish Ringer

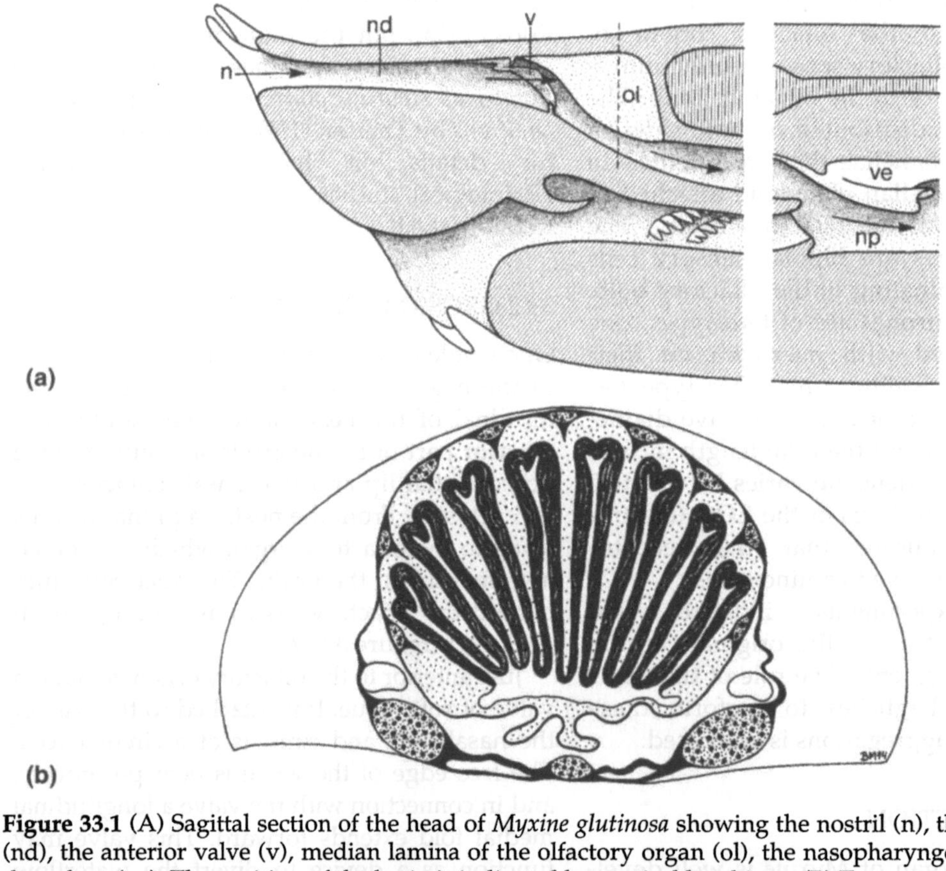

Figure 33.1 (A) Sagittal section of the head of *Myxine glutinosa* showing the nostril (n), the prenasal sinus (nd), the anterior valve (v), median lamina of the olfactory organ (ol), the nasopharyngeal duct (np) and the velum (ve). The position of the brain is indicated by vertical hatchings. (B) Transverse section through the olfactory organ at the position indicated by the broken line in Figure A. The seven olfactory lamellae are clearly seen.

solution and mechanically dissociating the cells. The cells are viable and show the following dimensions (n = 17): Cell somata: width and length (7.6 ± 1.4 µm, 12.0 ± 2.4 µm), range 4.5–10.7 µm. Total cell length (exclusive the axon) (77.5 ± 23.6 µm), range 40–135 µm.

The diameter of the dendrite varied from less than 2 µm to 5 µm, and the diameter of the olfactory knob could be up to 7 µm. The cells were found to keep well in an artificial seawater solution. Attempts to study their membrane properties by patch-clamp methods were hampered by poor seal resistance between the patch electrode and the cell membrane. The morphology of the isolated receptor cells corresponds well with those that Retzius observed in his studies of cyclostomes (Retzius, 1880).

33.3.1 Olfactory receptor neurones with microvilli

An extensive Golgi apparatus is found in the proximal part of the dendrite, and throughout the dendrite there are vesicles of agranular endoplasmatic reticulum. Profiles of granular endoplasmatic reticulum are most common just distal to the nucleus. Throughout the dendrite and oriented longitudinally to this are found neurotubules. Numerous mitochondria are found scattered throughout the receptor neurone (Figure 33.2).

Numerous microvilli extend from the terminal swelling, but no cilia are observed protruding from the same cell. The terminal swellings contain numerous microtubules, which appear to extend into the microvilli. Centrioles were not observed.

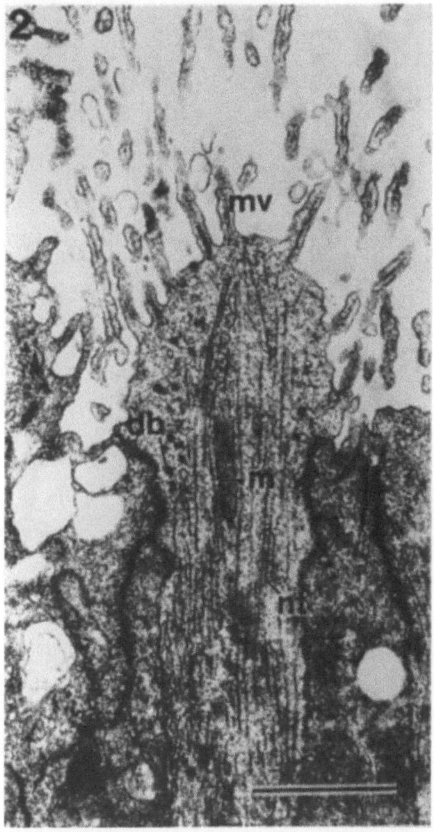

Figure 33.2 Transmission electron micrograph through the surface region of the olfactory epithelium of *Eptatretus deani* showing the apical part of a microvillous receptor neurone with microvilli (mv), dark bodies (db) and neurotubules (nt). Scale bar 1 µm (From Theisen, 1976.)

33.3.2 Olfactory receptor neurones with cilia

The fine morphology of the ciliated receptor neurones resembles those seen in microvillous receptor cells. The dark bodies are prominent (Figure 33.3). The terminal swelling (olfactory knob) contains a number of cilia varying between five and eleven. Cross-sections of the cilia show a 9+0 microtubular arrangement. At the basis of each cilium each doublet of microtubule is connected with the ciliary membrane by a Y-shaped structure. The doublets are not provided with dynein arms. The distal part of the cilia may contain a reduced number of doublets. The absence of dynein arms and a central doublet (Figure 33.4) indicate that the cilia are immobile, a property in line with observations from the olfactory receptor neurones of other vertebrates.

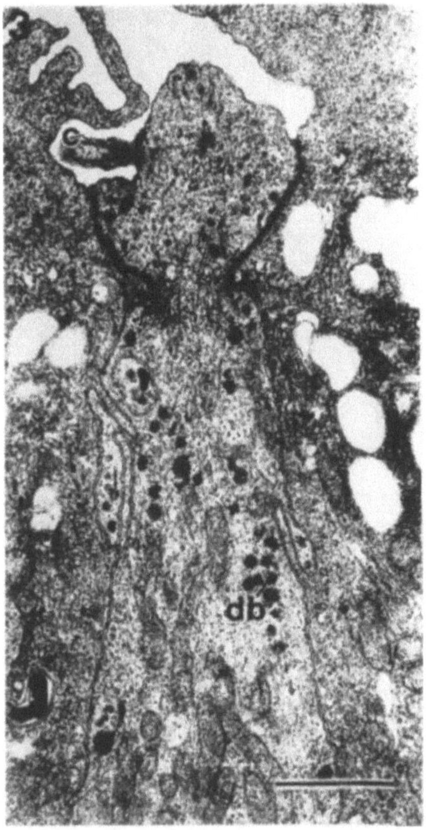

Figure 33.3 Transmission electron micrograph through the surface region of the olfactory epithelium of *Eptatretus deani* showing the apical part of a ciliated receptor neurone with a cilium (c) and dark bodies (db). Scale bar 1 µm. (From Theisen, 1976.)

33.3.3 Supporting cells

The supporting cell is nearly circular in appearance at the apical part, but in the more proximal part the outline becomes irregular with extensive extensions forming leaflike or finger-like protrusions and outgrowths. These extensions may surround single or groups of receptor neurones. The free apical part of the supporting cell bears microvilli, similar in diameter with that of the microvillous receptor neurones. The supporting cells surround as to isolate the receptor neurones as shown in Figure 33.5.

The supranuclear cytoplasm of the supporting cell contains prominent Golgi complexes. The cytoplasm also contains systems of granular endoplasmatic reticulum, free ribosomes and mitochondria. The most characteristic features are secretory granules and tonofilaments, both occurring abundantly. The secretions contain acid mucopolysaccharides. The tonofilaments traverse the supporting cell and apically a distinct layer is situated almost parallel to the free surface. In cells filled with secretory granules, this layer of tonofilaments is displaced basally. Tonofilaments also occur randomly distributed in the cytoplasm. Some tonofilaments converge on desmosomes. Desmosomes between supporting cells are found at various levels. *Maculae adherentes* are sometimes seen between supporting cells and receptor neurones. These structures are typically of asymmetrical appearance.

The lack of supporting cells with motile

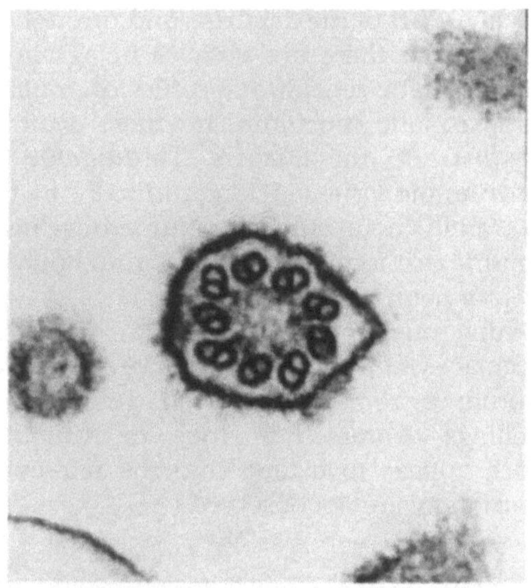

Figure 33.4 Transmission electron micrograph through the sensory cilium of a receptor neurone of *Myxine glutinosa*, demonstrating the lack of dynein arms of the outer microtubules and the missing central doublet of microtubules. (From Theisen, unpublished.)

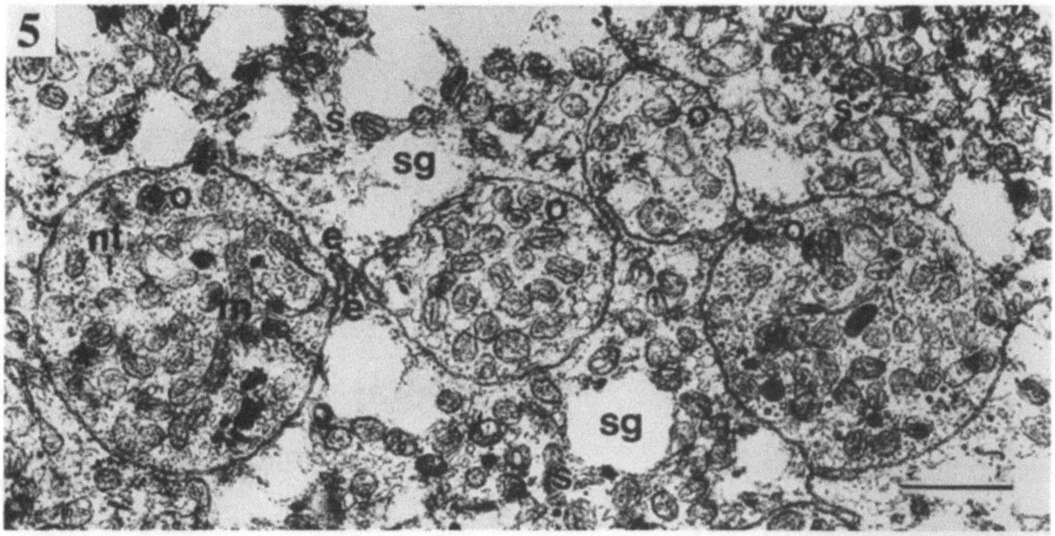

Figure 33.5 Horizontal section through the olfactory epithelium of *Eptatretus deani* showing the dendrites of the olfactory receptor neurones. Note that the olfactory receptor neurones (o) are separated from each other by supporting cells (s) and their extensions (e). In the receptor neurones are seen mitochondria (m) and neurotubuli (nt), and in the supporting cells secretory granules (sg). Scale bar 1 µm. (From Theisen, 1976.)

cilia in the sensory epithelium of hagfishes (Theisen, 1973, 1976) supports the idea that the waterflow is caused by the respiratory movements of the animal and not by ciliary beating of the supporting cells as found in many teleosts (Døving *et al.*, 1977). In the nomenclature adopted by these authors the hagfish will be classified as a cyclosmate.

33.4 PHYSIOLOGY

The electro-olfactogram, EOG (Ottoson 1956), was recorded from the olfactory organ of the Atlantic hagfish, *Myxine glutinosa* (Døving and Holmberg, 1974). Newly caught specimens were decapitated, and the dental plates, the ventral tissues and the skin overlying the dorsal surface of the olfactory organ were removed. Immobilization was secured by removing the brain. A plastic catheter was introduced into the nostril and a flow of seawater 1 ml s^{-1} was maintained through the olfactory organ. The electrical responses to stimulation with amino acids and other substances were recorded by two sintrated AG/AgCl electrodes. One electrode was connected to the flowing seawater. The other electrode, a micropipette filled with seawater, was placed on the dorsal surface of the preparation, just outside the olfactory cavity. Stimulation with the L-alanine, L-glutamine, glycine, gamma-amino butyric acid, glutathione and 4-hydroxy-L-proline caused a monophasic response similar to that exemplified in Figure 33.6A. The recording conditions were such that the surface of the olfactory epithelium was depolarized. The initial deflection was a transient peak of response. Following the peak there was a sustained depolarization that lasted for the period of stimulation. Upon cessation of stimulation the response gradually declined to the prestimulus level.

The responses varied with the position of the recording electrode placed on the dorsal surface. It was greatest in the medial part of the organ and decreased as the electrode was placed anterior or posterior to that region.

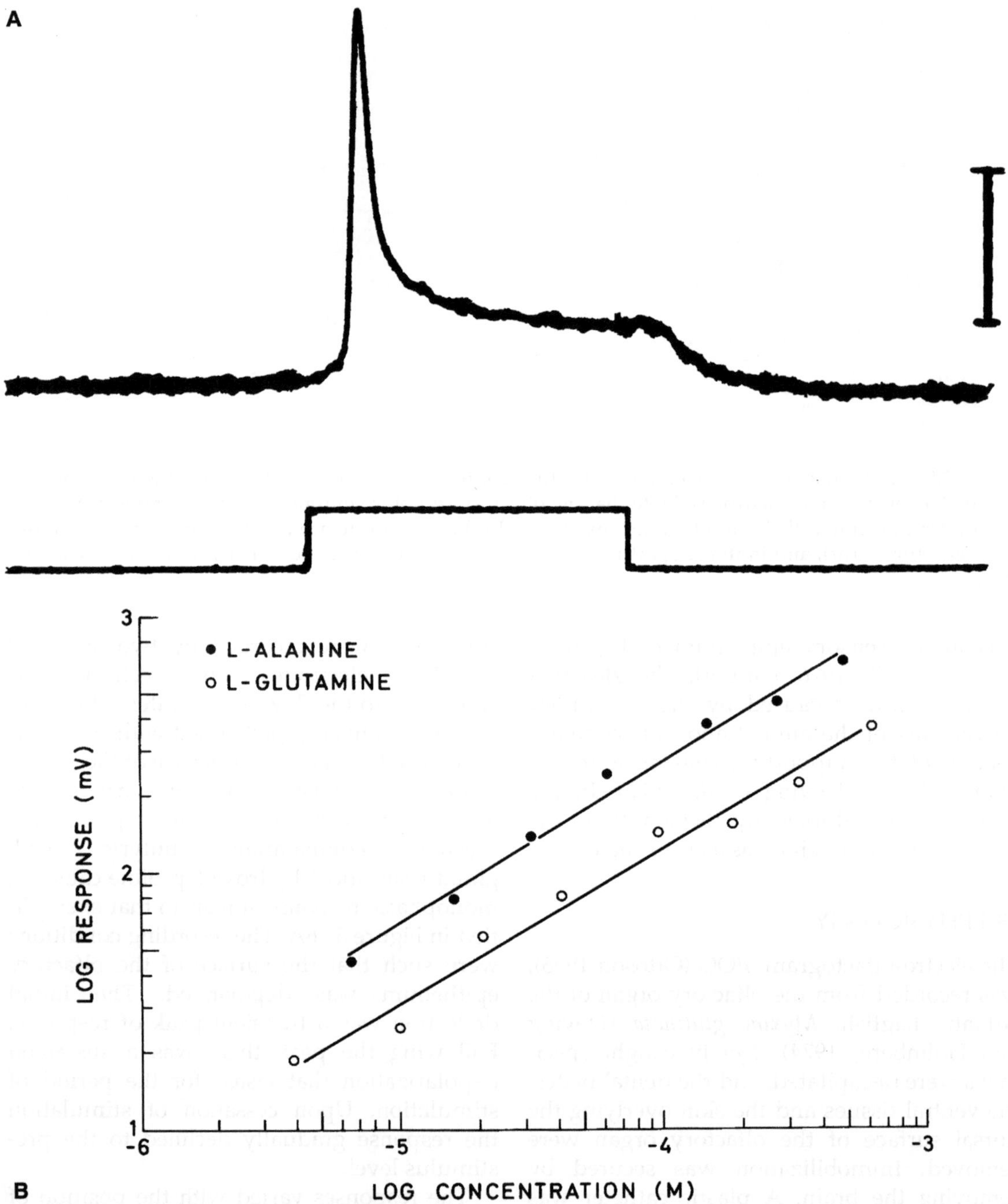

Figure 33.6 (A) The electro-olfactogram (EOG) recorded from the olfactory organ of *Myxine glutinosa* to a stimulation with 140 μM L-alanine. Stimulus duration indicated by the lower trace is 7.5 s. Vertical bar 200 μV. (B) Relation between the peak amplitude of the EOG and the stimulus concentration for L-alanine and L-glutamine. (From Døving and Holmberg, 1974.)

The responses also diminished in size as the electrode was placed lateral to the midline. A continuous flow of 3 M KCl caused a gradual decline in the amplitude of the response and eventually abolished it. It was concluded that the responses observed were analogous to the EOG first observed in frog and analysed in detail by Ottoson (1956). These were the first observations of underwater-EOG later described in other marine fishes (Silver *et al.*, 1976).

The amplitude of the response of the olfactory organ increased with increasing concentration of the stimulant. The dose–response curve for the initial peak amplitude in a log–log plot was close to a straight line with a slope of 0.64 both for L-alanine and L-glutamine (Figure 33.6B). Thus the relation between the response amplitude (*R*) and the stimulus concentration (C) could be described by a power function of the form:

$$R = k \cdot C^{0.64}$$

It is not known if the frequency of nerve impulses in the receptor neurones follow the same relation.

Of the many natural substances tested, the 'finger-rinse' was the most spectacular. One clean finger, thoroughly washed and rinsed for a long period, was dipped with two phalanges in artificial seawater for a period of 30 s. When this water was applied to the hagfish olfactory organ it caused an appreciable response from that organ. The active substances in this finger rinse are not known, but since bile salts are present in the skin and these are known as potent odorants for teleost olfactory organ (Døving *et al.*, 1980), a fair guess would be that bile salts could be responsible for a part of these responses.

33.5 FUNCTION OF THE OLFACTORY ORGAN

Hagfishes are easily caught in great quantities by simple baited traps. The short time needed to yield substantial quantities of hagfishes used for capture demonstrates that the hagfishes are abundant and that they probably are attracted to the trap by substances emanating from the bait (Martini, this volume). This means of catching the fish make them easily adaptable to tagging experiments.

Walvig (1967) developed methods to mark the hagfishes by indian ink (Black Rembrandt) and internal, subcutaneous, black plastic tags. He captured 53 hagfishes close to the biological station of the University of Oslo at Droebak, Oslofjord, in mid-December and released them 2.5 km southeast of the original site. The first recaptures were made in September the following year. A total of 18 recoveries were made over a period of 4 years. Two recaptures were made 9 and 12 years after release.

These experiments indicate that the hagfishes either migrate randomly around in the fjord or that they form discrete subpopulations that are recognized by the individuals. At present we do not know the role of the olfactory system in this return, but if it is similar to the return found in many salmonids it would be surprising if the olfactory system and specific odorants do not play a significant role in this behaviour. A return to the site of capture is an interesting observation that needs to be verified and extended.

ACKNOWLEDGEMENT

I am greatly indebted to Birgit Theisen for her kind help in providing illustrations and for correcting mistakes in an earlier version of this manuscript.

REFERENCES

Cole, J.F. (1913) A monograph on the general morphology of the myxinoid fishes, based on a study of *Myxine*. Part V. The anatomy of the gut and its appendages. *Trans. Roy. Soc. Edinburgh*, **49**, 293–344.

Døving, K.B., Dubois-Dauphin, M., Holley, A. and Jourdan, F. (1977) Functional anatomy of the olfactory organ of fish and the ciliary mechanisms of water transport. *Acta Zool.*, **58**, 245–55.

Døving, K.B. and Holmberg, K. (1974) A note on the function of the olfactory organ of the hagfish *Myxine glutinosa*. *Acta Physiol. Scand.*, **91**, 430–2.

Døving, K.B., Selset, R. and Thommesen, G. (1980) Olfactory sensitivity to bile acids in salmonid fishes. *Acta Physiol. Scand.*, **108**, 123–31.

Müller, J. (1834) *Vergleichende Anatomie der Myxinoiden, der Cyclostomen mit durchbohrtem Gaumen*, Berlin, K. Akad. der Wissenschaften, 108pp.

Müller, J. (1840) *Vergleichende Neurologie der Myxinoiden. Fortsetzung der vergleichende Anatomie der Myxinoiden*. Berlin, K. Akad. der Wissenschaften, 83pp.

Ottoson, D. (1956) Analysis of the electrical activity of the olfactory epithelium. *Acta Physiol. Scand.*, **35**, 1–83.

Parker, W.K. (1883) On the skeleton of the marsipobranch fishes. I. The myxinoids (*Myxine* and *Bdellostoma*). *Phil. Trans.* London, **174**, 373–457.

Peters, A. (1963) The peripheral nervous system, in *The Biology of Myxine* (eds A. Brodal and R. Fänge), Universitetsforlaget, Oslo, pp. 92–123.

Retzius, G. (1880) Das Riechepithel der Cyclostomen. *Arch. Anat. Entwicklungsgesch.*, 4, 9–21.

Roos, D.M. (1963) The sense organs of *Myxine glutinosa*, in *The Biology of Myxine* (eds A. Brodal and R. Fänge), Universitetsforlaget, Oslo, pp. 150–60.

Silver, W.L., Caprio, J., Blackwell, J.F. and Tucker, D. (1976) The underwater electro-olfactogram: a tool for studying the sense of smell of marine fishes. *Experientia*, **32**.

Theisen, B. (1973) The olfactory system in the hagfish *Myxine glutinosa*. I. Fine structure of the apical part of the olfactory epithelium. *Acta Zool.*, **54**, 271–84.

Theisen, B. (1976) The olfactory system in the Pacific hagfishes *Eptatretus stoutii*, *Eptatretus deani*, and *Myxine circifrons*. *Acta Zool.*, **57**, 167–73.

Walvig, F. (1967) Experimental marking of hagfish (*Myxine glutinosa* L.). *Nytt Mag. Zool.*, **15**, 35–9.

34
THE EYES OF HAGFISHES

N. Adam Locket and Jørgen Mørup Jørgensen

SUMMARY

Though probably functional light receptors, hagfish eyes are small, that of *Myxine glutinosa* only 500 μm diameter, and degenerate. Demonstrated extraocular photoreception may be more important for hagfish behaviour. *Eptatretus* species eyes are beneath an unpigmented skin patch, but *Myxine glutinosa* eyes are buried beneath muscle. All hagfishes have only an undifferentiated corneo-scleral layer, and extraocular muscles are absent. We found no lens in any hagfish examined. *Eptatretus* species have a vitreous cavity, with scattered collagen fibrils, some forming dense aggregates. Choroidal capillaries, but not pigment, occur in all species examined. *Eptatretus* retain a hollow optic cup, but at the margin epithelium and neuroretina are continuous, without extension to ciliary body or iris, both of which are absent. Developmental anomalies are common in peripheral retina in all. The *Myxine* optic cup has no lumen, the margins meeting at a fibrous plug. *Eptatretus* species retinas contain photoreceptors, with clear outer segments in the periphery, but few or none in the fundus. *Myxine* has few, degenerate outer segments, indenting the opposing epithelium. Receptor synapses are sessile. Synaptic bodies, like vertebrate ribbons, occur in *Eptatretus*, but only simple synapses in *Myxine*. *Myxine* optic nerve contains a few hundred thin axons only.

34.1 INTRODUCTION

Writing of the sense organs of hagfishes, Ross (1963) observed that the eye and ear, the most important sense organs of higher vertebrates, are small and simple in hagfishes. The olfactory organs, on the contrary, are well developed, and hagfishes are readily attracted to a bait by smell (Adams and Strahan, 1963). Few accounts of the vertebrate eye discuss the situation in hagfishes, for at least two good reasons. One is that hagfishes, though chordates, are not vertebrates, and the other is that their eyes are small and degenerate, and thus poorly suited to studies on vision. Even those of the most visually advanced species lack most components of an optical system, and those of *Myxine* are buried beneath muscle, with no trace of their existence visible from the surface. This is not to say that hagfishes are insensitive to light. Newth and Ross (1955) showed that destruction of *Myxine glutinosa* eyes had no effect on the light sensitivity, which was maximal in an area in front of the eyes. They studied the distribution of light-sensitive areas on the skin of *Myxine*, finding concentrations just behind the tentacles and in an area round the cloaca. Some light sensitivity was present over most of the rest of the animal. The reaction time of this dermal light sense was long, 2–3 min at threshold intensity, and about 10 s in maximal illumination. They found that removal of a portion of skin abolished the response, indicating that the sense organs

The Biology of Hagfishes. Edited by Jørgen Mørup Jørgensen, Jens Peter Lomholt, Roy E. Weber and Hans Malte. Published in 1998 by Chapman & Hall, London. ISBN 0 412 78530 7.

must reside in the skin. The animal's reaction was abolished anterior to a transection of the spinal cord, but behind the cut, reflex movements continued, showing that the afferents go more or less directly to the spinal medulla, rather than via a nerve, as they do in lampreys, which also have a cutaneous light sense.

The discussion of hagfish eyes which follows is based on findings reported in the literature, with our observations on new material, mostly fixed with a view to resin embedding and possible electron microscopy. We follow Bo Fernholm (this volume) for nomenclature of the hagfishes we have studied, but give authors' attributions, with Fernholm's equivalent in parentheses, where these are different in the literature.

Holmberg (1970) used the terms supranuclear and subnuclear to identify the portions of the receptors towards the outside and inside of the eye respectively, but we prefer the terms vitread and sclerad, by analogy with better developed vertebrate eyes, even though vitreous and differentiated sclera are lacking in *M. glutinosa*.

The hagfish eye has been examined by a number of workers, from Krause (1886), Kohl (1892) and Retzius (1893) in the nineteenth century to a succession of work more recently (Allen, 1905; Stockard, 1907; Dücker, 1924; Kobayashi, 1964). The important papers of Holmberg and colleagues from Stockholm, quoted *in situ*, were the first to provide ultrastructural information on the retinal cells, and form an important basis for this chapter.

Fernholm and Holmberg (1975) compared the structure of the eyes in three species of hagfish, *Myxine garmani*, *Polistotrema (Eptatretus) stoutii and M. glutinosa*. Of these Holmberg (1970) had already examined the retinal cells of *M. glutinosa*, with the least developed eyes. The same author (1971) had also compared the retinal cells in *Myxine* and *Polistotrema* (*Eptatretus*) *stoutii*, using electron microscopy in both cases, and giving excellent diagrams summarizing the features of epithelial and receptor cells. Fernholm and Holmberg's comparison showed that *M. garmani* had the most advanced structure, and that *P. stoutii* occupied an intermediate position. Holmberg (1971) quoted the results of Kobayashi (1964) on *M. garmani*, himself describing and illustrating the situation in *Polistotrema* (*E.*) *stoutii*.

34.2 *EPTATRETUS STOUTII, E. ATAMI, E. BURGERI* AND *E. CIRRHATUS*

Allen (1905) examined the eyes of 19 adult *Bdellostoma* (*Eptatretus*) *stoutii* from series sections of paraffin-embedded hagfishes which had been fixed entire in Müller's fluid. Material from the four species above, and from *Myxine glutinosa*, has been examined recently by the authors, and brief descriptions follow. The eyes of the *Eptatretus* species are all better developed than those of *Myxine glutinosa*, in that all have a vitreous cavity. None, however, has a lens. They are sufficiently similar that a single description will serve for all, with specific differences noted in context. Allen found that the eyes in his *E. stoutii* were variable in size between specimens, though these were all between 43 and 47 cm long, but were even more variable between right and left sides in a single individual. Approximately 1.3 mm diameter, the eyes were embedded in fat, just beneath a transparent patch in the skin on the side of the head; in some cases the eye abutted the surface tissues, when the external aspect of the eye was flattened, and in others it was wholly buried. He found no evidence of eye muscles, but did identify a slender optic nerve passing through the fat. His *Eptatretus* eyes had a thin and undifferentiated corneoscleral coat, in some cases fused with the deep surface of the overlying skin. The retinal portion of the eye was represented by an incompletely developed optic cup, of which the two layers remained separated and which enclosed a vitreous cavity. The outer layer of the optic cup was a simple single layer of cells, containing no pigment. The inner layer presented 'more or less clearly marked retinal elements', though the fixation was such that no further details could be seen. At the periphery, the inner layer of the optic cup was reduced to a single layer, continuous with the outer layer at the margin, and forming with that layer a two-layered epithelium, considered by Allen to be a rudimentary iris.

Allen found no sign of a lens, though quoting Price (1896) who had found a trace of a developing lens in early embryos, which disappeared in later ones. This course of lens regression was confirmed by Stockard (1907), who examined a series of *Bdellostoma* (*Eptatretus*) *stoutii* embryos. A 15 mm embryo showed a definite lens placode opposite the optic cup, but as the latter became buried beneath mesodermal tissues with development, the lens placode regressed until no trace of it remained. Stockard emphasized that the lens is induced by the contact of the optic cup with the overlying ectoderm, and that the cup is not invaginated by pressure from the lens. Kobayashi (1964) reported that *M. garmani* has a lens, though no other author has recorded one in a hagfish. In our specimens of *Eptatretus cirrhatus, E. burgeri, E. stoutii* and *Paramyxine* (*E.*) *atami* no trace of a lens was found, either by dissection or in sectioned material (Figure 34.1a).

None of the hagfishes examined appears to have a highly structured cornea, the light being admitted to the eye through a clear patch of skin at best, in *M. garmani,* while *M. glutinosa* does not even have a break in skin texture, the eye being buried beneath a layer of muscle (Figure 34.1b). Our findings confirm that the *Eptatretus* eye is located beneath skin, which in *E. burgeri, E. stoutii and E. cirrhatus* is non-pigmented over the eye. In all species the eye includes a fibrous coat, though this is not differentiated into an opaque sclera and transparent cornea. Microscopy suggests that the 'cornea' and sclera have similar optical properties; the ultrastructural regularity that characterizes the transparent corneas of, e.g., teleost fishes is not seen in the hagfish material, though the collagen fibrils are of even size, *c.* 52 nm diameter. The fibrils are present in sheets and bundles, which make large angles with each other, and amongst which fibroblasts are present. The corneo-scleral coat is thin, about 20 μm, in each case, and the collagen of the sclera is continuous with that of loose connective tissue surrounding the eye. This loose connective tissue enmeshes fat-containing cells, so that the eye is largely embedded in fat. Like previous authors, we have found no sign of extraocular muscles.

In no case is an epithelium present on the corneal surface, which is buried beneath the skin. In some sections from *E. cirrhatus, E. burgeri* and *E. stoutii* an irregular layer of cells lines the inner surface of the cornea, but this is not a true corneal endothelium. These cells, continuous with those of the choriocapillaris, but clearly separate from those of the retina, are absent in *E. atami.*

Lining the sclera is a tenuous layer of capillaries, representing the vascular component of the choroid, but no choroidal shielding pigment is present. A few choroidal nerve fibres are seen in some sections, but inconsistently, perhaps due to differences in the plane of section. The choroidal vessels all appear to be capillaries, consisting only of endothelial cells without fibrous or muscular components. Fine, irregularly arranged collagen fibres, but no elastin, are present between the capillary walls and the basement membrane of the underlying retinal epithelium; together they form an equivalent to Bruch's membrane in vertebrate eyes (Figure 34.1c).

The fibrous vesicle contains in each case a retina, clearly showing the two layers, epithelium and neuroretina, characteristic of vertebrate retinas and derived from the outer and inner layers of the optic cup respectively, and a vitreous cavity. In our *Eptatretus* species this retina lines half to two thirds of the globe, and is cup-shaped as in normal vertebrate eyes. The epithelium differs from that of almost all vertebrates in being devoid of the pigment granules that make the eyecup easily visible even in animals otherwise largely transparent, including many embryos, in which this is the first such pigment to appear. This absence of pigment is seen even in those hagfish species with darkly pigmented skin.

The cells of the retinal epithelium contain a more or less spherical nucleus, ribosomes and Golgi complex. Mitochondria are also present, including giant examples, c. 5 × 1.2 µm, in *E. cirrhatus*. The epithelial cells in some examples of *E. cirrhatus* and *E. atami* contain in addition aggregations, vitread or sclerad to the nucleus, which appear amorphous by light microscopy. Electron microscopy shows these to consist of masses of tortuous and interlacing tubular networks, probably derived from smooth endoplasmic reticulum (Figure 34.2a). They are reminiscent of the SER in the chloride cells in certain teleost gills (Conte, 1969) and of the situation in *Latimeria* retinal epithelium (Locket, 1973). That these arrays of ER are not present in other specimens of the same and related species suggests that they may be a labile feature. Electron microscopy shows that the epithelial cells are joined by tight junctions close to their vitread surfaces, though these are not prominent by light microscopy. The vitread surfaces of the cells, facing the ventricular space, show small, sparsely distributed microvilli extending into that space. In places outer segments are in contact with the epithelial cells, but do not indent the surface to be enclosed by the epithelial cell cytoplasm. Groups of membranous whorls are in some cases present within the cytoplasm, and are regarded as phagosomes, i.e. phagocytosed outer segment material. In all cases the ventricular space is clear and in some appears rather wide, the outer segments not bridging the gap between neuroretina and epithelium.

The neuroretina (Figure 34.1c) in all four species examined shows a layer of cells abutting the ventricular space, which, despite some doubt about the function of hagfish retina, certainly correspond to vertebrate photoreceptors. These cells adjoin each other at the outer limiting membrane, sclerad to which are low inner segments containing mitochondria, corresponding to the ellipsoids of vertebrate retinas. There is no sign of a glycogen-containing paraboloid, nor of an extensile myoid region. The ellipsoids contain a basal body, giving rise to a cilium with the 9+0 tubule pattern characteristic of the cilia of some invertebrate as well as vertebrate eyes. Unusually, this cilium is located centrally in the ellipsoid, where it lies free in an invagination; in a single instance in *E. burgeri* two cilia are present side by side in the same invagination. Sclerad to the ellipsoid in vertebrates there is typically an outer segment, showing the characteristic stack of bimembrane discs. These discs in *Eptatretus*, however, are not stacked closely, but in a loose and often disordered way; they appear to be fully enclosed within the outer segment membrane, so the receptors more approximate rods than cones of vertebrates. The outer segments are very loosely packed in the ventricular space, quite unlike those of most fishes, particularly those from the deep sea, in which as much space as physically possible seems to be occupied by outer segment material. In all the *Eptatretus* species examined, the outer segments are better formed and organised in the retinal periphery than in the centre. As with the inner segments, the cilium is located in the centre of the outer segment, not along one side as in most retinas, the basal discs projecting from it as a series of conical frills, with apices vitread. Some small groups of basal discs are enclosed within separate portions of the outer segment membrane, and others are within the cytoplasm of the inner segment (Figure 34.2b).

The junctions of the outer limiting membrane are clear, and in places cellular material intervenes between inner segments at this level. In some instances a cilium was observed arising from such a cell, and projecting free into the ventricular space. The cell

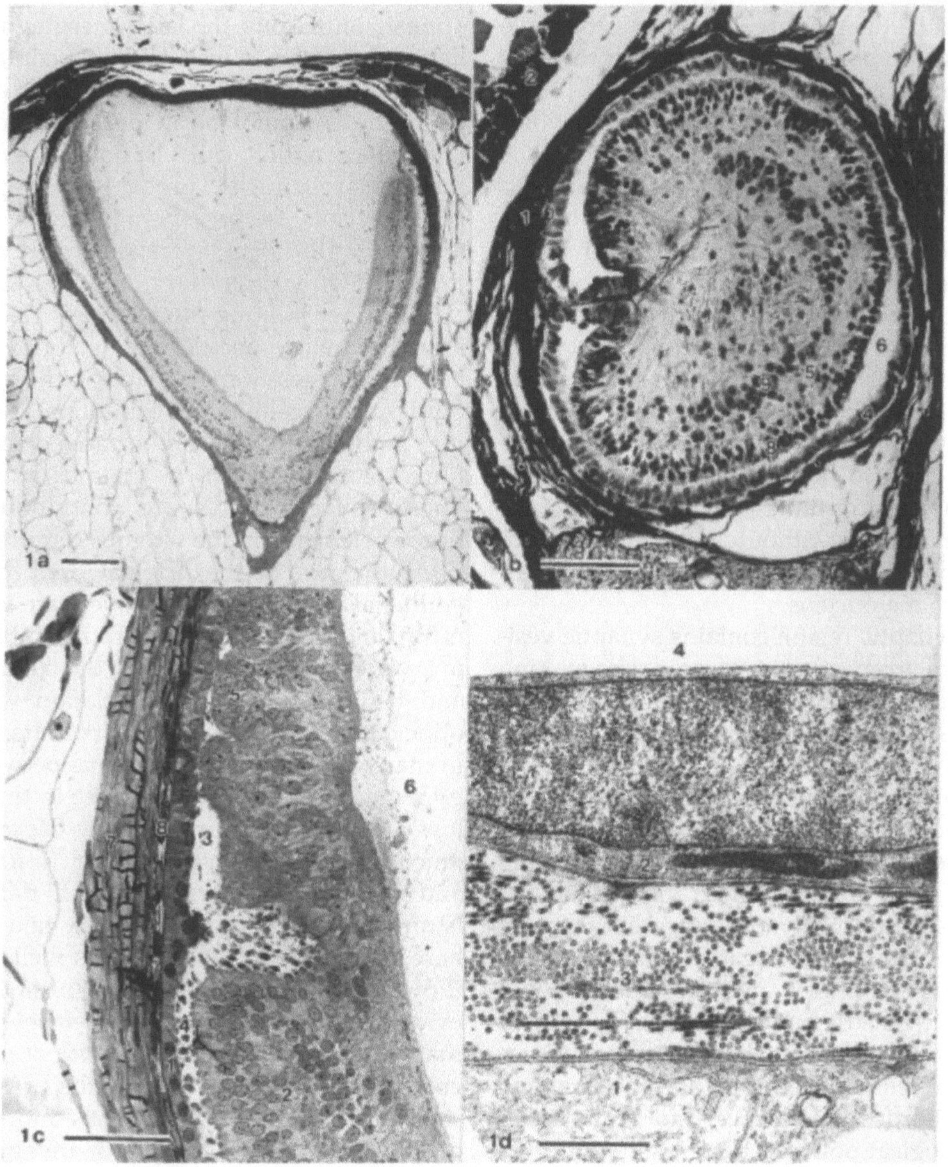

Figure 34.1 (a) Eye of *Eptatretus burgeri*. Scale bar 150 μm. (b) Eye, T.S., *M. glutinosa*. Fibrous corneo-sclera (1) lies beneath muscle bundle (2) and immediately dorsal to ophthalmic nerve (3). Retinal epithelium (4) lines whole capsule, and is separated from neuroretina (5) by ventricular space, artefactually wide at (6). Neuroretina occludes potential vitreous cavity, here represented by fibrous plug (7). Receptor nucleus (8) and inner nuclear layer (9) are recognizable. No outer segments visible in this micrograph. Scale bar 100 μm. (c) Retinal periphery, *E. atami*. Retinal epithelium (1) contains no pigment. Neuroretina (2) is separated from epithelium by ventricular space (3) which contains widely separated outer segments (4). Neuroretina shows folds in far periphery (5). Vitreous space (6) contains fine collagen fibrils. Sclera (7) and choroid (8) are thin and unpigmented. Scale bar 50 μm. (d) Retinal epithelium and choroid, *E. burgeri*. The epithelial cells (1) and endothelial cells of the choroidal capillaries (2) have basement membranes. Between the two are loose collagen fibres (3) but no elastin is present. Lumen of capillary (4). Scale bar 1 μm.

could not be traced vitreally, but it is possible that these represent Landolt's clubs, processes extending from a bipolar cell to the outer limiting membrane, or, in some taxa, beyond it (Locket, 1975; Figure 34.2c).

The visual cell nuclei are ellipsoid, and present in loose formation about two nuclei deep, though not forming a tight layer. Some ribosomes and a Golgi complex may be seen in the cytoplasm, there are scattered mitochondria apart from those concentrated to form the ellipsoid, but few microtubules are visible. Dense bodies, of unknown function and containing paracrystalline arrays, are present close to the nucleus in some visual cells. The cell continues unconstricted to the synaptic region, without any narrowing to form conducting fibres, either sclerad or vitread to the nucleus.

The synaptic region contains synaptic vesicles, often arranged close to processes making invaginating contacts with the receptor. Well-formed synaptic ribbons, with a halo of vesicles and an arciform density, are not seen in our material, though synaptic bodies as described by Holmberg and Öhman (1976) in *Eptatretus burgeri* and *E. stoutii* are certainly present. We confirm their view that the synapses are of the dyad type, rather than the triads of most vertebrates. Some of the contacting processes are also packed with synaptic vesicles, but the cells of origin of these processes cannot be determined, since the inner cells of the retina are not arranged in a well-ordered sequence, and do not offer morphological points of distinction. In particular the distinct layers of horizontal cells found in many elasmobranchs and teleosts are absent (Figure 34.2d).

Radial (Müller's) fibres are well seen in most fishes, but in the hagfishes examined the fibrous portions of these are hard to make out, and only in *E. cirrhatus* have clearly identifiable radial fibre nuclei been recognized. The inner nuclei are sparsely distributed, and do not form a well-defined layer; between them, the outer and inner synaptic layers appear continuous, the latter merging into the nerve fibre layer as well. Synaptic contacts are abundant among these neurites, but details of connections cannot be made out. A definite layer of ganglion cells is not recognizable, though large nuclei, probably belonging to ganglion cells, are present in the inner nuclear and inner plexiform layers; others are close to the retinovitreal boundary.

The inner limiting membrane is clear in all four *Eptatretus* species, and is formed by a basement membrane coating the vitread border of the cellular contributions, and varying amounts of loosely arranged collagen fibres embedded in a ground substance. These collagen fibres are continuous with fibrous elements in the vitreous (Figure 34.3a)

At the rim of the original optic cup the epithelial and neuroretinal layers are in continuity, and the formation here resembles that in a vertebrate embryo before the ciliary body and iris begin to form in relation with the optic cup. Where the two layers unite there is in many cases a peripheral sinus between the two, the cells of both layers close to this being low and simple in appearance. Close to the junction, the neuroretinal layer is irregular, and the synaptic layers thin and ill defined. It is probable that retinal growth takes place here, as it does in those teleosts well studied (Powers and Raymond, 1990), though no evidence of cell division has been seen in our material. The likelihood of growth at the optic cup margin is emphasized by the presence in several specimens of apparent abnormalities of growth. Some sections show invaginations of the ventricular space into the substance of the neuroretina, the invagination being lined with poorly ordered visual cells. In at least one example of *E. atami* an apparent cyst is present, the cavity of which contains disordered outer segment material, presumably formed by the lining cells and accumulated within the cyst. In other examples a close nest of undifferentiated cells is present, without evidence of their derivation.

The clear delineation of the vitreous from

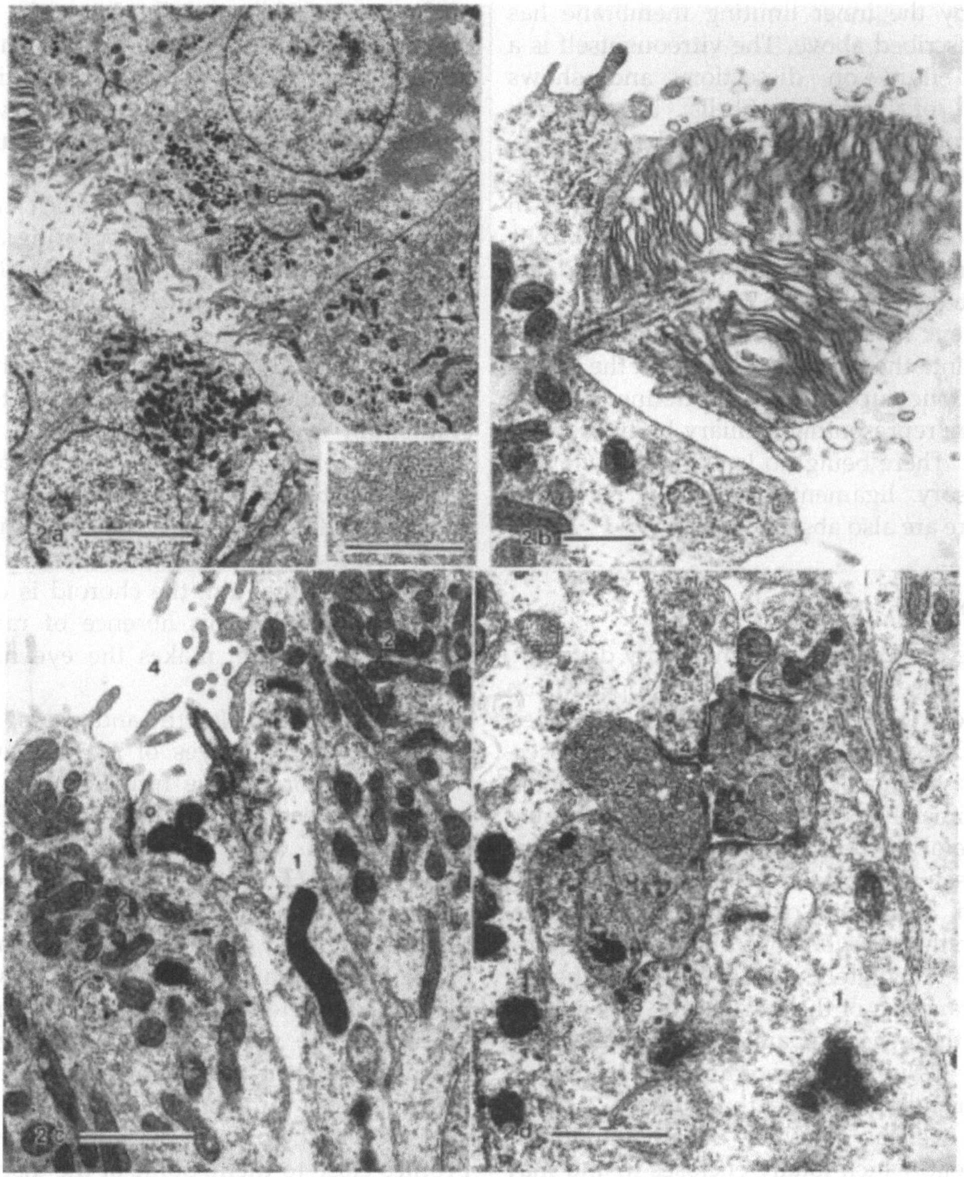

Figure 34.2 (a) Retinal epithelium cells and receptors, *E. atami*. Epithelial cells (1) and receptors (2) extend villous processes into the ventricular cavity (3). Receptor cells contain mitochondrial aggregate (4). In this field no outer segments are present. Epithelial cell contains dense bodies which are not pigment granules (5) and Golgi complex (6). Amorphous mass at (7) is tubular complex, shown in inset. Scale bar 5 µm, inset 0.5 µm. (b) Outer segment, *E. burgeri*. Irregular array of membranous discs are within outer segment membrane. Tangential sections show that cilium (1) is central in inner and outer segment. Scale bar 0.5 µm. (c) Probable Landolt's club, *E. burgeri*. Process (1) between receptor inner segments (2). Club ends at outer limiting membrane (3) and a cilium projects into the ventricular space (4). Scale bar 1 µm. (d) Receptor pedicle, *E. burgeri*. The pedicle (1) contains clear vesicles, which are even more numerous in some contacting processes (2). Dark synaptic bodies, but not ribbons (3), are surrounded by vesicles. Probable synapses between contacting processes are also seen (4). Scale bar 1 µm.

retina by the inner limiting membrane has been described above. The vitreous itself is a viscous fluid on dissection, and shows strands of collagen fibrils by electron microscopy. In some places on the inner retinal surface aggregations of such fibres appear, where they form blebs. Scattered in the vitreous are roundish cells, believed to be fibroblasts, which contain vacuoles and are associated with collagen fibrils (Figure 34.3b). Prominent loose bundles of collagen fibres extend into the vitreous space from the region of the optic cup rim, but no defined fibrous structure representing a ciliary body or iris is present. There being no lens, the muscle and suspensory ligament associated with this structure are also absent.

34.3 *MYXINE GLUTINOSA*

The eye of *M. glutinosa* is markedly different, in the sense of being less developed, from those of the *Eptatretus* species described above. Recognizable with difficulty at dissection as a small solid unpigmented body, approximately 600 μm in diameter, located just lateral to the junction of the capsules of the nasal organ and the brain, the eye is covered by a slip of longitudinal muscle, beneath the subcutaneous blood sinus (Figure 34.1b). Immediately posteroventral to the eye is the ganglion of the ophthalmic nerve, from which the eye is separated by some tenous fibrous tissue; the trigeminal nerve fibres continue forwards beneath the eye.

The eye has a poorly developed fibrous outer coat, which totally encloses it, and may be regarded as the sclera. There is no differentiation into a cornea, nor can any trace be found of eye muscles. The lens also is totally lacking. The solid contents of the sclera is almost entirely derived from the optic cup, of which the two layers derived from this cup, called here retina and epithelium, are continuous laterally, but no peripheral sinus is present between the layers at the cup margin, at least in resin sections. Where the margins of the cup would normally define the pupil, they are in contact, separated only by a small fibrous plug which peters out within the central mass of retinal tissue. The cavity of the cup, normally occupied by vitreous, and well shown in hagfishes with better developed eyes, is thus absent in *Myxine*. The retinal epithelium is irregular, appearing almost columnar in places, but low cubical elsewhere. The epithelial nuclei, in most retinas arranged in a regular single layer, are present in depth, though nowhere is the epithelium more than one cell thick. External to this epithelium is a vestigial uveal tract, represented merely by a tenuous choriocapillaris, containing blood cells, beneath the fibrous eye capsule (Figure 34.3c). The connections of these vessels have not been determined. Like the retinal epithelium, the choroid is devoid of dark pigment. This absence of melanin, with its small size, makes the eye hard to locate on dissection.

The work of Newth and Ross (1955) suggested that the eyes of *M. glutinosa* are without photoreceptive function; this might be due to defects in various parts of the visual system, from outer segments to the optic nerve fibres. Holmberg (1970) examined the retinal cells in detail, and found, as we have, that well-defined outer segments are lacking. Scattered globular membranous whorls about 3–5 μm diameter representing a disordered version of the vertebrate outer segment are however present (Figure 34.3d). Holmberg described the lamellar membranes as interspersed with cytoplasm, with many vesicular profiles close to them. Some of the discs have closed ends, clearly within the plasma membrane, suggesting an affinity with vertebrate rods. Such closed disc appearances could be due to a plane of section artefact; vertebrate cone discs are open only for part of their circumference, and if the section does not pass through the open region, the disc will appear closed. The degenerate outer segments usually indent the opposing face of the epithelium, but not every inner segment

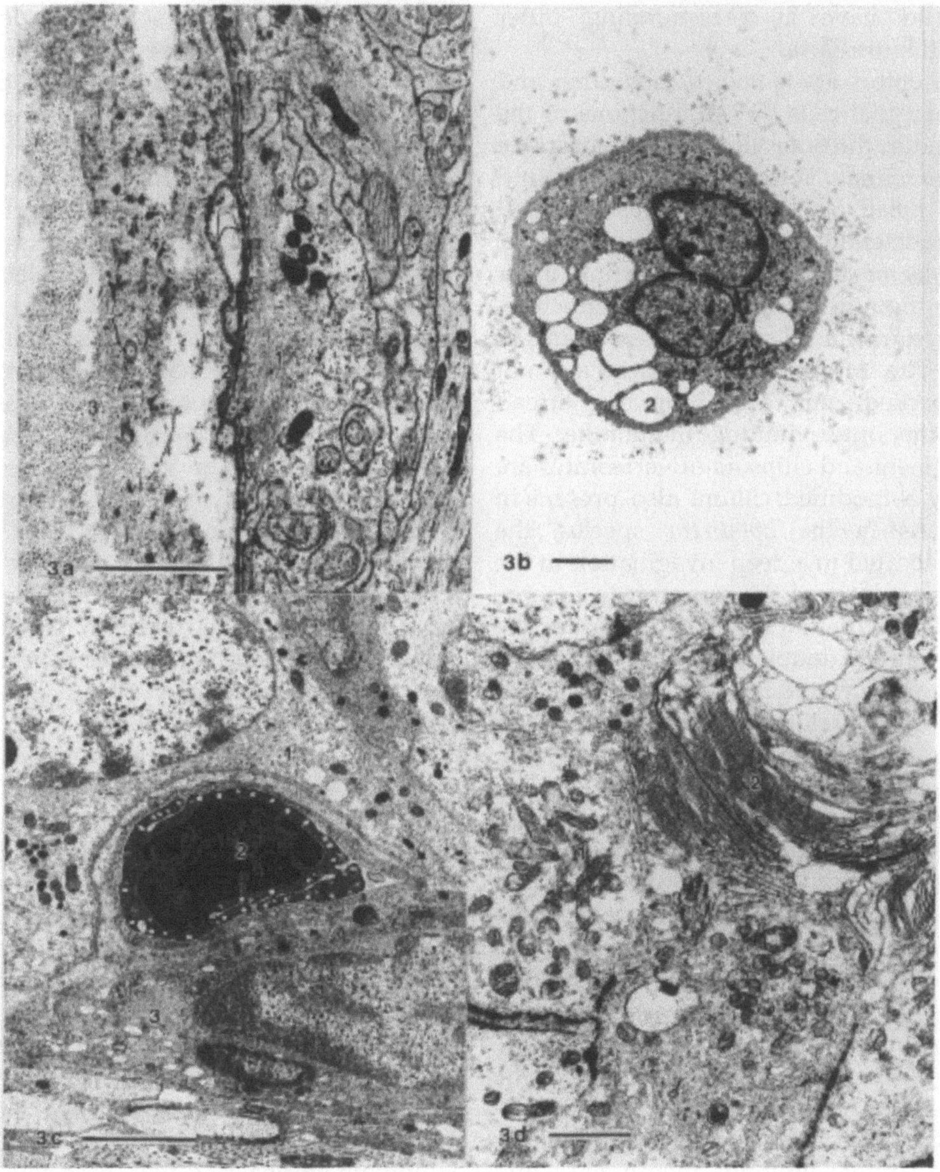

Figure 34.3 (a) Inner limiting membrane, *E. cirrhatus*. Cell processes, probably of radial fibres, contain fibrils just beneath the retinal surface (1). Cell surface is dense beneath the basement membrane (2). Matrix and collagen fibres are present in vitreous (3). Scale bar 2 µm. (b) Vitreous cell, *E. atami*. Cell contains double or bilobed nucleus (1) and clear vacuoles (2). Basement membrane (3) surrounds cell, which is associated with collagen fibrils. Scale bar 2 µm. (c) Epithelial cell and choroid, *M. glutinosa*. Epithelial cell (1) lies smoothly against choroid. Haemocyte (2) is in capillary, which indents epithelial cell. Choroidal fibroblast (3) lies against sclerad side of capillary. Scale bar 2 µm. (d) Receptor inner and outer segment, *M. glutinosa*. Inner segments (1) contain aggregates of mitochondria, and project slightly into ventricular space. This is hard to discern, as receptors and epithelium are in close apposition. Globular outer segment (2) contains whorls and stacks of disc membranes, and indents epithelial cell. A cilium (3) connects inner and outer segments. Scale bar 1 µm.

appears to have a corresponding outer segment (Figure 34.4a).

The receptors are joined to each other and to adjacent glial cells by cell junctions of the type found in the outer limiting membrane of vertebrate retinas. Vertebrate visual cells, and those of *Eptatretus*, have an inner segment projecting sclerally into the ventricular space, but in *Myxine* there is no projecting inner segment, though the aggregate of mitochondria that form an ellipsoid in vertebrates is present. The *Myxine* mitochondria are not tightly packed, and located on the vitread side of the outer limiting membrane. The outer segment and ellipsoid of vertebrates are joined by a modified cilium, also present in *Myxine*. As in the *Eptatretus* species, the cilium is located in a deep invagination in the sclerad cytoplasm of the receptor, to emerge at the level of the intercellular junctions. The 9+0 microtubule doublets extend as far as the outer segment, where their individual identities are lost. Occasional cilia are present with abnormal microtubule arrangements, e.g. 8+2. A second, obliquely oriented, centriole located within the receptor, to the vitread side of the mitochondria, is only occasionally seen in hagfishes, and may be an inconstant feature. No separate mass of smooth endoplasmic reticulum enmeshing glycogen and forming a paraboloid is present, though granules resembling glycogen are scattered throughout the cytoplasm, accounting for the diastase-sensitive PAS-positive staining of the receptor cells. A Golgi complex, with associated vesicles, is present between the nucleus and the outer limiting membrane, and patches of rough endoplasmic reticulum are scattered through the cytoplasm, mostly sclerad to the nucleus.

The glial cells present between the receptors are recognizable by the presence of filaments in their cytoplasm, though these become more sparse towards the sclerad end of the cell, where mitochondria, smooth endoplasmic reticulum and a Golgi complex are located. These cells contain numerous *c.* 40 nm vesicles close to the sclerad border; similar vesicles were also seen in receptors and epithelial cells. The vitread ends of the glial cells pass into the inner layers, where they cannot be well followed.

The synaptic apparatus is located at the vitread end of the receptor, as in vertebrates, but there is no constriction forming a conducting fibre between nucleus and synapse (Holmberg, 1970), which is thus sessile. Such synapses are not confined to hagfish; they also occur in certain teleost fishes, and were reported in alligators by Kalberer and Pedler (1963). The pedicles in *M. glutinosa* resemble in general those of the *Eptatretus* species, but, as reported by Holmberg and Öhman (1976), they do not contain synaptic ribbons or synaptic bodies. There are however patches of membrane densities, resembling those in conventional synapses between vertebrate neurites, and also found in the inner plexiform layer of vertebrate retina (Figure 34.4b).

Holmberg (1972), who examined the optic tract by electron microscopy, quoted the earlier works of J. Müller (1837), W. Müller (1874), Kohl (1892), Retzius (1893) and Holmgren (1919), of which the latter had described a thin and rudimentary tract, containing about 100 nerve fibres. Holmgren had also observed a small swelling close to the eye, which contained bipolar ganglion cells. Holmberg's investigations showed that the optic tract is indeed small, only some 30 μm diameter, and fibre counts of 810 and 1467 fibres were made on two tracts. Holmberg emphasized that these counts were a small sample, and that *Myxine* is in any case a variable animal; he also commented that there might be an unknown number of retinopetal fibres in the tract, which would further reduce the count of output fibres from the eye. The fibres themselves are all unmyelinated, with the largest about 1.4 μm diameter, ranging down to 0.1 μm. These fibres are in marked contrast with those of the nearby ophthalmic nerve, which range from

1 to 4 μm in diameter. The < 1 μm optic nerve fibres are mostly single, with some smaller fibres arranged in groups, and densely surrounded by glial cytoplasm where they leave the eye (Figure 34.4c). Cell bodies of two types are present at this part of the tract. One, packed with filaments, is considered to be glial, but the other type contains scattered rough ER, a Golgi apparatus and occasional annulate lamellae. In some cases Holmberg saw fibres containing 40 nm agranular vesicles in contact with these cells, sometimes with membrane densities with clusters of vesicles around them. He points out that Dücker (1924) found only about 150–180 cells in the inner fibrous layer, which would not account for the number of fibres in the tract, but that the cells of the second type above probably corresponded to ganglion cells in vertebrate retinas. From the normal appearance of the optic nerve fibres, and the presence of synapses at the vitread ends of the receptors and elsewhere in the retina and nerve head, Holmberg concluded that the *M. glutinosa* eye is probably a functional light receptor.

The retina vitread to the receptor synapses contains two fibrous layers, which may tentatively be equated with the inner and outer plexiform layers of vertebrate retinas. In both layers profiles of three types of fibre were described by Holmberg. One, with flocculent cytoplasm and scattered vesicles and mitochondria also contains agranular vesicles *c.* 50 nm diameter, which are associated with membrane densities, considered conventional synapses. A second type of fibre resembles the first, but contains scattered groups of ribosomes and a few filaments. The third fibre type is characterized by the presence of numerous filaments, among which are scattered a few mitochondria and some smooth membrane profiles. It has not been possible to assign these fibre types to particular cells of origin, though the third type are probably radial fibre. Holmberg described two types of cell bodies located between the inner and outer fibre layers. One has a deeply indented nucleus, the indentation containing numerous organelles. These include mitochondria, saccules and vesicles of the Golgi complex and rough endoplasmic reticulum continuous with annulated lamellae, and with expanded cisternae containing electron dense material. Smooth endoplasmic reticulum is usually prominent in these cells. The peripheral cytoplasm shows mitochondria, smooth and rough endoplasmic reticulum and filaments. The second cell type has a round nucleus, with even chromatin pattern and no peripheral chromatin condensation, while well-developed smooth endoplasmic reticulum is distributed throughout the cell body. The Golgi complex of these cells is located in a zone resembling a hillock. In some cases fibres containing 50 nm agranular vesicles have been observed making contact with the cell bodies of these cells.

Holmberg also described follicles, with a 20 μm lumen, located within the fibre layers of the retina, often close to the transitional zone between epithelial and receptor cells. These follicles contained two types of cell, one densely populated with 20–30 nm electron dense granules with mitochondria aggregated in the part of the cytoplasm nearest the lumen. The luminal surface was invaginated in places, with microvillous structures projecting into the invagination. Profiles of cilia, some associated with small whorls of parallel membranes, were also seen. These cilia had 8+1 or 7+2 microtubule patterns, not the 9+0 seen in the receptor cell cilia. The second cell type in the follicles usually contained large electron dense granules in a filamentous cytoplasm, and much vesicular endoplasmic reticulum associated with the Golgi complex in the luminal part of the cytoplasm. Vigh-Teichmann *et al.* (1984) also observed these follicle-like structures, concluding that they are tubular, and formed from invaginations of the receptor cell layer. They saw follicles close to the exit of the optic nerve, and traced a lumen into the nerve for some distance. Continuity was established between the lumen of the tubules and the ventricular space, and outer segments of photoreceptors were seen in the lumina of the tubules. Vigh-Teichmann *et al.* localized opsin by immunocytochemistry in retinal outer segments, and also in the outer segments in the tubules penetrating the optic nerve. Immunoelectron microscopy showed that some outer segments, particularly those in the tubules, did not react for opsin, though adjacent ones did. The opsin reactivity tends to reinforce the idea that the eyes, degenerate as they are, may still detect light. Examples of such cyst-like formations are also present in our material (Figure 34.4d).

DISCUSSION

The cells developed from the outer layer of the optic cup are pigmented in most chordates, and in many the pigment develops early, making the eye rudiment easy to see even in early embryos. In hagfishes this epithelium remains without pigmentation. In many vertebrates the bases of these cells, where they abut Bruch's membrane, are extensively folded, presenting a large surface area to the nearby choroid. In *M. glutinosa* the cell bases are smooth, though the lateral surfaces do interdigitate with those of adjacent cells. The nucleus of these cells is large and round, occupying much of the cell, but not specially located in the base of the cell. Close association of the epithelial cells with the receptor outer segments were not described by Holmberg, though dark staining inclusions about 0.4 μm diameter, which sometimes showed granular or pale internal structure, were present and could have represented lysosomes associated with the degradation of phagocytosed outer segments. We find that the outer segments, where present, indent the epithelial cells to lie in close contact with the vitread surface of those cells.

Compared with most vertebrate synaptic pedicles, those of hagfishes appear simple. No hagfish pedicles examined bear basal filaments, so prominent on some receptor cells, particularly cones, of many vertebrates (e.g. Ramón y Cajal, 1972). The synaptic specializations are not identical with those in vertebrates, and indeed are not consistent within the hagfish. The vitread plasma membrane of the receptor is invaginated, cell processes making contact within the invaginations. In all the hagfishes examined one type of these processes contain numerous 30–50 nm agranular vesicles, apparently of the same type as occur in the receptor cytoplasm, where they are abundant in the vitread cytoplasm. A second type of contacting fibre contains no vesicles, but scattered mitochondria are present in both types. Points of presumed synaptic contact are marked by membrane densities with a cluster of vesicles around them, but no synaptic ribbons of the type found in vertebrate retinal synapses have been observed in *M. glutinosa* (Holmberg and Öhman, 1976, present observations). The *Eptatretus* species reported have similar synaptic specializations, with the same two types of process making contact. Like vertebrates, these hagfishes have densities within the receptor synaptic region, surrounded by a halo of agranular vesicles, but unlike vertebrates, spherical not ribbon-shaped. Called by Holmberg synaptic bodies, these occur in relation to the type 2 fibres, or to adjacent types 1 and 2 (Holmberg's terms). A further study by Holmberg and Öhman (1976) showed marked differences between these bodies in the *Eptatretus* species they examined and those of the river lamprey, *Lampetra fluviatilis*, which has synaptic ribbons closely resembling those of vertebrates. They point out that the complexity of the synaptic bodies closely matches the degree of degeneration of the eyes; *M. glutinosa*, with the least organized eye having none, and the lamprey, with two populations of receptors, perhaps equivalent to rods and cones, with definite synaptic ribbons. They also speculate on the phylogenetic implications, with lamprey ribbons resembling those of vertebrates more closely

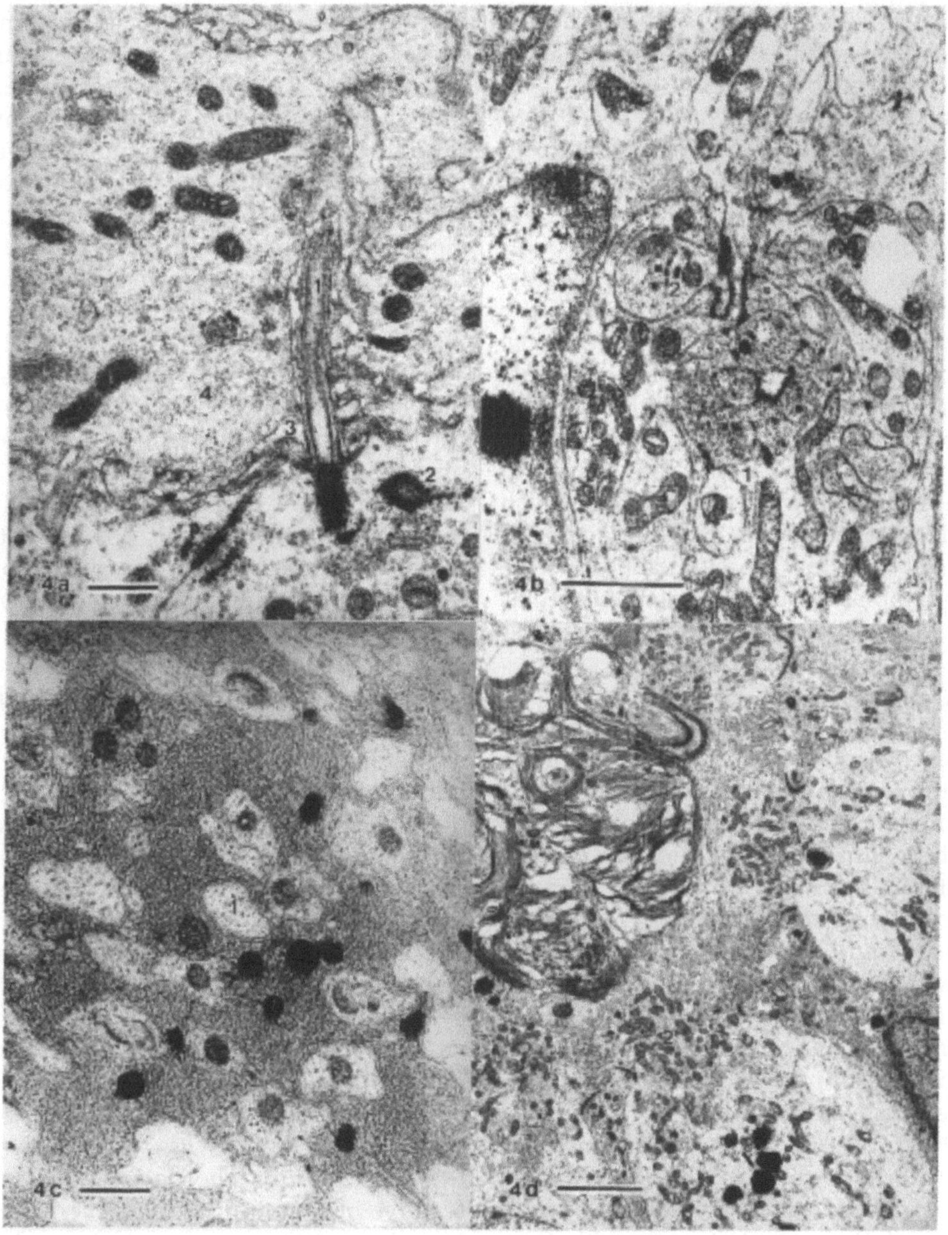

Figure 33.4 (a) Cilium without outer segment, *M. glutinosa*. The cilium (1) projects from its basal body; a second centriole is present (2). The cilium crosses the narrow ventricular space (3) and indents an epithelial cell (4). This cilium does not bear an outer segment. Scale bar 0.5 µm. (b) Receptor pedicle, *M. glutinosa*. Blunt ended pedicle (1) contains mitochondria and few synaptic vesicles. Contacting processes (2) contain more vesicles. Membrane densities are seen at some contact sites, but no synaptic bodies. Scale bar 1 µm. (c) Optic nerve fibres, *M. glutinosa*. Thin axons (1) are embedded in glial tissue densely packed with fine filaments. Scale bar 0.5 µm. (d) Cyst in retina, *M. glutinosa*. Cells resembling receptors are arranged around a cavity containing disorganized outer segment material (1). The cells contain aggregations of mitochondria (2). Scale bar 2 µm.

than do hagfish synaptic bodies, though lampreys and hagfishes are cyclostomes and vertebrates are gnathostomes.

Though most synaptic densities in *M. glutinosa* are related to single profiles of contacting processes, the synaptic bodies in the other hagfishes and ribbons in lampreys are frequently opposite places where two profiles make contact with the receptor, side by side. In this they resemble the dyad synapses in the inner plexiform layer of the vertebrate retina, where bipolar cell processes make contact with two postsynaptic processes, a synaptic ribbon with associated vesicles being present in the bipolar process. Gnathostome rods and cones usually have triad synapses, in which two horizontal cell processes flank a bipolar process opposite the synaptic ribbon in the receptor cytoplasm, but this arrangement has not been observed in hagfishes.

34.5 CONCLUSION

Hagfishes are undoubtedly sensitive to light. Newth and Ross (1955) explored the sensitivity of different parts of the body of *M. glutinosa* with light probes, finding marked sensitivity in the head, but in front of the eyes, and around the cloaca. They found that the animal would respond to light stimuli by swimming movements, preceded by movements of the head, or alterations in position from lying immobile on the side to dorsoventrally oriented, in preparation for swimming. They measured the delay of the response to different intensities, showing a reaction time of about 20 s even to bright illumination. They did not find definite evidence of light sensitivity of the eyes, but even with finer light probes than they used this would be difficult, due to the lack of shielding pigment round the retina. It was interesting that the light sense from the cloacal region was mediated through spinal nerves, not lateral line nerves as in lampreys (Young, 1935). Further responses of *Myxine* were studied by Steven (1955), using the same fish as Newth and Ross. Steven used the first movement in response to light, before swimming commenced, to measure responses, and found that reaction time was inversely proportional to stimulus intensity at low intensities, reaching a saturation value at intensities above c. 10 equivalent foot candles. He also explored the spectral sensitivity of these responses, finding maximal sensitivity between 500 and 520 nm, with little or no response above c. 600 nm. An analysis of bodies and livers of two hagfishes showed that they contained only Vitamin A1, the chromophore characteristic of rhodopsin. The immunocytochemical findings of Vigh-Teichmann and colleagues, i.e. reactivity of the *Myxine* receptors to bovine antiopsin, suggested to them that the *Myxine* pigment may be a rod pigment derived from Vitamin A1, i.e. rhodopsin, since the bovine retina contains predominantly rhodopsin.

Steven related the demonstrated light sensitivity to the amount of light penetrating seawater to the depths at which hagfishes are found, showing that illumination at those depths would be sufficient to stimulate the fishes to swim. The swimming response might cause the fishes to find dimmer conditions, and if burrowing resulted the hag would be concealed from sighted predators.

It is common, when dealing with chordate eye structure, to think in terms of vertebrates, amongst which there is a well-defined basic pattern of eye, albeit with many functional and phylogenetic adaptations. With hagfishes we are not dealing with vertebrates, though they are certainly chordates and have many features in common with true vertebrates. As Holmberg (1971) points out, the cyclostomes have been separate from the gnathostomes for many millions of years, and caution should be exercised when comparing the two groups.

The fact that there are many similarities, in ultrastructure as well as general organization, between cyclostome and gnathostome eyes shows the extreme conservatism of the arrangements of the chordate eye, in turn due to the common mode of eye development across the chordates. In all we see the development of the neuroretina from the inner layer of the optic cup, itself an evagination of the developing forebrain, while the outer layer gives rise to the surrounding epithelium, in most cases pigmented and called the pigment epithelium. This arrangement, and the development of visual cells from the lining of the ventricular space, corresponding to ependyma in the brain, result in the outer segments projecting sclerally into the ventricular space, incoming light having to pass through the rest of the neuroretina before arriving at the visual pigment. In teleost eyes studied (Powers and Raymond, 1990) the retina grows during life at the periphery. Such growth may well take place in hagfishes, or at least in *Eptatretus* species, as shown by the better formed outer segments nearer the periphery than at the centre. It is as if newly formed retina produces outer segments, which degenerate with time. The lens is induced from surface ectoderm by the underlying optic cup, and the uveal and corneoscleral layers form from mesoderm surrounding the optic cup. The cavity of the cup contains the vitreous. Those parts of the hagfish eye that remain still show these arrangements, though some, notably *Myxine glutinosa*, have lost all but traces of the retina. Clearly hagfish eyes, lacking components which occur in almost all vertebrates, are degenerate, as Fernholm and Holmberg (1975) showed. Even those parts still present show irregularities, probably of a degenerate kind. It is not entirely clear whether the various hollow structures described by authors, i.e. follicles (Holmberg, 1970), tubules (Vigh-Teichmann *et al.*, 1984), cysts and cell nests (our observations), are indeed tubular or cystic. Whatever their precise nature, these anomalous structures may all be accounted for by disordered growth of retinal elements, normally present in regular formations in fully functional retinas. Hagfishes are not alone in possessing degenerate eyes: various cave teleosts have degenerate or absent eyes, and the abyssal fish *Ipnops murrayi* has lost lens and vitreous, the retina being flattened out on the surface of the head, covered by a remnant of cornea (Munk, 1959).

Ross (1963) suggested that hagfishes had ancestors with functional eyes, and that their blindness and eye reduction is an example of the trend for reduction and elimination of eyes in cave-dwelling or burrowing species. That view was furthered by Fernholm and Holmberg (1975), who documented the degrees of degeneration in *Eptatretus cirrhatus*, *E. burgeri*, *Paramyxine* (*Eptatretus*) *atami* and *Myxine glutinosa*. They concluded that the extreme degeneration of the eyes and visual system is apomorphic, and that the plesiomorphic state was one of seeing eyes in shallow water organisms.

REFERENCES

Adam, H. and Strahan, R. (1963) Notes on the habitat, aquarium maintenance and experimental use of hagfishes, in *The Biology of Myxine* (eds A. Brodal and R. Fänge), Universitetsforlaget, Oslo, pp. 33–41.

Allen, B.M. (1905) The eye of *Bdellostoma stoutii*. *Anatomischer Anzeiger*, **26**, 208–11.

Conte, F.P. (1969) Salt secretion, in *Fish Physiology*, Vol. 1 (eds W.S. Hoar and D.J. Randall), Academic Press, New York, pp. 241–92.

Dücker, M. (1924) Über die Augen der Zyklostomen. *Jenaische Zeitschrift für Naturwissenschaft*, **60**, 471–528.

Fernholm, B. and Holmberg, K. (1975) The eyes in three genera of hagfish (*Eptatretus*, *Paramyxine* and *Myxine*) – a case of degenerative evolution. *Vision Research*, **15**, 253–9.

Holmberg, K. (1970) The hagfish retina: fine structure of retinal cells in *Myxine glutinosa* L., with special reference to receptor and epithelial cells. *Zeitschrift für Zellforschung*, **111**, 519–38.

Holmberg, K. (1971) The hagfish retina: electron microscopic study comparing receptor and epithelial cells in the Pacific hagfish, *Polistotrema stoutii*, with those in the Atlantic hagfish, *Myxine glutinosa*. *Zeitschrift für Zellforschung*, **121**, 249–69.

Holmberg, K. (1972) Fine structure of the optic tract in the Atlantic hagfish, *Myxine glutinosa*. *Acta Zoologica (Stockholm)*, **53**, 165–71.

Holmberg, K. and Öhman, P. (1976) Fine structure of retinal synaptic organelles in lamprey and hagfish photoreceptors. *Vision Research*, **16**, 237–9.

Holmgren, N. (1919) Zur Anatomie des Gehirns von *Myxine*. *Kungliga svenska Vetenskabs-Akademiens Handlingar*, **60**, 1–96.

Kalberer, M. and Pedler, C. (1963) The visual cells of the alligator: an electon microscopic study. *Vision Research*, **3**, 323–9.

Kobayashi, H. (1964) On the photo-perceptive function in the eye of the hagfish, *Myxine garmani* Jordan et Snyder. *Journal of the Shimonoseki University of Fisheries*, **13**, 67–83.

Kohl, C. (1892) Rudimentäre Wirbelthieraugen. *Zoologica (Stuttgart)*, **13**, 48–51.

Krause, C. (1886) Die Retina – II. Die Retina der Fische. Cyclostomata. *Internationale Monatschrift für Anatomie und Histologie*, **3**, 8–21.

Locket, N.A. (1973) Retinal structure in *Latimeria chalumnae*. *Philosophical Transactions of the Royal Society of London*, Series B, **266**, 493–521.

Locket, N.A. (1975) Landolt's club in some primitive fishes, in *Vision in Fishes* (ed. M.A. Ali), Plenum Press, New York, pp. 471–80.

Müller, J. (1837) Über den eigenthümlichen Bau des Gehörorgans bei den Cyclostomen, mit Bemerkungen über die ungleiche Ausbildung der Sinnesorgane bei den Myxinoiden. Fortsetzung der vergleichende Anatomie der Myxinoiden. *Abhandlungen der Kaiserlichen Akademie für Wissenschaft zu Berlin*.

Müller, W. (1874) Über die Stammesentwicklung des Sehorgans der Wirbelthiere, in *Beiträge zur Anatomie und Physiologie als Festgabe Carl Ludwig gewidment*. F.C.W. Vogel, Leipzig.

Munk, O. (1959) The eyes of *Ipnops murrayi* Günther 1887. *Galathea Report*, **3**, 79–87.

Newth, D.R. and Ross, D.M. (1955) On the reaction to light of *Myxine glutinosa* L. *Journal of Experimental Biology*, **32**, 4–21.

Powers, M.K. and Raymond, P.A. (1990) Development of the visual system, in *The Visual System of Fish* (eds R.H. Douglas and M.B.A. Djamgoz), Chapman & Hall, London, pp. 419–42.

Price, G.C. (1896) Some points in the development of a myxinoid (*Bdellostoma stoutii*). *Verhandlungen des Anatomisches Gesellschaft* (volume and page not given).

Ramón y Cajal, S. (1972) *The Structure of the Retina* (trans. S.A. Thorpe and M. Glickstein), Charles C. Thomas, Springfield.

Retzius, G. (1893) Das Auge von *Myxine*. *Biologische Untersuchungen*, **5**, 64–8.

Ross, D.M. (1963) The sense organs of *Myxine glutinosa*, in *The Biology of Myxine* (eds A. Brodal and R. Fänge), Universitetsforlaget, Oslo, pp. 150–60.

Steven, D.M. (1955) Experiments on the light sense of the hag, *Myxine glutinosa* L. *Journal of Experimental Biology*, **32**, 22–38.

Stockard, C.R. (1907) The embryonic history of the lens in *Bdellostoma stoutii* in relation to recent experiments. *American Journal of Anatomy*, **6**, 511–15.

Vigh-Teichmann, I., Vigh, B., Olsson, R. and van Veen, T. (1984) Opsin-immunoreactive outer segments of photoreceptors in the eye and in the lumen of the optic nerve of the hagfish, *Myxine glutinosa*. *Cell and Tissue Research*, **238**, 515–22.

Young, J.Z. (1935) The photoreceptors of lampreys. I. Light-sensitive fibres in the lateral line nerves. *Journal of Experimental Biology*, **12**, 229–38.

35 STRUCTURE OF THE HAGFISH INNER EAR

Jørgen Mørup Jørgensen

SUMMARY

The inner ear of the Atlantic hagfish *Myxine glutinosa* was examined by light microscopical serial sections as well as transmission and scanning electron microscopy. For comparison, ears from some Japanese and Pacific species have also been examined.

In all species, the labyrinth contains a large *macula communis* and two cristae. The macula is covered with numerous round statoconia, of different diameters. Various stages in the formation of statoconia can be seen in cells surrounding the macula.

The sensory epithelium of the *macula* may be divided in an anterior horizontal part, a middle vertical and a posterior horizontally positioned part. It consists of sensory hair cells and supporting cells. The hair cells are apically equipped with a bundle consisting of a kinocilium and 10–25 stereovilli, forming an organ pipe configuration. The ciliary axoneme consists of a ring of nine outer double microtubules, with no central microtubules.

Most of the apical cytoplasm of the hair cell is occupied by a cuticular plate. In some macular hair cells a filamentous structure can be seen in various parts of the cell. In the hair cell a roundish synaptic body is often seen, surrounded by vesicles, opposite to the afferent nerve endings. Cholinergic nerve fibres, supposed to be efferent nerve fibres, however, can be visualized in the sensory epithelia from cholinesterase-staining.

The cristae are ring-shaped and lack a proper cupula. Apically, the crista hair cells have very long kinocilia, up to 35 μm, and a bundle of stereovilli, arranged as on the macular hair cells. The presence of mitoses and apoptotic cells indicate a continuous turnover of the hair cells.

35.1 INTRODUCTION

The ear of the Atlantic hagfish *Myxine glutinosa* has previously been examined by a number of authors, the most prominent of which are Retzius (1881) and Lowenstein and Thornhill (1970). Accounts of the ear of other hagfishes, however, have not been published, and for that reason inner ears of the following species were also examined: *Eptatretus stoutii*, *Eptatretus burgeri*, *Eptatretus cirrhatus*, *Paramyxine atami*, and compared with the ear from *Myxine glutinosa*.

All hagfishes examined so far are unique among craniates because of the very simple structure of the ear. Only a single semicircular canal and a single macula are present. Lampreys have two semicircular canals and a single macula, which is more differentiated and can be divided in an anterior horizontal, a vertical and a posterior horizontal part (Lowenstein *et al.*, 1968). All gnathostomes have three semicircular canals and two or more maculae in their vestibular labyrinth (Retzius, 1881).

The Biology of Hagfishes. Edited by Jørgen Mørup Jørgensen, Jens Peter Lomholt, Roy E. Weber and Hans Malte. Published in 1998 by Chapman & Hall, London. ISBN 0 412 78530 7.

35.2 EAR CAPSULE

Hagfish ears consist of two kidney-shaped cartilaginous capsules, one on each side of the medulla oblongata (see McVean, this volume). The medial side of these capsules has a window consisting of connective tissue (Retzius, 1881). The two octaval nerves, *ramus anterior* and *ramus posterior*, as well as the endolymphatic duct and blood vessels penetrate this window.

35.3 MEMBRANOUS INNER EAR

In all examined hagfish species the inner ear is ring-shaped (Figure 35.1A). Their inner ears are sufficiently similar that a single description serves for all species examined. The diameter of the canal is wide compared to other chordates of similar size (McVean, 1991 and this volume).

Three sensory epithelia, a *macula communis* and two cristae are enclosed in the canal (Figure 35.1). They consist of two main cell types, the hair cells and supporting cells. The latter reach from the basal lamina to the surface of the epithelium, while the hair cells never contact the basal lamina.

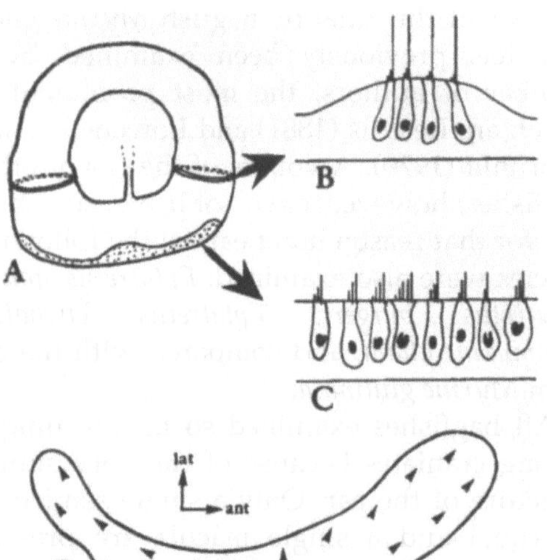

Figure 35.1 Outline of the hagfish inner ear (A), with the two ring-shaped cristae (B) and the macula (C). The orientation of the hair cells in the macula, viewed from above, is shown in D. The arrowhead points indicate the position of the kinocilium relative to the stereocilia.

The hair cells resemble hair cells described from lateral line and inner ear sensory organs in other craniates. The apical tuft consists of an eccentric kinocilium and a bundle of microvilli, termed stereovilli, or stereocilia (Figure 35.3A and B). The kinocilium is unique compared to kinocilia from vertebrate lateral line and inner ear cilia, because the two central microtubules are missing. The kinocilium arises from a basal body which has no basal foot (Figure 35.4B).

The stereovilli are unequal in length (Figure 35.3A, B and 35.4A). They form an organ pipe configuration, with the longest stereocilia standing next to the kinocilium. The stereovilli have a diameter of about 0.1 µm, and connect to the hair cells without the narrow neck, usually seen in gnathostome inner ears.

The stereovilli have an inner core of 60 Å microfilaments, which basally in the hair continue into the cuticular plate, a finely filamentous structure in the apical cytoplasm of the hair cell (Figure 35.4B). In some hair cells a peculiar structure consisting of filaments can be seen at different levels of the hair cells (Figure 35.4C).

Most hair cells have a lightly staining cytoplasm, with many widespread mitochondria, one or two Golgi bodies and a moderately developed endoplasmatic reticulum. Occasionally, dark staining hair cells with an irregular outline are found. These show many characteristics of apoptosis. Mitotic cells are rare. It is believed that there is a gradual replacement of hair cells in the hagfish inner ear, as presumed for octavolateralis organs of most vertebrates, with mammals as the main exception (see Rubel *et al.*, 1991, Corwin *et al.*, 1991, and Jørgensen, 1991, for a discussion of this subject).

In the basal part of the hair cell a round synaptic body may be found opposite to the

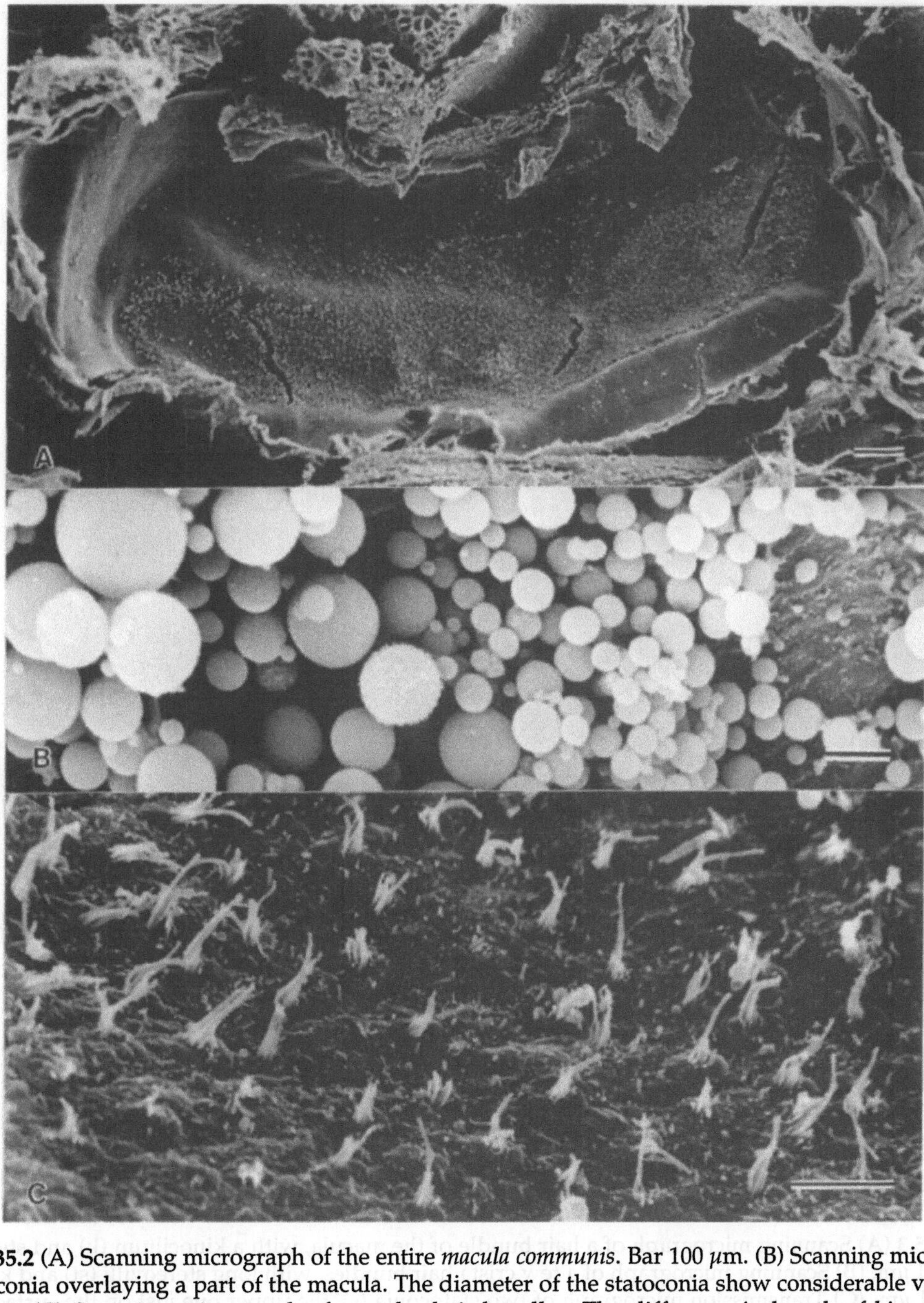

Figure 35.2 (A) Scanning micrograph of the entire *macula communis*. Bar 100 μm. (B) Scanning micrograph of statoconia overlaying a part of the macula. The diameter of the statoconia show considerable variation. Bar 4 μm. (C) Scanning micrograph of macular hair bundles. The difference in length of kinocilia and stereovilli is marked. Bar 5 μm.

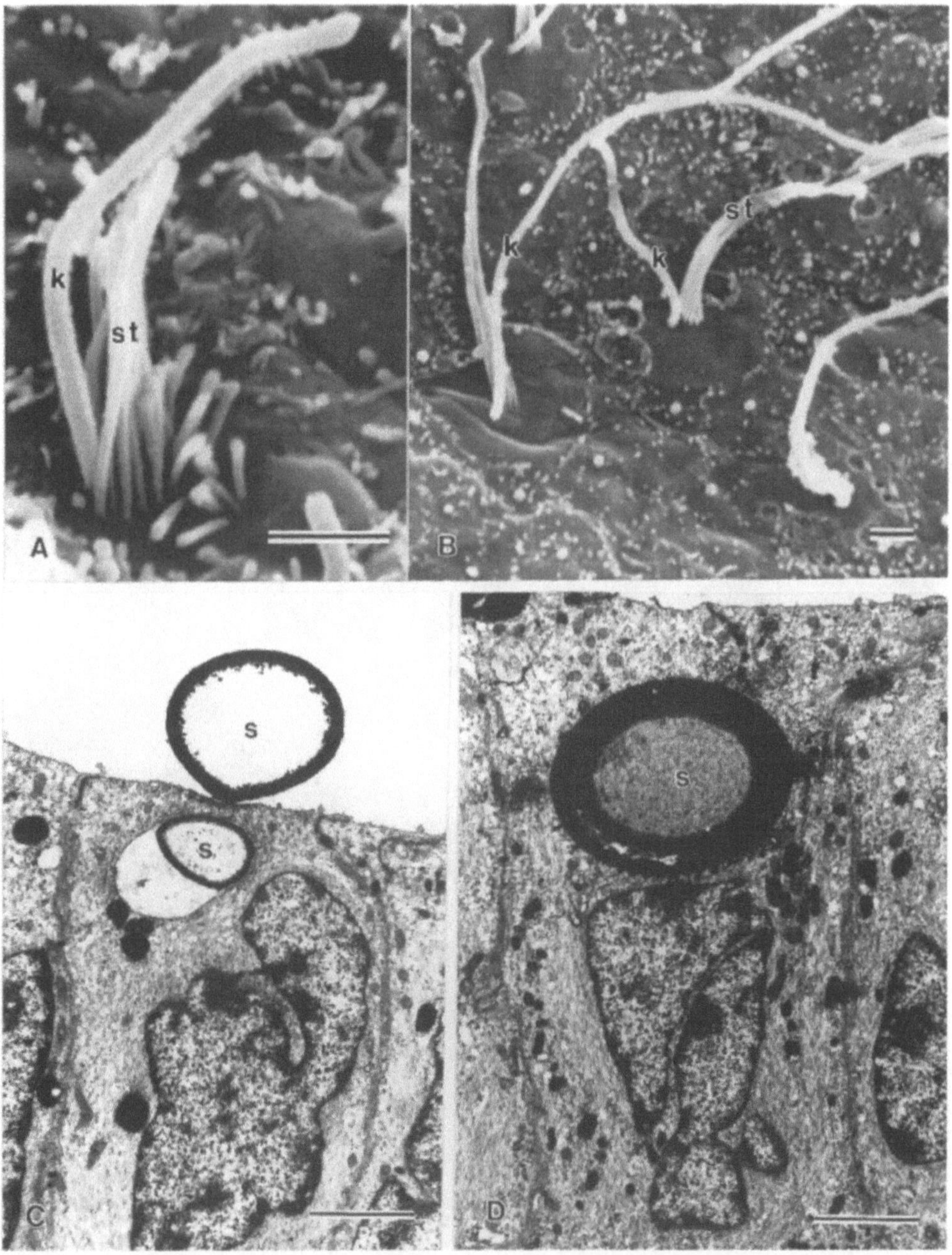

Figure 35.3 (A) Scanning micrograph of a hair bundle of the macula, with a kinocilium (k) and stereovilli (st). Bar 1 μm. (B) Scanning micrograph of a few crista hair bundles, with long stereovilli (st) and kinocilia (k). Bar 1 μm. (C) Transmission micrograph of two statoconia (s), the upper one just on the surface of an extramacular cell, the lower one intracellular in a vacuole. Bar 2 μm. (D) An intracellular statoconium with a thick electron dense surface layer. Bar 2 μm.

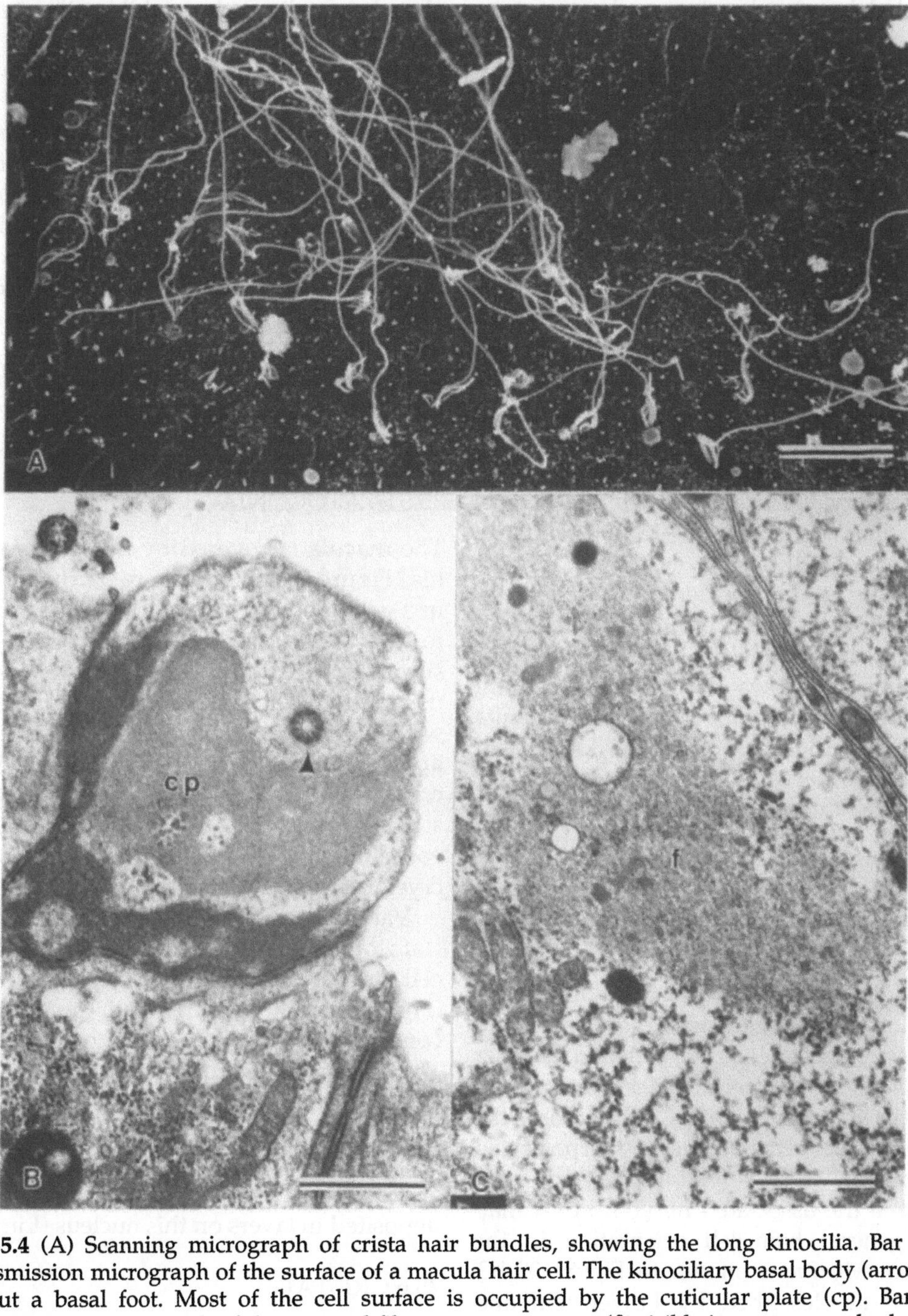

Figure 35.4 (A) Scanning micrograph of crista hair bundles, showing the long kinocilia. Bar 10 μm. (B) Transmission micrograph of the surface of a macula hair cell. The kinociliary basal body (arrowhead) is without a basal foot. Most of the cell surface is occupied by the cuticular plate (cp). Bar 1 μm. (C) Transmission micrograph of the unusual filamentous structure (f) visible in many macular hair cells. Bar 1 μm.

nerve endings. The synaptic body is often surrounded by vesicles, believed to contain synaptic transmitter substance.

35.4 *MACULA COMMUNIS*

Stretching along the ventral floor of the membranous ear is found a *macula communis* (Figure 35.2A). Transverse sections of the head show that the macula is divided in an anterior horizontal part, followed by a middle vertical and a posterior horizontal part (Jørgensen *et al.*, 1998). Also the lamprey macula can be divided in similar parts, which may reflect homology with the separate maculae in utriculus, sacculus and lagena of gnathostome fishes (Lowenstein *et al.*, 1968).

The sensory hairs are of moderate height, the kinocilia being about 8–12 µm, and the longest stereovilli about 4–10 µm (Figure 35.3A). The number of stereovilli supported by each hair cell can be counted by subjecting SEM-preparations to 30 min of ultrasound. This technique shows that 15–25 stereovilli protrude from each hair cell and also permits evaluation of the position of the kinocilium relative to the stereovilli.

The orientation of the kinocilium is believed to reflect the response polarity of the hair cell (Flock, 1965; Harris *et al.*, 1970). Electrophysiological experiments on lateral line and inner ear sensory organs of fishes and other vertebrates have shown that when the hair bundle is bent towards the kinocilium, the cell is depolarized and the neurone to which it connects sends an elevated number of impulses to the central nervous system. When the hairs are bent away from the kinocilium, the cell becomes hyperpolarized and none or only a few impulses can be detected in the associated nerve fibre. A map of the general hair cell orientation in the macula of *Myxine* is shown in Figure 35.1C. However, as described by Lowenstein and Thornhill (1971), some hair cells have an orientation that differs somewhat from their neighbours, or have hair bundles with different arrangements, in which the kinocilium is placed centrally in the middle of the stereovilli (Jørgensen, 1989) or there are two kinocilia.

The regional orientation of the hair cells within the macula should produce the following responses; if the hagfish lowers its head, in a nose-down position, the anterior hair cells will be excited, while the posterior cells will be inhibited. In a nose-up position the cells in the anterior part of the macula will be inhibited, while the cells in the posterior part will be excited. Rotational movements around the body axis should be sensed by hair cells in the middle part of the macula. The two maculae, in both ears, should thus detect rotation about all three axes.

35.5 STATOCONIA

The macula is covered by numerous statoconia (Figure 35.2B) which are partly embedded in a gelatinous structure. This can be followed to the apical tips of the microvilli of the supporting cells and is believed to be secreted from these cells. The statoconia are spherical with different diameters, and consist of apatite. Hagfishes and lampreys are the only craniates with apatitic statoconia; all gnathostomes examined so far have statoconia consisting of calcium carbonate in different crystal forms (Carlström, 1963).

Various stages in the formation of hagfish statoconia can be observed in parenchymatic cells in the immediate vicinity of the macula (Figure 35.3B and C). Thin sections show concentric rings in some, but not all statoconia. The reason for this difference is not known.

The formation of statoconia in hagfishes is also different from vertebrates. In vertebrates an organic nucleus is formed in supporting and other cells, and calcium carbonate is deposited in layers on this nucleus (Lim, 1973).

35.6 CRISTAE

The *crista anterior* and *crista posterior* are positioned close to the anterior and posterior ends of the macula. In *Myxine glutinosa* they form a belt, 5–10 hair cells wide, all the way round the inner circumference of the semicircular canal and are oriented perpendicularly to the canal wall. In *Eptatretus* species a more ordinary ridge-shaped crista is formed in one part of an ampulla-like widening, but a narrow zone containing a few hair cells does form an almost complete ring (Jørgensen *et al.*, 1998).

The crista sensory hairs are characterized by very long kinocilia, up to 35 μm (Figure 35.4A). The height of the stereovilli seems to vary from cell to cell. Some cells have long stereovilli (Figure 35.3A), while these are short in others. It is unclear whether this difference reflects different stages of development cells with different functions.

As stated by Retzius (1881) and Lowenstein and Thornhill (1970), a proper cupula is not present. In the scanning preparations, however, some sort of precipitation covers many of the crista sensory hairs. Staining or precipitation with fine powder in the living animal, or quick frozen ear preparations examined in the scanning electron microscope has hitherto failed to disclose the proper nature of this covering.

The orientation of the hair cells is generally away from the macula, but many cells deviate considerably from this arrangement.

35.7 INNERVATION

All hair cells examined receive afferent innervation. The boutons contain some mitochondria and a few vesicles. No traces of efferent boutons have been observed, but light microscopial staining for acetylcholinesterase (Hedreen *et al.*, 1985) revealed a few cholinergic fibres in *Myxine* (Jørgensen *et al.*, 1998), especially in the sensory epithelium of the cristae.

REFERENCES

Carlström, D. (1963) A crystalline study of vertebrate otoliths. *Biological Bulletin*, **125**, 441–63.

Corwin, J.T., Jones, J.E., Katayama, A., Kelley, M.W. and Warchol, M.E. (1991) Hair cell regeneration: the identities of progenitor cells, potential triggers and instructive cues, in *Regeneration of Vertebrate Sensory Receptor cells*. Ciba Foundation Symposium **160**, Wiley, Chichester, pp. 103–20.

Flock, Å. (1965) Electron microscopic and electrophysiological studies on the lateral line canal organ. *Acta Otolaryngologica (Stockholm), Suppl.* **199**, pp. 1–90.

Harris, G.G., Frishkopf, L.S. and Flock, Å. (1970) Receptor potentials from hair cells of the lateral line. *Science*, **167**, 76–9.

Hedreen, J.C., Bacon, S.J. and Price, D.L. (1985) A modified histochemical technique to visualize acetylcholinesterase-containing axons. *Journal of Histochemistry and Cytochemistry*, **33**, 134–40.

Jørgensen, J.M. (1989) Evolution of octavolateralis sensory cells, in *Neurobiology and Evolution of the Lateral Line System* (eds S. Coombs, P. Görner and H. Münz), Springer-Verlag, New York, pp. 118–45.

Jørgensen, J.M. (1991) Regeneration of lateral line and inner ear vestibular hair cells, in *Regeneration of Vertebrate Sensory Receptor Cells*. Ciba Foundation Symposium **160**, Wiley, Chichester, pp. 151–70.

Jørgensen, J.M., Shichiri, M., Geneser, F.A. (1998) Morphology of the hagfish inner ear. *Acta Zoologica (Stockholm)*, in press.

Lim, D.J. (1973) Formation and fate of the otoconia. *Annals of Otology, Rhinology and Laryngology*, **82**, 1–13.

Lowenstein, O., Osborne, M.P. and Thornhill, R.A. (1968). The anatomy and ultrastructure of the labyrinth of the lamprey (*Lampetra fluviatilis* L.). *Proceedings of the Royal Society of London*, Ser. B, **170**, 113–34.

Lowenstein, O. and Thornhill, R.A. (1970) The labyrinth of *Myxine*: anatomy, ultrastructure and electrophysiology. *Proceedings of the Royal Society of London*, Ser. B, **176**, 21–42.

McVean, A. (1991) The semicircular canals of the hagfish *Myxine glutinosa*. *Journal of Zoology, London*, **224**, 213–22.

Retzius, G. (1881) *Das Gehörorgan der Wirbelthiere. I. Das Gehörorgan der Fische und Amphibien*, Samson & Wallin, Stockholm, **222** pp.

Rubel, E.W., Oesterle, E.C. and Weisleder, P. (1991) Hair cell generation in the avian inner ear, in *Regeneration of Vertebrate Sensory Receptor Cells*. Ciba Foundation Symposium **160**, Wiley, Chichester, pp. 77–102.

36

PHYSIOLOGY OF THE INNER EAR

Alistair R. McVean

SUMMARY

Myxine glutinosa has a single semicircular canal on each side of the head. Each canal is oriented so that it projects onto all three major planes of rotation. The unusually large internal diameter of the canal enhances its mechanical sensitivity to rotation while the absence of a cupula enables the diameter of the torus to be kept relatively small without detracting from its function of a rotational velocity transducer. Each canal contains three sensory epithelia; two cristae and one macula communis. The hair cells of the cristae have their kinocilia directed away from the midline of the fish. Records obtained from SCC nerves generate velocity-dependent, phase-locked responses to sinusoidal oscillations in the horizontal plane between 0.25 and 2.0 Hz. The acuity of the response is less than that obtained from similar vertebrate preparations. The orientation of the kinocilia results in a situation in which only rotation in the horizontal plane generates unambiguous responses. This is also the plane in which mechanical sensitivity is greatest. Ambiguity of response from the cristae in rolling and pitching is resolved by the gravitational responses of the hair cells within the macula communis. Mole rats lack eyes which form a sharp visual image, yet have semicircular canals with enhanced sensitivity. Neither *Myxine* nor mole rats need a vestibular-ocular reflex (VOR) to stabilize the eyes, yet each has responded differently to visual deprivation. This suggests that the relativly poor acuity of the semicircular canals in *Myxine* is not dictated by the absence of a VOR but may be sufficient to orientate on the sea floor.

36.1 INTRODUCTION

Myxine, like every other free living animal, lives in a Newtonian world in which rotation about three axes and linear acceleration in another three are probable events. Unlike most craniates *Myxine* cannot substitute or supplement vestibular information with visual data from a sharp optical image. We might therefore expect to find a developed and mechanically sensitive vestibular system which, as in the eyeless world of submarines, provides accurate information on rotational and linear displacements. Such information is available to most other craniates from three orthogonally placed semicircular canals and associated maculae. It comes as a surprise to find that *Myxine* has to deduce positional information with an instrument that looks, at first sight, decidedly inferior; instead of three semicircular canals there is only one; instead of two or more maculae, there is only one. This poses the question of whether the vestibular apparatus is a poor design, providing less than adequate information though enough to furnish the limited requirements of a bottom dwelling marine fish which lives in mud burrows, or whether it is a simple but sophisticated instrument which achieves the performance displayed by vertebrate semicircular canals but with fewer working parts.

The Biology of Hagfishes. Edited by Jørgen Mørup Jørgensen, Jens Peter Lomholt, Roy E. Weber and Hans Malte. Published in 1998 by Chapman & Hall, London. ISBN 0 412 78530 7.

The truth, as is often the case, lies somewhere between these two extremes.

36.2 THE LABYRINTH CANAL AND ASSOCIATED SENSORY CELLS

Lowenstein and Thornhill (1970) have provided an excellent description of the anatomy of the sensory components of the *Myxine* labyrinth. The following is largely based on their account. The toroidally shaped membranous labyrinth is enclosed in cartilage which reaches to the centre. The labyrinth, as might be expected from a single canal that is attempting to measure rotation in three planes, is oriented at an angle that does not have an immediate reference to any one of the three major axes of the animal. Viewed from the front, the canal lies in a plane that is 60° to the dorsal–ventral axis; viewed from above the canal makes an angle of 30° with the anterior–posterior axis (Figure 36.1a).

36.2.1 Sensory epithelia

Each labyrinth is gently swollen by two ampullae, one placed anteriorly, one posteriorly. The anterior ampulla lies more laterally than the posterior ampulla and because of the canal orientation, is ventral to it. Lying between the ampullae, on the medial side of the canal, is the single macula communis (Figure 36.1c).

Each labyrinth is supplied by branches of the eighth cranial nerve. Closer to the labyrinth two ganglia supply the ampullae and macula (Figure 36.1b). In *Myxine* there is no sign of myelination but in *Eptatretus* some axons of the posterior ramus appear to be myelinated (Amemiya *et al.*, 1985). Size distinguishes two classes of cells in the anterior ganglion of *Eptatretus*. A ventral group has diameters of 20–40 μm, half the size of the cells of the group lying above them. In contrast, the cells in the posterior ganglion are uniformly small. The terminals of axons from both ganglia overlap in the medulla. In *Bdellostoma*, where the anterior ganglion also contains two cell sizes, the larger cells innervate the skin, the smaller cells the labyrinth (Ayers and Worthington, 1908). Amemiya *et al.* (1985) found no efferents supplying the labyrinth, unlike the situation in the lamprey (Lowenstein *et al.*, 1968). However Jørgensen (this volume) found a sparse number of cholinergic fibres in both cristae and the macula.

Each of the two ampullae contains two annular cristae, oriented at a compound angle to the canal. No cupula has been found associated with the cristae. The hair cells of each crista are polarized so that the kinocilia point away from the macula. The cells of the single macula (Figure 36.1c) form a sensory sheet on the median and ventral wall of the canal between and extending into the ampullae. The macula is covered by statoconia bound within a fibrous matrix. The hair cells are uniformly distributed over the macula with the kinocilia pointing away from the centre of the field.

36.3 PREDICTED RESPONSE OF THE HAIR CELLS IN THE AMPULLARY CRISTAE TO ROTATION

Within the plane of the canal the predicted response from each crista to rotation is clear enough but it can be difficult to picture the direction of fluid flow to rotation about the three major axes of the animal because of the unusual angle of the canals. If the canals were oriented in the horizontal plane the hair cells would detect yawing movements; clockwise yawing would excite the hair cells of the right hand side (RHS) posterior crista and those of the left hand side (LHS) anterior crista. The sensory cells in the other two cristae would be inhibited (Figure 36.2). This situation is comparable to the vertebrates where the cristae in orthogonal canals on each side of the head also complement each other. In the orthogonal situation the hair cells in the two

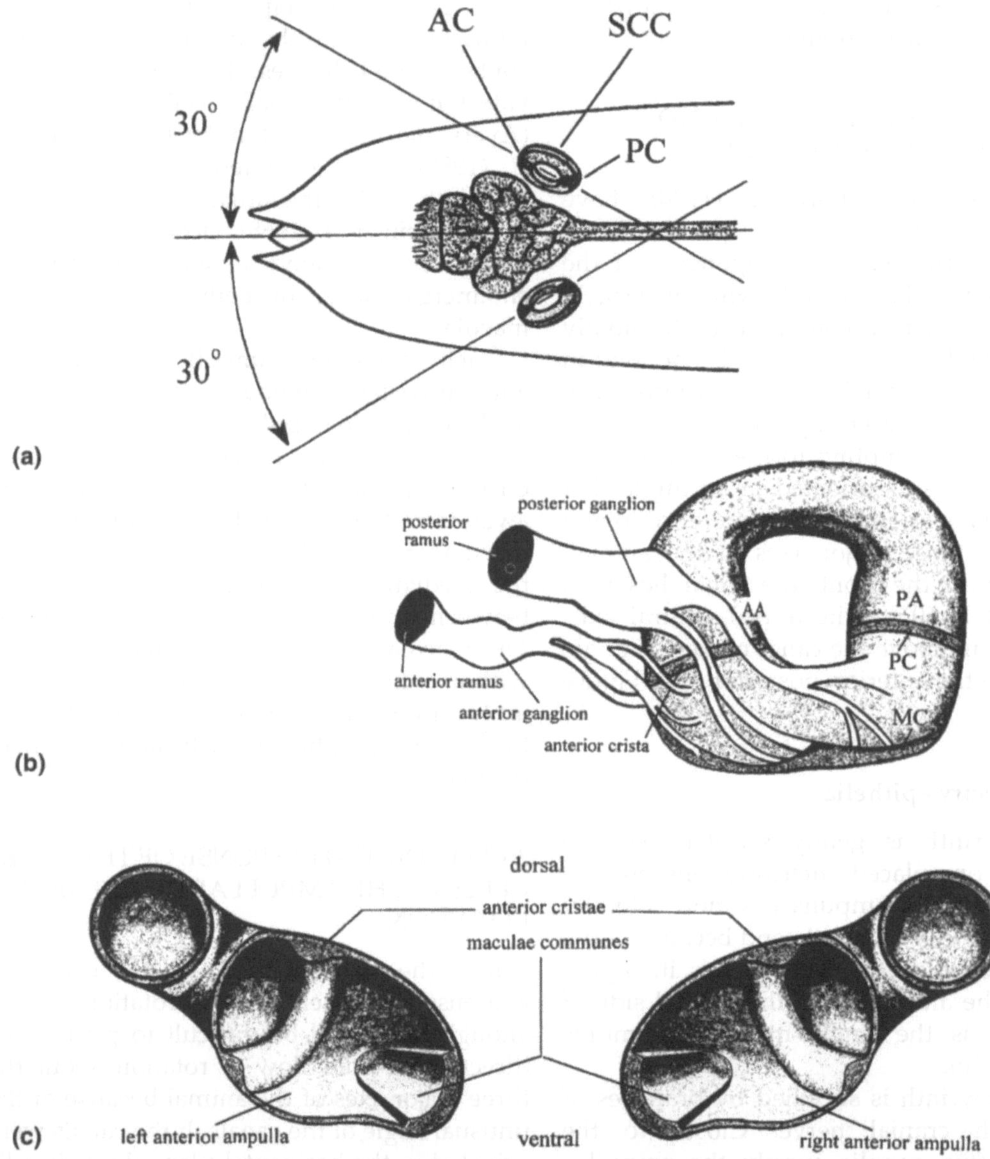

Figure 36.1 (a) Diagram shows the orientation of the single semicircular canal (SCC) lying on each side of the brain of *Myxine*. The canals are positioned so that the posterior cristae are dorsal and medial to the anterior cristae (from McVean, 1991 with permission). (b) Two ganglia serve the labyrinth, one on the anterior ramus of the eighth nerve, one on the posterior ramus. From each ganglion a number of branches divide to supply the torus, the largest going to the ampullae. (c) Vertical section through the left and right labyrinths of *Myxine glutinosa* showing the anterior ampullae and cristae as well as the macula communis. AA, anterior ampulla; AC, anterior crista; MC, macula communis; PA, posterior ampulla; PC, posterior crista. (a) reproduced with permission from McVean, A. (1991) The semicircular canals of the hagfish *Myxine glutinosa*. *Journal of Zoology, London*, **224**, 213–22. (b) and (c) redrawn with permission from Lowenstein, O. and Thornhill, R.A. (1970) The labyrinth of *Myxine*: anatomy, ultrastructure and electrophysiology. *Proc. Roy. Soc., London B.*, **176**, 21–42.

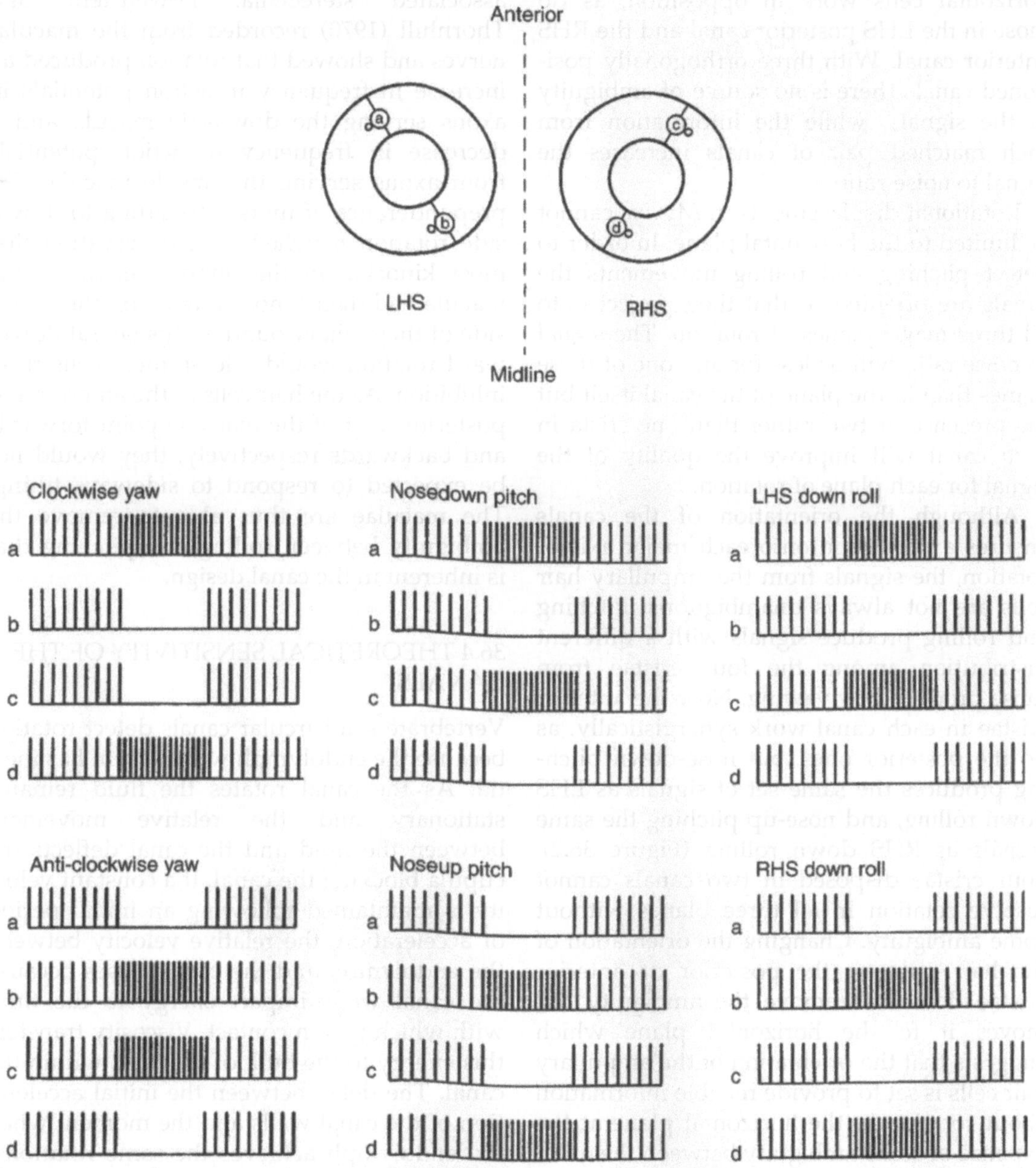

Figure 36.2 The hair cells of all four ampullary cristae are oriented with the kinocilia facing away from the midline. Each crista is represented by a line across the width of the canal and one hair cell in each (a,b,c,d) represents the orientation of all the hair cells in the crista. The kinocilium is represented by a small circle attached to the hair cell. In picturing the responses of the hair cells it is necessary to bear in mind the orientation of the canal as shown in Figure 36.1(a). The response of each cell to rotation is depicted as a change in its spontaneous frequency. The cell is shown as being either excited or inhibited, with a return to its spontaneous frequency at the end of the movement. Only yawing generates an unambiguous response.

horizontal cells work in opposition, as do those in the LHS posterior canal and the RHS anterior canal. With three orthogonally positioned canals there is no source of ambiguity in the signals, while the information from each matched pair of canals increases the signal to noise ratio.

Rotational displacements in *Myxine* cannot be limited to the horizontal plane. In order to detect pitching and rolling movements the canals are oriented so that they project onto all three major planes of rotation. The signal to noise ratio will be less for any one of these planes than in the plane of the canal itself but the presence of two rather than one crista in each canal will improve the quality of the signal for each plane of rotation.

Although the orientation of the canals ensures a projection onto each major axis of rotation, the signals from the ampullary hair cells are not always unambiguous. Pitching and rolling produce signals with a different distribution among the four cristae from those produced by yawing. Now the anterior cristae in each canal work synergistically, as do the posterior ones, but nose-down pitching produces the same set of signals as LHS down rolling, and nose-up pitching the same signals as RHS down rolling (Figure 36.2). Four cristae disposed in two canals cannot resolve rotation in all three planes without some ambiguity. Changing the orientation of the hair cells on the posterior or anterior cristae does not remove the ambiguity but moves it to the horizontal plane which suggests that the orientation of the ampullary hair cells is set to provide reliable information about rotation in the horizontal plane at the expense of some ambiguity between the other two planes.

However, information coming from the labyrinth includes signals coming from the maculae. When the dorsal surface of *Myxine* axis is vertical, the hair cells of the maculae point horizontally with the hair cells in the middle third of each macula oriented with their kinocilia lying above or below their associated stereocilia. Lowenstein and Thornhill (1970) recorded from the macular nerves and showed that rotation produced an increase in frequency in action potentials in axons serving the downside macula and a decrease in frequency in action potentials from axons serving the upside macula. The preponderance of units responding to downside rotation matched their observation that most kinocilia in the middle region of the macula had their kinocilia lying on the dorsal side of their ciliary bundles. Ipsilateral downward rotation would release these cells from inhibition. As the hair cells in the anterior and posterior field of the maculae point forwards and backwards respectively, they would not be expected to respond to sideways tilting. The maculae are thus able to remove the ambiguity between rolling and pitching that is inherent in the canal design.

36.4 THEORETICAL SENSITIVITY OF THE CANALS

Vertebrate semicircular canals detect rotation because the endolymph within them has inertia. As the canal rotates the fluid remains stationary and the relative movement between the fluid and the canal deflects the cupula blocking the canal. If a constant velocity is maintained following an initial period of acceleration, the relative velocity between the endolymph and the canal ceases because the canal walls impart energy to the fluid with which it is in contact. Viscosity transfers this energy to the bulk of the fluid within the canal. The delay between the initial acceleration of the canal walls and the moment when the endolymph achieves the same rotational velocity as the canal is characterized by the inertial time constant which, in the posterior canal of the squirrel monkey, is 3 ms (Fernandez and Goldberg, 1971). At this moment the signal from the cristae will have reached a maximum (or minimum if the flow has pushed the kinocilia towards their associated stereocilia). In vertebrates the hair cells of the cristae are embedded in a cupula which is attached along the length of its perimeter to the canal wall (Oman *et al.*, 1987). The pressure of the endolymph against the cupula makes the centre bulge in the opposite direction to the canal vector and in doing so deforms the cilia embedded in the cupula. The cupula has elasticity which, after the moment when the endolymph achieves the same velocity as the canal, pushes against the displaced endolymph and restores the cupula to its pre-acceleration shape. Only a further acceleration pulse or deceleration will deform the cupula for a second time. The ratio between the strength of this elastic restoring force and the moment of inertia of the endolymph determines a second time constant. This is the elastic time constant which, in the cat, is 4–5 s and 3 s in the frog (Carpenter, 1988).

The relationship between these two time constants is such that for abnormally slow rotations the hair cells in the cupulae measure acceleration; for extremely rapid changes they measure head position but for the normal spectrum of head rotational velocities the semicircular canals measure velocity (Carpenter, 1988).

The performance of a semicircular canal is determined by several aspects of its design. The sensitivity can be predicted from morphological data (Oman *et al.*, 1987; Lindenlaub *et al.*, 1995) according to the equation

$$\frac{X_c}{y} = \frac{P}{4\,\pi L A_c V(S/A^2)}$$

where X_c is the average displacement of the cupula face, y is head angular velocity, P is the plane area, L is the central streamline area, A_c is cupula cross-sectional area, V is endolymph kinemetic viscosity, S is the increase in drag flow in an elliptic segment due to increase in wall surface area relative to that of a canal with a circular cross-section and A is the cross-sectional area of the canal.

In order to compare the mechanical sensitivity of canals with different dimensions we can rewrite this equation in terms of the radius of curvature (R) and the radius of the canal (r) (see Figure 36.3a for L, P, R and r).

$$\frac{X_c}{y} = \frac{Rr^2}{8VS}$$

Thus the mechanical sensitivity is proportional to Rr^2.

In Figure 36.3b, Rr^2 is plotted for mammals and fishes against body mass using the data of Jones and Spells for R and r from different fishes and mammals. Figure 36.3b shows that Rr^2 increases with body mass in order to compensate for slower movements made by larger animals. It also shows that the semicircular canals of fishes are an order of magnitude more sensitive than those of mammals (Jones and Spells, 1963) possibly because of the absence of visual information available to supplement the vestibular reflexes (Montgomery and McVean, 1987). Figure 36.3b also shows that the value of Rr^2 for *Myxine* is comparable to fishes of the same size. However because there is no cupula in *Myxine* canals (Lowenstein and Thornhill, 1970; Jørgensen, this volume) the amplitude of crista hair cell deflections will be determined by the drag exerted by the endolymph directly on the cilia of the hair cells. In *Myxine* the crista hair cells project from a crest (Lowenstein and Thornhill, 1970) which will increase the velocity of the endolymph flowing over the hair cells both by decreasing the cross-sectional area of the canal at that point and by placing the hair cells out of the direct influence of drag exerted by the canal walls.

Where there is a cupula its long time elastic time constant ensures that it acts as a velocity detector. The time constant is determined by O/Δ where O = moment of friction per unit angular velocity and Δ = moment of the cupula restoring force per unit angular displacement of the endolymph (Jones and Spells, 1963). The amplitude of this ratio is governed by the ratio R^2/r^2. Jones and Spells (1963) point out that R

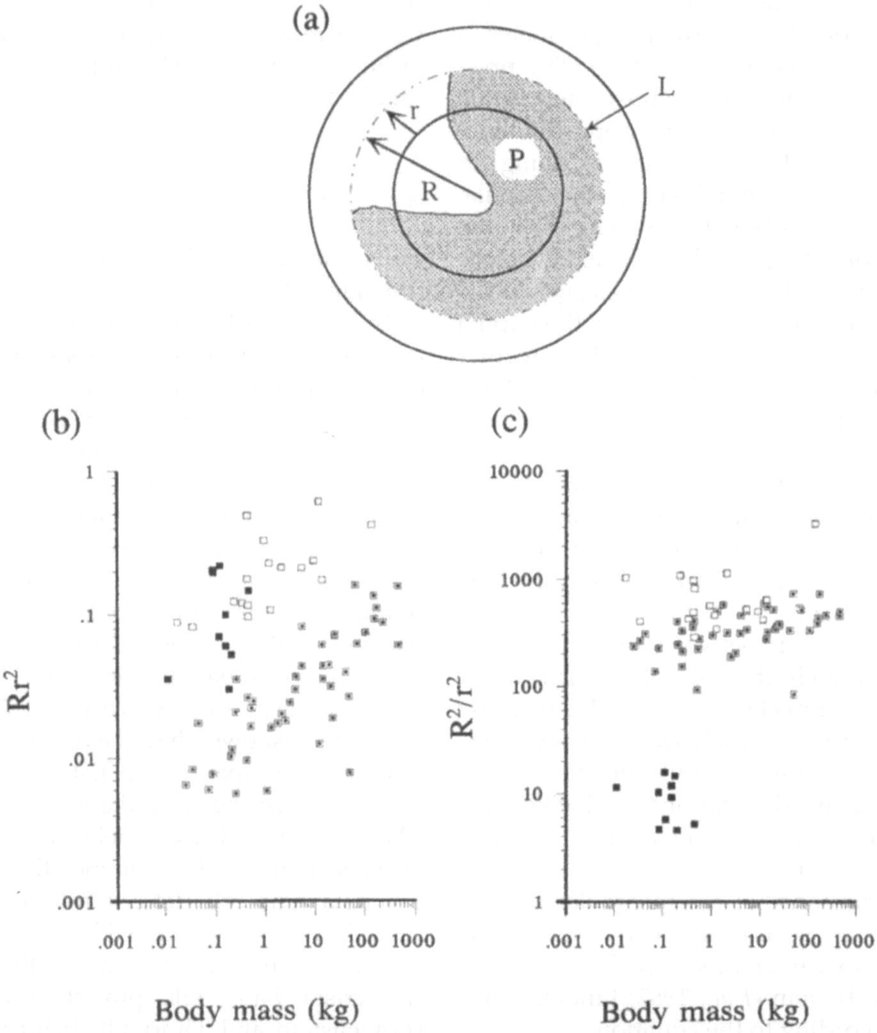

Figure 36.3 (a) Schematic drawing illustrating the plane area (*P* is circular but a portion has been cut away to reveal *R*), streamline length (*L*), radius of curvature (*R*) and internal radius (*r*) of a canal. (b) Rr^2 plotted against body mass for a variety of different mammals (▣), fishes (□) and 10 *Myxine glutinosa* (■) of different sizes. The Rr^2 values for *Myxine* are similar to those of the fishes. (c) R^2/r^2 plotted against body mass for the same animals as in (b). The values for *Myxine* are at least an order of magnitude smaller than those for fishes and mammals. Fish and mammal data from Jones, G.M. and Spells, K.E. (1963) A theoretical and comparative study of the functional dependence of the semicircular canal upon its physical dimensions. *Proc. Roy. Soc.*, London B., **157**, 403–19. *Myxine* data with permission from

must increase disproportionally with r^2 in order to keep the elastic time constant of the cupula in the region where the canal can continue to signal changes in velocity. Since *Myxine* does not have a cupula the elastic time constant will be governed by the ratio between elastic restoring force of the cilia and the viscosity of the endolymph. It will be virtually independent of canal dimensions, which explains the very low value for R^2/r^2 in

Myxine compared with fishes and mammals (Figure 36.3c). This independence from the requirement to satisfy the elastic time constant of a cupula allows *Myxine* to fit a canal with large internal diameter into a narrow head.

36.5 RESPONSES FROM THE CRISTA NERVES

Records obtained from a range of vertebrates confirm that the ampullary cristae detect the amplitude of rotational velocity (Melvill-Jones and Milsum, 1970; Fernandez and Goldberg, 1971; O'Leary *et al.*, 1976; Blanks and Precht, 1976; Montgomery, 1980; Hartmann and Klinke, 1980). By recording from nerves serving both the anterior and posterior cristae, Lowenstein and Thornhill (1970) showed that *Myxine* semicircular canals respond to rotation in a manner expected from the orientation of their kinocilia. Clockwise yaw excited the hair cells in the posterior RHS ampulla; anticlockwise rotation depressed its resting discharge. They cleverly oriented their preparations so that the nerve records would not be contaminated by macular responses. They were able to show that the cristae followed oscillations as slow as 0.2 Hz. Subsequently McVean (1991) showed that the response from the cristae to sinusoidal oscillations was 90° advanced with respect to head position, as would be expected in a velocity detector (Figure 36.4). A phase-locked response could be detected over the range 0.25–2 Hz which probably encompasses oscillation frequencies generated during swimming. McVean also showed that amplitude of the response from the crista hair cells was directly proportional to rotational velocity but that the average gain of four afferents (electronically isolated from records which included several units) to yawing with the head held in its normal position, was $0.042 \pm 0.047\ I\ s^{-1}\ deg^{-1}\ s^{-1}$ which is significantly less than has been recorded for canal afferents in other species (Montgomery and McVean, 1987). It would be interesting to measure gain with true single unit records from individual axons.

The projection of the *Myxine* canal onto the three rotational planes predicts that, in comparison to oscillation within the plane of the canal itself, the signal will be diminished to 82% for yawing, 52% for pitching and 43% for rolling. As in the case described above, where it was shown that only yawing was signalled without ambiguity, the relative prcjections suggest that signals associated with yawing are most important.

36.6 VESTIBULAR ORGANS IN ANIMALS IN A SENSORIALLY DEPRIVED ENVIRONMENT

McVean (1991) suggested that since one of the main functions of vestibular systems is to provide accurate and immediate information by which the eyes can be stabilized against potentially disruptive movements of the head and body, the absence of eyes in *Myxine* that generate a sharp image might relax the requirements for vestibular afferents with a high gain. Though *Myxine* ampullary cristae do appear to have a lower gain than other animals, a more direct test of this hypothesis is to examine the properties of the vestibular system in mole rats which live in a uniform underground environment, have eyes which cannot form a sharp image and whose predecessors undoubtedly possessed fully functional eyes and vestibular systems. Lindenlaub *et al.* (1995) compared the morphometrics of the semicircular canals in *Spalax ehrenbergi* (from Israel) and two *Cryptomys* spp. (from Zambia) with those of the laboratory rat. They also measured the area of the associated sensory epithelia and hair cell density in the cristae and the two maculae. They argued that either high gain vestibular information would no longer be needed in an underground tunnel system or, alternatively, that without eyes an enhanced sensory system would be needed to underpin

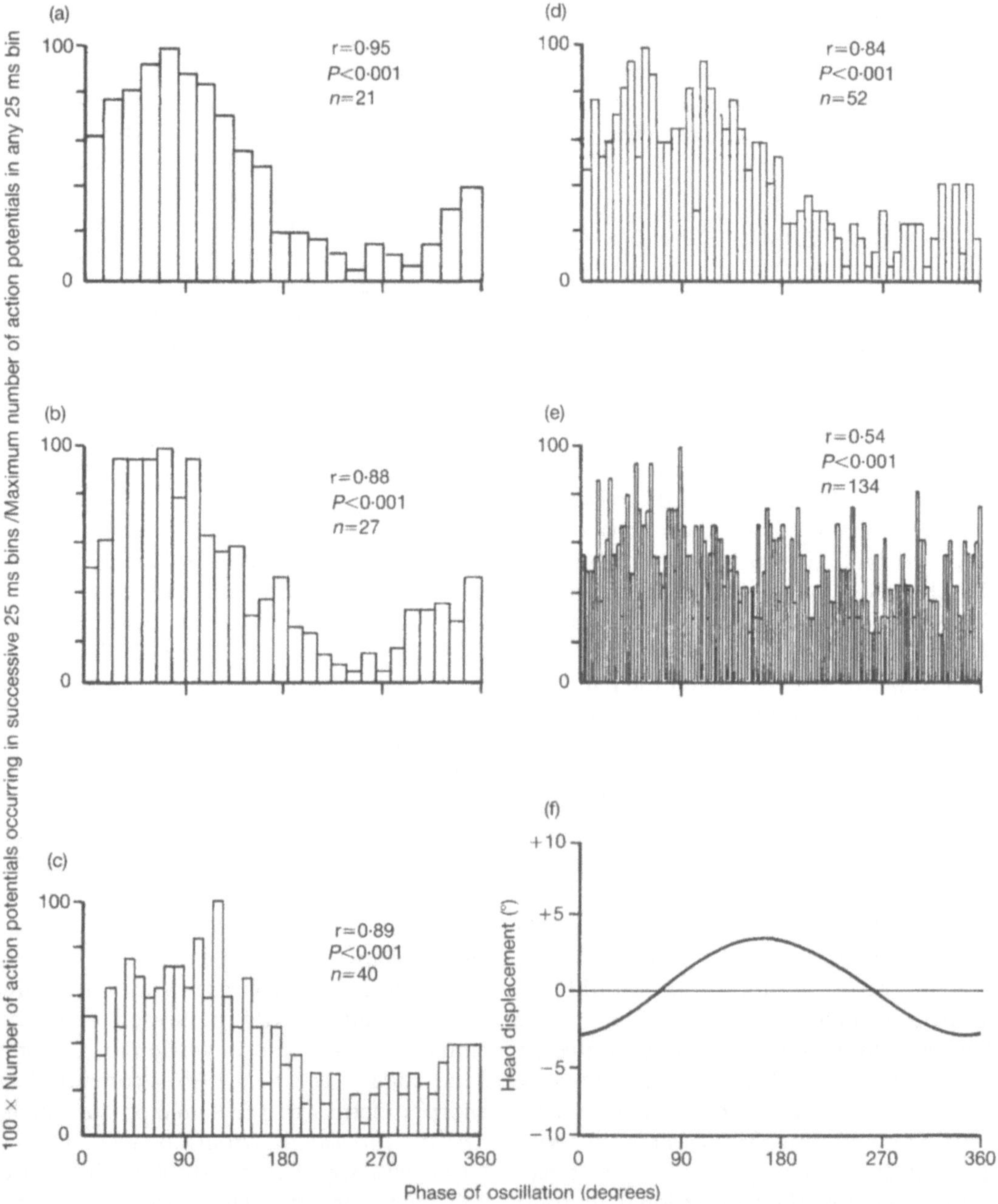

Figure 36.4 Response from the posterior crista of the right-hand side SCC to sinusoidal rotation at different frequencies. Several units have been recorded simultaneously. Nerve spike frequencies are given as the number occurring in successive 25 ms bins. Head position (f) is such that upwards on the graph represents movement of the head to the right. The discharge frequency is in the form a sine wave 90° in advance of the turntable position as would be predicted from a sense organ that measures rotational velocity. (a) 2 Hz, (b) 1.5 Hz, (c) 1 Hz, (d) 0.5 Hz, (e) 0.25 Hz. The correlation (*r*) between discharge frequency with head position displaced 90° in advance of head position is given for each record, using product-moment correlation. (From McVean, 1991, with permission.)

a navigational system that can no longer supplement locational cues with visual images. Using Oman's formula for assessing the theoretical mechanical sensitivity of the semicircular canals, they showed that the mechanical sensitivity of all three canals in both *Spalax* and *Cryptomys* were significantly ($P < 0.001$) more sensitive than those of the laboratory rat. Unlike *Myxine*, where discrimination in the horizontal plane appears to be most important, the sensitivity of the anterior vertical canal was better than in the other two canals in all three species. Both the maculae and the ampullary cristae had a relatively and absolutely larger surface area in the mole rats than in the laboratory rat. Because the density of the hair cells in all the sensory epithelia were similar, the total number of receptors in each sensory epithelium was significantly higher than in the laboratory rat.

The mole-rat data suggest that the selective forces which maintain a high gain vestibular system in the mole rat include factors separate from and additional to those moulding the vestibular-ocular system. This suggests that the (apparently) low gain of the hair cells in *Myxine* cristae may not have resulted from the loss of acute vision as the eye regressed, but may be adequate for stabilizing swimming.

REFERENCES

Amemiya, F., Kishida, R., Goris, R.C., Onishi, H. and Kusunoki, T. (1985) Primary vestibular projections in the hagfish, *Eptatretus burgeri*. *Brain Research*, **337**, 73–9.

Ayers, H. and Worthington, J. (1908) The finer anatomy of the brain of *Bdellostoma dombeyi*. 1. The acoustico-lateral system. *American Journal of Anatomy*, **8**, 1–33.

Blanks, R.H.I. and Precht, W. (1976) Functional characterisation of primary vestibular afferents in the frog. *Experimental Brain Research*, **25**, 369–90.

Carpenter, R.H.S. (1988) *Movement of the Eyes* (2nd edn), Pion, London.

Fernandez, C. and Goldberg, J.M. (1971) Physiology of peripheral neurons innervating semicircular canals of the squirrel monkey. II. Response to sinusoidal stimulation and dynamics of the peripheral vestibular system. *Journal of Neurophysiology*, **34**, 661–75.

Hartmann, R. and Klinke, R. (1980) Discharge properties of afferent fibres of the goldfish semicircular canal with high frequency stimulation. *Pflugers Archiv*, **388**, 111–21.

Jones, G.M. and Spells, K.E. (1963) A theoretical and comparative study of the functional dependence of the semicircular canal upon its physical dimensions. *Proceedings of the Royal Society, London, B*, **157**, 403–19.

Lindenlaub, T., Burda, H. and Nevo, E. (1995) Convergent evolution of the vestibular organ in the subterranean mole-rats, *Cryptomys* and *Spalax*, as compared with the aboveground rat, *Rattus*. *Journal of Morphology*, **244**, 303–11.

Lowenstein, O. and Thornhill, R.A. (1970) The labyrinth *Myxine*: anatomy, ultrastructure and electrophysiology. *Proceedings of the Royal Society, London, B*, **176**, **2** 42.

Lowenstein, O., Osborne, M.P. and Thornhill, R.A. (1968) The anatomy and ultrastructure of the labyrinth of the lamprey (*Lampetra fluviatilis* L.). *Proceedings of the Royal Society, London, B*, **170**, 113–34.

McVean, A. (1991) The semicircular canals of the hagfish *Myxine glutinosa*. *Journal of Zoology, London*, **224**, 213–22.

Melvill-Jones, G. and Milsum, J.H. (1970) Characteristics of neural transmission from the semicircular canal to the vestibular nuclei of cats. *Journal of Physiology, London*, **209**, 295–316.

Montgomery, J.C. (1980) Dogfish horizontal canal system: responses of primary afferent, vestibular and cerebellar neurons to rotational stimulation. *Journal of Neuroscience*, **5**, 1761–9.

Montgomery, J.C. and McVean, A.R. (1987) Brain function in Antarctic fish: activity of central vestibular neurons in relation to head rotation and eye movement. *Journal of Comparative Physiology A*, **160**, 289–93.

O'Leary, D.P., Dunn, R.F. and Honrubia, V. (1976) Analysis of afferent responses from isolated semicircular canal of guitarfish using rotational acceleration white-noise inputs. I. Correlation of response dynamics with receptor innervation. *Journal of Neurophysiology*, **39**, 631–44.

Oman, C.M., Marcus, E.N. and Curthoys, I.S. (1987) The influence of semicircular canal morphology on endolymph flow dynamics – an anatomically descriptive mathematical model. *Acta Oto-laryngolica (Stockholm)*, **103**, 1–13.

INDEX

MIX
Papier aus verantwortungsvollen Quellen
Paper from responsible sources
FSC® C105338

If you have any concerns about our products,
you can contact us on
ProductSafety@springernature.com

In case Publisher is established outside the EU,
the EU authorized representative is:
Springer Nature Customer Service Center GmbH
Europaplatz 3, 69115 Heidelberg, Germany

Printed by Libri Plureos GmbH
in Hamburg, Germany